# ORIGIN OF ELEMENTS IN THE SOLAR SYSTEM

# ORIGIN OF ELEMENTS IN THE SOLAR SYSTEM

## IMPLICATIONS OF POST-1957 OBSERVATIONS

Edited by

**O. Manuel**

*University of Missouri*
*Rolla, Missouri*

**Proceedings of the International Symposium Organized by Glenn T. Seaborg and Oliver K. Manuel**

Springer Science+Business Media, LLC

Library of Congress Cataloging-in-Publication Data

Origin of elements in the solar system: implications of post-1957 observations/edited by O.K. Manuel.

p. cm.

Includes bibliographical references and index.

ISBN 978-1-4757-7315-6 ISBN 978-0-306-46927-5 (eBook)

DOI 10.1007/978-0-306-46927-5

1. Cosmochemistry—Congresses. 2. Nucleosynthesis—Congresses. 3. Astrophysics—Congresses. I. Manuel, O. K., 1936–

QB450 .O78 2001

523′.02—dc21

2001018617

Proceedings of the American Chemical Society symposium: *Origins of Elements in the Solar System: Implications of Post 1957 Observations*, held August 22–26, 1999, in New Orleans, Louisiana, USA

ISBN 978-1-4757-7315-6



Originally published by Kluwer Academic / Plenum Publishers in 2000

Softcover reprint of the hardcover 1st edition 2000

http://www.wkap.nl/

10 9 8 7 6 5 4 3 2 1

A C.I.P. record for this book is available from the Library of Congress

# Contributors

**Ahmad**, Irshad, Physics, Argonne National Laboratory, Argonne, IL, USA

**Alexander**, E. Calvin, Jr., Geology and Geophysics, University of Minnesota, Minneapolis, MN, USA

**Anders**, Edward (retired), The Enrico Fermi Institute, University of Chicago, Chicago, IL, USA

**Anthony**, Don W., National Superconducting Cyclotron Laboratory, Michigan State University, East Lansing, MI, USA

**Arlandini**, Claudio, Forschungszenturm Karlsruhe Institut für Kernphysik III, Karlsruhe, GERMANY

**Armbruster**, Peter, GSI (Gesellschaft für Schwerionforschung), Darmstadt GERMANY

**Ballad**, Robert V., Knolls Atomic Power Laboratory, Schenectady, NY, USA*

**Bateman**, Nick, Simon Frasier University, Burnaby, B. C., CANADA

**Becker**, Victor J., Diatide, Inc., Londonderry, NH, USA*

**Bernas**, Monique, Institut de Physique Nuclèaire d'Orsay, Orsay, FRANCE

**Bildsten**, Lars, Physics and Astronomy, University of California, Berkeley, CA, USA

**Brown**, Wilbur K. (retired), Los Alamos National Laboratory, Los Alamos, NM, USA

**Browne**, Edgardo, Lawrence Berkeley National Laboratory, Berkeley, CA, USA

**Burbidge**, Geoffrey R., Center for Astrophysics and Space Sciences, University of California, San Diego, CA, USA

**Casey**, William H., L.A.W.R. and Geology Departments, University of California, Davis, CA, USA*

**Chevalier**, Roger A., Astronomy and Astrophysics, University of Virginia, Charlottesville, VA, USA
**Chiou**, George K. Y., IBM/FD Analytical Laboratory, IBM Corporation, San Jose, CA, USA*
**Cumming**, A., Physics and Astronomy, University of California, Berkeley, CA, USA
**D'Auria**, John, The DRAGON Collaboration, Simon Fraser University, Burnaby, B.C., CANADA
**De Laeter**, John R., Applied Physics, Curtain University, Perth, WESTERN AUSTRALIA
**Downing**, R. Greg, R. G. D. Research, Inc., Niskayuna, NY, USA*
**Ebihara**, Mitsuru, Chemistry, Tokyo Metropolitan University, Tokyo, JAPAN
**Fang**, Jiafu, Pennzoil Quaker State Company, Technology Division, The Woodlands, TX, USA*
**Fowler**, Malcolm M., Nuclear and Radiochemistry, Los Alamos National Laboratory, Los Alamos, NM, USA
**Frank**, Adam, Physics and Astronomy, University of Rochester, Rochester, NY, USA
**Ganapathy**, R., Bethlehem, PA, USA*
**Ghiorso**, Albert, Lawrence Berkeley National Laboratory, University of California, Berkeley, CA, USA
**Giesen**, U., Physics, University of Notre Dame, Notre Dame, IN, USA
**Goswami**, Jitendra Nath, Physical Research Laboratory, Ahmedabad, INDIA
**Greene**, John P., Physics, Argonne National Laboratory, Argonne, IL, USA
**Gregorich**, Ken, Lawrence Berkeley National Laboratory, University of California, Berkeley, CA, USA
**Grevesse**, Nicolas, Institut d'Astrophysique, Université de Liège, Ougrée-Liège, BELGIUM
**Guzik**, Joyce Ann, Applied Theoretical and Computational Physics Division, Los Alamos National Laboratory, Los Alamos, NM, USA
**Haight**, Robert C., Los Alamos Neutron Science Center (LANSCE), Los Alamos, NM, USA
**Hashimoto**, Yuji, Chemistry, Kanazawa University, Kanazawa, JAPAN
**Hoffman**, Robert D., Lawrence Livermore National Laboratory, Livermore, CA, USA
**Huss**, Gary R., Geological and Planetary Sciences, California Institute of Technology, Pasadena, CA, USA
**Hwaung**, Golden, Electrical and Computing Engineering, Lousiana State University, Baton Rouge, LA, USA
**Johnson**, Robert, Chemistry, University of Missouri, Rolla, MO, USA*

**Käppeler**, Franz, Forschungszentrum Karlsruhe Institut für Kernphysik III, Karlsruhe, GERMANY

**Koehler**, Paul E., Oak Ridge National Laboratory, Oak Ridge, TN, USA

**Kohman**, Truman (retired) Chemistry, Carnegie-Mellon University, Pittsburg, PA, USA*

**Kratz**, Karl L., Institut für Kernchemie, Universität Mainz, Mainz, GERMANY

**Kuroda**, Paul K. (retired), Chemistry, University of Arkansas, Fayetteville, AR, USA

**Kutschera**, Walter, Institut für Radiumforschung und Kernphysik, Universität Wien, Vienna, AUSTRIA

**Lal**, Devendra, Geosciences, Scripps Institution of Oceanography, La Jolla, CA, USA

**Larimer**, John W., Geology, Arizona State University, Tempe, AZ, USA*

**Lee**, Jauh T., Advanced Separation Technologies, Inc., Whippany, NJ, USA*

**Lewis**, Roy S., The Enrico Fermi Institute, University of Chicago, Chicago, IL, USA

**Lhersonneau**, Gerard, Physics, University of Jyvaskyla, Jyvaskyla, FINLAND

**Li**, Bin, Lunar and Planetary Laboratory, University of Arizona, Tucson, AZ, USA*

**Lietz**, Cara, Chemical Engineering, University of Missouri, Rolla, MO, USA

**Lofy**, P. A., National Superconducting Cyclotron Laboratory, Michigan State University, East Lansing, MI, USA

**Loss**, Robert D., Applied Physics, Curtin University, Perth, WESTERN AUSTRALIA

**Lu**, Qi, Chemistry, University of Tokyo, Tokyo, JAPAN

**Lugmair**, Günther W., Geosciences, Scripps Institute of Oceanography and Max-Planck Institut für Chemie, Mainz, GERMANY

**Maas**, Roland, Geology, Latrobe University, Victoria, AUSTRALIA

**Mantica**, P. F., National Superconducting Cyclotron Laboratory, Michigan State University, East Lansing, MI, USA

**Manuel**, Oliver K., Chemistry, University of Missouri, Rolla, MO, USA

**Masuda**, Akimasa (retired), Chemistry, University of Tokyo, Tokyo, JAPAN

**Miller**, Geoffrey G., Los Alamos National Laboratory, Los Alamos, NM, USA

**Möller**, Peter, Theoretical Division, Los Alamos National Laboratory, Los Alamos, NM, USA

**Morrissey**, David J., National Superconducting Cyclotron Laboratory, Michigan State University, East Lansing, MI, USA

**Morss**, Lester, Chemistry Division, Argonne National Laboratory, Argonne, IL, USA*

**Myers**, William A., Chemical Engineering, University of Arkansas, Fayetteville, AR, USA

**Nakanishi**, Takashi, Chemistry, Kanazawa University, Kanazawa, JAPAN

**Neuforge**, Corinne, Los Alamos National Laboratory, Los Alamos, NM, USA

**Ninov**, V., Lawrence Berkeley National Laboratory, Berkeley, CA, USA

**Nolte**, Adam, Chemical Engineering, University of Missouri, Rolla, MO, USA

**Norman**, Eric B., Lawrence Berkeley National Laboratory, University of California, Berkeley, CA, USA

**Ott**, Ulrich, Max-Planck Institut für Chemie, Mainz, GERMANY

**Oura**, Y., Chemistry, Tokyo Metropolitan University, Tokyo, JAPAN

**Ozaki**, Hiromasa, Chemistry, Tokyo Metropolitan University, Tokyo, JAPAN

**Palmer**, Phillip D., Los Alamos National Laboratory, Los Alamos, NM, USA

**Paul**, Michael, Racah Institiute of Physics, Hebrew University, Jerusalem, ISRAEL

**Pfeiffer**, Bernd, Institut für Kernchemie, Universität Mainz, Mainz, GERMANY

**Pilcher**, Carl, NASA Headquarters, Washington, DC, USA*

**Prisciandaro**, Joann I., Cyclotron Laboratory, Michigan State University, East Lansing, MI, USA

**Ramadurai**, Souriraja, TIFR/Astrophysics Group, Colaba, Mumbai, INDIA

**Rauscher**, Thomas, Physics, University of Basel, Basel, SWITZERLAND

**Rosman**, K. J. R., Applied Physics, Curtin University, Perth, WESTERN AUSTRALIA

**Rouse**, Carl A. (retired), GA Technologies, Inc., San Diego, CA, USA

**Rowe**, Marvin W., Chemistry, Texas A&M University, College Station, TX, USA

**Rundberg**, Robert S., Nuclear and Radiochemistry, Los Alamos National Laboratory, Los Alamos, NM, USA

**Sakamoto**, Koh, Chemistry, Kanazawa University, Kanazawa, JAPAN

**Sauval**, A. Jacques, Observatoire Royal de Belgique, Bruxelles, BELGIUM

**Schatz**, Hendrik, GSI (Gesellschaft für Schwerionenforschung), Darmstadt, GERMANY

**Seabury**, Edward H., Los Alamos Neutron Science Center (LANSCE), Los Alamos, NM, USA

**Shimamura**, Tadashi, School of Allied Health Sciences, Kitasato University, Sagamihara, Kanagawa, JAPAN

**Shinotsuka**, Kazunori, National Institute for Fusion Science, Gifu, JAPAN
**Steiner**, M., Cyclotron Laboratory, Michigan State University, East Lansing, MI, USA
**Sorlin**, Olivier, Institut de Physique Nucleaire, Orsay Cedex, FRANCE
**Tanaka**, T., Earth and Planetary Sciences, Nagoya University, Nagoya, JAPAN
**Tanimizu**, Masaharu, Earth and Planetary Sciences, Nagoya University, Nagoya, JAPAN
**Thielemann**, Friedrich-Karl, Department für Physik und Astronomie, University of Basel, Basel, SWITZERLAND
**Trimble**, Virginia, Astronomy and Astrophysics, University of Maryland, College Park, MD, USA
**Ullman**, John L., LANSCE-3, Los Alamos National Laboratory, Los Alamos, NM, USA
**Vahia**, Mayank N., Astronomy and Astrophysics, Tata Institute of Fundamental Research, Mumbai, INDIA
**Viola**, Vic, Chemistry, University of Indiana, Bloomington, IN, USA
**Voss**, Fritz, Forschungszentrum Karlsruhe Institut für Kernphysik III, Karlsruhe, GERMANY
**Ward**, Thomas E. (Consultant), U.S. Department of Energy, Washington, DC, USA
**Wiescher**, M., Physics, University of Notre Dame, Notre Dame, IN, USA
**Wilhelmy**, Jerry B., Los Alamos National Laboratory, Los Alamos, NM, USA
**Windler**, Ken, Chemistry, University of Illinois, Urbana-Champagne, IL, USA
**Wisshak**, Klaus, Forschungszentrum Karlsruhe Institut für Kernphysik III, Karlsruhe, GERMANY
**Woolsey**, Stanford E., Lawrence Livermore National Laboratory, University of California, Santa Cruz, CA, USA
**Xie**, Yixiang, Electronic Materials Applied Research Center, University of Missouri, Rolla, MO, USA*

*Participant or author of paper presented at the symposium but not published in these proceedings.

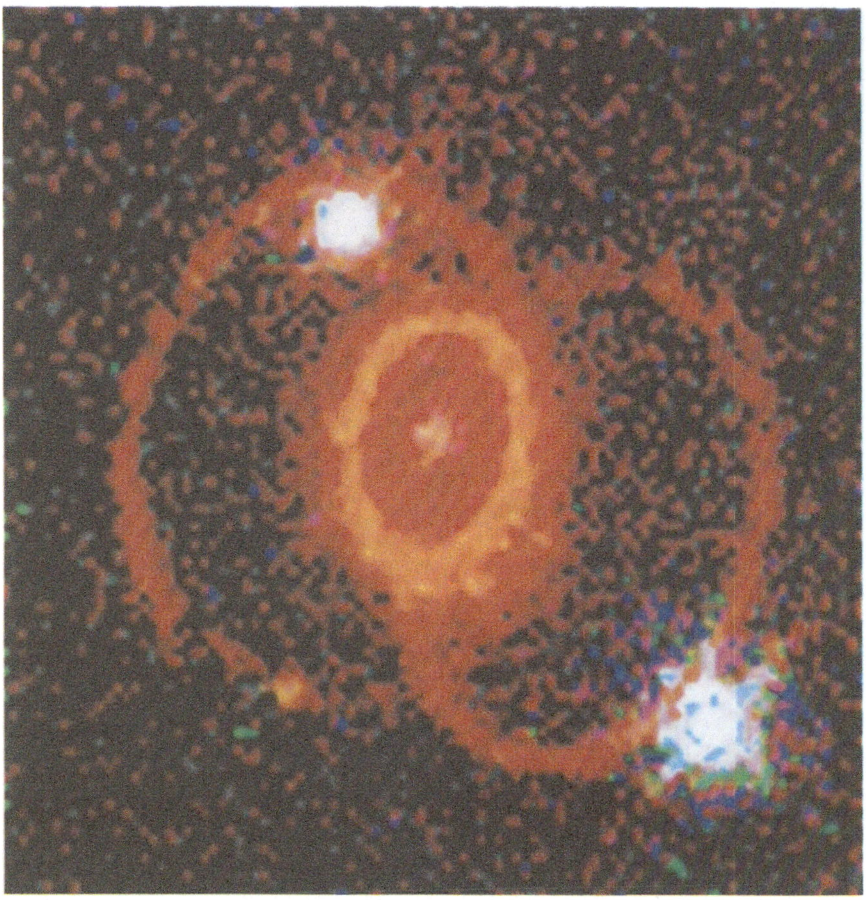

Supernova 1987A is a recent example of one of the stellar nucleosynthesis events that generated many of the nuclides in our solar system. (From NASA).

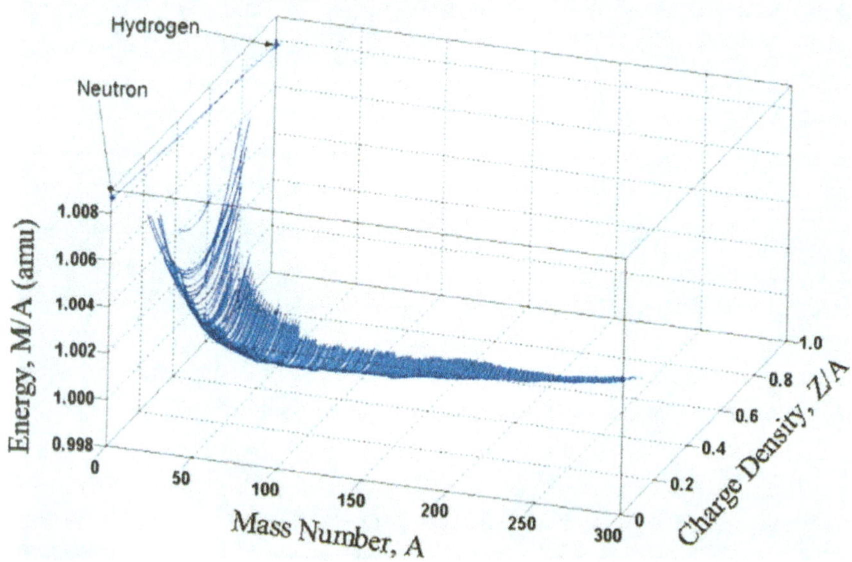


The cradle of the nuclides holds the products of nucleosynthesis. The more stable nuclides lie in the valley, those that are radioactive or readily consumed by fusion or fission occupy higher positions in the cradle. (From FCR, Inc.)

# Preface and Dedication

"Satyam eva jayate nanritan"
"Truth is victorious, never untruth"
Mundaka Upanishad, III.1.6

*"From where did all this come?"* has been one of the fundamental questions since the dawning of civilization. These proceedings record the diversity of opinions on this subject as the 20th century draws to an end.

When the landmark papers by Burbidge, Burbidge, Fowler and Hoyle ($B_2FH$) and Cameron on stellar nucleosynthesis appeared in 1957, the periodic table ended at element #101, the latest discovery of Ghiorso and co-workers. Experiments described here tell of the production of heavier elements in the intervening years, in labs at Dubna, GSI-Darmstadt and Berkeley. Element #114 was reported when this symposium was in the planning stages, and the Berkeley group reported the discovery of elements #118, #116 and #114 at the symposium. These proceedings also contain exciting new information from laboratory studies on the life-times and cross-sections of nuclides under extreme, stellar-like conditions, often far from the region of nuclear stability.

More stable nuclides are more abundantly formed in laboratory studies. "Fine-structure" elemental abundance peaks also occur in nature at multiples of the $^4$He nuclide, at iron (Fe), and at closed shells of nucleons. One issue debated here is whether this relationship between abundance and nuclear stability holds for the overall composition of the solar system. The answer depends if the solar interior ($\approx$ 99.8% of the mass of the solar system) is mostly $^1$H, the stable nuclide with the highest mass per nucleon, or $^{56}$Fe, the nuclide with the lowest mass per nucleon. Several papers here suggest that seismology, neutrinos and isotopic ratios from the sun may soon resolve differences of opinion on the sun's internal structure and composition.

Observations of the skies, particularly with the Hubble Space Telescope, have given us new details on the birth and death of stars and of planetary systems. The discovery of pulsar planets clearly illustrates one unexpected end product of a supernova explosion. The explosion of a nearby supernova, SN1987A, provides new details about the nuclear reactions that occur under these extreme conditions and the dynamics and composition of material ejected by supernovae. In these proceedings it is shown that the decay of doubly magic $^{56}Ni$, rather than a heavy element like $^{254}Cf$, dominated the light curve of SN1987A. Since $^{56}Ni$ is the parent of the most abundant iron isotope, $^{56}Fe$, SN1987A has clearly demonstrated one way to make isotopic anomalies, short-lived nuclides, and a region rich in iron.

As the 20th century was ending, Professor Glenn Seaborg and I sought to bring together a blend of experimentalists and theorists from astronomy, chemistry, geology, physics and related disciplines to summarize the current understanding of the origin of elements that comprise the solar system.

*Glenn Seaborg (Courtesy of Piermaria J. Oddone)*

This volume is dedicated to the memory of the late Professor Glenn T. Seaborg, who agreed in April of 1998 to give the keynote address and to co-chair this symposium. Drs. Al Ghiorso and Ken Gregorich knew Glenn much better than I did. Their papers in this volume give great insight into the genius of this man and his influence on science, education, and government policies in the 20th century.

Although I hardly knew Glenn personally, his influence on my scientific career has been tremendous since April of 1976 when one of his former students, Vic Viola, attended the annual AGU meeting in Washington, D.C. There, two students of Professor Paul K. Kuroda were trying to convince the audience that isotopically "strange" xenon in meteorites came from nuclear reactions in the supernova (SN) that made our elements, not from later fission of a superheavy element (SHE) within meteorite grains. Knowing that his mentor would want to hear the debate on SHE *vs* SN, Vic invited me to speak at a San Francisco ACS symposium on the origin of the elements in the fall of 1976. That was where I first met Professor Seaborg.

The next year, Glenn asked me to comment on SHE *vs* SN at the Robert Welch Conference on Cosmochemistry. Over the next 20 years, "strange" isotopic ratios were found in several elements of meteorites and great advances were made in producing SHEs in the laboratory (*e.g.*, the first three papers in this volume). Glenn was eager to co-chair the 1999 ACS symposium where these observations would be discussed and a cross section of opinions at the end of the 20th century recorded in these proceedings.

Glenn personally invited many distinguished speakers but Death, the greatest teacher of all, intervened shortly before the program was completed and SHEs # 118, 116, ... were discovered in Glenn's lab (Gregorich, this volume). Death itself teaches that, "*The all-knowing Self was never born, nor will it die. Beyond cause and effect, this Self is eternal and immutable. When the body dies, the Self does not die.*" (Katha Upanishad, I.2.18).

Following this tragedy, Professor Geoffrey Burbidge (this volume) graciously agreed to give the keynote address. Accompanied by his lovely wife, Margaret, these two towering figures firmly linked the discussion to the period when stellar nucleosynthesis was conceived, in the middle of the 20th century, to the time of the symposium, as the 20th century was drawing to an end.

In keeping with Glenn's spirit, we adopted the following motto for the symposium: *"Satyam eva jayate nanritan"* (Mundaka Upanishad, III.1.6). *"Truth is victorious, never untruth."*

O. Manuel

# Acknowledgements

Special thanks go to

- **Glenn T. Seaborg** for planning and organizing the meeting and for agreeing to co-chair the symposium.
- **Karl-Ludwig Kratz, Mitsuru Ebihara,** and **Mayank Vahia** served with Glenn as members of the Scientific Program Committee.
- **Doris Demerjian, Al Ghiorso,** and **Ken Gregorich** - close friends and associates of Glenn Seaborg - kept the forward momentum going during the difficult period of his illness and death and helped us bring the symposium to a successful completion.
- **Peter Armbruster** chaired the special session of the symposium in memory of Glenn Seaborg.
- **Al Ghiorso** prepared the video about Glenn that was the highlight of the special session in his memory.
- **Ken Gregorich** told us about the culmination of Seaborg's career with the discovery of elements 118, 116 and 114 in his laboratory.
- **Geoffrey Burbidge** presented an eloquent and memorable keynote address on the synthesis of elements.
- **Margaret Burbidge** and **Truman Kohman** were present to share memories of pioneering work on the origin of the elements.
- **Carl Pilcher, Jack Larimer, Bin Li,** and **Greg Downing** gave important papers at the meeting but were unable to submit papers for publication in this proceedings.
- **Virginia Trimble** and **Akimasa Masuda** were unable to attend the meeting but contributed papers to this proceedings.

- **Kittie Robertson** and **Sherry Adams** handled daily planning and arranged for housing and the reception at the symposium.
- **Phyllis Johnson** typed and organized the proceedings. She was assisted by **Patty Chism** and **Ami Willett**.
- **Robert Johnson** helped with the symposium and the proceedings.
- **Paul K. Kuroda** and **The Foundation for Chemical Research, Inc.**, provided financial and moral support, as did **Bill Casey**, Chair of the Geochemistry Division of ACS and **Lester Morss**, Chair of the Nuclear Chemistry Division of ACS. **Madalyn A. Hardy** and others at ACS Headquarters helped with publicity.
- **Ouyang Ziyuan** of the Guiyang Institute of Geochemistry, **Li Xibin** of the Beijing Astronomical Observatory, and other scientists around the globe were unable to attend the symposium despite great effort.
- Special friends, former and current students gave generously to the success of this undertaking: **R. Ganapathy, Marvin Rowe, Tom Ward, Ron Thompson, Koh Sakamoto, Calvin Alexander, Greg Downing, Golden Hwaung, George K.-Y. Chiou, Jiafu Fang, Vic Becker, Robert Ballad, Tadashi Shimamura, Uli Ott, Li Bin, Jauh T. Lee, Ken Windler, Yixiang Xie, Cara Lietz,** and **Robert Johnson**.
- **Susan Safren**, served as editor for Kluwer Academic/Plenum Publishers.

# Contents

Contributors v

Preface and Dedication xi

Acknowledgements xv

Contents xvii

**PART I: GLENN SEABORG AND THE QUEST FOR SUPERHEAVY ELEMENTS** 1

Ghiorso Remembers Seaborg 3
ALBERT GHIORSO

Superheavy Elements at Berkeley: The Culmination of Seaborg's Career 21
K.E. GREGORICH AND V. NINOV

The Discovery of Superheavy Elements 107-112 and of the Deformed Shell at N = 162 35
P. ARMBRUSTER

**PART II: THE NUCLEAR REACTIONS THAT MADE OUR ELEMENTS** 49

The Decay of $^{19}$N 51
D. W. ANTHONY, D. J. MORRISSEY, P. A. LOFY, P. F. MANTICA, J. I. PRISCIANDARO, M. STEINER, J. M. D'AURIA, AND U. GIESEN

Measuring the Astrophysics Rate of the $^{21}$Na(p,$\gamma$)$^{22}$Mg Reaction 63
JOHN M. D'AURIA FOR THE DRAGON COLLABORATION

Production and β-Decay Half-Lives of Very N-Rich Nuclei 71
MONIQUE BERNAS

The Role of the N = 28 and N = 40 Closed Shells in the Production of the Neutron-Rich Ca-Ti-Cr-Fe-Ni Elements in the Universe 81
O. SORLIN

Experimental Studies Related to s-Process Abundances 93
K. WISSHAK, F. VOSS, C. ARLANDINI, AND F. KÄPPELER

Neutron Capture Cross Section Measurements for the Analysis of the s-Process 103
ROBERT S. RUNDBERG, J. L. ULLMANN, J. B. WILHELMY, M. M. FOWLER, R. C. HAIGHT, G. G. MILLER, P. D. PALMER, F. KÄPPELER AND P. KOEHLER

About the Reliability of Extrapolation of Nuclear Structure Data for r-Process Calculations 111
G. LHERSONNEAU, B. PFEIFFER, AND K.-L. KRATZ

The Astrophysical r-Process 119
K.-L. KRATZ, P. MÖLLER, B. PFEIFFER, AND F.-K. THIELEMANN

Nuclear Aspects of Stellar and Explosive Nucleosynthesis 143
THOMAS RAUSCHER, FRIEDRICH-KARL THIELEMANN, ROBERT D. HOFFMAN, AND STANFORD E. WOOSLEY

Proton Captures in the Atmosphere of Accreting Neutron Stars 153
H. SCHATZ, L. BILDSTEN, A. CUMMING, AND M. WIESCHER

**PART III: A COSMOLOGICAL VIEW OF NUCLEO-SYNTHESIS** 165

The Origin of the Elements 167
G. BURBIDGE

Chemical Evolution Tomorrow 175
VIRGINIA TRIMBLE

LiBeB Nucleosynthesis and Clues to the Chemical Evolution of the Universe 189
V. E. VIOLA

Measurement of the 44Ti Half-life and its Significance for Supernovae 203
I. AHMAD, J. P. GREENE, W. KUTSCHERA, AND M. PAUL

On the Half-Life of $^{44}$Ti in Young Supernova Remnants 211
ERIC B. NORMAN AND EDGARDO BROWNE

Abundances in SN 1987A and other Supernovae 217
ROGER A. CHEVALIER

The Birth of Planetary Systems Directly from Supernovae 225
WILBUR K. BROWN

Bipolar Outflows in Stellar Astrophysics 241
ADAM FRANK

**PART IV: NUCLIDES IN THE SUN** 251

Mini-blackhole at the Solar Center and Isotopic Abundances in the Primitive Solar Nebula 253
S. RAMADURAI

Abundances of the Elements in the Sun 261
N. GREVESSE AND A. J. SAUVAL

Isotopic Ratios: The Key to Elemental Abundances and Nuclear Reactions in the Sun 279
O. MANUEL

Critical Evaluation of CI Chondrites as the Solar System Standard of Elemental Abundances 289
M. EBIHARA, K. SHINOTSUKA, H. OZAKI, AND Y. OURA

Sensitivity of Solar Oscillation Frequencies to Element Abundances 301
JOYCE A. GUZIK AND CORINNE NEUFORGE

Inverse and Forward Helioseismology: Understanding the Interior Composition and Structure of the Present Sun 317
CARL A. ROUSE

Heterogeneous Accretion of the Sun and the Inner Planets 345
GOLDEN HWAUNG

Interstellar Matter, Sun, and the Solar System 351
M. N. VAHIA AND D. LAL

**PART V: NUCLIDES IN THE SUN'S PLANETARY SYSTEM 359**

Isotope Anomalies in Tellurium in Interstellar Diamonds 361
R. MAAS, R. D. LOSS, K. J. R. ROSMAN, J. R. DE LAETER, U. OTT, R.S. LEWIS, G. R. HUSS, E. ANDERS, AND G. W. LUGMAIR

Isotope Abundance Anomalies in Meteorites: Clues to Yields of Individual Nucleosynthesis Processes 369
ULRICH OTT

Variation of Molybdenum Isotopic Composition in Iron Meteorites 385
QI LU AND AKIMASA MASUDA

Iron Meteorites and Paradigm Shifts 401
E. CALVIN ALEXANDER, JR.

Chronology of Early Solar System Events: Dating with Short-lived Nuclides 407
JITENDRA NATH GOSWAMI

Xenology, FUN Anomalies and the Plutonium-244 Story 431
P. K. KURODA AND W. A. MYERS

Extinct $^{244}$Pu: Chronology of Early Solar System Formation 501
MARVIN W. ROWE

A Search for Natural Pu-244 in Deep-Sea Sediment: Progress Report 511
K. SAKAMOTO, Y. HASHIMOTO AND T. NAKANISHI

Strange Xenon Isotope Ratios in Jupiter 519
K. WINDLER

Abundances of Hydrogen and Helium Isotopes in Jupiter 529
ADAM NOLTE AND CARA LIETZ

The Possible Role of PeP Weak Interactions in the Early History of the Earth 545
THOMAS E. WARD

Ce-Nd-Sr Isotope Systematics of Eucrites and Lunar Rocks 555
MASAHARU TANIMIZU AND TSUYOSHI TANAKA

**PART VI: THE ORIGIN OF THE SOLAR SYSTEM** 573

Abundance of $^{182}$Hf and the Supernova Model of the Solar System 575
S. RAMADURAI

Binary Origin of Solar System 581
M. N. VAHIA

Origin of Elements in the Solar System 589
O. MANUEL

**Author Index** 645

PART I

# GLENN SEABORG AND THE QUEST FOR SUPERHEAVY ELEMENTS

# Ghiorso Remembers Seaborg

*The text on which the video of the same name was based*

Albert Ghiorso
*Lawrence Berkeley National Laboratory, University of California, Berkeley, CA 94720*
*a_ghiorso@lbl.gov*

## 1. INTRODUCTION

Glenn T. Seaborg was a real pioneer in the transuranium field. A striking way of demonstrating this fact is shown by a glance at the chart showing the trans-Pb region in the upper part of the chart of the nuclides. Those with a pink background are the only nuclides known in 1942, the highest atomic number being $^{238,239}Pu$ which had just been discovered by Seaborg and his colleagues. Most of the rest were discovered and characterized by Seaborg and his associates and students in the next five decades.

*W. Billig, Glenn Seaborg and Albert Ghiorso in front of the birthplace of Madam Curie*

In this presentation my objective will be to provide a short retrospective of Glenn's long and exemplary scientific career among the transuranium elements. This will also provide a background for the recent exciting events with regard to the superheavy elements which will be reported on later.

Seaborg, who died on 24 February, 1999 will be remembered not only for his many accomplishments in science but also for his participation in the associated political structures. He was truly a giant who enjoyed the adventures that were a hallmark of the scientific research of his time. I think that it is fair to say that he was essentially the first of the fraternity of nuclear chemists for he went well beyond being a radiochemist. He and his dozens of students used the tools of radiochemistry to do front line research in nuclear physics that the physicists of his day could not do. I was very fortunate in that, after he hired me in 1942, he also became my mentor. I learned how to do research under his guidance, intensively for the first few years, and then at a more modest pace for the next five decades. I owe him a profound debt of gratitude for introducing me to the wonderful world of exploration.

## 2. MANHATTAN PROJECT

Seaborg was a very controlled person who always "kept his eye on the ball" and seemed to have the knack of doing the right thing at the right time. He was an excellent judge of the varying abilities of his people to perform the important tasks that were assigned to them and this ability became crucially important when he headed a large crew of PhD chemists on the Manhattan Project at the University of Chicago. His job was the determination of a suitable chemical procedure for isolating plutonium from the highly irradiated uranium that would be produced in the huge Hanford neutron reactors. This was a tremendous assignment to be carried out for it involved an element that had yet to be seen and he followed the work in every detail. During the war most of the scientists worked about 60 hours/week (the necessary meetings were always held at night so they would not interfere with the work!) so to relieve the stress Glenn developed the habit of playing golf occasionally with a few associates at the nearby Jackson Park Golf Course; inevitably, of course, the work would be discussed too. Although his tortured swing was something to behold somehow he managed to get on the green in three strokes on the average and often beat his opponents with his short game (I was often among them).

## 3. AMERICIUM AND CURIUM

His first love was the expansion of the Periodic Table and as soon as he felt that the chemical separation procedures were under control he decided that the next step was to look for elements beyond plutonium. He had two options: one was to use the newly-built neutron reactors (piles they were called in those days) to add neutrons to the plutonium until it beta-decayed to a new element with the next atomic number, 95; the other was to bombard Pu with helium ions from the 60" cyclotron at the Crocker Laboratory in Berkeley and fly the target to Chicago for the chemical separations that would isolate a new element with the atomic number 96. He used both methods and eventually was successful in both endeavors. There were two big problems. One was that the instruments were relatively primitive at that time so that their sensitivity was severely limited. The other was that, although the chemistry of plutonium was well-known by this time, that for these new elements was completely unknown and little progress had been made after months of effort. Finally, Seaborg made a big breakthrough when he formulated his theory that these elements were part of a Rare-Earth-like series, a theory that required a major realignment of the periodic table. For many years this idea was fought by other chemists because of the seeming ambiguity of the properties of the early actinide elements; however, his Actinide Theory *was* successful and it was this insight that enabled the chemical identification of elements up through element 103, lawrencium. I was a member of the team that worked on these elements and can testify to the dedication that he applied to this research.

## 4. RETURN TO BERKELEY

He returned to Berkeley in 1946 to take up once again his career as a Professor of Chemistry at the University of California. He had to start from scratch and build up an entirely new laboratory that would enable him to continue his heavy-element research program. It took some time to accomplish this but an important difference was that now much of the work could be done with the help of graduate students. Some of the veteran researchers at Chicago, I among them, also migrated to Berkeley with Seaborg to provide the starting cadre for the new laboratory. These people would provide the know-how that would allow it to function properly with known techniques; it became their job to supervise the students and develop new research tools and methods. To reach higher in Z our only course at that time in the late 40's was to learn how to bombard the very radioactive targets, americium and curium, that would eventually become available. This was not a simple task and it took a lot of preparation to learn how to do it safely. Because it would be a few years before the neutron reactors could produce enough Am and Cm to make suitable targets it would be another decade before we could actually make an attempt to produce elements 97 and 98.

## 5. 184" SYNCHROCYCLOTRON

In the meantime, to practice our skills we explored what we could do with the 184" Cyclotron using its 200-MeV proton and 400-MeV helium ion beams. This was an excellent training ground for we immediately ran into a host of new alpha emitters in the form of a number of alpha-decaying collateral series starting with uranium and protactinium. These decay chains tested all of our skills in chemistry, physics, and alpha particle ion chamber spectroscopy and the analyses of our results forced us to learn about the basic systematics of alpha decay.

## 6. BERKELIUM AND CALIFORNIUM

This research was important in its own right but it was especially important in honing our ability to detect small amounts of alpha activity because by Christmas of 1949 Stanley Thompson and Kenneth Street were ready to bombard Am with helium ions from the 60" Cyclotron on the nearby campus to make element 97. The very first experiment bombarding milligram amounts of $^{241}$Am was successful and much to our surprise we observed what turned out to be the unique 3-peak alpha spectra from 4.6-h $^{243}$Bk. A couple of months later early in 1950 we repeated the feat, this time to make element 98, bombarding microgram amounts of $^{242}$Cm and observing the distinctive alpha particles from 44-m $^{245}$Cf. Of course, Seaborg was delighted and the chemists lost no time in showing that these new elements behaved like actinide elements as he had expected.

## 7. THE NOBEL PRIZE

This work in the new elements beyond plutonium apparently enhanced Seaborg's chances for he received the Nobel Prize in 1951. This opened many doors for him and thus to service outside the Laboratory. But although he became highly involved with matters outside he always remained close to the research inside the Laboratory, especially that dealing with new elements.

## 8. MIKE!

In November of 1952 a most extraordinary event occurred at Bikini Lagoon in the South Pacific Ocean that was completely unexpected. In a test (code-named Mike) of the first thermonuclear device that was the predecessor of the hydrogen bomb, not only was there a tremendous 12 megaton explosion that made the large lagoon disappear, there were also several exotic new nuclides generated. In particular, in their normal analyses a new and very heavy isotope, $^{244}$Pu, was discovered by the Los Alamos and the Argonne Laboratories in the fallout from the blast. The $^{238}$U in the device had absorbed many neutrons almost simultaneously to make ultra heavy isotopes of uranium which had then undergone beta decay to elements of higher atomic number. When we were made aware of this clue it was enough for our Laboratory to involve itself and search for heavier elements in the debris using the same tools that had just been successful in finding elements 97 and 98 in the cyclotron bombardments.

And we were successful once again! In December, element 99 was discovered, and then element 100 in March of 1953. Seaborg had been naturally skeptical at first for what seemed to be a foolhardy chase, but soon became highly involved in the work, and kept careful records of the research as it progressed through the many new isotopes of elements 97-100 that we uncovered.

Armed with this new information it was relatively easy for us to discover some of these same isotopes in reactor-neutron-bombarded plutonium within the next couple of years. Moreover, it was only a few years before it was possible to make enough twenty-day einsteinium-253 to make even its bombardment productive as far as new elements were concerned. In 1955 a classic experiment was performed to make element 101, mendelevium, by bombarding $10^9$ atoms of 20-d $^{253}$Es with an intense helium ion beam to make $^{256}$Md. For the first time, recoil from a nuclear reaction was used to separate the transmutation products from the very radioactive target. In addition, a new ion exchange technique was used to isolate element 101 and show that its spontaneous-fissioning daughter, $^{256}$Fm, grew into that fraction.

## 9. THE HILAC

In the meantime it had been clear to us that at some point the approach to heavier elements would have to be through the use of heavier ions than helium and we had been persuaded by Luis Alvarez to build a big linear accelerator for this purpose. A design team was put together which included physicists from Yale University and by 1957 twin HILACs were built at Berkeley and Yale that enabled bombardments with ions up to neon. There followed a long learning period both for the accelerator and for the new type of experiments that had to be performed with heavy ions. That same year marked the beginning by Seaborg of a long range reactor program to produce gram amounts of californium and milligram amounts of einsteinium, essential building blocks for the future. At this point he could only propose this program but four years later he was in a position to direct its actual accomplishment.

## 10. THE ATOMIC ENERGY COMMISSION

Seaborg participated in the first experiments that showed that element 102 had been produced in 1958 and 1959 but it was not long before he felt the call to higher office. In 1961 President John F. Kennedy asked him to be the Chairman of the U.S. Atomic Energy Commission and this would launch his career into the international arena. This position was a very responsible one, both for science and for world politics, and he diligently applied himself to it for ten years through three presidents, Kennedy, Johnson, and Nixon, one of the most far reaching actions that he espoused being the comprehensive test ban treaty. Though the job as the Chairman was a very demanding one he always had time to talk about the science that was dearest to his heart, the transuranium elements, and I found that I always had access to him and his assistants. He followed our progress at Berkeley closely as we produced the next three elements, 103, 104, and 105 by using various heavy ions from the HILAC. The difference now was that he was able to guide the Transplutonium Program that he had launched by letter through the budget pitfalls of reality. It was this program that also provided the necessary kilogram amounts of curium for the satellite thermoelectric power supplies. At that time there was no other way to provide the vital power needed for the exploration of our planetary system. The Program was carried through to the point where the heaviest elements became readily available for various kinds of chemical and physical research and today our knowledge of these rare elements is very extensive indeed. One of the principal goals of this program was to manufacture large amounts of $^{244}$Pu and then isotopically separate it from the other isotopes of plutonium for we felt that it would be valuable for the future production of superheavy elements. This program was very expensive, so much so that no other country undertook to match it.

Seaborg's job as Chairman was a strenuous one that called upon him to travel all over the world and undergo the interminable speeches, dinners, etc. that go with that sort of thing. He had learned to cope with that sort of regime I found out when he invited me to be one of a group of ten scientists that for two weeks toured the laboratories of the Soviet Union in 1963 under his leadership. I found the trip exhausting and I was amazed that, at the end when Glenn gathered us all together in the American Embassy to assign our individual tasks in writing our report, he had an excellent grasp of most of the details that would go into it, details that would run the gamut from the investigations of small university laboratories to the mega-outputs of huge atomic energy installations.

He often would communicate with me by letter to acquaint me with bits of news that he thought would be important to me. Not long after he took over the job as Chairman of the AEC he told me that the Argonne National

Laboratory had come up with a proposal to build a very large cyclotron to accelerate very heavy ions. It was a very interesting idea with a lot of merit, I thought, but the implied message that I received was "You have the HILAC. What are your plans beyond that?" It so happened that we had already been thinking of how we would increase the energy of the ions from the HILAC to several hundred MeV/A. Our goal was to produce heavy particles with enough range to penetrate into the human body so that Ernest Lawrence's brother John could perform cancer therapy on people. At first we thought of a dual-use project: the HILAC beam would be injected into a very big cyclotron to do the high energy part of this job and the cyclotron by itself would be available for low energy heavy ion research.

## 11. THE OMNITRON

Over a period of weeks we progressed through various unsatisfactory designs until finally in 1964 Bob Main, Bob Smith, and I in the space of a concentrated half-hour's discussion brainstormed the idea of an entirely new type of accelerator. We called it "Omnitron" because it could accelerate all of the elements to either low or high energies. It was far ahead of its time and was one of the world's first complicated accelerators. This machine would have accomplished its purpose by the use of two large concentric synchrotron rings of magnets in which the particles were accelerated and/or stored. The particles could be passed easily from one ring to the other so that a cyclic regime could be set up in which particles would be accelerated first in an easily available low charge state in one ring and then stripped to a higher charge state and reaccelerated in the other. Though expensive, this was a tour de force among accelerators and some twenty five years ahead of its time. Since the concepts were new we needed some R & D funds to prove that our design was workable and Seaborg soon made those available. By 1964 we had a preliminary proposal ready to submit to the AEC. [It is worth noting that it was found to be still an excellently designed accelerator and not at all outdated twenty five years later by J.M. Nitschke in his work on the ill-fated ISL project.]

## 12. THE IDEA OF SUPERHEAVY ELEMENTS AND THE SUPERHILAC

At the last minute I heard about the famous thesis of Bill Myers working for his doctorate under the guidance of Wladek Swiatecki in which he suggested that an island of superheavy elements (SHE) might exist in the region of 126 protons and 184 neutrons. I immediately recognized that the Omnitron would be an ideal accelerator to explore this possibility so I included this thought in our very first proposal as an insert. Seaborg was immediately entranced by the idea and became a very strong supporter of our proposed machine. Though the scientific need was great and the political support excellent in the AEC, it was never to be constructed because it ran afoul of the tragic war in Vietnam. There the U.S. was destroying that country with an expenditure of funds equivalent to three Omnitrons/day! In case that the new machine was not approved, our fallback position (at one tenth the cost) was to build a much improved version of the HILAC which we dubbed SuperHILAC. It was this machine that we used for our initial SHE research and subsequently for the discovery of element 106.

## 13. THE BEVALAC

In 1971, as a substitute for the Omnitron for the high energy biomedical program, I came up with the idea of injecting our SuperHILAC beams into the nearby Bevatron. Thus was created the first coupled accelerator and I immediately christened it BevaLAC. This happened just before Seaborg's tenure as Chairman ended so that he was able to make available funds to start this program. This action prolonged the life of the Bevatron for another twenty years by demonstrating that high energy cancer therapy was a viable option.

The BevaLac also turned out to be an extremely important machine for high energy physics. It inspired the use of one of the major accelerators of the world, the CERN SPS, to take very heavy ions up to the highest collision energies in hopes that the primordial quark-gluon plasma could be observed. This, in turn, led to the construction of a new large accelerator called RHIC (Relativistic Heavy Ion Collider) at Brookhaven National Laboratory to fill the "hole-in-the-ground" left by the ill-fated Isabella project. This innovative machine will soon enable U-U collisions at high energy.

## 14. RETURN TO BERKELEY

When he returned to Berkeley after his ten-year stint in the Washington "jungle" it would not have been surprising if Glenn had settled down into a less hectic existence but that was not to be. He immediately applied himself to putting into book form what he had learned in these early years (his Journals are a marvel of detail). He continued to be an important influence on the national scene, and was often called upon by the Presidents to serve the country in various ways, particularly in education. He also continued to be a strong advocate for nuclear power, a very unpopular position to take. But for him, the growth of the civilian nuclear power program was the most significant achievement of his AEC tenure and he never wavered from his position that such a program would be best for the world in the long run.

By 1974 the SuperHILAC had been debugged and we were ready to look for the Magic Island of the Superheavy Elements. Everyone was disappointed when the "sure-fire" experiments did not uncover anything and the experiments settled down for the long haul. It would take another twenty five years before the promised land was reached. In the meantime, Glenn would tirelessly include in his many scientific talks the prospects for finding this SHE mythical island.

## 15. ELEMENT 106 (SEABORGIUM)

Disappointed by our initial failures to find the SHE I decided that at least it should be possible to produce element 106 by bombarding $^{249}$Cf with $^{18}$O ions from the SuperHILAC using the newest version of the "VW" system which had been so successful in the detection of elements 104 and 105. The equipment needed for us to discover the heaviest elements had become more and more complicated, of necessity, because the bombardment yields decreased steadily as the atomic number increased. Fortunately, the development of solid-state detectors had now made it possible to design experiments which were marvels of sensitivity. The apparatus had the designation VW. This stood for Vertical Wheel, a descriptive term for an apparatus with unparalleled sensitivity at the time. It was the culmination of a line of instruments that identified alpha particle activities by their alpha energies and by the genetic relationships that they had to other alpha activities. The last version of these, constructed especially for element 106, was even able to demonstrate the presence of the great granddaughter of $^{263}$106 by using a triple recoil technique. With this instrument we were able to refine our research and characterize isotopes with great accuracy, even when there were interfering activities present. In 1974 this experiment was performed and it was successful in finding 0.9-s $^{263}$106.

Within two weeks of the conclusion of the Berkeley experiment the famous head of the Soviet team, G.N. Flerov, just happened to come to Berkeley on a long-planned visit. In the seminar that he gave to us he presented evidence for what he and his colleagues thought was an isotope of element 106 with mass 259 or 260 that decayed by spontaneous fission. He was very much surprised when we told him of our findings for a heavier isotope of the same element. We were confident of our identification and they were confident of theirs so the two teams decided to withhold naming the element until it could be determined whether one or both laboratories should have that honor. It took twenty years until 1994 when another Russian team found out that Flerov's work had been confused by the $^{256}104$ spontaneous-fissioning ***daughter*** by alpha decay of $^{260}106$. Most of the SF activity that they had observed was from element 104 rather than from 106!

## 16. THE NAME SEABORGIUM

My long time colleague, Darleane Hoffman tells the story of what happened next:

"Following the Russian publication which showed that their early work did not constitute a valid discovery of element 106 the Hoffman/Gregorich Berkeley team at the 88-Inch Cyclotron set out to try and confirm the original SuperHILAC identification of $^{263}106$. After an arduous set of bombardments we managed to do so using a completely different technique so now there was no impediment to naming the element. It is worth noting that the long hiatus made a lot of difference in what the name would be! If it had not been for the priority conflict with the Dubna laboratory in 1974 it is highly unlikely at that time that the discoverers would have named element 106 after Seaborg. But the passage of twenty-years had made a tremendous difference and now that option became a very good choice. Without Glenn's participation the name seaborgium was chosen by the co-discoverers. Ghiorso was the one who informed him of their decision and for that occasion he had prepared a special cover page to a folder which contained pages from the Log Book that carried the experimental data. Seaborg was completely overwhelmed and said that this was the greatest honor that he had ever received, including the Nobel Prize!

The name Seaborgium for Element 106 was endorsed by the ACS Committee on Nomenclature and by the Board of Directors shortly thereafter. After some intervening initial disparate recommendations by the IUPAC Commission on Nomenclature of Inorganic Chemistry, approval finally came at the meeting of the IUPAC Council in Geneva, Switzerland on August 30, 1997.

It is worth noting that Glenn received another out-of-this-world honor in that an asteroid has also received his name. Minor Planet Seaborg was discovered in 1983 by C. S. and E. M. Shoemaker at Palomar. The name was proposed in 1995 by Truman Kohman one of Seaborg's colleagues on the Chicago Manhattan Project."

## 17. SHIP AND COLD FUSION

The GSI laboratory had been developing a very expensive and very elaborate recoil velocity separator for heavy ion reactions which they called SHIP. This was a very different approach from our "hot fusion" approach to producing new elements and a group of young scientists headed by veteran researcher Peter Armbruster soon showed that they had a winner by using a "cold fusion" reaction to produce element 107.

This extremely important reaction was first pointed out by Yuri Oganessian of Dubna. It takes advantage of the low Q values afforded by the double closed-shell of neutrons and protons in $^{208}$Pb to achieve much lower excitation energies. When this isotope is used as a target with medium weight accelerated ions, transuranium elements are produced in nuclear fusion reactions wherein fewer neutrons have to be evaporated to get rid of the excitation energy. This reduces the competition from fission and leads to higher cross sections. In addition, another important advantage is that the well-defined recoil velocity of the fusion products is high and forward focussed and this means that, in principle, a recoil can be passed through a suitable beam separator with high efficiency and implanted directly into a remote detector. This device, usually a position-sensitive detecting crystal, monitors the energy and arrival time of a recoil implantation and, subsequently at the same position, the genetically-related chain of decays and times that follows. This was the principal motivation for building SHIP.

That instrument turned out to be very successful and in a succession of brilliant experiments the teams of Armbruster, Gottfried Munzenberg, and Sigurd Hofmann over a period of years discovered elements one-by-one with atomic numbers from 107-112. Their work is not a subject for this paper except to note that it seemed to show very clearly that the cross sections to produce these cold fusion reactions decreased by a factor of 3-5 for each increase in Z. Extrapolation of this trend meant that the prospects for detecting elements with atomic numbers above element 112 were very dismal.

## 18. GAS-FILLED SEPARATORS

During the period that the Omnitron was under development by a corps of some of the best accelerator physicists in the world that had been recruited by Bob Main, I followed the work closely and as a result came up with an excellent idea for a new kind of separator. I was struck by the fact that a fast moving ion of a heavy element that emerged from a foil into vacuum had a higher mean charge state than when it emerged into a region of low gas pressure and furthermore, the charge state attained depended on the atomic number of the ion. Might this be a way of separating a heavy ion recoil from the beam that produced it?

I was struck by the simplicity of the idea so I did a simple experiment at the exit of the HILAC in April, 1967 before that machine was converted into the SuperHILAC. I mounted a thin target of $^{165}$Ho at the entrance to the steering magnet at the end of the accelerator and bombarded it with a beam of $^{40}$Ar ions. An aluminum catcher foil was mounted downstream at the end of the magnet so that an alpha radioautograph could monitor how much the 40-MeV alpha-emitting astatine recoil products of the bombardment were bent relative to the Ar beam by a given magnetic field when the pressure was about 1 torr of helium. A good separation was obtained. A later experiment using 67-MeV $^{12}$C ions to bombard a gold target worked extremely well with the 4-MeV At recoils also being completely separated from the beam. In addition, it was noted that a sharp image of a collimator was obtained and this indicated that charge-exchange oscillations occurred very frequently along the recoil trajectories so that the magnetic rigidity was essentially constant.

## 19. SASSY AND SASSY2

Unfortunately, this promising technique was not adequately followed up until much later because the successful experiments on elements 104, 105, and 106 took precedence and they used our more conventional techniques. It was not until 1972 that our first steps were taken to make a gas-filled recoil separator that would be suitable for work with heavy ions. A device called SASSY (Small Angle Separator System) was constructed using available magnets from the Bevatron. It had some good ideas, such as the measurement of the Time-of-Flight of the recoils, but suffered from the fact that the discrimination at the focal plane from beam particles was insufficient at high beam levels. In addition, the only available Si detector array was not large enough to catch all of the fusion recoils.

Some years later another version of the idea called SASSY2 was constructed based on a design by Saburo Yashita, one of Seaborg's graduate students, and Juris Kalnins. It had a larger acceptance for the recoils and bent them through a much larger angle. By this time I had "retired" and so had no access to funds which would have allowed the instrument to be constructed in the normal way. Faced with this situation, I had no choice, if I wanted to pursue the heavy element research in this way. I would have to build it myself. I had often built parts of the instruments that I used but now I would have to build the major parts of SASSY2. For instance, the pole pieces would have to be milled in a complicated way to provide the necessary double focussing. This was a daunting task but my son, Bill, persuaded me that it could be done and offered to help me with both advice and labor. Bill was right and we did succeed in making a viable instrument that was used for one important experiment.

## 20. ELEMENT 110

With this device we were able to bombard $^{209}$Bi with $^{59}$Co ions in the very last experiment at the SuperHILAC to look for the isotope, $^{267}$110. In a very difficult 40-day period we did find one event which we have attributed to element 110 but I hasten to point out that it is only one event and must be confirmed. Seaborg, himself, took an active part in the re-analysis of the data which disclosed the candidate event. Unfortunately, this discovery came too late to save the SuperHILAC from the budget ax that had cut off the expensive BevaLAC. Glenn tried valiantly to persuade the DOE administration to let the SuperHILAC keep running by itself but he was thwarted by people whom he could not influence. Without the SuperHILAC our heavy element program was dead.

## 21. LASSY

Or was it? We still had the 88-Inch Cyclotron which had undergone continuous improvements with the advent of the ECR ion source. It is interesting to note that this, the first of the isochronous heavy ion cyclotrons, was put into the budget back in 1957 in the Sputnik era by Seaborg when he appealed directly to very-influential Congressman Price of Illinois. But though the 88-Inch is an excellent cyclotron, without a good recoil separator it was clear to us that it would not be useful in the new element field. Probably the most important accomplishment of the two SASSY devices was that they had demonstrated that gas-filled separators did work and all they needed was an infusion of funds so, with the encouragement of Seaborg in 1994 George Kalnins and I began to think of the logical successor to SASSY2. We soon came up with a device called LASSY (Large Aperture Separator System) which used superconducting magnets. However, it was considered too expensive by the Division and we had to resort to our fallback position, a separator that was twice as long and made use of existing magnets from the old Bevatron Pool. Ken Gregorich and Victor Ninov became actively involved in the new design with Kalnins and after a couple of years the new separator came into being as the BGS (Berkeley Gas-filled Separator). Although the BGS is not as good as LASSY would have been, it was a considerable improvement over anything else in the world and, after a short period of testing, early in 1999 it was ready to enter the lists as Berkeley's contribution to the search for SHE.

## 22. THE SUPERHEAVY ELEMENTS AT LAST!

The Hoffman, Ghiorso, Seaborg book (The Transuranium People: the Inside Story) was finished a couple of days before Glenn suffered his terrible stroke on 24 August, 1998 and we assumed that all that was left to do was the proof reading. Then came a fantastic double climax to the search for the superheavy elements, a search which had been going on intermittently for such a long time, a search that had been a prime goal for such a large number of scientists throughout the world, including Seaborg. From Dubna in January, 1999 came the alert that a collaboration between Russian and American scientists were reporting a single chain of events in the bombardment of $^{48}$Ca on $^{244}$Pu, one that they felt could only be attributed to element 114! Now we knew that we would have to write an epilogue for the book and cover this important development.

Walter Loveland of Oregon State College, who has collaborated with Glenn and me for many years in many heavy element projects tells the story of the Dubna discovery:

> "A first glance at the Dubna findings seemed to indicate that indeed they had made the long-awaited SHE discovery. A putative chain of three $\alpha$-emitters: 30-s 9.71-MeV $^{289}$114 decaying to 15-m 8.67-MeV $^{285}$112 decaying to 1.6-m 8.83-MeV $^{281}$110 decaying to 17-m spontaneous-fission emitting $^{277}$108, was a text book example of what everyone expected for SHE decay. But as this event was examined in detail more critically by a number of experts in the field some doubts began to arise, principally because of the non-zero background that existed in the experiment. It was not quite perfect and shouldn't perfection be required of something so important as the first of the mythical superheavy elements?
>
> The Dubna experiment was performed in collaboration with the Lawrence Livermore National Laboratory. As well as some detector and

target expertise, that laboratory had also furnished the $^{244}$Pu target material that was used in the experiment. This plutonium, shared by Berkeley and Livermore, had come from the same batch that Glenn had husbanded through the AEC while he was Chairman so in a very real sense he had participated in their discovery. The device used to isolate the recoil products was based on SASSY, the original Berkeley gas-filled separator, but modernized by the use of a superior high efficiency detector "box". The Dubna team, headed by Yuri Oganessian, had measured their overall yield to be about 40% so they felt that they had a good chance to see a SHE even if its production cross section was as small as a picobarn, $10^{-36}$ cm$^2$, but it meant that they would have to run for a very long time at a high beam current. This they proceeded to do for about 40 days to the end of 1998. In the middle of December the famous single event showed up. One event of anything is never very satisfactory so the Dubna team started another series of long runs early in June, 1999 to make another."

My colleague, Ken Gregorich, at the 88-Inch continues the story:

"The experiment was the same as that which tentatively had been scheduled to begin in Berkeley about eight months later. Although we had been scooped in this most important experiment we felt happy that at last the Magic Island seemed to have been found and redoubled our efforts to get our own experiment under way so that we could try and confirm their important finding. The BGS under construction for about a year, was nearing completion and was about ready for its first tests with beam. This follow-on to SASSY2 had a much larger bending angle (77° instead of 55°) to obtain higher momentum resolution and we were anxious to find out how well it would work. But it was not an easy job to commission the BGS since it was a completely new machine and many new mechanical, electrical, and computer-related functions had to be built and checked out. Special attention had to be paid to the solid-state detector system and, though not yet complete, the device was put into preliminary operation late in 1998 by Victor Ninov and myself. The use of Rare Earth targets and relatively light ions soon showed that its performance exceeded expectations and that it should be a useful tool in the superheavy element region.

Since we were not yet ready to bombard radioactive targets it was decided to use a Pb target first in order to make alpha radioactivity that would calibrate the BGS in the heavy element region; a preliminary bombardment with $^{51}$V ions soon showed that all was well."

Darleane Hoffman describes a fortunate intervention:

"At about this time Robert Smolańczuk brought forward a suggestion that he had first made at GSI the year before. Robert is a theoretical physicist who, working as a student under Adam Sobiczewski in Poland, had calculated out the detailed energetics in the superheavy element region. After getting his PhD he went to GSI and there he went one step further and worked out an empirical correlation of production cross sections *vs.* reactions for many isotopes. To everyone's surprise, he found that the cross sections did not decrease in a uniformly catastrophic manner as had been predicted by other calculations. In particular, his method suggested that bombardment of $^{208}$Pb by $^{86}$Kr might proceed at a rate of a few hundred picobarns, larger by a factor of more than a thousand compared to the normal expectations; unfortunately, because GSI was engaged in a major program of modifying both the UNILAC and SHIP there was no possibility of trying out the experiment there at that time. When he was invited to work at Berkeley as a postdoctoral fellow by Darleane Hoffman and be closely associated with experimentalists he jumped at the chance to try and convince this different group of people that they should change their set program and try out his new ideas.

It was not that easy at Berkeley, either, because the theoretical evaluation of how nuclear reactions proceed had stagnated and it was not yet ready for a fresh new voice to liberate their thinking. Somewhat grudgingly, and pushed by Hoffman and Ghiorso, the theoretical contingent admitted that the experiment was worth trying because it was a new idea and one never knows, "maybe we are due for a miracle". Fortunately, the BGS program of bombarding $^{244}$Pu with $^{48}$Ca would not be ready to begin for many months and thus the schedule was not yet constricted by priorities for the "best" experiment. It was finally decided that "Robert's" experiment should be given a chance. After all, it was argued, the efficiency of the BGS was so high that it would only take a few days to find out that his prediction was wrong! Besides, it would give us some useful experience with the new system."

Victor Ninov describes the consequences of the decision to proceed:

"In April the experiment started with lots of things going wrong as might be expected and it was a continuing fight for Ken Gregorich, Walter Loveland, and me to find and fix problems. Notable was the use of PPAC counters to differentiate the legitimate recoil particles from background. And another Si detector downstream from the focal plane was used to cancel particles that went completely through the focal plane detector to eliminate an annoying background. It was a good "dry run" for a while. But then there was an intriguing event, a series of six alpha particles, each with more than 10 MeV of energy that occurred in the short space of less than a minute, something that had not been seen in any of the previous runs! Could it be that this was the chain of alpha decays that had been predicted? And then later it occurred again! Now I devoted myself to analyzing the data ever more carefully to make sure that there was not some obscure "glitch" and I soon satisfied myself that the data from the experiment fit the predictions of Smolańczuk and that element 118 had been found! And of course, elements 116 and 114 to boot."

It was such a startling discovery that strenuous efforts were made to find out what had gone wrong. But there was nothing wrong that could be found by several other experienced experimenters, and a few days later after a new PPAC had been added to cover all of the focal plane and some other minor improvements had been made, a new run was started with a new set of targets. This time only one chain was observed, but it was an excellent one because it confirmed the previous ones by showing that the same alpha energies were produced. Altogether, four atoms of element 118 were produced and three of these were included in the paper that was soon prepared for publication. Needless to say, the startling news was an enormous surprise to the scientific circles throughout the world. Now there was no question, we had reached the shore of the mythical Magic Island. It was actually out there and it looked as if it should be possible to go even further!

My immediate reaction to this exciting event was sorrow that Glenn had missed it. He would have been immensely pleased that the superheavy element concept had been proved to be valid in Berkeley, the home of its birthplace. He would have enjoyed the miraculous way in which it was found and I am sure that his next reaction would have been to ask two questions: What are your plans for performing some kind of chemistry on the new elements? and How soon will you be looking for element 119 (eka-francium) and its daughters? It was this spirit of inquiry that endeared him to me and the world of science.

I will miss him!

# Superheavy Elements at Berkeley: The Culmination of Seaborg's Career

K.E. Gregorich and V. Ninov
*Lawrence Berkeley National Laboratory, MS-88, Berkeley, CA 94720*
*KEGregorich@lbl.gov*

**Abstract:** Through many years and many heavy element research projects, Seaborg always brought the authors' thoughts and discussions back to the superheavy elements (SHE). This guidance led to the construction of the Berkeley Gas-filled Separator (BGS) with its primary program being a search for superheavy elements. In April and May of 1999, the BGS was used to search for the production and decay of element 118 from the $^{86}Kr + ^{208}Pb$ reaction according to Smolańczuk's predictions of relatively large production rates. Three decay chains, each consisting of an implanted heavy ion, followed by a rapid (ms) succession of high-energy (> 10 MeV) alpha-particle decays were detected. These chains are consistent with the production and decay of element 118 with mass number 293. These results a) show experimental evidence for the existence of shell-stabilized SHE, b) provide experimental values for refinement of nuclear mass models in the SHE region, and, most importantly, c) present a "new" reaction pathway for the production of SHE.

Since the first predictions of the possible existence of long-lived superheavy element (SHE) isotopes, Glenn Seaborg was always very excited about the possibility of producing, identifying, and measuring the chemical properties of the superheavy elements. At that time in the early 1970s, the experimental trends in nuclear stability indicated that with increasing Z and N (beyond the heaviest known isotopes at the time with Z ~ 106 and N ~ 158) spontaneous fission (SF) would dominate the decay and result in ever-decreasing half-lives. However, according to the SHE predictions (Seaborg and Loveland, 1990), closed proton and neutron shells near Z = 114 and N = 184 should result in increased stability.

Traditionally, the best nuclear reactions for production of SHE with particle accelerators were thought to involve $^{48}Ca$ projectiles with actinide targets, and several attempts to produce SHE were made (Otto *et al.*, 1978; Armbruster *et al.*, 1985; Lougheed *et al.*, 1985) but ultimately proved to be unsuccessful. Recently, very exciting results from experiments with $^{48}$Ca projectiles and $^{238}$U, $^{242}$Pu, and $^{244}$Pu targets have been reported (Oganessian *et al.*, 1999a,b,c) with indications that element 114 isotopes may have been produced in the $^{48}$Ca bombardments of Pu targets.

Heavy element production rates in experiments (Ghiorso *et al.*, 1995; Hoffman, 1998; Lazarev, 1996) to synthesize elements 110-112 indicate that it will be difficult to produce elements beyond Z = 113 by using the "cold fusion" approach (Oganessian *et al.*, 1975) to bombard Pb or Bi target nuclei to produce heavy compound nuclei at low excitation energies. Standard extrapolations of these production cross sections indicate that large increases in beam currents would be needed to approach experimentally accessible production rates of at least one atom per month.

However, the recent predictions of Smolańczuk (1999a,b,c) indicate that the cold fusion of projectiles near $^{86}$Kr with $^{208}$Pb or $^{209}$Bi targets should lead to relatively large SHE production rates. According to his calculations, the $^{208}$Pb($^{86}$Kr,1n)$^{293}$118 reaction should proceed with a cross section of 670 pb. In addition, Smolańczuk predicted that these SHE isotopes would decay by a rapid series of high-energy α-particles. The predicted decay properties of $^{293}$118 and the resulting daughter activities are presented in Table I.

Based on these predictions, a search was made for SHE production in the interaction of $^{86}$Kr projectiles with $^{208}$Pb targets by using the recently completed Berkeley Gas-filled Separator (BGS) (Ninov and Gregorich, 1998) at the 88-Inch Cyclotron of the Lawrence Berkeley National Laboratory (LBNL). Results from these first experiments have been published by (Ninov *et al.*, 1999). A schematic of the BGS is presented in Figure 1. The $^{86}$Kr$^{19+}$ beam was produced in the Advanced Electron Cyclotron Resonance Source (Xie and Lyneis, 1996) and accelerated by the 88-Inch Cyclotron to an energy of 459 MeV (ΔE = 2.3 MeV FWHM).

During the experiments, the beam current averaged 0.3 particle micro-Amperes (2.0 × 10$^{12}$ ions/s). At the BGS, the beam first passed through a

*Table 1.* Predicted (Smolańczuk, 1999a,b) decay sequence from $^{293}$118.

| $^{A}Z_{N}$ | $Q_\alpha$(MeV) | $T_{1/2}$ |
|---|---|---|
| $^{293}118_{175}$ | 12.23 | 31 μs – 310 μs |
| $^{289}116_{173}$ | 11.37 | 960 μs – 9.6 ms |
| $^{285}114_{171}$ | 11.18 | 800 μs – 8.0 ms |
| $^{281}112_{169}$ | 11.00 | 610 μs – 6.1 ms |
| $^{277}110_{167}$ | 10.77 | 620 μs – 6.2 ms |
| $^{273}108_{165}$ | 9.69 | 120 ms – 1.2 s |
| $^{269}Sg_{163}$ | 8.35 | 8.0 min – 80 min |
| $^{265}Rf_{161}$ | SF | 41 min |

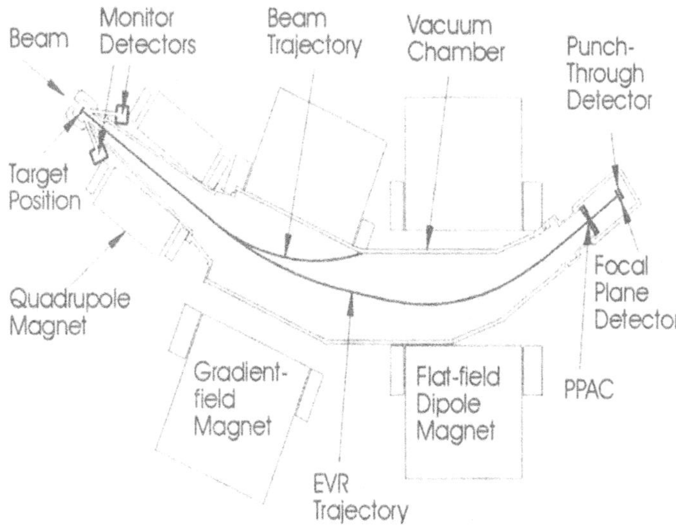


*Figure 1.* Schematic of the Berkeley Gas-filled Separator

0.1 mg/cm$^2$ carbon vacuum window before entering the 1.0 Torr mounted at the periphery of a 14-inch diameter rotating wheel positioned 0.5 cm downstream of the vacuum window. The targets consisted of 300-400 μg/cm$^2$ of $^{208}$Pb metal, evaporated onto the downstream side of 40 μg/cm$^2$ carbon foils. The Pb layer was covered by an additional 10 μg/cm$^2$ layer of carbon to prevent sputtering of the $^{208}$Pb and to aid in the blackbody radiation cooling of the targets. The $^{86}$Kr energy at the center of the $^{208}$Pb target material was 449 MeV (Hubert *et al.,* 1990). The compound nucleus evaporation residues (EVRs) recoiled out of the target at near the beam momentum ($E_{EVR}$ = 131 MeV). Upon passing through the He gas, the EVRs experience many charge-changing collisions, leading to a well-defined average charge state. This average charge state is nearly proportional to the EVR velocity, resulting in a large charge and velocity acceptance in the BGS. The EVRs then passed through a vertically-focusing quadrupole magnet, a 25° dipole magnet with a strong horizontally focusing field gradient and a 45° flat-field dipole magnet, before entering the detector chamber. Other nuclear reaction products, unreacted beam, and scattered beam particles take on a lower magnet rigidity in the He gas and are directed toward the beamstop along the upper edge of the vacuum chamber shown in Figure 1. The efficiency of the BGS for focusing element 118 EVRs onto the focal plane detector was estimated to be 75% by measuring the α-decay of Po EVRs from a $^{86}$Kr + $^{116}$Cd test reaction and comparing it with the expected Po production rates (Reisdorf, 1981). This efficiency agrees well with Monte Carlo simulations of ion trajectories through the separator.

Upon entering the detector chamber, the EVRs passed through a parallel plate avalanche counter (PPAC) (Swan *et al.*, 1994) where time and position were recorded. The Si-strip focal plane detector was located 30 cm downstream of the PPAC. This Si-strip detector was 80 mm wide × 35 mm high and was divided into 16 vertical strips. Along each of these strips, the vertical position was measured by resistive charge division. The energy and position resolution of the Si-strip detector were determined by using implanted recoil atoms from a $^{86}Kr$ + $^{116}Cd$ test reaction. The energy resolution for 5-9 MeV α particles was 30 keV FWHM, and the vertical position resolution for these EVR-α correlations was ± 0.58 mm. Directly behind the focal plane detector was a second Si-strip punch-through detector, used to reject particles passing through the focal plane detector.

Events registering signals in both the Si-strip focal plane detector and either the PPAC or punch-through detector were interpreted as reaction products passing through the separator, while events leaving only a signal in the Si-strip detector were assumed to be decays of nuclei previously implanted in the Si-strip detector. The focal plane detector had a calculated efficiency of 60% for detection of full-energy 12-MeV α-particles from nuclei implanted to a depth of 13 μm. The other 40% of the time, α particles escaped out the front of the detector, "escape alphas", depositing a partial energy ($1.5 \leq E(MeV) \leq 4.0$). For both the focal plane detector and the punch-through detector, the energy, position and time of all events were recorded. The Si strip detector events were recorded over two amplifier gain settings: a low-energy setting covering 0.5-15 MeV and a high-energy setting covering energies up to 200 MeV.

Two separate $^{86}Kr$ + $^{208}Pb$ experiments were carried out. In the first experiment, the overall rate of events registering greater than 0.5 MeV in the focal plane detector was 50 $s^{-1}$. In this experiment, a 5 cm × 5 cm Si-strip detector, which did not cover the entire focal plane detector, was used as the punch-through detector. The integrated beam flux for this experiment was $0.7 \times 10^{18}$ ions. For the second experiment with an integrated beam flux of $1.6 \times 10^{18}$ ions, an 8 cm × 3.5 cm Si-strip detector was used as the punch-through detector, which provided complete coverage of the focal plane detector. Improvements to the beamstop within the BGS reduced the overall count rate in the Si-strip focal plane detector to less than 20 $s^{-1}$. The low-energy spectrum recorded in the focal plane detector for the entire second experiment is presented in Figure 2. Figure 2(a) contains the ungated spectrum. Figure 2(b) shows the remaining events after applying a veto based on all events recorded in the PPAC. Figure 2(c) includes the additional sensitivity of applying a veto for all events registering in the punch-through detector, thereby giving a spectrum of events satisfying the gating conditions for "α-particles" or "escape alphas". The detection rate for "α-particles" with 4.0 $\leq$ E(MeV) $\leq$ 13.0 was 0.5 $s^{-1}$. Figure 2(d) shows the spectrum of all events from Figure 2(c) with 8.1 $\leq$ E(MeV) $\leq$ 13.0, which are correlated in position and time (within 1 s) with an implanted recoil. Note that the three events labeled with arrows are members of a single decay chain.

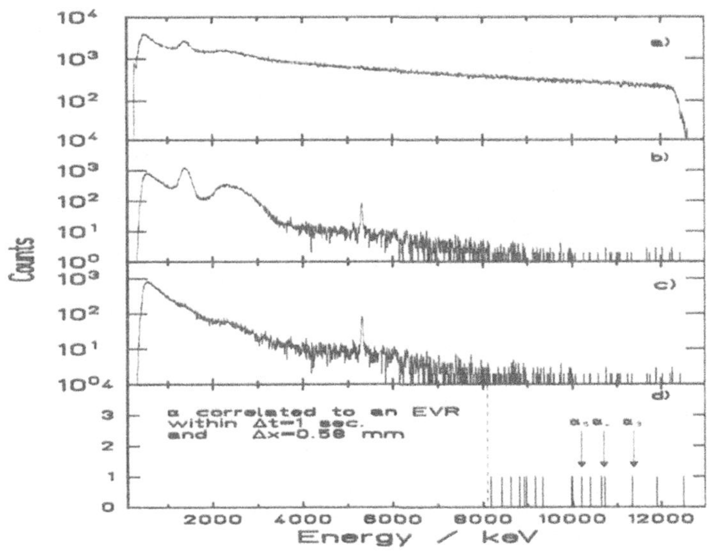


*Figure 2.* The low-energy spectrum in the focal plane detector for the entire second experiment. The gating conditions for the four panels are explained in the text.

Three decay chains were observed, each consisting of an implanted heavy ion recoil (with implantation energies consistent with that expected for an element 118 recoil) correlated in position and time with chains of 6 to 7 subsequent α-particle decays. This corresponds to a production cross section of $2.2^{+2.6}_{-0.8}$ pb. The observed correlation chains are shown in Figure 3, interpreted as the decay of $^{293}$118. The first two chains were observed in the first experiment, and the third was observed in the second experiment. For the third chain, we have indicated that the α-decay of the $^{293}$118 was not detected because it occurred during the 120 μs dead time of the acquisition system while recording the recoil implantation event.

An analysis of the positions of the members of each chain was carried out. The FWHM position resolution for EVR-α correlations was 0.58 mm. It was assumed that the position resolution for correlations between pairs of escape alphas is the same, and for pairs of full energy α-particles, it is half as large. From this, the 1-sigma position resolutions for the event types were calculated: $\sigma_{recoil}$ = 0.230 mm, $\sigma_{\alpha}$ = 0.087 mm, and $\sigma_{escape}$ = 0.174 mm. The calculated, and are presented in Figure 4.

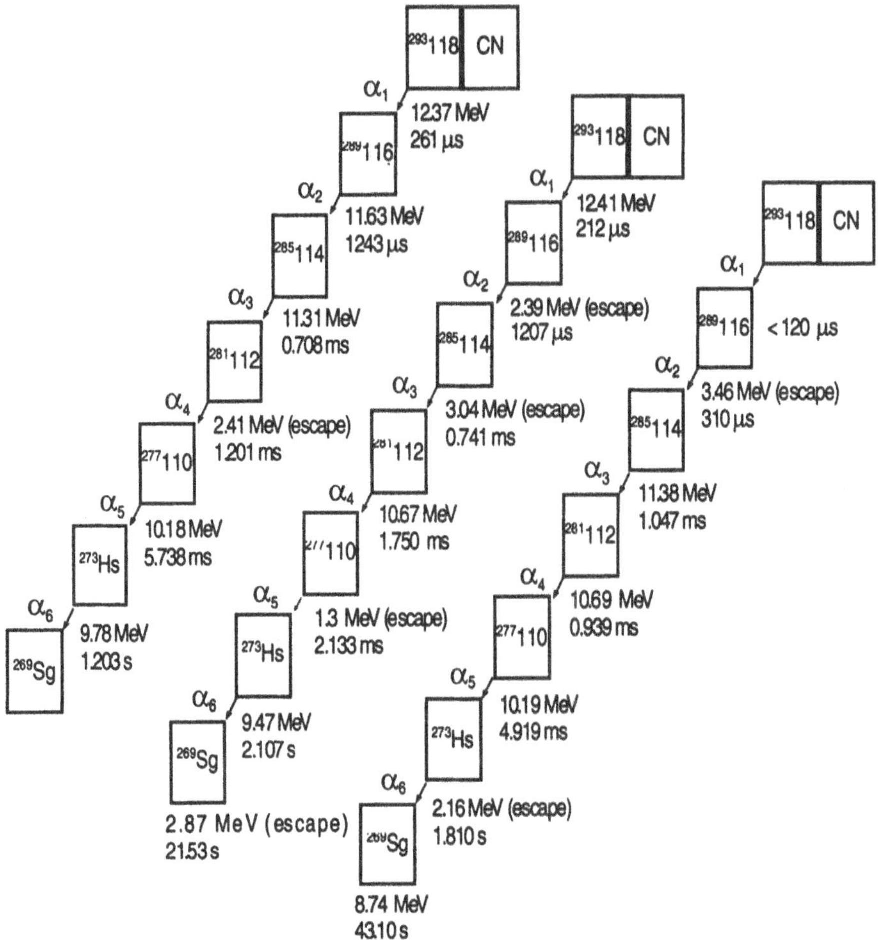


*Figure 3.* Observed decay chains from the reaction of 449-MeV $^{86}$Kr with $^{208}$Pb.

The large $X^2$ for the second chain weighted average and the reduced $X^2$ for the positions in each chain were results from the error in determining the positions near the bottom of the strip. (The positions were determined by comparing the pulse-heights from the top of the strips with the total pulse-height. At a 3.5 mm vertical position, the signal for escape alphas measured at the top of the strip is as small as 1% of the full scale values for the amplifier and analog to digital converter (ADC). In future experiments, position signals will be recorded from both the top and bottom of the detector strips to ensure accurate position measurements over the full vertical height of the strips.)
During the entire experiment, one additional chain was recorded, which could possibly be interpreted as being due to the implantation and decay of an atom of $^{293}$118. It occurred in strip 14 during the first experiment.

A recoil implantation event was recorded: followed 190 μs later by a 12.0 MeV α particle, then followed 936 μs later by a 11.4-MeV α particle, then followed at 1.125 ms, 1.740 ms, 1.503 ms, and 13.879 s later by escape alpha events with energies of 1.3, 3.1, 2.3, and 1.2 MeV, respectively. The vertical positions (in mm) in strip 14 of the implantation, the two α events, and the four escapes were: 18.3, 18.6, 18.5, 19.1, 19.4, 19.3, and 19.8. Since strip 14 was not backed by the punch-through detector in the first experiment, we cannot be sure of the validity of the escape alpha events. In fact, the positions of the escape events are significantly different from those for the recoil and the full energy alphas. We have, therefore, decided not to include this chain in any analysis of decay properties or cross section.

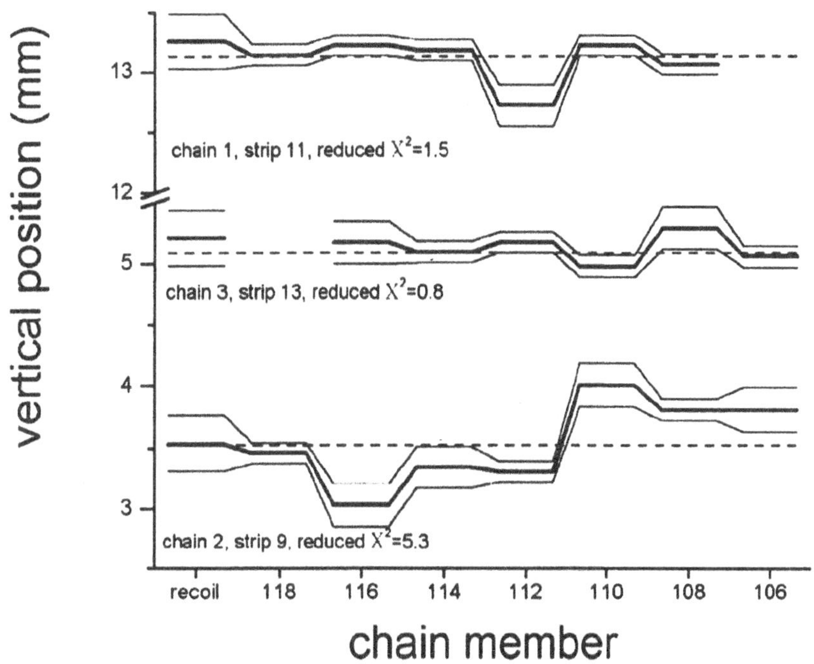


*Figure 4.* Analysis of the vertical positions of the three observed chains from the implantation and decay of $^{293}$118. The dashed lines are the weighted average positions of the chains. The heavy lines show the measured positions of the chain members, and the thin lines indicate the 1-σ limits on the measured positions.

Based on the sequences shown in Figure 3, the half-lives of the decay chain members were calculated by the method of Gregorich (1991). These half-lives are presented together with the calculated α-decay hindrance factors based on α-decay systematics for unhindered transitions (Hatsukawa *et al.,* 1990) in Table 2 showing that the observed α-transitions have hindrance factors in the range expected for odd-N nuclei.

*Table 2.* Measured α-particle energies, half-lives, and hindrance factors (Hatsukawa *et al.*, 1990) for decay chain members.

| $^{A}Z_{N}$ | $E_{\alpha}$(MeV) | Measured $T_{1/2}$ | H.F.[a] |
|---|---|---|---|
| $^{293}118_{175}$ | 12.39 | $120^{+180}_{-60}$ μs | 8.7 (4.3 – 14) |
| $^{289}116_{173}$ | 11.63 | $600^{+860}_{-300}$ μs | 3.6 (1.8 – 8.9) |
| $^{285}114_{171}$ | 11.34 | $580^{+870}_{-290}$ μs | 3.0 (1.5 – 7.4) |
| $^{281}112_{169}$ | 10.68 | $890^{+1300}_{-450}$ μs | 0.5 (0.2 – 1.2) |
| $^{277}110_{167}$ | 10.18 | $3.0^{+4.7}_{-1.5}$ ms | 0.4 (0.2 – 1.0) |
| $^{273}108_{165}$ | 9.78<br>9.47 | $1.2^{+1.7}_{-0.6}$ s | 110 (56 – 270)[b]<br>15 (8 – 37)[b] |
| $^{269}Sg_{163}$ | 8.74 | ~ 22 s[c] | ~ 4.2 |

[a]HF for the most probable half life with HF for lower and upper half-life limits in parentheses.
[b]Hindrance factors assuming each α-energy has 50% abundance.
[c]This half-life is approximate because maximum search time is poorly defined.

The observation of long decay chains of rapid, high-energy α-particles indicates the formation of SHE, as there are no known nuclei that could exhibit a similar decay pattern. The energies of the observed α-particles agree well with the predictions of Smolańczuk, supporting the proposed $^{293}$118 assignments for $^{A}$Z. However, it is possible that we have produced other SHE via different exit channels, for example, $^{208}$Pb($^{86}$Kr, 1α)$^{290}$116 or $^{208}$Pb($^{86}$Kr, 1p)$^{293}$117. Since the excitation energy of the compound nucleus is ~13 MeV (Smolańczuk, 1999a,b,c), emission of two particles (*i.e.*, 2n, αn, or pn, etc.) is energetically forbidden. Statistical model considerations suggest that the ratio $\Gamma_n/\Gamma_\alpha$ is proportional to $\exp\{-[S_n - (B_\alpha - Q_\alpha)]/T\}$, where $S_n$ is the neutron separation energy, $B_\alpha$ is the Coulomb barrier for α emission, $Q_\alpha$ is the energy released in removing an α particle from the nucleus, and T is the nuclear temperature. Substituting appropriate values for the binding energies (Smolańczuk, 1999a,b,c) and barriers (Parker *et al.*, 1991) gives $\Gamma_n/\Gamma_\alpha \sim 60$ and similarly, $\Gamma_n/\Gamma_p \sim 2000$, indicating that neutron emission is the most probable deexcitation path. In addition, extrapolation of the α-decay energies from known isotopes to those at the ends of the chains indicate that assignment of our observed chains to production of $^{290}$116 by the $^{208}$Pb($^{86}$Kr, 1α)$^{290}$116 reaction is extremely unlikely. This is demonstrated in Figure 5, where the observed α-decay energies of the chain members are plotted along with Smolańczuk's predicted decay energies under three assumptions: 1) that the head of the chain is $^{293}$118 from the $^{208}$Pb($^{86}$Kr, 1n) reaction, 2) that the head of the chain is $^{293}$117 from the $^{208}$Pb($^{86}$Kr, 1p) reaction, and 3) that the decay chain originates with $^{290}$116 from the $^{208}$Pb($^{86}$Kr, 1α) reaction. It should be noted that in this last case, the decay energy observed for the last member of the chain (which would be $^{266}104$) is as high as the experimental values for the more neutron-deficient assignments are not absolutely ruled out, we are confident in the assignment of these chains to the decay of $^{293}118$.

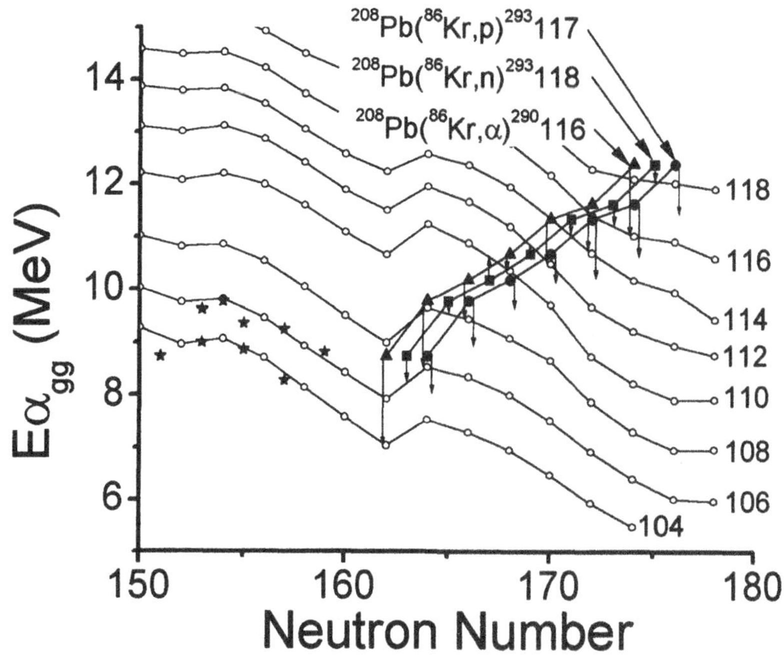


*Figure 5* Comparison of the our observed α-decay energies with those predicted by Smolańczuk (open circles) and three assumptions: 1) that the head of the chains is $^{293}118$ from the $^{208}Pb$ ($^{86}Kr$, 1n) reaction (squares); 2) that the head of the chain is $^{293}117$ from the $^{208}Pb$ ($^{86}Kr$, 1p) reaction (filled circles); and 3) that the decay chain originates with $^{290}116$ from the $^{208}Pb$ ($^{86}Kr$, 1α) reaction (triangles). Arrows from the our experimental points show the difference between measured and predicated α-decay energies. Previously measured values for α-decay energies are included as stars, indicating the accuracy of the predictions closer to the actinide region.

In Figure 6, we show the comparison of our measured α–particle energies with predictions of several modern nuclear mass models. The best agreement with our observations is obtained with Smolańczuk's prediction. The finite range droplet model (Möller *et al.,* 1995) and the Thomas-Fermi model (Myers and Swiatecki, 1996) are macroscopic-microscopic models, which both use Möller's shell effects. It appears that their predicted Z = 114 shell effect near N = 175 is too strong. The $Q_\alpha$ values from the SLy4 Skyrme-Hartree-Fock Bogoliubov method (Ćwiok *et al.,* 1999) show extra stability near Z = 116, which does not appear in the experimental data. The empirical mass model of Liran and Zeldes (Liran and Zeldes, 1976) may not be appropriate for extrapolation into this region.

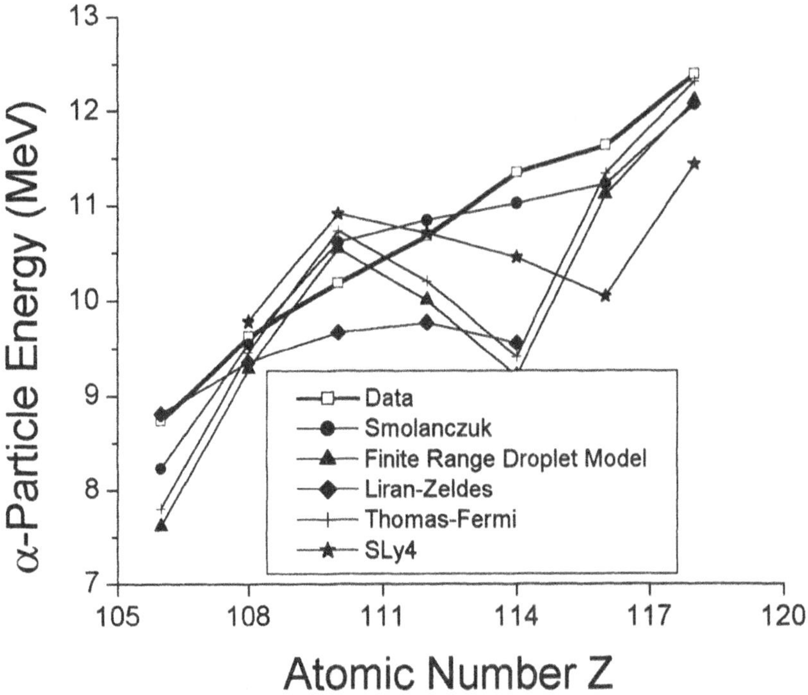


*Figure 6.* Comparison of the α-particle energies observed in this work with the predictions of various mass models for the N – Z = 57 nuclei.

We have presented evidence for the first synthesis of new superheavy elements ($^{293}$118 and its decay products $^{289}$116, $^{285}$114, $^{281}$112, $^{277}$110, $^{273}$Hs, and $^{269}$Sg). The measured α-decay energies can be used directly for refinement of nuclear mass and shell models. Our results show an unexpected viability of the cold fusion approach to the synthesis of superheavy nuclei by using projectiles heavier than $^{70}$Zn with targets near $^{208}$Pb. The 2.2 pb cross section, although more than 300 times smaller than that predicted by Smolańczuk, is orders of magnitude larger than expected from extrapolations of cold fusion reactions with lighter projectiles (Ghiorso *et al.*, 1995; Hoffman, 1998; Lazarev *et al.*, 1996). This relatively large cross section may be explained by the idea of "unshielded fusion" (Swiatecki, 1999) where, with heavier projectiles, the optimal bombarding energy for the 1n deexcitation channel is above the Coulomb barrier so that the Coulomb barrier is no longer the first thing stopping the fusion process.

## ACKNOWLEDGMENTS

We gratefully acknowledge the operations staff of the 88-Inch Cyclotron for providing intense, steady beams of $^{86}$Kr. We thank B. Lommel and W. Talheimer of Gesellschaft für Schwerionenforschung for providing the entrance windows and lead targets, N. Kurz and H. Essel for help in setting up the data acquisition system, and G. Münzenberg and S. Hoffman for their support. R. Smolańczuk graciously assisted with helpful discussions and continuous theoretical support. M. Steiner, J. Yurkon, and D. J. Morrissey at Michigan State University generously loaned us the PPACs. Financial support was provided by the Office of High Energy and Nuclear Physics, Nuclear Physics Division and the Office of Energy Research, Office of Basic Energy Sciences, Chemical Sciences Division of the U.S. Department of Energy under Contract No, DE-AC03-76SF0098 and Grant No., DE-FG06-88ER40402.

## REFERENCES

Armbruster, P., Leino, M., Lee, D., Lemmertz, P. Schmidt, K.H., Schadel, M., Yashita, S., Vonguten, H.R., Welch, R.D., Wilmarth, P,. Wirth, G., Reisdorf, W., Poppensieker, K., Trautmann, M., Vermeulen, D., Summerer, K., Seaborg, G.T., Schneider, W.F.W., Schneider, J.H.R., Agarwal, Y.K., Frink, C., Ghiorso, A., Gaggeler, H., Fowler, M.M., Brugger, M., Bruchle, W., Dufour, J.P., Dornhofer, H., Greulich, N., Gregorich, K.E., Daniels, W.R., Moody, K.J., Munzenberg, G., Herrmann, G., Hessberger, F.P., Hickmann, U., Hildebrand, N., Hoffman, D.C., Hofmann, S. and Kratz, J.V.: 1985, "Attempts to produce superheavy elements by fusion of Ca-48 with CM-248 in the bombarding energy-range of 4.5-5.2 MeV/U", *Phys. Rev. Lett.* **54**, 406-409.

Ćwiok, S., Nazarewicz, W. and Heenan, P.H.: 1999, "Structure of odd-n superheavy elements", *Phys. Rev. Lett.* **83**, 1108-1111.

Ghisorso, A., Lee, D., Somerville, L.P., Loveland, W., Nitschke, J.M., Ghiorso, W., Seaborg, G.T., Wilmarth, P., Leres, R., Wydler, A., Nurmia, M., Gregorich, K., Czerwinski, K., Gaylord, R., Hamilton, T., Hannink, N.J., Hoffman, D.C., Jarzynski, C., Kacher, C., Kadkhodayan, B., Kreek, S., Lane, M., Lyon, A., Mcmahan, M.A., Neu, M., Sikkeland, T., Swiatecki, W.J., Turler, A., Walton, J.T. and Yashita, S.: 1995, "Evidence for the possible synthesis of element-110 produced by the Co–9 + Bi–209 reaction", *Phys. Rev.* **C51**, R2293-R2297.

Gregorich, K.E.: 1991, "Maximum-likelihood decay curve fits by the simplex-method", *Nucl. Instrum. Methods Phys. Res. Section* **A302**, 135-142.

Hatsukawa, Y., Nakahara, H. and Hoffman, D.C.: 1990, "Systematics of alpha-decay half-lives", *Phys. Rev.* **C42**, 674-682.

Hofmann, S.: 1998, "Sputter depth profile analysis of interfaces", *Rep. Prog. Phys.* **61**, 827-888.

Hubert, F., Bimbot, R. and Gauvin, H.: 1990, "Range and stopping-power tables for 2.5-500 MeV nulleon heavy-ions in solids", *At. Data Nucl. Data Tables* **46**, 1-213.

Lazarev, Y., Lobanov, Y.V., Oganessian, Y.T., Utyonkov, V.K., Abdullin, F.S., Polyakov, A.N., Rigol, J., Shirokovsky, I.V., Tsyganov, Y.S., Iliev, S., Subbotin, V.G., Sukhov, A.M., Buklanov, G.V., Gikal, B.N., Kutner, V.B., Mezentsev, A.N., Subotic, K., Wild, J.F., Longheed, R.W. and Moody, K.J.: 1996, "Alpha-decay of (273)110 – Shell closure at N = 162", *Phys. Rev.* **C54**, 620-625.

Liran, S. and Zeldes, N.: 1976, "A semiempirical shell-model formula", *At. Data Nucl. Data Tables* **17**, 431-441.

Lougheed, R.W., Wild, J.F., Schadel, M., Seaborg, G.T., Landrum, J.H., Gregorich, K.E., Dougan, R.J., Dougan, A.D., Gaggeler, H., Hulet, E.K. and Moody, K.J.: 1985, "Search for superheavy elements using the Ca–48 + ES–254 reaction", *Phys. Rev.* **C32**, 1760-1763.

Möller, P., Nix, J.R., Myers, W.D. and Swiatecki, W.J.: 1995, "Nuclear ground-state masses and deformations", *At. Data Nucl. Data Tables* **59**, 185-381.

Myers, W.D. and Swiatecki, W.J.: 1996, "Nuclear properties according to the Thomas-Fermi model", *Nucl. Phys.* **A601**, 141-167.

Ninov, V. and Gregorich, K.E.: 1998, "The Berkley gas-filled separator", ENAM98, eds., Sherrill, B.M., Morrissey, D.J. and Davids, C.N., (AIP, Woodbury, NY,1999) pp. 704-707 and http://bgsmc01.lbl.gov/.

Ninov, V., Gregorich, K.E., Loveland, W., Ghiorso, A., Hoffman, D.C., Lee, D.M., Nitsche, H., Swiatecki, W.J., Kirbach, U.W., Laue, C.A., Adams, J.L., Patin, J.B., Shaughnessy, D.A., Strellis, D.A. and Wilk, P.A.: 1999, "Observation of superheavy nuclei produced in the reaction of $^{86}$Kr with $^{208}$Pb", *Phys. Rev. Lett.* **83**, 1104-1107.

Oganessian, Y.T., Iljinov, A.S., Demin, A.S. and Tretyakova, S.P.: 1975, "Experiments on the production of fermium neutron-deficient isotopes and new possibilities of synthesizing elements with Z > 100", *Nucl. Phys.* **A239**, 353-364.

Oganessian, Y.T., Yeremin, A.V., Gulbekian, G.G., Bogomolov, S.L., Chepigin, V.I., Gikal, B.N., Gorshkov, V.A., Itkis, M.G., Kabachenko, A.P., Kutner, V.B., Lavrentev, A.Y., Malyshev, O.N., Popeko, A.G., Rohac, J., Sagaidak, R.N., Hofmann, S., Munzenberg, G., Veselsky, M., Saro, S., Iwasa, N. and Morita, K.: 1999a, "Searching for new isotopes of element-112 by irradiation of U-238 with Ca-48", *Eur. Phys. J.* **A5**, 63-68.

Oganessian, Y.T., Yeremin, A.V., Popeko, A.G., Bogomolov, S.L., Buklanov, G.V., Chelnokov, M.L., Chepigin, V.I., Gikal, B.N., Gorshkov, V.A., Gulbekian, G.G., Itkis, M.G., Kabachenko, A.P., Lavrentev, A.Y., Malyshev, O.N., Rohac, J., Sagaidak, R.N., Hofmann, S., Saro, S., Giardian, G. and Morita, K.: 1999b, "Synthesis of nuclei of the superheavy element 114 in reactions induced by $^{48}$Ca", *Nature* **400**, 242-245.

Oganessian, Y.T., Utyonkov, V.K., Lobanov, Y.V., Abdullin, F.S., Polyakov, A.N., Shirokovsky, I.V., Tsyganov, Y.S., Gulbekian, G.G., Bogomolov, S.L., Gikal, A.N., Nezentsev, A.N., Iliev, S., Subbotin, V.G., Sukhov, A.M., Buklanov, G.V., Subotic, K., Itkis, M.G., Moody, K.J., Wild, J.F., Stoyer, N.J., Stoyer, M.A. and Lougheed, R.W.: 1999c, "Synthesis of superheavy nuclei in the $^{48}$Ca + $^{244}$Pu reaction", *Phys. Rev. Lett.* **83**, 3154-3157.

Otto, R.J., Morrissey, D.J., Lee, D., Ghiorso, A., Nitschke, J.M., Seaborg, G.T., Fowler, M.M. and Silva, R.J.: 1978, "A search for superheavy elements with half-lives between a few minutes and several hundred days, produced in the $^{48}$Ca + $^{248}$Cm reaction", *J. Inorg. Nucl. Chem.* **40**, 589-595.

Parker, W.E., Kaplan, M., Moses, D.J., Larana, G., Logan, D., Lacey, R., Alexander, J.M., Rizzo, D.M.D., Deyoung, P., Welberry, R.J. and Boger, J.T.: 1991, "Charged-particle evaporation from hot composite nuclei – evidence over a broad Z-range for distortions from cold nuclear profiles", *Phys. Rev.* **C44**, 774-795.

Reisdorf, W.: 1981, "Analysis of fissionability data at high-excitation energies. 1. The level density problem", *Z. Phys.* **A300**, 227-238.

Seaborg, G.T. and Loveland, W.: 1990, "The Elements Beyond Uranium", Wiley, New York, NY, 359 pp.

Smolańczuk, R.: 1999a, "Production mechanism of superheavy nuclei in cold-fusion reactions", *Phys. Rev.* **C59**, 2634-2639.

Smolańczuk, R.: 1999b, "Production of superheavy elements", *Phys. Rev.* **C60**, 021301-1 – 021301-3.

Smolańczuk, R.: 1999c, "Production of even-even superheavy nuclei in cold fusion reactions", *Phys. Rev. Letter* **83**, 4705-4708.

Swan, D., Yurkon, J. and Morrissey, D.J.: 1994, "A simple two-dimensional PPAC", *Nucl. Instrum. Methods Phys. Res. Sect.* **A348**, 314-317.

Swiatecki, W.J.: 1999, private communication.

Xie, X.Q. and Lyneis, C.M.: 1996, "Production of high charge state ions with the advanced electron cyclotron resonance ion source at LBNL", *Rev. Sci. Instrum.* **67**, 886-888.

# The Discovery of Superheavy Elements 107-112 and of the Deformed Shell at N = 162

P. Armbruster
*GSI-Darmstadt, Germany*
*p.armbruster@clri6a.gsi.de*

**Abstract:** The production of radioactive elements at GSI Darmstadt during the last 20 years is reviewed. The development of highly sensitive separation and identification methods lead to the synthesis of elements Z = 107-112. A single alpha-decay chain was shown to be sufficient to identify an element, as was demonstrated first for $^{266}$Mt in 1982. The new elements are stabilized by the nuclear structure of the nuclear ground state and its increased binding energy. They are the first superheavy elements. Not the instability of the SHE's, but their small production cross sections were found to be the limiting factor to proceed toward still higher elements. A one-step reaction mechanism is discussed for the 1n-reaction channel responsible for the cold fusion reactions observed.

## 1. OUR CHANCE TO ENTER THE FIELD

The element seaborgium (Z = 106) is the last element for which the chemistry is studied (Schädel *et al.*, 1997). It was discovered in 1974 at LBL-Berkeley by A. Ghiorso and collaborators. An intensive period of competition between LBL-Berkeley and JINR-Dubna ended with the discoveries of elements 102-106. It was at that time, when our group at GSI-Darmstadt was preparing to start into the field of heavy element synthesis. The shell-model introduced in 1948 shows that nuclei have a more refined structure than the fermion-liquids of neutrons and protons underlying the earlier liquid drop model of the nucleus. The idea of superheavy elements (SHE), stabilized by microscopic order of the many-body nucleonic system beyond the vanishing stability of the macroscopic nuclear liquid drops, was born in 1966. An extrapolation of shell structure beyond $^{208}$Pb gave the next doubly magic nucleus at N = 184 and Z = 114, 126. The shell stabilization energy was large, and high fission barriers protecting these nuclei against spontaneous fission were predicted. An island of long-lived isotopes was supposed to be separated from the known elements by a swamp of spontaneously fissioning isotopes, as uranium and thorium are separated from lead and bismuth by a region of less stable elements. A world wide effort to discover the SHE's started in many laboratories in the seventies. But already in the beginning of the eighties the enthusiasm cooled down. There was no doubt that SHE's were not found in nature, nor made in the laboratory. Their production cross sections were not in the 100mb-region, but smaller by at least 6 orders of magnitude, and their half-lives were shorter than the age of the solar system. Under these discouraging auspices for making SHE's, began the work of our group at Darmstadt, where exciting possibilities were going to be opened.

The rush for SHE's triggered the decision in 1969 to build a laboratory for heavy ion research, the GSI-Darmstadt. Christoph Schmelzer and his team had the chance to construct the UNILAC, Universal Linear Accelerator, the design of which was thoroughly prepared at Heidelberg in the sixties. The UNILAC became fully operational with the world-wide first U-beams for nuclear physics in 1976.

In the sixties we developed techniques to separate fission products making use of the recoil obtained in the nuclear reaction. "Lohengrin" was built at the ILL, Grenoble. This method, which is universal for all elements, is fast (1 $\mu$ sec), efficient ($\varepsilon$ = 50%), and selective, and it could also be applied to fusion reactions. Together with the Physics Institute of the University of Giessen, the two-stage velocity filter SHIP was built at GSI (Figure 1) and it was ready to start work in 1975 (Münzenberg *et al.*, 1979).

To use the recoil of a nuclear reaction for separation goes back to the work of O. Hahn and L. Meitner, who separated isotopes using the recoil of $\alpha$-decay already in 1909 (Hahn and Meitner, 1909). To apply this principle to fusion, and its kinematics of emitting all recoils in the beam direction at a well defined velocity, gave, as we know now, the experimental breakthrough towards a new period of heavy element research (Table 1). Velocity filters and gas-filled magnetic separators became the method, and it is now used world-wide in research of SHE's. SHIP had to be equipped with detector systems allowing the identification of single decaying atoms. Luminosities of 3 pb$^{-1}$ month$^{-1}$ are reached today. Chains of $\alpha$-decays and spontaneous fission of a single decaying nucleus allow one to identify unambiguously, via one atom, an element produced with a cross section of $10^{-36}$ cm$^2$ at a rate of 1 atom every 10 days.

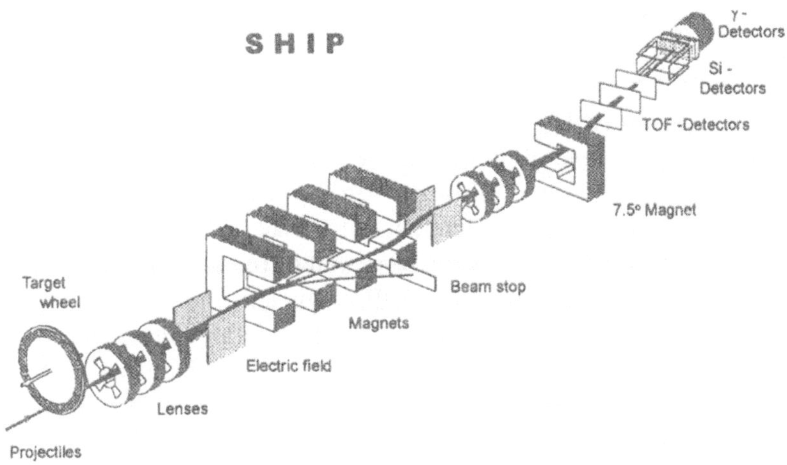


*Figure 1.* SHIP (Separator for Heavy Ion Reaction Products) a two stage, velocity filter to separate fusion products. Conservation of mass and momentum in the collision force fusion products to fly in the beam direction with a well-defined velocity. This allows their separation kinematically from the projectile beam by a $0^0$-velocity filter. The separated fusion products are implanted into position-sensitive Si-detectors and identified by a chain of correlated radioactive decays.

*Table 1.* The period of "Cold Fusion"

Six elements synthesized in 1n-reactions, $E^x$ = (10 - 16) MeV, using UNILAC. Irradiation of $^{208}$Pb and $^{209}$Bi with medium weight stable ions (Z ~ 24-30) and separation by SHIP

| Researchers | Year | Achievement |
|---|---|---|
| Y. Oganessian *et al.* | (1973) | "Cold fusion", $^{208}$Pb($^{40}$Ar, 2n)$^{246}$Fm |
| Ch. Schmelzer | (1976) | Heavy Ion Accelerator UNILAC |
| G. Münzenberg *et al.* | (1980) | 1n-channel established in $^{208}$Pb($^{50}$Ti, 1n)$^{257}$Rf |

| Atomic Number | Elements | Symbol | Research Group | Year |
|---|---|---|---|---|
| Z = 107 | Bohrium | (Bh) | G. Münzenberg *et al.* | (1981) |
| Z = 108 | Hassium | (Hs) | G. Münzenberg *et al.* | (1984) |
| Z = 109 | Meitnerium | (Mt) | G. Münzenberg *et al.* | (1982) |
| Z = 110 | unnamed | | S. Hofmann *et al.* | (1994) |
| Z = 111 | unnamed | | S. Hofmann *et al.* | (1994) |
| Z = 112 | unnamed | | S. Hofmann *et al.* | (1996) |

In 1973, Y. Oganessian and A. Demin found spontaneous fission activities from $^{246}$Fm at Dubna in an irradiation of $^{208}$Pb and $^{40}$Ar. Shortly afterwards, the production of isotopes of heavier elements using Pb targets was announced by Y. Oganessian: "Cold Fusion" was born (Oganessian,

1974). He was the first to propose that the reduction of excitation energy of the compound system, due to the large shell correction of $^{208}$Pb, might be responsible for the unexpected gift of a new method in heavy element research. We learned later, that the stability of $^{208}$Pb in deep penetration stages of the amalgamation process is the second decisive factor, changing the dynamics of fusion and allowing "Cold Fusion" to become the most successful method in SHE-production. The triad, UNILAC, SHIP and its detectors, and "Cold Fusion", is the base which finally allowed our team to discover the next 6 elements. The UNILAC is still the best accelerator to produce high beam intensities ($3 \times 10^{12}$ s$^{-1}$) at excellent beam quality. The SHIP is the best adapted separation scheme for "Cold Fusion" reactions, and its detector, approaching 100% efficiency for correlated chains, is the most efficient method to identify single new isotopes. $^{208}$Pb- and $^{209}$Bi-based "Cold Fusion" reactions decouple our work from the supply of actinide-targets not produced in non-nuclear weapons states. The enriched, expensive n-rich isotopes used as projectiles could be purchase on the free market in the U.S. or in Russia. The patient, persistent, and continuous effort of our group during a time of more than 20 years is another highly important factor, but the lucky triad at our beginning is certainly the main reason why a new-comer in the game could take the lead.

As we will discuss, our research lead to the discovery of a region of shell-stabilized deformed nuclei at Z = 108 (Armbruster, 1985; Münzenberg, 1988; Hofmann, 1998). Deformed SHE's were foreseen by A. Bohr already in 1974. The center of stability of the deformed region was since fixed theoretically at $^{270}$Hs (Ćwiok *et al.*, 1983; Leander *et al.*, 1984). Now the deformed shell at N = 162 could be established experimentally (Hofmann *et al.*, 1996).

## 2. THE FIRST SHE'S: Bh, Hs, Mt, 110, 111 and 112

The 6 elements discovered since 1981 at GSI (Table 1) are all well documented by a number of at least two complete chains (Münzenberg *et al.*, 1981, 1982, 1984, 1988; Hofmann *et al.*, 1995a,b, 1996). The birthday chains of each element are presented in Figure 2. All chains show correlated α-decays to known isotopes. There is no doubt of the isotope assigned to the new element, of its decay mode, and of its half-life. The unambiguous assignment demonstrates the experimental progress achieved since the previous period, which gave rise to so many fights and destruction of good will among scientists. To have introduced a technique, which really proves that a new element has been synthesized is, beside the pleasure of discovery, the most awarding achievement of our work.

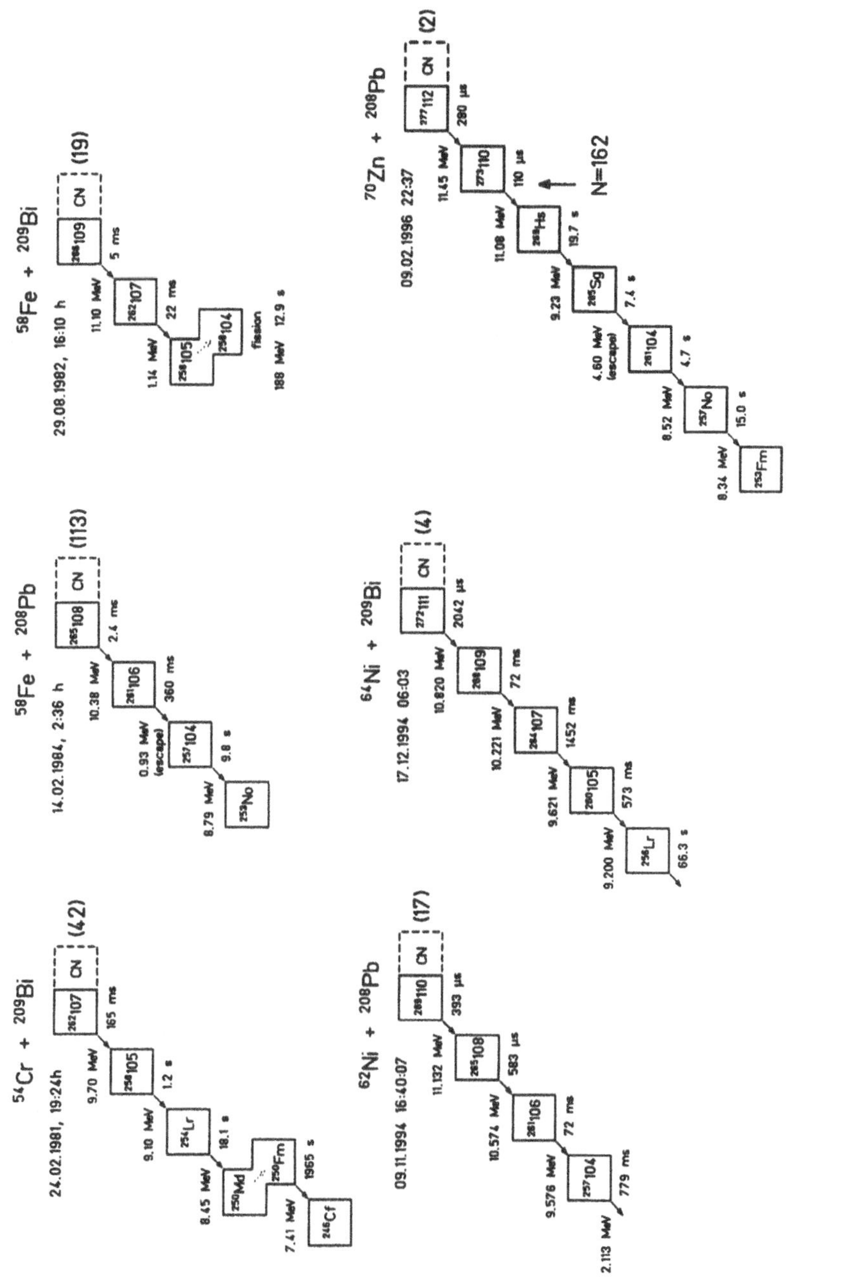


*Figure 2.* The birthday chains observed for the elements, bohrium, hassium, meitnerium, Z = 110, Z = 111 and Z = 112. The time of discovery and the reactions are indicated. The first box indicates the fused system, which decays by emission of one neutron to an isotope of the new element. The number in brackets gives the total of observed chains for the element. The measured decay energies and correlation time between the decays are given. All the new elements decay by α-decay, none by fission. This indicates high protecting fission barriers.

The correlation technique after implantation into an active Si-detector with its low random rates is a clock working from a few microseconds to several minutes. It covers 8 orders of magnitude of the time scale. Longer half-lives, as expected for the 3h-isotope $^{268}$Sg at the N = 162-shell, are a new challenge for experimentalists. This isotope could be reached by the reaction, $^{232}$Th ($^{48}$Ca, 4n)$^{276}$110, as the third member of the decay chain. The experiment proposed by Y. Oganessian is beyond the present possibilities of existing separator facilities, and it was performed without success recently at Dubna. Even though $^{268}$Sg, the longest lived isotope with N = 162 had not yet been identified, the predicted shell closure (Hofmann *et al.*, 1996) could be established in one of our recent experiments (Hofmann *et al.*, 1996). In the decay chain of $^{277}$112 (Figure 2) we observe an acceleration and subsequent slowing down of α-decay half-lives before and after passing the shell at N = 162. The first two members of the chain with N = 165 and N = 163 show short correlation times in the 0.1 ms-range; whereas the isotope $^{269}_{161}$Hs shows a correlation time of 20 s, which is longer by more than 5 orders of magnitude. This slowing down of half-lives at N = ($N_{shell}$ −1) is a well-known observation for N = 125-isotopes, where similar delays are observed passing $^{216}_{126}$Th. The $^{277}$112-chain is not only the longest chain ever observed, moreover, it allows one to fix a shell closure at N = 162. Having seen this shell closure at the neutron number predicted is a great success for both theory and experiment.

Analyzing the shell corrections to the binding energy of heavy nuclei, we compare to the smoothly varying liquid-drop barriers. The shell corrections of Möller and Nix (1988) and liquid-drop barriers are underlying the chart of nuclides shown in Figure 3. The outer contour-lines are estimated lines for fission barriers of 4 MeV, and for proton binding energies of −1.5 MeV corresponding to half-lives of about 1 μs, our experimental detection limit. There are different regions indicated in the figure. In region (1) the liquid-drop barrier has fallen below the zero-point energy ($B_f$ = 0.5 MeV) and shell corrections alone dominate. This is the SHE-region and there we find the isotopes of elements 107 to 112 that have been discovered. Going further down to the region (2) from Sg (Z = 106) and Fm (Z = 100), the shell correction energies become weaker, but the liquid drop barriers start to increase. Spontaneous fission becomes a dominant decay mode at Rf (Z = 104). Region (3) shows shell correction energies and liquid drop barriers of about equal height, starting with a ratio of 2 and ending with a ratio of 0.5 for these quantities. We find two sub-regions. First, at N = 126 there are strongly shell-stabilized spherical nuclei for elements above radium, which have small liquid drop fission barriers (Schmidt *et al.*, 1980).

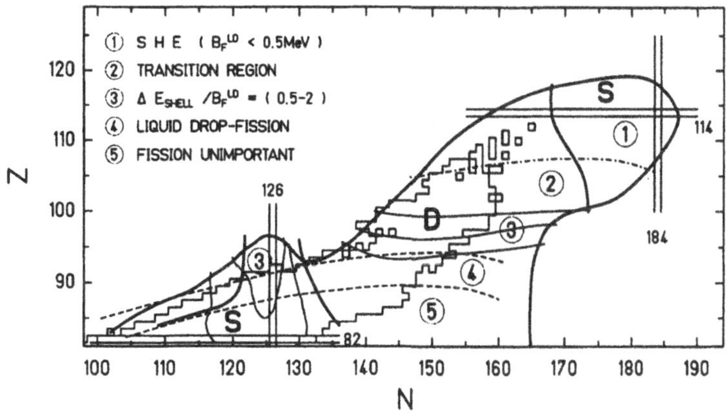


*Figure 3.* The region of shell-stabilized nuclei may be divided into five sub-regions presented by their liquid drop fission barriers: (1) The region of superheavy nuclei is defined by a liquid drop barrier smaller than the zero-point energy ($B_f^{LD}$ < 0.5 MeV). With the discovery of elements 107-112, this region was entered for the first time. (2) The hatched-dotted line at $B_f^{LD}$ = 0.5 MeV and the line $\Delta E_{shell}/B_f^{LD}$ = 2 define a region where liquid drop barriers gradually disappear, Z = 100-106. (3) The regions with the ratio of shell-correction energies to liquid drop fission barriers equal to 0.5; 1; 2 are indicated. (4) The protection against spontaneous fission is mainly given by the liquid drop barrier in the region between the two hatched lines. (5) Below the heavy hatched line ($B_f^{LD} = B_n$), nuclei are stable against fission. In the regions indicated by S, nuclei are spherical. In the region D, they are deformed.

These nuclei are the best approximation to the elusive spherical superheavy nuclei at N = 184. Second, between Es (Z = 99) and Pu (Z = 94) we find the well-studied region of deformed nuclei and fission isomers characterized by the interplay of shell correction energies and liquid drop barriers. In this region, the most complex nuclear structure is expected. The region (4) is dominated by liquid drop barriers (Z = 89-93). Below the line of equal neutron binding energy and liquid drop barriers (Z < 88), region (5), fission is not important for the ground state properties of nuclei.

We have determined fission barriers and barrier curvatures for a sequence of heavy nuclides with N - Z = 48 running up to $^{264}$Hs (Armbruster, 1984; Patyk *et al.*, 1989). The analysis of the data corroborates the new elements to be purely shell-stabilized and in this sense to be the first SHE's. In agreement with calculations we showed that the heaviest nuclei have a single, high and narrow fission barrier. *E.g.*, for $^{260}$Sg calculations predict a barrier exit point configuration close to the 2:1 superdeformation with a deformation of only about 3 fm, as compact as the second minimum in fission isomers. We expect these barrier exit points as well for the deformed nuclei around $^{270}$Hs, as for the spherical nuclei around N = 184. We are confident that theory will correctly predict the spherical SHE's. But the experimental confirmation of a shell at N = 184 is still in the future. We may expect about 12 SHE's beyond Sg and up to Z = 118, with about as many isotopes as the number of stable isotopes. At GSI we discovered six SHE's with 13 isotopes since 1981, the rest is elusive. It is not the ground-state properties of SHE's, but the vanishing stability of shell corrections to heating and the dynamics of the production mechanism that prevents us from penetrating deeper into the region of purely shell-stabilized nuclei.

## 3. PRODUCTION CROSS SECTIONS – THE LIMITING FACTOR

Going to higher atomic numbers of synthesized heavy elements, the production cross sections decrease with the steadily growing, repelling Coulomb-forces between the collision partners. Each additional atomic number is paid by a penalty, a decrease in cross section by a factor of 3.2. Figure 4 shows the 1 n-cross sections for fusion of $^{208}$Pb/$^{209}$Bi and projectiles between $^{48}$Ca and $^{70}$Zn giving the elements nobelium to Z = 112. There are small variations from the generally observed decrease by gain factors, in case the neutron number of the projectile allows an increase in the isospin, or the projectile has a closed-shell configuration. A penalty is paid replacing $^{208}$Pb by $^{269}$Bi as a target atom. The larger cross section of $\sigma = 15$ pb for $^{208}$Pb($^{64}$Ni, n)$^{271}$110, compared to $\sigma = 8$ pb for $^{209}$Bi ($^{58}$Fe, n) $^{266}$Mt, is the only case of an inversion observed. The colliding partners and their nuclear structure, together with the atomic number of the synthesized element, determine the cross section. The nuclear structure of the fused element does not help to increase the observed cross sections. This is a most amazing observation, which cools the optimism that shell-corrections to the binding energy of the end product might increase not only the stability of SHE's, but also their production yields. The 4n- and 5n-reactions using $^{238}$U-targets and lighter projectiles between $^{22}$Ne and $^{34}$S (See Figure 4) show the same slope to smaller cross sections. To reach the same element, there is no advantage using actinide-targets up to Z = 112. For Z = 110, the gap in favor of 1n-reactions is 40. For element 112, we reached production rates of 1 detected atom a week at a cross section of 1 pb. In the range of excitation energies of (10-15) MeV, there is a narrow window of 5 MeV width in the projectile energy, which allows one to make the elements presented in Figure 2 by the emission of one neutron. De-excitation by two neutron- or $\gamma$-emission was not observed. The reaction mechanism is sketched in Figure 5 (Armbruster, 1989; Armbruster, 1999a). The length scale chosen for the distance between the center of mass of the partners is in units of the radius of the fused nucleus.

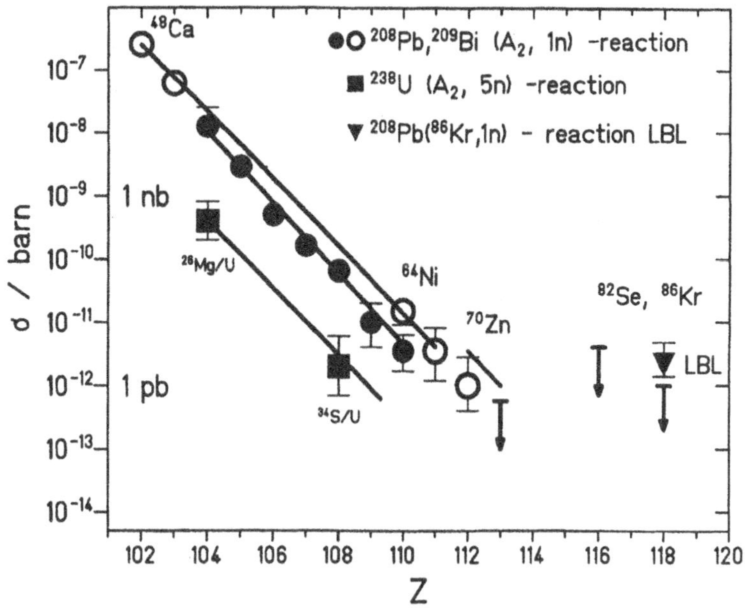


*Figure 4.* The production cross sections for $^{208}$Pb- and $^{209}$Bi- based 1 n-reactions from GSI (cold fusion) in comparison to actinide-based reactions from Dubna (hot fusion). The cross section of an experiment aiming for Z = 118 (Ninov *et al.*, 1999) is given together with lower limits for Z = 113, Z = 116 and Z = 118 from latest experiments at GSI (Hofmann, 1999).

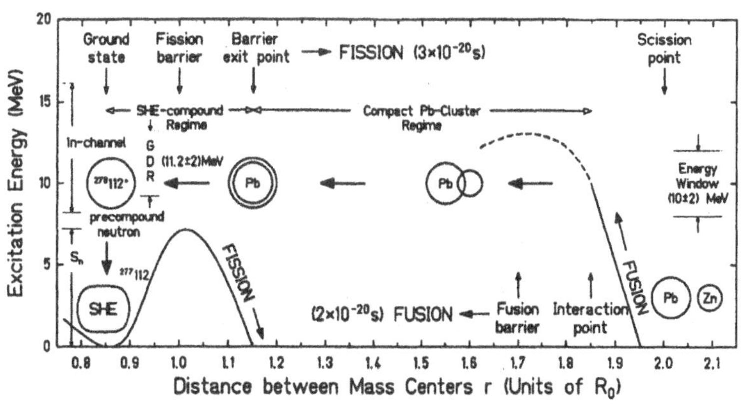


*Figure 5.* Stages corresponding to the fusion path toward element 112 for the $^{70}$Zn/$^{208}$Pb collision system. The excitation energy of the collision system above the ground state of the final superheavy element is presented as a function of the distance between the center-of-mass values of the two collision partners in units of the radius of the final product. The difference between the interaction distance of the two partners and the center-of-mass distance of the two halves of the final product is 6 fm, equal to the radius of the latter. Only $10^{-10}$ of the collisions succeed to pass the 6 fm and to cool down to the ground state of the SHE.

A very rare ($10^{-12}$ of the total cross section), cold (~ 12 MeV), one-step rearrangement process exploiting the microscopic order in all stages of the process gives us the SHE's. Besides the nuclear structure of closed shell collision partners cooling the process, and the superheavy end product stabilized by a shell closure, it is important that the doubly magic $^{208}$Pb nucleus still maintains its cluster-structure in deep penetration stages and amalgamates with the nucleons of the projectile in a late stage, close to the fission barrier exit point. Before the equilibration of a compound system is reached, the system is cooled by emission of a neutron (8 MeV). It is then protected by the fission barrier of the nascent SHE toward immediate (2 x $10^{-20}$ s) reseparation. The probability to reach the compact shape of a weakly excited fused system at $r \leq R_0$ for the system $^{208}$Pb/$^{70}$Zn is 3 x $10^{-6}$, and the probability to emit the neutron in the good moment is 3 x $10^{-5}$. With a reduced cross section for central collisions of 7 mb, a cross section for the formation of element 112 of 1 pb follows. Thus, 2 x $10^{18}$ atoms of $^{70}$Zn (0.1 mg) mixing with about the same number of atoms of $^{208}$Pb (0.5 mg) produce 1 atom of element 112 during the $10^{-4}$ s passage time through the target, in case their velocity (0.11c) was well defined within 1%.

## 4. PERSPECTIVES

We should never forget, while hunting for higher atomic numbers, the trivial truth that the number of elements is finite. N. Bohr stated the world ofnuclear liquid drops would be restricted to about 100 elements. The stability gained by increasing the falling liquid drop fission barriers by microscopic nuclear structure gave us 6 elements more, and seaborgium was reached and considered the end in the seventies. The pure stabilization by nuclear structure on an unstable liquid drop potential energy surface, the idea of SHE's, gave us for sure until now another 6 elements, and experiments going up to Z = 118 were published recently (Ninov *et al.*, 1999). Whether we will succeed to synthesize the SHE's with spherical nuclei has to be assured. We have to envisage staying with fusion measurements cross sections well below the pb-limit, or a new reaction must be found, as indicated with the recent "Cold Fusion" of Z = 118. Extrapolating a next doubly closed-shell, spherical nucleus following the known series of n-rich spherical nuclei of $^{48}$Ca, $^{78}$Ni, $^{132}$Sn and $^{208}$Pb for some seems to be the most exciting. For me, the unexpected discovery of a strong deformed shell at N = 162, gives as much excitement. A phenomenon was established in the region of the heaviest nuclei that was never seen anywhere with such a strength before. To study this nuclear structure effect thoroughly is as important as to reach still higher elements. At least three even-even nuclei, on the shell of N = 162, are accessible $^{272}110$, $^{270}Hs$ and $^{268}Sg$. It is a challenge to the field to measure their binding energies by finding more α-decay bridges through the region of spontaneously fissioning nuclei of nobelium and rutherfordium. The decay chain of $^{276}112$ in the reaction $^{70}Zn(^{207}Pb, n)^{276}112$ with an estimated cross section close to the values of the reactions $^{70}Zn(^{209}Bi, n)^{278}113$ leading to the next element, would fix the strength of the N = 162-shell, if also for $^{260}Rf$, as was found for $^{256}Rf$ (Hessberger, 1985), a small α-decay branch would be detected. There is interesting physics to be done for many years on the elements up to $Z \leq 114$ and $N \leq 170$ (Armbruster, 1999b), even if running for spherical superheavy nuclei in these years is accompanied by general applause (Oganessian *et al.*, 1999a,b,c). We will learn how many of these experiments will stand the checks of the next century. The program at GSI now under the responsibility of Sigurd Hofmann will explore also in the next years the until now so successful "Cold Fusion" reactions, eventually up to the challenge of Z = 118 (Hofmann, 1999).

## REFERENCES


Armbruster, P.: 1984, "Discovery of element 108. The island of α-active nuclei beyond 105", *Proc. Int. School of Phys., "Enrico Fermi"*, Course **91**; 1986, North-Holland **91**, 222-240.

Armbruster, P.: 1985, "On the production of heavy-elements by cold fusion – The elements 106 to 109", *Ann. Rev. Nucl. Part. Sci.* **35**, 135-194.

Armbruster, P.: 1989, "Neutron-rich clusters and the dynamics of fission and fusion", *J. Phys. Soc. Jpn.* **58**, 232-248.

Armbruster, P.: 1999a, "Nuclear-structure in cold rearrangement processes in fission and fusion", *Rep. On Progress in Physics* **62**, 465-525.

Armbruster, P.: 1999b, "The exponential break-down of production cross-sections: Self-termination of heavy element making at Z = 115", GSI Annual Report 1998, *GSI Report-1-99*, 8-10.

Ćwiok, S., Dudek, J., Nazarewicz, W. and Pashkevich, W.: 1983, "Fission-barriers of transfermium elements", *Nucl. Phys.* **A410**, 254-270.

Hahn, O. and Meitner, L.: 1909, "Eine neue Methode zur Herstellung radioaktiver Zerfalls-produkte; Th D, ein kurzlebiges Produkt des Thoriums", *Verh. D. Phys. Ges.* **11**, 55-62.

Hessberger, F.P., Schtt, H.J., Reisdorf, W., Thuma, B., Vermeulen, D., Schmidt, K.H., Armbruster, P., Hofmann, S., Hingmann, R. and Munzenberg, G.: 1985, "Study of evaporation residues produced in reactions of Pb-207, Pb-208 with Ti-50", *Z. Phys.* **A321**, 317-327.

Hofmann, S., Ninov, V., Hessberger, F.P., Armbruster, P., Floger, H., Munzenberg, G., Schott, H.J., Popeko, A.G., Yeremin, A.V., Andreyev, A.N., Janik, R., Leino, M.: 1995a, "Production and decay of (269)110", *Z. Phys.* **A350**, 277-280.

Hofmann, S., Ninov, V., Hessberger, F.P., Armbruster, P., Folger, H., Munzenberg, G., Schott, H.J., Popeko, A.G., Yeremin, A.V., Saro, S., Janik, R. and Leino, M.: 1996, "The new element 112", *Z. Phys.* **A354**, 292-230.

Hofmann, S., Ninov, V., Hessberger, F.P., Armbruster, P., Folger, H., Munzenberg, G., Schott, H.J., Popeko, A.G., Yeremin, A.V., Andreyev, A.N., Saro, S., Janik, R., and Leino, M.: 1995b, "The new element-111", *Z. Phys.* **A350**, 281-282.

Hofmann, S.: 1998, "New element – Approaching Z = 114", *Rep. On Progress in Physics* **61**, 639-689.

Hofmann, S.: 1999, "Studies of superheavy elements – status and prospects", *Experimental Nuclear Physics in Europe (ENPE 99): Facing the next millenium,* Proceedings of the Conference in Sevilla, Spain, June 1999, eds., Rubio, B., Lozano, M. and Gelletly, W., *AIP Conf. Proc.* **495**, pp. 137-144.

Möller, P. and Nix, J.R.: 1988, "Nuclear-masses from a unified macroscopic microscopic model", *At. Data Nucl. Data Tables* **39**, 213-223.

Münzenberg, G., Armbruster, P., Folger, H., Hessberger, F.P., Hingmann, R., Hofmann, S., Keller, J., Leino, M.E., Poppensieker, K., Reisdorf, W., Schmidt, K.H. and Schott, H.J.: 1984, "The identification of element-108", *Z. Phys.* **A317**, 235-236.

Münzenberg, G., Armbruster, P., Hessberger, F.P., Hofmann, S., Reisdorf, W., Sahm, C.C., Schmidt, K.H. , Schneider, J.H.R. and Thuma, B.: 1981, "Identification of element 107 by alpha-correlation chains", *Z. Phys.* **A300**, 107-108.

Münzenberg, G., Armbruster, P., Hessberger, F.P., Poppensieker, K., Reisdorf, W., Sahm, C.C., Schmidt, K.H., Schneider, J.H.R. and Schneider, W.F.W.: 1982, "Observation of one correlated alpha-decay in the reaction Fe-58 on Bi-209", *Z. Phys.* **A309**, 89-90.

Münzenberg, G., Faust, W., Hofmann, S., Armbruster, P., Guettner, K. and Ewald, H.: 1979, "The velocity filter SHIP, a separator of unslowed heavy ion fusion products", *Nucl. Instr. Meth.* **161**, 65-82.

Münzenberg, G., Hofmann, S., Hessberger, F.P., Folger, H., Ninov, V., Poppensieker, K., Quint, A.B., Residorf, W., Schott, H.J., Summerer, K., Armbruster, P., Leino, M.E., Achermann, D., Gollerthan, U., Hanelt, E., Morawek, W., Fujita, Y., Schwab, T. and Turler, A.: 1988, "New results on element-109", *Z. Phys* **A330**, 435-436.

Münzenberg, G.: 1988, "Recent advances in the discovery of trans-uranium elements", *Rep. On Progress in Physics* **51**, 57-104.

Ninov, V., Gregorich, K.E., Loveland, W., Ghiorso, A., Hoffman, D.C., Lee, D.M., Nitsche, H., Swiatecki, W.J., Kirbach, U.W., Laue, C.A., Adams, J.L., Patin, J.B., Shaughnessy, D.A., Strellis, D.A. and Wilk, P.A.: 1999, "Observation of superheavy nuclei produced in the reaction of $^{86}$Kr with $^{208}$Pb", *Phys. Rev. Lett.* **83**, 1104-1107.

Oganessian, Y.T., Yeremin, A.V., Gulbekian, G.G., Bogomolov, S.L., Chepigin, V.I., Gikal, B.N., Gorshkov, V.A., Itkis, M.G., Kabachenko, A.P., Kutner, V.B., Lavrentev, A.Y., Malyshev, O.N., Popeko, A.G., Rohac, J., Sagaidak, R.N., Hofmann, S., Munzenberg, G., Veselsky, M., Saro, S., Iwasa, N. and Moritak, K.: 1999a, "Searching for new isotopes of element-112 by irradiation of U-238 with Ca-48", *Eur. Phys. J.* **A5**, 63-68.

Oganessian, Y.T., Utyonkov, V.K., Lobanov, Y.V., Abdullin, F.S., Polyakov, A.N., Shirokovsky, I.V., Tsyganov, Y.S., Gulbekian, G.G., Bogomolov, S.L., Gikal, B.N., Mezentsev, A.N., Iliev, S., Subbotin, V.G., Sukhov, A.M., Buklanov, G.V., Subotic, K., Itkis, M.G., Moody, K.J., Wild, J.F., Stoyer, N.J., Stoyer, M.A. and Lougheed, R.W.: 1999b, "Synthesis of superheavy nuclei in the $^{48}$Ca + $^{244}$Pu reaction", *Phys. Rev. Lett.* **83**, 3154-3157.

Oganessian, Y.T., Yeremin, A.V., Popeko, A.G., Bogomolov, S.L., Buklonov, G.V., Chelnokov, M.L., Chepigin, V.I., Gikal, B.N., Gorshkov, V.A., Gulbeklan, G.G., Itkis, M.G., Kabachenko, A.P., Lovrentev, A.Y., Malyshev, O.N., Rohac, J., Sagaldak, R.N., Hofmann, S., Saro, S., Giardinas, G. and Morita, K.: 1999c, "Synthesis of nuclei of the superheavy element 114 in reactions induced by $^{48}$Ca", *Nature* **400**, 242-245.

Oganessian, Y.T.: 1974, "Fission and fusion induced by heavy ions", in *Lecture Notes in Physics* **33**, 221-252.

Patyk, Z., Sobiczewski, A., Armbruster, P. and Schmidt, K.H.: 1989, "Shell effects in the properties of the heaviest nuclei", *Nucl. Phys.* **A491**, 267-280.

Schädel, M., Brüchle, W., Dressler, R., Eichler, B., Gäggeler, H.W., Gunther, R., Gregorich, K.E., Hoffman, D.C., Hübner, S., Jost, D.T., Kratz, J.V., Paulus, W., Schumann, D., Timokhin, S., Trautmann, N. and Türler, A.: 1997, "Chemical-properties of element-106 (Seaborgium)", *Nature* **388**, 55-57.

Schmidt, K.H., Faust, W., Münzenberg, G., Reisdorf, W., Clerc, H.-G., Vermeulen, D. and Lang, W.: 1980, Proc. Symp. Chem of Fission, Jülich 1979, IAEA Vienna, vol. **I**, 409-420.

PART II

# THE NUCLEAR REACTIONS THAT MADE OUR ELEMENTS

# The Decay of $^{19}N$

D. W. Anthony[1], D. J. Morrissey[1], P. A. Lofy[1], P. F. Mantica[1], J. I. Prisciandaro[1], M. Steiner[1], J. M. D'Auria[2], and U. Giesen[3]
*[1]National Superconducting Cyclotron Laboratory, 164 S. Shaw Lane, East Lansing, Michigan, 48824, U.S.A., [2]Department of Chemistry, Simon Fraser University, 8888 University Dr., Burnaby, British Columbia, V5A 1S6, Canada, [3]Department of Physics, University of Notre Dame, 225 Nieuwland Science Hall, Notre Dame, IN, 46556, U.S.A.*
*anthony@nscl.msu.edu*

**Abstract:** An intense secondary radioactive ion beam of $^{19}N$ was produced at the National Superconducting Cyclotron Laboratory and quantitative measurements of the decay properties of this nuclide were made. A spectroscopic measurement of delayed neutrons emitted during the decay has yielded a total delayed neutron branching ratio of 51 (11) %. Gamma-rays observed suggest a total beta feeding strength of 44 (10) % to bound excited states of $^{19}O$. The half-life of $^{19}N$ has been re-measured to be 299 (19) milliseconds.

## 1. INTRODUCTION

Recent studies of inhomogeneous models (IMs) of the Big Bang have suggested that mass production above A = 20 depends on the decay and reaction properties of the nuclides with mass numbers between 15 and 18 (Meissner *et al.*, 1996; Applegate *et al.*, 1987; Rauscher *et al.*, 1994). For example, mass build up through the $^{18}O$ seed nuclide will be enhanced if the $^{18}O(n,\gamma)^{19}O$ rate is higher than the $^{18}O(p,\alpha)^{15}N$ rate (Rauscher *et al.*, 1994). It is therefore important to make quantitative measurements of these reaction rates in order to refine the network calculations, not only for these particular IMs, but for this mass region in general as it may be important to other astrophysical processes (Käppeler *et al.*, 1990; Raiteri *et al.*, 1991).

A recent experiment performed at the National Superconducting Cyclotron Laboratory (NSCL) was designed to provide insight into the resonant part of the $^{18}O$(n,() reaction by making a quantitative and complete measurement of the decay properties of the $^{19}N$ nuclide. The beta-delayed neutron emission probability and the spin and parity for the populated levels of $^{19}O$ were measured. From these spin and parity assignments, and the energy at which the resonance occurs above the ($^{18}O$ + n) threshold, the resonant part of the reaction strength can be calculated (Rolfs and Rodney, 1988).

The $^{19}N$ beta-decay Q-value is about 12.5 MeV to the ground state of the $^{19}O$ daughter and the energy available for delayed neutron emission is about 8.5 MeV (Tilley *et al.*, 1995). There are eight excited levels above the $^{19}O$ ground state that are bound with respect to neutron emission and more than 40 known excited levels of $^{19}O$ that are both above the neutron emission threshold and below the ground state energy of $^{19}N$. The spin and parity are known for about twenty of these levels (Tilley *et al.*, 1995).

Prior measurements of the $^{19}N$ decay properties have produced disparate results. Table (1) shows the sum of the experimental data available in the literature pertinent to the decay of $^{19}N$.

There have been five prior measurements of the decay half-life, however the measurements with the better statistical accuracy do not agree to within two standard deviations (Reeder *et al.*, 1991; Samuel *et al.*, 1988; Mueller *et al.*, 1988; and Dufour *et al.*, 1988, 1986). There have been three prior measurements of the total neutron emission probability (Reeder *et al.*, 1991; Mueller *et al.*, 1988; and Reeder *et al.*, 1994) but no spectroscopic measurement of the neutrons emitted following the beta decay. The only spectroscopic information for $^{19}N$ delayed neutrons was published by Ozawa and collaborators (Ozawa *et al.*, 1995) where $^{19}N$ appeared as the beta daughter in an experiment designed to measure the decay properties of $^{19}C$. In that study, one peak in their neutron spectrum was assigned to $^{19}N$ on the basis of the variation of the peak's intensity as a function of time.

*Table 1.* Summary of published experimental data pertinent to the decay of $^{19}N$.

| QUANTITY | VALUE | REFERENCE |
|---|---|---|
| $Q_\beta$ (MeV) | 12.54 | Tilley *et al.*, 1995 |
| Half-Life (milliseconds) | 329 (19) | Reeder *et al.*, 1991 |
| | 235 (32) | Samuel *et al.*, 1988 |
| | 210 (+200) (-100) | Mueller *et al.*, 1988 |
| | 300 (80) | Dufour *et al.*, 1988 |
| | 320 (100) | Dufour *et al.*, 1986 |
| Delayed Neutron Emission (%) | 48.7 (21) | Reeder *et al.*, 1994 |
| | 62.4 (26) | Reeder *et al.*, 1991 |
| | 33 (+34) (-11) | Mueller *et al.*, 1988 |
| Decay γ-Rays (keV) | 96, 709, 3138 | Dufour *et al.*, 1986 |

In addition to the half-life and total neutron emission probability, an experiment by Dufour and collaborators found three gamma-ray transitions characteristic of bound state transitions in $^{19}O$ (Dufour *et al.*, 1986). They ascribed these gamma-rays to the beta decay of $^{19}N$. However, due to the low statistics of the measurement, no specific beta branching strengths to individual levels in $^{19}O$ were reported.

Due to the shortcomings of the published experimental decay data on $^{19}N$ and the insight a quantitative measurement of its decay properties would provide into the astrophysically important $^{18}O(n,\gamma)$ reaction rate, the decay of $^{19}N$ was re-measured at the NSCL. We report here the techniques used for this study and present the preliminary experimental results.

## 2. RESULTS AND DISCUSSION

A secondary beam of $^{19}N$ was produced at the NSCL by fragmenting a $^{22}Ne$ primary beam (80MeV/A) in a Be target. The product of interest was separated from other reaction products using the A1200 fragment separator and transported to a low-background experimental end-station. The yield of $^{19}N$ at the focal plane of the A1200 was about 400 nuclei per second, per particle nano-amp of primary beam. This production rate yielded between 3000 and 5000 $^{19}N$ nuclei per second at the end-station.

To measure effectively the decay properties of $^{19}N$, the end-station consisted of four, loosely independent detection systems. The first, designated the implantation system (Harkewicz, 1991), is shown in Figure 1. The purified $^{19}N$ radioactive beam entered the implantation system from the left in Figure 1, passed through a thin surface barrier silicon detector (Delta E), and slowed down by a series of Aluminium foils prior to implantation at the approximate mid-point of a plastic scintillator designated the "Implantation" or "START" detector. In addition to these, there was a second silicon "VETO" detector situated immediately downstream of the START detector.

The Delta E detector provided useful online data regarding the purity and yield of $^{19}N$ as it entered the end-station, while the DeltaE in conjunction with the VETO detector provided information on the composition and yield of the secondary beam that was actually deposited within the START scintillator. For this experiment, the secondary beam was delivered to the experimental end-station in a pulsed fashion.

The implantation system was placed at the geometric center of the second detector array known as the Neutron Bar Array (Harkewicz, 1991) (Figure 2). The neutron bar array consists of sixteen plastic scintillators cast to a constant radius of curvature such that the distance is one meter from the Implantation or START detector to any point on the array. Spectroscopic measurements of neutrons emitted following beta-decay are made using a time-of-flight (TOF) technique. The initial beta-decay is detected by the implantation scintillator, giving a start signal, while the subsequent neutron emission detected in the neutron bar array provides a stop signal. The time difference, taken together with the known flight distance, yields the energy of the emitted neutron.

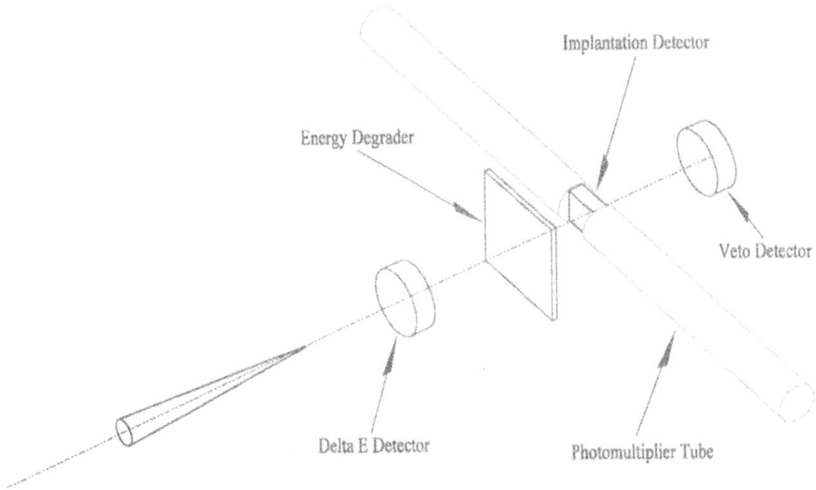


*Figure 1.* The implantation system showing the diagnostic silicon detectors and the START scintillator. The secondary beam enters from the left.

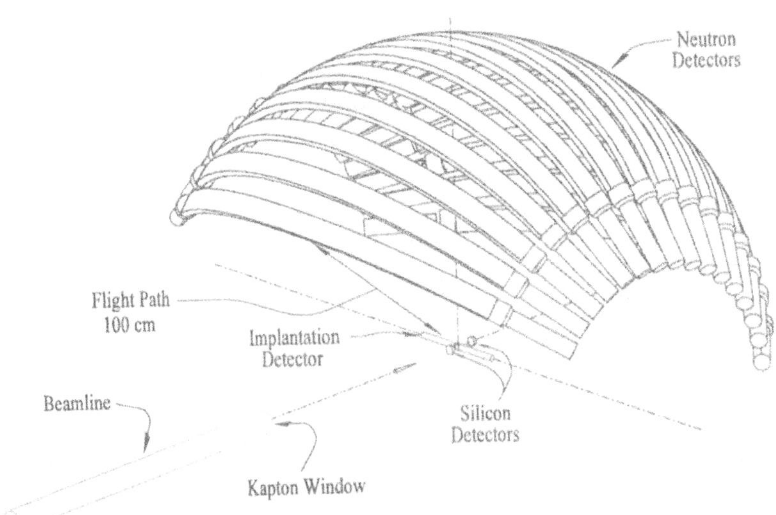


*Figure 2.* The Neutron Bar Array shown in relation to the implantation system and START scintillator.

The neutron bar array was selected for its ability to efficiently detect neutrons in the energy range of about 800 keV through 10 MeV. Below 800 keV, the neutron detection efficiency is greatly reduced. However, the ability to quantitatively record lower energy neutrons depends not only on the detection efficiency, but also on the strength of the beta feeding and on the total number of decay events observed. The true lower level threshold of the array during this experiment was about 350 keV.

In addition to the neutron bar array, the set-up included three $^{6}Li$ glass scintillators for neutron detection. The $^{6}Li$ glass detectors were selected to record neutron events in the energy range of 50 keV through 1 MeV. These devices also work on a TOF technique and were placed at distances between 12 and 30 centimeters from the implantation position.

In order to record gamma-rays emitted during the decay of $^{19}N$, two high purity germanium detectors were included in the set-up. They were 80% and 120% efficient relative to standard NaI crystals and were placed as close as possible to the source activity, given the physical constraints of the neutron bar array and its supporting apparatus.

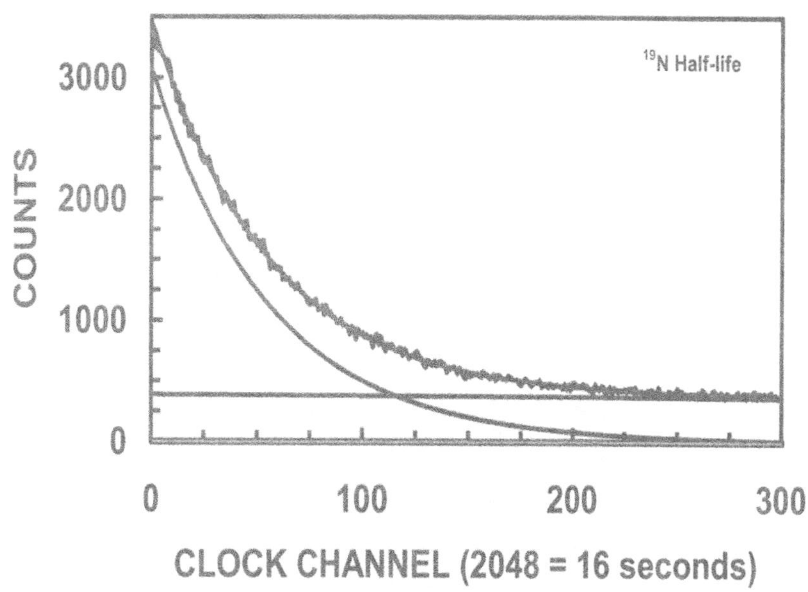


*Figure 3.* Counts versus clock channel for the $^{19}N$ measurement showing the experimental data (histogram), the sum of the fitted components (curve), and the background (flat line).

On line calibration of the implantation system and neutron detectors was accomplished from the known half-lives and delayed neutron emission probabilities of $^{16}C$ and $^{17}N$ (Alburger and Wilkinson, 1976 and Tilley *et al.*,

1993). The germanium detectors were calibrated using Standard Reference Materials and other standard sources of gamma-ray activity.

The first experimental result for the decay of $^{19}N$ was a re-measurement of the half-life. For this measurement the intensity of the primary beam was attenuated by a factor of about 30 to minimize error associated with electronic dead time. Figure 3 shows the data from this run, plotted as counts versus clock channel. The fit to the data includes parameters for the decay constant of $^{19}N$, the initial activities of $^{19}N$ and the beta daughter $^{19}O$, and a constant background term. The decay constant of $^{19}O$ was fixed to the well known value from the literature (Tilley *et al.,* 1995). The result of the fit indicates that the half-life of $^{19}N$ is 299 (19) milliseconds.

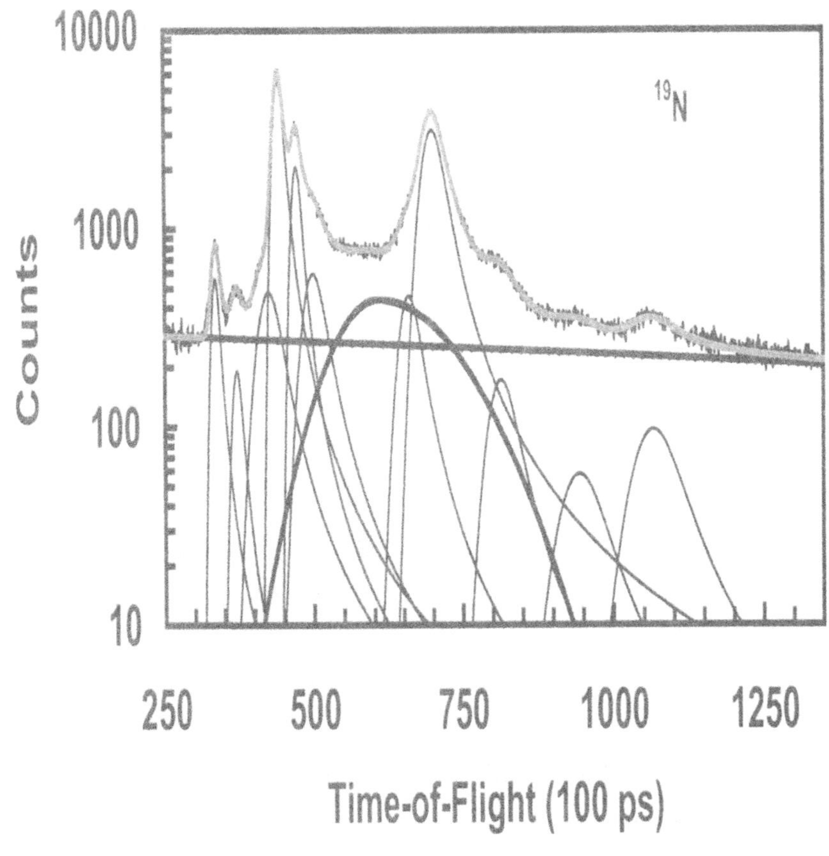


*Figure 4.* Delayed neutron TOF spectrum recorded by the neutron bar array during the decay of $^{19}N$. Shown are the raw data (histogram), the aggregate fit (uppermost curve), peak components and background terms (horizontal line and large peak near 600).

Figure 4 shows the aggregate neutron TOF spectrum accumulated by fifteen of the sixteen neutron array bars. The figure includes the total fit to the experimental data, the component fits to each of the eleven observed peaks and shoulders, and the two component background including a linear term to account for random noise and a broad, asymmetric gaussian to model for scattered neutrons off the concrete walls and floor of the expermental vault. The neutron peaks were fit to a functional form that is gaussian below the centroid and Breit-Wigner above the centroid. This functional form was selected as it provided the best approximation for the response function of the neutron array bars to neutron events. The peak shapes were restricted such that the ratio of the gaussian width to the Breit-Wigner width for all peaks was fixed to a constant value derived from the fits to the calibration data. Also, a smoothing condition was included such that the gaussian and Breit-Wigner functions matched at the centroid. These two conditions guaranteed that no additional parameters were necessary to the peak fits beyond those prescribed by the gaussian component.

*Table 2.* Delayed neutrons observed in the neutron bar array during the decay of $^{19}N$. Branching ratios and log ft values for the deduced beta transitions from the ground state of $^{19}N$ to unbound levels in $^{19}O$. The error in the neutron energy is about 40 keV.

| NEUTRON ENERGY (keV) | BRANCHING RATIO (%) | Log ft |
|---|---|---|
| 460 | 7.7 (16) | 4.7 |
| 590 | 0.9 (3) | 6.0 |
| 800 | 1.6 (4) | 4.2 |
| 1070 | 16.5 (32) | 4.8 |
| 1210 | 1.4 (3) | 5.2 |
| 2100 | 2.4 (5) | 4.6 |
| 2350 | 3.6 (8) | 5.1 |
| 2680 | 9.9 (18) | 3.7 |
| 3980 | 1.4 (4) | 5.3 |
| 3870 | 1.0 (2) | 5.0 |
| 4600 | 1.4 (3) | 4.6 |

The branching ratios for the delayed neutrons were calculated from the integrated area of the neutron peaks and the response of the implantation detector to the initial beta-decay. Table 2 lists the neutron peaks observed by their energy in keV and the calculated branch for that transition as a percent of the total decay strength. The last column gives the log ft value for the corresponding beta transition from the ground state of $^{19}N$ to the deduced excited state in $^{19}O$. These were derived from the measured composite half-life of $^{19}N$, the branching ratio for the transition, and log f values given in (Gove and Martin, 1971). It should be noted that a number of different functional forms were tested during the analysis to approximate the neutron peak shapes, ranging from simple gaussian to the compound function described above. For the branching ratio determination, the functional form of the peak did not significantly affect the result, provided that the function chosen was applied consistently to both the calibration and $^{19}N$ data. The compound peak shape reported here gave the lowest reduced chi-squared in the aggregate fit. The errors quoted for the branching ratios are mainly due to uncertainty in the detection efficiency and the shape of the background components of the fit.

Taking the results for the individual branches in concert, the total neutron emission probability for neutrons above threshold is found to be 47.8 (93) % of the decay strength. Due to the fact that the quoted individual errors are ascribed to the efficiency and background subtraction, they are necessarily highly correlated. Because of this, the reported error in the total delayed neutron emission probability is the straight sum of the individual errors and not an addition in quadrature.

The $^{6}Li$ glass scintillators were used to look for delayed neutrons below the threshold of the neutron bar array. Preliminary analysis shows an additional neutron transition at about 300 keV. The branching ratio for this transition appears to be about three percent of the total decay strength, with an error of about two percent. Recent unrelated experimental studies on the response of the $^{6}Li$ glass scintillators to neutrons (Massey, 1999) have demonstrated significant uncertainty in the intrinsic efficiency of these devices relative to that quoted by the manufacturer (Levy Hill Laboratories, LTD, 1997). Consequently, further analysis is required for this component of the study. However, taking the result of the $^{6}Li$ glass scintillator data with the neutron bar array analysis indicates that the total delayed neutron emission probability of $^{19}N$ is 51 (11) % of the overall decay.

To determine the $^{19}N$ beta branching to bound excited levels in $^{19}O$, the data recorded by the germanium detectors were analysed. Figure 5 shows the gamma-ray spectrum collected by one of the two germanium detectors, the one with the higher efficiency. Also in the figure are markers indicating those gamma-rays that are at energies characteristic of transitions in $^{19}O$. These peaks were confirmed by their half-lives as belonging to the decay of $^{19}N$. From the intensity of the lines and the known cascades originating from specific levels in $^{19}O$ (Tilley *et al.*, 1995), the strength of beta feeding from the decay of $^{19}N$ to specific bound excited levels of $^{19}O$ has been deduced.

The data support at least two branches to bound excited states of $^{19}O$. One, a very strong branch having about 36% of the total decay strength to the first 3/2- state at 3.95 MeV and another having about 5% of the total decay strength to the unconfirmed 1/2- state at 3.23 MeV. From the cascade data and the observed intensities, the fraction of $^{19}N$ beta decay populating bound excited states of $^{19}O$ has been found to be 44 (10) % of the total decay strength.

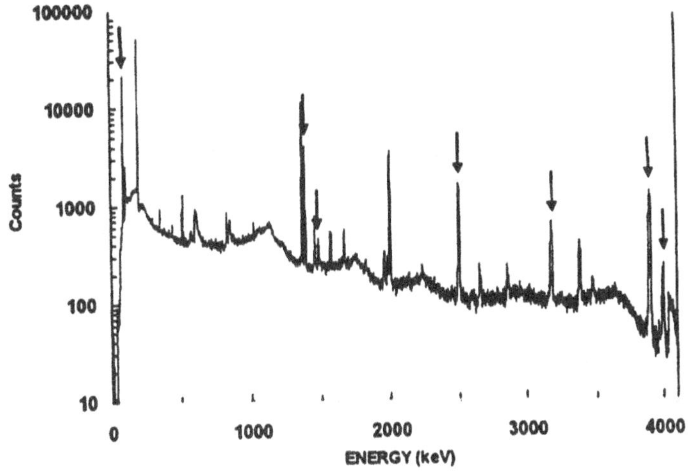


*Figure 5.* Gamma-ray spectrum collected by one of two Germanium detectors. The markers indicate peaks due to transitions in $^{19}O$.

# 3. CONCLUSION

The decay properties of $^{19}N$ were monitored using a variety of detection devices. The half-life of this nuclide was re-measured to be 299 (19) milliseconds, which is in close agreement with that reported in (Reeder *et al.*, 1991). The first spectroscopic measurement of delayed neutrons emitted during the decay of $^{19}N$ has been made yielding branching ratios for twelve transitions recorded by the neutron bar array and the $^{6}Li$ glass scintillators. Taken in concert, the total delayed neutron emission probability has been found to be 51 (11) % of the total decay strength.

Beta feeding to bound states of $^{19}O$ has been deduced from the observation of seven gamma-rays characteristic of $^{19}O$. The data indicate that there are at least two, and probably three, bound excited states of $^{19}O$ populated during the decay of $^{19}N$, accounting for 44 (10) % of the total decay strength. Within error, the (beta, neutron) and (beta, gamma) data account for the decay in its entirety. However, this experiment was only sensitive by difference to beta decay directly to the ground state of $^{19}O$ and, furthermore, was not sensitive to delayed alpha emission which is energetically possible by about 3.5 MeV (Tilley *et al.*, 1995).

From the calculated log ft values, the observed transitions are most likely allowed. The known spin and parity of some of the levels that are populated (3/2- and 1/2-) suggest a ground state spin of and parity for $^{19}N$ as (3/2-) or (1/2-), consistent with an assumed (1/2-) assignment based on the known (1/2-) ground state spin of the other odd-mass nitrogen isotopes.

# REFERENCES

Alburger, D.E. and Wilkinson, D.H.: 1976, "Beta decay of $^{16}$C and $^{17}$N", *Phys. Rev.* **C13**, 835-846.

Applegate, J., Hogan, C. and Scherrer, R.: 1987, "Cosmological baryon diffusion and nucleosynthesis", *Phys. Rev.* **D35**, 1151-1160.

Dufour, J.P., Del Moral, R., Hubert, F., Jean, D., Pravikoff, M.S., Fleury, A., Delagrange, H., Mueller, A.C., Schmidt, K.J. and Hanelt, E.: 1988, "Spectroscopic measurements with a new method: The projectile fragments isotopic separation", *AIP Conf. Proc. 164*, p. 344.

Dufour, J.P., Delagrange, H., Del Moral, R., Fleury, A., Geissel, H., Hubert, F., Jean, D., Pravikoff, M.S. and Schmidt, K.H.: 1986, "Beta decay of $^{17}$C, $^{19}$N, $^{22}$O, $^{24}$F, $^{26}$Ne, $^{32}$Al, $^{34}$Al, $^{35,36}$Si, $^{36,37,38}$P, $^{40}$S", *Z. Phys.* **A324**, 487-488.

Gove, N.B. and Martin, M.J.: 1971, "Log f tables for beta decay", *Nucl. Data Tables* **10**, 205-317.

Harkewicz, R.: 1991, "The beta-delaying branching rations of the neutron-rich nucleus 15B", Ph.D. Thesis, NSCL, p. 127.

Käppeler, F., Gallino, R., Busso, M., Picchio, G. and Raiteri, C.M.: 1990, "s-Process nucleosynthesis: Classical approach and asymptotic giant branch models for low-mass stars", *Ap. J.* **354**, 630-643.

Levy Hill Laboratories, LTD: 1997, private communication.

Massey, T.N.: 1999, private communication.

Meissner, J., Schatz, H., Gorres, J., Herndl, H., Wiescher, M., Beer, H. and Käppeler, F: 1996, "Neutron capture cross section of $^{18}$O and its astrophysical implications", *Phys. Rev.* **C53**, 459-468.

Mueller, A.C., Bazin, D., Schmidt-Ott, W.D., Anne, R., Guerreau, D., Guillemaud-Mueller, D., Saint-Laurent, M.G., Borrel, V., Jacmart, J.D., Pougheon, F. and Richard, A.: 1988, "Beta-delayed neutron emission of $^{15}$B, $^{18}$C, $^{19,20}$N, $^{34,35}$Al, $^{37}$P", *Z. Phys.* **A330**, 63-68.

Ozawa, A., Raimann, G., Boyd, R.N., Chloupek, F.R., Fujimaki, M., Kimura, K., Kitagawa, H., Kobayashi, T., Kolata, J.J., Kubono, S., Tanihata, I., Watanabe, Y. and Yoshida, K.: 1995, "Study of the beta-delayed neutron emission of $^{19}$C", *Nucl. Phys.* **A592**, 244-256.

Raiteri, C.M., Busso, M., Gallino, R., Picchio, G. and Palone, L.: 1991, "s-Process nucleosynthesis in massive stars and the weak component. I. Evolution and neutron captures in a 25 [solar mass] star", *Ap. J.* **367**, 228-238.

Rauscher, T., Applegate, J., Cowan, J., Thielemann, F.K. and Wiescher, M.: 1994, "Production of heavy elements in inhomogeneous cosmologies", *Ap. J.* **429**, 499-530.

Reeder, P.L., Kim, Y., Hensley, W.K., Miley, H.S., Warner, R.A., Vieira, D.J., Wouters, J.M., Zhou, Z.Y. and Seifert, H.L.: 1994, "Beta decay data for neutron rich Li-Cl nuclides", Int. Conf. Nucl. Data for Science and Tech., Gatlinburg.

Reeder, P.L., Warner, R.A., Hensley, W.K., Vieira, D.J. and Wouters, J.M.: 1991, "Half-lives and delayed neutron emission probabilities of neutron rich Li-Al nuclides", *Phys. Rev.* **C44**, 1435-1453.

Rolfs, C.E. and Rodney, W.S.: 1998, "Cauldrons in the cosmos", University of Chicago Press, Chicago, IL, pp. 169-178.

Samuel, M., Brown, B.A., Mikolas, D., Nolen, J., Sherrill, B., Stevenson, J., Winfield, J.S. and Xie, Z.Q.: 1988, "Measurement of the beta-decay half-life of $^{17}$B", *Phys. Rev.* **C37**, 1314-1317.

Tilley, D.R., Weller, H.R. and Cheves, C.M.: 1993, "Energy levels of light nuclei A=16-17", *Nucl. Phys.* **A564**, 1-183.

Tilley, D.R., Weller, H.R., Cheves, C.M. and Chasteler, R.M.: 1995, "Energy levels of light nuclei A = 18-19", *Nucl. Phys.* **A595**, 1-170.

# Measuring the Astrophysics Rate of the $^{21}Na(p,\gamma)^{22}Mg$ Reaction

John M. D'Auria for the DRAGON Collaboration
*Department of Chemistry, Simon Fraser University, Burnaby, Canada*
*dauria@popserver.sfu.ca*

**Abstract:** The $^{21}Na(p,\gamma)^{22}Mg$ reaction is believed to play an important role determining the amount of the long-lived, radioisotope $^{22}Na$ in the universe. This nuclide is important as a target of current and future generations of gamma ray telescopes, since its production is part of the pathway of a nova, and because its daughter, $^{22}Ne$, has been found in pre-solar grains in meteorites. Unfortunately the rate of this reaction has never been measured primarily due to the fact that $^{21}Na$ is radioactive ($T_{1/2} = 23$ s). This report describes an experimental program aimed at measuring this rate using inverse kinematics at the new radioactive beams facility, ISAC. A new recoil mass separator, DRAGON, is being constructed to perform this and similar reaction rate studies; the status of this new facility is presented.

## 1. INTRODUCTION

### 1.1 Overview

Novae occur on the surfaces of white dwarf stars that accrete matter from a companion star. The accreted material gains a large amount of energy as it falls into the gravitational field of the white dwarf, and this energy heats the surface layer. When the temperature is high enough to ignite thermonuclear reactions, an explosion occurs. Various reaction networks are believed to occur during the hydrogen burning of the outburst and the NeNa cycle could play a dominant role for a certain class of novae. This cycle consists of:

$$^{20}Ne(p,\gamma)\,^{21}Na(\beta^+)^{21}Ne(p,\gamma)^{22}Na(p,\gamma)^{23}Mg(\beta^+)^{23}Na(p,\alpha)^{20}Ne \qquad (1)$$

and

$$^{20}\mathrm{Ne}(p,\gamma)^{21}\mathrm{Na}(p,\gamma)^{22}\mathrm{Mg}(\beta^+)^{22}\mathrm{Na}(p,\gamma)^{23}\mathrm{Mg}(\beta^+)^{23}\mathrm{Na}(p,\alpha)^{20}\mathrm{Ne} \qquad (2)$$

As the warm NeNa cycle (Equation 1) includes beta decay, the rate of the energy generation is limited by this ($T_{1/2}$=22 s) decay. By bypassing this decay, the hot NeNa cycle (Equation. 2) can dramatically increase the energy generation of the explosion. As the rate of the $^{21}$Na(p,γ)$^{22}$Mg determines when the transition between cycles takes place, knowledge of the actual rate of this reaction plays a key role in understanding this type of nova.

The radioactive nucleus, $^{22}$Na, is of interest because it decays with the emission of a 1.28 MeV gamma ray, and because its daughter, $^{22}$Ne, has been found in meteorites. Ne-E was first found in carbonaceous meteorites (Black, 1972). This gaseous component consists of neon, enriched in $^{22}$Ne by factors up to $10^3$, or in other words, it is nearly pure $^{22}$Ne. Ne-E is carried by grains of silicon carbide, probably pre-solar in nature. While the origin of the grains is unclear, clearly one production path for $^{22}$Ne is through decay of $^{22}$Na. There is speculation that understanding of how and why Ne-E exists could play an important role in expanding our understanding of nucleo-synthesis in the present day Galaxy and in the Galaxy of 4.5 billion years ago. The production of $^{22}$Na in the universe is governed by the NeNa cycle, but the $^{21}$Na(p,γ)$^{22}$Mg reaction has not yet been measured.

Another important role that the isotope $^{22}$Na plays is as a target of gamma ray astronomy. The emitted gamma ray of 1.28 MeV should be observable by gamma ray observatories and can be a useful probe of recent nova activity; the half-life is 2.6 years. Present estimates of the flux of $^{22}$Na (Politano *et al.*, 1995) in a nova using estimates of the $^{21}$Na(p,γ)$^{22}$Mg rate (Wiescher and Langanke, 1986) indicate that it should be observable by the COMPTEL telescope on the NASA Gamma Ray Observatory. However, it has not yet been observed in several nova, and only upper limits have been set (Iyudin *et al.*, 1995) for the novae Her 1991 and Cyg 1992. Again, better knowledge of the rate of the $^{21}$Na(p,γ)$^{22}$Mg reaction is needed to understand the lack of this observation in these complex, rapid explosions.

## 1.2 Current Knowledge of the $^{21}$Na(p,γ)$^{22}$Mg Reaction

It is believed that the rate of this reaction is dominated by narrow, isolated resonances and few of the key parameters are known to a reasonable degree of accuracy. The level scheme for $^{22}$Mg and its isobaric analogue, $^{22}$Ne are shown in Figure 1.

The key levels lie above 5.5 MeV, and it is presently believed that only three resonances at 212.4, 336 and 464 keV are important at nova temperatures, although new information (D'Auria *et al.*, 1999) may effect this as described below. The resonance strengths of these states have not been measured but using estimates (Wiescher and Langanke, 1986), the expected thick target yields of a p, γ reaction can be estimated (See Table 1). Given the uncertainty in both the strengths and the energy of the states, however, the availability of a $^{21}Na$ beam offers an excellent opportunity to study these.

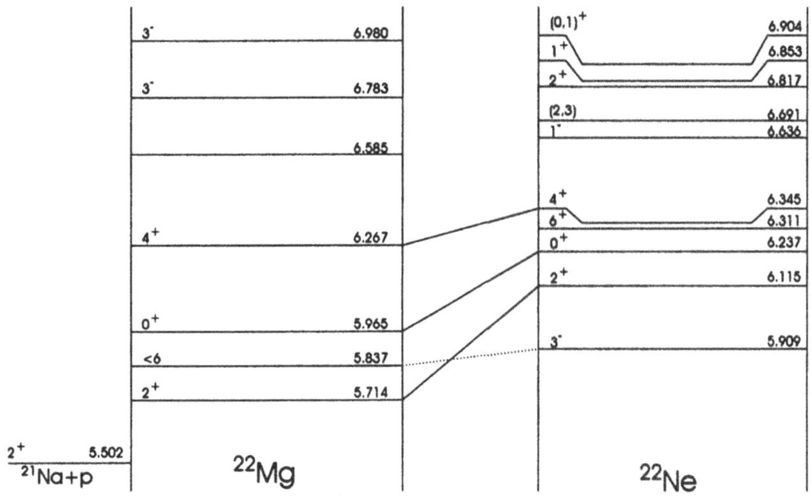


*Figure 1.* Levels of $^{22}Ne$ and $^{22}Mg$.

*Table 1.* Parameters for the $^{1}H(^{21}Na,^{22}Mg)\gamma$ reaction

| $E_r(\Delta E)$ (keV) | $E_x$ (MeV) | $J^\pi$ | $\omega\gamma$ (meV) | Yield* ($x10^{-12}$) | $E_{beam}$ (MeV/u) | $E_{recoil}$ (MeV/u) | Recoil Cone (mrad) | Count Rate** (cph) |
|---|---|---|---|---|---|---|---|---|
| 212.4(1.9) | 5.714 | $2^+$ | 2.2 | 6 | 0.22 | 0.20 | 13.3 | 10 |
| 336(5) | 5.84 | < 6 | 11.3 | 20 | 0.35 | 0.33 | 10.9 | 33 |
| 464(25) | 5.97 | $0^+$ | 2.5 | 3 | 0.486 | 0.46 | 9.5 | 5 |

* Assumes a gas target of $2 \times 10^{18}$ atoms/cm² in a 10 cm long cylinder; P = 3 torr.
** Assumes a $^{21}Na$ beam intensity of $10^9$ pps and a DRAGON transmission of 40%.

## 2. EXPERIMENTAL

In order to perform this study will require a radioactive beam of $^{21}Na$, a windowless gas target of hydrogen gas, a high transmission and acceptance recoil mass separator in which all beam particles are eliminated, and a detection system of the recoiling reaction products. The study must be performed using inverse kinematics since the heavy reactant is radioactive. In addition preliminary studies must be performed to obtain more precise information on properties of $^{22}$Mg. These will be described below.

## 2.1 Production of the $^{21}$Na beam

The $^{21}$Na beam will be produced at the new accelerated radioactive beam facility, ISAC, located at the TRIUMF National Nuclear Laboratory in Vancouver, Canada. This facility started operations in 1998 and accelerated beams are expected by the end of 2000. The energy of the beam is variable from 0.15 to 1.5 MeV/u; Table 1 shows the required beam energies for the $^{21}$Na study. Figure 2 displays the planned layout of the new experimental hall.

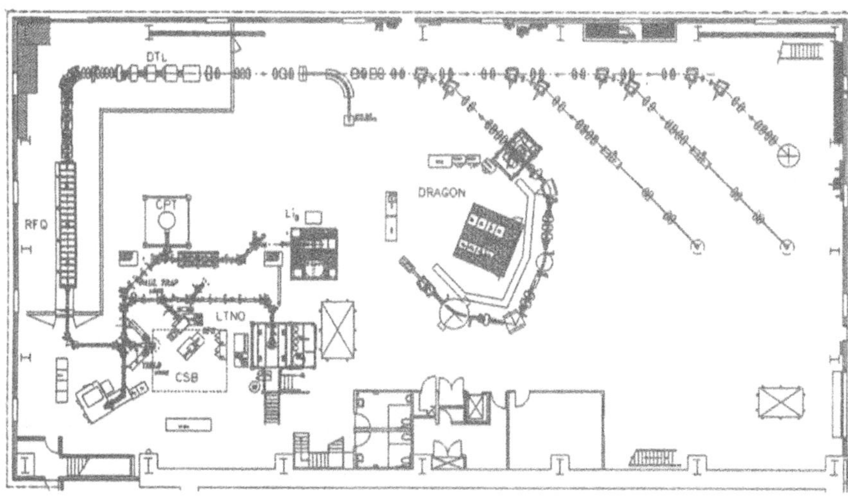


*Figure 2.* Plan View of ISAC Experimental Hall.

The $^{21}$Na beam will be produced using 500 MeV protons from the TRIUMF cyclotron incident onto a thick target (probably either SiC or MgO), and the $^{21}$Na atoms extracted through diffusion into a heated surface ion source. This source will only ionize alkali and alkaline elements, with ions extracted at a total energy of < 60 keV. Subsequent mass analysis of this ion beam will result in a pure beam of $^{21}Na^+$ ions. These will be directed into the RFQ, a stripper, and then a DTL LINAC acceleration system. Previous studies indicate that the final, accelerated beam incident onto the gas target should be of the order of > $10^9$ p/s.

## 2.2 Windowless Gas Target

A windowless gas target system for both He and $H_2$ gas has been designed, built, and is now in the testing phase. It consists primarily of pumps (roots blowers, turbos and backing) in order to maintain a gas pressure of the order of 3 torr and yet keep the pressure levels at both ends of the target ~ $10^{-6}$ torr. The inner target chamber is 10 cm in length with a circular opening of 6 mm. The target thickness is estimated to be 2 x $10^{18}$ atoms/cm$^2$. Initially a flow-through system will be used but ultimately, a gas re-circulation and cleaning system will be introduced. Initial tests of the outer components of the gas target have been completed and it is to be installed at TRIUMF early in the fall of 1999.

## 2.3 The Separator

Following the target will be a recoil mass separator with 100 % acceptance of all radiative proton and alpha capture reactions of interest. This separator consists of two separate sections with a momentum analyzer (magnetic dipole) and an energy analyzer (electric dipole) in each. The momentum analyzer will transmit one charge state of the beam and reaction products, and then the beam is rejected at the energy analyzer. With these two sections a beam rejection factor of > $10^{10}$ has been estimated assuming various beam scattering and beam charge exchange scenarios. This system also contains 10 quadrupoles, 4 steerers and 4 sextupoles and all are being constructed with high precision. All units have been designed, and in the construction phase. Installation is expected to commence by the end of 1999. Figure 3 displays a conceptual view of the DRAGON facility.

## 2.4 Detection Systems

At the end of the separator there will be several detectors. Initially the ions will pass through a carbon foil with the resulting electrons detected in a multichannel plate detector to generate a fast start signal. These ions may then pass into a gaseous detection chamber (ionization detector) through a thin polymer/grid supported window. At the front of the chamber, they pass through a PGAC (parallel gridded avalanche counter) positioned 0.5 m from the start detector, and this would provide a fast stop signal. The t-o-f would be the primary measure of mass when combined with the energy of the recoil. An alternate possibility is to use a second MCP detector system to provide a stop signal, without the PGAC. The total energy of the ion would be provided by the multichamber ionization chamber. It may also be possible in some instances to provide some Z discrimination based upon the small ΔE in the sub-chambers. The ionization chamber has been built and tested but has only demonstrated an energy resolution of within a factor of 2 of the desired value of 1%. It should also be noted that a secondary t-o-f signal would also be generated using the pulsing structure of the accelerator as a start signal. A further beam rejection factor of at least a factor of 10 or optimistically, $10^2$ is expected when using the recoil detection system.

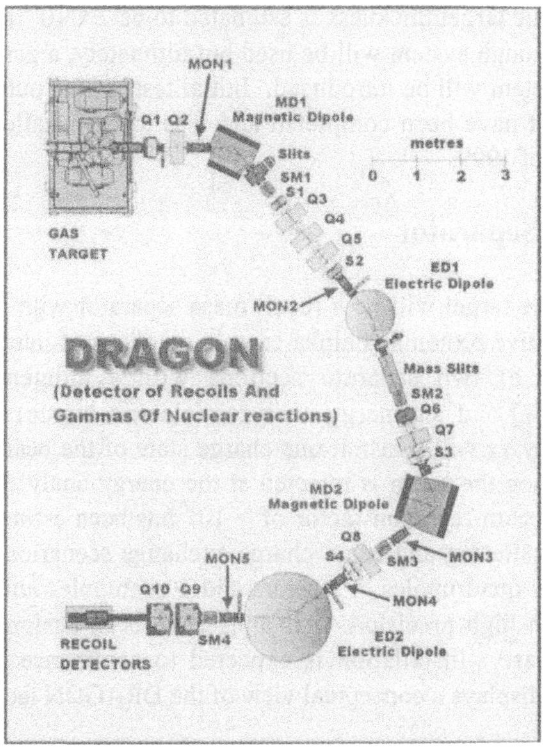


*Figure 3.* Conceptual representation of the DRAGON facility.

## 2.5 Gamma Array

It is also proposed to surround the gas target with a gamma array to provide an additional beam rejection trigger when the prompt gamma from the reaction is used in coincidence with the recoil detector. The design of the array however does face several challenges. The proton-rich ion beam itself is radioactive and upon passing through the target with its slits will deposit approximately 0.1% of its activity or about $10^7$ cps. A highly segmented, fast scintillator system is under detailed review to detect with good energy resolution, the 1-8 MeV gammas expected from the various reactions under consideration for study with DRAGON in the field of the

0.511 MeV annihilation gammas. The latest design and one for which funding may be sought, utilizes a BGO based, modular system with a total detection efficiency of about 65%. It is believed use of this system will provide a further beam rejection factor of $>10^2$.

## 2.6 Experimental Estimates

The last column in Table 1 are estimates of the expected counting rates that would be observed for the different resonances given the assumptions indicated. These may have to be modified depending upon a final analysis of the study mentioned below on the key levels of $^{22}$Mg.

## 2.7 Related Studies

### 2.7.1 Studies of the $^{22}$Mg levels

As indicated above additional information is needed on the energies of the levels in $^{22}$Mg in order to attempt to perform this study. The resonance at 464 keV is only known to an accuracy of 25 keV, which is higher than the expected energy loss (10 keV) in the gas target. It is very difficult to attempt to find narrow resonances with the relatively weak radioactive beams. Therefore studies were performed to measure this energy more accurately. Table 2 presents some of the preliminary data obtained in a (p,t) reactions performed at the CNS cyclotron in Tokyo, Japan. Also indicated are the accepted energies in the literature (Endt, 1990) for $^{22}$Mg. It is clear that additional levels have been observed in the energy range, which may play a role in the astrophysical rate of this reaction. A theoretical analysis is underway to understand the nuclear structure involved (D'Auria *et al.*, 1999).

*Table 2.* Preliminary data from $^{24}Mg(p,t)^{22}Mg$ reaction

| $E_r$ (keV) (Endt) | 8° | 16° | 20.5° |
|---|---|---|---|
| 4400.9 (1.4) | | 4399.0 (5.3) | 4400.5 (5.2) |
| 5037.0 (1.4) | 5037.0 | 5037.0 | 5037.0 |
| | 5089.5 (1.9) | 5090.9 (1.8) | 5089.1 (1.6) |
| 5292 (3) | 5296.5 (1.3) | 5296.5 (1.3) | 5295.0 (1.2) |
| 5464 (5) | 5454.8 (1.3) | 5454.3 (1.3) | 5454.5 (1.3) |
| 5713.9 (1.2) | 5713.9 | 5713.9 | 5713.9 |
| 5965 (25) | 5961.1 (2.4) | 5964.4 (2.6) | |
| | 6044.6 (2.9) | 6048.6 (3.0) | 6046.3 (3.0) |
| 6267 (15) | 6244.3 (4.9) | 6251.0 (5.2) | 6246.2 (5.0) |
| | 6321.6 (5.8) | 6325.7 (6.1) | 6322.7 (5.9) |
| 6585 (35) | 6613.5 (10.2) | 6621.9 (10.8) | 6604.7 (10.3) |
| 6783 (19) | 6787 (14) | | |

# 3. SUMMARY

An experimental program is underway to measure the astrophysical rate of the $^{1}H(^{21}Na,\gamma)^{22}Mg$ reaction. This involves a new accelerated radioactive beams facility, ISAC, and a new detection facility, the DRAGON at the TRIUMF laboratory in Vancouver, Canada. It is expected that this study will be ready to receive beam by the end of 2000. New results have also been obtained concerning the levels of $^{22}Mg$ and further analysis and possibly other indirect experiments will be performed. More details on many aspects of ISAC and DRAGON can be found at http://www.triumf.ca and http://www.sfu.ca/triumf.

# ACKNOWLEDGEMENTS

Financial support of the Natural Sciences and Engineering Research Council of Canada and infrastructure support of TRIUMF are gratefully acknowledged. The efforts of all members of the DRAGON collaboration are also acknowledged.

# REFERENCES

Black, D.C.: 1972, "On the origin of trapped helium, neon and argon isotopic variations in meteorites-II. Carbonaceous meteorites", *Geochim. Cosmochim. Acta,* **36**, 377-394.

D'Auria, J. M.: 1999, in preparation.

Endt, P.M.: 1990, "Energy levels of A = 21-44 Nuclei (VII)", *Nucl. Phys.* **A521**, 1-830

Iyudin, A.F., Bennett, K., Bloemen, H., Diehl, R., Hermsen, W., Lichti, G.G., Morris, D., Ryan, J., Schonfelder, V., Steinle, H., Strong, A., Varendorff, M. and Winkler, C.: 1995, "COMPTEL search for $^{22}Na$ line emission from recent novae", *Astron Astrophys.* **300**, 422-428.

Politano, M., Starrfield, S. , Truran, J.W., Weiss, A. and Sparks, W.M.: 1995, "Hydrodynamic studies of accretion onto massive white dwarfs: One Mg-enriched nova outburst. I. Dependence on white dwarf mass", *Ap. J.* **448**, 807-821.

Wiescher, M. and Langanke, K.-H.: 1986, "The proton capture reactions on $^{21}Na$ and $^{22}Na$ under hydrogen burning conditions", *Z. Phys.* **A325**, 309-315.

# Production and β-Decay Half-Lives of Very N-Rich Nuclei

Monique Bernas
*Institut de Physique Nuclèaire d'Orsay, IN2P3-CNRS, F-91406 Orsay Cedex, France*
*bernas@ipno.in2p3.fr*

**Abstract:** Investigations of extremely neutron-rich nuclei are bringing significant information; neutron halos, neutron skin and other surprises already appear on the frontier of the nuclear binding. We have initiated the experimental study of fission of U-projectiles at relativistic energies with the FRS (GSI) in which 117 new n-rich isotopes have been separated and unambiguously identified. They will be further investigated starting with half-life ($T_{1/2}$) and mass-excess (m.e.) measurements. The first or the second of these quantities has already been measured for 42 new n-rich isotopes of elements between Cu and Ti using different production techniques at the MP-tandem (IPN-Orsay), the on-line mass-separator (GSI-Darmstadt), the nuclear reactor (ILL-Grenoble), and the FRS (GSI).

## 1. INTRODUCTION

The r-process is one of the best established mechanisms for stellar nucleosynthesis as its properties have been understood for more than 40 years (Burbridge *et al.*, 1957). It requires a high neutron density at high temperature, which produces the combination of neutron-captures and beta-decays that synthesizes the r-process nuclei. The neutron captures are rapid, and occur so successively to the point where they become unlikely. This happens when the binding energy of an extra-neutron drops below a limiting value for the capture cross section. The further progression of the r-process requires a beta-decay, the time constant of which is much longer than for n-capture or γ-dissociation. Therefore the abundance tends to build up at closed neutron-shells, and the isotones involved become 'waiting points'

*Origin of Elements in the Solar System: Implications of Post-1957 Observations.*
Edited by O. Manuel, Kluwer Academic/Plenum Publishers, 2000. 

nuclei. The r-process proceeds through nuclei close to the n-drip line, for which very little is known. To reproduce the solar abundances, binding energies are derived from mass formulae. $Q_\beta$ values are deduced and the structure of excited states of the final nucleus is approximated from a distinct theory. Finally the half-life, essentially an integrated nuclear quantity, is calculated. We have tried to extend the domain of measured half-lives toward more and more n-rich species with the hope to constrain on experimental basis the astrophysical parameters and potential sites of the r-process.

## 2. PRODUCTION MODES

During the eighties, kinematics of two-body transfer reactions were successfully used to study the mass excess (m.e.) and spectroscopy of the neutron-rich isotopes of Fe, Ni and Zn (See Bernas *et al.*, 1984 and references therein). In case of Ni for example, we measured the m.e. of $^{67,68,69}$Ni isotopes (Dessagne *et al.*, 1984) which are still the heaviest known-mass Ni isotopes as well as the first spectroscopic features of $^{67,68}$Ni (Girod *et al.*, 1988).

Then in our quest for new species we moved to deep inelastic reactions as a production mode. At the on-line separator at GSI, a number of new isotopes were discovered and their ($\beta$-$\gamma$ X ion) delayed coincidences gave their $\beta$-half-lives and decay schemes; for example the decay schemes of $^{67,69}$Ni were accurately measured (Runte *et al.*, 1985).

The next step lead us to use thermal fission at the nuclear reactor of the ILL-Grenoble. A very asymmetric fission mode was discovered using the Lohengrin separator of fission fragments (Sida *et al.*, 1989). Neutron-rich fragments down to $^{68}$Fe were identified (Bernas *et al.*, 1991). We developed then an experimental method in order to measure $\beta$-decay half-life of short-lived rare isotopes. Identified ions are implanted in Si pin-diode detectors and their subsequent $\beta$-decay detected in the same device. Signals from an identified implant and from beta-decay are analyzed with double sensitive preamplifiers. Even with low statistics, the half-life of a number of new isotopes was evaluated; in case of Ni, measurements of $T_{1/2}$ were conducted up to $^{71,72,73,74}$Ni (Bernas *et al.*, 1990).

At GSI, with the relativistic ion beam of $^{86}$Kr, projectile fragments were separated and selectively implanted in a similar series of pin-diodes. A number of new values of $T_{1/2}$ were determined, and for Ni isotopes the frontier was carried up to $^{76}$Ni.

Finally, the r-process path was recently reached by a new production method named the U-projectile fragmentation. I will report here on this last discovery since it opens a new large field to experiments.

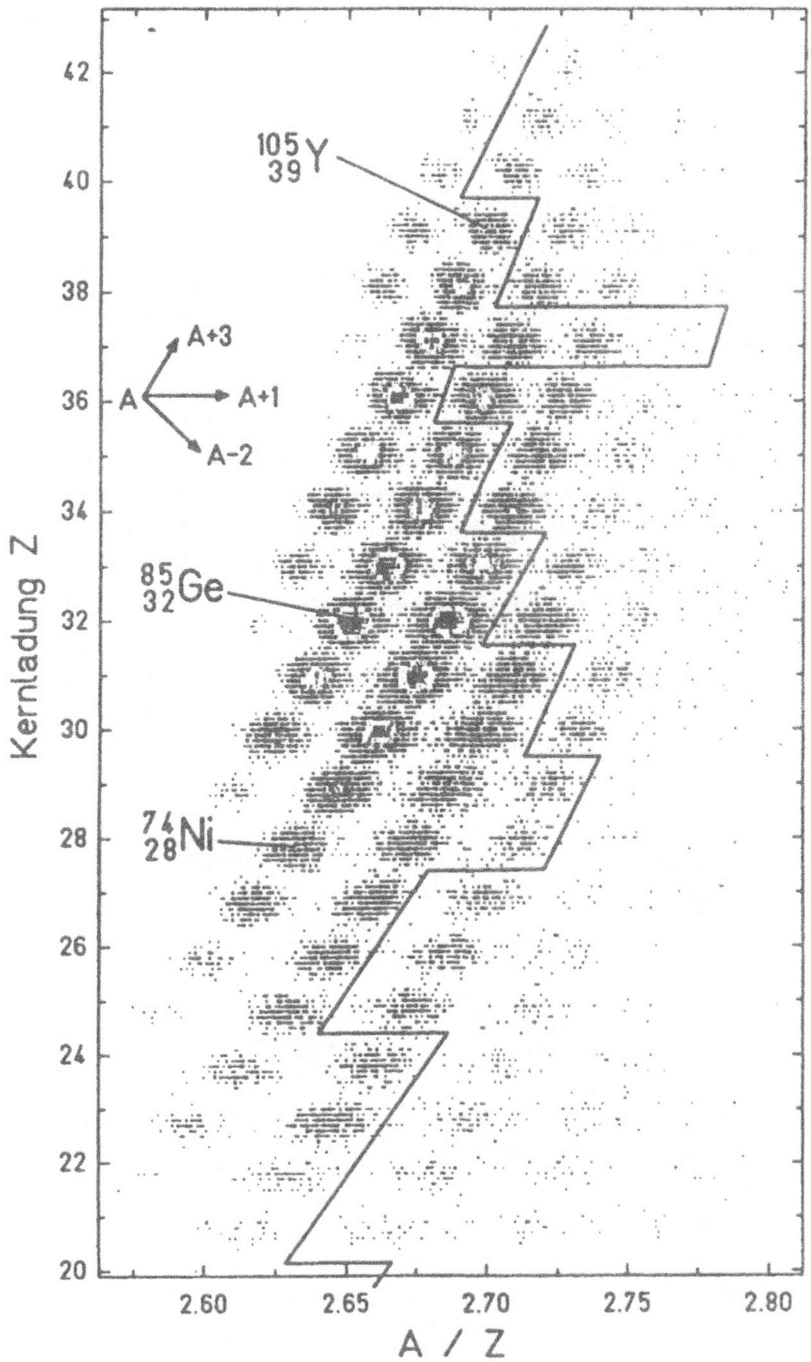


*Figure 1.* A/Z versus Z scatter-plat of fission fragments observed at large magnetic rigidity.

In inverse kinematics, fission fragments from U-projectiles are emitted forward and effectively transmitted to a separator. The transmission is multiplied by more than 4 orders of magnitude as compared with thermal fission conditions. At large velocities ($\beta = 0.83$ for E = 750 A•MeV, [MeV per nucleon]) fragments are totally stripped of their electrons up to Yb and their magnetic separation is much more easy and efficient. The resolving power of the fragment separator equipped with time of flight scintillation detectors and ionization chamber for energy loss measurements reaches $A/\Delta A = 500$ and $Z/\Delta Z = 130$. For the first time, each of the 2000 fragments produced by fission and fragmentation can be identified unambiguously. Fission fragments are produced with similar velocities. The rigidity of the dipole magnets of the FRS is related to the mass over charge ratio by $B\rho = 3.107\ A/Z\ \beta\gamma$, where $\gamma$ is the Lorenz factor. Thus, when the rigidity is increase, the A/Z ratio of selected fragments increases linearly. An example is shown in Figure 1 of the scatter plot of the fragments transmitted at a rigidity larger than the value leading to the maximum of the fission yield. Each cloud corresponds to an isotope defined in A and Z. New isotopes appear on the right of the line, the current limit of the known species.

Measurements were performed by tuning the FRS to higher and higher rigidities and using Pb and Be targets. Altogether, 117 new isotopes were identified and their cross sections measured (Bernas *et al.*, 1997; Engelmann *et al.*, 1995, 1999). These are shown in Figure 2.

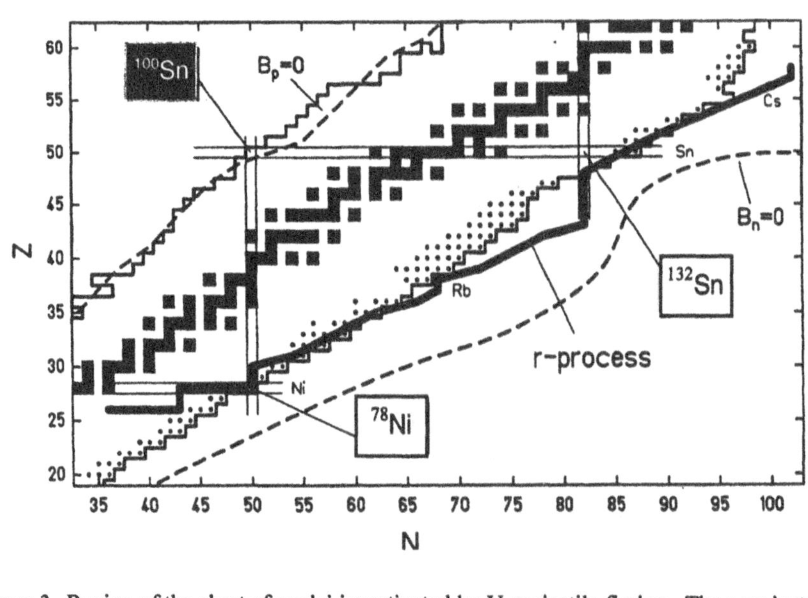


*Figure 2.* Region of the chart of nuclei investigated by U-projectile fission. The new isotopes are indicated with dots while the r-process path is illustrated with the thick full line.

## 3. HOW TO MEASURE $T_{1/2}$ WITH LOW PRODUCTION RATES

In thermal fission the low-energy fragments are emitted isotropically. Given their small range, they are implanted on the very surface of the detector. In projectile-fission with the FRS tuned in a mono-energy mode, samples of a given isotope come out with the same range. They are selectively implanted in the 0.3 mm thick Si pin-diodes. The next more abundant isotone simultaneously transmitted comes out with an equal range; in the frame of event-by-event analysis, this contamination provides a good test for the extraction of the value of $T_{1/2}$. Time correlation between implant and β signals are the basic data from which to deduce beta-decay half-lives. With this method, 22 new isotopes were investigated for the first time (Ameil *et al.*, 1998).

To extract the decay constant λ from time correlation, we encounter two limitations:

- The β-detection efficiency is only around 50%.
- The β signal is very small and not well characterized. It cannot be distinguished from background signals.

Therefore reliable information requires a few conditions:

- To keep the background level as low as possible. This is the reason for the division of the implanting device into 20 independent cells.
- The time structure of the beam must leave an appropriate time window for observing the β within a low background environment.
- The test on an implant with a known β-decay half-life must be systematically performed.
- The statistical analysis of the time correlation must be appropriate.

The ratio between the decay constant λ and the background rate b should also be sufficiently large. In order to make this tangible, we plot the time correlation between the implant and the first β detected in terms of N(t) = f [1n(t)] (Schmidt *et al.*, 1984). Let us consider the Poisson law for the background:

$$\Delta N = N_F b \exp[-(bt)]\Delta t$$

or

$$\Delta N = N_F\, bt \exp[-(bt)]\Delta t / t$$

with dt/t = d[1n (t)] we obtain the standard curve u exp –u. Adding the contribution of the decay with a constant λ, with ε the detection efficiency of a β-particle the formula becomes:

$$\Delta N = N_F\,[(1-\varepsilon)bt\,\exp-(bt)+\varepsilon(b+\lambda)t\,\exp-(b+\lambda)t]\,\Delta t/t \qquad (1)$$

$$i.e.,\ [(1-\varepsilon)u\exp-u+\varepsilon v\exp-v]d[\ln(t)].$$

This is shown in Figure 3.

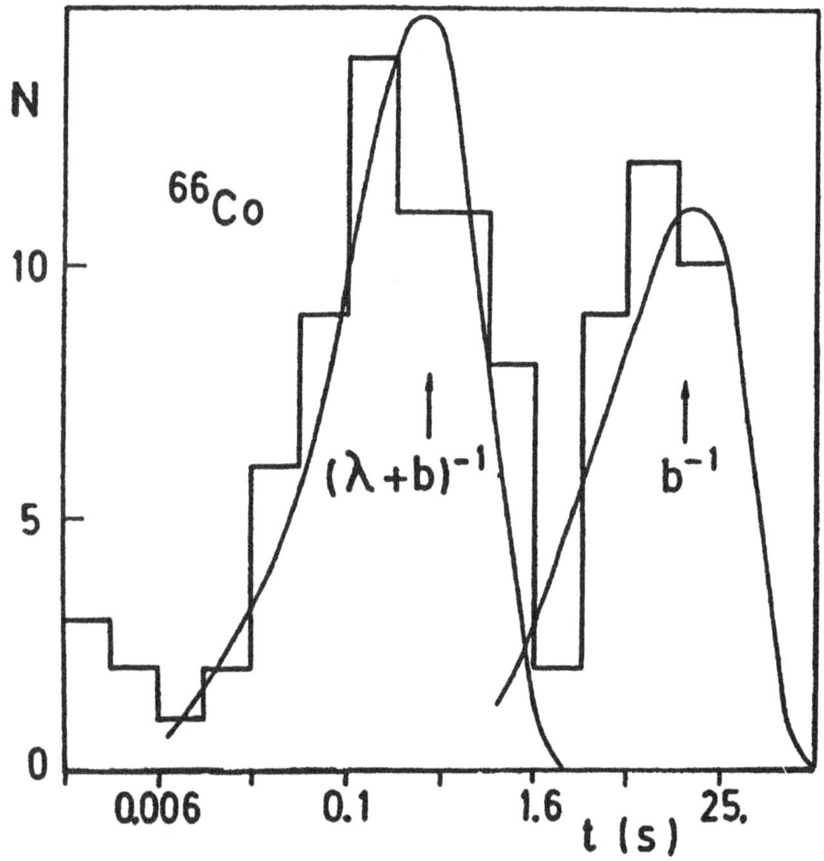


*Figure 3.* Distribution of time-gaps between identified implants and first β signals. The choice of coordinates emphasizes a clear cut between real and background contributions.

We clearly see that each contribution must be seperate with a ratio of at least 20 between λ and b+λ. This ratio depends on ε and could be less in case of a larger ε or better statistics.

To safely extract the decay constant λ, we determine independently the background b (/s/det) and ε, the β-detection-efficiency. The value of b was obtained by constructing the random time-reversed correlation, between implants and the β detected *previously*. The value of ε was extracted as the unknown parameter when reproducing the decay-curve of the isotonic implant. It was cross-checked by different ways. The least square fit and the maximum likelihood methods were used to extract λ (Bernas *et al.*, 1990). The daughter decay was included when necessary, *i.e.* when it was comparable to the primary decay constant.

Being far apart from stability, the nuclei undergo a cascade of fast β-decays. If the time-gap is much less than 1/b sec, we can observe a sequence of 2 and eventually 3 decays. The first decay is identified with an efficiency of ε, the sequence of 2 β-decays with efficiency $\varepsilon^2$, etc. We select the faster ones among the β-signals following the implant, those occurring in less than twice the presumed half-life. Among these events we choose the cases where a second and then a third β signals are recorded in a time smaller than twice the respective daughter half-lives. For each step the half-lives are deduced from a small number of valid correlation. In the case of the $^{65}$Mn decay chain (Figure 4) the Co half-life, already known, confirms the validity of the criteria and confirms Mn and Fe half-lives. The frequency of the next β, from $^{65}$Ni, is dominated by b (Ameil *et al.*, 1998).

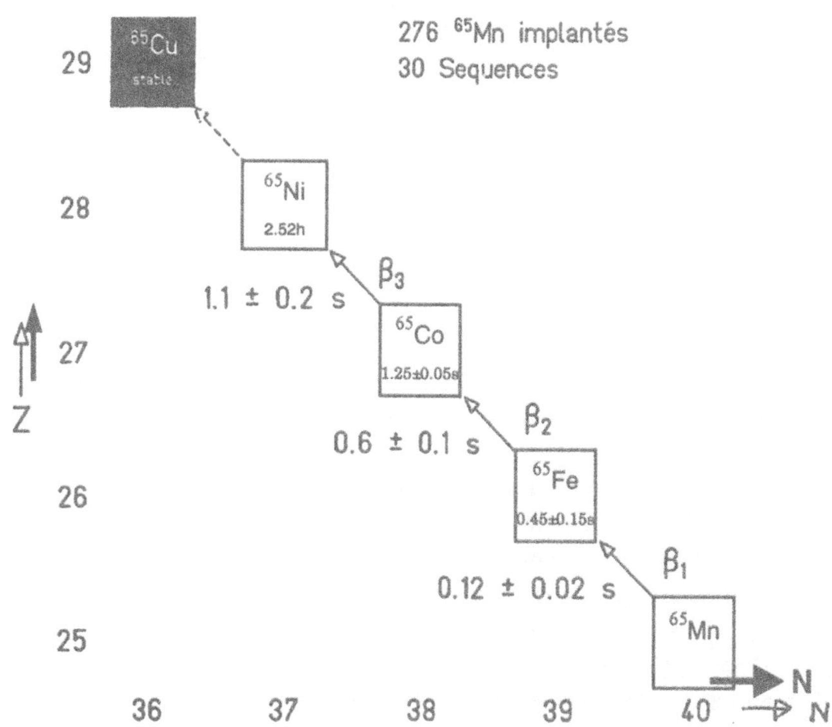


*Figure 4.* Fast sequence of β-decay.

# 4. RESULTS

Figure 5 illustrates the present results obtained for n-rich isotopes in deep-inelastic reactions (Runte *et al.*, 1985), in thermal fission (Bernas *et al.*, 1990), in projectile fragmentation (Ameil *et al.*, 1998) and in a recent experiment by Franchoo *et al.* (1998) where ions were extracted using selective ionization. The experimental half-lives are compared with theoretical expectations. Values obtained by J. Zylicz *et al.* (1998) for Ni agree rather well with the experimental values. In case of the more general calculation of Möller *et al.* (1997), there is a discrepancy of about a factor of 5.

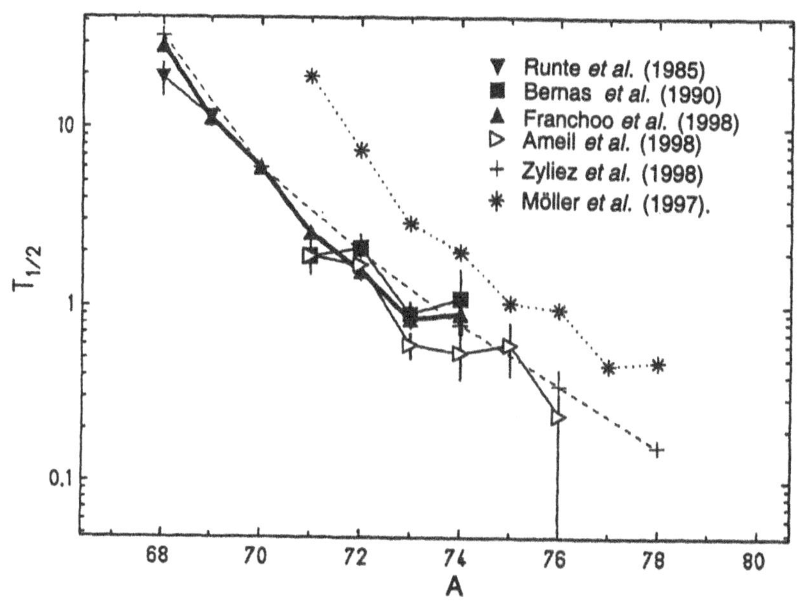


*Figure 5.* Comparison of half-lives of Ni-isotopes obtained experimentally with recent calculations.

A wider comparison is performed over the 7 elements investigated by Ameil *et al.* (1998). A mean discrepancy of the order 5 is found, except for V isotopes where it is larger. The more n-rich the nuclei, the better the models reproduce the data mainly with the Gross Theory (Tachibana *et al.*, 1988). For V, the value of the $T_{exp}/T_{theor}$ ratio increases abruptly with the n-excess, whatever the model.

Compared with the predictions of P. Möller *et al.* (1997), we observe: a) a staggering of this ratio for even-odd elements, and b) a smaller value of this ratio ($T_{theor}$ too large) for isotopes with even number of neutrons, mostly for Ti, Cr and Fe.

# 5. CONCLUSION AND PROSPECTS

We have shown that β-decay half-lives of very n-rich isotopes can be measured even with a limited number of selected ions unambiguously separated and implanted in a Si detector. Our method does not rely on any physico-chemical properties of the element investigated. All β half-lives are accessible.

The same method can be applied to the many new isotopes discovered in U-projectile fission (Bernas *et al.,* 1997). Those rare species are produced daily in nuclear plants and our achievement consists of separation and identification of the isotopes. The values of $T_{1/2}$ can be precisely measured at Isolde, Leuven or Jyväskylä as soon as the element is efficiently extracted. The mass excess remains, however, the very important parameter to know for the r-process. In the future, both $T_{1/2}$ and mass excesses will be simultaneously measured for a large number of n-rich isotopes combining projectile fission at FRS and the ESR in isochronous mode.

# REFERENCES

Ameil, F., Bernas, M., Armbruster, P., Czajkowski, S., Dessagne, P., Geissel, H., Hanelt, E., Kozhuharov, C., Miehe, C., Donzaud, C., Grewe, A., Heniz, A., Janas, Z., Dejong, M., Schwab, W. and Steinhauser, S.: 1998, "β-decay half-lives of very neutron-rich isotopes of elements from Ti to Li", *Eur. Phys. J.* **A1**, 275-283 and Ameil, F., Ph.D. thesis report IPNO-T-97-02 Orsay.

Bernas, M., Armbruster, P., Bocquet, J.P., Brissot, R., Faust, H., Kozhuharov, C. and Sida, J.L.: 1990, "Beta-decay half-lives of neutron-rich Cu and Ni isotopes produced by thermal fission of $^{235}$U and $^{239}$Pu", *Z. Phys.* **A336**, 41-51.

Bernas, M., Armbruster, P., Czajkowski, S., Faust, H., Bocquet, J.P. and Brissot, R.: 1991, "Discovery of neutron-rich Co and Fe isotopes in $^{239}$Pu($n_{th}$,f) - yields and half-lives", *Phys. Rev. Lett.* **67**, 3661-3664.

Bernas, M., Dessagne, P., Girod, M., Langevin, M., Payet, J., Pougheon, F., Roussel, P., Schmidtott, W.D. and Tidemandpetersson, P.: 1984, "Mass and excited levels of the neutron-rich nuclei $^{73}$Zn and $^{74}$Zn", *Nucl. Phys* **A413**, 363-374.

Bernas, M., Englemann, C., Armbruster, P., Czajkowski, S., Ameil, F., Bockstiegel, C., Dessagne, P., Donzaud, C., Geissel, H., Heinz, A., Janas, Z., Kozhuharov, C., Miehe, C., Munzenberg, G., Pfutzner, M., Schwab, W., Stephan, C., Summerer, K., Tassangot, L., and Voss, B.: 1997, "Discovery and cross-section measurement of 58 new fission products in projectile-fission of 750 A.MeV U-238", *Phys Lett.* **B415**, 111-116.

Burbidge, G.R., Burbidge, E.M., Fowler, W.A. and Hoyle, F.: 1957, "Synthesis of the elements in stars", *Rev. Mod. Phys.* 29, 547-650.

Dessagne, P., Bernas, M., Langevin, M., Morrison, G.C., Payet, J., Pougheon, F. and Roussel, P.: 1984, "The complex transfer reaction ($^{14}$C, $^{15}$O) on Ni, Zn and Ge targets; existence and mass of $^{69}$Ni", *Nucl. Phys.* **A426**, 399-412.

Engelmann, C., Ameil, F., Armbruster, P., Bernas, M., Czajkowski, S., Dessagne, P., Donzaud, C., Geissel, H., Heinz, A., Janas, Z., Kozhuharov, C., Miehe, C., Munzenberg,

G., Pfutzner, M., Rohl, C., Schwab, W., Stephan, C., Summerer, K., Tassangot, L. and Voss, B.: 1995, "Production and identification of heavy Ni isotopes - evidence for the doubly magic nucleus $Ni^{78}_{28}$" *Zeit. Phys.* **A352**, 351-352.

Franchoo, S., Bruyneel, B., Huyse, M., Koester, U., Kratz, K.-L., Krugov, K., Kudryavtsev, Y., Mueller, W.F., Pfeiffer, B., Raabe, R., Reusen, I., Thirolf, P., Van Duppen, P., Van Roosbroeck, J., Vermeeren, L., Walters, W.B., Weissman, L. and Woehr, A.: 1998, "β-decay of neutron-rich cobalt and nickel isotopes", in *ENAM '98 Int. Conf.*, Bellaire, MI (USA), pp 757-760.

Girod, M., Dessagne, P., Bernas, M., Langevin, M., Pougheon, F. and Roussel, P.: 1988, "Spectroscopy of neutron-rich nickel isotopes: Experimental results and microscopic interpretation", *Phys. Rev.* **37**, 2600-2612.

Möller, P., Nix, J.R. and Kratz, K.L.: 1997, "Nuclear properties for astrophysical and radioactive-ion-beam applications", *At. Data Nucl. Data tables* **66**, 131-343.

Runte, E., Tidemandpetersson, P., Roeckl, E., Schardt, D., Schmidtott, W.D., Rykaczewski, K., Peuser, P., Ziegeler, L., Dessagne, P., Bernas, M., Gippert, K.L., Kaffrell, N., Klepper, O., Kirchner, R., Langevin, M. and Larsson, P.O.: 1985, "Decay study of neutron-rich isotopes of Mn, Fe, Co, Ni, Cu and Zn", *Nucl. Phys.* **A441**, 237-260.

Schmidt, K.H., Clerc, H.G., Pielenz, K. and Sahm, C.C.: 1984, "Some remarks on the error analysis in the case of poor statistics", *Zeit. Phys.* **A316**, 19-26.

Sida, J.L., Armbruster, P., Bernas, M., Bocquet, J.P., Brissot, R. and Faust, H.R.: 1989, "Mass, charge and energy distributions in very asymmetric thermal fission of $^{235}U$", *Nucl. Phys.* **A502**, C233-C241.

Tachibana, T., Uno, M., Yamada, M. and Yamada, S.: 1988, "Empirical mass formula with proton neutron interaction", *At. Data Nucl. Data Tables* **39**, 251-258.

Zylicz, J., Dobaczewski, J. and Szymanski, Z.: 1998, "Gamov-Tellar decay of even isotopes $^{68}Ni$ to $^{78}Ni$", in *ENAM '98 Int. Conf.*, pp 813-815.

# The Role of the N = 28 and N = 40 Closed Shells in the Production of the Neutron-Rich Ca-Ti-Cr-Fe-Ni Elements in the Universe

O. Sorlin
*Institut de Physique Nucléaire, IN2P3-CNRS, 91406 Orsay Cedex, France*
*sorlin@ipno.in2p3.fr*

**Abstract:** The nuclear structure of neutron-rich elements, as for instance their beta-decay times, may considerably influence the neutron-capture paths in explosive stellar events. Special care is given in this contribution to the links between nuclear and astrophysics for the synthesis of the stable neutron-rich Ca-Ti-Cr-Fe-Ni isotopes.

## 1. INTRODUCTION

The interpretation of the observed elemental and isotopic composition in the universe is intimately related to the description of their originating nucleosynthesis processes. Each of these processes is characterized by different conditions of temperature, neutron or proton fluxes which result in a wide variety of nuclear reactions involved. By direct comparison between the calculated and observed abundances of the elements for given conditions, a probable astrophysical site can be assessed. The origin of the neutron-rich stable isotopes $^{48}$Ca, $^{50}$Ti, $^{54}$Cr, $^{58}$Fe, $^{64}$Ni is still unclear. These isotopes are found in correlated overabundances, as compared to the solar system ones, in certain refractory inclusions of meteorites. It is still questionable how and where all these neutron-rich isotopes can be produced in a self-consistent way. Under a neutron flux of $n_n \sim 4.10^{19}$ $cm^{-3}s^{-1}$ and a temperature of $T \sim 10^9$K, it is possible to produce neutron-rich unstable nuclei of atomic number A = 48, 50, 54, 58 and 64 by successive neutron-captures from stable seed nuclei. These neutron-rich isotopes will be plausible progenitors of the stable $^{48}Ca$, $^{50}Ti$, $^{54}Cr$, $^{58}Fe$ and $^{64}Ni$ after β-decay. These progenitors are found in the neutron-capture path when the β-decay time in a given isotopic chain is shorter than the neutron-capture time. The location of these "turning-points" requires the knowledge of their β-decay half-lives. These short-lived nuclei (few tens of milliseconds) have to be synthesized artificially in laboratories in order to measure their half-lives. An evaluation of the neutron-capture β-decay path under different stellar conditions can subsequently be deduced using calculated neutron-capture cross-sections.

After an introduction on the main motivations for studying neutron-rich nuclei, experimental results of astrophysical motivation on nuclei "south to" $^{48}Ca$ will be shown. A typical experiment will then be presented, focusing on nuclei in the Ti-Cr region. The way to produce these neutron-rich species in terrestrial laboratories, and the experimental set-up to determine their half-lives will be described. Some results will be presented and discussed both in the frame of nuclear and astro-physics.

## 2. EXOTIC NUCLEI AND MAGICITY

The field of nuclear research has been drastically widened for the last 15 years due to the recent access to species which have either a large proton or neutron enrichment as compared to stable species observed on earth. One of the main goal of nuclear physics is to modelize an "ensemble" of nucleons (neutrons and protons) in strong and short-range interactions. From the study of stable nuclei, it has been demonstrated experimentally that some edifices of nucleons are more tightly bound than others. This occurs for given numbers of protons (Z) or neutrons (N) which are equal to 8, 20, 28, 50, 82 and 126. These numbers have been coined as magic numbers. The strength of this magicity can be characterized by the energy that is required for a nucleus to be excited in its first excited state. The higher this energy is, the harder it is to be excited. When making an analogy with atomic physics, the nuclear magic nuclei have similar properties (with respect to their ability to be excited) as the rare-gas atoms at the very right of the Mendeleev periodic table of the elements. The main difference arises from the fact that nuclei contain two fluids (neutrons and protons) whereas atomic clouds contain only electrons. Also, of course, the energy-range for exciting a nucleus or atom is very different due to the nature of the forces which govern nuclear or atomic realms. In Figure 1, the energy required to excite $_{20}Ca$, $_{16}S$ and $_{12}Mg$ isotopes is shown as a function of their neutron enrichment.

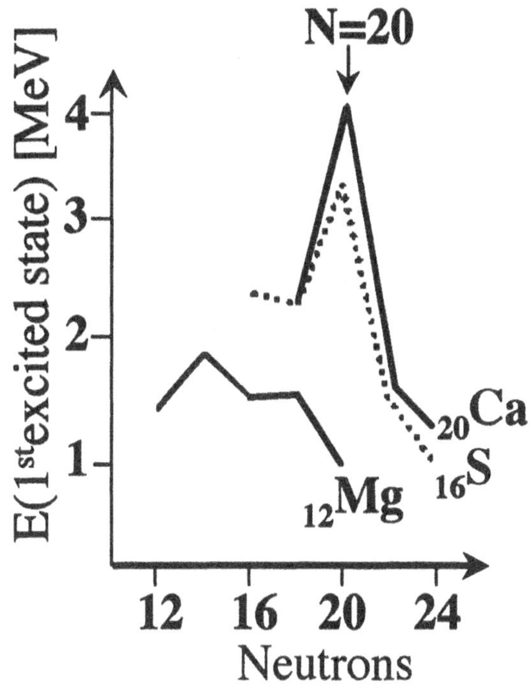


*Figure 1.* First excited states energies of different nuclei as a function of their neutron richness.

The $^{40}Ca$ contains Z = 20 protons and N = 20 neutrons and is thus a doubly magic nucleus. The sudden rise in energy at N = 20 clearly establishes the difficulty to excite $^{40}Ca$ to its first excited state. A similar large energy is required to excite $^{36}S$ which contains four protons less than $^{40}Ca$. The stable $^{36}S$ is already a neutron-rich nucleus. Its abundance is 2% of the total sulfur isotopes. When removing again four protons, the nucleus $^{32}Mg$ exhibits a large neutron excess and is thus unstable by β-decay. It is remarkable to see that it is more easily excited than the other N = 20 isotones even if it contains fewer nucleons and thus fewer degrees of freedom. This reveals a drastic change in the structure of this nucleus which has to be understood. Some theoretical interpretations suggest that this nucleus minimizes its intrinsic potential energy by adopting a strong quadrupole deformation. This small energy of 885 keV (Guillemaud-Mueller, 1984) would correspond to that required to make the nucleus rotate. As in molecules, the rotational mode is the easiest way to excite a deformed structure. The other N = 20 isotones are spherical and rotational modes are inoperable in these nuclei due to the symmetry of the nucleus. The first excited mode for these even Z – even N spherical species corresponds to a quadrupolar surface-vibration about their spherical shape. It is interesting to note that the doubly magic nucleus $^{28}O$, which corresponds to removal of four protons from $^{32}Mg$, is most probably unbound (Tarasov *et al.*, 1997). This feature, together with the low excitation energy in $^{32}Mg$, supports a disappearance of magicity when a large N/Z ratio is encountered. Hence, the well established concept of nuclear magicity must be reconsidered by taking into account more specifically the role of proton or neutron enrichment. Experimental investigations are still being made in order to interpret the possible decoupling of these two fluids which may result in the appearance of new deformed regions of nuclei or new excitation modes between protons and neutrons. Also, the surface of these neutron-rich nuclei may be composed essentially of a diffuse neutron surface, which may result in a modification of the structure of the nucleus and in the appearance of new magic numbers.

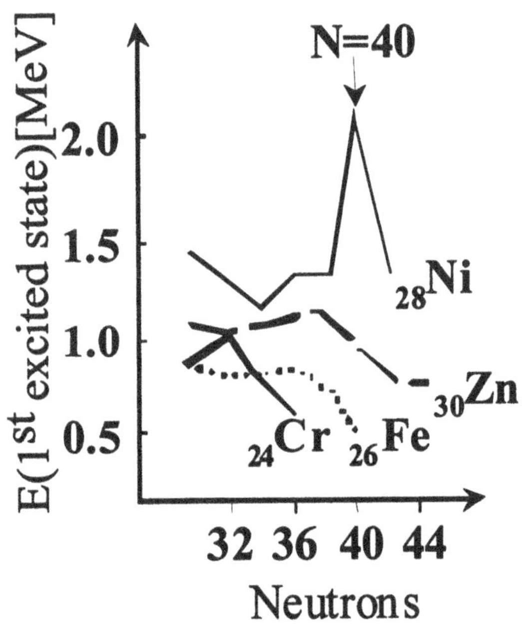


*Figure 2.* First excited states energies of different nuclei as a function of their neutron richness around the neutron number, N=40.

As shown in Figure 2, when removing two protons from the stable $^{70}Zn$, the energy required to excite the N = 40 nucleus, $^{68}Ni$, is substantially higher (Girod *et al.*, 1988; Broda *et al.*, 1995). This effect could be interpreted as the emergence of a new magic number at N = 40 which is magnified by the fact that $^{68}Ni$ is also magic in protons with Z = 28. Another experimental study has shown that this nucleus is spherical (Leenhardt, 2000). When decreasing the number of protons, and thus providing a larger neutron enrichment to the nucleus, this excitation energy decreases drastically in $^{66}$Fe (Hannawald *et al.*, 1999). This surprising result shows that a nucleus is extremely sensitive to the removal or addition of a few nucleons. As we will see in the following section, the N = 40 nucleus $^{64}$Cr is thought to play an important role in the nucleosynthesis of the stable $^{64}$Ni (Figure 3). The energy of its first excited state has not been measured so far, due to the difficulty of producing this neutron-rich nucleus in laboratory. But, from the experimental trend of Figure 2, the N = 40 isotope $^{64}$Cr is expected to have a very low excitation energy, and thus probably a very deformed shape as a minimum of its potential energy. It is therefore very important to measure key nuclear properties in this region of mass, and not to rely too much on theoretical calculations which cannot always predict very accurately these sudden changes in nuclear structure (as masses, deformations, half-lives).

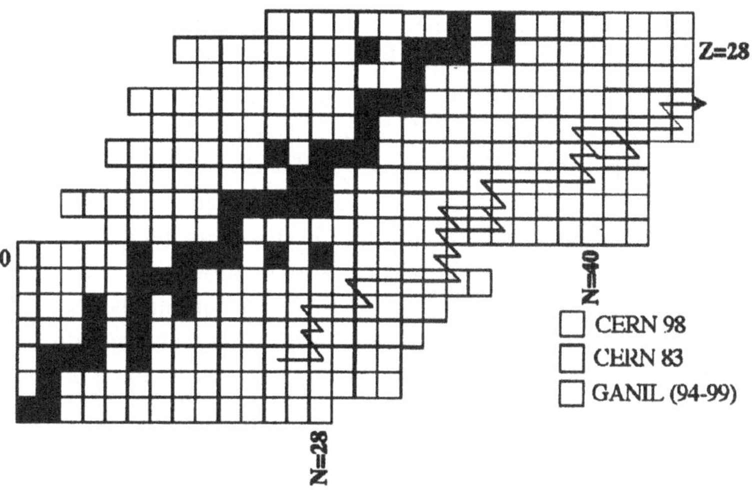


*Figure 3.* r-Process path in the region of sulfur to chromium nuclei for a neutron flux of $4.10^{19}$ cm$^{-3}$ s$^{-1}$ and for a temperature of $10^{9}$ K. Under such conditions, neutron-captures proceed until β-decay occur. Measurements referenced as CERN 98, CERN 83 and GANIL 94-99 are published in Hannawald *et al.* (1999), Langevin *et al.* (1983) and (Sorlin *et al.*, 1993, 1995, 1999), respectively.

## 3. THE Ca-Ti-Cr-Fe-Ni ISOTOPIC "ANOMALIES"

Correlated overabundances of neutron-rich stable isotopes $^{48}$Ca, $^{50}$Ti, $^{54}$Cr, $^{58}$Fe and $^{64}$Ni are found in certain refractory inclusions of meteorites (Harper *et al.*, 1990; Volkening and Papanastassiou, 1989; Loss and

Lugmair, 1990; Birck and Allègre, 1984). This feature emphasizes that these nuclei should be produced in the same astrophysical environment. Since all these nuclei are neutron-rich, it is plausible to think that these isotopes are produced within a neutron-rich stellar environment. Under moderate conditions of neutron flux $n_n \sim 4 \times 10^{19}$ $cm^{-3}s^{-1}$ and temperature of $T \sim 10^9$ K, it has already been suggested that the large $^{48}Ca/^{46}Ca = 270$ abundance ratio observed in the EK-1-4-1 inclusion of the Allende meteorite (Lee *et al.*, 1978) can be explained (Kratz *et al.*, 1991). It is therefore interesting to see whether all other neutron-rich isotopes found in correlated abundances can be produced within the same neutron-density and temperature conditions. Under such a neutron-rich environment, neutron-captures proceed from stable seed nuclei until β-decay time is shorter than the neutron-capture time. At this location, the nucleosynthesis path is depleted to higher Z nuclei at a rate which depends on the β-decay time (See Figure 3).

Beta-decay studies south to $^{46,48}Ca$ region have been investigated in order to find the location of the branching points in the neutron-capture path and to explain why $^{46}Ca$ is so underabundant in the solar system and in this inclusion of meteorite. The neutron-rich nuclei around the N = 28 shell-closure have been produced and studied at the Grand Accélerateur National d'Ions Lourds (GANIL). Beta-decay half-lives of $^{43}P_{28}$, $^{43-45}S_{27-29}$ , $^{44-46}Cl_{27-29}$, has been determined in two experiments (Sorlin *et al.*, 1993, 1995). It has been found that all nuclei with 28 neutrons (expected to be semi-magic nuclei) have half-lives shorter than predicted by factors of 3 to 10. This was interpreted as the vanishing of this magic shell N = 28 when a large neutron enrichment is encountered, as that mentioned in the section II for a neutron number N = 20. As a consequence of these shorter half-lives, the neutron-capture path in the Cl-chain is depleted already at $^{45}Cl_{28}$, without producing the main progenitors of $^{46}Ca$, $^{46}Cl_{29}$ and $^{47}Cl_{30}$, too much. This, hitherto unexpected nuclear structure at the neutron magic number N = 28, could account for the underproduction of $^{46}Ca$ in certain inclusions of meteorites. Such a neutron-flux could be found in the outermost layers of an exploding SNII. For the bulk composition of $^{46}Ca$ and $^{48}Ca$ in the solar system, it has been demonstrated by Meyer *et al.* (1996) that their isotopic ratio can be explained by α-particle captures in SNI. Thus, the bulk solar material of $^{46}Ca$, $^{48}Ca$, $^{50}Ti$, $^{54}Cr$, $^{58}Fe$ and $^{64}Ni$ would be produced from a SNI star, which are much more abundant than SNII ones. However, small amounts of exotic composition, such as that in EK-1-4-1, can be well produced in a SNII scenario. This suggestion would be reinforced if all correlated over-abundances can be produced under the same neutron density and temperature conditions. It is therefore important to determine the half-lives of neutron-rich isotopes of heavier elements in order to find the possible progenitors of $^{58}Fe$ and $^{64}Ni$. The way to produce these nuclei and to determine their half- lives are described in the next section.

# 4. PRODUCTION AND BETA-DECAY MEASUREMENTS OF NEUTRON-RICH Ti-Co ISOTOPES

The neutron-rich $^{57-59}$Ti, $^{59-62}$V, $^{61-64}$Cr, $^{63-66}$Mn, $^{65-68}$Fe, $^{67-70}$Co have been produced at GANIL via interactions of a 60.4 MeV/u $^{86}$Kr beam of $10^{12}$ pps with a $^{58}$Ni target of 140 mg cm$^{-2}$ thickness. During these collisions of $10^{12}$ particles per seconds at about one third of the speed of light, many fragments of different N/Z ratios are produced. For more peripherical collisions, the incident projectile looses few nucleons only, whereas for central collisions it is completely destroyed. Among all the species produced in this process, some, with large neutron enrichment, are of astrophysical interest. These neutron-rich unstable isotopes have to be selected among the $10^{12}$ pps outgoing particles from the target. Due to the kinematics of the reaction, the fragments originating from the break-up of the primary beam are focused at very forward angles. This ensures the possibility of guiding these fragments in a spectrometer of 1 degree angular acceptance with a reasonable efficiency of about 15%. At these velocities v, the fragments are fully stripped of their electrons in their interactions in the target and their total charge Q equals to Z.

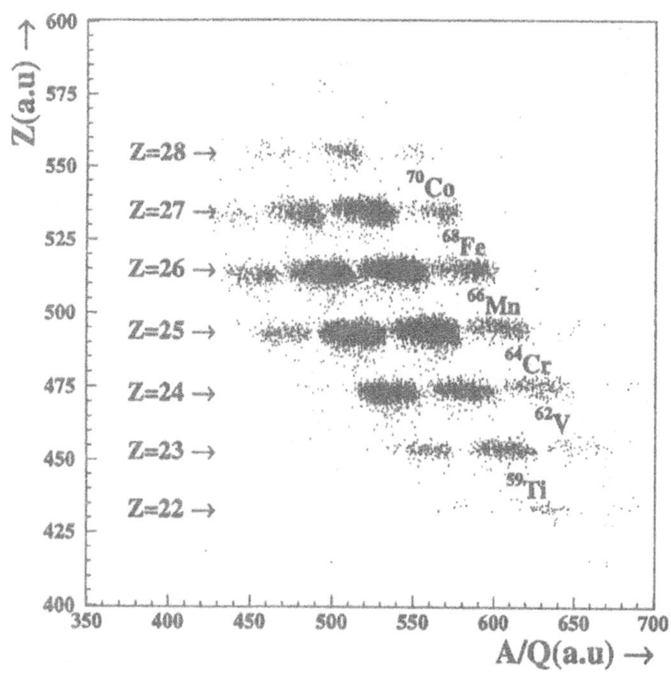


*Figure 4.* Identification matrix of the nuclei produced in the fragmentation of a $^{86}$Kr beam for a given setting of the spectrometer.

The nuclei of interest with mass A and charge Z have been separated by the doubly achromatic spectrometer LISE3 (Anne and Mueller, 1992). This spectrometer is composed of two dipoles which sort out nuclei according to their $A \times v/Z$ ratios according to the laws of magnetic forces. An additional selection is achieved by the use of a thin Be-foil in between the two dipoles of LISE3. The magnetic field of the second dipole is then tuned to match the energy of the chosen fragments after their energy loss in the foil. With this device, a tremendous selection has been achieved, resulting in the optimal transmission of nuclei of interest which are produced at a rate of 1 nucleus per minute and the rejection of the unwanted fragments and remnants of the primary beam. The produced fragments are identified event by event by using Si-detectors in which the energy-loss of a nucleus is proportional to $Z^2/v$. The measured time-of-flight along the spectrometer, after a flight length of 44 meters, provides a complementary identification of each nucleus. An example of an identification matrix is shown in Figure 4. The fragments with the larger neutron-richness are found in the right part of the figure for each isotopic chain Z.

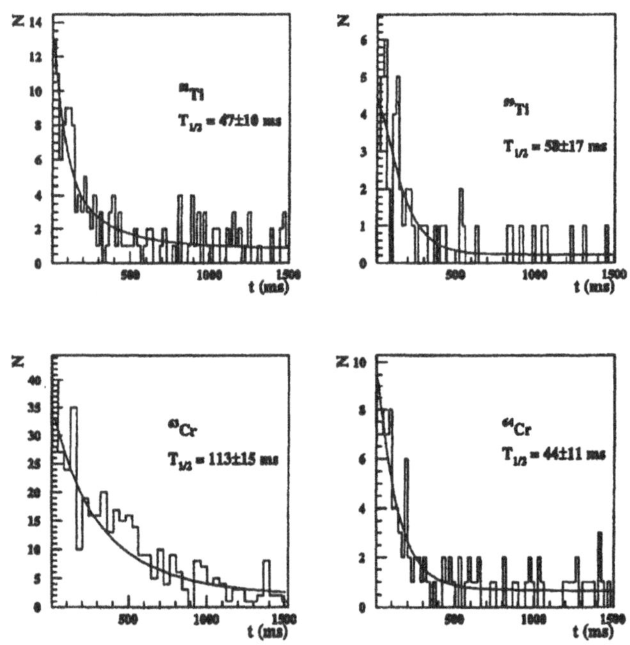


*Figure 5.* Decay curves of Ti and Cr isotopes. Numbers refer to the half-lives.

The selected nuclei were implanted in a 500 μm silicon detector placed behind the first one. This detector was divided in twelve 2 mm-wide, 24 mm-height vertical strips. This detector served both for the detection of the heavy nuclei implanted and for the subsequent β-particles coming from their decay. Each time a nucleus was implanted, the primary beam was switched off during 1.5 seconds, in order to avoid implanting any other nucleus which may also emit a β-particle.

In the case of light-Z neutron-rich nuclei, this duration was long enough to observe the decay of the mother and daughter nuclei. A β-event was considered as valid if occuring in the same strip #i as the precursor nucleus or in one of the neighboring strips, #i-1 and #i+1. Beta-events detected in other strips were counted as β-background and were mainly coming from the decay of long-lived nuclei produced by filiations. By constructing a time histogram between the implantation of a nucleus and the detection of its correlated β, it is possible to deduce the half-lives of the species implanted in this experiment. Figure 5 shows the results obtained for four nuclei of astrophysical interest $^{58,59}$Ti and $^{63,64}$Cr. Other results obtained in this experiment are published in the reference (Sorlin *et al.*, 1999).

## 5. ASTROPHYSICAL IMPLICATIONS

In an astrophysical context, the half-lives of $^{58}$Ti (47(10) ms) and $^{64}$Cr (44(11) ms) are of key importance since these nuclei are expected to be r-progenitors of the stable neutron-rich $^{58}$Fe and $^{64}$Ni respectively. Half-lives of $^{58}$Ti and $^{64}$Cr are about four times shorter than predicted by Möller *et al.* (1997). From these experimental values, the β-decay times are shorter than the 70 ms neutron-capture times for $^{58}$Ti and $^{64}$Cr assuming a neutron flux of 3.6 x $10^{19}$ $cm^{-3}s^{-1}$. The nucleosynthesis path in this region of mass can therefore be modified, both nuclei acting as turning points in the neutron-capture path, subsequently feeding by β-decays the $^{58}$Fe and $^{64}$Ni nuclei, after the freeze-out of the neutron captures. In Figure 6, the expected turning points are shown in the titanium (upper) and chromium (lower) chain. By extrapolation of the $T_{1/2}$ experimental trend, the turning points occur, for this neutron flux, at masses A = 58, 60 and A = 64, 66 in Ti and Cr chains respectively.

It is interesting to note that the N = 40 nucleus $^{64}$Cr, which corresponds to the removal of four protons from the nearly doubly-magic nucleus $^{68}$Ni, would then play an important role in the nucleosynthesis due to its short half-life. This is "analog" to the case of $^{44}$S, which plays an important role in the understanding of the large ratio $^{48}$Ca/$^{46}$Ca = 270 observed in the EK-1-4-1 meteoritic inclusion (Lee *et al.*, 1978). These two cases at N = 40 and at N = 28 confirm that the appearance or vanishing of magicity far from stability should be considered carefully when trying to reproduce the abundance of the neutron-rich stable elements in refractory inclusions of meteorite or in the bulk solar-system.

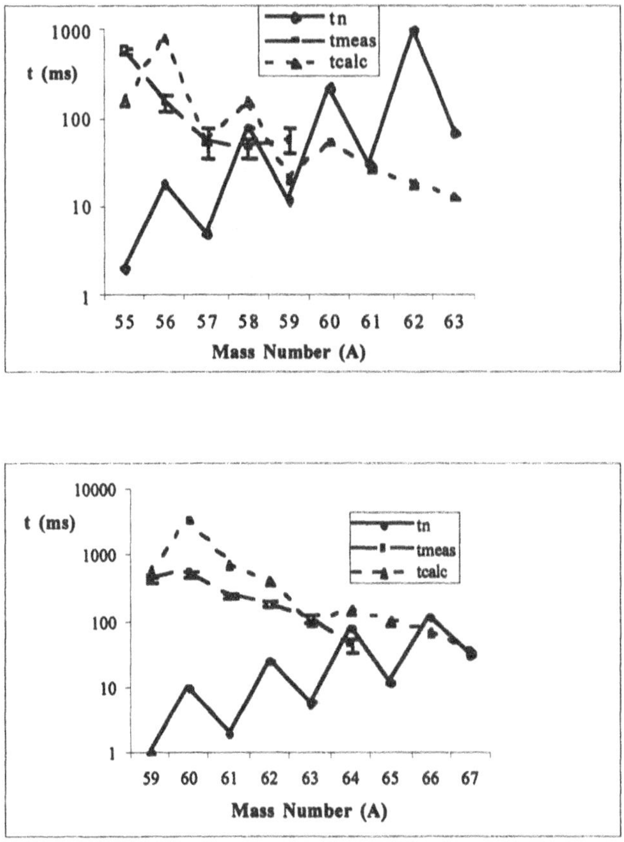


*Figure 6.* Comparison between neutron-capture times $t_n$ and measured β-decay times $t_{meas}$ in the Ti isotopic chain (upper) and Cr chain (lower). Turning points are located where the β-decay times are shorter than the neutron-capture times. Values of $t_n$ are calculated by Thielemann *et al.* (1986) using a simple Hauser Feshbach formalism and a neutron flux of 3.6 x $10^{19}$ $cm^{-3}$. They are indicated by thick lines. These times are also compared to calculated half-lives of Möller *et al.* (1997).

## 6. CONCLUSION AND PERSPECTIVES

Experimental beta-decay studies around the N = 28 and N = 40 shell or subshell closures have demonstrated unexpected patterns as compared to theoretical predictions. In particular, the half-lives of $^{45}Cl$, $^{58}Ti$, and $^{64}Cr$ are shorter than predicted. These three nuclei are thus considered as branching points in the neutron-capture path, assuming a neutron flux of 3.6 x $10^{19}$ $cm^{-3}s^{-1}$. It is interesting to see that under this parameter-study condition, both the underabundance of $^{46}Ca$ and the overabudance of $^{58}Fe$ and $^{64}Ni$

could be explained. More realistic time-dependant calculations should be achieved, including these recent experimental informations, in order to check the above statement. A better treatment of the neutron-captures should be also taken into account when nuclear structure information is available. Rauscher *et al.* (1995) has demonstrated that the disappearance of the magicity at N = 28 has an important influence on the calculated neutron-capture times. Eventually, a realistic exploding stellar site can be deduced from the comparison of the abundance pattern found in the calculation and that found in the EK1-4-1 inclusion of meteorite.

## REFERENCES


Anne, R. and Mueller, A.C.: 1992, "LISE-3: A magnetic spectrometer wien filter combination for secondary radioactive beam production", *Nuc. Instr. Meth.* **B70**, 276-285.

Birck, J.-L. and Allègre, C.J.: 1984, "Chromium isotope anomalies in Allende refractory inclusions", *Geophys. Res. Lett.* **11**, 943-946.

Broda, R., Fornal, B., Krolas, W., Pawlat, T., Bazzacco, D., Lunardi, S., Rossi Alvarez, C., Menegazzo, R., De Angelis, G., Bednarczyk, P., Rico, J., De Acuna, D., Daly, P.J., Mayer, R.H., Sferrazza, M., Grawe, H., Maier, K.H. and Schubart, R.: 1995, "N = 40 neutron subshell closure in the Ni-68 nucleus", *Phys. Rev. Lett.* **74**, 868-871.

Girod, M., Dessagne, P., Bernas, M., Langevin, M., Pougheon, F. and Roussel, P.: 1988, "Spectroscopic of neutron-rich nickel isotopes – Experimental results and microscopic interpretation", *Phys. Rev.* **C37**, 2600-2612.

Guillemaud-Mueller, D., De Saint-Simon, M., Détraz, C., Epherre, M., Langevin, M., Naulin, F.: 1984, "Beta-decay schemes of very neutron-rich sodium isotopes and their descendants", *Nucl. Phys.* **A426**, 37-76.

Hannawald, M., Kautzsch, T., Wohr, A., Walters, W.B., Kratz, K.L., Fedoseyev, V.N., Mishin, V.I., Bohmer, W., Pfeiffer, B., Sebastian, V., Jading, Y., Koster, U., Lettry, J. and Ravn, H.L.: 1999, "Decay of neutron-rich Mn nuclides and deformation of heavy Fe isotopes", *Phys. Rev. Lett.* **82**, 1391-1394.

Harper, C.L., Nyquist, L.E., Shih, C.-Y. and Wiesmann, H.: 1990, "The isotopic astronomy of correlated $^{96}$Zr-$^{50,49,47}$Ti anomalies in the Allende meteorite", in *Nuclide in the Cosmos*, Proceedings of the 1st International Symposium on Nuclear Astrophysics, Baden, Vienna, eds., Oberhummer, H. and Hillebrandt, W., MPI für Physik und Astrophysik, Garching, pp. 138-144.

Kratz, K.-L., Anne, R., Bazin, D., Borcea, C., Borrell, V., Détraz, C., Dogny, S., Gabelmann, H., Guillemaud-Mueller, D., Hillebrandt, W., Lewitowicz, M., Lukyanov, S.M., Möller, P., Mueller, A.C., Penionzhkevich, Yu, E., Pfeiffer, B., Pougheon, F., Saint-Laurent, M.G., Salamatin, S.M., Schäfer, F., Sohn, H., Sorlin, O., Thielemann, F.-K. and Wöhr, A.: 1991, "Exotic nuclide – The nuclear structure origin of isotopic anomalies in the Allende meteorite", in *Proceedings of the 1st Europ. Biennal Workshop on Nucl. Physics*, Megeve, France, 1991, eds., Guinet, D. and Pizzi, J.P., World Scientific, Singapore, pp. 218-226.

Langevin, M., Détraz, C., Epherre, M., Guillemaud-Mueller, D., Klotz, G., Miehe, C., Mueller, A.C., Richard Serre, C., Thibault, C., Tochard, F. and Walter, G.: 1983, "K-53, K-54 and Ca-53: Three neutron-rich isotopes", *Phys. Lett.* **B130**, 251-253.

Lee, T., Papanastassiou, D.A. and Wasserburg, G.J.: 1978, "Calcium isotopic anomalies in the Allende meteorite", *Ap. J. Lett.* **220**, L21-L25.

Leenhardt, S.: 2000, "Etude du caractere doublement magique du $^{68}$Ni", thesis work, University of Orsay, France, to be published.

Loss, R.D. and Lugmair, G.W.: 1990, "Zinc isotope anomalies in Allende meteorite inclusions", *Ap. J. Lett.* **360**, L59-L62.

Meyer, B.S., Krishnan, T.D. and Clayton, D.D.: 1996, "Ca-48 production in matter expanding from high temperature and density", *Ap. J.* **462**, 825-838.

Möller, P., Nix, J.R. and Kratz, K.-L.: 1997, "Nuclear properties for astrophysical and radioactive-ion-beam applications", *At. Data Nucl. Data Tables* **66**, 131-343.

Rauscher, T., Böhmer, W., Kratz, K.-L., Balogh, W. and Oberhumer, H.: 1995, "Theoretical neutron-capture cross-sections for r-process nucleosynthesis in the $^{48}$Ca region", in *Proceedings of the Int. Conf. On Exotic Nuclei and Atomic masses*, ENAM95, Arles, France, eds., de Saint-Simon, M. and Sorlin, O., Editions Frontières, Gif sur Yvette, France, pp. 683-688.

Sorlin, O., Donzaud, C., Axelsson, L., Belleguic, M., Béraud, R., Borcea, C., Canchel, G., Chabanat, E., Daugas, J.M., Emsallem, A., Guillemaud-Mueller, D., Kratz, K.-L., Leenhardt, S., Lewitowicz, M., Longour, C., Lopez, M.J., Saint-Laurent, M.G. and Sauvestre, J.E.: 1999, "Beta-decay half-lives of neutron rich Ti-Co isotopes", *Nucl. Phys.* **A**, in press.

Sorlin, O., Guillemaud-Mueller, Anne, R., Axelsson, L., Bazin, D., Böhmer, W., Borrel, V., Jading, Y., Keller, H., Kratz, K.-L., Lewitowicz, M., Lukyanov, S.M., Mehren, T., Mueller, A.C., Penionzhkevich, Y.E., Pougheon, F., Saint-Laurent, M.G., Salamatin, V.S., Shoedder, S. and Wöhr, A.: 1995, "Beta-decay studies of far from stability nuclide near N = 28", *Nucl. Phys.* **A583**, C763-C768.

Sorlin, O., Guillemaud-Mueller, D., Mueller, A.C., Borrel, V., Dogny, S., Pougheon, F., Kratz, K.-L., Gabelmann, H., Pfeiffer, B., Salamatin, V.S., Anne, R., Borcea, C., Fifield, L.K., Lewitowicz, M., Saint-Laurent, M.G., Bazin, D., Détraz, C., Thielemann, F.-K. and Hillebrandt, W.: 1993, "Decay properties of exotic N-similiar-or-equal-to-28-S and Cl nuclide and the Ca-48/Ca-46 abundance ratio", *Phys. Rev.* **C47**, 2941-2953.

Tarasov, O., Allatt, R., Angélique, J.C., Anne, R., Borcea, C., Dlouhy, Z., Donzaud, C., Grévy, S., Guillemaud-Mueller, D., Lewitowicz, M., Lukyanov, S., Mueller, A.C., Nowacki, F., Oganessian, Yu., Orr, N.A., Ostrowski, A.N., Page, R.D., Penionzhkevich, Yu., Pougheon, F., Reed, A., Saint-Laurent, M.G., Schwab, W., Sokol, E., Sorlin, O., Trinder, W. and Winfield, J.S.: 1997, "Search for 280 and study of neutron-rich nuclei near N = 20 shell closure", *Phys. Lett.* **B409**, 64-70.

Thielemann, F.-K., Arnould, M. and Truran, J.W.: 1986, "Thermonuclear reaction rates from statistical model calculations", in *Advances in Nuclear Astrophysics*, eds., Vanghioni-Flam, E. Audouze, J., Cassè, M., Chieze, J.P. and Tran Thanh Van, J., Editions Frontières, Gif sur Yvette, France, 611 pp.

Volkening, J. and Papanastassiou, D.A.: 1989, "Iron isotope anomalies", *Ap. J. Lett.* **347**, L43-L46

# Experimental Studies Related to s-Process Abundances

K. Wisshak, F. Voss, C. Arlandini, and F. Käppeler
*Forschungszentrum Karlsruhe, Institut für Kernphysik, P.O. Box 3640,D-76021 Karlsruhe, Germany*
*wisshak@ik3.fzk.de*

**Abstract:** The accurate determination of the neutron capture cross sections at stellar temperatures of kT = 10-100 keV is an essential prerequisite for investigations of the nucleosynthesis of heavy elements. For this purpose a $4\pi$ Barium Fluoride Detector was constructed at Karlsruhe and implemented at the 3.75MV Van de Graaff accelerator. This detector combines high efficiency with good energy resolution for the detection of gamma-rays, and allows to register capture events with a probability of ~ 98%. Neutrons are produced by the $^7Li(p,n)^7Be$ reaction. Over the last decade the cross sections of 57 isotopes have been measured in the mass range from Cd to Ta with typical uncertainties of 1-2%, an improvement of factors 5-10 compared to other methods. These data were used in extensive s-process studies based on phenomenological and stellar models for deriving reliable sets of s- and r-process abundances.

## 1. INTRODUCTION

Neutron reactions are responsible for the formation of all elements heavier than iron. The corresponding scenarios relate to helium burning in Red Giant stars and to supernova explosions. In the first case, moderate neutron fluxes are produced in the slow neutron capture process (s-process) by $(\alpha,n)$ reactions on $^{13}C$ and $^{22}Ne$, which imply neutron capture times much longer than typical $\beta$-decay times. Thus the reaction path follows the valley of beta stability and the respective neutron capture reactions are accessible to laboratory experiments. In the second case, neutron fluxes are more than ten order of magnitudes larger, giving rise to the rapid neutron capture process (r-process). The correlated synthesis path is shifted by 10 to 20 mass units

*Origin of Elements in the Solar System: Implications of Post-1957 Observations.*
Edited by O. Manuel, Kluwer Academic/Plenum Publishers, 2000.

to the neutron-rich side of the stability valley. Since experiments in this region are extremely difficult, by far most of the required nuclear data have to be obtained by model calculations.

On average, the observed solar abundances of the heavy elements, $N_{\odot}$, are produced by both processes in equal parts. The reliable separation of the two components represents a key problem for the comparison with the abundance distributions predicted by nucleosynthesis studies. Since the s-process part can be determined for all isotopes (Käppeler *et al.*, 1989) the r-process contribution follows simply from the difference to the solar values, $N_r = N_{\odot} - N_s$. An essential feature of the s-process results from the rather steady neutron flux which leads to reaction equilibrium over large mass regions. The produced abundances $N_s$ are inversely proportional to the respective stellar (n,γ) cross sections $\langle\sigma\rangle$ thus yielding the product $N_s \langle\sigma\rangle$ as a smooth function of mass number A. This function can be modeled by a phenomenological approach known as the canonical s-process (Käppeler *et al.*, 1989). The few parameters of that prescription can be determined by fitting the empirical $N_s \langle\sigma\rangle$ values of those isotopes which are of pure s-process origin ($N_s = N_{\odot}$) since they are shielded by stable isobars against β-decays from the r-process region. Important nuclei in that respect are $^{110}$Cd, $^{124}$Te, $^{136}$Ba, $^{150}$Sm, or $^{160}$Dy.

Once the function $N_s \langle\sigma\rangle$ (A) is determined, the reliability of the isotopic s-process abundances depends essentially on the accuracy of the respective cross sections and solar abundances. In general, both data sets exhibit uncertainties of 5 to 10% (Käppeler, 1999), except for the important region of the rare earth elements, where the solar abundances are defined to ~ 2% because of the chemical similarity. In this region between $^{139}$La and $^{176}$Lu, the accuracy of the $N_s \langle\sigma\rangle$ values is clearly limited by the cross sections.

In view of this situation a new effort was started in Karlsruhe some 10 years ago to set up an experiment which allows to determine the required (n,γ) cross sections at keV energies with uncertainties of ~ 1%. Measurements started in the lanthanide region but were meanwhile extended to lower and higher mass numbers as well. Accurate cross sections are also required for analyses of branchings in the s-process path which give information on the physical parameters like temperature and neutron density during the s-process. These analyses are important constraints for stellar models, for which considerable progress was achieved in the past few years. With the new set of accurate cross sections the s-abundances can now be determined with uncertainties of 5% (Arlandini *et al.*, 1999).

The discovery of isotopic anomalies in acid-resistant residues found in a certain class of meteorites has opened a completely new field of nucleosynthesis studies. The carriers of these anomalies were identified as presolar grains enriched in s-process material that have formed in the atmospheres of mass-loosing red giants. With modern spectroscopic methods these anomalies can be precisely measured even for individual grains. Hence, their interpretation in terms of an s-process origin requires correspondingly accurate neutron capture rates as well.

## 2. EXPERIMENTAL

A schematic sketch of the experimental set-up is shown in Figure 1. Continuous neutron spectra in the energy range from 3 to 225 keV were produced via the $^{7}Li(p,n)^{7}Be$ reaction using the pulsed proton beam of the Karlsruhe Van de Graaff accelerator. The average beam characteristics are a repetition rate of 250 kHz, a pulse width of 700 ps, and an average current of 2 μA. Compared to other sources, neutron production by nuclear reactions leads to relatively low γ-ray backgrounds. Hence, a rather compact shielding of the neutron target is sufficient, allowing for short flight paths and correspondingly favorable neutron fluxes. In the present experiment, the neutron energy is determined by the time-of-flight technique using a flight path of only 77 cm.

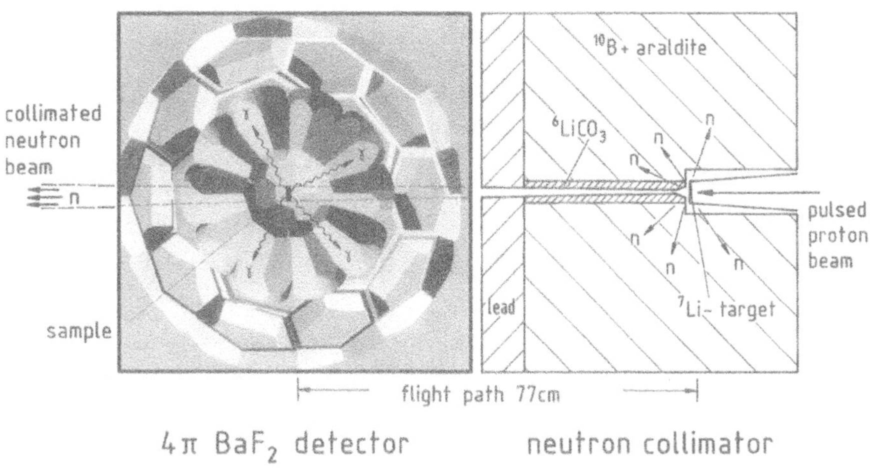


*Figure 1.* Schematic view of the experimental set-up at the accelerator.

The complete γ-cascade from neutron capture events is registered by the $4\pi BaF_2$ detector (Wisshak *et al.*, 1990), which consists of a spherical shell of barium fluoride with 20 cm inner diameter and 15 cm thickness subdivided into 42 hexagonal and pentagonal crystals. Each crystal is supplied with a reflector and a photomultiplier tube, thus representing an independent γ-ray detector. The 4π array is characterized by an energy resolution of 14% at

662 keV and 7.1% at 2.5 MeV. The time resolution is ~ 500 ps and the efficiency exceeds 90% in the γ-ray energy range up to 10 MeV.

Up to 9 samples can be mounted on a sample changer, three positions being reserved for the gold reference sample, a graphite or $^{208}$Pb sample for measuring the background due to scattered neutrons, and an empty sample can. This means that six samples can be investigated in an experiment, corresponding to a complete isotope sequence along the s-process chain in practically all elements.

## 3. MAIN FEATURES

With respect to accurate neutron capture cross section measurements the present set-up is characterized by several relevant features:

The main advantage results from the unique combination of good energy resolution and high γ-ray detection efficiency. Accordingly, most capture events fall in a sharp line at the neutron binding energy (See left part of Figure 2), and 98% of all capture events are observed above the detection threshold at 1.6 MeV. This implies that the efficiency is independent of the multiplicity of the capture cascades, thus avoiding the most crucial correction in other types of experiments.

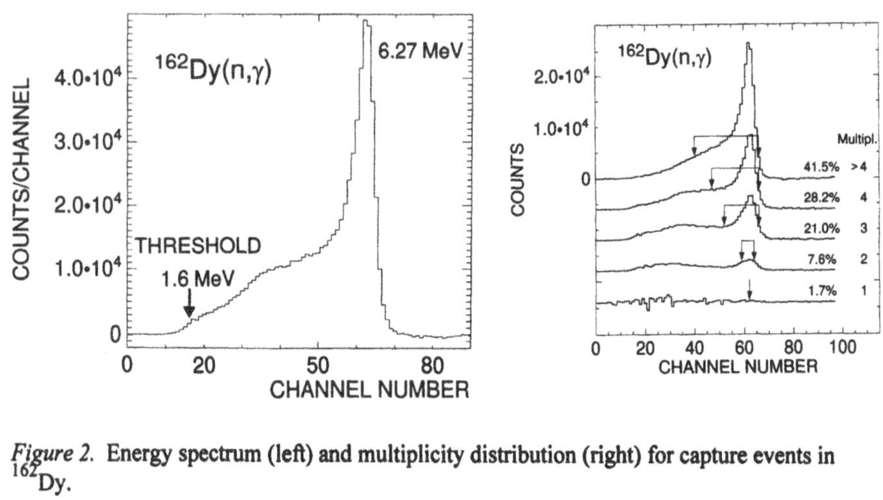


*Figure 2.* Energy spectrum (left) and multiplicity distribution (right) for capture events in $^{162}$Dy.

The neutron flux at the sample position is determined via the gold standard used in the same experiment. This eliminates many of the systematic uncertainties correlated with electronics, accelerator stability, and detector efficiency.

To a large extent, backgrounds from natural radioactivity and from the neutron beam, which are concentrated at low energies, can be discriminated by the good energy resolution, whereas true capture events are found at higher γ-ray energies. A further background discrimination is provided by the multiplicity, defined as the number of detector modules contributing to an event. This is illustrated in the right part of Figure 2 which shows that typically 90% of the true capture events are recorded with multiplicities ≥ 3, while background events are mainly restricted to multiplicities one or two.

Backgrounds from isotopic impurities and scattered neutrons can be discriminated by γ-ray energy as can be seen at the example of $^{164}$Dy in Figure 3. More than 60% of the capture events in the $^{163}$Dy impurity are located above the $^{164}$Dy peak, a region which is excluded from the evaluation of the cross section. The respective background is subtracted by a normalized spectrum measured with the $^{163}$Dy sample (middle part in Figure 3). The remaining background due to capture of sample scattered neutrons in the barium isotopes of the scintillator is eliminated using a normalized spectrum recorded with a scattering sample (graphite or $^{208}$Pb). The final spectrum in the right part of Figure 3 shows good but not perfect compensation of both backgrounds. It is to be emphasized that these corrections can be calculated versus neutron energy and that the good resolution allows to check the result in each neutron energy interval. Thus reliable results are obtained even at low neutron energies where the correction for scattered neutrons is larger.

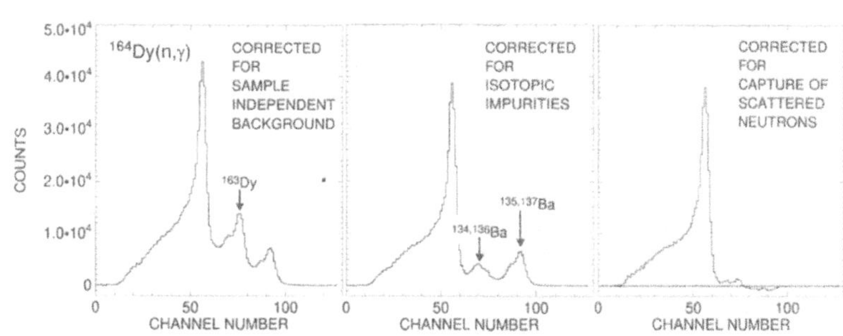


*Figure 3.* Correction for isotopic impurities and capture of scattered neutrons.

An important feature of the present set-up is the short primary flightpath. Consequently, events due to capture of scattered neutrons are significantly delayed in time compared to true capture events. Due to the small capture cross section of the barium isotopes, neutrons are scattered in the scintillator 20 times on average before they are captured. Therefore, most of the corresponding background is sufficiently delayed and does not interfere with the time window of true events. The left part of Figure 4 shows the primary capture events in $^{162}$Dy and the remaining background from scattered neutrons. Due to the delay of the scattering background a reasonable signal/background ratio is obtained although the total cross section exceeds the capture cross section at 30 keV by a factor of 35.

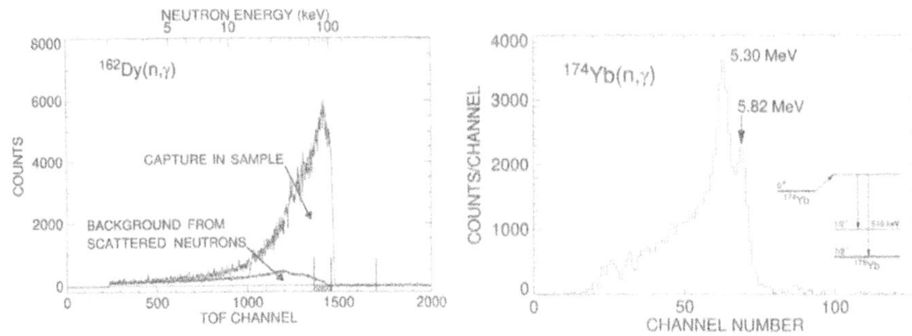


*Figure 4.* Left: Time of flight distribution of capture events and background from scattered neutrons. Right: γ-ray spectrum from neutron captures in $^{174}$Yb feeding ground and isomeric states in $^{175}$Yb.

As demonstrated in the right part of Figure 4, partial capture cross sections to ground and isomeric states can be separated by the sum energy difference of the prompt capture γ-ray cascades. Such cases are particularly troublesome to deal with in experiments based on conventional techniques where the binding energy enters directly as a parameter in the cross section evaluation.

# 4. RESULTS FOR S- AND R-PROCESS ABUNDANCES

With the described technique, the neutron capture cross sections of 57 isotopes between $^{110}$Cd and $^{181}$Ta have been measured in the energy range from 3 to 225 keV. This set contains 16 out of 21 s-only isotopes in the mass range from A = 100-180. From these data stellar cross sections were derived by averaging over a Maxwellian temperature distribution for temperatures between kT = 10-100 keV. These data were used in extensive s-process studies using the canonical approach and a stellar model. The results on the neodymium isotopes (Wisshak *et al.*, 1998) may be considered as a typical example.

In the canonical s-process the differential equations for production and destruction of isotopes under neutron irradiation are solved with certain assumptions. This allows to describe the $N_s$ $<\sigma>$ systematics using in addition to the capture cross sections only two free parameters, representing the average strength of the neutron flux and the number of $^{56}$Fe seed nuclei (Käppelar *et al.*, 1989).

In stars, the s-process in the mass region A > 90 occurs during the helium shell burning of low mass stars (M = 1.5-3M$_{\odot}$) (Gallino *et al.*, 1998). This so called asymptotic giant branch (AGB) phase is characterized by successive thermal instabilities where neutrons are released in two different ways. A long pulse of low neutron density is produced under radiative conditions between subsequent thermal instabilities via the $^{13}$C(α,n) reaction, followed by a short burst with high neutron density during the next thermal instability when the $^{22}$Ne(α,n) reaction is ignited in the convectively burning He shell. After each thermal instability there is significant overlap of the He shell (where the s-process material is produced) and the convective H-rich envelope. The s-processed material is efficiently engulfed by the convective envelope and transported to the surface from where it is remixed into the interstellar medium by strong stellar winds. In this way the s-abundances from many different stars contributed to the material from which the protosolar cloud formed, the original composition being still preserved in the solar photosphere and in primitive meteorites.

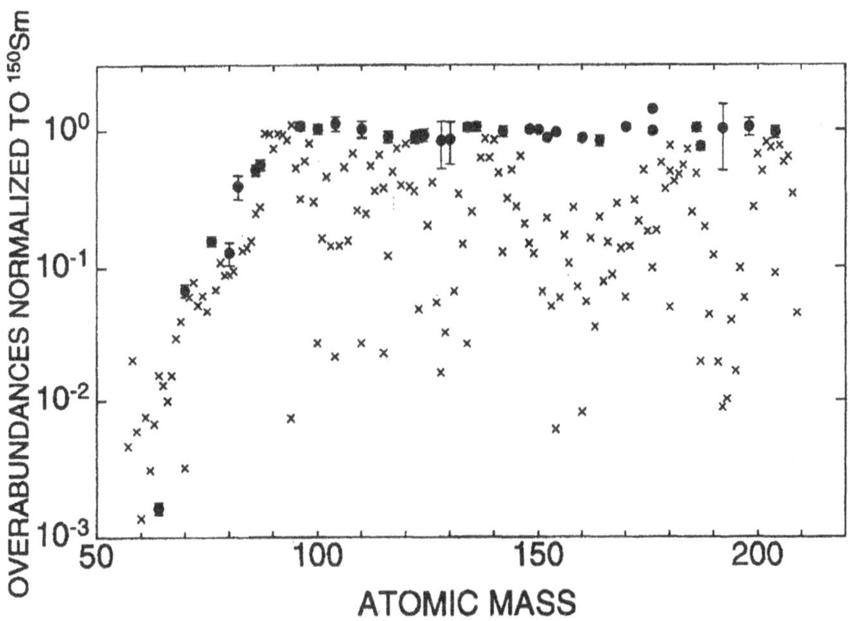


*Figure 5.* Overabundances of heavy elements in the envelope of a Red Giant star.

In stellar model calculations each thermal instability is followed in short time steps, the amount of s-process material mixed into the envelope after each episode is calculated and the individual contributions are added

(Arlandini *et al.*, 1999). The resulting overabundance factors are defined as the final abundance of each isotope in the stellar envelope divided by its solar abundance. These values, arbitrarily normalized to 1 for $^{150}$Sm, are shown in Figure 5. Black symbols represent the important s-only isotopes. Obviously, in the mass range from A = 100 to 180 all s-only isotopes fall on a horizontal line which means that they are produced in solar proportions. The error bars are dominated by the respective cross section uncertainties but are hardly visible for the isotopes measured with the Karlsruhe 4πBaF$_2$ detector. Significant uncertainties remain for a few cases which have not yet been studied experimentally, *e.g.*, for $^{128,130}$Xe and $^{192}$Pt.

On average the abundances of the important s-only nuclei are reproduced within ~ 5%. All other isotopes (crosses in Figure 5) fall below that level, which means that they are underproduced compared to solar. This behavior is expected since their abundances contain also the respective r-process contributions. In fact, this difference can be converted into an r-process abundance distribution shown in the lower part of Figure 6 where the open symbols are the r-process abundances derived from the stellar model via the relation $N_r = N_\odot - N_s$. This pattern fits exactly to the abundances of the r-only isotopes (black symbols) which are totally independent of this model. An almost identical distribution is derived from the canonical model (upper part of Figure 6). Numerical values for the s- and r-process abundances derived with both models can be found in (Arlandini *et al.*, 1999).

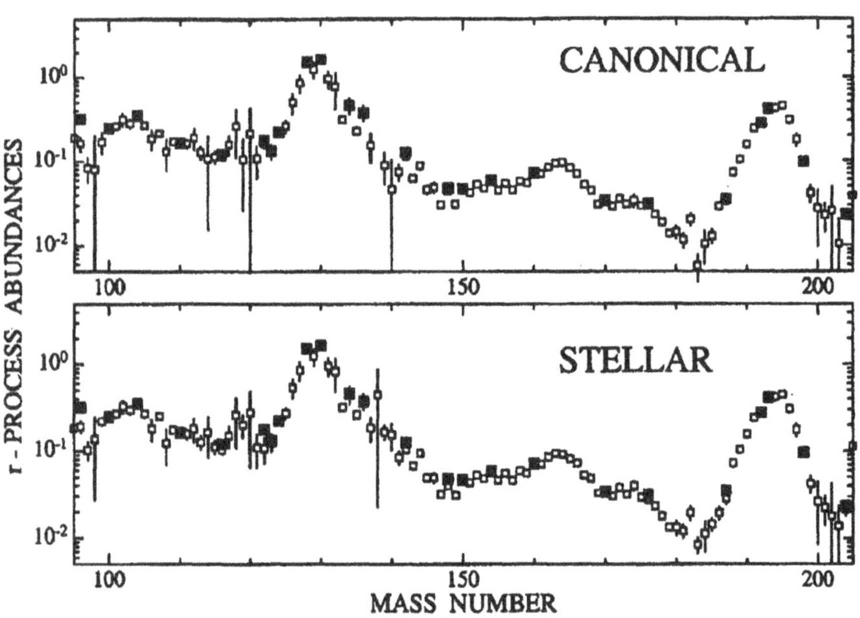


*Figure 6.* The r-process residuals as calculated by the canonical and a stellar model.

## REFERENCES

Arlandini, C., Käppeler, F., Wisshak, K., Gallino, R., Lugaro, M., Busso, M. and Straniero, O.: 1999, "Neutron capture in low mass asymptotic giant branch stars: Cross sections and abundance signatures", *Ap. J.* **525**, 886-900.

Gallino, R., Arlandini, C., Busso, M., Lugaro, M, Travaglio, C., Straniero, O., Chieffi, A. and Limongi, M.: 1998, "Evolution and nucleosynthesis in low-mass asymptotic giant branch stars. II. Neutron captures and the s-process", *Ap. J.* **497**, 388-403.

Käppeler, F., Beer, H. and Wisshak, K.: 1989, "s-Process nucleosynthesis - Nuclear physics and the classical model", *Rep. Prog. Phys.* **52**, 945-1013.

Käppeler, F.K.: 1999, "The origin of the heavy elements: The s-process", *Progress in Nucl. and Particle Phys.* **43**, 419-484.

Wisshak, K., Guber, K., Käppeler, F., Krisch, J., Muller, H., Rupp, G. and Voss, F.: 1990, "The Karlsruhe 4-pi barium fluoride detector", *Nucl. Instr. Meth. A* **292**, 595-618.

Wisshak, K., Voss, F., Käppeler, F., Kazakov, L. and Reffo, G.: 1998, "Stellar neutron-capture cross-sections of the Nd isotopes", *Phys. Rev. C* **57**, 391-408.

# Neutron Capture Cross Section Measurements for the Analysis of the s-Process

Robert S. Rundberg[1], J. L. Ullmann[1], J. B. Wilhelmy[1], M. M. Fowler[1], R. C. Haight[1], G. G. Miller[1], P. D. Palmer[1], F. Käppeler[2] and P. Koehler[3]
*[1]Los Alamos National Laboratory, [2]Forschungszentrum Karlsruhe, Germany, [3]Oak Ridge National Laboratory*
*rundberg@lanl.gov*

**Abstract:** The pioneering work of Burbidge *et al.* (1957) outlined the basic mechanism for heavy element nucleosynthesis in the stellar environment. However, some specifics of the s-process remain poorly determined. Central to these are an understanding of the "branching point" nuclei. These are isotopes that have half-lives such that there is a competition between beta-decay and neutron-capture. Such isotopes have a critical role in understanding the stellar temperature and neutron density and thus the dynamics of nucleosynthesis. At Los Alamos we have begun a program to irradiate radioactive targets and measure the differential capture reactions that are required to unfold the details of stellar evolution. This program utilizes the high fluence available at the LANSCE neutron spallation source to investigate targets having mass on the order of 1 mg. The radioactive target material is obtained via spallation reactions at high intensity accelerators or from irradiations at nuclear reactors. For purification and preparation of the target material, we have dedicated chemical and isotope separator capabilities housed in hot cell environments. A 162 element $4\pi$ $BaF_2$ detector array that will permit highly segmented, calorimetric measurements on capture gamma-rays is being procured for this program. The project status and preliminary results from radioactive target measurements on $^{171}$Tm are presented.

## 1. INTRODUCTION

Understanding the origins of elements and isotopes has challenged scientists for centuries. Now, the situation is more complex in that we need

*Origin of Elements in the Solar System: Implications of Post-1957 Observations.*
Edited by O. Manuel, Kluwer Academic/Plenum Publishers, 2000.

to account for quantitative data of abundances as found on the Earth, in meteorites, in inclusions in meteorites, and as deduced from astronomical emission and absorption spectra measured both from the sun and distant stars and galaxies. Although a detailed picture is emerging, several of the crucial details depend on theoretically calculated nuclear quantities. Here we focus on heavy elements, that is those beyond iron where the mechanisms for synthesis were proposed by Burbidge *et al.* (1957) and Cameron (1955). These mechanisms consist chiefly of a combination of neutron-capture and beta- decay. The main processes are a rapid ("r-") process, and a slow ("s-") process, combined with two additional processes of lesser importance. The r-process takes place in the explosive environment of a supernova, where extremely high neutron fluxes can result in neutron-capture to nuclei far off the line of stability. It is difficult to calculate the r-process, both because of the extreme hydrodynamics of a stellar explosion and because of the lack of knowledge of the properties of nuclei far from stability. The s-process, on the other hand, takes place in the "asymptotic giant branch" stage of evolution in low to medium mass stars, or in red-giant stars, and consists of sequential neutron-capture along the valley of beta stability. A very simple model of the s-process has been shown to work quite well in reproducing solar and cosmic isotopic abundances. Because the basic process is understood, precise measurements can be used to test the details of realistic stellar models. In particular, the dynamics of the stellar models must be included in the understanding, and here especially knowledge of neutron-capture cross sections over the range of relevant energies is essential.

Of critical importance for understanding the dynamics of the s-process are the reactions on branching point nuclei. These are unstable isotopes that have half-lives comparable to the s-process production times – generally believed to be in the 1-100 year time frame (Wallerstein *et al.*, 1997; Käppeler *et al.*, 1989, 1990; Toukan *et al.*, 1995; Wisshak *et al.*, 1993) in the asymptotic giant branch evolution of small to medium mass stars. The predominate neutron producing nucleosynthesis reactions are from pulses of a multiyear occurring $^{22}$Ne($\alpha$,n) reaction with a 30 keV thermal energy distribution coupled with a slower $^{13}$C($\alpha$,n) reaction occurring between pulses with a temperature less than 8 keV. Therefore, isotopes that have lifetimes comparable to the production period will be sensitive to the detailed dynamics of the stellar environment. If the product $\Phi \bullet \sigma \bullet t >> \lambda \bullet t$ then neutron capture will predominate and the nucleosynthesis will proceed to the next heaviest isotope of the element. If on the other hand $\lambda \bullet t >> \Phi \bullet \sigma \bullet t$ then the isotope will predominately beta-decay to the next heavier element. This is illustrated in Figures 1 and 2 for two regions of lanthanide nuclei.

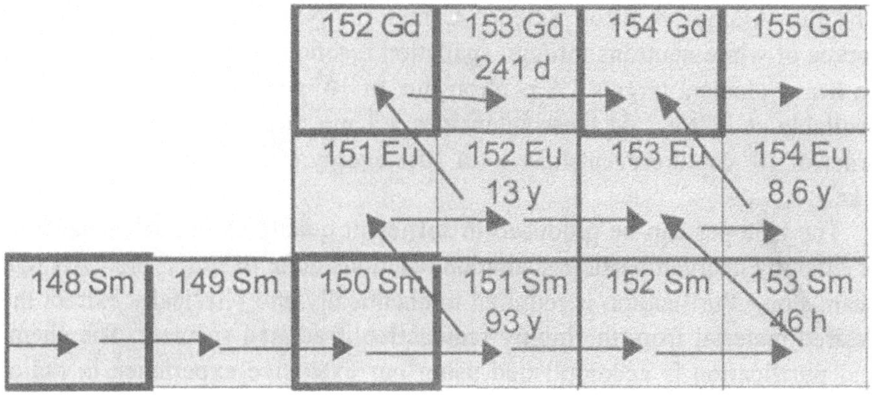


*Figure 1.* s-Process nucleosynthesis in the Sm region. Branching points for radioactive isotopes are indicated.

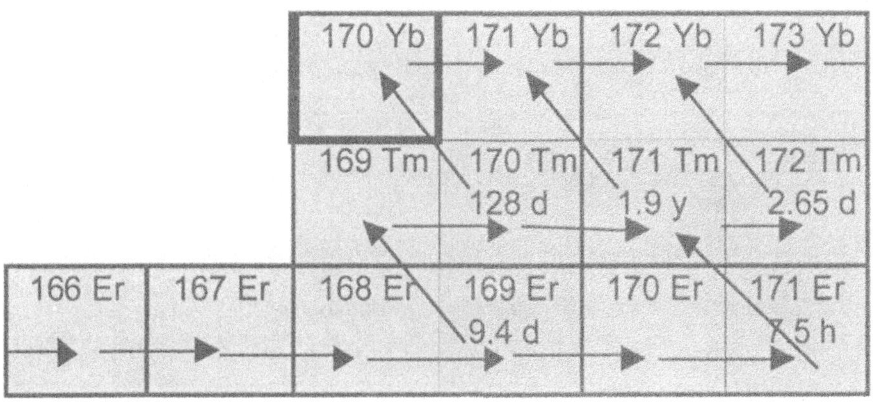


*Figure 2.* s-Process nucleosynthesis in the Tm region. Branching points for radioactive isotopes are indicated.

## 2. EXPERIMENTAL APPROACH

While neutron-capture cross sections have been measured for most stable species, very little information has been obtained on unstable species in the multi keV range of importance to s-process nucleosynthesis. The challenges for such measurements are formidable: a high intensity neutron source in the keV energy range is required, sufficient isotope has to be obtained in sufficient purity, and a detector system that can efficiently operate in the high radiation field of the experiment and target is required. At Los Alamos we have, or are obtaining, all of the ingredients necessary for this effort.

The Los Alamos Neutron Science Center (LANSCE) provides an intense source of white neutrons through spallation reactions with 800 MeV protons on heavy element targets. A neutron flux of $10^8$ ncm$^{-2}$sec$^{-1}$steradian$^{-1}$eV$^{-1}$ is available at 1 keV. At these intensities a 1 mg target will undergo on the order of $10^5$ capture events/day for a 10% energy bin at a neutron energy of 1 keV.

The isotopes can be produced in sufficient quantities in nuclear reactors or often by charged particle spallation reactions done *in situ* at the LANSCE beam stop. Purification is required to chemically and physically extract the desired material from the highly radioactive irradiated species. The chemical purification is accomplished using our extensive experience in radiochemical procedures and is carried out in our hot cell facilities. Most often the desired species is not obtained in an isotopically pure state. An isotopic enrichment technique is needed. We have constructed a dedicated Radioactive Species Isotope Separator (RSIS) (Figure 3) that is completely housed in a hot cell and designed for remote manipulator operation and service.

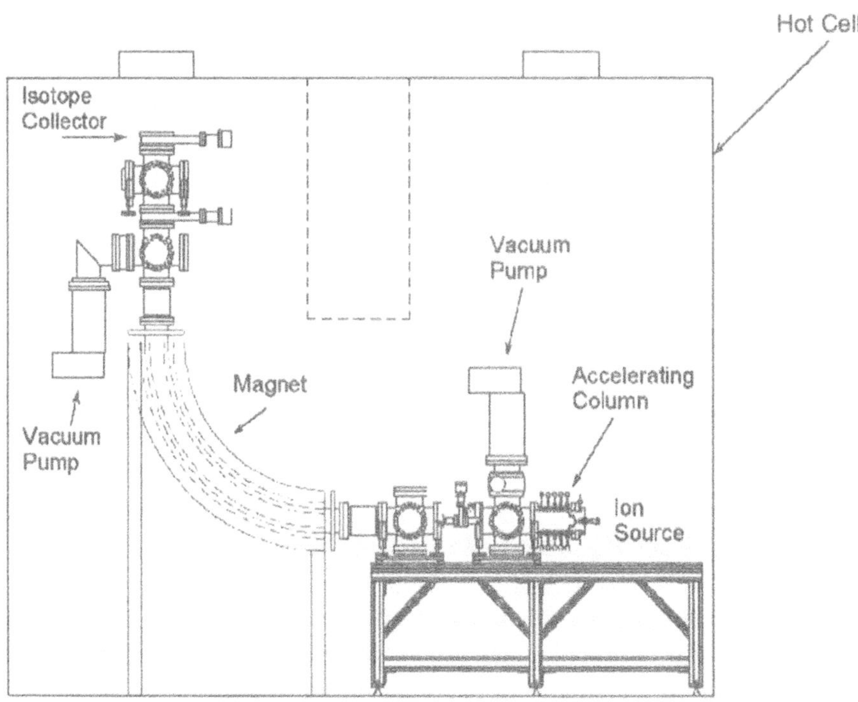


*Figure 3.* Radioactive Species Isotope Separator (RSIS) configuration.

The third component required is a high efficiency, high granularity detector array that can record capture gamma-rays in the highly radioactive environment associated with the experimental accelerator facility and the intrinsic radiation coming from the target. For this purpose we are constructing the Device for Advanced Neutron Capture Experiments (DANCE) array. This will be a $4\pi$ detector array (Figure 4) consisting of 162 $BaF_2$ elements each 15 cm thick. $BaF_2$ is chosen for several reasons: it has relatively low neutron-capture probabilities and is, therefore, fairly insensitive to scattered neutrons, it has a fast scintillation component that can be used to minimize radiation pile up, and it is a high Z material that has a high photo-electric interaction probability. The components of this array are currently being procured and construction, and full scale operation are expected by the end of the year 2000.

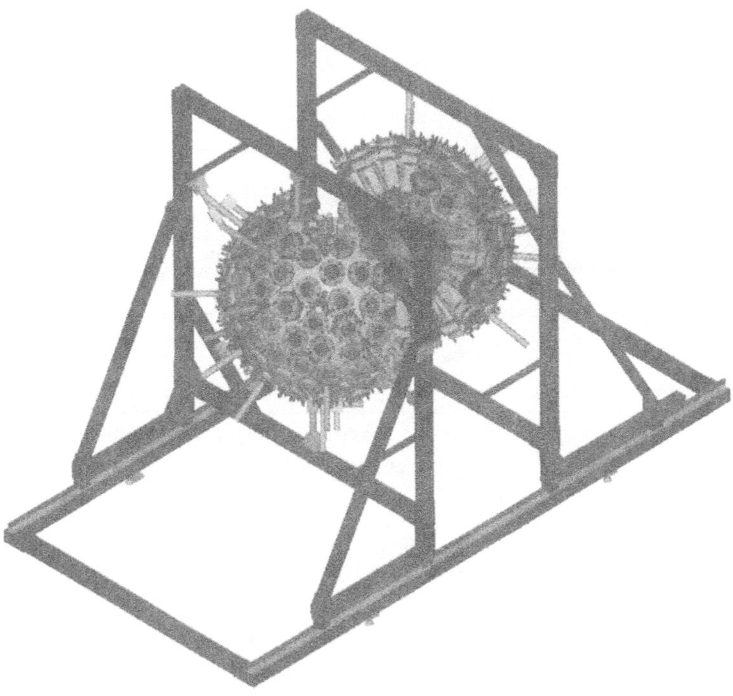

*Figure 4.* Device for Advance Neutron Capture Experiments (DANCE).

# 3. RESULTS

The first proof of principal experiment was performed on the 1.9 yr. radioactive target $^{171}$Tm. This isotope was chosen for its ease of production and its radiation properties. It could be produced in an isotopically pure manner by irradiation of $^{170}$Er in a nuclear reactor. Tm was chemically extracted from the radioactive target material and electrodeposited on a 0.5 mil (12.7 μm) Be foil onto which 70 nm of Ti had been vapor deposited. A second 0.5 mil Be foil was used to cover the foil to ensure radiation encapsulation. This material is also relatively benign having only a low energy, 97 keV β and a 66 keV γ-ray in its decay sequence. The experiment was done at an 8 meter irradiation station at LANSCE. Since the DANCE array was not constructed, the measurements were done using two $C_6D_6$ scintillators that were 5 inches in diameter and 3 inches thick.

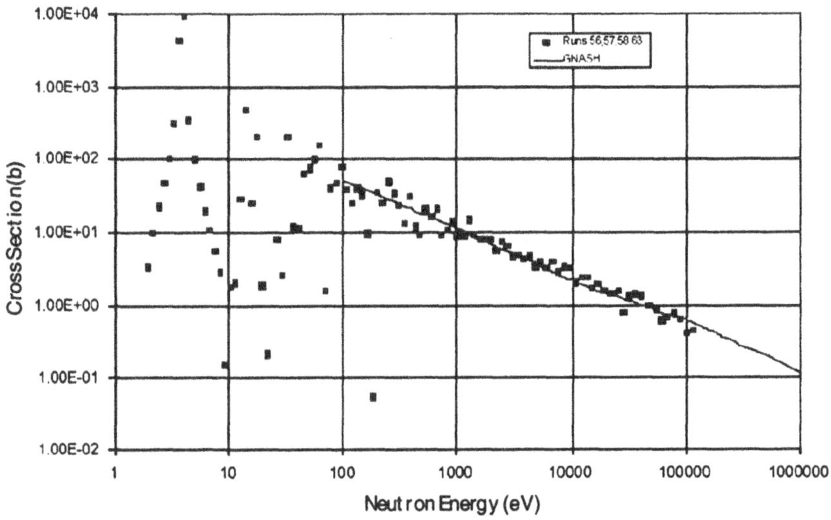


*Figure 5.* Comparison of Tm-169 cross sections measured at LANSCE with a GNASH calculation. The GNASH calculation had been adjusted to fit the data of Macklin *et al.* (1982).

In addition to the radioactive $^{171}$Tm, several stable targets were also irradiated to test and calibrate the system. A stable $^{169}$Tm target was prepared in an identical manner to the radioactive $^{171}$Tm. Figures 5 and 6 show comparisons of the results obtained with the $^{169}$Tm and the $^{171}$Tm target, respectively, with the respective calculated capture cross-sections obtained using the GNASH code. The code agrees very well with the results from the stable $^{169}$Tm but differs substantially (by up to a factor of 5) from the results with the radioactive $^{171}$Tm. It should be noted that the GNASH calculation for $^{169}$Tm was adjusted to fit the measurements of Macklin *et al.* (1982).

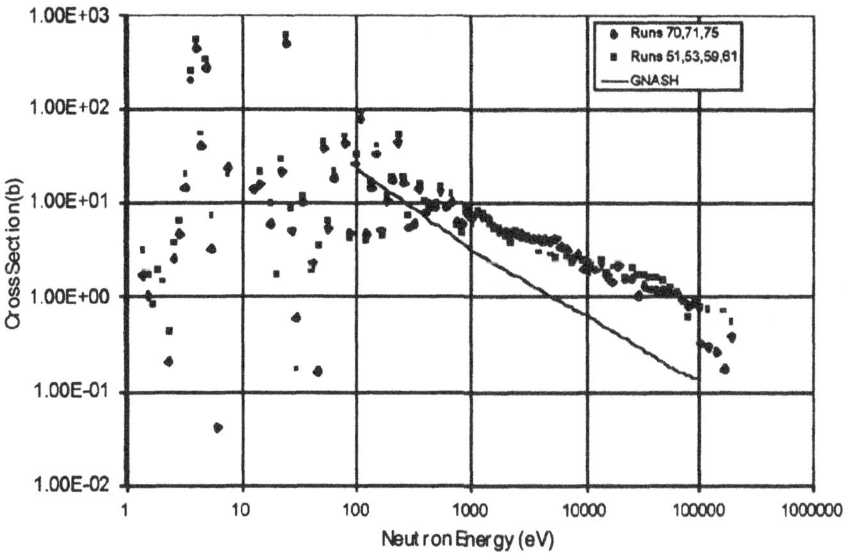


*Figure 6.* Comparison of Tm-171 cross sections measured at LANSCE with a GNASH calculation.

The neutron cross-sections for the important s-process branching nuclide, $^{151}$Sm, will be measured later this year when the Manuel Lujan Center at LANSCE reopens after a lengthy safety shut down. The target material for this experiment is 550 micrograms of 70 percent enrichment $^{151}$Sm. Future experiments will measure the cross sections of s-process branching nuclei listed in Table 1. This list was derived from lists compiled by Matthews (1998) and Koehler and Käppeler (1998).

*Table 1.* Radioactive and rare isotopes for which neutron capture cross sections are needed that are feasible at Los Alamos.

| Isotope | Half-Life | Units | Priority |
|---|---|---|---|
| $^{79}$Se | 1.0E + 04 | y | 1 |
| $^{85}$Kr | 10.7 | y | 1 |
| $^{95}$Zr | 64 | d | 1 |
| $^{134}$Cs | 2 | y | 1 |
| $^{135}$Cs | 3.0E + 06 | y | 2 |
| $^{147}$Pm | 2.6 | y | 1 |
| $^{151}$Sm | 90 | y | 1 |
| $^{155}$Eu | 5 | y | 1 |
| $^{153}$Gd | 241 | d | 1 |
| $^{163}$Ho | 33 | y | 1 |
| $^{170}$Tm | 128 | d | 1 |
| $^{171}$Tm | 1.9 | y | 1 |
| $^{185}$W | 75 | d | 1 |

| Isotope | Half-Life | Units | Priority |
|---|---|---|---|
| $^{186}$Re | 1.0E + 05 | y | 1 |
| $^{192}$Ir | 74 | d | 1 |
| $^{193}$Pt | 50 | y | 1 |

# REFERENCES


Burbidge, E.M., Burbidge, G.R., Fowler, W.A. and Hoyle, F.: 1957, "Synthesis of elements in stars", *Rev. Mod. Phys.* **29**, 547-650.

Cameron, G. W.: 1955, "Origin of anomalous abundances of the elements in giant stars", *Ap. J.* **121**, 144-160

Käppeler, F., Beer, H. and Wisshak, K.: 1989, "s-Process nucleosynthesis - nuclear physics and the classical model", *Rep. Prog. Phys.* **52**, 945-1013.

Käppeler, F., Gallino, R., Busso, M., Picchio, G. and Raitreri, C.M.: 1990, "s-Process nucleosynthesis: Classical approach and asymptotic giant branch models for low-mass stars", *Ap. J.* **354**, 630-643.

Koehler, P. and Käppeler, F. 1998, private communication.

Macklin, R.L., Drake, D.M., Malanify, J.J., Arthur, E.D., and Young, P.G.,: 1982, "Cross-sections of the $^{169}$Tm(n,γ) reaction from 2.6 keV to 2 Mev", *Nucl. Sci. Eng.* **82**, 143-150.

Matthews, G.J.: 1998, private communication.

Toukan, K.A., Debus, K., Käppeler, F. and Reffo, G.: 1995, "Stellar neutron-capture cross sections of Nd, Pm, and Sm isotopes", *Phys. Rev. C* **51**, 1540-1550.

Wallerstein, G., Iben, I., Parker, P., Boesgaard, A.M., Hale, G.M., Champagne, A.E., Barnes, C.A., Käppeler, F., Smith, V.V., Hoffman, R.D., Timmes, F.X., Sneden, C., Boyd, R.N., Meyer, B.S. and Lambert, D.L.:1997, "Synthesis of elements in stars: Forty years of progress", *Rev. Mod. Phys.* **69**, 995-1084.

Wisshak, K., Guber, K., Voss, F., Käppeler, F., Reffo, G.: 1993, "Neutron capture in $^{148,150}$Sm: A sensitive probe of the s-process neutron density", *Phys. Rev. C* **51**, 1401-1419.

# About the Reliability of Extrapolation of Nuclear Structure Data for r-Process Calculations

G. Lhersonneau[1], B. Pfeiffer[2], and K.-L. Kratz[2]
[1]*Accelator Laboratory, University of Jyvaskyla, P.O. Box40351, Jyvaskyla, Finland*
[2]*Institut for Nuclear Chemistry, University of Mainz, D-55099, Mainz, Germany*
*ihers@phys.jyu.fi*

**Abstract:** Gross decay properties are the nuclear part of the input for calculations of elemental abundances. They depend, sometimes very sensitively, on details of nuclear structure. Models for predictions of nuclear masses and shapes have to be used for isotopes very far from stability. The reliability of extrapolations far from experimentally reachable nuclei is, however, not always granted due to singularities in the nuclear landscape. We review data on the region of the neutron-rich isotopes near A = 100, which is a region of especially dramatic changes.

## 1. INTRODUCTION

Knowledge of properties of neutron-rich nuclei situated between the r-process path and further down β-stability is essential in order to calculate nuclear abundances. The deficiencies of models which reproduce the observed abundances for certain mass numbers could be removed if the strength of standard nuclear magic numbers far from stability would be decreased (Kratz *et al.*, 1993 and Chen *et al.*, 1995). On the other hand new magic numbers emerge far from stability (Walters, 1998 and Grzywacz *et al.*, 1998). As a matter of fact, strong shell closures at non-standard magic numbers are known to occur also close to stability. In this contribution a survey of our systematical spectroscopic investigations in such a region will be presented. First, we review briefly how nuclear structure affects quantities of astrophysical relevance.

*Origin of Elements in the Solar System: Implications of Post-1957 Observations.*
Edited by O. Manuel, Kluwer Academic/Plenum Publishers, 2000. 

In the r-process calculation the relevant nuclear properties are the β-decay lifetime, $T_{1/2}$, and the probability for β-delayed neutron emission, $P_n$. Both are intimately related to nuclear masses, or more exactly to the difference in binding energies of the parent and daughter nuclei. The strong energy dependence of the Fermi integral, roughly proportional to the 5$^{th}$ power of the decay energy, makes the decay $Q_{\beta^-}$ value crucial in determining $T_{1/2}$. The nuclear binding energy is affected by the underlying structure. For instance, it is increased by a shell closure or deformation. Fortunately, $T_{1/2}$ and $P_n$ are experimentally accessible for very neutron-rich nuclei produced in minute amounts owing to the sensitive detection of β-delayed neutrons (Mehren *et al.,* 1996). Statistical considerations about the feeding pattern provide a simple guide for an estimate of $P_n$ (Kratz and Herrmann, 1973).

However, details of nuclear structure play a non-neglectible role in decay properties via the transition matrix element. Thus, we need to know the wave functions of initial and final states. It may happen, especially near shell closures, that configurations are very pure and a strong single branching is observed. In that case, the decay lifetime is very sensitive to the excitation energy of the final level, *i.e.*, on single-particle energies in the daughter nucleus. The situation is less critical for deformed nuclei for which the wave function is a mixture of spherical states with a selected $K$ value (the spin projection on the nuclear symmetry axis). This often results in fragmentation of the decay strength, becoming more evenly distributed. Experimentally, wave functions are extracted from transition rates or moments. This requires detailed γ-spectroscopic studies which cannot be performed on the most exotic nuclei due to their modest production rates. Models have to be used for these nuclei.

Thus, properties of nuclei relevant for r-process calculations are calculated using parameters extrapolated from regions where enough detailed experimental information is available. There could be sometimes a large range in nucleon number where no detailed experimental data are available, *e.g.*, for refractory elements. Calculations using a global approach like the Gross-Theory (Nakahata *et al.,* 1997), or the microscopic QRPA (Möller and Randrup, 1990) are surprisingly successful, considering the huge number of nuclei calculated. Nevertheless, their reliability might be questionable in regions where nuclear structure is not as smooth as expected, as we show in the following.

## 2. THE NEUTRON-RICH A = 100 REGION

We present the specific features of the so called A = 100 region, where neutron-rich nuclei show dramatic changes of structure within a narrow range of Z and N.

We first discuss nuclei in the vicinity of $^{96}Zr$. The simultaneous occurrence of closures at Z = 40 and N = 56 results in large gaps above the $g_{9/2}$ proton and below the $d_{5/2}$ neutron subshells, respectively. We note that in the Ni region, N = 40 is also associated with a local increase of magicity (Grazywacz *et al.*, 1998). Limits of the magic $^{96}Zr$ region are in the Nb isotopes (Z = 41) and the N = 59 isotones (Lhersonneau *et al.*, 1997, 1998). Still we are lacking data for lower-Z nuclei which are very difficult to produce. A first step was made in a spectroscopic study of bromine decays to krypton isotopes up to A = 93 (Woehr, 1992).

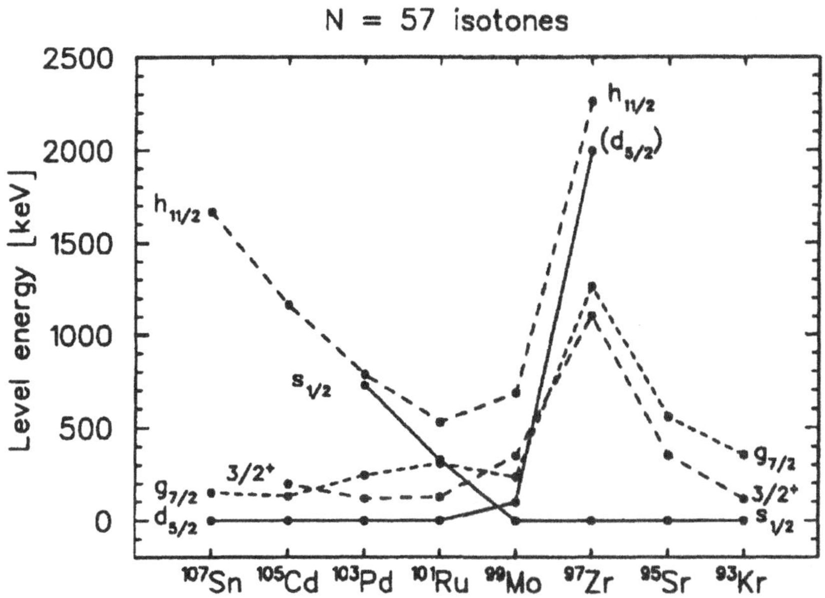


*Figure 1.* Level systematics of N = 57 isotones. Note the $d_{5/2}$ shell closure effect for $^{97}Zr$ (Z = 40) and the level sequence for the neutron-rich nuclei $^{97}Zr$, $^{95}Sr$ and $^{93}Kr$ for which N = 56 is the relevant closure. Conventional calculations are based on the levels in the Sn region, of very different character with a high-lying $s_{1/2}$ level and nearly degenerate $d_{5/2}$ and $g_{7/2}$ levels.

Figure 1 shows a cut through the N = 57 isotones. The shell effects at $^{97}Zr$ are clearly visible, as well as the fact that the neutron levels change their nature depending whether the $g_{9/2}$ proton shell is occupied or not. In global calculations, parameters are adjusted to the Sn region with a characteristic near degeneracy of the $d_{5/2}$ and $g_{7/2}$ shells. It is obvious that problems arise when describing the region of $^{97}Zr$ and the more neutron-rich nuclei beyond it. Correct placement and occupation of the $g_{7/2}$ neutron shell, since the Gamow-Teller decay in this region is of $\nu g_{7/2} \rightarrow \pi g_{9/2}$, are very important.

Next, we present two level systematics leading to the very neutron-rich nucleus $^{100}Sr$. Figure 2 shows the N = 62 isotones. The proximity of the neutron midshell (N = 66) is favorable for the development of collective features by increasing the number of proton-neutron interactions. The evolution of levels looks very smooth, but it is misleading to conclude a similar character of these nuclei. The stable $^{108}$Pd, neighbor of one of the best known vibrators, $^{110}$Cd, is interpreted in the interacting boson language as vibrational with an admixture of $\gamma$-softness (Kim *et al.*, 1996). The latter dominates in Ru (Lu *et al.*, 1995) before a new change occurs towards strong axial deformation ($\beta$ = 0.4) with $^{100}$Sr as a textbook example of a rigid rotor (Lhersonneau *et al.*, 1990; Hamilton *et al.*, 1995). This region exhibits identical transitions, first noticed between the g.s. bands of $^{98}$Sr and $^{100}$Sr, and recently between $^{99}$Sr and $^{100}$Sr (Lhersonneau *et al.*, 1999b). This phenomenon, else known in connection with superdeformation, is here unique since these transitions are at very low excitation energies.

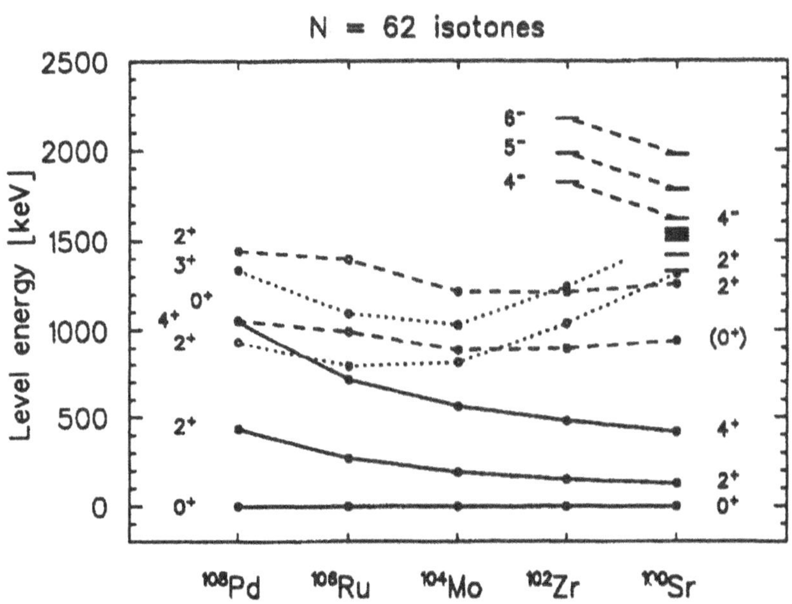


*Figure 2.* Systematics of levels in N = 62 isotones. In spite of a smooth behavior of their level energies, these neighboring nuclei have different structures. They span the three limits of the interacting-boson model U5 (Pd), O6 (Ru) and SU3 (Sr), corresponding to the vibrator, triaxial and axially deformed rotor, respectively.

Differences in structure of these even-even nuclei are reflected by the excitation energies and density of the two-neutron quasiparticle states populated by $\beta$-decay of their odd-odd parents. A single and very fast (logft = 4.9) branch is observed for Rh decays to excited $I = 5$ states near 2.5 ~ MeV in various Pd isotopes (Lhersonneau *et al.*, 1999a). In other decays, feeding is rather fragmented. Nevertheless, there are strong branches in terms of intensity (but weaker than the previous ones in terms of matrix elements since logft are about 6) in both $^{100}$Rb and $^{102}$Y decays to $4^-$ states at 1619 and 1821 keV, respectively. The low excitation energy of these deformed states is a consequence of a local pairing reduction (Capote *et al.*, 1998).

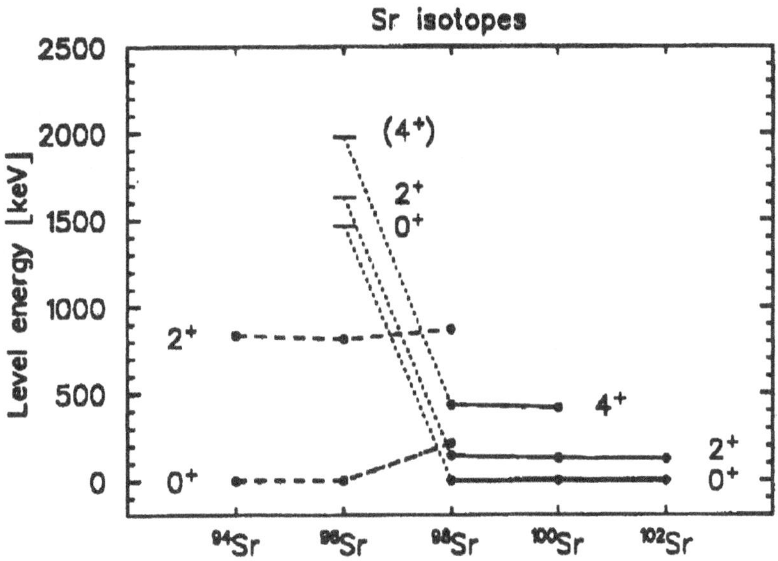


*Figure 3.* Systematics of levels in neutron-rich even-even Sr isotopes. The dramatic drop of the $2^+$ energy from $^{96}$Sr to $^{98}$Sr, *i.e.*, between N = 58 and 60, is due to the lowering of the very deformed minimum below the spherical one under the proton-neutron interaction. At neutron number N = 59, the $^{97}$Sr ground state still is spherical while deformed states have been identified near 600 keV.

As the exotic nucleus $^{100}$Sr is approached, the Sr systematics display the most dramatic changes, see Figure 3. Near stability, the levels look rather vibrational although with little collectivity due to the N = 56 closure. A deformed structure shows up in $^{96}$Sr at 1.5 MeV (Hamilton *et al.*, 1995; Lhersonneau *et al.*, 1994). This is lowered with increasing N by the increased proton-neutron interaction, becoming the strongly deformed ground state in $^{98}$Sr and higher-mass Sr isotopes. A similar phenomenon occurs also for Zr isotopes at same neutron numbers. However, for higher-Z even-even nuclei the shape transition is smooth. Sharp transitions are also displayed by the odd-proton Y and Nb nuclei at the same neutron numbers. As shown by our decay study of $^{97}$Rb, allowed decay of deformed Rb parent nuclei to the spherical and lowest-lying states in $^{97}$Sr is rather slow (Lhersonneau *et al.*, 1990). Thus, in the nuclei with shape coexistence in a narrow excitation energy range, *i.e.*, $^{96}$Sr, $^{97}$Sr, $^{98}$Sr and their neighboring isotones, the prediction of the decay pattern and consequently of the lifetime is sensitive to the calculated positions of the spherical and deformed minima of the potential.

So far, many calculations have turned out to be able to reproduce a shape transition in the A = 100 region. See for instance the recent work by Skalski *et al.* (1997). However, the magicity of $^{96}$Zr, the exact location of the first deformed ground states and the coexistence of potential minima remain a real challenge even to calculations of local character.

## 3. CONCLUSION

The occurrence of accidents in the nuclear landscape such as depicted above is rare, but it is unlikely that there are no other such cases among the numerous nuclei yet to discover far from stability. The need for improved experimental data for a better understanding of nuclear structure in general is obvious. In this context, studies of neutron-rich nuclei far from stability are compulsory in order to obtain a more reliable basis for extrapolations of nuclear properties to the nuclei relevant for astrophysical calculations.

## ACKNOWLEDGMENTS

Part of this work was performed in collaboration between the Institute for Nuclear Chemistry in Mainz and the Department of Physics in Jyväskylä and was supported by the German DAAD and the Academy of Finland.

## REFERENCES

Capote, R., Mainegra, E. and Ventura, A.: 1998, "Quantum Monte Carlo study of pairing interaction of the neutron-rich A ≅ 100 nuclei", *J. Phys. (London) G* **24**, 1113-1123.

Chen, B., Dobaczewski, J., Kratz, K.-L., Langanke, K., Pfeiffer, B., Thielemann, F.-K. and Vogel, P.: 1995, "Influence of shell-quenching far from stability on the astrophysical r-process", *Phys. Lett.* **355B**, 37-44.

Grzywacz, R., Beraud, R., Borcea, C., Emsallem, A., Glogowski, M., Grawe, H., Guillemaud-Mueller, D., Hjorth-Jensen, M., Houry, M., Lewitowicz, M., Mueller, A.C., Nowak, A., Plochocki, A., Pfutzner, M., Rykaczewski, K., Saint-Laurent, M.G., Sauvestre, J.E., Schaefer, M., Sorlin, O., Szerypo, J., Trinder, W., Viteritti, S. and Winfield, J.: 1998, "New island of Mu-s isomers in neutron-rich nuclei around the Z = 28 and N = 40 shell closures", *Phys. Rev. Lett.* **81**, 766-769.

Hamilton, J., Ramayya, A.V., Zhu, S.J., Ter-Akopian, G.M., Oganessian, Yu.Ts., Cole, J.D., Rasmussen, J.O. and Stoyer, M.A.: 1995, "New insights from studies of spontaneous fission with large detector arrays", *Progress in Particle and Nuclear Physics* **35**, 635-704.

Kim, K.H., Gelberg, A., Mizusaki, T., Otsuka, T. and Von Brentano, P.: 1996, "IBM-2 calculations of even-even Pd nuclei", *Nucl. Phys. A* **604**, 163-182.

Kratz, K.-L. and Herrmann, G.: 1973, "Systematics of neutron emission probabilities from delayed neutron precursors", *Z. Phys.* **263**, 435-442.

Kratz, K.-L., Bitouzet, J.-P. Thielemann, F.-K., Möller, P. and Pfeiffer, B.: 1993, "Isotopic r-process abundances and nuclear structure far from stability: Implications for the r-process mechanism", *Ap. J.* **403**, 216-238.

Lhersonneau, G., Gabelmann, H., Kaffrell, N., Kratz, K.-L., Pfeiffer, B., Heyde, K. and the ISOLDE Collaboration: 1990, "Saturation of deformation at N = 60 in the Sr isotopes", *Z. Phys. A* **337**, 143-148.

Lhersonneau, G., Pfeiffer, B., Kratz, K.-L., Ohm, H., Sistemich, K., Brant, S. and Paar, V.: 1990, "Structure of the N = 59 nucleus $^{97}$Sr: Coexistence of spherical and deformed states", *Z. Phys. A* **337**, 149-159.

Lhersonneau, G., Pfeiffer, B., Kratz, K.-L., Enqvist, T., Jauho, P.P., Jokinen, A., Kantele, J., Leino, M., Parmonen, J.M., Penttila, H., Aysto, J. and the ISOLDE Collaboration: 1994, "Evolution of deformation in the neutron-rich Zr region from excited intruder state to the ground state", *Phys. Rev. C* **49**, 1379-1390.

Lhersonneau, G., Dendooven, P., Honkanen, A., Huhta, M., Jones, P.M., Julin, R., Juutinen, S., Oinonen, M., Penttila, H., Persson, J.R., Perajarvi, K., Savelius, A., Wang, J.C. and Aysto, J.: 1997, "New interpretation of shape coexistence in $^{99}$Zr", *Phys. Rev. C* **56**, 2445-2450.

Lhersonneau, G., Suhonen, J., Dendooven, P., Honkanen, A., Huhta, M., Jones, P., Julin, R., Juutinen, S., Oinonen, M., Penttila, H., Persson, J.R., Perajarvi, K., Savelius, A., Wang, J.C., Aysto, J., Brant, S., Paar, V. and Vretenar, D.: 1998, "Level structure of $^{99}$Nb", *Phys. Rev. C* **57**, 2974-2990.

Lhersonneau, G., Wang, J.C., Hankonen, S., Dendooven, P., Jones, P., Julin, R. and Aysto, J.: 1999a, "Decays of $^{110}$Rh and $^{112}$Rh to the near neutron midshell isotopes $^{110}$Pd and $^{112}$Pd", *Phys. Rev. C* **60**, 014315-20.

Lhersonneau, G., Pfeiffer, B. and Kratz, K.-L.: 1999b, "Identical transitions in ground-state bands of $^{99}$Sr and $^{100}$Sr", in *Eur. Phys. J.A*.

Lu, Q.H., Butler-Moore, K., Zhu, S.J., Hamilton, J.H., Ramayya, A.V., Oberacker, V.E., Ma,W.C., Babu, B.R.S., Deng, J.K., Kormicki, J., Cole, J.D., Aryaeinejad, R., Dardenne, Y.X., Drigert, M., Peker, L.K., Rasmussen, J.O., Stoyer, M.A., Chu, S.Y., Gregorich, K.E., Lee, I.Y., Mohar, M.F., Nitschke, J.M., Johnson, N.R., McGowan, F.K., Ter-Akopian, G.M., Oganessian, Yu.Ts. and Gupta, J.B.: 1995, "Structure of $^{108,110,112}$Ru: Identical bands in $^{108,110}$Ru", *Phys. Rev. C* **52**, 1348-1354.

Mehren, T., Pfeiffer, B., Schoedder, S., Kratz, K.-L., Huhta, M., Dendooven, P., Honkanen, A., Lhersonneau, G., Oinonen, M., Parmonen, J.-M., Penttila, H., Popov, A., Rubchenya, V. and Aysto, J.: 1996, "Beta-decay half-lives and neutron-emission probabilities of very neutron-rich Y to Tc isotopes", *Phys. Rev. Lett.* **77**, 458-461.

Möller, P. and Randrup, J.: 1990, "New developments in the calculation of beta-strength functions", *Nucl. Phys. A* **514** 1-48.

Nakahata, H., Tachibana, T. and Yamada, M.: 1997, "Semi-gross theory of nuclear beta-decay", *Nucl. Phys. A* **625**, 521-553.

Skalski, J., Mizutori, S. and Nazarewicz, W.: 1997, "Equilibrium shapes and high-spin properties of the neutron-rich A ≅ 100 nuclei", *Nucl. Phys. A* **617**, 282-315.

Walters, W.B.: 1998, "Monopole shifts of single-particle energy levels and the positions of magic nuclides for very neutron-rich nuclides", *Proc. Int. Workshop on Nuclear Fission and Fission-Product Spectroscopy,* Seyssins, France **AIP 447**, 196 pp.

Woehr, A.: 1992, "Kernstrukturuntersuchungen an neutronenreichen Brom-isotopen (Nuclear structure studies of neutron-rich bromine isotopes)", *PhD dissertation, University of Mainz.*

# The Astrophysical r-Process
*From $B^2FH$ to Present*

K.-L. Kratz[1], P. Möller[2], B. Pfeiffer[1], and F.-K. Thielemann[3]
*[1]Institut für Kernchemie, Universität Mainz, D-55128 Mainz, Germany, [2]P. Moller Scientific Computing and Graphics, Inc., Los Alamos, NM 87545, USA, [3]Departement für Physik und Astronomie, Universität Basel, CH-4056 Basel, Switzerland*
*klkratz@vkcmzd.chemie.unimainz.de*

**Abstract:** In 1957, Burbidge, Burbidge, Fowler and Hoyle ($B^2FH$) provided a basis for forty years of research in various aspects of nucleosynthesis in stars. We will focus in this paper on progress in r-process nucleosynthesis, with emphasis on the most recent developments in nuclear physics. In 1986, the first experimental data on two crucial, neutron-magic "waiting-point" nuclei provided valuable clues to the astrophysical conditions and the nature of the r-process site. Beginning in the 1990's, our group ($FK^2L$) presented considerably improved r-abundance calculations, which were for the first time based on a modern, internally consistent nuclear-theory input. The phenomenon of shell-quenching far from β-stability required at that time seems to become evident by now. This new nuclear-structure phenomenon leads to improved predictions of astrophysical observables like the solar-system isotopic r-abundances ($N_{r,\odot}$) or the recently observed abundances of ultra-metal-poor halo stars.

## 1. INTRODUCTION

Summarizing the knowledge about nucleosynthesis in stars in the mid 1950's, in their famous review about the *"Synthesis of the Elements in Stars"*, Burbidge, Burbidge, Fowler and Hoyle (1957; in the following referred to as $B^2FH$) combined progress from stellar evolution and solar-system abundances with nuclear-physics data to explain how stars can produce all known nuclear species in eight separate processes, some of them still being denoted today by the well-known lettering notation α, e, s, r, and p. At the same time, several groups were working along similar lines, (Coryell, 1956, 1961; Cameron, 1957; Suess and Urey, 1956). Their papers

*Origin of Elements in the Solar System: Implications of Post-1957 Observations.*
Edited by O. Manuel, Kluwer Academic/Plenum Publishers, 2000. 

provided the basis for *Nuclear Astrophysics,* a field that has now been active for more than four decades.

In the present paper, we will focus on the rapid neutron-capture process, the *r-process*, which is responsible for the nucleosynthesis of about half of the abundance of elements heavier than iron. We will first recall the general features for an r-process that $B^2FH$ suggested, and then briefly summarize progress in confirming and extending their initial ideas. Comprehensive reviews by Seeger *et al.* (1965), Hillebrandt (1978), Schramm (1982), Mathews and Ward (1985), and Cowan *et al.* (1991) provide thorough and complete background material until beginning of the 1990's. The main part will then be devoted to recent experimental and theoretical progress of nuclear physics far from β-stability during the last decade, to a considerable part originating from our collaboration (See, *e.g.* F.-K. Thielemann and K.-L. Kratz, 1992) denoted as $FK^2L$ by E. Sheldon at the 1991 Masurian Lake Summer School; or (Kratz *et al.*, 1993; Thielemann *et al.*, 1994). Summaries of the present nuclear-physics and astrophysics status of explosive nucleosynthesis can be found in Sherrill *et al.* (1998), Forkel-Wirth and Bollen (1999) and Wallerstein *et al.* (1997).

## 2. $B^2$ FH – THE NUCLEAR ASTROPHYSICS "BIBLE"

The possibility that neutrons may play an important role in astrophysics had been recognized already very soon after Chadwick's discovery of the neutron. However, it was not until 1956, when an improved solar-system abundance ($N_\odot$) curve, together with data on fast neutron-capture cross sections, became available that the essential idea for neutron-capture nucleosynthesis could be outlined in the review article by $B^2FH$. The atomic abundance curve of Suess and Urey (1956) showed two important features (See Figure 1): the association of magic neutron numbers with $N_\odot$ peaks, and a splitting of these peaks corresponding to a *slow* ("s-process") and a *rapid* neutron-capture process ("r-process"), which are naturally explained by nuclear-structure aspects of the two capture processes.

These ideas (Burbidge *et al.*, 1957; Coryell, 1956; Cameron, 1957; and Suess and Urey, 1956) still form the basis of the contemporary viewpoint of nucleosynthesis of the elements beyond Fe. Focussing on the r-process in this paper, we will first recall the basic nuclear-physics and astrophysics input of $B^2FH$, and then briefly review what has remained true and what has changed over the last forty years.

$B^2FH$ showed that *"the essential feature of the r-process is that a large flux of neutrons becomes available in a short time interval for addition to*

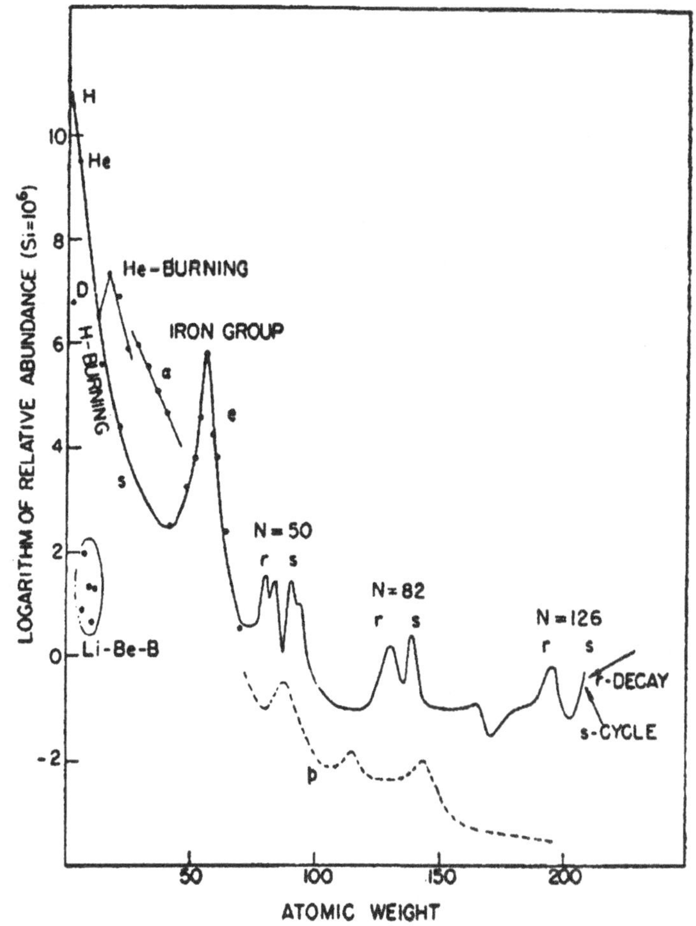


*Figure 1.* Schematic curve of solar-system abundances as a function of atomic weight, based on the 1956 data of Suess and Urey (1956).

*seed elements of the Fe group".* This mode of nucleosynthesis was suggested to be *"responsible for a large number of isotopes in the range 70 ≤ A ≤ 209, and also for synthesis of uranium and thorium"*, an assumption which, in principle, has remained true up to present. Stating that *"the precise source of the neutrons is not important"* (although they later suggest $^{21}Ne(\alpha,n)$ as the main neutron source) for their description of details, $B^2FH$ associate the r-process with supernova explosions (they in fact favor SNs of type I), where a *"neutron density of $10^{24}/cm^3$ and a flux of $4\times10^{32}/cm^3$ might be produced at temperatures $\approx 10^9$ degrees"*. They furthermore require *"that neutrons be added extremely rapidly, so that the total time-scale for the addition of a maximum of about 200 neutrons per iron nucleus is ≈ 10 - 100 s"*. Apart from the much too long r-process time-scale, (which is mainly due to their–at that time understandably very simple–nuclear-physics input to calculate β-decay half-lives, as will be discussed below), again the general aspects represent the current viewpoint.

Also their ideas about the *"path of the r-process"* along neutron-rich, unstable nuclei with neutron binding energies ($B_n$) of approximately 2MeV, where further *"neutron addition is limited not by the (n,γ) cross sections but by (γ,n) competition and by "waiting" for $\beta^-$ processes to take place"*, are still valid today. And their assumption of an (n,γ)⇔(γ,n) equilibrium, *i.e.* the well-known "waiting-point" concept, is still currently used in many "canonical" r-process calculations. It is interesting to note in this context, that even the detailed behavior at the magic neutron numbers N = 50, 82 and 126, was predicted correctly by $B^2FH$. The *"climb up the staircases"* at these closures, the major waiting-point nuclei involved, as well as the *"break-through pairs"* (A = 81, Z = 31), (A = 131, Z = 49) and (A = 196, Z = 70), and their *"association with the rising sides of major peaks in the abundance curve for the r-process"* are still today important properties to be studied experimentally and theoretically, in order to establish progress in the understanding of r-process nucleosynthesis.

As far as nuclear-physics input is concerned, understandably $B^2FH$ had to rely on the scarce experimental data basis of the mid-1950's, and on rather simple model predictions of nuclear masses and β-decay half-lives. For example, *"the smooth Weizsäcker mass formula"* with empirical corrections (*e.g.*, near the shell closures) was applied to calculate neutron binding and β-decay energies at the waiting-points; the β-decay lifetimes were deduced from the Fermi expression assuming **one** major allowed transition with a log(ft) value of 3.85 to a hypothetical low-lying excited state in the daughter nucleus. It is again interesting to note in this context, that $B^2FH$ admit that their simple treatment will *"lead in general to an underestimate"* of their effective β-decay energies at the waiting points, *"so that our values of $\tau_\beta$ could be too large by a factor of as much as 10"*.

A very good estimate at those times! We can check this point in the context of their assumption that the total time for the r-process is approximately the sum of the β-decay half-lives of all waiting-point nuclei. Under conditions of a steady flow starting from $^{56}Fe$ *"as the source nucleus"*, $B^2FH$ predict the first waiting point to be $^{68}Fe$, for which *"beta decay occurs with a lifetime ($\tau_\beta$) of ≈ 21 sec. This is ... in fact ≈ 25% of the total time for the r-process"*. Today, the half-life of $^{68}Fe$ is experimentally known to be $T_{1/2} = \tau_\beta/\ln 2 \cong 200$ ms; hence, a factor of roughly 70 shorter than initially predicted. In consequence, with today's 10 to 100 times shorter half-lives (some of the most important, neutron-magic ones being experimentally known by now Thielemann and Kratz, 1992), also the total time of a steady-flow r-process has become much shorter, *i.e.* of the order 1–3.5 sec.

There are a number of additional very interesting aspects discussed by B$^2$FH, such as effects of spheroidal deformation on the r-process abundances, the role of β-delayed neutron emission, the termination and freezing of the r-process, or Th–U cosmochronometry, which we cannot discuss in this paper, which are, however, worth looking at in the original paper.

We rather want to summarize the highly creative work of B$^2$FH by showing their Figure VII,3 (Here Figure 2) in which the authors present their calculated abundances, *"arbitrarily taking* $_{52}Te_{76}^{128}$ *near the N–82 peak as standard at 1.48 on the Suess and Urey scale"*. B$^2$FH conclude, that a *"reasonable but not exact agreement with observed abundances is obtained"*. This is, indeed, a very modest statement, in particular when considering that a **substantial** improvement of the detailed fit to the isotopic solar-system r-process abundances of Anders and Grevesse (1989) was, in fact, only achieved 35 years later by our collaboration (Thielemann and Kratz, 1992; Kratz *et al.*, 1993; and Thielemann *et al.*, 1994).

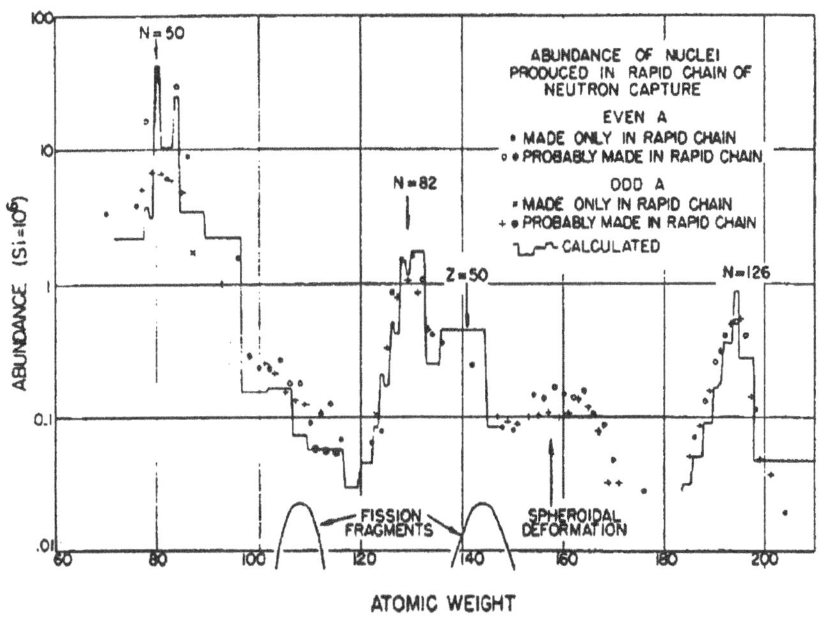


*Figure 2.* Classical static r-process calculations of B$^2$FH (Burbidge *et al.*, 1957) compared to observed abundances of Suess and Urey (1956).

## 3. FROM B$^2$FH TO MID 1980'S

Since the pioneering work of B$^2$FH outlined in some detail above, great effort has been put into a better understanding of r-process nucleosynthesis.

The ideas were following a gradual evolution of more sophisticated stellar models combined with improved nuclear-physics input and computational technology. But, although some promising candidate environments have been suggested in the following three decades, the site of the r-process has remained an unsolved puzzle.

As mentioned above, it was initially suggested that the r-process could be associated with supernovae (Burbidge *et al.*, 1957). Quite a number of models were developed (See, *e.g.*, Seeger *et al.*, 1965; Cameron *et al.*, 1970; Kodama and Takahashi, 1973; Schramm, 1973; Sato, 1974; Hillebrandt *et al.*, 1976) to describe how heavy nuclei might be produced in the regions just outside neutron-rich supernova cores. Unfortunately, none of these models were able to account for the production and ejection of r-process elements with the correct distribution to explain the solar-system abundances in a realistic and self-consistent manner. Nevertheless, in terms of the definitive conclusions (Thielemann and Kratz, 1992; Kratz *et al.*, 1993) drawn about 15 years later by $FK^2L$ about the necessity of "a superposition of r-process components", at least the initial ideas of Seeger, Fowler and Clayton (Seeger *et al.*, 1965) along this line are worth mentioning here. From their abundance calculations, they conclude that *"all three (r-abundance) peaks cannot have been produced under the same set of circumstances"*. In their Figures 12 and 13, they show calculated abundance patterns for short-time ($\tau_r$=0.44 s and 1.77 s) and long-time ($\tau_r$=3.54 s and 4.9 s) solutions which cover different mass ranges of r-process nuclei. However, the authors were not (yet) able to draw more definite conclusions, although they believed that *"the general features of the r-process abundance curve seem to be well understood in terms of nuclear physics"* (known in 1965). They had to admit that *"the uncertainty of the form of the semi-empirical atomic mass law when extrapolated to very neutron-rich nuclei introduces a fundamental impedient to the detailed calculations of abundances"*.

Because of the difficulties with SN-core models, other possibilities have been studied with varying degrees of success. A nice review of these scenarios is given, *e.g.*, in the article of Mathews and Ward (1985). We shortly summarize their discussion here: The proposed r-process sites can be divided into two categories: those in which the r-process elements are produced as **primary** products directly in a star; and those in which the r-elements are **secondary**, thus requiring pre-existing heavy "seed" nuclei which capture neutrons to produce r-process elements. In the former category are high-temperature, neutron-rich environments such as shock (Colgate, 1971) or jet (Leblanc and Wilson, 1970; Schramm and Barkat, 1972; Meier *et al.*, 1976; Symbalisty *et al.*, 1985) ejection of material from neutron-rich SN cores, inhomogeneous Big Bang cosmologies (Applegate *et al.*, 1988; Kajino *et al.*, 1990), or ejecta from binary neutron-star tidal disruption (Lattimer and Schramm, 1974, 1976; Symbalisty and Schramm, 1982) or coalescence (Evans and Mathews, 1988). The latter category includes less neutron-rich and lower-temperature environments, such as nova outbursts (Hoyle and Clayton, 1974), shock-induced explosive helium (Hillebrandt and Thielemann 1977; Truran *et al.*, 1978) or carbon (Lee *et al.*, 1979) burning, the core helium flash in low-mass stars (Cowan *et al.*, 1982), super-heated proton-rich helium burning in a pre-collapse 11-$M_\odot$ star (Woosley and Weaver, 1986), neutron-star accretion disks (Hogan and Applegate, 1987), or neutrino inelastic scattering in the outer envelopes of core-bounce supernovae (Epstein *et al.*, 1988).

# 4. FROM MID 1980'S TO FK$^2$L

Still in the early 1980's, astrophysicists believed that nuclear-structure information on r-process waiting-point isotopes would **never** become available from terrestrial experiments. However, only shortly after in 1986, a new area in nuclear astrophysics began with the identification of two of the most important "classical", neutron-magic waiting-point isotopes at different on-line mass-separator facilities; *i.e.* $^{130}Cd_{82}$ at ISOLDE Kratz *et al.*, 1986), and $^{80}Zn_{50}$ at OSIRIS (Lund *et al.*, 1986) and Tristan (Gill *et al.*, 1986).

These measurements immediately improved the prospect for understanding the conditions for the r-process. From the β-decay properties of the above two isotopes, together with information on the "break-through" pairs (Burbidge *et al.*, 1957) $^{131,133}$In and $^{81,83}$Ga, respectively, where the r-process branches off from the neutron closed shells, strong evidence has been obtained that the – just at that time improved – solar-system r-abundances ($N_{r,\odot} \cong N_\odot - N_{s,\odot}$) of their stable isobars (Käppeler *et al.*, 1989) result from a β-flow equilibrium *e.g.*, (Kratz, 1988; Kratz, *et al.* 1988a). This concept implies approximate equality of the progenitor abundance ($N_{r,prog}$) times the β-decay rate ($\lambda_\beta$). An additional test was performed for the – at that time unknown – N = 50 and 82 r-process isotopes $^{76}$Fe to $^{79}$Cu and $^{127}$Rh to $^{129}$Ag, respectively. From the known $N_{r,\odot}$ and predicted $P_n$-values, values for the $N_{r,prog}$ of the waiting-point nuclei were calculated; and with this also values for the $T_{1/2}$ for these isotopes were "predicted" – provided that a β-flow equilibrium existed. Although these astrophysical requirements were found to agree closely with the nuclear-physics values (QRPA-predictions; See Table 1 in (Kratz, 1988)), it was concluded at that time, that a final decision whether or not the steady-flow picture holds for the $A \cong 80$ and $A \cong 130$ $N_{r,\odot}$-peaks would only be possible from further experimental data on waiting-point nuclei. In any case, Kratz, (1988) and Kratz *et al.* (1988a) pointed out that within the $N_r(Z)\lambda_\beta(Z) \cong$ constant assumption, apart from neutron density ($n_n$) and stellar temperature

($T_9$), the knowledge of nuclear masses (respectively neutron binding energies $B_n$) and β-decay half-lives ($T_{1/2}$) as well as β-delayed neutron branching ratios ($P_n$) alone would be sufficient to predict the whole set of $N_{r,\odot}$ as a function of A. And this success strongly motivated further experimental and theoretical nuclear-structure investigations, as well as astrophysical r-process studies. For example, the second N = 50 waiting-point isotope $^{79}$Cu could be identified (Kratz *et al*, 1991). Experimentally known strong $P_n$ branches were shown to be the nuclear-structure origin of the odd-even staggering in the A ≅ 80 $N_{r,\odot}$ peak region (Kratz *et al.*, 1990); and from the interpretation of the $^{80}Zn_{50}$ decay scheme, evidence for a vanishing of the spherical N = 50 shell closure far from stability was obtained (Kratz *et al.*, 1988b).

Following this concept, in the early 1990's Thielemann and Kratz tried – without assuming a particular astrophysical site or model – to deduce stellar conditions which produce the total r-abundance pattern and to check which of the many r-process scenarios were still consistent with the improved knowledge of the $N_{r,\odot}$ (Käppeler *et al.*, 1989) and the status of nuclear-structure theory and data very far from β-stability at that time. These systematic studies led to the well recognized FK$^2$L papers (Thielemann and Kratz, 1992; Kratz *et al.*, 1993; Thielemann *et al.*, 1994). Their main results will be summarized in the following.

Let us first remember that, apart from the above first measurements in the r-process path, another major advancement in the understanding of the nuclear-physics input to r-process calculations was the development of a *unified macroscopic-microscopic approach* by Möller and collaborators in the early 1990's (Möller *et al.*, 1990; Möller *et al.*, 1995; Möller *et al.*, 1997; Möller and Randrup, 1990), within which all nuclear properties could be studied for the first time in an internally consistent way: A microscopic (folded-Yukawa) single-particle model with extensions is combined with a macroscopic (finite range droplet) model which includes Coulomb redistribution effects and an improved formulation of the (Lipkin-Nogami) pairing model. As a first step, nuclear ground-state masses and shapes are calculated. Once these quantities are known, nuclear wave-functions are derived for the appropriate shapes. Matrix elements giving β-decay rates and other quantities of interest may then be determined. $T_{1/2}$ and $P_n$-values are deduced from theoretical Gamow-Teller (GT) strength functions calculated within the QRPA (Möller and Randrup, 1990).

This consistent nuclear-data set certainly yielded more reliable predictions than earlier models. Nevertheless, being aware that this new approach must also have its deficiencies, FK$^2$L have further improved the data set by also including into the GT strength-function calculations first-forbidden transitions (Takahashi *et al.*, 1973), and by taking into account all available experiments on $Q_\beta$, $B_n$, $T_{1/2}$ and $P_n$ as well as known nuclear-structure properties, either model-inherently not contained (*e.g.*, p-n residual interactions) or not properly described (*e.g.*, the onset of deformation at A ≅ 100) by the above – still too simplistic – global approach. For a detailed discussion, see *e.g.*, Kratz *et al.* (1993).

By applying this new experimental – theoretical nuclear-physics data base, FK²L (Thielemann and Kratz, 1992; Kratz *et al.*, 1993; Thielemann *et al.*, 1994) have tried to generalize the earlier fits to selected mass regions (Kratz, 1988; Kratz *et al.*, 1988a) to the whole $N_{r,\odot}$ distribution by assuming a *global* steady flow. The result of such a calculation for freeze-out conditions, normalized to the A ≅ 80 peak $[N_{r,\odot}(^{80}Se)\lambda_\beta(^{80}Zn) \cong 29.5\ s^{-1}]$, is displayed in Figure 3. Their calculation did produce three r-abundance maxima; however, the A ≅ 130 and 195 peaks were much too tall and were also shifted to higher A, which means that in these regions $n_n$ is too low. Thus, FK²L definitively concluded that the $N_{r,\odot}$ distribution, with declining peak heights as a function of A, *cannot* be explained by a global steady flow with an (n,γ)⇔(γ,n)-equilibrium.

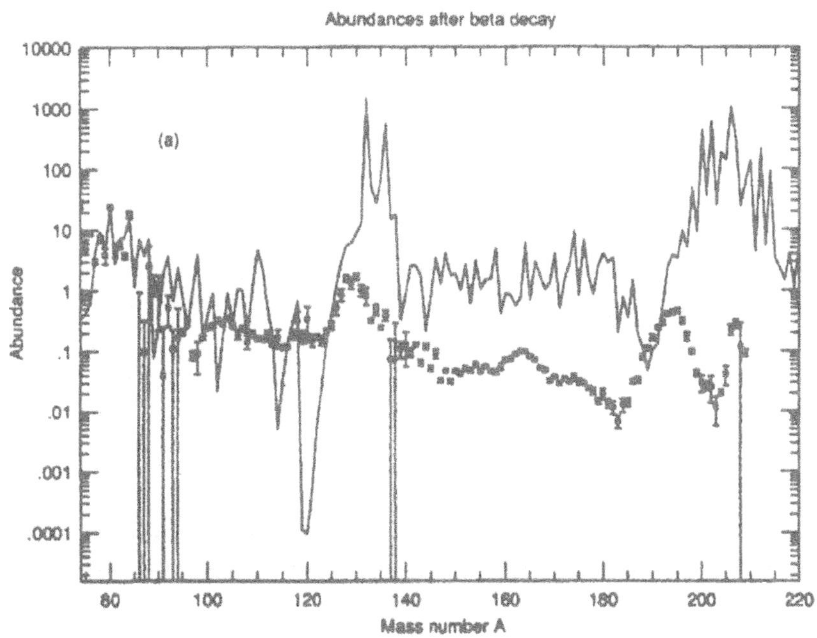


*Figure 3.* Global r-abundance curve, using the waiting-point and steady-flow approximation with $T_9$=13 and $n_n=10^{20}\ cm^{-3}$ and the unified nuclear-data set for $B_n$, $T_{1/2}$ and $P_n$ (See Kratz *et al.*, 1993), in comparison with the $N_{r,\odot} = (N_\odot - N_s)$ of Käppeler *et al.* (1989).

In a next step, FK²L have focussed on reproducing the details of individual mass regions, and finding the break points between steady-flow areas.

The results of these *static* calculations can be summarized as follows: The r-process, indeed, has reached a steady-flow equilibrium which is, however, no longer global but only local in between the r-abundance peaks. It breaks down at the top of each peak, *i.e.* at the N = 50, 82 and 126 closed shells. Similar to the $N_s\sigma$-curve for the s-process, $FK^2L$ obtained a three- (or four-) step r-process $N_r\lambda_\beta$-curve with different neutron number densities. The statistical weights of the first three $N_r\lambda_\beta$-components were approximately 10:3:1, being roughly proportional to the $T_{1/2}$ of the respective neutron-magic waiting-point nuclei $^{80}Zn_{50}$ (550 ms), $^{130}Cd_{82}$ (195 ms) and $^{195}Tm_{126}$ (≅ 55 ms from QRPA). The "best-fit" $T_9$-$n_n$ conditions for the different mass ranges are shown in Figure 4.

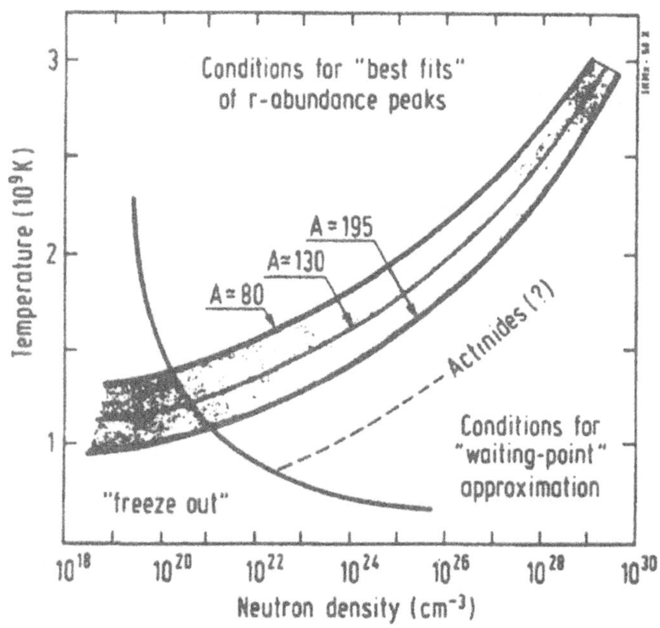


*Figure 4.* $T_9$-$n_n$ conditions for the waiting-point and steady-flow approximation for different mass ranges of $N_{r,\odot}$ (Käppeler *et al.*, 1989). Since r-process nuclei beyond Bi decay via β- and α-chains, no clear features emerge for a "best fit"; hence, the dashed line indicates but a first guess for the fourth component (Kratz *et al.*, 1993).

There remained, however, an uncertainty for the conditions which produce the $N_{r,\odot}$ beyond A ≅ 200. Since their progenitors decay via β- and α-decay chains, there is no simple shape to fit, and the 1993 study was just exploratory. Nevertheless, their results were able to rule out explosive He-burning as a possible solution, and supported the high-entropy neutrino wind scenario of SN II, which just came up at that time, *e.g.*, Meyer *et al.* (1992), Woosley and Hoffman (1992), Woosley *et al.* (1994), Takahashi *et al.* (1994).

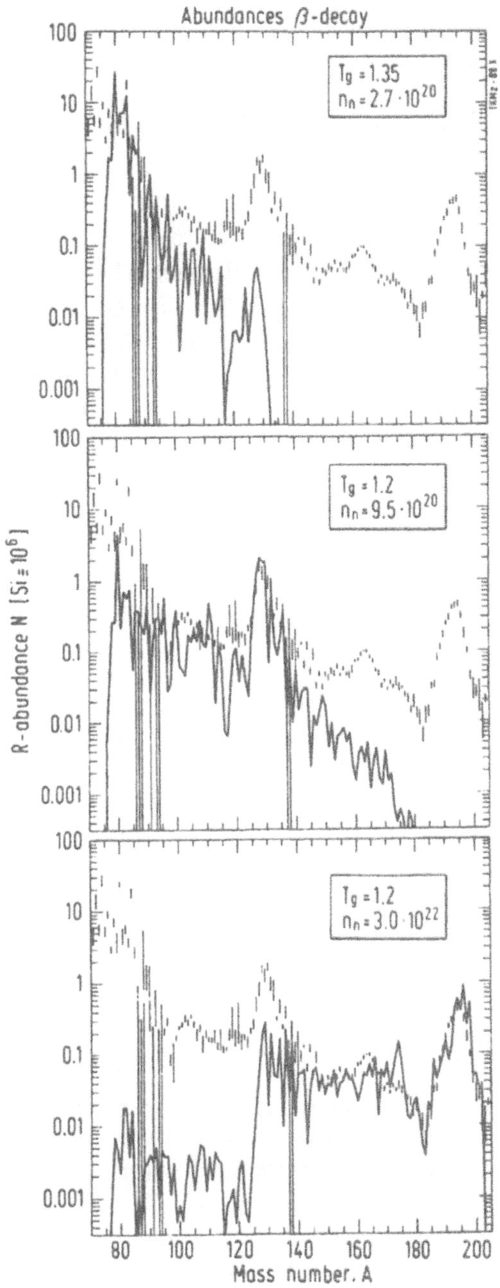


*Figure 5.* Time dependent calculations with "best-fit" conditions from static steady-flow calculations for a seed nucleus with Z = 26. Upper part: Formation of nuclei in the A ≅ 80 peak at 1.5 s; middle part: Formation of the 85 ≤ A ≤ 130 region at 1.5 s; lower part: Formation of nuclei beyond A ≅ 130 up to the A ≅ 195 abundance peak after 2.5 s.

The final question FK$^2$L asked (Kratz *et al.*, 1993) was whether the $N_{r,\odot}$ curve can be the result of a superposition of three (or more) *local* steady flows, where each component dominates one peak. They concluded that in the "static extreme" this is not possible, as already indicated by their *global* steady-flow calculations (see Figure 3). Such a superposition would only make sense when the abundances for a specific $T_9$-$n_n$ condition, appropriate to a region $A \leq A_{peak}$, would be set to zero for $A > A_{peak}$. This situation can, however, occur in time-dependent calculations, where the r-peaks containing neutron-magic nuclei with the longest $T_{1/2}$ will act as "bottle-necks", over which only small amounts of matter will pass. When starting their time-dependent calculations with a $Z = 26$ seed, FK$^2$L found that the time scales for reaching the appropriate peaks were comparable with 1.5 to 2.5 s, and were consistent with the expected duration of an r-process in a SN (See Figure 5).

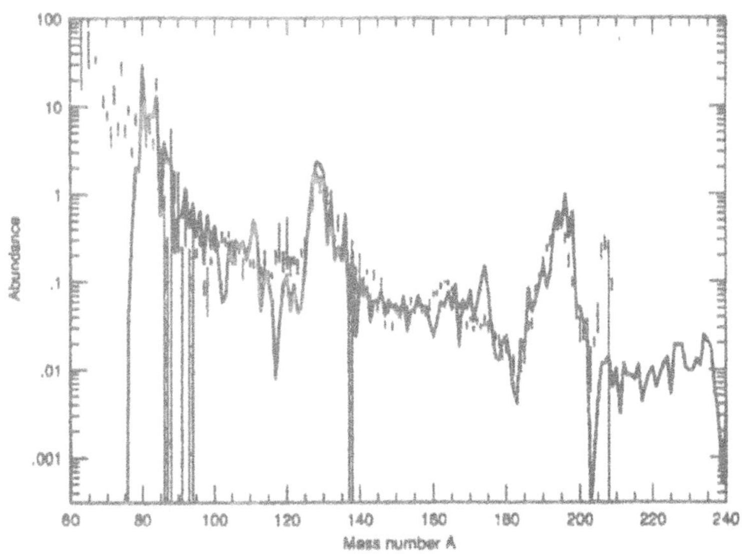


*Figure 6.* R-abundance distribution obtained from a superposition of four time-dependent calculations with the "best-fit" $T_9$-$n_n$ values for the $A \cong 80$ peak and the $90 \leq A \leq 130$ and $135 \leq A \leq 195$ mass ranges, and a first guess for the $A > 195$ region. From Kratz *et al.* (1993).

A typical result for a superposition of a four-component time-dependent calculation obtained by Kratz *et al.* (1993) and Thielemann *et al.* (1994) is shown in Figure 6. Clearly, the main features of the $N_{r,\odot}$ distribution were well reproduced much better than in earlier attempts of other authors (See, *e.g.*, the respective figures in (Mathews and Ward, 1985), and in certain regions even *isotopic abundances* became meaningful. However, there remained several deficiencies which – apart from $A \leq 78$ where the $(N_\odot - N_s)$ residuals are not of genuine r-origin – were interpreted to indicate nuclear-structure effects very far from stability not accounted for properly in the 1993 *unified* macroscopic-microscopic approach.

The solution to the A ≅ 120 abundance trough was attributed to the nuclear masses along the r-process path. In the FRDM mass model (Möller *et al.*, 1995), the decrease of the $B_n$'s occurs too slowly when approaching N = 82. This implies that – in an extreme view – there would exist *not a single* $B_n \cong 2$ MeV waiting-point isotope between $^{112}Zr_{72}$ and $^{125}Tc_{82}$. This model-deficiency presumably has its origin in the neglect of the p-n residual interaction which seems to manifest itself in an overestimation of the N = 82 shell strength below $^{132}Sn_{82}$ (Thielemann and Kratz, 1992; Kratz *et al.*, 1993; Thielemann *et al.*, 1994).

In the 103 ≤ A ≤ 115 and 160 ≤ A ≤ 180 mass regions, sinusoidal deviations show up in Figure 6. As is discussed in detail by FK$^2$L (Kratz *et al.*, 1993; Thielemann *et al.*, 1994), these deviations are due to model-deficiencies in the development of quadrupole deformation around neutron mid-shells, N = 66 and 104, which are clearly correlated with the other nuclear properties $B_n$, $T_{1/2}$ and $P_n$.

In summary, FK$^2$L concluded that both the overall success in the reproduction of the r-abundance peaks, as well as the local failures discussed above, indicate nuclear-structure signatures of extremely neutron-rich isotopes, the vast majority of them not accessible in terrestrial but only in stellar laboratories.

This interpretation of the overall $N_{r,\odot}$ fits – *"learning neutron-dripline physics from astrophysical observables"* – on the one hand, was greeted with enthusiasm by the majority of the nuclear-physics community. On the other hand, however, it has led to repeatedly pessimistic statements by some astrophysicists, such as: *"...with the numerous uncertainties remaining...,the reliability of the r-process calculations comes naturally to the mind...These problems should call for a much deeper study...before rushing into...premature comparisons of these results with the observed r-abundances."* (Goriely and Bouquelle, 1993). But, in most cases – with a delay of a few years – these authors also came themselves to quite similar conclusions (about the effect of shell quenching on the A ≅ 115 r-abundances, *e.g.*, Pearson *et al.* (1996), or the validity of the "waiting-point" concept, Goriely and Clerbaux, (1999), as our collaboration had deduced from detailed, systematic studies before.

## 5. FROM FK$^2$L TO PRESENT

Since the FK$^2$L papers (Thielemann and Kratz, 1992; Kratz *et al.*, 1993; Thielemann *et al.*, 1994), considerable progress has been achieved in the study of nuclear-structure systematics of neutron-rich, medium – to heavy –

mass nuclei at different laboratories using various production, separation and detection methods, with CERN-ISOLDE always playing a leading role in this field. See, *e.g.*, Springer (1988), IOP Conference (1993), Frontières (1995), Kratz *et al.* (1999). However, due to the generally very low production yields, the restriction to chemically non-selective ionization modes, or the application of non-selective detection methods, no further information on isotopes lying **on** the r-process path could be obtained for quite some time. Only in recent years, further progress in the identification of r-process nuclides was achieved by considerably improving the **selectivity** in production and detection methods of rare isotopes, *e.g.* by applying Z-selective laser ion-source systems, or projectile-fragmentation TOF techniques. Details about recent experimental information on isotopes relevant to r-process nucleosynthesis can be found, *e.g.* in Sherrill *et al.* (1998), Forkel-Wirth and Bollen (1999) and Kratz *et al.* (1999).

We will briefly summarize here the main results. The first mass region, where at present nuclear-structure information is of astrophysical interest, is the *r-process seed region* of very neutron-rich Fe-group nuclei. Recent spectroscopic studies around the double (semi-) magic $^{68}$Ni have revealed that N = 40 obviously is **not** a good closed subshell (For details, see *e.g.*, Sherrill *et al.* (1998) and Forkel-Wirth and Bollen (1999), and references therein), in disagreement with most (but not all) theories. The observed collectivity in this region, presumably extending towards a rather "weak" magic N = 50 shell closure around $^{78}Ni_{50}$, may therefore change the astrophysical picture of the development of the r-process matter flow from the Fe-group seed to the N = 50, A ≅ 80 first $N_{r,\odot}$ peak.

The second region of interest is that of far unstable nuclei around A ≅ 115. Here, most r-process calculations using standard *global* mass models (Möller *et al*, 1995; Hilf *et al.*, 1976; Aboussir *et al*, 1995; Meyers and Swiatecki, 1994, 1996) show a pronounced *r-abundance trough*, which – as mentioned above – had been interpreted by FK$^2$L (Thielemann *et al*, 1994) as being due to nuclear model deficiencies. Recent $Q_\beta$ measurements and γ-spectroscopic studies on very neutron-rich isotopes of refractory Zr to Pd elements have, indeed, indicated that predicted quadrupole deformation is obviously too strong in the N = 66 mid-shell region, and is maintained over a too wide mass range towards N = 82. Hence, as concluded by FK$^2$L from astrophysical and nuclear-structure considerations, the (correlated) decrease of the $B_n$ values with increasing neutron number occurs too slowly when approaching the N = 82 magic shell. See, *e.g.*, Thielemann *et al.* (1994), Pfeiffer *et al.* (1997), Kratz *et al.* (1998) and references therein. These deficiencies in the above mass models seem to have their origin in an overestimation of the N = 82 shell strength below $^{132}$Sn.

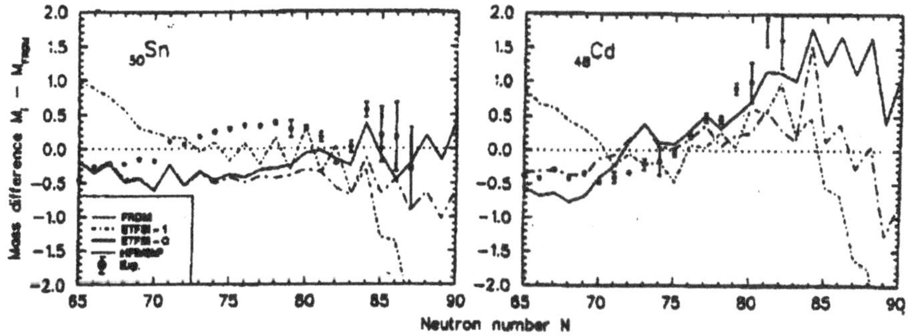

*Figure 7.* Experimental and theoretical mass differences for $_{50}$Sn and $_{48}$Cd isotopes normalized to FRDM values ($M_i - M_{FRDM}$). For discussion, see text.

Consequently, the third region of present interest is that around the double magic isotope $^{132}$Sn. Apart from astrophysical importance (formation of the A ≅ 130 peak of the $N_{r,\odot}$ (Thielemann *et al.*, 1994; Kratz *et al.*, 1998), this area is of considerable shell-structure interest. The isotope $^{132}$Sn itself, together with the properties of the nearest-neighbor single-particle ($^{133}{}_{51}$Sb and $^{133}$Sn$_{83}$) and single-hole ($^{131}{}_{49}$In and $^{131}$Sn$_{81}$) nuclides are essential for tests of the shell model and as input for any reliable future microscopic nuclear-structure calculations towards the neutron drip-line. Recent spectroscopic data, mainly from the new PS-booster ISOLDE facility at CERN, together with earlier β-decay studies at OSIRIS have revealed a number of interesting features that none of the potentials currently used in *ab initio* shell-structure calculations are able to reproduce properly. Among the highlights of these results are, (i) the identification of the second N = 82 waiting-point isotope 46-ms $^{129}$Ag; (ii) the level systematics of even-even $_{48}$Cd isotopes which show an extension of their vibrational character up to the N = 78 and 80 isotones $^{126,128}$Cd with deduced B(E2) values of ≅ 4,000 $e^2fm^4$; and (iii) the trend in mass measurements in the $^{132}$Sn shown in Figure 7. As can be seen from this figure, between the sequence of $_{50}$Sn and $_{48}$Cd isotopes there is a significant change in the slope of the experimental and theoretical mass differences (here normalized to the FRDM predictions, $M_i$–$M_{FRDM}$). Clearly, for the Cd isotopic chain, the best agreement with the measured masses is obtained with a mass model which contains the effect of *shell quenching,* such as HFB/SkP (Dobaczewski, 1999) or ETFSI-Q (Pearson, 1999) chosen here.

Summarizing the recent results from the $^{132}$Sn region, we conclude that by now there is sufficient experimental evidence to confirm our earlier predictions concerning a weaker strength of the N = 82 shell closure far from stability than described by the standard FRDM and ETFSI-1 mass models For a more detailed discussion, see *e.g.*, Kratz *et al.* (1999) and related papers (Sherrill *et al.*, 1998; Forkel-Wirth and Bollen, 1999). Therefore, these new nuclear-structure signatures should no longer be ignored in astrophysical applications. Consequently, we prefer to use "quenched" mass models, and short-range extrapolations of the new nuclear-structure signatures in our QRPA calculations of β-decay properties. On the other hand, it is difficult to understand why one tries to compensate for these obvious deficiencies in the nuclear-physics input by including astrophysically questionable parameters in an 8,000-component iterative inversion procedure without any predefined limit to the parameter space to fit the $N_{r,\odot}$ pattern to highest precision (See, *e.g.*, (Goriely and Clerbaux, 1999; Arnould and Takahashi, 1999, and references therein to their earlier papers).

Although also on the astrophysical side there has been considerable improvement since $FK^2L$ in 1993. For a recent review, see *e.g.*, (Wallerstein *et al.*, 1997). Even today none of the presently favored r-process models, in particular (I) the supernova type II neutrino wind and (ii) neutron star mergers [See *e.g.*, Woosley *et al.* (1994), Takahashi *et al.* (1994), Freiburghaus *et al.* (1999), and paper by Kratz *et al.* in Forkel-Wirth and Bollen (1999)] were able to account for the production and ejection of r-process material with the correct distribution to explain the solar-system r-abundances in a fully realistic and self-consistent manner. Therefore, our collaboration has continued to use the more deductive $FK^2L$ approach to obtain further insight into the r-process mystery (see, *e.g.* (Pfeiffer *et al.*, 1997; Kratz *et al.*, 1998)). In our new attempts, which still are independent of a specific stellar model or site, the above modern nuclear data base has been combined with most recent astrophysical observables (Beer *et al.*, 1997; Arlandini *et al.*, 1999; Sneden *et al.*, 1996; McWilliam, 1997; Cowan *et al.*, 1994, 1999; Westin *et al.*, 1999) in order to derive the necessary conditions required to observe these features.

As an example of the effect of the *new neutron-dripline physics* on the calculated isotopic r-abundance pattern, in Figure 8 we show global $N_r$ distributions from a superposition of 16 components with different, *correlated* neutron densities, process durations and weights at constant $T_9$ = 1.35 (Pfeiffer *et al.*, 1997; Kratz *et al.*, 1998; Freiburghaus *et al.*, 1999) for the two versions of the ETFSI nuclear mass model (Aboussir *et al.*, 1995; Pearson, 1999). In both cases, identical conditions for the stellar parameters were used. With the ETFSI-1 mass model, apart from pronounced A ≅ 115 and A ≅ 175 abundance troughs, too little r-process material is observed in the whole region beyond the A ≅ 130 peak. As mentioned above, the solution to the A ≅ 115 r-abundance deficiency clearly lies in the prediction of nuclear properties along the r-process path in this mass range up to the N = 82 shell closure below $^{132}Sn$. Analoguous deficiencies in describing shape-transitions around N ≅ 104 midshell and the N = 126 shell strength seem to be the origin of the deep A ≅ 175 r-abundance trough (Pfeiffer *et al.*, 1997; Kratz *et al.*, 1998).

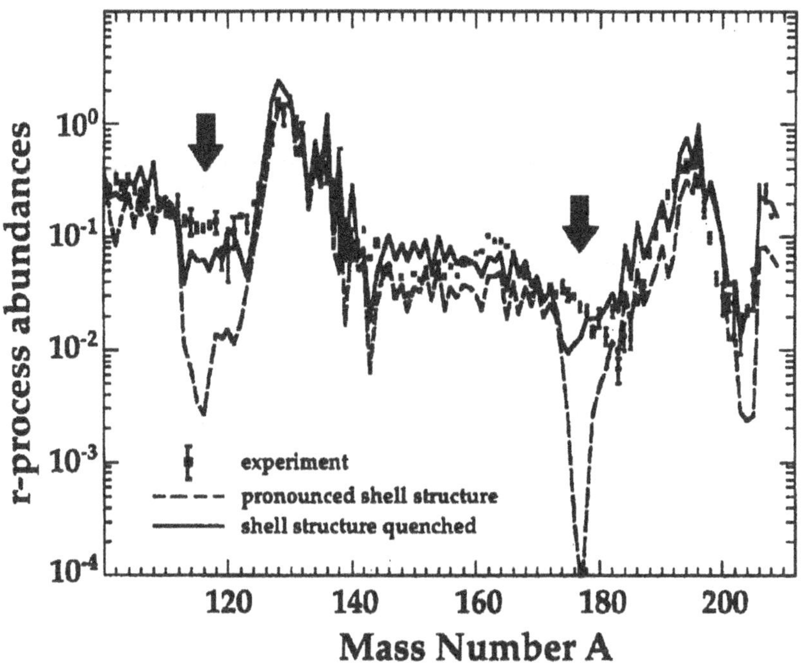


*Figure 8.* Global r-abundance fits for the ETFSI-1 mass model (Aboussir *et al.*, 1995) with pronounced shell structure and the ETFSI-Q formula (Pearson, 1999) with quenched shell gaps. For discussion, see text and (Pfeiffer *et al.*, 1997; Kratz *et al.*, 1998).

With the "quenched" mass model ETFSI-Q, however, a considerable improvement of the overall $N_{r,\odot}$ fit is observed. In particular, the prominent abundance troughs in the A ≅ 115 and 175 regions are eliminated to a large extent, and the r-matter flow is speeded up at A ≥ 130. Hence, with our new experimental data the application of a "quenched" mass model (Dobaczewski, 1999; Pearson, 1999; Satpathy and Nayak, 1998; Brown, 1998) for r-abundance calculations beyond A ≅ 150 seems to be highly recommended, in particular if predictions for the $^{203}$Tl to $^{209}$Bi region are required. Moreover, also extrapolations up to the progenitors of the long-lived actinides $^{232}$Th and $^{235,238}$U should now be more reliable (Pfeiffer *et al.*, 1997), since they are for the first time based on an internally consistent, modern nuclear-physics input.

Recently, Sneden and coworkers (Sneden *et al.*, 1996; Cowan *et al.*, 1994, 1999; Westin *et al.*, 1999) – in a new category of astrophysical observables – have determined *elemental* abundances of up to 9 medium-heavy elements between $_{38}$Sr and $_{48}$Cd, and 17 heavy elements between $_{56}$Ba and $_{90}$Th for the ultra-metal-poor, neutron-capture-rich halo stars CS 22892-052, HD 115444, HD 122563, and HD 126238. Similar abundance patterns have also been observed in other halo stars (McWilliam, 1997).

In Figure 9, the values of CS 22982-052 are compared to the solar r-element distribution and to our ETFSI-Q predictions for the entire Z-range from $_{38}$Sr to $_{90}$Th, after adjustment to the star's actual metallicity of [Fe/H] ≅ -3.1.

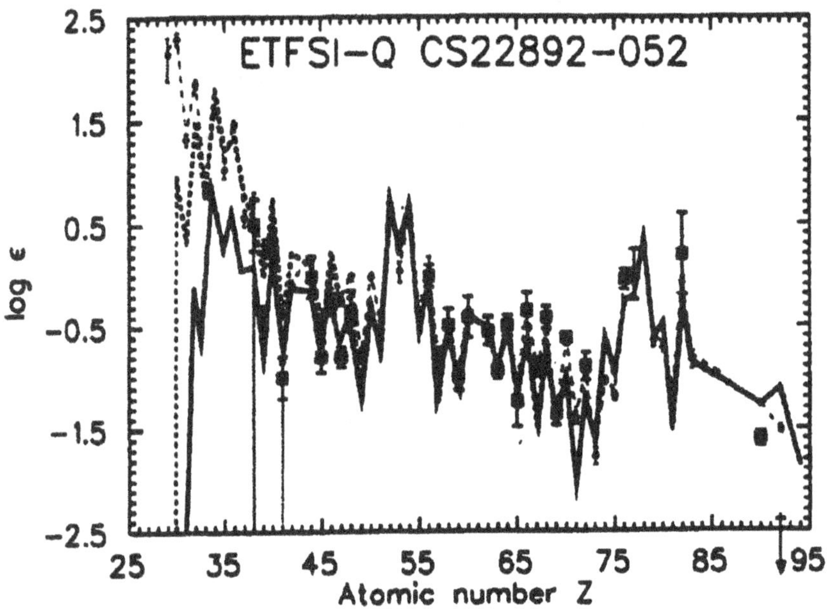


*Figure 9.* Comparison between observed (filled squares) and calculated elemental r-abundances from the ultra-metal-poor halo star CS 22892-052, 0 after normalization to solar metallicity [Fe/H] ≅ -3.1. The solar-system ($N_{r,\odot}$) distribution is shown as grey dashed curve with filled circles, and the respective theoretical distribution obtained with the ETFSI-Q masses is superimposed as black dotted curve (For details, see Cowan *et al.*, 1999). A calculation reproducing the observed abundances in CS 22892-052 is shown as a full black curve. This might correspond to the main r-process component proposed in Kratz *et al.* (1999), whereas the "residuals", $N_{r,\odot}$, - $N_{r,main}$, could correspond to a second, *weak* r-process component.

For the Z ≥ 40 region, four important conclusions can be drawn from this figure. First, from the *observed* Th/Eu abundance ratio in CS 22892-052, combined with our ETFSI-Q prediction of the *initial* Th/Eu ratio, a lower limit of its "decay age" of about 14.5 Gyrs is obtained (Cowan *et al.*, 1999), which overlaps with recent Galactic ages on the young side (See, *e.g.*, Salaris and Weiss (1997; Takahashi (1998). Second, the observed abundance pattern beyond $_{56}$Ba clearly indicates that the heavy elements in CS 22892-052 (as in the three other halo stars mentioned above) are of pure r-process origin, and must have been synthesized as a *primary* process already very early in the Galactic evolution, prior to the onset of main s-process nucleosynthesis (Cowan *et al.*, 1999; Travaglio *et al.*, 1999). Third, as for the abundances of the elements between Z = 38 and 48, an interesting feature is their pronounced odd-even-Z staggering, which clearly reflects nuclear-structure properties (mainly masses and delayed-neutron branchings) of the progenitor isotopes involved. All odd-Z elements from $_{39}$Y to $_{47}$Ag are clearly under-abundant compared to the (metallicity-scaled) solar pattern, whereas the even-Z elements ($_{38}$Sr – $_{48}$ Cd) are closer to solar (See Figure 9). These new data seem to support the recent evidence for a two-source nature of the r-process concluded from extinct radioactivities in meteorites (Wasserburg *et al.*, 1996). And fourth, the excellent agreement between the heavy r-element abundances of all four first-generation stars and the global $N_{r,\odot}$ pattern (which – in fact – represents a summation over the whole Galactic evolution) indicates that there is probably a *unique* scenario for the r-process at least, above A ≅ 130.

Taken together, this indicates that it is very unlikely, that neither the elemental abundances in the halo stars, nor the isotopic solar abundances resulted from a random superposition of varying r-components or abundance patterns from different nucleosynthesis sites (Goriely and Clerbaux, 1999; Arnould and Takahashi, 1999).

## 6. SUMMARY

We have shown in this paper, how the nuclear-physics and astrophysics basis of the rapid neutron-capture process have changed from B$^2$FH in 1956 (where the basic ideas were obtained) via FK$^2$L in 1993 (with the impact of the first waiting-point experiments) up to present. Although our knowledge in both fields has considerably improved over the past 40 years, there are still many open questions, such as the magnitude of shell quenching at N = 50, 82 and 126 when approaching the neutron dripline, or the nature and stellar site of the r-process – or the sites of several r-processes. In any case, we believe that continued progress in our understanding of the coupling between nuclear structure far from β-stability and explosive nucleosynthesis scenarios will remain an exciting challenge also in the next millenium.

## ACKNOWLEDGEMENTS

We thank J. Dobaczewski and J.M. Pearson for making available to us their HFB/SkP, respectively, ETFSI-Q mass tables. Discussions with H. Beer, T.C. Beers, A. Brown, F. Corvi, R. Gallino, K. Heyde, F. Käppeler, U.

Ott, P.G. Reinhard, C. Sneden, J.W. Truran and W.B. Walters are gratefully acknowledged. Support for this work was provided by various grants from the German BMBF, DFG and DAAD, and from the Swiss Nationalfonds.

## REFERENCES


Aboussir, Y., Pearson, J.M., Dutta, A.K. and Tondeur, F.: 1995, "Nuclear-Mass formula via an approximation to the Hartree-Fock method", *At. Data Nucl. Data Tables* **61**, 127-176.

Anders, E. and Grevesse, N.: 1989, "Abundances of the elements – meteoritic and solar", *Geochim. Cosmochim. Acta* **53**, 197-214.

Applegate, J.H., Hogan, C.J. and Scherrer, R.J.: 1988, "Cosmological quantum chromodynamics, neutron diffusion, and the production of primordial heavy-elements", *Ap. J.* **329**, 572-579.

Arlandini, C., Käppeler, F., Wisshak, K., Gallino, R., Lugaro, M., Busso, M. and Straniero, O.: 1999, "Neutron capture in low mass asymptotic giant branch stars: Cross sections and abundance signatures", *Ap. J.* **525**, 886-900.

Arnould, M. and Takahashi, K.: 1999, "Nuclear astrophysics," *Rep. Progr. Phys.* **62**, 395-464.

Beer, H., Corvi, F. and Mutti, P.: 1997, "Neutron-capture of the bottleneck isotopes Ba-138 and Pb-208, s-process studies, and the r-process abundance distribution", *Ap. J.* **474**, 843-861.

Brown, B.A.: 1998, "New Skyrme interaction for normal and exotic nuclei", *Phys. Rev.* **C58**, 220-231.

Burbidge, E.M., Burbidge, G.R., Fowler, W.A., and Hoyle, F.: 1957, "Synthesis of the elements in stars", *Revs. Mod. Phys.* **29** , 547-650.

Cameron, A.G.W., Delano, M.D. and Truran, J.W.: 1970, "The dynamics of the rapid neutron capture process", *Int. Conf. on the Properties of Nuclei Far from the Region of β-stability*, Leysin, Switzerland, *CERN Rep. 70-30*, 735–768.

Cameron, A.G.W.: 1957, "Nuclear reactions in stars and nucleosynthesis,", *Report Atomic Energy of Canada, Ltd., CRL-41*; See also 1957 paper, "A revised semiempirical atomic mass formula", *Can. J. Phys.* **35**, 1021-1032.

Colgate, S.A.: 1971, "Neutron star formation, thermonuclear supernovae, and heavy-element reimplosion", *Ap. J.* **163**, 221-230.

Coryell, C.D.: 1956, "Neutron and proton shells in the cosmic abundance of the elements", *Prog. Report MIT Lab. for Nucl. Sci.* **42**, 7-27.

Coryell, C.D.: 1961, "The chemistry of creation of the heavy elements", *J. Chem. Educ.* **38**, 67-72.

Cowan, J.J., Cameron, A.G.W. and Truran, J.W.: 1982, "The thermal runaway r-process:", *Ap. J.* **252**, 348-355.

Cowan, J.J., Pfeiffer, B., Kratz, K.L., Thielemann, F.K., Sneden, C., Burles, S., Tytler, D. and Beers, T.C.: 1999, "r-Process abundances and cosmochronometers in metal-poor stars", *Ap. J.* **521**, 194-205.

Cowan, J.J., Sneden, C., Ivans, I., Burles, S., Beers, T.C. and Fuller, G.: 1999, "Abundances in the ultra-metal-poor halo giant CS 22892-052: Implications for the production of neutron-capture elements in the early galaxy", *194th AAS Meeting*, Chicago, #67.04.

Cowan, J.J., Thielemann, F.K. and Truran, J.W.: 1991, "The r-process and nucleochronology", *Phys. Rep.* **208**, 267-394.

de Saint Simon, M. and Sorlin, O., eds.: 1995, *Exotic Nuclei and Atomic Masses - ENAM95*, Editions Frontières, Gif sur Yvette, France, 1072 pp.

Dobaczewski, J.: 1999, HFB/SkP Mass Table, private communication.

Eberth, J., Meyer, R.A. and Sistemich, K., eds.: 1988, "Nuclear Structure of the Zr Region", *Research Reports in Physics*, Springer, New York, NY, 424 pp.

Epstein, R.I., Colgate, S.A. and Haxton, W.C.: 1988 "Neutrino-induced r-process nucleosynthesis", *Phys. Rev. Lett.* **61**, 2038-2041.

Evans, C.R. and Mathews, G.J.: 1988, in *Origin and Distribution of the Elements*, World Scientific, Singapore, p. 619.

Forkel-Wirth, D. and Bollen, G.: 1999, "The ISOLDE Laboratory Portrait", *Hyperfine Interaction*, in press.

Freiburghaus, C., Rosswog, S. and Thielemann, F.-K.: 1999, "r-Process in neutron star mergers", *Ap. J.* **525**, L121-L124.

Gill, R.L., Casten, R.F., Hill, J.C., Mach, H. and Moreh, R.: 1986, "Half-life of Zn-80 – the 1st measurement for an r-process waiting-point nucleus", *Phys. Rev. Lett.* **56**, 1874-1877.

Goriely, S. and Bouquelle, V.: 1993, "The r-process: At last some results with microscopic nuclear predictions", in *Nuclei in the Cosmos II Second Int. Symp. on Nuclear Astrophysics*, IOP, pp. 595-600.

Goriely, S. and Clerbaux, B.: 1999, "Uncertainties in the Th cosmochronometry", *Astron. Astrophys.* **346**, 798-804.

Hilf, E.R., Van Groote, H. and Takahashi, K.: 1976, "Gross theory of nuclear masses and radii", in *Proc. 3rd Int. Conf. Nuclei Far from Stability*, Geneva, CERN-Rep. **76-13**, 142-148.

Hillebrandt, W. and Thielemann, F.K.: 1977, "The production of r-process seeds at low densities", *Astron. Astrophys.* **58**, 357-362.

Hillebrandt, W., Kodama, T. and Takahashi, K: 1976, "r-Process nucleosynthesis – A dynamic model", *Astron. Astrophys.* **52**, 63-68.

Hillebrandt, W.: 1978, "The rapid neutron-capture process and the synthesis of heavy and neutron-rich elements", *Space Sci. Rev.* **21**, 639-702.

Hogan, C.J. and Applegate, J.H.: 1987, "Neutron tori and the origin of r-process elements", *Nature* **330**, 236-238.

Hoyle, F. and Clayton, D.D.: 1974, "Nucleosynthesis in white-dwarf atmospheres", *Ap. J.* **191**, 705-710.

Kajino, T., Mathews, G.J., Fuller, G.M.: 1990, "Primordial nucleosynthesis of intermediate–mass elements in baryon-number inhomogeneous big-bang models – observational tests:", *Ap. J.* **364**, 7-14.

Käppeler, F, Beer, H. and Wisshak, K: 1989, "s-Process nucleosynthesis – Nuclear-physics and the classical-model", *Rep. Prog. Phys.* **52**, 945-1013.

Kodama, T. and Takahashi, K.: 1973, "On the delayed neutrons at the final state of the r-process", *Phys. Lett.* **B43**, 167-169.

Kratz, K.L., Bitouzet, J.P., Thielemann, F.K., Möller, P. and Pfeiffer, B.: 1993, "Isotopic r-process abundances and nuclear-structure far from stability – Implications for the r-process mechanism", *Ap. J.* **403**, 216-238.

Kratz, K.L., Gabelmann, H., Hillebrandt, W., Pfeiffer, B., Schlosser, K. and Thielemann, F.K.: 1986, "The beta-decay half-life of $^{130}_{48}Cd_{82}$ and its importance for astrophysical r-process scenarios", *Z. Phys.* **A325**, 489-490.

Kratz, K.L., Gabelmann, H., Möller, P., Pfeiffer, B., Ravn, H.L., and Wohr, A.: 1991, "Neutron-rich isotopes around the r-process waiting-point nuclei $^{79}Cu_{50}$ and $^{80}Zn_{50}$", *Z. Phys.* **A340**, 419-420.

Kratz, K.L., Harms, V., Hillebrandt, W., Pfeiffer, B., Thielemann, F.K. and Wohr, A.: 1990, "Origin of the odd-even staggering in the A $\simeq$ 80 solar r-abundance peak", *Z. Phys.* **A336**, 357-358.

Kratz, K.L., Harms, V., Wohr, A. and Möller, P.: 1988b, "Gamow-Teller decay of Zn-80 – Shell structure and astrophysical implications", *Phys. Rev.* **C38**, 278-284.

Kratz, K.L., Moeller, P., Pfeiffer, B. and Walters, W.B.: 1999, "New information on r-process nuclei", in *Proc. 10$^{th}$ Capture Gamma-Ray Spectroscopy and Related Topics - CGS10*, Santa Fe, NM, AIP Conf. Proc., in press.

Kratz, K.L., Pfeiffer, B. and Thielemann, F.K.: 1998, "Nuclear-structure input to r-process calculations", *Nucl. Phys.* **A630**, C352-C367.

Kratz, K.L., Thielemann, F.K., Hillebrandt, W., Möller, P, Harms, V., Wohr, A. and Truran, J.W.: 1988a, "Constraints on r-process conditions from beta-decay properties far off stability and r-abundances", *J. Phys.* **G14**, S331-S342.

Kratz, K.L.: 1988, "Nuclear physics constrains to bring the astrophysical r-process to the 'waiting point' ", *Revs. Mod. Astr.* **1**, 184-209.

Lattimer, J.M. and Schramm, D.N.: 1974, "Black-hole – neutron-star collisions", *Ap. J.* **192**, L145-L147.

Lattimer, J.M. and Schramm, D.N.: 1976, "The tidal disruption of neutron stars by black holes in close binaries", *Ap. J.* **210**, 549-567.

Leblanc, J.M. and Wilson, J.R.: 1970, "A numerical example of the collapse of a rotating magnetized star", *Ap. J.* **161**, 541-551.

Lee, T., Schramm, D.N., Wefel, J.P. and Blake, J.B.: 1979, "Isotopic anomalies from neutron reactions during explosive carbon burning", *Ap. J.* **232**, 854-862.

Lund, E., Ekstrom, B., Fogelberg, B. and Hoff, P.: 1986, "Decay properties of $^{80}Zn_{75}$ and beta-values of neutron-rich Zn and Ga isotopes", *Phys. Scr.* **34**, 614-623.

Mathews, G.J. and Ward, R.A.: 1985, "Neutron-capture processes in astrophysics", *Rep. Prog. Phys.* **48**, 1371-1418.

McWilliam, A: 1997, "Abundance ratios and galactic chemical evolution", *Ann. Rev. Astron. Astrophys.* **35**, 503-556.

Meier, D.L., Epstein, R.L., Arnett, W.D. and Schramm, D.N.: 1976, "Magnetohydrodynamic phenomenamil in collapsing stellar cores", *Ap. J.* **204**, 869-878.

Meyer, B.S., Mathews, G.J., Howard, W.M., Woosley, S.E. and Hoffman, R.D.: 1992, "r-Process nucleosynthesis in the high-entropy supernova bubble", *Ap. J.* **399**, 656-664.

Meyers, W.D. and Swiatecki, W.J.: 1994, "Table of nuclear masses according to the 1994 Thomas-Fermi model", *Report LBL-36803*, 181 pp.

Meyers, W.D. and Swiatecki, W.J.: 1996, "Nuclear properties according to the Thomas-Fermi model", *Nucl. Phys.* **A601**, 141-167.

Möller, P. and Randrup, J.: 1990, "New developments in the calculation of beta-strength functions", *Nucl. Phys.* **A541**, 1-48.

Möller, P., Nix, J.R. and Kratz, K.L.: 1997, "Nuclear properties for astrophysical and radioactive-ion-beam applications", *At. Data Nucl. Data Tables* **66**, 131-343.

Möller, P., Nix, J.R., Kratz, K.L. and Howard, W.M.: 1990, "Nuclear structure calculations for astrophysical applications" in *Nuclei in the Cosmos Proc. Int. Symp. on Nuclear Astrophysics*, MPA/P4, 226-275.

Möller, P., Nix, J.R., Myers, W.D. and Swiatecki, W.J.: 1995, "Nuclear ground state masses and deformations", *At. Data Nucl. Data Tables* **59**, 185-381.

Neugart, R. and Woehr, A., eds.: 1993, *Nuclei Far From Stability/Atomic Masses and Fundamental Constants 1992*, IOP Conference Series, R7-R8.

Pearson, J.M.: 1999, ETFSI-Q Mass Table, private communication.

Pearson, J.M., Nayak, R.C. and Goriely, S.: 1996, "Nuclear-mass formula with Bogolyubov-enhanced shell-quenching – Application to r-process", *Phys. Lett.* **B397**, 455-459.

Pfeiffer, B., Kratz, K.L. and Thielemann, F.K.: 1997, "Analysis of the solar-system r-process abundance pattern with the new ETFSI-Q mass formula", *Z. Phys.* **A357**, 235-238.

Salaris, M. and Weiss, A.: 1997, "Chronology of the halo globular-cluster system formation", *Astron. Astrophys.* **327**, 107-120.

Sato, K.: 1974, "Formation of elements in neutron rich ejected matter of supernovae II.", *Prog. Theor. Phys.* **51**, 726-744.

Satpathy, L. and Nayak, R.C.: 1998, "Masses of atomic-nuclei in the infinite nuclear-matter model", *At. Data Nucl. Data Tables* **39**, 241-249.

Satpathy, L. and Nayak, R.C.: 1998, "Study of nuclei in the drip-line regions", *J. Phys.: Nucl. Part. Phys.* **G24**, 1527-1533.

Schramm, D.N. and Barkat, Z.: 1972, "The neutron-proton ratio in stars exploding from dense, high-temperature states", *Ap. J.* **173**, 195-204.

Schramm, D.N.: 1973, "Explosive r-process nucleosynthesis", *Ap. J.* **185**, 293-301.

Schramm, D.N.: 1982, "The r-process and nucleocosmochronology" in *Essays in Nuclear Astrophysics*, Cambridge Univ. Press, Cambridge, UK, p. 325.

Seeger, P.A., Fowler, W.A. and Clayton, D.D.: 1965, "Nucleosynthesis of heavy elements by neutron capture", *Ap. J. Suppl.* **11**, 121-166.

Sherrill, B.M., Morrissey, D.J. and Davids, C.N., eds.: 1998, *Exotic Nuclei and Atomic Masses - ENAM98*, AIP Conf. Proc. **455**.

Sneden, C., McWilliam, A., Preston, G.W., Cowan, J.J., Burris, D.L. and Armosky, B.J.: 1996, "The ultra-metal-poor, neutron-capture-rich giant star CS 22892-052", *Ap. J.* **467**, 819-840.

Suess, H.E. and Urey, H.C.: 1956, "Abundances of the elements", *Rev. Mod. Phys.* **28**, 53-74.

Symbalisty, E. and Schramm, D.N.: 1982, "Neutron star collisions and the r-process", *Ap. J. Lett.* **22**, 143-145.

Symbalisty, E.M.D., Wilson, J.R. and Schramm, D.N.: 1985, "An expanding vortex site for the r-process in rotating stellar collapse", *Ap. J* **291**, L11-L14.

Takahashi, K., Witti, J. and Janka, H.T.: 1994, "Nucleosynthesis in neutrino-driven winds from proto-neutron stars . 2. The r-process", *Astron. Astrophys.* **286**, 857-869.

Takahashi, K., Yamada, Y. and Kondoh, T.: 1973, "Beta-decay half-lives calculated on gross theory", *At. Data Nucl. Data Tables* **12**, 101-142.

Takahashi, K.: 1998, in *Tours Symp. On Nuclear Physics III*, AIP Conf. Proc. **425**, 616.

Thielemann, F.K. and Kratz, K.L.: 1992, "Synthesis of nuclei in astrophysical environments" in *Frontier Topics in Nuclear and Astrophysics,* Proc. 22[nd] Masurian Lakes Summer School, 1991, IOP, 187-226.

Thielemann, F.K., Kratz, K.L., Pfeiffer, B., Rauscher, T., Vanwormer, L. and Wiescher, M.C.: 1994, "Astrophysics and nuclei far from stability", *Nucl. Phys. A* **570**, C329-C343.

Travaglio, C., Galli, D., Gallino, R., Busso, M., Ferrini, F. and Straniero, O.: 1999, "Galactic chemical evolution of heavy elements: From barium to europium", *Ap. J.* **521**, 691-702.

Truran, J.R., Cowan, J.J. and Cameron, A.G.W.: 1978, "The helium driven r-process in supernovae", *Ap. J.* **222**, L63-L67.

Wallerstein, G., Iben, I., Parker, P., Boesgaard, A.M., Hale, G.M., Champagne, A.E., Barnes, C.A., Käppeler, F., Smith, V.V., Hoffman, R.D., Timmes, F.X., Sneden, C., Boyd, R..N., Meyer, B.S. and Lambert, D.L.: 1997, "Synthesis of the elements in stars – 40 years of progress", *Rev. Mod. Phys.* **69**, 995-1084.

Wasserburg, G.J., Busso, M. and Gallino, R.: 1996, "Abundances of actinides and short-lived nonactinides in the interstellar-medium – diverse supernova sources for the r-process:, *Ap. J.* **466**, L109-L113.

Westin, J., Sneden, C., Gustafsson, B. and Coqan, J.J.: 1999, "The r-process-enriched low metallicity giant HD 115444", submitted to *Ap. J.*

Woosley, S.E. and Hoffman, R.: 1992, "The alpha-process and the r-process", *Ap. J.* **395**, 202-239.

Woosley, S.E. and Weaver, T.A.: 1986, in *Nucleosynthesis and Its Implications for Nuclear and Particle Physics*, eds., Audouze, Jean and Mathieu, Nicole, NATO ASI Ser. C, Kluwer Academic Publishers, Dordrecht, Holland, vol. **163**, p. 145.

Woosley, S.E., Wilson, J.R., Mathews, G.J., Hoffman, R.D. and Meyer, B.S.: 1994, "The r-process and neutrino-heated supernova ejecta:, *Ap. J.* **433**, 229-246.

# Nuclear Aspects of Stellar and Explosive Nucleosynthesis

Thomas Rauscher[1], Friedrich-Karl Thielemann[1], Robert D. Hoffman[2], and Stanford E. Woosley[2,3]
*[1]Department of Physics and Astronomy, University of Basel, 4056 Basel, Switzerland, [2]Lawrence Livermore National Laboratory, Livermore, CA 94550, USA, [3]Department of Astronomy and Astrophysics, University of California, Santa Cruz, CA 95064, USA*
*tommy@quasar.physik.unibas.ch*

**Abstract:** The majority of nuclear reactions in astrophysics involve unstable nuclei which are not yet fully accessible by experiments. Therefore, there is high demand for reliable predictions of cross sections and reaction rates by theoretical means. The majority of reactions can be treated in the framework of the statistical model (Hauser-Feshbach). The global parameterizations of the nuclear properties needed for predictions far off stability probe our understanding of the strong force and take it to its limit. The sensitivity of astrophysical scenarios to nuclear inputs is illustrated in the framework of a detailed nucleosynthesis study in type II supernovae. Abundances resulting from calculations in the same explosion model with two different sets of reaction rates are compared. Key reactions and required nuclear information are identified.

## 1. INTRODUCTION

The investigation of stellar and explosive nuclear burning in astrophysical environments is a challenge for both theoretical and experimental nuclear physicists. Highly unstable nuclei are produced in such processes which can again be targets for subsequent reactions. Cross sections and astrophysical reaction rates for a large number of nuclei are required to perform complete network calculations which take into account all possible reaction links and do not postulate *a priori* simplifications. Most of the involved nuclei are currently not accessible in the laboratory and therefore theoretical models have to be invoked in order to predict reaction rates.

In astrophysical applications, usually different aspects are emphasized than in pure nuclear physics investigations. Many of the latter in this long and well established field were focused on specific reactions, where all or most "ingredients" were deduced from experiments. As long as the reaction mechanism is identified properly, this will produce highly accurate cross sections. For the majority of nuclei in astrophysical applications, such information is not available. The real challenge is thus not the application of well-established models, but rather to provide all the necessary ingredients in as reliable a way as possible, also for nuclei where no such information is available.

Considering the still remaining uncertainties in the prediction of nuclear reaction rates, it is of great interest to investigate the sensitivity of abundance yields to variations in the rates. This is also important if one wants to disentangle stellar physics and nuclear effects in the comparison of models which differ in both aspects. We compared two sets of reaction rates by employing them in the stellar model of (Woosley and Weaver, 1995).

## 2. NUCLEAR REACTION RATES

To compute the number of reactions, r, per volume and time, the velocity (energy) distribution of the interacting particles has to be folded with the cross section, σ, giving the probability for an interaction. Nuclei in an astrophysical plasma follow a Maxwell-Boltzmann distribution (MBD) and the thermonuclear reaction rates will have the form (Fowler *et al.*, 1967).

$$r_{j,k} = \langle \sigma v \rangle n_j n_k$$
$$\langle \sigma v \rangle := \left(\frac{8}{M\pi}\right)^{1/2} (kT)^{-3/2} \int_0^\infty E\sigma(E) exp(-E/kT) dE \tag{1}$$

Here $M$ denotes the reduced mass of the target-projectile system and $n_{j,k}$ is the number of projectiles and target nuclei, respectively. In astrophysical plasmas with high densities and/or low temperatures, electron screening becomes highly important, which reduces the Coulomb repulsion.

In general, the cross section will be the sum of the cross sections resulting from compound reactions via an average over overlapping resonances (HF) and via single resonances (BW), direct reactions (DI) and interference terms:

$$\sigma(E) = \sigma HF(E) + \sigma BW(E) + \sigma DC(E) + \sigma int(E) \tag{2}$$

Depending on the number of levels per energy interval in the system, projectile + target, different reaction mechanisms will dominate (Rauscher *et al.*, 1997). Since different regimes of level densities are probed at the various projectile energies, the application of a specific description depends on the energy. In astrophysics, one is interested in energies in the range from a few tens of MeV down to keV or even thermal energies (depending on the charge of the projectile).

## 3. THE STATISTICAL MODEL

It has been shown (Rauscher *et al.*, 1997) that the majority of nuclear reactions in astrophysics can be described in the framework of the statistical model (HF) (Hauser and Feshback, 1952). This description assumes that the reaction proceeds via a compound nucleus which finally decays into the reaction products. With a sufficiently high level density, average cross sections

$$\sigma^{HF} = \sigma_{form} b_{dec} = \sigma_{form} \frac{\Gamma_{final}}{\Gamma_{tot}} \tag{3}$$

can be calculated which can be factorized into a cross section $\sigma_{form}$ for the formation of the compound nucleus and a branching ratio $b_{dec}$, describing the probability of the decay into the channel of interest compared with the total decay probability into all possible exit channels. The partial widths $\Gamma$ as well as $\sigma_{form}$ are related to (averaged) transmission coefficients, which comprise the central quantities in any HF calculation.

Many nuclear properties enter the computation of the transmission coefficients: mass differences (separation energies), optical potentials, giant dipole resonance widths, and level densities. The transmission coefficients can be modified due to pre-equilibrium effects which are included in width fluctuation corrections (Tepel *et al.*, 1974) (see also Rauscher *et al.* (1997) and references therein) and by isospin effects. It is in the description of the nuclear properties where the various HF models differ. A choice of what is thought of being the currently best parametrizations is incorporated in the new HF code NON-SMOKER (Rauscher and Thielemann, 1998), which is based on the well-known code SMOKER (Thielemann *et al.*, 1987).

## 4. A REACTION RATE LIBRARY

Utilizing the NON-SMOKER code, cross sections and reaction rates for reactions with nucleons, $\alpha$ particles or $\gamma$ rays in entrance and exit channels, respectively, were calculated for all targets between proton and neutron drip line in the range 9 < Z < 84. Tabulated cross sections and rates can be found at *http://quasar.physik.unibas.ch/~tommy/reaclib.html*. Analytic fits to these rates, along with further information, can be obtained as an electronic file from the authors or on-line from Atomic Data and Nuclear Data Tables. A selection from these fits is published in (Rauscher and Thielemann, 1999).

In all applications, these rates should be supplemented or replaced with experimental rates as they become available. Such a combination of theoretical and experimental rates is provided, *e.g.*, in the REACLIB compilation. Currently, a new version is being compiled, in which the theoretical rates presented here will be included. Latest information on REACLIB can be found on the WWW at *http://ie.lbl.gov/astro.html*. Further details on the NON-SMOKER code are presented at *http://quasar.physik.unibas.ch/~tommy/reaclib.html*.

# 5. REACTION RATE SENSITIVITY OF NUCLEOSYNTHESIS IN TYPE II SUPERNOVAE

When comparing results from different supernova models, one faces the difficulty caused by the fact that it is hard to differentiate between influences of differing reaction rate sets and different stellar physics. We tried to segregate the abundance differences between the two models of Woosley and Weaver (1995) and Thielemann *et al.* (1996) existing because of the dichotomy of stellar models from those reflecting purely the choice of nuclear physics. For that purpose, hybrid calculations were performed, using the same stellar evolution code as in Woosley and Weaver (1995) but with rates from both models. In addition to helping to understand why calculations of the two groups differ, the use of independent rate sets in identical stellar models helps determine the nuclear physics portion of the error bar one should assign to nucleosynthesis studies of this sort. Furthermore, the reason for the differences in the theoretical rates was investigated, pointing to possibilities for future improvements of rate predictions. In the following, the findings are briefly summarized. A very detailed account of the work can be found in Hoffman *et al.* (1999).

## 5.1 The Rate Sets

The reaction rates utilized by Woosley and Weaver (1995) were those of (Woosley *et al.*, 1978) (WFHZ), and Thielemann *et al.* (1996) used (Thielemann *et al.*, 1987) (TAT). As examples, Figures 1 and 2 show a comparison of the two sets to each other and to experimental values for 30 keV neutron capture and proton capture at $T_9 = 3$. The dotted lines give deviations of a factor 2.

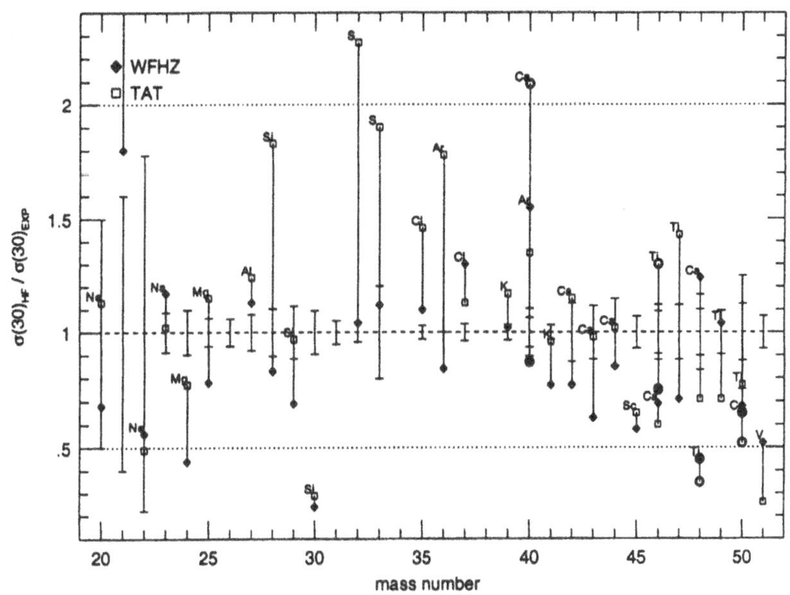


*Figure 1.* 30 keV neutron capture cross sections.

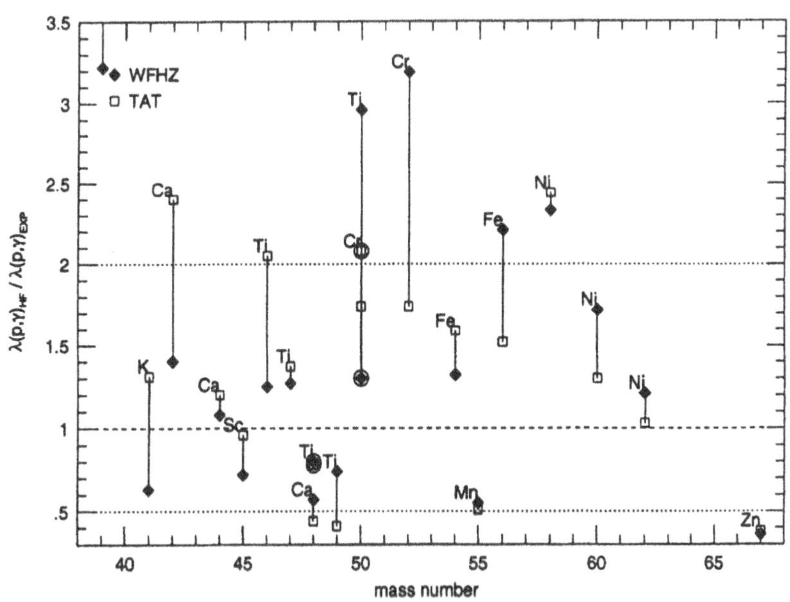


*Figure 2.* Proton capture reaction rates at $T_9 = 3$.

Typical differences at astrophysically interesting temperatures are less than a factor of two. There are individual cases, however, where the difference exceeds a factor of 10. Some of the larger differences occur for reactions where scarce experimental information is available and different assumptions were made regarding the photon transmission function, for example, $(\alpha,\gamma)$ reactions on $Z = N$ nuclei. Different assumptions were also made about the particle transmission functions, nuclear partition functions, and level densities. More modern and complete data used in the TAT rates makes them superior in cases where the partition function is important. WFHZ used an equivalent square well with empirical reflection factors; TAT used a more detailed optical model. Given the quite different values, *e.g.*, the neutron and proton transmission function, it is perhaps surprising that the rates differ so little. This is because the relevant temperatures for explosive burning are high. For incident particles in the Gamow window, the deviations in the particle transmission functions are typically smaller than a factor of two. In addition, higher partial waves contribute. A comparison of rates at a lower temperature would have revealed larger discrepancies.

Compared to experiment, both sets of theoretical rates give similar agreement, typically to a factor of two. The standard deviations between the two theoretical sets and cross section data are almost identical. In summary, the two rate sets have comparable merit when compared to experiment.

All the authors of this paper agree that the new rate set, the "NON-SMOKER" set, will be preferable to both TAT and WFHZ and will be adopted by both groups (Woosley and Weaver, 1995; Thielemann *et al.*, 1996) for future work.

## 6. RESULTS AND CONCLUSIONS

The comparison of the yields obtained with the two reaction rate sets in the model of Woosley and Weaver (1995) is shown in Figures 3 and 4 for a 15 and a 25 $M_{\odot}$ supernova, respectively. The dotted lines give deviations of a factor 2.

When the two current rate sets are included in otherwise identical stellar models we find that the nucleosynthesis, with some interesting exceptions, is not greatly changed. For example, only about a dozen (out of 70) stable isotopes in the mass range 12 to 70 have nucleosynthesis that differ by over 20% in two supernovae of 15 $M_{\odot}$ that use the same rate for $^{12}C(\alpha,\gamma)\,^{16}O$. It can, however, be noticed that most of these isotopes - with the exception of $^{44}Ti$ - are products of hydrostatic burning where individual reaction flows are governed by the cross sections involved. Nevertheless, none differ by more than a factor of 1.7. Given the significantly larger differences that exist in individual reaction rates, one may wonder at the robust nature of the final nucleosynthesis. We see three major causes.

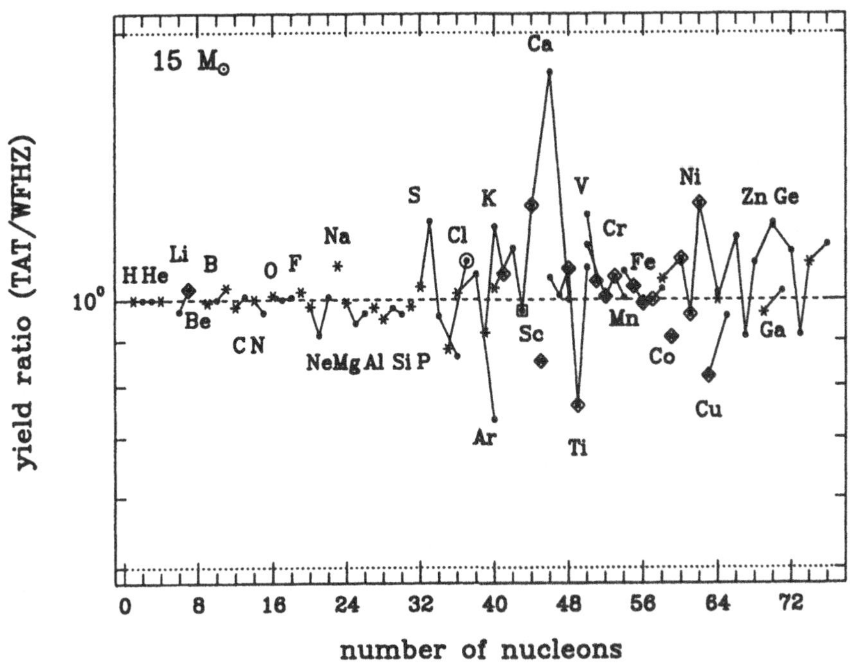


*Figure 3.* Yield ratios for a 15 $M_\odot$ supernova.

First, as the star burns and becomes hotter, the nuclear flow follows the valley of β stability making heavier nuclei as it goes. In doing so, it follows the path of least resistance – those reactions having the largest cross section for a nucleon or α particle reacting with a given nucleus. These large cross sections are reasonably well replicated by any calculation, normalized to experiment, that treats the Coulomb barrier and photon transmission function approximately correctly. Large differences may exist in rate factors for reactions that are in competition, especially a small channel in the presence of a large one, but these small channels are frequently negligible, at least for the major abundances, while they can cause larger differences when one is interested in the abundances of trace isotopes.

If one is however interested in accurate abundance predictions resulting from these smaller flows in hydrostatic burning stages, this can in most cases only be obtained by improving the reliability of the cross sections (and reaction rates) that determine these weak flows on light and intermediate mass nuclei. As new experimental information becomes available, a continuous improvement is therefore highly warranted.

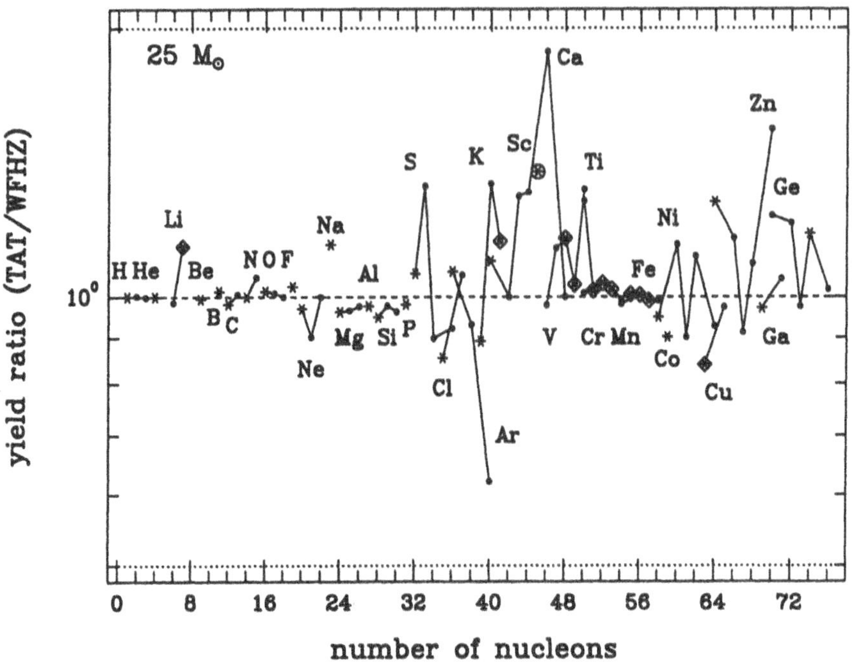


*Figure 4.* Yield ratios for a 25 $M_\odot$ supernova.

Second, beyond oxygen burning, which is to say for nuclei heavier than calcium, nucleosynthesis increasingly occurs in a state of full or partial nuclear statistical equilibrium. There the abundances are given by binding energies and partition functions. So long as the "freeze-out" is sufficiently rapid, individual rates are not so important.

Third, the reaction rates varied here were only those theoretical values from Hauser-Feshbach calculations for intermediate mass nuclei, *i.e.*, nuclei heavier than magnesium. The really critical reaction rates are, for the most part, those below magnesium. These reactions, like *e.g.*, $^{12}C(\alpha,\gamma)^{16}O$, govern the energy generation, major nucleosynthesis, and neutron exposure in the star. The rest are perturbations on these dominant flows.

This is not to say, however, that the nuclear and stellar details of heavy element synthesis are now well understood. Differences in the *stellar model* may account not just for 20% variation, but *orders of magnitude*. That is, uncertainty in stellar physics – especially the treatment of convection and how it is coupled (or not coupled) to the nuclear network – accounts for most of the differences in current nucleosynthesis calculations – provided such calculations use the same nuclear reaction rates below magnesium.

Even in a perfect stellar model though, there will still be interesting nuclear physics issues. Stellar nucleosynthesis is becoming a mature field rich with diverse and highly detailed observational data. The "factor of two" accuracy that was adequate in the past may not do justice to the observations of the future. There are many individual cases where the nuclear physics uncertainty is still unacceptably large. We point out just two of them.

The suppression of radiative capture reactions into self-conjugate (isospin zero) nuclei is very uncertain. Past Hauser-Feshbach calculations have adopted empirical factors for this suppression. The new NON-SMOKER rates include a significantly improved treatment (Rauscher *et al.*, 1999). The $\alpha$-capture reactions, like $^{24}$Mg($\alpha$,$\gamma$)$^{28}$Si, $^{28}$Si($\alpha$,$\gamma$)$^{32}$S,. . ., $^{44}$Ti($\alpha$,$\gamma$)$^{48}$Cr, are very important to nucleosynthesis in oxygen and silicon burning. The reaction $^{40}$Ca($\alpha$,$\gamma$)$^{44}$Ti also directly affects the synthesis of $^{44}$Ti. Modern accurate determinations of most of the reaction rates are missing (as well as (p,$\gamma$) reactions into the same nuclei). Measurements here would be most welcome.

The Hauser-Feshbach rates are also only as good as the local experimental rates to which the necessary parameters of the calculation are calibrated. In that regard, we would point out the near absence of charged particle reaction rate data for $A > 70$. Charged particle reactions are important, especially on unstable nuclei, at significantly higher atomic weights in the r-process and in the p-process.

## ACKNOWLEDGEMENTS

This work has been supported by the U.S. NSF (AST 96-17494, AST 96-17161, AST 97-31569, PHY-74-07194) and the Swiss NSF (20-47252.96, 2000-53798.98). T.R. is a PROFIL fellow of the Swiss NSF.

## REFERENCES

Fowler, W.A., Caughlan, G.E. and Zimmermann, B.A.: 1967, "Thermonuclear reaction rates", *Ann. Rev. Astron. Astrophys.* **5**, 525-570.

Hauser, W. and Feshbach, H.: 1952, "The inelastic scattering of neutrons", *Phys. Rev.* **A87**, 366-373.

Hoffman, R.D., Woosley, S.E., Weaver, T.A., Rauscher, T. and Thielemann, F.K.: 1999, "The reaction rate sensitivity of nucleosynthesis in type II supernovae", *Ap. J.* **521**, 735-752.

Rauscher, T. and Thielemann, F.K.: 1998, "Global statistical model calculations and the role of isospin", in *Stellar Evolution, Stellar Explosions and Galactic Chemical Evolution*, ed., Mezzacappa, A., IOP Publishing, Ltd., Bristol, UK, pp. 519-523.

Rauscher, T. and Thielemann, F.K.: 1999, "Astrophysical reaction rates from statistical model calculations", *Atomic Data Nucl. Data Tabl.*, in press.

Rauscher, T., Görres, J. and Wiescher, M.C.: 1999, "Capture of α-particles by isospin-symmetric nuclei", *Phys. Rev.* **C**, to be submitted.

Rauscher, T., Thielemann, F.K. and Kratz, K.L.: 1997, "Nuclear level density and the determination of thermonuclear rates for astrophysics", *Phys. Rev.* **C56**, 1613-1625.

Tepel, J., Hoffmann, H. and Weidenmüller, H.: 1974, "Hauser-Feshbach formulas for medium and strong absorption", *Phys. Lett.* **49B**, 1-4.

Thielemann, F.K., Arnould, M. and Truran, J.W.: 1987, "Thermonuclear reaction rates from statistical model calculations", in *Advances in Nuclear Astrophysics*, eds., Vangioni-Flam, E., Audouze, J., Cassè, M., Chieze, J.P. and Tran Thanh Van, J., Editions Frontières, Gif sur Yvette, France, pp. 525-540.

Thielemann, F.K., Nomoto, K. and Hashimoto, M.: 1996, "Core-collapse supernovae and their ejecta", *Ap. J.* **460**, 408-436.

Woosley, S., Fowler, W., Holmes, J. and Zimmerman, B.: 1978, "Semiempirical thermonuclear reaction-rate data for intermediate-mass nuclei", *Atomic Data Nucl.Data Tabl.* **22**, 371-441.

Woosley, S.E. and Weaver, T.A.: 1995, "The evolution and explosion of massive stars. 2. Explosive hydrodynamics and nucleosynthesis", *Ap. J. Suppl.* **101**, 181-235.

# Proton Captures in the Atmosphere of Accreting Neutron Stars

H. Schatz[1], L. Bildsten[2], A. Cumming[2] and M. Wiescher[3]
*[1]Gesellschaft für Schwerionenforschung GSI, Planckstr. 1, D-64291 Darmstadt, Germany; [2]ITP, University of California, Santa Barbara, CA 93106, USA; [3]Department of Physics, University of Notre Dame, Notre Dame IN 46556, USA*
*h.schatz@gsi.de*

**Abstract:** We calculate the ashes of combined hydrogen and helium burning during steady state burning in x-ray pulsars and during x-ray bursts. We find that in both cases, the rp process can produce elements as heavy as $A = 100$ and beyond.

## 1. INTRODUCTION

Neutron stars in x-ray binaries accrete hydrogen and helium rich material from their companion and burn it within their atmosphere into heavier nuclei. The nature of this nuclear burning depends on the local accretion rate per unit area $m$ (Bildsten, 1998a,b). The local accretion rate can be much larger than the global accretion rate when the material is funneled by a strong magnetic field. In this paper we discuss two burning regimes: At lower local accretion rates between 0.03 and 1 $\dot{m}_{\rm Edd}$, (with $\dot{m}_{\rm Edd}$ being the Eddington accretion rate), the accreted material forms after a few hours a hydrogen and helium rich layer on the neutron stars surface, which at some point becomes unstable and burns explosively within tens of seconds. This cycle of accretion and explosive burning repeats continuously and the thermonuclear explosions are observed as x-ray bursts. For $\dot{m} > \dot{m}_{\rm Edd}$ the burning of hydrogen and helium becomes time independent (steady state).

This happens on the neutron stars with strong magnetic fields in x-ray pulsars and in some of the highest accretion rate weakly magnetized neutron stars.

Understanding of the reaction sequences during nuclear burning on accreting neutron stars is important as they determine the nuclear energy release and the composition of the ashes. While the nuclear energy release defines the observed x-ray light curve, knowledge of the composition of the ashes is important for a variety of reasons:

1. The ashes of the nuclear burning replaces, after some time, the original composition of the liquid ocean and the solid crust, which form the outer layers of the neutron star. Knowledge of this composition is crucial to calculate a variety of (possible) observables including gravity wave frequencies on the oceans (Bilsten and Cumming, 1998; Bilsten and Cutler, 1995), the thermal structure affecting the off-state luminosity in some transient neutron stars (Brown *et al.*, 1998) and the possible emission of gravitational radiation (Bildsten, 1998a,b) and the evolution of magnetic fields (Brown and Bildsten, 1998).
2. If the ashes contains unburned fuel like hydrogen or carbon, then this fuel would burn in deeper layers - hydrogen via electron capture (deep hydrogen burning) and carbon via a thermonuclear carbon flash. This would strongly affect the thermal structure and the gravity wave frequencies of the neuton stars surface (Bildsten and Cumming, 1998) and could also explain some observational features in x-ray bursts (Taam *et al.*, 1993; Taam *et al.*, 1996; Ayasli and Joss, 1982; Woolsey and Weaver, 1984) and x-ray pulsars (Brown and Bilsten, 1998).
3. During very luminous x-ray bursts it is an open question whether a small fraction of the burned material can escape the strong gravitational field of the neutron star. If burned material escapes, it might contribute to the solar system isotopic abundances.

For x-ray bursts a variety of authors has performed nucleosynthesis calculations, but the calculations were mostly limited to reaction networks ending at $^{56}$Ni (Taam *et al.*, 1993; Taam *et al.*, 1996; Woolsey and Weaver, 1984), Se (Hanawa *et al.*, 1983), Kr (Koike *et al.*, 1999) or Y (Wallace and Woosley, 1981). Wallace and Woolsey (1984) performed a pioneering calculation up to $^{96}$Cd, but used a very crude 16 nuclei network. Here we will discuss the first calculation of the nucleosynthesis in an x-ray burst based on a complete reaction network up to $A$ = 100. The reaction network used in this study includes 631 nuclei between H and Sn and is described in Schatz *et al*, (1998).

The nucleosynthesis in steady state burning has been calculated for the first time in the work reported here (Schatz *et al.*,1999). Previously, the composition of the neutron stars ocean and crust in x-ray pulsars was simply assumed to be pure iron.

## 2. EXPLOSIVE NUCLEAR BURNING IN X-RAY BURSTS

To calculate the burning processes during an x-ray burst, we used a selfconsistent 1 zone model coupled to our large reaction network. In this model we assume a constant pressure of $10^{22}$-$10^{23}$ dyne/cm$^2$. The opacities are calculated as described in (Schatz *et al.*, 1999). The calculated temperaturc is shown in Figure 1. The temperature raises within seconds to 2 GK, and then cools down with a much longer timescale of roughly 100 s. The density is around $10^6$ g/cm$^3$.

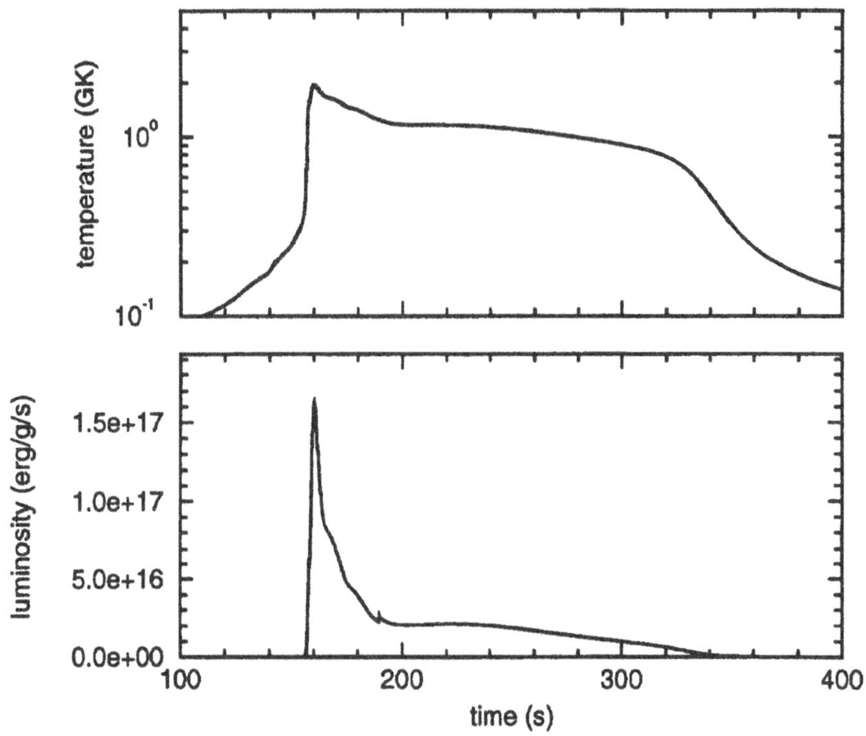


*Figure 1.* Calculated temperature and luminosity during an x-ray burst.

We find that the major reaction flows are carried by three basic processes: (1) Helium is burned via the 3$\alpha$ reaction into $^{12}$C. (2) Then, the capture of two protons produces $^{14}$O, which is the starting point of the $\alpha$p process, a sequence of ($\alpha$,p) and (p,$\gamma$) reactions up to $^{41}$Sc. The $^{18}$Ne($\alpha$,p) $^{21}$Na reaction represents the breakout of the hot CNO cycles which do not play a role during the burst (but might be important during the accretion phase). In the $\alpha$,p process hydrogen serves as catalyst and only helium is burned. (3) The helium burning processes provide the seed nuclei for the hydrogen burning via the rp process, a sequence of rapid proton captures and slow β decays along the proton drip line up to $A \approx 96$ (Wallace and Woolsey, 1981). Clearly, the rp process does not end at $^{56}$Ni as assumed by some authors before, and even when it reaches the $A \approx 96$ region, it is not terminated by hydrogen exhaustion (in fact, 16% of the hydrogen remains unburned). The rp process ends when the process timescale given by the sum of the slow β decay lifetimes along the reaction path, exceeds the limited burst timescale of roughly 100 s. The fact that the rp process does not exhaust the hydrogen can be understood from the extended α,p process, which is a result of the high temperatures: in a simple approximation the mass number $A_{rp}$ which the rp process would have to reach to burn all hydrogen is related to the endpoint of the α,p process $A_{\alpha p}$ (Schatz *et al.*, 1998) via.

$$A_{rp} \approx A_{\alpha p}\left(1+\frac{X}{Y_{burn}}\right), \tag{1}$$

where X and Y are the burned hydrogen and helium mass fractions respectively. Note that one additional (α,p) reaction in the α,p process moves $A_{rp}$ by 13 mass units. This is because an extended α,p process does not only move the startpoint of the rp process but also increases the hydrogen to seed ratio for the rp process. For our x-ray burst calculation we obtain $A_{\alpha p} = 41$, X = 0.65, and Y = 0.28, which gives $A_{rp} = 136$. Therefore, hydrogen is left unburned because the rp process cannot reach A = 136 nuclei during the relatively short burst.

The final abundance pattern is shown in Figure 2. Besides hydrogen and helium, the main products are a mixture of isotopes between mass $A$ = 68 - 100. To estimate the possible contribution to the solar system abundances, Figure 3 shows the overproduction factors, which represent the ratio of the produced abundances to the observed solar system abundances.

For some isotopes the overproduction factors are very large and simple estimates show that the escape of a small fraction of the burned material could be sufficient to make x-ray bursts contribute to galactic nucleosynthesis (Schatz *et al.*, 1998). Especially interesting are the large overproduction factors for light p-nuclei, of which some are severely underproduced in the standard p-process scenarios (Rayet *et al.*, 1995; Howard and Meyer, 1993).

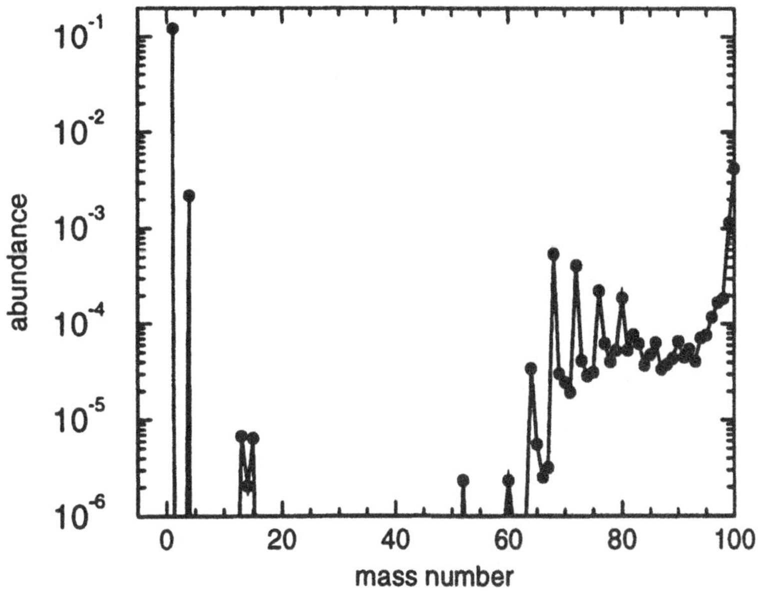


*Figure 2.* Final abundance distribution in the ashes of an x-ray burst.

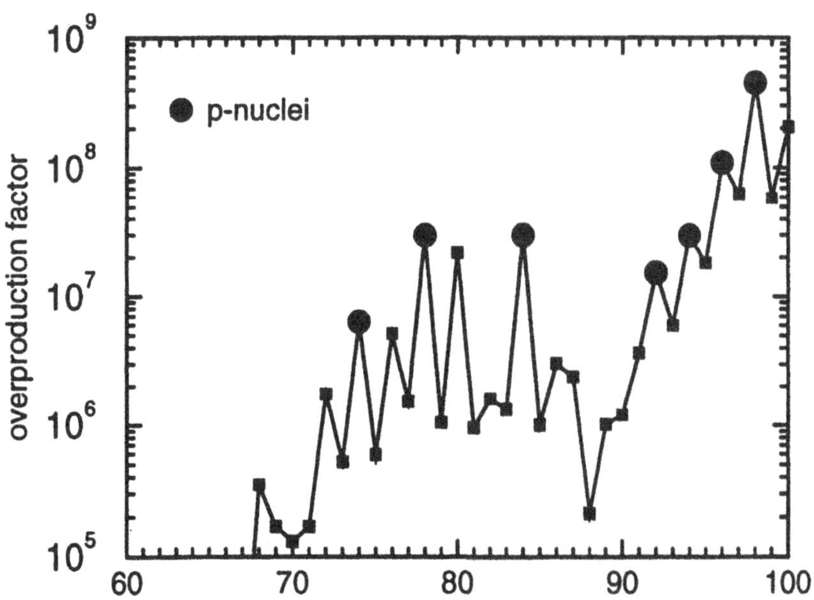


*Figure 3.* Overproduction factors of isotopes in the ashes of an x-ray burst.

Whether x-ray bursts could be the solution to this longstanding problem depends on the escape factor and on the co-production of some s-nuclei like $^{80}$Kr, which limits the possible contribution from this scenario. Concerning the latter point it has to be taken into account that there are still large uncertainties in the nuclear physics input parameters, especially for the calculation of the rp process path between $A$ = 68 and $A$ = 100. The major uncertainty comes from the experimentally unknown location of the proton drip line and the β-decay half-lives of the even-even N = Z nuclei. Recently, there have been considerable experimental efforts to improve this situation by measuring the β-decay half life of $^{80}$Zr (Ressler *et al.*, 1998) and of other even-even N = Z nuclei (Wefers *et al.*, 1999). For example, the new $^{80}$Zr half life of Ressler *et al.*, (1998) is somewhat smaller (4.2 s) than the one used in our calculation (6.8 s), which would reduce the $A$ = 80 abundance peak somewhat thus easing the constraint from the co-production of the s-only nucleus $^{80}$Kr. If in improved calculations the problem of the overproduction factors for s-nuclei remains, then the arguments from above could be reversed, using the production of s-nuclei to constrain the escape factor.

The production of nuclei heavier than $^{56}$Ni also affects the light curve of x-ray bursts, as has been already pointed out for a smaller network by (Koike *et al.*, 1999). We find that isotopes above $^{56}$Ni are produced during the cooling phase of the burst, and that the rp process up to $A$ = 100 leads to the extended 100 s temperature plateau of 1 GK shown in Figure 1. As can be seen from the corresponding x-ray light curve, this leads to a long burst tail following the short x-ray burst itself. This agrees with x-ray burst observations, showing in fact two timescales – a burst of the order of 10 s and a longer tail with a timescale of 100 s (Fox, 1999).

## 3. THE ASHES OF NUCLEAR STEADY STATE BURNING IN X-RAY PULSARS

The thermal and continuity equations describing the hydrostatic evolution of the atmosphere of a constantly accreting neutron star (1.4 $M_\odot$, R = 10 km) (Brown and Bildsten, 1998) were solved in connection with the reaction network providing energy generation and nuclear abundances. We find that for all accretion rates in this regime the conditions for helium ignition via the 3α-reaction are reached before a significant amount of hydrogen has been consumed by the CNO cycles. Helium therefore ignites in a hydrogen rich environment and the released energy is sufficient to trigger a break out of the hot CNO cycles via $^{18}$Ne(α,p). As a consequence, nuclear burning in steady state takes place via the same three processes as in x-ray bursts: the 3α reaction, the α,p process, and the rp process. The main difference to x-ray bursts is that there is more time for the nucleosynthesis (the only limit being the time it takes to reach the depth, where hydrogen starts to capture electrons), so the rp process always ends when hydrogen is consumed. Therefore, the average endpoint of the rp process is directly given by equation 1 and depends therefore mainly on the α,p process. Nevertheless, the timescale for the rp process plays still a minor role, as it determines how much helium can be burned together with hydrogen, thus affecting $Y$ in equation 1 (which denotes only the fraction of helium actually burned during hydrogen burning).

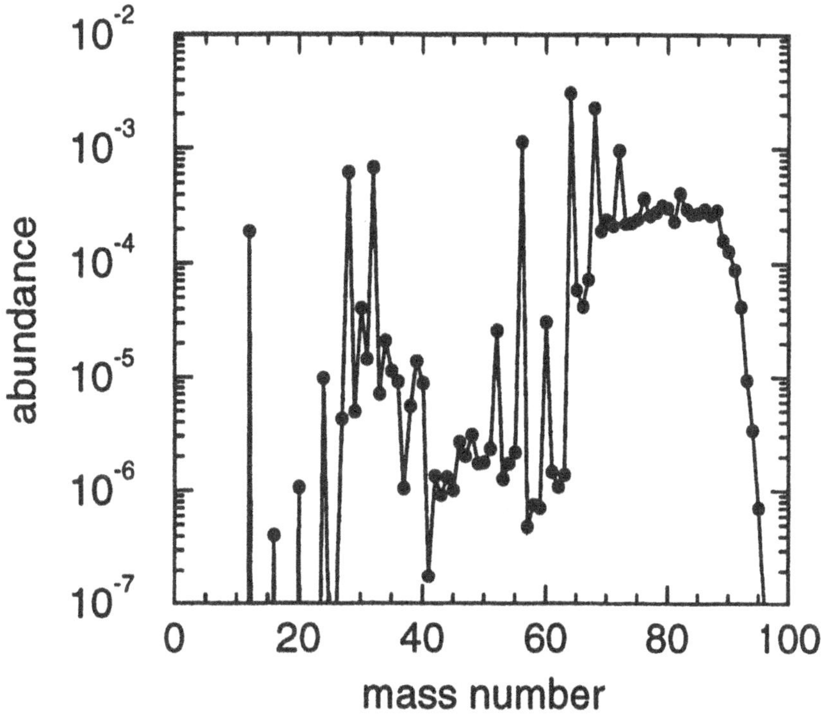


*Figure 4.* Final abundance distribution for steady state burning at $\dot{m} = 5\,\dot{m}_{\rm Edd}$.

Figure 4 shows the final abundance distribution for an accretion rate of $\dot{m} = 5\,\dot{m}_{\rm Edd}$. Obviously the rp process produces nuclei up to $A = 88$. However, the final composition is strongly dependent on the accretion rate. This is illustrated in Figure 5, which shows the last isotope produced in the rp process as a function of accretion rate. For increasing accretion rates heavier and heavier nuclei are produced. The reason for this behavior is the accretion rate dependence of the burning conditions: while the density is roughly the same for all cases (rising from $10^5$ to $10^6$ g/cm$^3$ during burning), the burning temperature ranges from 0.5 to 1.7 GK for accretion rates between 1 and 50 $\dot{m}_{Edd}$. At the higher temperatures the α,p process proceeds to heavier elements ($A_{\alpha p}$ = 18 - 38 for $\dot{m}/\dot{m}_{Edd}$ = 1 - 50) which allows the rp process to synthesize still heavier nuclei (equation 1). For accretion rates above 20 $\dot{m}_{Edd}$, the rp process reaches the end of our network at $A$ = 100. For higher accretion rates we can use equation 1 to estimate the endpoint of the rp process from the calculated endpoint of the α,p process. The result is also shown in Figure 5. For an accretion rate of 45 $\dot{m}_{Edd}$ the rp process is expected to reach $A$ = 150 nuclei. This estimate assumes that there are no loops occuring in the reaction flow. For even higher accretion rates the burning temperature gets so high that the rp process nuclei are destroyed by photodisintegration and the material is driven into nuclear statistical equilibrium. The result are then iron peak isotopes, and this explains the drop in $A_{rp}$ in Figure 5 at very high accretion rates.

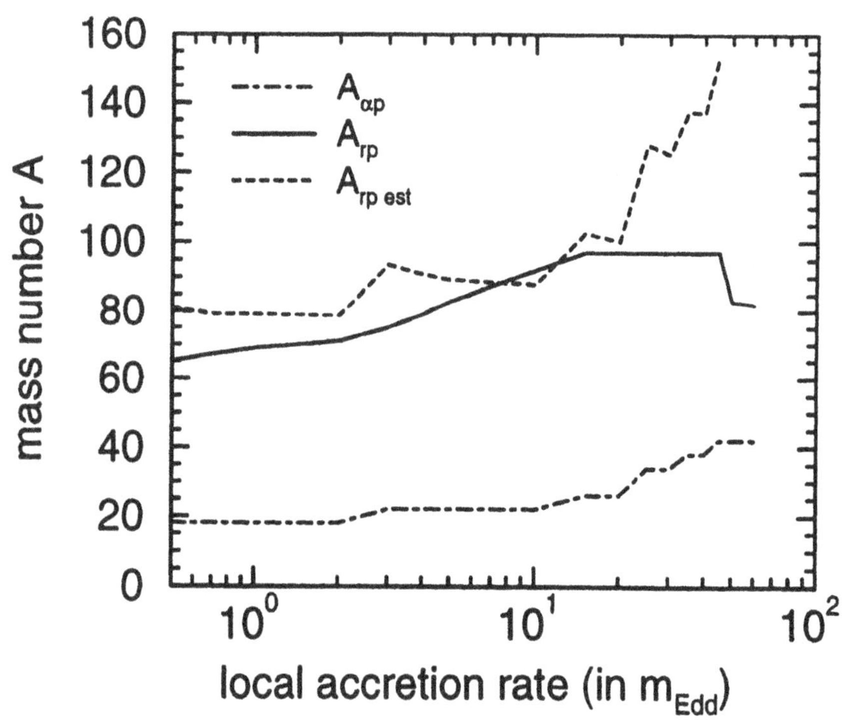


*Figure 5.* Mass number of last isotope produced in the α,p process ($A_{\alpha p}$), in the rp process ($A_{rp}$), and estimated end of the rp process in steady state burning.

For all accretion rates, hydrogen is burned completely via the rp process and after hydrogen consumption the remaining amount of helium burns via 3 α-reaction and α-captures into $^{12}C$ and other α-nuclei. However, the mass fraction of $^{12}C$ in the ashes never exceeds a few percent, which is not sufficient to fuel an observable carbon flash.

## 4. CONCLUSIONS

The same hydrogen and helium burning processes occur in steady state and in explosive burning on the surface of accreting neutron stars, when the global accretion rate exceeds $0.03\,\dot{m}_{Edd}$. These are the 3α-reaction, the α,p process and the rp process. The final composition depends in steady state on the local accretion rate, in x-ray bursts on the burst conditions and timescales. In any case we can conclude that the surface of a neutron star accreting hydrogen and helium for a sufficiently long time will not be composed of pure iron as has been assumed previously, but of a mixture of nuclei that can be as heavy as $A = 100$ or beyond. This will most definitely have a strong impact on calculations of the crust conductivity, since the simple picture of a single species lattice with few impurities is clearly invalid.

In x-ray bursts, large amounts of light p-nuclei are produced, but it is an open question whether any fraction of the burned material can escape the neutron star. Calculations taking into account new experimental nuclear physics data will have to be performed to determine, whether the co-production of s-nuclei puts a severe constraint on the possible contribution of x-ray bursts to galactic nucleosynthesis of p-nuclei.

No significant amount of hydrogen or carbon can reach deeper layers beneath the zone of steady state hydrogen and helium burning. Therefore, electron capture on hydrogen or explosive burning of carbon will not occur in these scenarios. On the other hand we find that in x-ray bursts significant amounts of hydrogen remain unburned. This confirms the results of previous studies with much smaller reaction networks and contradicts speculations (Fujimoto *et al.*, 1987) that the use of a complete reaction network will yield complete hydrogen burning.

## ACKNOWLEDGEMENTS

The authors would like to thank T. Rauscher, F. Rembges, and F.-K. Thielemann for providing us with the network part of the code and latest theoretical reaction rates as well as for many fruitful discussions.

# REFERENCES

Ayasli, S. and Joss, P.C.: 1982, "Thermonuclear processes on accreting neutron stars – a systematic study", *Ap. J.* **256**, 637-665.

Bildsten, L. and Cumming, A.: 1998, "Hydrogen electron-capture in accreting neutron-stars and the resulting g-mode oscillation spectrum", *Ap. J.* **506**, 842-862.

Bildsten, L. and Cutler, C.: 1995, "Nonradial oscillations in neutron-star oceans-a source of quasi-periodic x-ray oscillations", *Ap. J.* **449**, 800-812.

Bildsten, L.: 1998a, "Gravitational-radiation and rotation of accreting neutron-stars", *Ap. J.* **501**, L89-L93

Bildsten, L.: 1998b, "Thermonuclear burning on rapidly accreting neutron stars", in *The Many Faces of Neutron Stars*, eds., Alpar, A., Bucceri, L. and VanParadijs, J., Dordrecht, Kluwer Academic Publishers, Dordrecht, The Netherlands, pp. 419-449.

Brown, E.F. and Bildsten, L.: 1998, "The ocean and crest of a rapidly accreting neutron-star - implications for magnetic-field evolution and thermonuclear flashes", *Ap. J.* **496**, 915-933.

Brown, E.F., Bildsten, L. and Rutledge, R.E.: 1998, "Crustal heating and quiescent emission from transiently accreting neutron-stars", *Ap. J. Lett.* **504**, 95-98.

Fox, D.: 1999, private communication.

Fujimoto, M.Y., Sztajno, M., Lewin, W.H.G. and Vanparadijs, J.: 1987, "On the theory of type-I x-ray-bursts – the energetics of bursts and the nuclear-fuel reservoir in the envelope", *Ap. J.* **319**, 902-915.

Hanawa, T., Sugimoto, D. and Hashimoto, M.: 1983, "Nucleosynthesis in explosive hydrogen burning and its implications in 10-minute interval of x-ray bursts", *Pub. Astr. Soc. Japan* **35**, 491-506.

Howard, W.M. and Meyer, B.S.: 1993, "The p-process in type Ia supernovae", in *Nuclei in the Cosmos*, eds., Käppeler, F. and Wisshak, K., Institute of Physics, Bristol, UK, pp. 575-580.

Koike, O., Hashimoto, M., Arai, K. and Wanajo, S.: 1999, "Rapid proton capture on accreting neutron-stars – effects of uncertainty in the nuclear process", *Astron. Astrophys.* **342**, 464-473.

Rayet, M., Arnould, M., Hashimoto, M., Prantzos, N. and Nomoto, K.: 1995, "The p-process in type-II supernovae", *Astron. Astrophys.* **298**, 517-527.

Ressler, J.J., Peichaczek, A., Walters, W.B., Aprahamian, A., Batchelder, J.C., Bingham, C.R., Brenner, D.S., Ginter, T., Gross, C.J., Grzywacz, R., Kulp, D., MacDonald, B., Reviol, W., Rykaczewski, K., Rikovska, J.R., Wiescher, M., Winger, J.A. and Zganjar, E.F.: 1998, "Measurement of the Zr-80 half-life by delayed gamma tagging", Technical report, ORNL-6957, *Physics Division Progress Report* RIB039.

Schatz, H, Bildsten, L., Cumming, A. and Wiescher, M.: 1999, "Rapid proton captures in the atmosphere of accreting neutron-stars", *Book of Abstracts, 217th ACS National Meeting*, NUCL 054.

Schatz, H. Aprahamian, A., Gorres, J., Wiescher, M., Rauscher, T., Rembges, J.F., Thielemann, F.K., Pfeiffer, B., Moller, P., Kratz, K.L., Herndl, H., Brown, B.A. and Rebel, H.: 1998, "rp Process nucleosynthesis at extreme temperature and density conditions", *Phys. Rep.* **294**, 168-263.

Taam, R.E., Woosley, S.E. and Lamb, D.Q.: 1996, "The effect of deep hydrogen burning in the accreted envelope of a neutron-star on the properties of x-ray-bursts", *Ap. J.* **459**, 271-277.

Taam, R.E., Woosley, S.E., Weaver, T.A. and Lamb, D.Q.: 1993, "Successive x-ray-bursts from accreting neutron-stars", *Ap. J.* **413**, 324-332.

Wallace, R.K. and Woosley, S.E.: 1981, "Explosive hydrogen burning", *Ap. J.* **45**, 389-420.

Wallace, R.K. and Woosley, S.E.: 1984, "Nuclear-physics problems for accreting neutron stars", in *High Energy Transients in Astrophysics*, ed., Woolsey, S.E., *AIP Conference Proceedings* **115**, 319-324.

Wefers, E., Faestermann, T., Münch, M., Schneider, R., Stolz, A., Sümmerer, K., Friese, J., Geissel, H., Hellström, M., Kienle, P., Körner, H.-J., Münzenberg, G.M., Schlegel, C., Thirolf, P., Weick, H. and Zeitelhack, K.: 1999, "Halflives of rp process waiting points", in *ENPE99: Facing the Next Millennium*, Sevilla, Spain.

Woosley, S.E. and Weaver, T.A.: 1984, "Repeated thermonuclear flashes on a accreting neutron star", in *High Energy Transients in Astrophysics*, ed., Woolsey, S.E., *AIP Conference Proceedings* **115**, 273-297.

PART III

# A COSMOLOGICAL VIEW OF NUCLEOSYNTHESIS

# The Origin of the Elements

G. Burbidge
*Department of Physics and Center for Astrophysics and Space Science, University of California, San Diego*
*gburbidge@ucsd.edu*

**Abstract:** The current state of our understanding of the origin of the elements in stars is reviewed. Only some selected topics are discussed.

## 1. EARLY HISTORY

Modern theories of the origin of the chemical elements based on what was known about relative abundances and nuclear physics date from about 1946. Very well known physicists in that era, Enrico Fermi, Maria Mayer, Edward Teller, Rudolf Peierls, George Gamow and others argued that the elements must all have been built up by neutron capture starting with primordial nucleons, electrons and radiation. For a prolific source of particles and radiation, they suggested a very hot early universe by using time reversal of the expansion of the universe to explain the high density. This did not work because there are no stable nuclei with mass 5 or 8 in the periodic table. Also because the elements are spread around with too much spatial irregularity to be attributed to a universal process. (This latter point was known to astronomers at the time but not to physicists).

With the exception of George Gamow and his students, the leading physicists dropped the problem. However, Gamow had become intrigued with the physics of a hypothetical early universe with a hot dense radiation and particle field and that was the beginning of the hot big bang universe hypothesis which is now so popular. Only Fred Hoyle in 1946 took a different approach to the problem. He noticed that there is a peak in the solar system abundance curve centered around iron, and deduced that this indicated that an equilibrium process requiring a temperature ~ $3 \times 10^9$ degrees was responsible. He concluded that this must have something to do with supernovae and hence, stars. This was the real beginning of the stellar nucleosythesis story.

By 1957, we ($B^2FH$) and Al Cameron had shown that nearly all of the isotopes in the periodic table were most likely made in the stars (Burbidge *et al.*, 1957; Cameron, 1957). The abundances of the elements that we used were based on stellar and solar data, and the isotope ratios were largely terrestrial or meteoritic except for a few elements like carbon. We relied very heavily on the abundances given by Hans Suess and Harold Urey (1956).

Forty years later, the question of whether or not one believes that *all* of the isotopes were made in stars, to some extent depends one's view of cosmology. The key isotopes are $^4$He and deuterium.

A view that has been widely held was originally due to Gamow and his colleagues. It is that many of the lightest isotopes, $^2$D, $^3$He, $^4$He, $^6$Li, $^7$Li, $^9$Be, $^{10}$B, $^{11}$B, were made in a hot big bang. Indeed, the correct abundances at least of $^2$D and $^4$He used in the model is one of the two major observational arguments supporting it.

However, the very accurate measurements of the microwave background radiation give a black body temperature of $T = 2.728^{\circ}$ K or an energy density of ~ $4.19 \times 10^{-13}$ ergs/cm$^3$. Also, the matter density in the universe is about $3 \times 10^{-31}$ gm/cm$^3$. Assuming that the matter has the normal He/H ratio by mass of about 0.24, if it has been produced in stars by hydrogen burning, the energy released is about $6 \times 10^{18}$ ergs/gm. Thus, the energy density from this process will be about: $3 \times 10^{-31} \times 0.24 \times 6 \times 10^{18} = 4.32 \times 10^{-13}$ ergs/cm$^3$. *This is almost exactly the same as the energy density of the cosmic black body radiation.* This agreement strongly suggests that the helium was produced in stars and not in a hot big bang. If this is the case, it is very likely that the other light isotopes were also produced by stars (Burbidge and Hoyle, 1998). This takes place in a quasi steady state cosmological model where the timescale for a cycle is about $10^{11}$ years (Burbidge *et al.*, 1999). It is clear that if all of the isotopes were made in the stars, most of the evidence that there ever was a hot big bang has gone away.

## 2. 1957-1999

The development in stellar nucleosynthesis since 1957 is a huge subject. A good update is contained in an article by Wallerstein *et al.* (1997). These authors review all of the eight processes originally proposed: H-burning, He-burning, the $\alpha$-process, the *e*-process, and the *s*-process, *r*-process, *p*-process and *x*-process.

Seven of them are still considered to be operating. In our original work ($B^2FH$), we used the term *x*– to designate processes that we were very uncertain about. However, over the last thirty years, as just stated, it has been argued that many of the light isotopes originally put into the category of the *x*-process were made in primordial nucleosynthesis, and this is part of the observational evidence now used to support the big bang hypothesis.

Since it has just been shown that there is strong evidence from the energy density of cosmic microwave background that this is a result of hydrogen burning in stars giving rise to the cosmic helium, which is by far the most abundant of the light isotopes, there is no longer good reason to suppose that any of the perishable light isotopes were made in a hypothetical early universe. Burbidge and Hoyle (1998) have recently discussed very plausible stellar origins for these isotopes. The details are as follows:

## 3. $^6$Li, $^7$Li, $^9$Be, $^{10}$B, and $^{11}$B

Much work has been done on these nuclides in recent years. It is generally accepted that $^6$Li, $^9$Be, $^{10}$B, and $^{11}$B were produced in spallation reactions of high-energy protons on $^{12}$C and $^{16}$O with the energy ultimately coming from galactic processes as we originally proposed ($B^2FH$). Reeves, Fowler and Hoyle (1970) showed that galactic cosmic rays are an important ingredient. The most modern work shows that it is the C and O that bombard the protons and $\alpha$-particles. The Be and B abundances are proportional to the Fe/H ratio in subdwarfs, and Vangioni-Flam *et al.* (1996) have shown that spallation by high-energy C and O can account for this. The high-energy C and O nuclei are ejected in the winds from massive stars and supernovae. It is now clear from the observations that there may be at least three possible effects that have contributed to the observed Li abundance. They are (*a*) stellar processing, which tends to deplete Li, (*b*) galactic production, which tends to build Li, and (*c*) big bang nucleosynthesis. From the observations, the relative importance of *(a), (b),* and *(c)* is not yet clear. However, in view of my earlier argument concerning the origin of $^4$He, I consider it likely that *(c)* is not operating. Thus, I believe that *(a)* and *(b)* alone can explain the Li abundance and that further observational investigations will show this.

## 4. D and $^3$He

The light isotope $^3$He is produced in large quantities in dwarf stars in which the masses are not large enough for it to be destroyed by $^3$He ($^3$He,

2p) $^4$He. It is also the case that there is a class of stars in which it has been shown from measurements of the isotope shift that most of the helium in their atmospheres is $^3$He. These stars include 21 Aquilae, 3 Centaurus A and several others (Burbidge and Burbidge, 1956; Sargent and Jugaku, 1961; Hartoog and Cowley, 1979; Stateva *et al.*, 1998). The stars are peculiar A, F, and B stars having He/H abundances that are ~ 1/10 of the normal helium abundance. The $^3$He/$^4$He ratio can range from 2.7 to 0.5. These stars occupy a narrow strip in the (log g-$T_{eff}$)-plane between the B stars with strong helium lines and those with weak helium lines, which show no evidence for the presence of $^3$He. However, the detection of $^3$He from the isotope shift will fail if the $^3$He/$^4$He ratio is $\lesssim 0.1$. Thus many of the weak helium-line stars may well have $^3$He/$^4$He abundance ratios far higher than the abundance ratio that is normally assumed to be present, namely $^3$He/$^4$He $\approx 2 \times 10^{-4}$. The high abundance of $^3$He in these stars has been attributed by G. Michaud and his colleagues to diffusion (Michaud *et al.,* 1979 and earlier references). Whether or not this is the correct explanation, results do tell us that stellar winds from such stars will enrich the interstellar gas with $^3$He in large amounts. This $^3$He is in addition to the $^3$He that will be injected from dwarf stars. The final abundance required is $^3$He/H $\approx 2 \times 10^{-5}$. It has been argued by those who believe that $^3$He is a product of big bang nucleosynthesis that there has not been time to build up the required abundance by astrophysical processes. However, not only do we not know what the rate of injection from stars is, but in the quasi steady state cosmology, the timescale for all of this stellar processing is ~ $10^{11}$ years rather than ~ $1.5 \times 10^{10}$ yr., as it is in the big bang cosmology. Thus, we believe that $^3$He may very well have been produced by stellar processes.

I turn now to the production of deuterium. In view of the fact that the $^3$He/H and D/H ratios are very similar, and because I believe that the $^3$He is likely to be produced by low-mass stars, I believe that the most likely source of the cosmic deuterium is the dwarf stars.

It is known that the dwarf M stars are a major constituent of normal galaxies. They have extensive convective envelopes, and thus they are likely to have outer layers in which extensive flaring activity takes place. In my view, it is the cumulative effect of stellar winds and flares from these low-mass stars that has led to the build up of the deuterium.

Deuterium is known to be produced in solar flares (Chupp *et al.*, 1973; Anglin *et al.*, 1973), and early work by Coleman and Worden (1976) has shown how much mass can be ejected from the dwarf stellar component. They estimated that for a typical galaxy containing $10^{11}$ – $10^{12}$ dwarf M stars, the mass-loss rate will amount to about 0.1 $M_{\odot}$ $yr^{-1}$ from the dwarfs. If I add to this the fact that the programs now underway to detect faint stars through microlensing are now showing that the number of dwarf stars is very large and the additional fact that in the QSSC cosmology, the timescale for the build-up of D in the interstellar gas is much greater than $10^{10}$ yr, a large amount of interstellar gas enriched in deuterium will be produced in a timescale corresponding to a cycle of oscillation $Q$ in the QSSC, i.e., in about $10^{11}$ yr.

Of course, in the same period, the deuterium contained in the gas that is recondensed into stars will be destroyed, so that the final abundance will depend on how much uncondensed gas remains. More measurements are required of D/H both in the gas our Galaxy (Linsky *et al.,* 1993, 1995) and elsewhere. Much has been made recently of the D/H ratio determined in the absorption line spectra of QSOs with large redshifts. The value obtained by Tytler and his colleagues (Tytler *et al.*, 1996; Burles and Tytler, 1996), D/H $\leq$ approximately 2 x $10^{-5}$, is the best estimate that has been made so far for extragalactic material, and this has been discussed in the context of big bang cosmology. In the QSSC, the absorbing clouds giving rise to the absorption spectrum may also have existed at an earlier epoch in the cycle. However, as discussed elsewhere (Hoyle and Burbidge, 1996), there is independent evidence that many QSOs may not be at the distances indicated by their redshifts, so the epoch to which these values of D/H correspond is not clear.

My prediction is that with the deuterium made largely in stellar flares, there will be a range of values of the D/H ratio with values of D/H ~ $10^{-5}$ at the high end. I do not expect that the D/H ratio will have a constant value throughout an individual galaxy or throughout a cycle of the QSSC. Thus, a possible test is to look for differences in the D/H ratio both inside and outside our Galaxy.

I now touch on one or two other results out of a huge number of advances on all fronts. For a full account, I refer you to Wallerstein *et al.* (1997).

## 5. ABOUT SUPERNOVAE

A tremendous amount of work has been done on supernovae models, and there are observations of large numbers of new supernovae.

The detection of neutrinos from Supernova 1987a is observational proof of the basic physics of the collapse and explosion. Exponentially decaying light curves in supernovae are due to $^{56}\mathbf{Ni} \longrightarrow {}^{56}\mathbf{Co} \longrightarrow {}^{56}\mathbf{Fe}$, and not $^{254}$Cf as we ($B^2$FH) originally supposed. The decay half-life of $^{56}$Co is 78.5 days, but the observed half-life from the light curve is 53-40 days depending on color. This is thought to be due to the leakage of gamma rays from the envelope. While our original suggestion was wrong, it actually encouraged us to believe in a nuclear explosion as a key element in a supernova outburst.

Supernovae do explode after imploding, but while it is believed that the bounce is due to neutrinos, no really successful models demonstrating this have yet been made.

## 6. ABOUT THE SOLAR SYSTEM

Many researchers are probably more interested in the abundances in the solar system than in the abundances in the cosmos. It is still the case that most of the isotopic ratios that they measure come from the Earth or from the meteorites and do not know what they are in stars in general.

However, the abundances of the more predominant elements are not too different from solar values in many stars, even those in distant galaxies.

To understand the abundances in the solar system, one has to find a way to incorporate isotopes made in very different conditions in the interiors of stars. It is known that star formation occurs in our Galaxy in dense molecular clouds and that multiple condensations are common. Because the solar system condensed long after the first stars in the galactic disk, one realizes that the material out of which the solar nebula condensed will be made up of an admixture of several stellar processes.

Recently Busso *et al.* (1999) have discussed in detail how one can understand the solar system abundances in terms of nucleosynthesis in the asymptotic giant branch (AGB) of stellar evolution.

The s-process (slow neutron capture) has as its source of neutrons $^{13}C(\alpha,n)^{16}O$ and at higher temperatures $^{22}Ne(\alpha,n)^{25}Mg$. These neutrons are captured in stars on the AGB branch at different stages.

Busso *et al.* (1999) find that a good reproduction of the solar system's main component is obtained through Galactic chemical evolution that mixes the outputs of AGB stars of different stellar generations, which are born with different metallicities and produce different patterns of *s*-process nuclei. The main solar *s*-process pattern is thus not considered to be the result of a standard archetypal *s*-process, occurring in all stars. Concerning the $^{13}C$ neutron source, its synthesis requires penetration of small amounts of protons below the convective envelope, where they are captured by the abundant $^{12}C$ forming a $^{13}C$-rich pocket. The models are still not sophisticated enough to explain the observations of CNO isotopes in red giants and in presolar dust grains. These observations require some circulation of matter between the bottom of convective envelopes and regions close to the H-burning shell.

Busso *et al.* (1999) have also shown that the pollution of the protosolar nebula by a close-by AGB star may account for concordant abundances of $^{26}$Al, $^{41}$Ca, $^{60}$Fe, and $^{107}$Pd. The AGB star must have undergone a very small neutron exposure and be of small initial mass ($M \leq 1.5\ M$). There is still a shortage of $^{26}$Al in such models.

This presentation is only a very brief review of the status of stellar nucleosynthesis.

## REFERENCES


Anglin, J.D., Dietrich, W. and Simpson, J.: 1973, "Deuterium and tritium from solar flares at 10 MeV per nucleon", *Ap. J.* **186**, L41-L46.

Burbidge, E.M. and Burbidge, G.: 1956, "On the possible presence of He$^3$ in the magnetic star 21 Aguilae", *Ap. J.* **124**, 655-659.

Burbidge, E.M., Burbidge, G., Fowler, W.A. and Hoyle, F. (B$^2$FH): 1957, "Synthesis of elements in stars", *Rev. Mod. Phys.* **29**, 547-650.

Burbidge, G. and Hoyle, F.: 1998, "The origin of helium and the other light-elements", *Ap. J.* **509**, L1-L3.

Burbidge, G., Hoyle, F. and Narlikar, J.V.: 1999, "A different approach to cosmology", *Physics Today* **52**, 38-44.

Burles, S. and Tytler, D.: 1996, "The cosmological density and ionization of hot gas - O-VI absorption in quasar spectra", *Ap. J.* **460**, 584-600.

Busso, M., Gallino, R. and Wasserburg, G.: 1999, "Stellar nucleosynthesis and extinct radio-activities", *Ann. Rev. Astron. Astrophys.* **37**, 239-309.

Cameron, A.G.W.: 1957, "Stellar evolution, nuclear astrophysics and nucleosynthesis", *Chalk River Report CRL-41*, Atomic Energy of Canada, Ltd.

Chupp, E., Forrest, D., Higbie, P., Suri, A., Tsai, C. and Dunphy, P.: 1973, "Solar gamma-ray lines observed during the solar activity of August 2 to August 11, 1972", *Nature* **241**, 333-335.

Coleman, G. and Worden, P.: 1976, "Mass loss from dwarf M stars through solar flaring", *Ap. J.* **205**, 475-481.

Hartoog, M. and Cowley, A.: 1979, "The helium-3 stars", *Ap. J.* **228**, 229-239.

Hoyle, F. and Burbidge, G.: 1996, "Anomalous redshifts in the spectra of extragalactic objects", *Astron. Astrophys.* **309**, 335-344.

Linsky, J.L., Brown, A., Gayley, K., Diplas, A., Savage, B.D., Ayres, T.R., Landsman, W., Shore, S.N. and Heap, S.R.: 1993, "Goddard high-resolution spectrograph observations of the local interstellar-medium and the deuterium hydrogen ratio along the line of sight toward Capella", *Ap. J.* **402**, 694-709.

Linsky, J.L., Diplas, A., Wood, B.E., Brown, A., Ayres, T.R. and Savage, B.D.: 1995, "Deuterium and the local interstellar-medium – Properties for the Procyon and Capella lines of sight", *Ap. J.* **451**, 335-351.

Michaud, G., Martel, A., Montmerle, T., Cox, A.N., McGee, N.H., and Hudson, S.W.: 1979, "Helium abundance anomalies and radiative forces in stellar envelopes", *Ap. J.* **234**, 206-216.

Reeves, H., Fowler, W.A. and Hoyle, F.: 1970, "Galactic cosmic-ray origin of Li, Be and B in stars", *Nature* **226**, 727-729.

Sargent, W. and Jugaku, J.: 1961, "The existence of $^3$He in Centauri A", *Ap. J.* **134**, 777-796.

Stateva, I., Ryabchikova, T. and Iliev, I.: 1998, "Search for the $^3$He isotope in the atmospheres of HgMn stars", in *Proc. of the 26th Workshop of the Europeon Working Group on CP Stars*, eds., North, P., Schnell, A. and Ziznovsky, J., *Contrib. Astr. Obs. Skalnate Pleso*, **27**, No. 3, in press.

Suess, H.E. and Urey, H.C.: 1956, "Abundances of the elements", *Rev. Mod. Phys.* **28**, 53-74.

Tytler, D., Fan, X.M. and Burles, S.: 1996, "Cosmological baryon density derived from the deuterium abundance at redshift z = 3.57", *Nature* **381**, 207-209.

Vangioni-Flam, E., Casse, M., Fields, B. and Olive, K.A.: 1996, "LiBeB production by nuclei and neutrinos", *Ap. J.* **468**, 199-206.

Wallerstein, G., Iben, I., Parker, P., Boesgaard, A.M., Hale, G.M., Champagne, A.E., Barnes, C.A., Kappeler, F., Smith, V.V., Hoffman, R.D., Timmes, F.X., Sneden, C., Boyd, R.N., Meyer, B.S. and Lambert, D.L.: 1997, "Synthesis of the elements in stars - 40 years of progress", *Rev. Mod. Phys.* **69**, 995-1084.

# Chemical Evolution Tomorrow

Virginia Trimble
*Physics Department, University of California, Irvine, CA 92697 and Astronomy Department, University of Maryland, College Park, MD 20742*

**Abstract:** The solar system is just one sample, though of course the most thoroughly studied sample, of cosmic chemistry. Some perspective on it can be gained from an overview of the chemical evolution of the Galaxy and Universe, how our knowledge of this evolution has developed, and what problems remain to be solved. The very long-range future, comparable with the age of the universe, is also potentially of interest.

## 1. HISTORICAL INTRODUCTION

It is reasonable to begin by asking whether the standard picture of nucleosynthesis and galactic chemical evolution is on at least roughly the right track. Thomas Gold is supposed to have said, "If we are all going the same direction, it must be forward." This can be taken as either a definition or a criticism of progress in science, making it salutary to look at where we have come from.

A century and a half ago, we had no data of any kind on the chemical composition of anything outside the earth, and indeed a philosopher (Auguste Compte) had used the composition of the planets and stars as an example of real knowledge that we could never acquire. Then, in 1858-59 Bunsen and Kirchhof demonstrated that absorption lines in the sun had the same wavelengths as emission lines from gaseous sodium and iron on earth, and the unknowable began to be known.

A century ago, it was widely supposed that the lack of spectral features due to iron, titanium, and other relatively heavy elements in the light from the hottest, blue stars meant that those elements had been broken up into smaller atoms, like oxygen, helium, and neon. An intermediate step, about

*Origin of Elements in the Solar System: Implications of Post-1957 Observations.*
Edited by O. Manuel, Kluwer Academic/Plenum Publishers, 2000. 

75 years ago, was the assumption that the stars and sun had about the same mix of elements as the earth, dominated by oxygen, silicon, iron, magnesium, and so forth. Input of some new physics, the Boltzman and Saha equations, and of a creative mind (that of Cecilia H. Payne, later Payne Gaposchkin) were need to demonstrate that all stars have about the same mix of elements, and that the mix is heavily dominated by hydrogen and helium (Payne, 1925).

Fifty years ago, the sun seemed to be running on the CN cycle, and no one had ever seriously attempted to evolve a galaxy. Credit for the first models of galactic luminosity and chemical evolution belongs unambiguously to Beatrice M. Tinsley (1968). Credit for recognizing that only somewhat more massive, hotter stars than the sun draw more energy from the CN cycle than from the proton-proton chain can be claimed by many. I tend to think first of J. Beverley Oke, because when he first tried to pass on the information to Martin Schwarzschild, he was not believed.

Even 25 years takes us back to a time when no quantitative abundance information existed either for the emission line gas in quasars or for the clouds responsible for introducing absorption lines into their spectra (most of which are distant from both the QSOs and us). Indeed, no normal galaxies further away than $z = 0.5$ had any spectral information at all, and record "highest redshifts" have appeared erratically ever since, reaching 6.48 as I write.

As for $t = 0$, you are there, and presentations in New Orleans addressed many aspects of current understanding. Detailed (or at least numerous) references to the historical topics and to many of the items discussed below can be found in Trimble (1975, 1991, 1996).

"Tomorrow" in the sections that follow will include, first, time scales of $10^{10\pm1}$ years, over which the composition of galaxies can be expected to change significantly (mostly in the direction of an increased fraction of heavy elements at the expense of hydrogen; helium also gains) and, second, time scales of $10^{1\pm1}$ yr, over which our understanding of cosmochemistry should improve, particularly, I hope, in the direction of being able to calculate accurately many items that must now be derived heuristically (like star formation rates and initial mass functions) or treated as variable parameters (like gas inflow and outflow and ratios of types of supernovae in the past). The first time scale can be described whimsically as Z(z(z)), meaning mass fraction of heavy elements *vs.* position in a galaxy as a function of how far back in time we look to see it. The second is "the curse of the adjustable parameter," of which there were only one or two in 1950 and about ten today.

Many of these "adjustable parameters" were first invoked to try to solve what we now call the G dwarf problem, meaning that the fraction of nearby stars with less than 20% or so of solar metallicity is considerably smaller than the simplest models predict. The apparent complete absence of population III stars (ones with no heavy elements at all, but just the 77% hydrogen and 23% helium left by Big Bang nucleosynthesis) is closely related. It has occurred to me that, if intergalactic communication is ever established, the first thing you might want to ask the entity at the other end of the phone cord is "Do you guys have a G dwarf problem?"

## 2. THE LONG RANGE FUTURE

How long a future we have to explore depends, first, on whether the universe will expand forever or recontract and, second, on the lifetime of the proton. A universe that recontracts sometime between $10^{11}$ yr and $10^{33}$ yr from now (limits set by the present non-decreasing expansion and likely limit to proton decay timescale) could have a spectacular future. A second epoch of mergers of galaxies and the gas now found between them (probably more than the gas in galaxies) will produce enormous bursts of star formation and, probably, as much heavy element production as occurred when the universe was young and galaxies first formed. The fireworks will, however, be short-lived. Things will begin merging when the microwave background temperature has been heated back up to 5K, and, when it reaches 3000 K, gas will begin to evaporate from stellar surfaces, shutting off the nuclear reactions.

In a universe that expands forever, we can expect, first, that the average metallicity will increase, though probably not by much more than a factor of two. This may mean an increased number of habitable planets, based on the observation that both the sun and the hosts of known extra-solar-system planets are richer in heavy elements than the general run of nearby stars and the local interstellar medium. Causality could go either way: the stars might be metal-rich on their surfaces (only) because they have accreted cometary, meteoritic, and planetary material; or they could have planets because these are easier to form out of metal-rich gas, which cools more readily. My prejudice inclines to the latter. Next, there will be changes in the ratios of amounts of the various elements and isotopes and in the details of the supernovae (and preceding stellar nucleosynthesis) responsible for them.

Dynamical evolution will accompany chemical changes. Thus galaxies will get rounder, more massive, and fewer. Looking inside them, we will find new sorts of stellar populations (meaning correlations of age, composition, location, and kinematics of stars) developing. The Milky Way, for instance, now has a flat, rotating distribution of metal-rich stars (disk or population I) and a round, non-rotating distribution of old, metal-poor stars (halo or population II). A major merger with the Andromeda galaxy would, for instance, probably result in round, non-rotating distributions made up of the current Pop I and Pop II stars and a new, young, metal-rich population whose formation was triggered in the merger.

The fraction of mass in heavy elements (and helium, which is co-produced) could become very large in a few special environments where gas is retained through many generations of supernova recycling. The gas responsible for the emission lines from the centers of quasars may be a case where this has already happened.

Several major uncertainties remain. Some of these might be addressed by simply letting existing models of galactic evolution run on past the point where they best match the Milky Way or other observed galaxies. (It is not clear whether anyone has done this.) Others will require additional understanding. Many rich clusters of galaxies have X-ray emitting gas whose cooling time is comparable to or less than the age of the universe. This must eventually flow to the cluster center (They are called cooling flows.) and presumably form stars. Because we don't see bright, blue stars at cluster centers, it is generally supposed that the products are low-mass objects in which little nucleosynthesis will occur, but this is not certain. Even less clear is the very long terms future of gas now in diffuse filaments and pancakes at a temperature of $10^{5\text{-}6}$ K in intergalactic space. Recent simulations of galaxy formation suggest that this may comprise more baryonic material than is currently in galaxies and clusters. If so, then whether it ever cools and gathers into units that make stars will make an enormous difference to the long-range future.

## 3. SHORTER TIME SCALES

You will not be surprised to hear that our picture of the total sweep of things chemical in the universe is still incomplete and "more work is needed" on many aspects. The items discussed below include (3.1) laboratory and other data on fundamental atomic and molecular properties, (3.2) measured abundances, where the goal is to have all stable elements and isotopes in a full range of stars, interstellar clouds, galaxies, intergalactic clouds, and so forth, (3.3) some red herrings that need to be cleared out of the way to reveal the chemical story we are interested in (Some of them include very interesting parts of astronomy and astrophysics, but they are noise rather than signal in the present context), (3.4) processes, sites, and the scheme of stellar evolution in which they are embedded (This topic was the core of the enormously important work of Cameron (1957) and Burbidge, Burbidge,

Fowler and Hoyle (1957), $B^2FH$), and (3.5) galaxy formation and evolution as both dynamical and chemical processes and the coupling of the two.

## 3.1 Basic Physical Data

We sometimes tend to think that we know all there is to know about the structure and transitions of atoms, molecules, and nuclei. This turns out not to be true even for gases (or gas-dust mixtures) in thermodynamic equilibrium, and is even more false out of equilibrium, when for instance, the amount of an ion that has to be present to make a given line strength depends on the balance between excitation and de-excitation due to photons (not in a Planck distribution) and collisions with electrons and several kinds of atoms and other ions.

A random sweep of a few weeks' worth of major astronomical journals uncovered the following examples: (1) transition probabilities (gf values) for Eu III, needed to decide just how over-abundant europium is in certain peculiar stars, (2) nuclear rates like $C^{12}(\alpha,\gamma)O^{16}$, where the stellar process is dominated by a sub-threshold resonance and so cannot be directly probed with laboratory data, but the answer determines the ratio of carbon to oxygen made in helium fusion (important if you want to end up with habitable planets), (3) the branching ratios for the slow and rapid capture of neutrons (s- and r-processes), especially when the most likely capture takes place on an excited nuclear level or on a long-lived but unstable nuclide, (4) radioactive and collisional cross-sections for ionization, recombination, excitation and de-excitation for all sorts of atoms and molecules, and (5) "missing opacity" – the observation that, especially for ultraviolet light, when you add up all the known lines and bands that will try to stop photons from getting out, the star still knows about more than you do.

## 3.2 Observed Abundances

Within the solar system, there are relatively few remaining discrepancies. Meteoritic and solar values for the amount of iron differ by 0.1-.2 dex; but it was a factor of nearly 10 within living memory (The meteorites were largely right.) Another solar system topic where an enormous amount is going on is the measurement of assorted isotopic anomalies in individual meteorite grains, including those associated with fossil radioactivities, like $Mg^{26}$ remaining from the decay of $Al^{26}$. The goal is to associate these with particular kinds of supernovae or supernova ejecta and so learn in detail about the balance of nuclides produced from, for instance, carbon burning in a 15 $M_\odot$ star. When the task has been completed, the results will probably be comprehensible to anybody with a periodic table on his desk or disk. We are a good long ways from that now.

Among the heaviest nuclides, we have not observed the products of the p-process anywhere outside the solar system (and inside only for earth, meteorites, and solar wind). The process produces the rare neutron-poor isotopes of elements beyond the iron peak and does not dominate any element. Products of the s- and r-processes have been studied in many galactic stars and a few gas clouds, but we have almost no information on them in other galaxies, even nearby.

The unstable elements probe relatively recent nuclear reactions and the time elapsed since then. Technetium is famously present in many highly-evolved, carbon-rich stars. One report of promethium in a similar star has never been confirmed. Uranium and thorium live long enough to tell us ages in the 1-20 Gyr range, if you can figure out how much was present to begin with. Two halo stars have Th/Eu ratios at the present time that indicate (if the production ratio was what you expect from a normal r-process) ages near 15 Gyr, rather higher than is generally now coming from globular cluster studies. U has not been seen in these or other stars (except the sun), but might be equally interesting.

Moving to the lightest element, we would like to know whether there are real variations in the ratio of deuterium to hydrogen in either the interstellar or intergalactic unprocessed medium. The former would tell us about how much of the gas in various places has been through stars and had its deuterium destroyed (called astration by the modellers). The latter is vital to understand if you want to use D/H to learn the baryon density in the universe. Real variations could result from either local destruction or local production that is not associated with the formation of much in the way of heavier elements.

The amounts of both helium and heavy elements increase as more stars throw out their reaction products, but the observed ratio of the enhancements, $\Delta Y/\Delta Z \approx 3\text{-}4$ are rather higher than what you expect from a typical stellar population under present conditions ($\Delta Y/\Delta Z \approx 1\text{-}2$). Insufficient knowledge about nucleosynthesis in stars of initially low metallicity may be part of the problem.

Still other issues where additional observations are needed include (1) Do young stars agree in composition with the interstellar gas they leave behind, or pick out more (or less) than their fair share of heavies, perhaps in the form of dust? (2) How much super metal rich stuff is really around, where (planet hosts, quasar gas, in cores of giant elliptical galaxies), and why? (3) What is the real range of variability in globular cluster stars of elements that really could not have been made in those stars themselves (aluminum, magnesium, silicon, etc.), and if it is large, how did this happen? and (4) What are the various correlations or abundance patterns, for instance of O/Fe and the alpha nuclei (Mg, Si, Ti, Ca) to Fe *vs.* Fe/H, the CNO isotopic ratios, and lithium with various stellar properties, since each of these constrains some important aspect of overall chemical evolution.

On larger scales and in stranger places, we do not know enough about (1) the real composition of the ejecta from various sorts of supernovae, nova explosions, stellar winds, and planetary nebulae (the correct bridge between calculated nucleosynthesis and the resulting average abundance), (2) the efficiency and time scale with which new ejecta are mixed into interstellar material, (3) gradients in abundances with radius and distance from galactic planes (remember Z(z(z))?!), and their correlations with galaxy types and masses; particularly one would like more detailed information than just the ratio of iron or oxygen to hydrogen, (4) compositions of stars and gas in strange galaxies like starbursters and the broad absorption line gas in QSOs, (5) "intergalactic" abundances, meaning the gas in X-ray emitting clusters (It is definitely not pristine.) and clouds responsible for narrow emission lines in quasars (and just where are those clouds anyhow, so that we know what it is whose composition we are measuring!), and (6) of course, everything as a function of redshift (with the additional difficulty of needing a good set of cosmological parameters so that data observed *vs.* redshift can be compared with calculations, which necessarily operate in ordinary time).

## 3.3 Red Herrings

These are the heart of some astronomical subfields, but mostly a nuisance to students (and teachers) of chemical evolution. A classic example is abundance of isotopes in molecules. Chemical fractionation is interesting and important, but it makes a CO a poor probe of $C^{13}/C^{12}$, and of $O^{18}/O^{17}/O^{16}$, unless you know a great deal about the conditions under which the molecules formed. Other examples include what drives the shocks out of type II supernovae and what are the progenitors of Type Ia's (though this is needed statistically to track chemical enrichment with time).

There is a whole constellation of places where the abundances we see have been modified by accretion from something odd (white dwarfs with metals; Lambda Boo stars without them), by gravitational settling, radiative uplift and so forth (peculiar A stars, helium in various contexts, and all "first ionization potential" effects and their inverses as in AR Lac), or partial ionization (like He/H in HII regions that are not completely HeII regions).

Sometimes you have to sort out which is which to make progress; occasionally the answer makes sense. A nice case is that of barium-rich stars, which could have made it for themselves or acquired it from binary companions that shed enriched outer layers and have since become white dwarfs. There is a signature: the self-polluters also show Tc. But this does not help us much in using these stars to follow the growth of barium abundance in the galaxy as a whole. Traditional carbon stars are evolved giants and thought to be polluted by carbon made from helium fusion in the stars themselves. On this basis, there "should" be no dwarf carbon stars. There are however, and they too are the victims of material deposited from evolved close binary companions (now white dwarfs).

## 3.4 Processes, Sites, and Stellar Physics

The set of processes and sites identified by Cameron (1957) and by $B^2FH$ (1957) as capable of producing the full range of elements and isotopes has withstood the forces of time remarkably well. The following short paragraphs describe the processes they identified (some of which had been recognized earlier, starting with hydrogen fusion in the 1920's and 30's) and what has become of them since.

Hydrogen fusion, by either the proton-proton chain or CNO cycle. It produces helium and converts $C^{12}$ and $O^{16}$ into the full range of CNO isotopes in normal stars. Hot hydrogen burning also occurs, in nova explosions and probably late in the lives of massive stars. Its products continue up the periodic table from oxygen to fluorine, neon, sodium, and magnesium and are visible in certain nova ejecta.

Helium fusion, or the triple-alpha process. Neither of the two body reactions H + He or He + He has a stable product. Thus three helium nuclei (alpha particles) must come together to form carbon. Capture of a fourth makes $O^{16}$, and the two elements are produced together in roughly equal amounts. This happens because of the details of the excited levels, their spins and parities, of the $C^{12}$ and $O^{16}$ nuclei.

Alpha process: Additional alpha particles were supposed by the pioneers to be captured by $O^{16}$ yielding the dominant isotopes of neon, magnesium, silicon, and sulfur. In fact, the necessary excited nuclear states are not present. Thus we get a series of heavy element burning processes that gradually synthesize elements up to the iron peak. The sequence is carbon burning, neon burning, oxygen burning, and silicon burning (which works largely through photodisintegration of some of the silicon nuclei, with the products being captured by remaining ones). There are loose single protons and neutrons in all of the stages, so that all stable elements and isotopes from oxygen up to the iron peak are produced (but have their relative abundances fine-tuned by heating during supernova explosions). As stellar cores get hotter and denser, the processes occur out of equilibrium, and the full chain of nuclear reactions begins to strain current computing power. In the future, it will be possible to track everything simultaneously, and the discrete burning phases may begin to blur together.

$B^2FH$ and Cameron both recognized that three separate processes would be needed to account for all the nuclides heavier than the iron peak, from roughly germanium up to uranium. First, the most tightly bound (most stable) isotopes, along what is called the valley of beta stability (in a map of neutron number *vs.* proton number) could be made by adding neutrons to iron peak nuclei on time scales longer than those of typical beta decays. This s-process will occur in stars of moderate mass during the stage in their lives when both helium and hydrogen are burning in thin shells around a carbon-oxygen core. Material gets carried back and forth between the two shells, resulting in liberation of neutrons from reactions like $C^{13}(\alpha,n)O^{16}$ and others more complex. The neutrons are captured by heavy nuclei already present (thus s-process elements like barium are secondary products–as is nitrogen–made only in second generation and later stars that begin with some heavy elements).

The most neutron-rich nuclides are attributed to the r-process, rapid capture. A nucleus sweeps up as many neutrons as it can bind, then later decays. Because you need lots of free neutrons and lots of heavy elements at the same time to make this happen, the r-process is generally supposed to occur in Type II supernovae, while the iron-peak core is collapsing to a neutron star. Ejection of neutron rich material when binary systems containing one neutron star and one black hole merge is also possible. Only the r-process reaches up to U and Th.

Finally, the neutron-poor isotopes, which are all of low abundance, arise from the p-process, in which neutrons are removed, probably by photo-ejection, or perhaps protons added. Supernovae are also the most likely site for this.

The initial compilations had some nuclides left over–the isotopes of lithium, beryllium, and boron. Later it was recognized that stars cannot have as much deuterium as we see. These left-overs were blamed on an x-process. We now recognize that the deuterium, the helium-3, and most of the normal helium around us (along with a small amount of lithium-7) are left over from the hot, dense phase early in the life of the universe (big bang), and their abundances are clues to the physics of that phase. Most of the lithium, beryllium, and boron are secondary products from the break-up (spallation) of carbon and oxygen in interstellar gas when they are hit by cosmic rays.

Some of the continuing questions in this area of reactions and how they fit into stellar struction and evolution are: (1) Can you make deuterium anywhere except the big bang? (Solar flares make a bit, but far too little in relation to Li, Be, and B.), (2) Is there a contribution from very massive or supermassive objects (100-$10^5$ $M_{\odot}$) that formed before the first generations of stars?, (3) Just where do the r- and p-processes happen? (The problem with the most promising zones in supernovae is the difficulty in getting the products out without exposing them to further reactions), (4) Just how many types of supernovae are there from a nucleosynthetic point of view, and what does each contribute?, (5) Does star formation take a fair sample of the gas it starts with? (6) How do mixing and mass loss in stars (the amount of which is quite variable and perhaps dependent on rotation and magnetic fields) affect the range of nuclear products?, and (7) Given that most stars occur in pairs (binary stars) with some interaction between the two members, what does nucleosynthesis in interacting binaries look like?

## 3.5 Galaxy Formation and Evolution

It is here that we must say most strongly that "some assembly is required." A handful of major unknowns remain. First, we need to know what the dark matter is made of and how it contributes, besides gravitationally, to formation of galaxies and larger scale structures. That basic formation process has many uncertainties, beginning with its very direction: Do large-scale lumps first acquire their identities and then break up into galaxy-sized pieces that collapse as a whole, spinning up and forming stars as they go (a scenario associated with the names of Eggen, Lynden-Bell, and Sandage, 1962)? Or, alternatively, do subgalactic structures separate out first and later merge? This latter is called hierarchical formation and is currently favored, but it is required to match some of the same observed properties of the Milky Way that originally inspired Eggen, Lynden-Bell, and Sandage.

If mergers are important, then how do they affect the rate of star formation, and the spectrum of stellar masses (binaries, etc.) that will be formed? These are essential inputs to models of galaxy evolution. We generally think of discrete stellar populations, defined by the ages, metallicities, location, and kinematics of stars, but some of these may be artificial slices cut out of continua. If they are discrete, then it makes sense to ask whether an old, thick disk population can solve the G-dwarf problem in the younger, thin disk. If not, probably not. It is worth remembering that Baade's (1944a,b) definition of populations I and II involved only location and appearance of color-magnitude diagrams. The correlations with metallically and age were discovered later (though he is often given credit).

Next, when we come to compare models with data, we find a partial degeneracy in age, metallicity, and initial mass function. A stellar population can collectively look red because it is old, or metal rich, or dominated by low-mass stars (or some combination) and blue because it is young, or metal poor, or dominated by massive stars. Breaking this degeneracy requires more detailed spectral information than is generally available outside the Local Group of galaxies.

## 3.6 The Curse of the Adjustable Parameter

Many of the things that you need to know to put together a complete simulation of galactic chemical evolution are, in principle, causal, calculable processes, for instance the number of stars of each mass that should form from a dense molecular cloud of particular temperature, magnetic energy, turbulence, and so on. In practice, we get approximations to them by looking at clusters of young stars, etc. to find out the possible range of the parameters and then, in our models, choosing plausible values until we like the results (meaning, usually, that they agree with observations of some galaxy or stellar population). The items typically treated this way include:

1. The initial mass function, N(M), that forms out of a gas cloud (power laws with cut-offs and log normal distributions are population). The characterization of binary stars formed belongs here, too. How many compared to the single stars, and what is the distribution of mass ratios and orbit periods or sizes?

2. The star formation rate as a function of time, or of local gas density and whatever other factors may enter. Delta functions and declining power laws are popular.

3. Infall of unprocessed gas from the surroundings of the galaxy. So-called high velocity clouds of neutral hydrogen approaching the disk of the Milky Way indicate that this is an on-going process but don't much help in finding out the amount of the infall or how it has changed with time.

4. Conversely, many galaxies seem to have winds blowing (sometimes perpendicular to their disks) with large enough velocities to remove the gas completely. The existence of metals in the intracluster gas of X-ray clusters says that outflows after the onset of nucleosynthesis must be widespread. If the drivers of the flows are supernova shock waves, then metal-rich material may be lost preferentially. How much is another parameter to play with.

5. The actual composition of the gas involved in processes 3 and 4 is an issue separate from the total amount of gas. Another adjustable is the amount of gas that just moves from one place in a galaxy to another, carrying its metals with it. Inward and outward are both possible

6. The fraction of gas that goes into a generation of stars that eventually comes out as gas of some kind gets a symbol like $f$ in the equations for

chemical evolution. Much of the returned material is still hydrogen and helium.

7. The fraction of gas that goes into a generation of stars and eventually comes out as newly-synthesized heavy elements is called y, the yield. It depends on the full assortment of details of stellar structure and evolution. Assuming that it is constant, at 1% or so, is common but surely wrong.

8. "Instantaneous recycling" is the approximation that the metals come out as soon as new stars are made. Type II supernova progenitors indeed have short lifetimes (a few x $10^7$ yr is typical), but they are not the whole story, and the high ratio of oxygen to iron in some metal-poor stars says that instantaneous recycling is not good enough. Unfortunately, we don't really know the lifetimes of progenitors of Type Ia supernovae (the ones with no hydrogen lines in their spectra that are generally blamed for making lots of iron).

9. The rates of the various kinds of supernovae in different sorts of galaxies as a function of time is only very partially constrained by observations (and perfectly honest people get estimates that differ by factors of two or three, even for the common SN types and normal galaxies like ours).

10. Finally there are other inputs of heavy elements from stellar winds, planetary nebulae, novae, and other events. These will not dominate total metallicity anywhere but may be important for specific elements like carbon and nitrogen and isotopes like $^{21,22}$Ne from novae on massive white dwarfs.

## 3.7 A Case Study: Empirical Determination of Star Formation Rate

The gold standard for measured rates of star formation would come from counting all the stars in some volume (cluster, population, whatever), measuring the mass of each, and determining its age from the extent to which it has evolved away from the main sequence. This is not entirely impossible. We can, for instance, say on evolutionary grounds that the sun is more than a billion and less than 10 billion years old. But this rigorous method is quite impossible if you cannot resolve all the individual stars (that is, anywhere outside the Local Group), and it provides no information about stars born longer ago than their lifetimes.

We have, instead, a number of "indicators" of star formation rate in various contexts, most of them sensitive only to stars of the largest masses, highest luminosities, and shortest lives (except for measuring total masses in clusters and such, which tells you about the lower masses preferentially). When more than one indicator is applied to the same population, results tend to be correlated but not identical, and it is sometimes obvious why this should be so. Examples (This is meant to be exhaustive, but I may have forgotten something.), in order from most to least straightforward include:

1. Direct radiation. This will be mostly blue and ultraviolet light, which is easily absorbed by dust and lost to the inventory, though it then comes out as

2. Reradiated photons, in the full range of infrared wavelength bands and the sub-millimeter (depending on the temperature of the dust doing the reradiating and on whether the galaxy is nearby or at appreciable redshift).

3. Radio emission from supernova remnants and electrons accelerated in them which then radiate in a general interstellar magnetic field (non-thermal radiation) plus the thermal radiation from HII regions ionized by young stars and in older supernova remnants. The thermal and non-thermal components can often be separated and used independently.

4. X-ray emission. Once again, this will come from a combination of various kinds of very hot gases, some diffuse (supernovae and their remnants), some compact, like the emission from X-ray binaries and very young white dwarfs. The sum should at least be correlated with the number of massive stars formed over the past $10^8$ yr or so.

5. Supernova rates. These have the advantage that the events can be seen in quite distant galaxies, and, at least for Type II (core collapse) SNs probe a fairly definite mass range.

6. Heavy element production. This is the primary tracer used at the largest redshifts, where we indeed see that the gas clouds responsible for producing QSO absorption lines are very metal poor by solar system standards. The strongest correlations are, however, probably with where, relative to large galaxies, the gas clouds lie rather than with look-back time.

7. Gamma ray burst rate. These are even remotely useful only if you are sure that the bursters are in distant galaxies (demonstrated by data in the last few years) and that you can associate them with some definite event in the lives of massive stars, for instance failed supernova explosions or the merger of neutron star binaries (plausible but not certain). Some very preliminary evidence from statistics of the brightnesses and spectra of GRBs has suggested that their rate might track the star formation rate, with a peak at moderate redshift ($z \approx 2$) and lower rates before and after.

8. Ionization of QSO absorption line clouds. Because the main source of ionizing photons is thought to be leakage from star forming-galaxies, this provides yet another indicator, and again correlations have been claimed. But if the correlation should be poor, I think we would claim that other sources, like active galaxies, dominated the UV background, not that star formation wasn't a well-defined process.

## ACKNOWLEDGEMENTS

I am grateful to Dr. Oliver K. Manuel for the invitation to give a talk at New Orleans (though it was impossible to accept) and to Drs. Richard Kron, James Truran, and Craig Wheeler for the invitation to give a real (rather than virtual) talk on chemical evolution at the Chicago Meeting of the American Astronomical Society in June 1999.

## REFERENCES

Baade, W.: 1944a, "The resolution of Messier 32, NGC 205, and the central region of the Andromeda Nebula", *Ap. J.* **100**, 137-146.

Baade, W.: 1944b, "NGC 147 and NGC 185, two new members of the local group of galaxies", *Ap. J.* **100**, 147-150.

Burbidge, E.M., Burbidge, G.R., Fowler, W.A. and Hoyle, H. ($B^2FH$): 1957, "Synthesis of the elements in stars", *Rev. Mod. Phys.* **29**, 547-650.

Cameron, A.G.W.: 1957, "Stellar evolution, nuclear astrophysics and nucleogenesis", Chalk River Report CRL-41; "Nuclear reactions in stars and nucleogenesis", *Publ. Astron. Soc. Pacific* **69**, 201-222.

Eggen, O.J., Lynden-Bell, D. and Sandage, A.R.: 1962, "Evidence from the motions of old stars that the galaxy collapsed", *Ap. J*. **136**, 748-766.

Payne, C.H.: 1925, *Stellar Atmospheres; A Contribution to the Observational Study of High Temperature in the Reversing Layers of Stars*, Harvard Observatory Monographs, No. 1, Cambridge, MA, 215 pp

Tinsley, B.M.: 1968, "Evolution of the stars and gas in galaxies", *Ap. J.* **151**, 547-565.

Trimble, V.: 1975, "The origin and abundances of the chemical elements", *Rev. Mod. Phys.* **47**, 877-976.

Trimble, V.: 1991, "The origin and abundances of the chemical elements", *Astron. Astrophys. Rev.* **3**, 1-46.

Trimble, V.: 1996, "Cosmic abundances: Past, present, and future", in *Cosmic Abundances: Past, Present and Future*, eds., Holt, S. S. and Sonneborn, G., Astron. Soc. Pacific Conf. Ser. **99**, pp. 3-35.

# LiBeB Nucleosynthesis and Clues to the Chemical Evolution of the Universe

V. E. Viola
*Indiana University, Department of Chemistry and IUCF, Bloomington, IN 47405*
*viola@indiana.edu*

**Abstract:** Theories of LiBeB nucleosynthesis are reviewed in the context of experimental nuclear reaction cross section data. These fragile elements disintegrate readily in the hot dense conditions present in the Big Bang and stellar evolution. Thus, their synthesis must occur in a cold, dilute cosmological environment. Reactions of H and He ions with CNO nuclei during the passage of galactic cosmic rays through the interstellar medium can account for most of the LiBeB abundances. The isotope $^7$Li requires production in the Big Bang and other sources. Because Be and B formation requires the existence of CNO formed during stellar evolution, their presence in old halo stars serves as a signal for nucleosynthesis. Measurements of cross sections for LiBeB production can now be combined with highly sensitive new measurements of the elemental abundances of these elements to gain new insights into the chemical evolution of the universe.

## 1. INTRODUCTION

Big Bang nucleosynthesis and stellar evolution have been remarkably successful theories for describing the observed abundances of the chemical elements. However, the elements lithium, beryllium and boron are notable exceptions. As is apparent in Figure 1, in the interstellar medium the abundances of the stable isotopes $^6$Li, $^7$Li, $^9$Be, $^{10}$Be and $^{11}$B (LiBeB) appear to be suppressed by more than a factor of $10^6$ relative to their immediate neighbors. In fact, calculations indicate that even these abundances are much larger than expected. In the most optimistic scenarios for Be and B synthesis in an inhomogenous Big Bang , these elements are $\leq 10^{-14}$ smaller than the abundance of hydrogen, although some $^7$Li is produced at near the observed levels. In stellar interiors , LiBeB are burned up rather than being synthesized. Thus, a unique mechanism is required to account for LiBeB synthesis, a fact that Burbidge *et al.* (1957) recognized when they labeled this mechanism as the *x*-process.

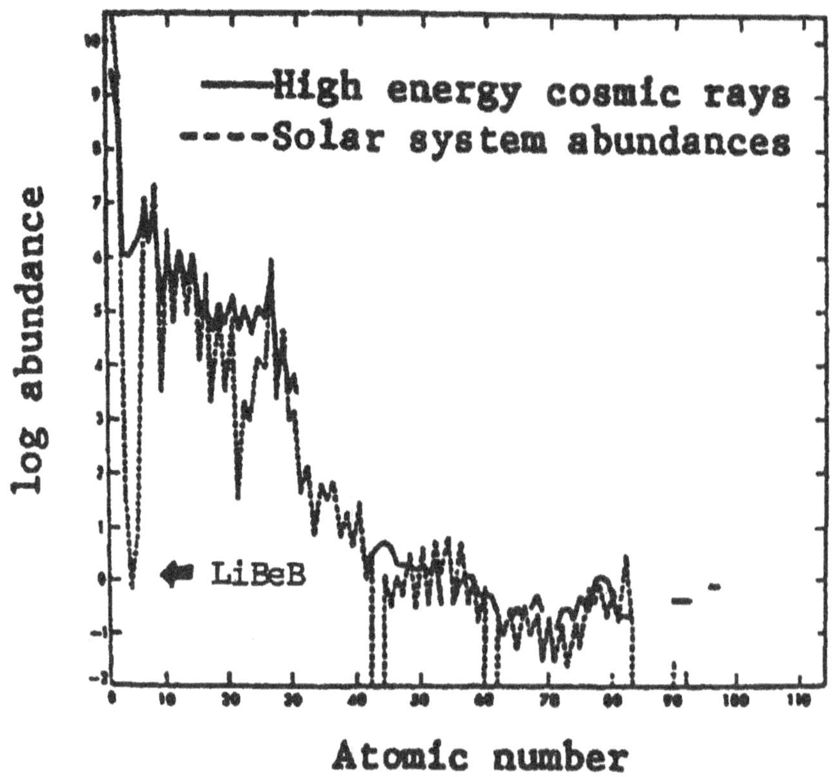


*Figure 1.* Comparison of interstellar medium abundances (dashed line) with abundances in the galactic cosmic-ray flux (solid line) as a function of atomic number.

Any successful model of LiBeB synthesis must take into account two basic nuclear constraints on the astrophysical site of their production. First, these nuclei are very loosely bound, a fact that strongly inhibits their production via thermonuclear fusion chains such as those that occur in the Big Bang or stellar burning. Second if LiBeB are to be produced in spallation-like reactions on heavier nuclei, one must consider the high energetic thresholds for these processes, ~ 10-20 MeV. As a consequence, the synthesis of LiBeB requires a flux of high energy particles and a cold, dilute environment that will permit them to survive, once formed. Based on the very high enrichment of LiBeB in the GCR flux (solid line in Figure 1), in 1970 Reeves *et al.* (1970) proposed the interaction of galactic cosmic rays

(GCR) with the interstellar medium (ISM) as one possible source. Subsequently, Austin (1981) stressed the inevitability of this mechanism, given that all the major components of the model are known or subject to direct experimental verification. Austin also identified the critical reaction cross sections necessary to test the model: $\alpha + \alpha$, and p, $\alpha$ + CNO, over all energies from threshold to several hundred MeV. Over the past 30 years several groups – primarily Michigan State, the University of Washington, Orsay and Lawrence Berkeley Laboratory, as well as our own at Indiana and Maryland – measured these cross sections, which are compiled in (Read and Viola, 1984). With these results in hand, Walker *et al.* (1985) examined the GCR + ISM model, as well as other proposed mechanisms such as flare production and synthesis in supernova shock waves. The results are discussed in Section 2.

More recently, there has been a rebirth of interest in LiBeB as possible indicators of chemical evolution in the early galaxy, made possible by the greatly improved resolution provided by the new generation of ground-based observatories such as Keck (See for example Boesgaard *et al.*, 1999). The object of this paper is to review briefly the basic models of LiBeB nucleosynthesis and the implications of the results for several important astrophysical questions, *e.g.*, the Big Bang, cosmology implications, the chemical evolution of the galaxy, and galactic cosmic rays.

## 2. TESTS OF LiBeB SYNTHESIS MODELS

In order to test models of nuclide formation one must solve the production equation:

$$\frac{dN_L}{dt} = \sum_i N_i \int_{E_o}^{\infty} \sum_j \phi_j(E)\sigma_{ijL}(E)S_{ijL}(E)dE \qquad (1)$$

Here, $N_i$ is the abundance of each target nucleus, usually the ISM abundances from Figure 1; $\phi_j(E)$ is the flux of the incident particles, as shown schematically in Figure 2 for the GCR, stellar flares, and supernova shock waves; $\sigma_{ijL}(E)$ is the cross section for formation of nuclide L from the reaction i + j, and $S_{ijL}(E)$ contains the physical parameters of the model – *e.g.*, stopping powers, medium thickness, etc. Both normal (p + A) and inverse (A + p) reactions must be considered. One then compares the formation rate with experimental values, or conversely, the data can be used in an effort to unfold the flux function $\phi_j(E)$.

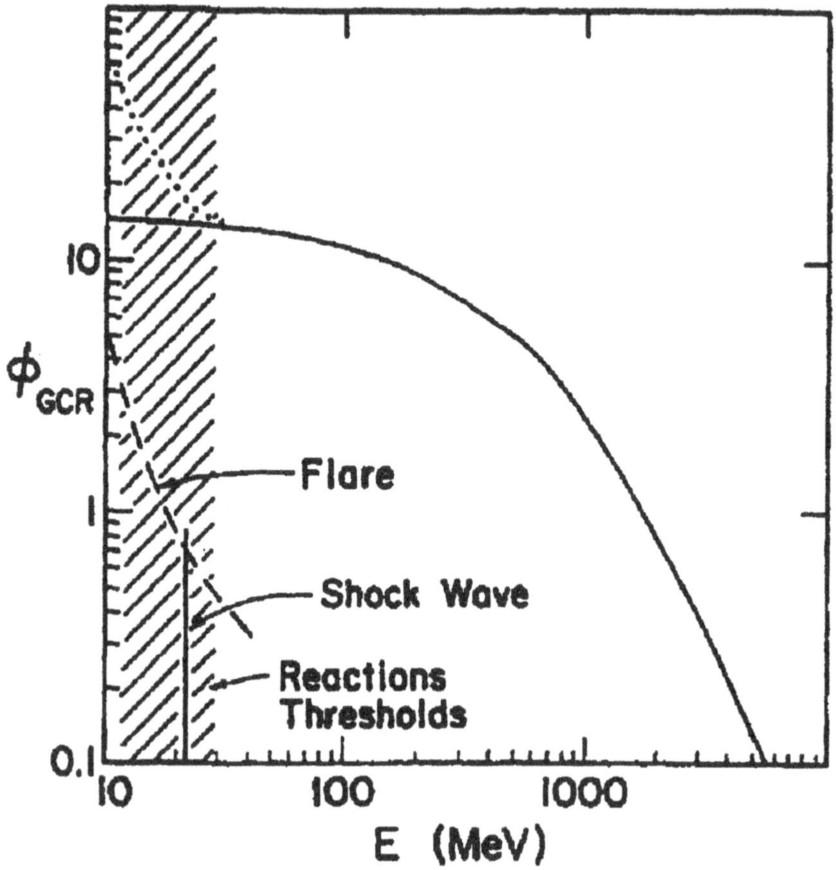


*Figure 2.* Schematic diagram of spectral shapes for the GCR flux, energetic flares and supernova shockwaves. Reaction thresholds for LiBeB synthesis are indicated by shaded area. Dotted line indicates assumed low energy component of the cosmic ray flux.

The essential nuclear cross-section information has now been measured by various techniques (Read and Viola, 1984). In our studies at Indiana, the experimental configuration employed electronic techniques. The basic concept involved the use of a detector telescope with a channel-plate fast-timing detector (Moyle *et al.*, 1978), or a beam rf pickoff signal, as a time-of-flight (TOF) start signal and a thin, totally depleted silicon detector as a TOF stop and energy (or energy-loss) detector to identify all fragments with $A \geq 6$. Cross sections for all fragments with $E/A \geq 0.05$ MeV can be determined with this technique. Analysis of data for target nuclei up to $^{28}Si$ has demonstrated that target nuclei heavier than oxygen are not important in LiBeB nucleosynthesis (Woo *et al.*, 1993).

In order to evaluate models for the synthesis of LiBeB in nature, it is necessary to compare calculated abundance ratios based on equation 1 with experimentally observed values (Austin, 1981; Reeves and Meyer, 1978; Cameron, 1982; Anders and Ebihara, 1982). Natural abundance ratios (based on $^1H = 10^{12}$) are summarized in Table 1 using the average values of Austin (1981).

Since we are concerned with light element production in global astrophysical environments, we initially take $N_i$ to be interstellar medium abundances. In our work the cross-section systematics, $\sigma_{ij}(E)$, of Reed and Viola (1984) have been incorporated into the calculations for the above reactions. These include all p and $^4He$ cross sections for both normal and inverse reactions with $^4He$, $^{12}C$, $^{13}C$, $^{14}N$ and $^{16}O$ from threshold to the high-energy cross section saturation region. The range-energy relations of Northcliffe and Schilling (1970) have been used as a basis for the determination of stopping powers. For production by galactic cosmic rays (Mitler, 1972; Meneguzzi *et al.*, 1971), the spectra are given by

$$\phi_i(E) = \alpha_i(E + E_o)^{-2.6} \tag{2}$$

with $E_o = M_oc^2$ and the set of $\alpha_i$ normalized to the observed galactic cosmic ray proton flux and $\phi_p = 12.5\ cm^{-2}\ sec^{-1}\ GeV^{-1}\ nucleon^{-1}$.

*Table 1.* Abundances of LiBeB isotopes*

| Abundance Ratio | Experiment | GCR | GCR + f $E^{-7}$ |
|---|---|---|---|
| $^6Li/H$ | 70. (2) | 69 | 83 |
| $^7Li/H$ | 900 (2) | 99 | 180 |
| $^9Be/H$ | 14 (1.6) | 15 | 15 |
| $^{10}B/H$ | 30 (2) | 71 | 82 |
| $^{11}B/H$ | 120 (2) | 180 | 320 |
| $^7Li/^6Li$ | 12.6 ± 0.2 | 1.4 | 2.2 |
| $^{11}B/^{10}B$ | 4.05 ± 0.05 | 2.5 | 4.05 |

*Abundances are based on $^1H = 10^{12}$ and are compared with experimental values from Austin (1981). Parentheses indicated estimated error factors. Calculated abundances are based on (1) a GCR + ISM model (equation 2) and (2) a GCR model that includes an $E^{-7}$ flare component, normalized to $^{11}B/^{10}B = 4.05$.

The energy loss term $S_L(E)$ takes a form corresponding to the leaky-box model of GCR propagation (Cesarsky, 1980). For present purposes, a value of 5 g/cm$^2$ is chosen, which adequately reproduces the light element-to-CNO ratio observed in high energy cosmic rays (Mitler, 1972; Meneguzzi *et al.*, 1971). Finally, in order to compare with absolute abundances, we introduce the first parameter into equation 1: the time scale for nucleosynthesis. We have used $\Delta t \approx 10^{10}$ years for the galactic-cosmic-ray calculations. Implicit in these calculations is the assumption that the higher cosmic-ray activity presumably present in the early universe is compensated by the lower abundances of the ISM at that time due to limited galactic chemical evolution.

In Table 1 the results of calculations based on the galactic-cosmic-ray nucleosynthesis model are compared with the deduced abundances. Based on the analyses of GCR and ISM abundance errors, uncertainties in the nuclear reaction cross sections, and the function $S_L(E)$, we estimate the errors in these calculations are about 30% for all isotopes.

The GCR nucleosynthesis calculations of Walker *et al.* (1985), which include complete experimental cross sections, are found to reproduce successfully the observed abundances of $^6$Li, $^9$Be, $^{10}$B and $^{11}$B, within the quoted limits of error. It is further reassuring that the calculations reproduce the absolute $^9$Be abundance quite well, since this isotope is subject to the smallest uncertainties, both in terms of experimental and calculated values. However, $^7$Li remains underproduced by nearly an order of magnitude and the $^{11}$B/$^{10}$B ratio is also unsatisfactory. As a second parameter, it is reasonable to introduce a low-energy component of the cosmic ray flux, which can enhance the $^{11}$B abundance via the $^{14}$N(p,$\alpha$)$^{11}$C($\beta^+$)$^{11}$B reaction and some $^7$Li via $^4$He($\alpha$,p)$^7$Li, while at the same time leaving the $^6$Li, $^9$Be, and $^{10}$B abundances relatively undisturbed. In Table 1 the results of such a low-energy component for a power-law spectrum $E^{-\gamma}$ with $\gamma = 7$ is combined with the GCR component and normalized so that $^{11}$B/$^{10}$B = 4.05. Thus our GCR flux function becomes:

$$\phi_i(E) = \alpha_i(E + E_o)^{-2.6} + f_i\, E^{-7} \tag{3}$$

where $f_i$ is the normalization factor for the low-energy component. We have taken the flux ratios of the various low energy cosmic ray nuclei to be the same as their high energy counterparts. Since $^4$He + $^4$He reactions dominate the production of Li, the He/H ratio is particularly relevant to these studies, but not so important when viewed in comparison with observational uncertainties in light element abundances. Note also that such low-energy fluxes must not over-ionize the ISM at low energies.

From Table 1, we see that a high-energy GCR model supplemented with a softer low-energy component with spectral index $\gamma \approx 7$ can reproduce the $^{11}$B/$^{10}$B isotopic abundance ratio without overproducing $^6$Li, $^9$Be, $^{10}$B, and $^{11}$B. An additional virtue of this spectrum is the rather low total energy requirements placed on the accelerating environment in order to produce the additional $^{11}$B. We conclude that GCR nucleosynthesis with a judiciously chosen low-energy component is a reasonable model for production of $^6$Li, $^9$Be, $^{10}$B, and $^{11}$B observed in the galaxy. However, $^7$Li cannot be produced in adequate amounts by this mechanism and another source is therefore required to account for the natural abundance of $^7$Li.

Other plausible mechanisms involving energetic nuclear reactions may also account for LiBeB synthesis. Two cases that are representative of such mechanisms are synthesis in flare spectra (Fowler *et al.*, 1962) and in supernovae shock waves (Colgate, 1973). In Table 2 the calculated abundance ratios for a flare source, normalized to the $^9$Be spectra, are presented as a function of the power law index $\gamma$. The present results argue rather decidedly against a flare mechanism as a significant source of LiBeB synthesis. In all cases the calculated abundance ratios significantly exceed the experimental values. The best agreement for the LiBeB ratios is found for low values of $\gamma$, which, it should be noted, begin to approximate the GCR flux of equation 2. However, low values of $\gamma$ imply an accelerating mechanism that utilizes a significant fraction of total gravitational energy of the source. Under any circumstances, the $^7$Li/$^6$Li and $^{11}$B/$^{10}$B ratios are unsatisfactory at all values of $\gamma$.

*Table 2.* Calculated* abundances of LiBeB isotopes for flares with power-law spectra $E^{-\gamma}$, normalized to $^9$Be

| Abundance Ratio | Experimental | $\gamma = 3$ | $\gamma = 5$ | $\gamma = 7$ |
|---|---|---|---|---|
| $^6$Li/$^9$Be | 5.0 | 20 | 26 | 34 |
| $^7$Li/$^9$Be | 64 | 57 | 130 | 180 |
| $^9$Be (≡ 1) | 1 | 1 | 1 | 1 |
| $^{10}$B/$^9$Be | 2.1 | 6.1 | 10 | 26 |
| $^{11}$B/$^9$Be | 8.7 | 32 | 99 | 360 |
| $^7$Li/$^6$Li | 12.6 | 2.9 | 4.8 | 5.2 |
| $^{11}$B/$^{10}$B | 4.05 | 5.2 | 9.9 | 14 |

*Calculated abundance ratios based on production in a transported flare spectrum $E^{-\gamma}$, where E is in energy per nucleon. All values are normalized to $^9$Be ≡ 1. Experimental values are based on (Austin, 1981).

In summary, while the flare mechanism may be important in accounting for small anomalies in isotopic ratios (*e.g.*, $^{11}$B/$^{10}$B), the present analysis appears to rule out this mechanism as a primary source of LiBeB synthesis.

Results of the shock wave abundance calculations (not shown) yield strong variations in the LiBeB ratios and demand that the energy of the shockwave be very finely tuned in order to reproduce the observed abundances, a constraint that would appear to limit the usefulness of this mechanism as a global model. Furthermore, even allowing for a source spectral shape defined precisely enough to satisfy the observed ratios, a severe limitation is imposed on this mechanism due to the tendency of such a shock wave to disperse by energy-loss when passing through any given medium. However, recently a variation on the GCR + ISM mechanism has been proposed (Duncan *et al.*, 1997) in which the propagation of GCRs is associated with type II supernovae. In this model LiBeB are formed as the

GCRs are propagated in the explosion and interact with the α CNO products present in the final stages of stellar evolution. This mechanism is discussed further in the following section.

In summary, with two plausible parameters, the GCR + ISM mechanism is able to account for the absolute abundances of $^6$Li, $^9$Be, $^{10}$B, and $^{11}$B satisfactorily. The only free parameters are the time over which nucleosynthesis has occurred and the presence of a low-energy component of the GCR transported flux of the form $\phi(E) \alpha E^{-\gamma}$ where $\gamma \approx 5\text{-}7$. Because all the major ingredients of the model are experimentally determined, this mechanism must certainly exist and under any circumstances must form the background for discussing LiBeB nucleosynthesis in any astrophysical setting.

## 3. THE BIG BANG AND CHEMICAL EVOLUTION

The failure of the GCR + ISM mechanism to account for $^7$Li signals the need for an additional source of this isotope. Based on the production rates for $^7$Li predicted by calculations of primordial light-element nucleosynthesis (See for example Walker *et al.*, 1993), the Big Bang becomes an immediate candidate for the third parameter needed to describe the origins of LiBeB fully. From the difference between the observed $^7$Li abundance (Austin, 1981) and GRC + ISM model calculations of Walker (1985), a Big Bang abundance of [$^7$Li/H] ≈ 7 × $10^{-10}$ is inferred. As shown in Figure 3, this value is consistent with primordial abundance ratios for $^2$H, $^3$He and $^4$He for a value of the universal mass density ratio of $\Omega \leq 0.03$.

Here $\Omega_B = \rho_B/\rho_{crit}$, where $\rho_B$ is the observed baryon density of the universe and $\rho_{crit}$ is the critical density required to produce a flat universe; *i.e.*, one that is expanding, but at a rate that asymptotically approaches zero. As pointed out by Mathews and Viola (1979), even more stringent limits can be placed on $\Omega_B$ by taking ratios of [$^7$Li/$^2$H], which are observed to have opposite slopes in Figure 3. Hence, the $^7$Li abundance also carries broader cosmological significance, supporting an open universe ($\Omega < 1$) as opposed to one that is closed ($\Omega > 1$). Of course, the question of the "missing mass" that would give $\Omega = 1$, necessary for models of inflation, remains one of major scientific interest.

What appeared to be a self-consistent picture of LiBeB origins broke down in the mid 1980s when studies of LiBeB abundances in old halo stars became possible due to greatly improved ground-based detection techniques. Low metallicity halo stars outside the galactic disk are believed to represent the oldest stars in the galaxy, and thus are most likely to reflect the primordial Big Bang light-element abundances.

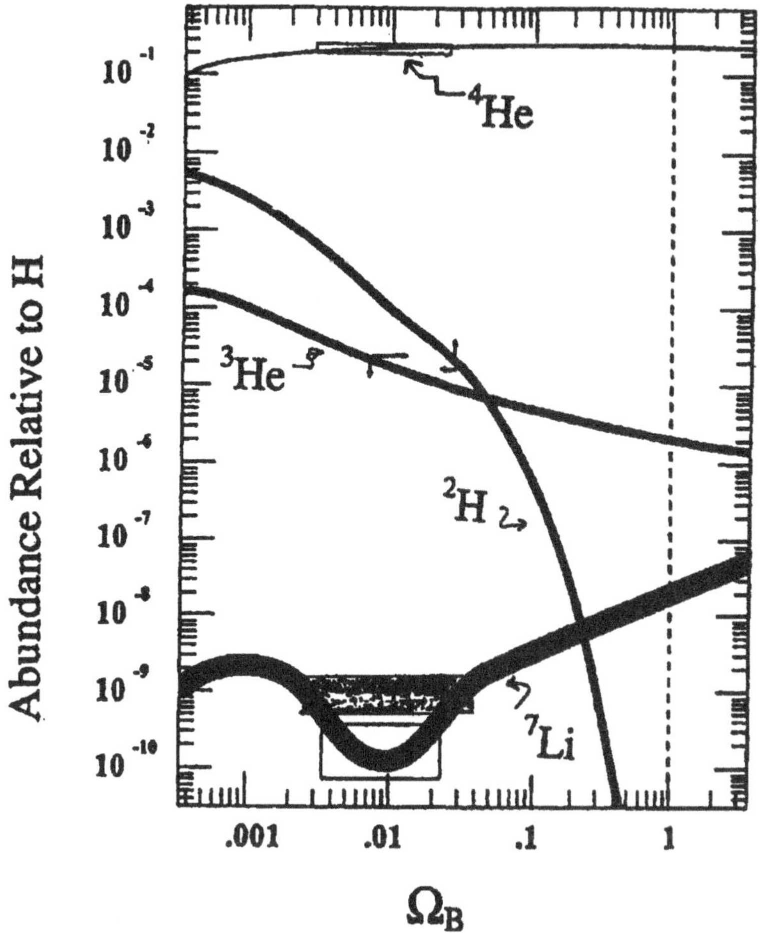


*Figure 3.* Abundance ratios for $^{2}H$, $^{3}He$, $^{4}He$, and $^{7}Li$ from primordial nucleosynthesis calculations (lines) compared with observed abundances (boxes or limits defined by arrows). For $^{7}Li$ the upper shaded box is the Spite plateau value (Spite and Spite, 1982, 1983). Abundances are plotted as a function of the universal baryon density parameter Ω.

The measurements of Spite and Spite (1982, 1983), shown in Figure 4, demonstrated for the first time what appears to be evidence for the Big Bang $^{7}Li$ abundance. Figure 4 plots the log [Li/H] ratio as a function of log [Fe/H] in these stars. Since Fe can only be produced in the supernova stage of stellar evolution, the [Fe/H] ratio serves as a chronometer for galactic chemical evolution; *i.e.*, the lower the [Fe/H] ratio, the older the star. The independence of [Li/H] on [Fe/H], the "Spite plateau", is taken as evidence that the Li in these stars is of Big Bang origin. Otherwise, if it came from the GCR + ISM mechanism, one would expect an increase with increasing iron content, since CNO also increases with Fe (Boesgaard *et al.*, 1999).

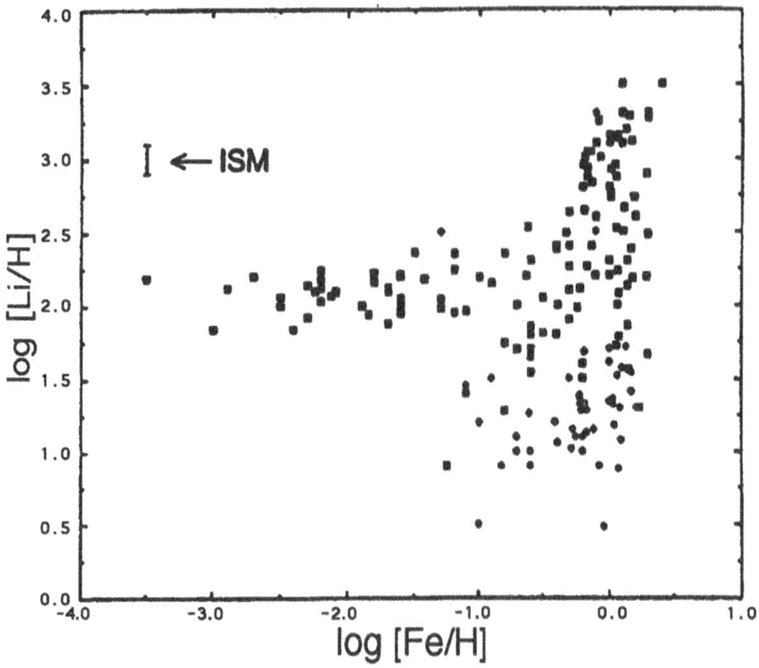


*Figure 4.* Plot of derived Li abundances as a function of Fe. From Spite and Spite (1982, 1983).

While the Spite and subsequent measurements strongly suggest that the Li in these stars is derived from the Big Bang, the amount, $[Li/H] \approx 1 \times 10^{-10}$, is far below the observed Li abundance in Table 1. This is the minimum value consistent with the Big Bang nucleosynthesis calculations (at any value of $\Omega$) and further constrains values of $\rho_B$. Unless $^7Li$ has been highly depleted by astration, the Spite result leads to the conclusion that the astrophysical sites of $^7Li$ are probably ubiquitous – stemming from the Big Bang, GCR + ISM reactions, and additional sources such as the surface of Red Giant stars. As is noted in Figure 4, for normal stars (log[Fe/H] = 0), a wide range of [Li/H] abundance ratios is observed.

The Spite and Spite result subsequently spurred interest in the possibility of observing similar behavior for Be and B. If such behavior existed at levels well above the calculated predictions, it would suggest that significant synthesis of heavier elements might be possible in the Big Bang – and call into question the standard model assumptions. Measurements have now been made, as shown in the Be results of Boesgaard *et al.* (1999) in Figure 5. In this case a direct dependence on metallicity (Fe) is observed. Since Be can only be formed in the presence of CNO nuclei, the behavior observed in Figure 5 must be attributed to the growth of both the GCR flux and chemical evolution in the early galaxy. Thus, the Be and B abundances in these stars may provide insight as to the level of GCR activity in the early universe. However, the standard model of the Big Bang appears to remain valid.

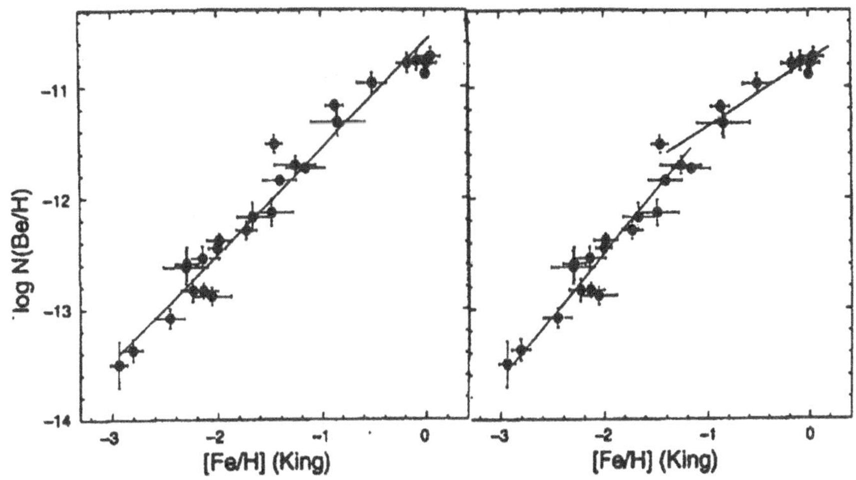


*Figure 5.* Derived Be abundances plotted versus [Fe/H] according to King (1993). Left frame shows linear fit to the data and right frame shows two-component fit, one for [Fe/H] ≥ -1.0 and one for [Fe/H] ≤ -1.1. Plots are from Boesgaard *et al.* (1999).

The Be results of Boesgaard *et al.* (1999) have also led to speculation as to whether GCR-induced LiBeB synthesis in the early universe takes place in the ISM or in the vicinity of supernovae. In the former case one would expect a quadratic growth curve for Be relative to Fe due to the need to produce CNO in one event, followed by LiBeB synthesis in a separate environment. In the latter case Be should exhibit a linear dependence on Fe, since both the GCR and CNO nuclei would be associated with the same event. As seen in the two identical plots in Figure 5, the present data do not distinguish clearly between the one- and two-component assumptions. At any rate, it has been shown (Boesgaard *et al.*, 1999) that [Be/H] increases with [O/H] much more rapidly than with [Fe/H], a result that favors the GCR + ISM scenario.

# 4. CONCLUSIONS

In summary, the study of LiBeB abundances and questions concerning their origin remains a stimulating subject. The isotopes $^{6}$Li, $^{9}$Be, $^{10}$B and $^{11}$B can be accounted for via the GCR + ISM mechanism, with two reasonable parameters. Whether the site of these reactions is isolated or in the vicinity of a supernova remains an open question that promises to be addressed as measurement techniques become increasingly sophisticated. The isotope $^7$Li is most likely produced via many sources, but its Big Bang component implies consistency with the standard model and $\Omega_B \leq 0.03$.

## ACKNOWLEDGMENTS

The author wishes to thank Constantine Deliyannis and Grant Mathews for many enlightening conversations. This work was supported by the U.S. Department of Energy.

## REFERENCES

Anders, E. and Ebihara, M.: 1982, "Solar-system abundances of the elements", *Geochim. Cosmochim. Acta* **46**, 2363-2380.

Austin, S.M.: 1981, "The creation of the light elements: Cosmic rays and cosmology", *Prog. Part. Nucl. Phys.* **1**, 1-456.

Boesgaard, A.M., Deliyannis, C.P., King, J.R., Ryan, S.G., Vogt, S.S. and Beers, T.C.: 1999, "Beryllium abundances in halo stars from Keck/Hires observations", *Astron. J.* **117**, 1549-1562.

Boesgaard, A.M., King, J.R., Deliyannis, C.P. and Vogt, S.S.: 1999, "Oxygen in unevolved metal-poor stars from Keck ultraviolet hires spectra", *Astron. J.* **117**, 492-507.

Burbidge, E.M., Burbidge, G.R., Fowler, W.A. and Hoyle, F.: 1957, "Synthesis of the elements in stars", *Rev. Mod. Phys.* **29**, 547-650.

Cameron, A.G.W.: 1982, "Elemental and nuclidic abundances in the solar system", in *Essays in Nuclear Astrophysics*, eds., Barnes, C.A., Clayton, D.D. and Schramm, D.N., Cambridge Univ. Press, Cambridge, UK, pp. 23-43.

Cesarsky, C.J., 1980, "Cosmic-ray confinement in the galaxy", *Ann. Rev. Astron. Astrophys.* **18**, 289-319.

Colgate, S.A.: 1973, "The production of deuterium in supernova shocks", *Ap. J.* **181**, L53-L54.

Duncan, D.K., Primas, F., Rebull, L.M., Boesgaard, A.M., Deliyannis, C.P., Hobbs, L.M., King, J.R. and Ryan, S.G.: 1997, "The evolution of galactic boron and the production site of the light-elements", *Ap. J.* **488**, 338-349.

Fowler, W.A., Greenstein, J.L. and Hoyle, F.: 1962, "Nucleosynthesis during the early history of the solar system", *Geophys. J. Royal Astron. Soc.* **6**, 148-220.

King, J.R.: 1993, "Stellar oxygen abundances. 1. A resolution to the 7774-angstron-Oi abundance discrepancy", *Astron. J.* **106**, 1206-1221.

Mathews, G.J. and Viola, V.E.: 1979, "On the light-element abundances, galactic evolution and the universal bargon density", *Ap. J.* **228**, 375-378.

Meneguzzi, M., Audoze, J. and Reeves, H.: 1971, "The production of the elements Li, Be and B by galactic cosmic rays in space and its relation to stellar observations", *Astron. Astrophys.* **15**, 337-359.

Mitler, H.E.: 1972, "The formation of Li, Be and B in cosmic-ray induced reactions", *Astrophys. Space Sci.* **17**, 186-218.

Moyle, R.A., Glayola, B.G., Mathews, G.J., Viola, V.E.: 1978 "Nucleosynthesis of Li, Be and B contributions from the p + $^{16}$O reactions at 50-90 MeV", *Phys. Rev.* **C19**, 631-640.

Northcliffe, L. and Schilling, R.F.: 1970, "Range and stopping-power tables for heavy ions", *At. Nucl. Data Tables* **7A**, 233-463.

Read, S.M. and Viola, V.E. 1984, "Excitation functions for A ≥ 6 fragments formed in $^{1}$H- and $^{4}$He-induced reactions on light nuclei", *At. Nucl. Data Tables* **31**, 359-397.

Reeves, H. and Meyer, J.P.: 1978, "Cosmic-ray nucleosynthesis infall rate of extragalactic matter in the solar neighborhood", *Ap. J.* **226**, 613-631.

Reeves, H., Fowler, W.A. and Hoyle, F.: 1970, "Galactic cosmic-ray origin of Li, Be, and B in stars", *Nature* **226**, 727-729.

Spite, M. and Spite, F.: 1982, "Lithium abundance at the formation of the galaxy", *Nature* **297**, 483-485.

Spite, M. and Spite, F.: 1983, "Abundance of lithium in unevolved-halo-stars and old-disk-stars – Interpretation and consequences", *Astron. Astrophys.* **115**, 357-366.

Walker, T.P., Mathews, G.J. and Viola, V.E.: 1985, "Astrophysical production-rates for Li, Be, and B-isotopes from energetic H-1 and He-4 reactions with HeCNO nuclei", *Ap. J.* **299**, 745-751.

Walker, T.P., Steigman, G., Schramm, D.N., Olive, K.A. and Fields, B.: 1993, "The boron-to-beryllium ratio in halo stars – A signature of cosmic-ray nucleosynthesis in the early galaxy", *Ap. J.* **413**, 562-570.

Woo, L.W., Kwiatkowski, K., Wilson, W.G., Viola, V.E., Breuer, H. and Mathews, G.J.: 1993, "Cross-sections for A = 6-30 fragments from the He-4 + Si-28 reaction at 117-MeV and 198-MeV", *Phys. Rev.* **C47**, 267-276.

# Measurement of the $^{44}Ti$ Half-life and its Significance for Supernovae

I. Ahmad[1], J. P. Greene[1], W. Kutschera[2], and M. Paul[3]
*[1]Argonne National Laboratory, Argonne, IL 60439, USA; [2]University of Vienna, Vienna, Austria; [3]Hebrew University, Jerusalem, Israel*
*ahmad@anlphy.phy.anl.gov*

**Abstract:** In 1998, we reported the three-laboratory measurements of the $^{44}Ti$ half-life which was determined relative to the well known value (5.2714 ± 0.0005 yr) of the $^{60}Co$ half-life. We have continued the measurement at Argonne and Jerusalem and inclusion of data points for additional two years does not change our published value of 59.0 ± 0.6 yr.

Supernova events are one of the main processes which generate galactic radioactivities. The light curves of the supernova are initially powered (Suntzeff *et al.*, 1992) by the radioactive decays of $^{56}Ni$ (5.9 d) and its daughter $^{56}Co$ (77.3 d), and later by the 272-d $^{57}Co$ and 59-y $^{44}Ti$. Theoretical models have, over the years, been developed to explain the production of these elements in supernova. $^{44}Ti$ is one of the nuclei produced in the supernova ejecta. Its synthesis requires exceptionally high temperatures, at least $5 \times 10^9$ K. This implies that it is produced in the deepest layers ejected from a supernova (Woosley and Diehl, 1998), near the so-called "mass cut" that separates the remnant from the ejected supernova. Thus a knowledge of the amount of $^{44}Ti$ produced in a supernova provides very critical information to check the validity of theoretical models. An accurate value of the $^{44}Ti$ half-life is also needed to determine the amount of $^{44}Ti$ produced by cosmic rays in meteorites (Bonino *et al.*, 1995).

The nucleus $^{44}Ti$ decays to $^{44}Sc$, which in turn decays to $^{44}Ca$ (Figure 1) and a gamma ray of 1157.0 keV is emitted in this decay. The amount of $^{44}Ti$ in a sample can be determined by measuring the 1157.0 keV gamma ray intensity. The half-life of $^{44}Ti$ can be determined either by specific activity method or by following the decay of the 1157.0 keV gamma line. Until the early nineties,

*Origin of Elements in the Solar System: Implications of Post-1957 Observations.*
Edited by O. Manuel, Kluwer Academic/Plenum Publishers, 2000. 

there was no consensus among the published values of the $^{44}$Ti half-life. The earliest three values (Wing *et al.*, 1965; Moreland and Heymann, 1965; Frekers *et al.*, 1983) were determined by the specific activity method, where the number of atoms in a sample and its decay rate were measured.

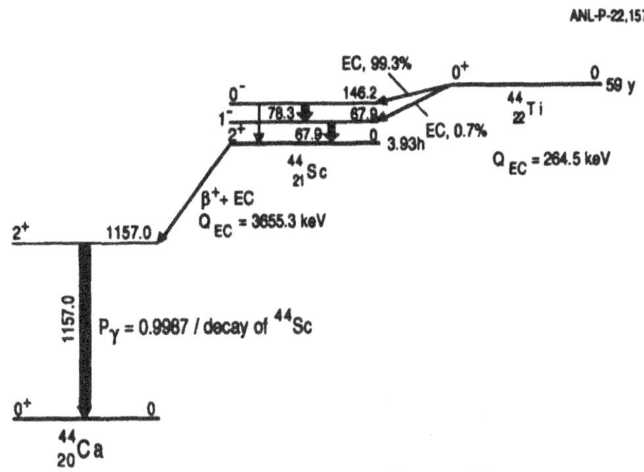


*Figure 1.* Decay schemes of $^{44}$Ti and $^{44}$Sc.

The last two values (Alburger and Harbottle, 1990; Norman *et al.*, 1997) shown in Figure 2 were obtained from the direct decay method. Alburger and

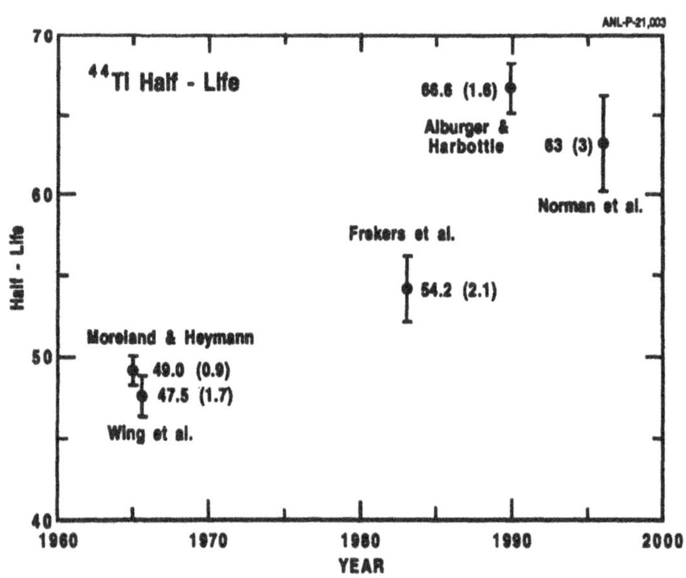


*Figure 2.* Values of $^{44}$Ti half-life published before 1998.

Harbottle (1990) measured the activity of a $^{44}$Ti sample with a proportional counter at regular intervals over a period of 3 years. It is obvious that the disagreement among the five values is caused by systematic uncertainties which, by their nature, are difficult to evaluate. Another impetus for a remeasurement of the $^{44}$Ti half-life came from the launching of space-based gamma ray spectrometers COMPTEL (Schönfelder *et al.*, 1993) and OSSE to observe $^{44}$Ti gamma ray lines (Figure 3).

The characteristic γ-ray of 1157.0 keV has been observed in the supernova Cassiopeia A both by the COMPTEL (Iyudin *et al.*, 1994; Schönfelder *et al.*, 1996; Dupraz *et al.*, 1997) and by OSSE (The *et al.*, 1996) detectors. This supernova was first sighted in 1680. More recently, the $^{44}$Ti gamma line has been identified (Iyudin *et al.*, 1998) in a previously unknown supernova in the Vela region. The age of this supernova is estimated to be about 680 years, its distance ~ 200 pc (pc = parsec = 3.27 light years), and it is the nearest known young supernova remnant. The abundance of $^{44}$Ti in a supernova can be determined from the known gamma ray flux from the supernova, its distance from earth and the $^{44}$Ti half-life. The half-life enters the calculations twice: once to apply decay correction to determine the flux at the time when the supernova was first seen and secondly to convert the decay rate to the number of atoms (Figure 4).

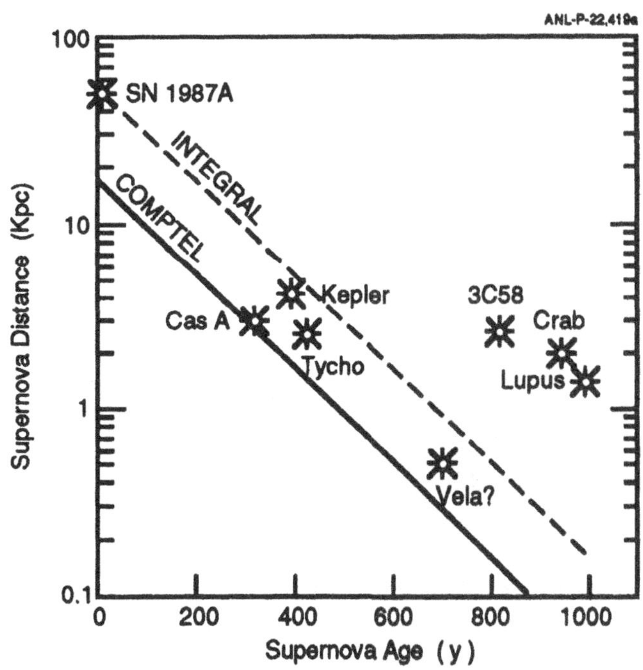


*Figure 3.* Sensitivities of the COMPTEL and INTEGRAL spectrometers. The instruments will be able to detect supernova explosions with distance and age below the line.

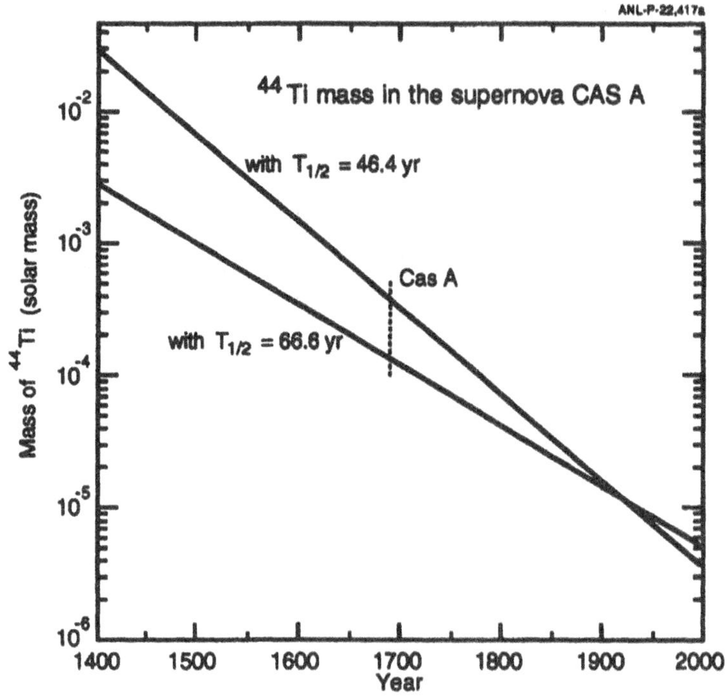


*Figure 4.* Effect of the $^{44}$Ti half-life on the mass of $^{44}$Ti in supernova remnants.

We started our measurement of the $^{44}$Ti half-life in 1992 at three laboratories-Argonne, Jerusalem and Torino. The aim of our experiment was to measure the $^{44}$Ti half-life relative to the half-life of $^{60}$Co. The reference isotope $^{60}$Co was chosen because its γ-ray of 1173.2 keV is quite close to the $^{44}$Ti gamma line of 1157.0 keV and its half-life is quite accurately known (King, 1993) (5.2714 ± 0.0005 yr). Because of the small difference in the $^{44}$Ti (1157.0 keV) and $^{60}$Co (1173.2 keV) γ-ray energies, small variations in the Ge detector efficiency will have a negligible effect.

The $^{44}$Ti activity was produced by the $^{45}$Sc(p,2n) reaction. Details of its production have been described in Frekers *et al.* (1983). The purity of $^{44}$Ti and also $^{60}$Co was checked by measuring their gamma-ray spectra with a Ge detector placed in a very low-background shield. Three samples containing 0.3 μCi $^{44}$Ti and 0.3 μCi $^{60}$Co were prepared at Argonne and were distributed to the three laboratories. A sample of pure $^{44}$Ti and a sample of pure $^{60}$Co were also prepared and their spectra were measured at Argonne at regular intervals. At Argonne, the γ-ray spectra of the mixed $^{44}$Ti + $^{60}$Co source and the two pure sources of $^{44}$Ti and $^{60}$Co were measured with a 25% Ge spectrometer, which had a resolution [FWHM] of 1.80 keV at 1.3 MeV. No shield was built around the Ge spectrometer because low background count rate was expected above the 1-MeV region. The total background rate was 110 Hz. A plastic holder was placed on the detector end cap and held with plastic screws. The source in a Lucite disk was placed in an aluminium holder, taped, and placed in one of the slots of the plastic holder such that the source-to-detector distance was 10.2 cm. In this arrangement, the source material was facing the Ge crystal. Each sample was counted for 48 h live time. In addition, a background spectrum with the source removed, but all holder materials in placc, was also measured for 48 h. After measuring a set of four spectra, the source-to-detector distance was changed to 5.2 cm, and another set of four spectra were taken. These sets of spectra were measured approximately every six months between February 1993 and June 1997, covering a period of 4 years. The γ-ray spectra of a mixed $^{44}$Ti + $^{60}$Co source, recorded at the start of the half-life measurement and at the present time, are displayed in Figure 5.

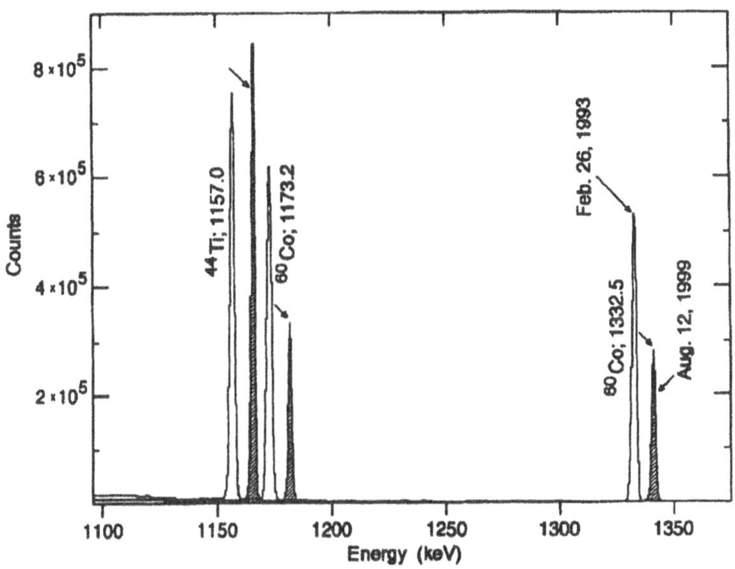


*Figure 5.* Gamma-ray spectra of the $^{60}$Co+$^{44}$Ti sample measured with a 25% Ge detector at Argonne at the start of the measurement and at present. The 1999 spectrum is offset to the right.

For the analysis, the respective background spectra were subtracted from the three sample spectra and the counts in the 1157.0, 1173.2, and 1332.5 keV peaks were determined. The peak counts were obtained by drawing a straight line between counts on the left side and counts on the right side of the peak and subtracting the background counts from the peak counts. The ratios of the counts in the 1157.0 keV peak to the counts in the 1173.2-keV peak were fitted as a function of time with an exponential function. The slope of the line (Figure 6) gave the difference between the decay constants $\lambda(^{60}Co) - \lambda(^{44}Ti)$. Using the known half-life of $^{60}Co$, 5.2714 ± 0.0005 yr, the half-life of $^{44}Ti$ was deduced. The half-life was also determined from the 1157.0/1332.5 peak area ratios. In this way eight values of half-life were obtained from the Argonne data set and a weighted average gave a value of 59.0 ± 0.8 yr.

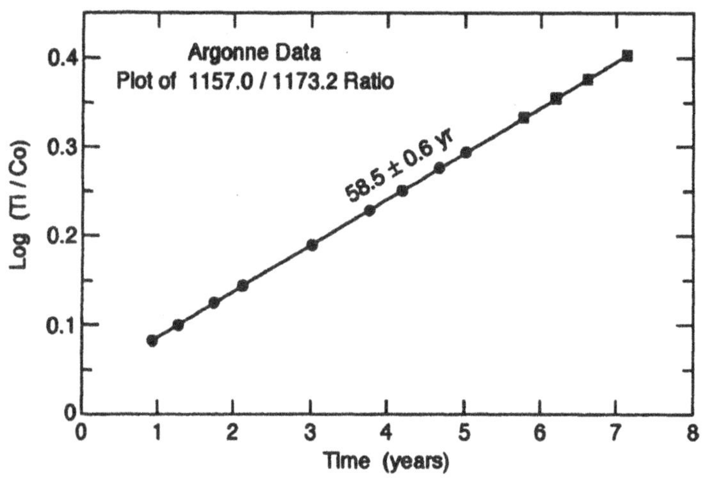


*Figure 6.* A semilogarithmic plot of the ratio of the counts in the 1157.0 keV peak to the counts in the 1173.2 keV peak measured as a function of time. The points with square symbol were measured after the publication of the paper (Ahmad *et al.*, 1998) in 1998. This line represents one of eight decay curves.

At Jerusalem, sets of spectra of a mixed source and a pure $^{60}Co$ source, and background spectra were measured with a 35% Ge(Li) spectrometer at a source-to-detector distance of 6.7 cm. Again, the detector was not shielded. Thirteen sets of spectra were measured over a period of 5 years. Each spectrum was analyzed with a peak fitting routine, using as a peak model for the shape of the 1157.0 keV line the shape of the 1173.2 keV peak measured in the same data set with a pure $^{60}Co$ source. A 60-keV region around the 1157.0 and 1173.2 keV lines, measured with the mixed source, was fitted to the sum of the two peak shapes scaled in energy and intensity and a linear component representing the background spectrum. The procedure gave excellent fits, and the ratio of the peak areas of the 1157.0 and 1173.2 keV lines was obtained directly from the scaling factors. The ratios were corrected for the very small contribution of the 1155.0 keV background line of $^{214}Bi$ (decay product of natural occurring $^{238}U$) which sits under the 1157.0 keV peak. The ratios of the peak areas were analyzed in the same way as the Argonne data to deduce $^{44}$Ti half-life. Half-life values were obtained both from the 1157/1173 and 1157/1332 ratios. The Jerusalem data gave a value of 58.9 ± 1.0 yr.

At Torino, only two sets of spectra, separated by ~3 yr, were measured which gave a half-life of 59.4 ± 1.4 yr. The weighted average of the Argonne, Jerusalem and Torino values gave a final value of 59.0±0.6 yr (1σ error) which was published in 1998 (Ahmad *et al.*, 1998). Three more measurements of $^{44}$Ti half-life were reported between 1998 and 1999. All three values - 60.3 ± 1.3 yr (Görres *et al.*, 1998), 62 ± 2 yr (Norman *et al.*, 1998), 60.7 ± 1.2 yr (Wietfeldt *et al.*, 1999) - are in good agreement with our value.

In all the decay measurements, the decay of $^{44}$Ti has been followed for only a small fraction of the half-life. For measurements over such a short interval, it is difficult to observe systematic errors. A measurement over a longer period is more likely to display systematic errors. For this reason, we have continued the measurement of the $^{44}$Ti half-life by measuring the spectra of the $^{60}$Co+$^{44}$Ti mixed samples at Argonne and Jerusalem using the same set-up which was used to determine the published value of the half-life. We have data points for an additional 2 years; at Argonne we have measured four points and at Jerusalem two points. As discussed in our previous publication (Ahmad *et al.*, 1998), we have eight half-life values at Argonne and two half-life values at Jerusalem. One of the decay curves from Argonne is displayed in Figure 6. We obtain a value of 58.5 ± 0.6 yr from this decay curve which is in excellent agreement with our published value of 59.0 ± 0.6 yr. We have obtained a weighted average value, which agrees with the published value within one standard deviation. We plan to publish the new value of the $^{44}$Ti half-life in the year 2000.

## ACKNOWLEDGMENTS

This work was supported by the U.S. Department of Energy, Nuclear Physics Division, under contract No. W-31-109-ENG-38.

## REFERENCES

Ahmad, I., Bonino, G., Cini Castagloni, G., Fischer, S.M., Kutschera, W. and Paul, M.: 1998, "Three-laboratory measurement of $^{44}$Ti half-life", *Phys. Rev. Lett.* **80**, 2550-2553.

Alburger, D.E. and Harbottle, G.: 1990, "Half-lives of $^{44}$T and $^{207}$Bi", *Phys. Rev.* **C41**, 2320-2324.

Bonino, G., Cini Castagnoli, G., Bhandari, N. and Taricco, C.: 1995, "Behavior of the heliosphere over prolonged solar quiet periods by $^{44}$Ti measurements in meteorites", *Science* **270**, 1648-1650.

Dupraz, C., Bloemen, H., Bennett, K., Diehl, R., Hermsen, W., Iyudin, A.F., Ryan, J. and Schönfelder, V.: 1997, "COMPTEL three-year search for galactic sources of $^{44}$Ti gamma-ray line emission at 1.157 MeV", *Astron. Astrophys.* **324**, 683-689.

Frekers, D., Henning, W., Kutschera, W., Rehm, K.E., Smither, R.K., Yntema, J.L., Santo, R., Stievano, B. and Trautmann, N.: 1983, "Half-life of $^{44}$Ti", *Phys. Rev.* **C28**, 1756.-1762.

Görres, J., Meiβner, J., Schatz, H., Stech, E., Tischhauser, P., Wiescher, M., Bazin, D., Harkewicz, R., Hellström, M., Sherrill, B., Steiner, M., Boyd, R.N., Buchmann, L., Hartmann, D.H. and Hinnefeld, J.D.: 1998, "Half-life of $^{44}$Ti as a probe for supernova models", *Phys. Rev. Lett.* **80**, 2554-2557.

Iyudin, A.F., Diehl, R., Bloemen, H., Hermsen, W., Lichti, G.G., Morris, D., Ryan, J., Schönfelder, V., Steinle, H., Varendorff, M., de Vries, C. and Winkler, C.: 1994, "COMPTEL observation of $^{44}$Ti gamma-ray line emission from Cas A", *Astron. Astrophys.* **284**, L1-L4.

Iyudin, A.F., Schönfelder, V., Bennett, K., Bloemen, H., Diehl, R., Hermsen, W., Lichti, G.G., van der Meulen, R.D., Ryan, J. and Winkler, C.: 1998, "Emission from $^{44}$Ti associated with a previously unknown galactic supernova", *Nature* **396**, 142-144.

King, M.M.: 1993, "Nuclear data sheets update for A = 60", *Nucl. Data Sheets* **69**, 1-67.

Moreland, Jr., P.E. and Heymann, D.: 1965, "The $^{44}$Ti half-life", *J. Inorg. Nucl. Chem.* **27**, 493-496.

Norman, E.B., Browne, E., Chan, Y.D., Goldman, I.D., Larimer, R.-M., Lesko, K.T., Nelson, M., Wietfeldt, F.E. and Zlimen, I.: 1997, "On the half-life of $^{44}$Ti", *Nucl. Phys.* **A621**, C92-C95.

Norman, E.B., Browne, E., Chan, Y.D., Goldman, I.D., Larimer, R.-M., Lesko, K.T., Nelson, M., Wietfeldt, F.E. and Zlimen, I.: 1998, "Half-life of $^{44}$Ti", *Phys. Rev.* **C 57**, 2010-2016.

Schönfelder, V., Aarts, H., Bennett, K., Deboer, H., Clear, J., Collmar, W., Connors, A., Deerenberg, A., Diehl, R., Vondordrecht, A., Denherder, J.W., Hermsen, W., Kippen, M., Kuiper, L., Lichti, G., Lockwood, J., Macri, J., McConnell, M., Morris, D., Much, R., Ryan, J., Simpson, B.N., Snelling, M., Stacy, G., Steinle, H., Strong, A., Swanenburg, B.N., Taylor, B., Devries, C., Winkler, C.: 1993, "Instrument description and performance of the imaging gamma-ray telescope COMPTEL aboard the Compton Gamma-ray Observatory", *Ap. J. Supp.* **86**, 657-692.

Schönfelder, V., Bennett, K., Bloemen, H., Diehl, R., Hermsen, W., Lichti, G.G., McConnell, M., Ryan, J., Strong, A. and Winkler, C.: 1996, "COMPTEL overview: Achievements and expectations", *Astron. Astrophys. Suppl.* **120**, C13-C21.

Sonzogni, A.A.: 1999, private communication.

Suntzeff, N.B., Phillips, M.M., Elias, J.H., DePoy, D.L. and Walker, A.R.: 1992, "The energy sources powering the late-time Bolometric Evolution of SN 1987A", *Ap. J.* **384**, L33-L36.

The, L.-S., Leising, M.D., Kurfess, J.D., Johnson, W.N., Hartmann, D.H., Gehrels, N., Grove, J.E. and Purcell, W.R.: 1996, "CGRO/OSSE observations of the Cassiopeia A SNR", *Astron. Astrophys. Suppl.* **120**, 357-360.

Wietfeldt, F.E., Schima, F.J., Coursey, B.M. and Hoppes, D,D.: 1999, "Long-term measurement of the half-life of $^{44}$Ti", *Phys. Rev.* **C59**, 528-530.

Wing, J., Wahlgren, M.A., Stevens, C.M. and Orlandini, K.: 1965, "Carrier-free separation of $^{44}$Ti and half-life determination of $^{44}$Ti", *J. Inorg. Nucl. Chem.* **27**, 487-491.

Winkler, C.: 1996, "INTEGRAL, The International Gamma-ray Astrophysical Laboratory", *Astron. Astrophys. Suppl. Series* **120**, 637-640; "INTEGRAL", *website page http://sci.esa.int/integral/*

Woosley, S. and Diehl, R.: 1998, "Titanium-44 gets a lifetime", *Phys. World 11*, No. **7**, 22-23.

# On the Half-Life of $^{44}$Ti in Young Supernova Remnants

Eric B. Norman and Edgardo Browne
*Nuclear Science Division, Lawrence Berkeley National Laboratory, Berkeley, CA 94720*
*ebnorman@lbl.gov*

**Abstract:** The electron-capture decay rate of $^{44}$Ti strongly depends on the number of atomic electrons that are bound to the nucleus. Recent x-ray observations of the Cas A and RXJ0852.0-4622 supernova remnants suggest that conditions of high temperatures and low densities exist in these objects. Under such conditions, the half-life of $^{44}$Ti would be significantly longer than its laboratory value. This effect implies that only an upper limit on the mass of $^{44}$Ti ejected by supernovae can be deduced from gamma-ray observations of these two supernova remnants.

The long-lived radioisotope $^{44}$Ti is currently of considerable interest in astrophysics. The relevant portions of its decay scheme and that of its daughter, $^{44}$Sc, are shown in Figure 1. $^{44}$Ti decays via electron-capture to $^{44}$Sc emitting γ-rays of 68, 78, and a very weak one of 146 keV. $^{44}$Sc subsequently decays via electron capture and positron emission with a 3.9-hour half-life to $^{44}$Ca emitting an 1157-keV γ-ray.

$^{44}$Ti is one of the few long-lived γ-ray emitting nuclides expected to be produced in substantial amounts during a supernova explosion (Clayton, 1982). Its characteristic 1157-keV γ ray was observed from the young supernova remnant Cassiopeia A [Cas A] (Iyudin *et al.*, 1994) and more recently from supernova remnant, RXJ0852.0-4622 (Aschenbach, 1998; Iyudin *et al.*, 1998). In order to deduce the mass of $^{44}$Ti ejected in these explosions using the γ-ray fluxes measured from these supernova remnants, one needs to know their ages and distances as well as the half-life of $^{44}$Ti.

For Cas A, there are reasonably good historical records that this supernova exploded in about 1680. For RXJ0852.0-4622, Aschenbach (1998) has estimated an age of less than 1500 years and Iyudin *et al.* (1998) estimated about 680 years. Until last year, there was great uncertainty in the half-life of $^{44}$Ti because published values ranged from 39.0 years (Meissner, 1996) to 66.6 years (Alburger and Harbottle, 1990). The results of four recent experimental studies (Norman *et al.*, 1998; Ahmad *et al.*, 1998, Gorres *et al.*, 1998; Wietfeldt *et al.*, 1999) however, have yielded a consistent value of 60 $\pm$ 1 years for the laboratory half-life of $^{44}$Ti.

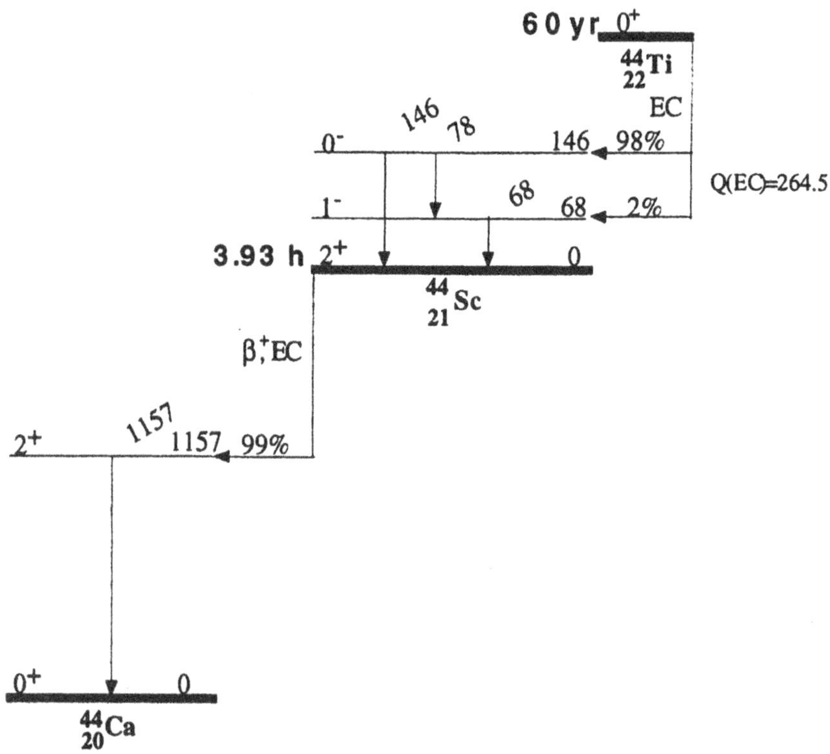


*Figure 1.* Decay schemes of $^{44}$Ti and $^{44}$Sc. All energies are given in keV.

In the laboratory the electron-capture decay of $^{44}$Ti takes place with neutral atoms. Thus, 22 bound atomic electrons surround the $^{44}$Ti nucleus. The binding energy of a 1s electron in a neutral Ti atom is 4.966 keV, and that of a 2s electron is 0.564 keV. For a neutral $^{44}$Ti atom, the probability of electron capture from the K (1s) shell is 0.8891 and from the L-shell (2s) is 0.0960. Therefore, neglecting electron screening, for a charge-state $19^+$ $^{44}$Ti ion (*i.e.*, one electron in the 2s shell) its half-life would be (60yr/0.9371) = 64 years. For a charge-state $20^+$ ion (zero electrons in the 2s shell) its half-life would be (60yr/0.8891) = 67.5 years, and for a charge-state $21^+$ ion (one electron in the 1s shell) its half-life would be (60yr/0.4446) = 135 years. Finally, for a charge-state $22^+$ $^{44}$Ti ion (no bound electrons), electron capture decay would not be possible and the nucleus would become stable.

The question thus arises as to how many electrons would be bound to a $^{44}$Ti nucleus under various conditions of temperature and density found in a young supernova remnant. Assuming thermal equilibrium has been reached, at a temperature $T$ and electron density $N_e$, the average number of bound atomic electrons in the atomic orbital with principal quantum number $n$ is (Chandrasekhar, 1967):

$$\overline{N}_n = \frac{2n^2}{1+\frac{G(T)}{N_e}exp(-\chi_n/(kT)} \qquad (1)$$

$$\text{where} \quad G(T) = \frac{2(2\pi m_e kT)^{3/2}}{h^3} \quad \text{and} \quad \chi_n = \frac{2^2\alpha^2 m_e c^2}{2n^2}$$

X-ray observations of the Cas A supernova remnant have provided estimates for both the temperature and the electron density in this object. Shull (1982) analyzed the x-ray data obtained by the High Energy Astrophysical Observatory (HEAO2) Solid State Spectrometer to obtain: $N_e \approx 1$ cm$^{-3}$; $T_1 = 6 \times 10^6$ K and $T_2 = 5 \times 10^7$ K, where $T_1$ was deduced from the x-ray line spectrum and $T_2$ from a fit to the x-ray continuum. Tsunemi *et al.* (1986) inferred from Tenma satellite observations that $T_{\text{continuum}} = 4.4 \times 10^7$ K and that $N_e \approx 20$ cm$^{-3}$. Vink *et al.* (1996) analyzed ASCA satellite data and deduced two temperature components of $7.7 \times 10^6$ K and $4.9 \times 10^7$ K for Cas A. From x-ray observations of RX J0852.0-4622, Aschenbach (1998) deduced a temperature in this remnant greater than $3 \times 10^7$ K. Thus it appears that conditions of high temperature and low density occur in both of these supernova remnants.

Inserting these measured temperatures and densities into equation 1 leads to the conclusion that $^{44}$Ti should be completely ionized (and thus stable) under the conditions that exist in these supernova remnants. However, the decay of $^{44}$Ti in these remnants has been observed. Furthermore, very strong evidence (Nittler *et al.*, 1996) for the existence of $^{44}$Ti in presolar grains suggests that at least some $^{44}$Ti ejected from a supernova cooled off sufficiently fast to become incorporated into solid objects. Consequently, not all of the $^{44}$Ti present in these objects is subject to such extremes of temperature and density. By fixing one of the parameters (*e.g.*, mean electron density) at a measured value, equation (1) may be used to calculate the maximum temperature at which, for example, there would be two K electrons and one L electron bound to a $^{44}$Ti nucleus. The results of these calculations are shown in Table 1.

*Table 1.* Average number of K and L electrons bound to a $^{44}$Ti nucleus as a function of temperature, *T*, and electron density, $N_e$, calculated using equation 1. The numbers shown in bold italics were calculated using fixed values for the other parameters listed in this table.

| $T(10^6 K)$ | $\overline{N_e}(cm^{-3})$ | $\overline{N_K}$ | $\overline{N_L}$ |
|---|---|---|---|
| ***0.35*** | 1 | 2.0 | 1.0 |
| ***0.375*** | 20 | 2.0 | 1.0 |
| ***1.34*** | 1 | 1.0 | 0.0 |
| ***1.4*** | 20 | 1.0 | 0.0 |
| 6 | ***4.2 x 10^23*** | 2.0 | 1.0 |
| 50 | ***1.6 x 10^36*** | ***0.6*** | 1.0 |

These results suggest that $^{44}$Ti would be highly ionized even at the lower range of temperatures inferred for the CasA and RX J0852.0-4622 supernova remnants. In order to estimate the mass of $^{44}$Ti ejected by these two supernovae, one must take the present-day observed gamma-ray fluxes and extrapolate them backward in time to the date of the supernova explosion.

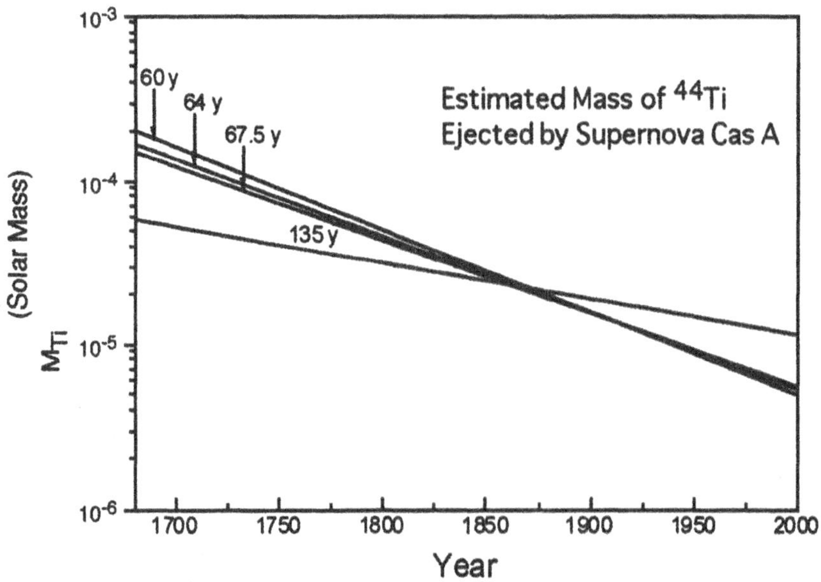


*Figure 2.* Extrapolated mass of $^{44}$Ti ejected by the Cas A supernova calculated by using four different half-lives for $^{44}$Ti: (i) the laboratory value of 60 years, (ii) the value estimated for an ion with two K electrons and only one L electron bound to the nucleus - 64 years, (iii) the value estimated for an ion with two K electrons but no L electrons bound to the nucleus – 67.5 years, and (iv) the value estimated for an ion with only one K electron bound to the nucleus – 135years.

Figures 2 and 3 show these extrapolations for the values of the $^{44}$Ti half-life previously calculated for four ionization states. For Cas A there is a factor of three range in the inferred mass of $^{44}Ti$ due to the variation in half-life with ionization state. RX J0852.0-4622 is much older, therefore this effect is much greater. Thus, the possible $^{44}Ti$ mass ejected by this supernova varies by a factor of 33 depending on the ionization state. Note that using the laboratory value for the $^{44}Ti$ half-life leads to the largest possible mass of this radioisotope ejected by a supernova. Thus, we conclude that as the result of a possible lengthening of the $^{44}Ti$ half-life due to ionization effects, the mass of $^{44}Ti$ ejected from a supernova and deduced from a gamma-ray measurement constitutes only an upper limit.

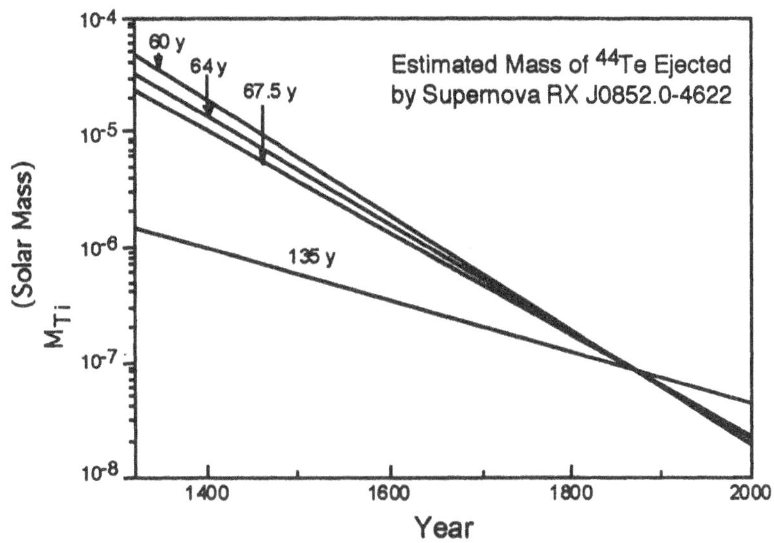


*Figure 3.* Mass of $^{44}Ti$ ejected by the RXJ0852.0-4622 supernova calculated using the four different half-lives for $^{44}Ti$ described above.

## ACKNOWLEDGMENTS

This work was supported by the Director, Office of Science, Division of Nuclear Physics of the U.S. Department of Energy under Contract No. DE-AC03-76SF00098.

## REFERENCES

Ahmad, I., Bonino, G., Cini Castagloni, G., Fischer, S.M., Kutschera, W. and Paul, M.: 1998, "Three-laboratory measurement of $^{44}Ti$ half-life", *Phys. Rev. Lett.* **80**, 2550-2553.

Alburger, D. and Harbottle, G.: 1990, "Half-lives of $^{44}$Ti and $^{207}$Bi", *Phys. Rev.* **C41**, 2320-2324.

Aschenbach, B.: 1998, "Discovery of a young nearby supernova remnant", *Nature* **396**, 141-142.

Chandrasekhar, S.: 1967, "An introduction to the study of stellar structure", Dover Publications, Dover, NY, p. 258.

Clayton, D.D.: 1982, "Cosmic radioactivity: A gamma-ray search for the origins of atomic nuclei", in *Essays in Nuclear Astrophysics*", eds., Barnes, C.A. Clayton, D.D. and Schramm, D.N., Cambridge University Press, Cambridge, UK, pp. 401-426.

Gorres, J., Meissner, J., Schatz, H., Stech, E., Tischauser, P., Wiescher, M., Bazin, D., Harkewicz, R., Hellstrom, M., Sherrill, B., Steiner, M., Boyd, R.N., Buchmann, L., Hartmann, D.H. and Hinnefeld, J.D.: 1998, "Half-life of $^{44}$Ti as a probe for supernova models", *Phys. Rev. Lett.* **80**, 2554-2557.

Iyudin, A.F., Diehl, R., Bloemen, H., Hermsen, W., Lichti, G.G., Morris, D., Ryan, J., Schonfelder, V., Steinle, H., Varendorff, M., de Vries, C. and Winkler, C.: 1994, "COMPTEL observations of $^{44}$Ti gamma-ray line emission from Cas A", *Astron. Astrophys.* **284**, L1-L4.

Iyudin, A.F., Schonfelder, V., Bennett, K., Bloemen, H., Diehl, R., Hermsen, W., Lichti, G. G., van der Meulen, R.D., Ryan, J. and Winkler, C.: 1998, "Emission from $^{44}$Ti associated with a previously unknown galactic supernova", *Nature* **396**, 142-144.

Meissner, J.: 1996, "Activity measurements of importance to stellar nucleosynthesis", *Ph.D. Thesis*, Univ. of Notre Dame.

Nittler, L.R., Amari, S., Zinner, E., Woosley, S.E. and Lewis, R.S.: 1996, "Extinct $^{44}$Ti in presolar graphite and SiC: Proof of a supernova origin", *Ap. J.* **462**, L31-L34.

Norman, E.B., Browne, E., Chan, Y.D., Goldman, I.D., Larimer, R.-M., Lesko, K.T., Nelson, M., Wietfeldt, F.E. and Zlimen, I.: 1998, "Half-life of $^{44}$Ti", *Phys. Rev.* **C57**, 2010-2016.

Shull, J.M.: 1982, "X-ray emission from young supernova remnants: Nonionization equilibrium abundances and emissivities", *Ap. J.* **262**, 308-314.

Tsunemi, H., Yamashita, K., Masai, K., Hayakawa, S.and Koyama, K.: 1986, "X-ray spectra of the Cassiopeia A and Tycho supernova remnants and their element abundances", *Ap. J.* **306**, 248-254.

Vink, J., Kaastra, J.S. and Bleeker, J.A.: 1996, "A new mass estimate and puzzling abundances of SNR Cassiopeia A", *Astron. Astrophys.* **307**, L41-L44.

Wietfeldt, F.E., Schima, F.J., Coursey, B.M. and Hoppes, D.D." 1999, "Long-term measurement of the half-life of $^{44}$Ti", *Phys. Rev.* **C59**, 528-530.

# Abundances in SN 1987A and Other Supernovae

Roger A. Chevalier
*Department of Astronomy, University of Virginia, P.O. Box 3818, Charlottesville, VA 22903*
*rac5x@virginia.edu*

**Abstract:** Although the cosmic abundances of elements point to a supernova origin, abundance measurements in individual supernovae are difficult. Abundances of radionuclides with relatively short half-lives are the best determined, especially for SN 1987A. These radionuclides excite emission from the surrounding heavy elements, enabling their study in the late spectra of supernovae. Shock interactions with the ambient medium provide another means of study, which will become facilitated in the next few years with the launch of new X-ray space observatories.

## 1. INTRODUCTION

The general picture for the origin of the elements heavier than helium proposed by Burbidge *et al.* (1957) is that they are synthesized in stars and then ejected into the interstellar medium. Supernovae play an important role in the ejection process. Computations of massive star evolution and the accompanying nucleosynthesis and supernova explosions have shown that the cosmic abundance pattern for most of the elements can be reproduced (*e.g.*, Timmes *et al.*, 1995). However, the problem of observationally determining what elements are ejected by supernovae has been developing slowly. Although there is excellent evidence that the expected heavy elements are ejected in supernovae, the determination of abundances is complicated in many cases. My aim here is to briefly review the present situation.

My emphasis will be on SN 1987A, which was relatively nearby and is the best observed supernova. Observations of other supernovae and their remnants will also be considered. Supernovae are divided into two major

*Origin of Elements in the Solar System: Implications of Post-1957 Observations.*
Edited by O. Manuel, Kluwer Academic/Plenum Publishers, 2000. 

observational types depending on whether hydrogen lines are absent (Type I) or present (Type II) in their spectra. The Type II supernovae are thought to result from core collapse in a massive star that has retained its H envelope. The Type I events are divided into Type Ia, which are probably thermonuclear explosions of white dwarfs, and Types Ib and Ic, which are probably the result of core collapse in a massive star that has lost its H envelope. Here, the term "massive" indicates a star that has an initial main sequence mass greater than about 8 $M_\odot$ (solar masses). SN 1987A belonged to the Type II category and had an initial progenitor mass of ~20 $M_\odot$.

In section 2, I discuss evidence for radioactive isotopes synthesized in the explosions. The amounts of the parent nuclei can be estimated directly from γ-ray line observations or indirectly from the strength of reprocessed radiation. Abundances estimated from optical emission lines are treated in section 3. Opportunities for abundance studies from the interaction of supernovae with their surroundings are discussed in section 4. The conclusions are in section 5.

## 2. RADIOACTIVITY

The explosion of the nearby supernova SN 1987A in the Large Magellanic Cloud has given us our best opportunity to study radioactivity in a supernova. Starting about 120 days after the supernova and continuing to 900 days, the infrared/optical/ultraviolet luminosity of the supernova approximately followed an exponential decay with a half-life of 77 days, the half-life of radioactive $^{56}$Co. $^{56}$Co is expected to be synthesized as $^{56}$Ni, which decays with a half-life of 6 days to $^{56}$Co, which in turn decays to $^{56}$Fe. The dominant isotope of Fe in the universe is $^{56}$Fe so the abundance of the isotope formed in a supernova is of considerable interest. For SN 1987A, the luminosity of the supernova implied that 0.07 $M_\odot$ of $^{56}$Ni was initially ejected (McCray, 1993). The presence of $^{56}$Co in SN 1987A was confirmed directly by the detection of $^{56}$Co γ-ray lines with balloon and satellite experiments (*e.g.*, McCray, 1993). The γ-ray lines, as well as continuum X-rays from downscattered γ-rays, emerged more rapidly than had been predicted in one-dimensional models. This can be attributed to Rayleigh-Taylor instabilities that occur when inner, dense layers are slowed down by outer, lower density layers.

The other radioactive species to be expected have increasing half-lives. After $^{56}$Co, $^{57}$Co, with a half-life of 272 days, is expected to become the dominant power source. Direct evidence for the isotope came with the detection of the gamma-ray line with the Compton Gamma-Ray Observatory (Kurfess *et al.*, 1992; Clayton *et al.*, 1992). A mass of 0.003 $M_\odot$ of $^{57}$Co in

SN 1987A was deduced to be present, which implies that the initial $^{57}Ni/^{56}Ni$ ratio in SN 1987A was about 1.5 times the solar ratio of $^{57}Fe/^{56}Fe$.

The isotope $^{57}Co$ became the dominant power source in SN 1987A at an age of about 1200 days, but was overtaken by $^{44}Ti$ at an age of 1500 days. A mass of $^{44}Ti$ of (1-2) x $10^{-4}$ $M_\odot$ has been deduced both on the basis of the light curve of SN 1987A and from analysis of the late optical spectrum (Chugai *et al.*, 1997; Kozma and Fransson, 1998). $^{44}Ti$ has a half-life of 60 years (Ahmad *et al.*, 1998) and can be expected to dominate the power to the inner ejecta nebula for future decades.

At a distance of 50 kpc, the $^{44}Ti$ $\gamma$-ray line from SN 1987A is not accessible with current experiments. Even the upcoming *INTEGRAL* observatory is not assured of detecting the $\gamma$-ray line. However, the $^{44}Ti$ line is accessible from nearby young Galactic supernova remnants. It has been observed from Cas A (Iyudin *et al.*, 1994), with an age of about 300 years and a distance of about 3 kpc, and from a previously undetected supernova superposed on the Vela supernova remnant (Iyudin *et al.*, 1998). In both cases, the inferred mass is (1-2) x $10^{-4}$ $M_\odot$, as in SN 1987A. This amount is at the high end of expectations for explosive nucleosynthesis in a massive star (Woosley and Weaver, 1995; Thielemann *et al.*, 1996).

Because most supernovae are too distant to detect $\gamma$-ray lines, supernova luminosities are the primary means of estimating amounts of radioactive isotopes. In most supernovae, $^{56}Co$ is thought to be the primary power source for the optical light starting 10's of days after the explosion. In most core collapse supernovae (Types II, Ib, Ic), the amount of $^{56}Fe$ resulting from $^{56}Co$ and $^{56}Ni$ decays is estimated at ~ 0.1 $M_\odot$, although some supernovae are inferred to have very little, *e.g.*, 0.002 $M_\odot$ of $^{56}Co$ in SN 1997D, (Turatto *et al.*, 1998). Most Type Ia supernovae are thought to eject about 0.2-0.4 $M_\odot$ of $^{56}Ni$. However, SN 1991T was unusually bright and SN 1991bg was unusually faint, and Arnett (1996) estimates that these supernovae synthesized 0.7 $M_\odot$ and 0.075 $M_\odot$ of $^{56}Ni$, respectively. Type Ia supernovae appear to produce more $^{56}Fe$ than do core collapse supernovae and, given that both types occur at a comparable rate in our Galaxy, the Type Ia's probably dominate the production of Fe. However, they have smaller ejected masses of heavy elements and core collapse supernovae probably are the dominant contributors to most heavy elements.

## 3. ABUNDANCES FROM SUPERNOVA SPECTRA

The lines observed in supernova spectra clearly contain information about the element abundances. Near maximum light, most of the supernova mass is optically thick at optical wavelengths and only the outer layers of the star can be observed. In the case of Type II supernovae, the outer layers are H rich and the processed gas is not visible. For Type I supernovae, processed material is visible, although only high velocity layers are seen. Studies of these photospheric lines have successfully identified most of the lines, but reliable estimates of element abundances have not been forthcoming. The problem is that the gas is far from LTE (local thermodynamic equilibrium) and the treatment of the radiative transfer taking into account the many possible line transitions is a complicated problem. In many cases, there are uncertainties in the needed atomic physics data.

The prospects for abundance determinations are better at late (approximately $\geq$ 100 days) times when the supernova becomes optically thin to continuum emission and it is possible to observe emission from the deep layers. Late spectra of SN 1987A have been extensively discussed (McCray, 1993; Wang *et al.*, 1996; Chugai *et al.*, 1997; Kozma and Fransson, 1998) and here I deal with some of the points made by Chugai *et al.* (1997). The spectrum in 1995, at an age of 8 years, was determined by power input from radioactive $^{44}$Ti, which is expected to be mixed in with Fe-rich gas. Decays of $^{44}$Ti produce both $\gamma$-rays and positrons; in order to explain the strengths of Fe lines, most of the positrons must be retained within the Fe rich gas by the magnetic field. The $\gamma$-rays can pass through the supernova gas and are mostly lost. The cooling of the heavy element rich gas is very strong, and the equilibrium temperature of the gas is at most several 100 K; the exact conditions depend on which composition region is under consideration. At these low temperatures, most of the cooling occurs in thermal emission at infrared wavelengths. The Fe-rich gas should primarily radiate in the [Fe II] 26$\mu$m line. Unfortunately, this spectral region is not very accessible, and even *ISO* apparently did not have the sensitivity necessary to detect the line. The ultraviolet/optical lines of Fe result from nonthermal excitation and carry about 10% of the power from the Fe rich region.

The regions outside of the Fe rich parts are illuminated by $\gamma$-rays coming from the Fe rich regions. A determination of the heavy element abundances depends on an understanding of the physical conditions in the enriched gas and on the process of nonthermal excitation of the optical lines. One problem is that the emission from some elements is from the H rich gas as well as from the heavy element gas (Kozma and Fransson, 1998). The two regions can be distinguished to some extent in the line profiles because they have different velocities. Chugai *et al.* (1997) estimated from the late spectra of SN 1987A that 1.5 $M_\odot$ of O was ejected in the explosion, with an uncertainty related to the amount of $^{44}$Ti. Kozma and Fransson (1998) estimated 1.9 $M_\odot$ of O, with an uncertainty of a factor of 2; the theoretical expectation is near the upper end of this range. Kozma and Fransson also found an abundance of C in the He zone that was below theoretical expectations.

## 4. EMISSION FROM SHOCK INTERACTIONS

If a supernova occurred in a vacuum, it would cool and become unobservable. However, supernova ejecta collide with both mass lost before the explosion and the interstellar medium. The interaction gives rise to shock waves which heat the gas and the radiation from the gas is potentially subject to spectroscopy and abundance analysis. In the case of SN 1987A, the radio and X-ray fluxes have been increasing since 1990 and part of the bright surrounding ring has recently started to increase in flux (Sonneborn *et al.*, 1998).

However, it is only the high velocity, H rich layers that are becoming visible by the interaction. It is on a longer timescale (100's of years) that one can expect the inner, heavy element layers to become visible in the interaction. To observe this sort of event, it is necessary to turn to the young Galactic supernova remnants.

A particularly interesting source in this regard is Cas A, the remnant of a supernova that occurred in the late 1600's. There is optical emission from fast moving knots that appear to be pieces of the core of a massive star. The emission is probably from shock waves that are driven into the knots as they interact with their surroundings, although the details of the emission process are not known. Chevalier and Kirshner (1979) found that the knots showed different types of emission that clearly depended on their abundance properties, not atomic physics. One knot showed only emission of O (OI, OII, and OIII), but another knot showed emission from S, Ar, and Ca, in addition to O. The three additional elements are all products of O burning, so there was a signature of different amounts of nuclear processing in the two knots. Although interesting, the emission from these knots is representative of only a small fraction of a $M_{\odot}$ of material; most of the mass of ejecta does not radiate at optical wavelengths.

A better chance to observe the ejecta is presented at X-ray wavelengths. The optical knot emission in Cas A is probably just from denser parts of the ejecta. Most of the ejecta are expected to be heated to a high temperature (1-10 keV) and do not cool rapidly. The analysis of X-ray line emission from supernova remnants is another source of abundance information. In the case of Cas A, Borkowski *et al.* (1996) performed one of the best studies of the X-ray emission because they included the fact that the supernova appears to be colliding with a shell from presupernova mass loss. The X-ray emission depends on the hydrodynamic model because the gas is not expected to be in ionization equilibrium and the ionization histories of individual elements of gas must be followed. Uncertainties in the evolution because of uncertainties in the supernova structure and in the structure of the surrounding medium make abundance estimates unreliable.

Another approach to the problem is to undertake high resolution spectroscopy of a remnant so that the line ratios for one element can reveal both the temperature and the product of density and time (the "ionization time"). If several elements can be observed, there is the chance to determine reliable abundance ratios. So far, this has been feasible only for a very high surface brightness knot in the remnant Puppis A (Canizares and Winkler, 1981).

## 5. CONCLUSIONS AND FUTURE PROSPECTS

The subject of supernova abundance determinations is still not well developed. Abundances derived directly from radioactivity are probably the most reliable, especially for SN 1987A. The number of observable radionuclides is small, but the radioactive decays serve to illuminate other heavy element gas that can be observed in the late time spectra of supernovae. The heavy elements also become observable when the gas collides with the surrounding medium.

The future prospects for abundance determination work are promising. The *Compton Gamma-Ray Observatory* has been able to detect γ-ray lines from only a few sources. The situation will be improved with the launch of the *INTEGRAL* mission and other future missions. Supernova spectra at late times are of special interest because the ejecta become optically thin and it is possible to observe the inner heavy element layers. The supernovae are faint at these times, but will be accessible to the new generation of telescopes with approximately ≥ 8 m mirrors. The situation regarding spectral computations and the needed atomic physics is also improving. The new generation of X-ray missions (*CHANDRA*, *XMM*, and *ASTRO-E*) will be able to carry out spatially resolved spectroscopy of supernova remnants. The *CHANDRA* observatory has Cas A as one of its initial targets. All of these studies will be providing tests of the well-developed theory of stellar nucleosynthesis.

## ACKNOWLEDGMENT

This work was supported in part by NASA grant NAG5-8232.

## REFERENCES

Ahmad, I., Bonino, G., Castagnoli, G.C., Fischer, S.M., Kutschera, W. and Paul, M.: 1998, "Three-laboratory measurement of the $^{44}$Ti half-life", *Phys. Rev. Lett.* **80**, 2550-2553.

Arnett, D.: 1996, *Supernovae and Nucleosynthesis, An Investigation of the History of Matter, from the Big Bang to the Present*, Princeton University Press, Princeton, NJ, 598 pp.

Borkowski, K., Szymkowiak, A.E., Blondin, J.M. and Sarazin, C.L.: 1996, "A circumstellar shell model for the Cassiopeia A supernova remnant", *Ap. J.* **466**, 866-870.

Burbidge, E.M., Burbidge, G., Fowler, W.A. and Hoyle, F.: 1957, "Synthesis of the elements in stars", *Rev. Mod. Phys.* **29**, 547-650.

Canizares, C.R. and Winkler, P.F.: 1981, "Evidence for elemental enrichment of Puppis A by a type II supernova", *Ap. J.* **246**, L33-L36.

Chevalier, R.A. and Kirshner, R.P.: 1979, "Abundance inhomogeneities in the Cassiopeia A supernova remnant", *Ap. J.* **233**, 154-162.

Chugai, N.N., Chevalier, R.A., Kirshner, R.P. and Challis, P.M.: 1997, "Hubble space telescope spectrum of SN 1987A at an age of 8 years: Radioactive luminescence of cool gas", *Ap. J.* **483**, 925-940.

Clayton, D.D., L.-S., Leising, M.D., Johnson, W.N. and Kurfess, J.D.: 1992, "The $^{57}$Co abundance in SN 1987A", *Ap. J.* **399**, L141-L144.

Iyudin, A.F., Diehl, R., Bloemen, H., Hermsen, W., Lichti, G.G., Morris, D., Ryan, J., Schönfelder, V., Steinle, H., Varendorff, M., de Vries, C. and Winkler, C.: 1994, "COMPTEL observations of $^{44}$Ti gamma-ray line emission from Cas A", *Astro. Astrophys.* **284**, L1-L4.

Iyudin, A.F., Schönfelder, V., Bennett, K., Bloemen, H., Diehl, R., Hermsen, W., Lichti, G.G., van der Meulen, R.D., Ryan, J. and Winkler, C.: 1998, "Emission from $^{44}$Ti associated with a previously unknown galactic supernova", *Nature* **396**, 142-144.

Kozma, C. and Fransson, C.: 1998, "Late spectral evolution of SN 1987A. II. Line emission", *Ap. J.* **497**, 431-457.

Kurfess, J.D., Johnson, W.N., Kinzer, R.L., Kroeger, R.A., Strickman, M.S., Grove, J.E., Leising, M.D. and Clayton, D.D.: 1992, "Oriented scintillation spectrometer experiment observations of $^{57}$Co in SN 1987A", *Ap. J.* **399**, L137-L140.

McCray, R.: 1993, "Supernova 1987A revisited", *Ann. Revs. Astr. Ap.* **31**, 175-216.

Sonneborn, G., Pun, C.S.J., Kimble, R.A., Gull, T.R., Lundqvist, P., McCray, R., Plait, P., Boggess, A., Boweres, C.W., Danks, A.C., Grady, J., Heap, S.R., Kraemer, S., Lindler, D., Loiacono, J., Maran, S.P., Moos, H.W. and Woodgate, B.E.: 1998, "Spatially resolved STIS spectroscopy of SN 1987A: Evidence for shock interaction with circumstellar gas", *Ap. J.* **492**, L139-L142.

Thielemann, F.-K., Nomoto, K. and Hashimoto, M.: 1996, "Core-collapse supernovae and their ejecta", *Ap. J.* **460**, 408-436.

Timmes, F.X., Woosley, S.E. and Weaver, T.A.: 1995, "Galactic chemical evolution: Hydrogen through zinc", *Ap. J. Suppl.* **98**, 617-658.

Turatto, M., Mazzali, P.A., Young, T.R., Nomoto, K., Iwamoto, K., Benetti, S., Cappellaro, E., Danziger, I.J., Demello, D.F., Phillips, M.M., Suntzeff, N.B., Clocchiatti, A., Piemonte, A., Leibundgut, B., Covarrubias, R., Maza, J. and Sollerman, J.: 1998, "The peculiar type II supernova 1997D: A case for a very low $^{56}$Ni mass", *Ap. J.* **498**, L129-L133.

Wang, L., Wheeler, J.C., Kirshner, R.P., Challis, P.M., Filippenko, A.V., Fransson, C., Panagia, N., Phillips, M.M. and Suntzeff, N.: 1996, "Hubble space telescope spectroscopic observations of the ejecta of SN 1987A at 2000 days", *Ap. J.* **466**, 998-1010.

Woosley, S.E. and Weaver, T.A.: 1995, "The evolution and explosion of massive stars. II. Explosive hydrodynamics and nucleosynthesis", *Ap. J. Suppl.* **101**, 181-235.

# The Birth of Planetary Systems Directly from Supernovae

Wilbur K. Brown
*5179 Eastshore Drive, Lake Almanor, CA 96137*
*wkbrown@inreach.com*

**Abstract**: The hypothesis explored in this article focuses on the outward-racing accretion shock produced in the course of a Type II supernova explosion. The shock arises in a massive giant star after the core collapse and hydrodynamic bounce. The piling up of material at the shock forms a shell of material which is subjected to an extremely high flux of neutrons. It is therefore rich in elements heavier than Iron – produced by r-process nucleosynthesis. The relatively small amount of material falling back from the volume inside the shell produces a planetary nebula around the central neutron star. The shell itself is ejected, and soon breaks up into gravitationally-bound fragments. Each fragment subsequently proceeds outward, sweeping up additional mass, to form a nebula surrounding a dominant, but relatively small central mass. It is shown that this scenario illustrates how stellar nebulae can be produced directly from Type II supernovae to form all known types of solar systems.

## 1. INTRODUCTION

In this article, I examine the possibility that solar systems of all types can arise directly from the shell of material at the outward-racing accretion shock wave in a Type II supernova explosion. Mass falling back toward the center is assumed to result in planets orbiting a central neutron star. The shell of mass propelled outward, rich in elements heavier that Fe, is assumed to fragment, whereupon each spinning fragment later produces the stellar nebula and the central star of a new solar system.

Although individual references will be indicated in the text, it is not too much to say that Arnett's excellent book (1996), "Supernovae and Nucleosynthesis", is the primary reference for the present article, and has been a great help in the writing of it.

*Origin of Elements in the Solar System: Implications of Post-1957 Observations.*
Edited by O. Manuel, Kluwer Academic/Plenum Publishers, 2000. 

## 2. THE MANUEL-WOLSZCZAN PLANETARY SYSTEM

The idea that a planetary system might be formed at what had previously been the center of a supernova was first advanced by Manuel and Sabu (1975) and Manuel (1981). In 1996, Brush, a highly respected historian of science called the idea "*bizarre*" (1996, p. 125) – and he was right! However, nature, as it often does, found a way to confound us: Wolszczan and Frail (1992) discovered the first extrasolar planetary system – around the pulsar PSR B1257 + 12. This was later confirmed by Wolszczan (1994), who found the system to comprise three roughly earth-mass planets orbiting the pulsar in a disk about one-half A.U. in radius. The parameters of the system are shown in Table 1.

*Table 1.* The planetary system of pulsar PSR B1257+12 (Wolszczan, 1994)

| Planet | A | B | C |
|---|---|---|---|
| $M_p/M_{earth}$ | 0.015/sin i | 3.4/sin i | 2.8/sin i |
| $r_2$ (AU) | 0.19 | 0.36 | 0.47 |
| T (days) | 25.43 | 66.54 | 98.22 |



I note that all of the three planets appear to be in resonance with each other. B,C (3:2); A,B (8:3); A,C (4:1). Because the millisecond pulsar is quite old (perhaps $10^9$ years), there has been plenty of time for the resonances to occur. It seems that the orbits of the three planets must be essentially co-planar.

Lin, Woosley, and Bodenheimer (1991) proposed that a small fraction of mass falls back to form planets after the supernova explosion, and I endorse this mechanism in the present article.

I have calculated a planetary disk mass distribution, involving the conservation of mass and specific angular momentum (Manuel *et al.*, 1998) (angular momentum per unit mass) and assuming uniform density in the material in the volume behind the moving shock wave (Arnett, 1996, p. 419). The spherical mass is mapped into a planetary disk. The proposed mass distribution both with and without rotational shear in the fall-back mass is shown in Figure 1.

If I assume that the edge of the planetary disk is at $r_{2m} = 0.6\ A.U.$, as I have in (Manuel *et al.*, 1998), then the three planets would be plotted at $r_2 / r_{2m} = 0.32,\ 0.60$ and $0.78$. Because the ordinate in Figure 1, $m(r_2)$, has dimensions of mass per unit radius, and only the mass of the three planets has been determined (not the mass per unit radius), plotting the ordinates of the planets would only be a guess, at best.

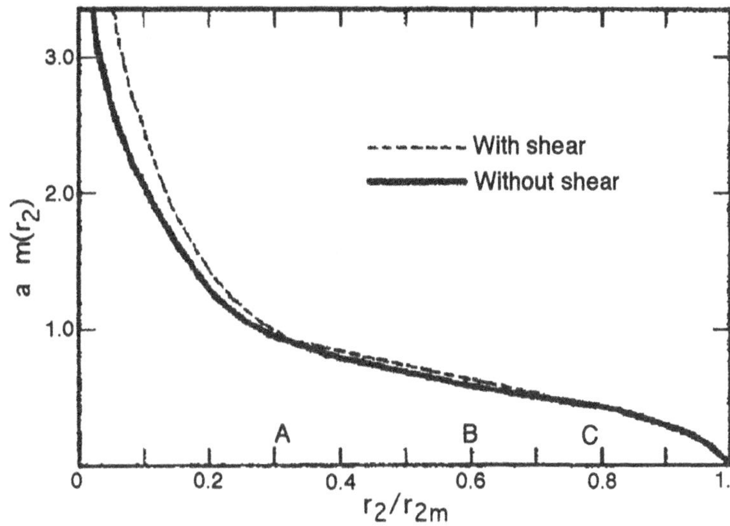


*Figure 1.* The mass distribution in the planetary disk of the Manuel-Wolszczan system, with and without shear. The two curves are quite similar. The ordinate, $m(r_2)$, has dimensions of mass per unit radius. The above graph is from Manuel *et al.* (1998).

## 3. ORDINARY SOLAR SYSTEMS

The category "ordinary" solar systems now includes recently-discovered extrasolar systems as well as our own. I propose that such solar systems arise from the expelled, fragmenting shell that consists of mass previously piled up at the outward-racing accretion shock wave formed in the course of a Type II supernova explosion. The material is subjected to an extremely high flux of neutrons giving rise to the heavy element inventory (heavier than Fe) through r-process nucleosynthesis. The expanding shell soon breaks up into gravitationally-bound fragments (Brown and Gritzo, 1986), each of which proceeds to produce a small central mass surrounded by a proto-planetary disk (Brown and Gritzo, 1986; Brown, 1992). I have examined the fragmentation process elsewhere (Brown, 1989; Brown and Wohlety, 1995). The calculated mass distribution of the fragments describes the Initial Mass Function of stars quite well (Brown, 1989, and see Appendix A and B).

Just as in the case of the Manuel-Wolszczan planetary system, I have calculated the mass distribution in ordinary systems conserving mass and specific angular momentum in each particular fragment. Because there are several parameters (the angle of ejection from the supernova axis of rotation, the ratio of bulk to shear rotation (see below), and the size and shape of the fragment) we have written a brief computer program (Brown and Gritzo, 1986), "SOLSYS", to calculate the resulting mass distribution in the nebula.

It is probable that there exist opposing flows in the disk of any particular nascent system: one produced by the bulk rotation of the material in the shell, and a second arising from Couette shear flow within the shell at the moment of fragmentation (Brown and Gritzo, 1986; Brown 1992). A large variety of stellar nebulae can be produced by these opposing flows (Brown, 1987). Included in these possible configurations is that of our own system, shown in Figure 2 (Brown and Gritzo, 1986, reprinted with kind permission from Kluwer Academic Publishers).

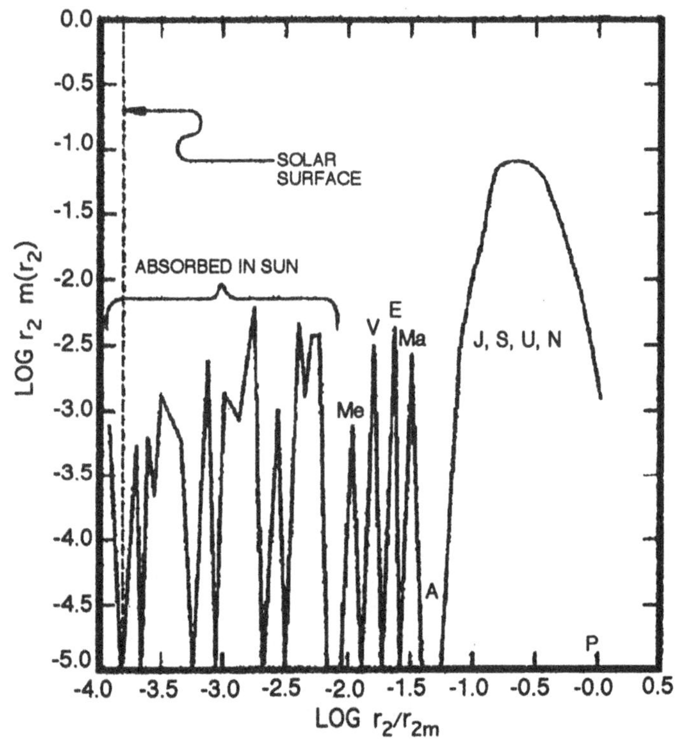


*Figure 2.* An example of the post-T Tauri wind solar nebula mass distributions calculated from the supernova shell fragmentation model of solar system formation. This particular pattern closely resembles our Solar System, and was found over a wide range of model parameters. The ordinate $r_2m(r_2)$, is the mass distribution at $r_2$ in terms of mass per natural logarithmic interval in radius, calculated using the computer program "SOLSYS". The location of the surface of our Sun is shown at the left.

The present model does not attempt to treat the actual mechanisms of planet formation. This subject is being ably investigated by others, *e.g.* Goldreich and Ward (1973), Bodenheimer and Pollack (1986), and Wetherill

(1980, 1989). In fact, an entire volume was recently devoted to this subject (Weaver and Danly, 1989).

Several auspicious circumstances follow from the opposing flow scenario. The colliding flows preferentially eject the light elements (*i.e.*, H and He) from the vicinity of disk center and form the T Tauri wind. The heavier elements tend to remain in the disk center, and form small, inner, rocky planets. The angular momentum of virtually all of the colliding light particles is nullified, and this matter eventually finds its way to the still relatively small, slowly-rotating central mass without adding angular momentum to it. In the outer disk, beyond the opposing flows, the light elements are retained, and there, gas giant planets, containing almost all of the angular momentum of the system, will form. Subsequently, the new star, gathering mass as it moves away from the supernova site, accretes its full mass and begins normal fusion burning. (See the widely-distributed Hubble Space Telescope photograph of the Eagle Nebula. It is plain to see that the new stars are emerging *in motion* from the nebula.)

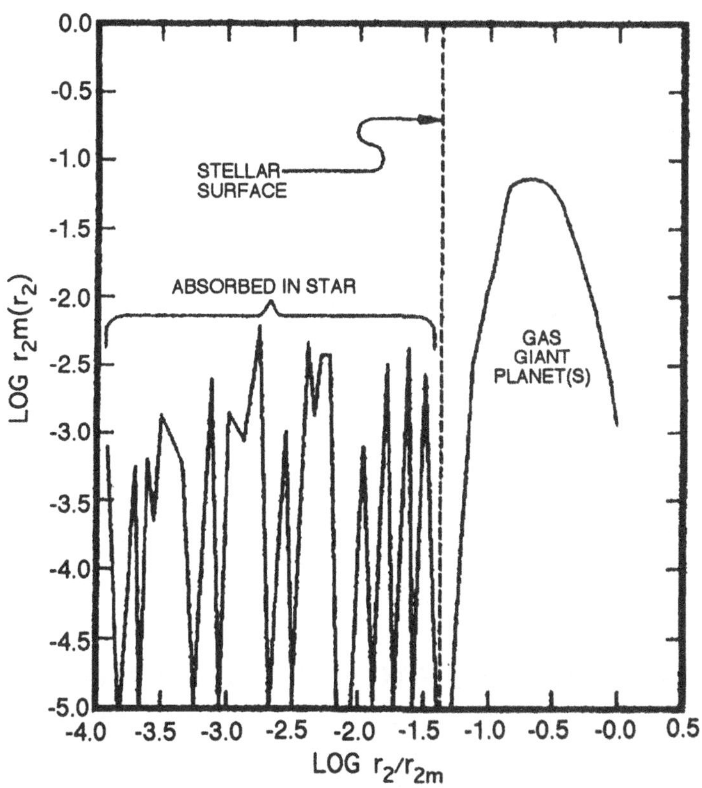


*Figure 3.* An example of the mass distribution in one of the recently discovered systems with one or more gas giant planets forming quite near the central star (calculated using "SOLSYS").

The recently observed extrasolar planetary systems can also be described by the present model (Brown, 1987). If the radius of the edge of the nebular disk is small (dictated by the original angular momentum input and the radial extent of the opposing flows) one or more gas giants may form quite near the central star. An example of this circumstance is shown in Figure 3 (Brown and Gritzo, 1986, reprinted with kind permission from Kluwer Academic Publishers).

## 4. CONCLUSION

It is probable that planetary systems are produced at or in the immediate neighborhood of a Type II supernova explosion. In the present article I have focused attention on the accretion shock wave that races out from the hydrodynamic core bounce. The mass piled up against this shock wave forms a shell. A high density flux of neutrons, produces the heavy elements via r-process nucleosynthesis, in the shell.

The relatively small amount of material falling back from the shock wave and the enclosed volume is hypothesized to result in a planetary system around the central neutron star – as in the Manuel-Wolszczan system.

The shell, ejected outward, soon breaks up. Each spinning fragment proceeds to form a nebula around a relatively small, slowly-rotating central star. Calculations of the mass distribution in the nascent stellar nebulae describe that of our own solar system, as well as those recently discovered, quite well.

## ACKNOWLEDGMENTS

I thank K. R. Alrick for checking references for me at the Los Alamos National Laboratory library.

## APPENDIX A

## SEQUENTIAL FRAGMENTATION THEORY

The sequential fragmentation theory has been published elsewhere (Brown, 1989; Brown and Wohlety, 1995), and here I will only outline the results, (including recent improvements). The theory begins with the solution of the integral equation:

$$n(m) = c\int_m^\infty n(m')f(m' \to m)dm' , \tag{A1}$$

where $m'(>m)$ is the mass of the fragmenting particle, and $f(m' \to m)$ represents the mass distribution in $m$ resulting from a single fragmentation event of $m'$. $n(m)$ is the number of particles per unit mass and $c$ is a constant $\equiv m_1^{-1}$.

I chose

$$f(m' \to m) = \left(\frac{m}{m_1}\right)^\gamma , \tag{A2}$$

where $-1 < \gamma \le 0$, signifying that lighter particles are preferentially produced. We have shown (Brown and Wohlety, 1995), that Equation (A2) implies a branching tree of cracks that can be described by a fractal, where the fractal dimension, $D_f = -3\gamma$. When Equation (A2) is inserted into Equation (A1), there results the integral equation

$$n(m) = \left(\frac{m}{m_1}\right)^\gamma \int_m^\infty n(m')d\left(\frac{m'}{m_1}\right) , \tag{A3}$$

for which the solution is the Weibull distribution in particle number per unit mass:

$$n(m) = \frac{N_T}{m_1}\left(\frac{m}{m_1}\right)^\gamma \exp\left[-\frac{(m/m_1)^{\gamma+1}}{\gamma+1}\right]. \tag{A4}$$

Equation (A4) has been normalized such that

$$N_T = \int_0^\infty n(m)dm , \tag{A5}$$

so that $N_T$ is the total number of particles in the distribution.

Several authors have derived the Weibull distribution as discussed in Brown and Wohlety (1995), but one derivation has heretofore been overlooked (Stauffer, 1979).

The ramifications of Equation (A4) have been discussed at length in Brown (1989), Brown and Wohlety (1995), Stauffer (1979) and Wohlety *et al.* (1989) and references therein.

Setting the derivative of Equation (A4) equal to zero, we find that

$$\frac{m_p}{m_1} = \gamma^{\frac{1}{\gamma+1}}, \text{and} \therefore \gamma > 0 , \tag{A6}$$

where $m_p$ is the peak value (= "most probable", or "mode"). For $\gamma > 0$, larger particles would be preferentially produced, and this may describe agglomeration rather than fragmentation. Inserting Equation (A6) into Equation (A4) yields.

$$n(m) = \frac{N_T}{m_p}\gamma\left(\frac{m}{m_p}\right)^{\gamma} \exp\left[-\frac{\gamma}{\gamma+1}\left(\frac{m}{m_p}\right)^{\gamma+1}\right], (\gamma > 0). \tag{A7}$$

The mass distribution corresponding to Equation (A4) is $mn(m)$, and has dimensions of the mass of fragments per unit mass:

$$mn(m) = N_T\left(\frac{m}{m_1}\right)^{\gamma+1} \exp\left[-\frac{(m/m_1)^{\gamma+1}}{\gamma+1}\right]. \tag{A8}$$

For this distribution

$$\frac{m_p}{m_1} = (\gamma+1)^{\frac{1}{\gamma+1}}, \tag{A9}$$

and

$$mn(m) = N_T(\gamma+1)\left(\frac{m}{m_p}\right)^{\gamma+1} \exp\left[-\left(\frac{m}{m_p}\right)^{\gamma+1}\right], \ (\gamma > -1) . \tag{A10}$$

It should be mentioned that for $\gamma = 1$, Equation (A10) yields the same shape as the Maxwell-Boltzmann distribution (but, of course, each describes quite different phenomena).

We may find the average mass of the distribution from

$$\overline{m} = \frac{M_T}{N_T} = \frac{\int_0^\infty mn(m)dm}{\int_0^\infty n(m)dm}, \tag{A11}$$

where $M_T$ is the total mass of fragments in the distribution, and

$$\frac{\overline{m}}{m_p} = \Gamma\left(\frac{\gamma+2}{\gamma+1}\right). \tag{A12}$$

Here, $\Gamma(\alpha)$ is the complete gamma function:

$$\Gamma(\alpha) = \int_0^\infty t^{\alpha-1} e^{-t} dt \; . \tag{A13}$$

The cumulative form of Equation (A10) is

$$M(>m) = \int_m^\infty mn(m)dm \; , \tag{A14}$$

and inserting Equation (A10) into Equation (A14), we find that

$$\frac{M(>m)}{M_T} = \frac{\Gamma\left(\frac{\gamma+2}{\gamma+1}, x\right)}{\Gamma\left(\frac{\gamma+2}{\gamma+1}\right)} \; . \tag{A15}$$

Here

$$x \equiv \frac{(m/m_1)^{\gamma+1}}{\gamma+1} = \left(\frac{m}{m_p}\right)^{\gamma+1}, \tag{A16}$$

and

$$\Gamma(\alpha, x) \equiv \int_x^\infty t^{\alpha-1} e^{-t} dt \; , \tag{A17}$$

*i.e.*, the complementary incomplete gamma function.

Another way of setting forth the mass distribution is in dimensions of mass of particles per unit ln m interval:

$$m^2 n(m) = N_T m_1 \left(\frac{m}{m_1}\right)^{\gamma+2} \exp\left[-\frac{\left(m/m_1\right)^{\gamma+1}}{\gamma+1}\right] \tag{A18}$$

In this case

$$\frac{m_p}{m_1} = (\gamma+2)^{\frac{1}{\gamma+1}}, \tag{A19}$$

so that

$$m^2 n(m) = N_T m_p (\gamma+2) \left(\frac{m}{m_p}\right)^{\gamma+2} \exp\left[-\left(\frac{\gamma+2}{\gamma+1}\right)\left(\frac{m}{m_p}\right)^{\gamma+1}\right]. \tag{A20}$$

It has been shown that the shape of the curve denoted by Equation (A20) is quite similar to the lognormal distribution (Brown and Wohlety, 1995).

I find that

$$\frac{\overline{m}}{m_p} = \left(\frac{\gamma+1}{\gamma+2}\right)^{\frac{1}{\gamma+1}} \frac{\Gamma\left(\frac{\gamma+3}{\gamma+1}\right)}{\Gamma\left(\frac{\gamma+2}{\gamma+1}\right)}. \tag{A21}$$

For the cumulative form of Equation (A20), Equation (A15) still holds, but now

$$x = \frac{\gamma+2}{\gamma+1}\left(\frac{m}{m_p}\right)^{\gamma+1}. \tag{A22}$$

Another useful form of the Weibull distribution (Brown, 1989; Brown and Wohlety, 1995), is

$$mn(m) = \frac{M_T}{m_2}\frac{k}{3}\left(\frac{m}{m_2}\right)^{\frac{k}{3}-1} \exp\left[-\left(\frac{m}{m_2}\right)^{\frac{k}{3}}\right], \tag{A23}$$

where $m_2$ is a mass related to the average and peak masses.
Finally, the mass per unit ln m interval for Equation (A23) is given by

$$m^2 n(m) = M_T \frac{k}{3}\left(\frac{m}{m_2}\right)^{\frac{k}{3}} \exp\left[-\left(\frac{m}{m_2}\right)^{\frac{k}{3}}\right], \tag{A24}$$

where $D_f = 3\left(\frac{k}{3}\right) = k$.

# APPENDIX B

## SEQUENTIAL FRAGMENTATION AS A COSMOGONICAL MECHANISM

My hypothesis is that following the Big Bang, about at the time of recombination (~ 300,000 years) the universe – driven by expansion – separated into what later became superclusters of galaxies, that these supercluster fragments subsequently separated into what later became galactic clusters, and that these cluster fragments, in turn, separated into what eventually became galaxies. Further, the galaxy fragments separated into volumes that collapsed into massive stars that died as supernovae. The ejected supernova shells separated into volumes that collapsed into lower-mass stars surrounded by stellar nebulae – the subject of this article.

I have modeled this top-down cosmogonical scenario using the present model of sequential fragmentation (see Appendix A). As we have shown, the model, in any of its various formulations, is related to a fractal-based Weibull distribution (Brown and Wohlety, 1995).

There has been minimal amount of work done on cosmogonical fragmentation, but three early papers, two by Bird (1968, 1969) and one by Layzer

(1975) are pertinent. I have been told that in the laboratory, under the proper flow conditions, gas does indeed separate into discrete volumes, *i.e.*, the gas becomes fragmented. In the remainder of this appendix, I will refer to the separation of gas into discrete volumes simply as fragmentation.

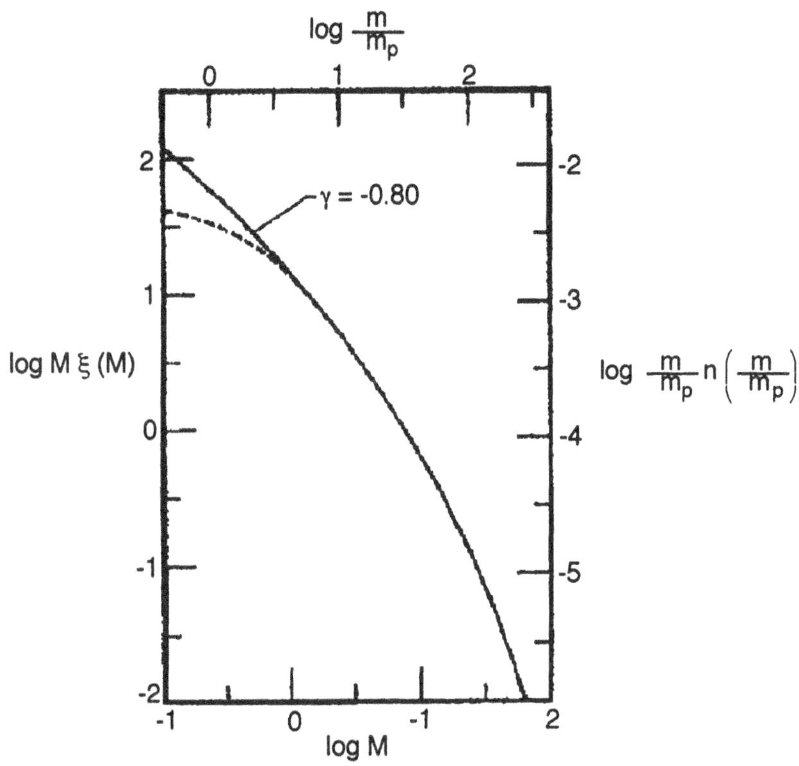


*Figure 4.* A comparison of Miller and Scalo's Initial Mass Functions of stars (Miller and Scalo, 1979), (dashed line) with the Weibull-based sequential fragmentation distribution of fragment masses given by Equation (A10) (solid line) for $\gamma = -0.80$. For stellar masses above $M = 1$ (solar mass units), the two curves are coincident (Brown, 1986a, reprinted with kind permission from Kluwer Academic Publishers).

Regarding fragmentation into solar systems, the Initial Mass Function of stars $\xi(M)$, was given by Miller and Scalo (1979). Their mass distribution, $M\xi(M)$, is compared to the fractal-based Weibull distribution (Equation (A10)) in Figure 4. The comparison yields $\gamma = -0.80$. The corresponding fractal dimension is $D_f = -3\gamma = 2.4$. It can be seen that the agreement is excellent for $M > 1M_{\odot}$. There are two reasons for the discrepancy below $1M_{\odot}$: (1) below ~ $8M_{\odot}$, Type II supernovae cannot occur, and thus stars of $<8M_{\odot}$ do not contribute to the distribution of stars at smaller masses, and (2) the lifetimes of low-mass stars are relatively long – giving an effect opposite to that of the first reason. Both of these factors conflict with the formulation of the sequential fragmentation theory in which it is assumed that fragmentation progresses steadily at all masses.

I have compared Equation (A10) to several of the conditions given by Miller and Scalo (1979), (differing galactic ages, and differing stellar birthrates) and found that with a small adjustment of $\gamma$, the match is as good as that shown in Figure 4.

The agreement between Miller and Scalo's Initial Mass Function and Equation (A10) supports the hypothesis that fragmentation of supernovae to form lower mass stars is a credible cosmogonical mechanism on this scale.

The collapse of a supernova shell fragment to a planetary disk with a small central star can be approximated in a manner identical to the collapse of a cluster fragment to a galaxy (Brown, 1985a).

Interestingly enough, it is possible to fit the mass distribution of galaxies with Equation (A23). To do this, however, I must assume that the mass of each galaxy is proportional to its total luminosity – a risky assumption, owing to the presence of dark matter. I have, nevertheless, made this assumption.

The distribution of galactic total luminosities is well-represented by the empirical Schechter number distribution (Schechter, 1976),

$$n\left(\frac{L}{L^*}\right) = n^*\left(\frac{L}{L^*}\right)^{\alpha} \exp\left[-\frac{L}{L^*}\right] \tag{B1}$$

where $L$ is the total luminosity of a galaxy, and $n\,(L/L^*)$ is the distribution, *i.e.*, the number of galaxies of total luminosity $L/L^*$ per unit luminosity between $L/L^*$ and $L/L^*$ plus $d\,(L/L^*)$. The quantity $n^*$ is a constant with dimensions of number of galaxies per unit luminosity. Strictly speaking, both $L^*$ and $\alpha$ are free parameters, but because I will be comparing only distribution shapes, I will use $L/L^*$ as the independent variable. Thus $\alpha$ is the single free parameter. $\alpha$ dictates the log-log slope of the faint end of the distribution. A rather wide range of $\alpha$ has been found $(-1.6 < \alpha < 0)$, depending on the sample of galaxies studied. (See, *e.g.*, Sandage *et al*, 1979; and Tammann *et al*, 1979).

Having made the assumption that the mass of each galaxy is proportional to its total luminosity, Schechter's number distribution, Equation (B1), is proportional to a mass distribution:

$$\frac{L}{L^*}n\left(\frac{L}{L^*}\right) = n^*\left(\frac{L}{L^*}\right)^{\alpha+1} \exp\left[-\frac{L}{L^*}\right]. \tag{B2}$$

In Figure 5, the mass distribution of Equation (B2) is compared to the fractal-based Weibull mass distribution of Equation (A23).

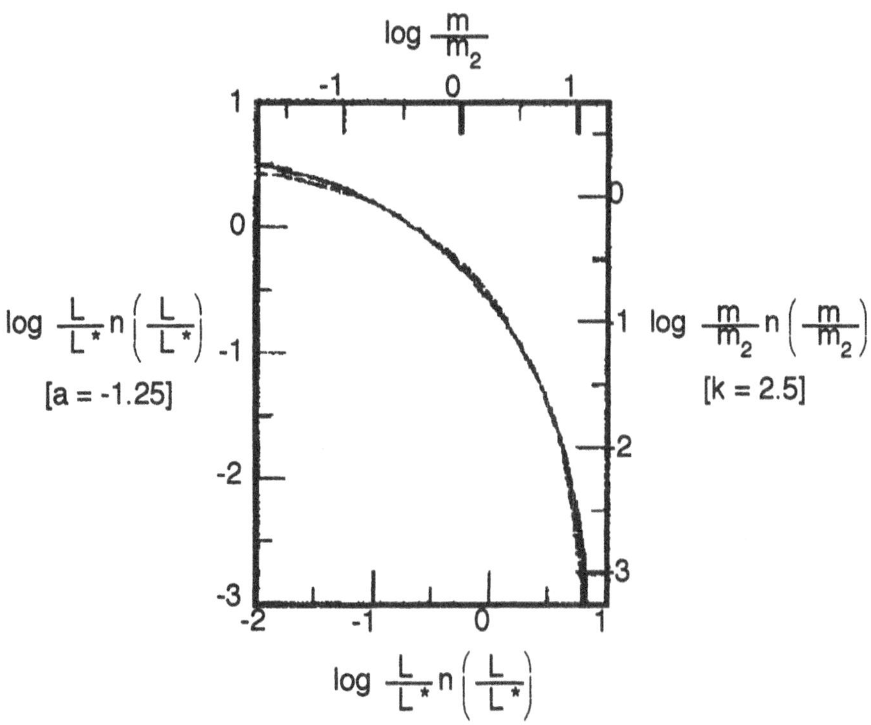


*Figure 5.* A comparison of the Schechter distribution of total galactic luminosities (Schechter, 1976), of Equation (B2) (dashed line) with the Weibull distribution of fragment masses given by Equation (A23) (solid line) for the specific parameter values $\alpha$ = -1.25 and k = 2.5 (Brown, 1986b, reprinted with kind permission from Kluwer Academic Publishers).

It can be seen from Figure 5, that the comparison of Equation (B2) with $\alpha = -1.25$ ( a representative value) is in excellent agreement with Equation (A23) with $k = 2.5$. I have described the collapse of cluster fragments to galaxies (Brown, 1985a), and the resulting galactic rotation curves in spiral galaxies (Brown, 1985b).

I conclude from Figure 5 that sequential fragmentation is a credible cosmogonical mechanism on this scale as well. Further, from Figure 4, $Df = -3\gamma = 2.4$, and from Figure 5, $D_f = k = 2.5$ – the fractal dimensions on both scales are quite comparable.

Elmegreen and Efremov (1998) studied the hierarchical groupings of stars, and obtained a fractal dimension of 2.3.

I have outlined elsewhere my conception of a universe that would be compatible with sequential framentation (Brown, 1994).

If I could but summon the courage, I would claim that Sequential Fragmentation is a valid cosmogonical mechanism on *all* scales, and that we live in a fractal-based universe with a fractal dimension of about two and one-third.

## REFERENCES


Arnett, D.: 1996, "Supernovae and nucleosynthesis, an investigation of the history of matter, from the big bang to the present", Princeton University Press, Princeton, NJ, 598 pp.

Bird, J.F.: 1968, "Cosmogonic fragmentation", *Nature* **217**, 1239-1240.

Bird, J.F.: 1969, "Gravitational instability of spheroidal expansions: A cosmogonic fragmentation mechanism", *Astrophys. Space. Sci.* **3**, 312-329.

Bodenheimer, P. and Pollock, J.B.: 1986, "Calculations of the accretion and evolution of giant planets: The effects of solid cores", *Icarus* **67**, 391-408.

Brown, W.K.: 1985a, "A model of protogalactic cloud collapse", *Astrophys. Space. Sci.* **113**, 143-153.

Brown, W.K.: 1985b, "Approximate rotation curve solutions for the evolution of a viscous protogalactic disk", *Astrophys. Space Sci.* **111**, 139-155.

Brown, W.K.: 1986a, "Comparison of a theory of sequential fragmentation with the initial mass function of stars", *Astrophys. Space. Sci.* **122**, 287-298.

Brown, W.K.: 1986b, "Universal fragmentation", *Astrophys. Space. Sci.* **121**, 351-355.

Brown, W.K.: 1987, "Possible mass distributions in the nebulae of other solar systems", *Earth Moon, and Planets* **37**, 225-239.

Brown, W.K.: 1989, "A theory of sequential fragmentation and its astronomical applications", *J. Astrophys. Astr.* **10**, 89-112.

Brown, W.K.: 1992, "The supernova as a genesis site of solar systems", *Speculat. Sci. Tech.* **15**, 149-160.

Brown, W.K.: 1994, "A thick, rotating universe", *Speculat. Sci. Tech.* **17**, 186-190.

Brown, W.K. and Gritzo, L.A.: 1986, "The supernova fragmentation model of solar system formation", *Astrophysics and Space Science* **123**, 161-181.

Brown, W.K. and Wohlety, K.H.: 1995, "Derivation of the Weibull distribution based on physical principles and its connection to the Rosin-Rammler and lognormal distributions", *J. Appl. Phys.* **78**, 2758-2763.

Brush, S.G.: 1996, *Fruitful Encounters: The Origin of the Solar System and of the Moon from Chamberlain to Apollo*, Cambridge University Press, Cambridge, UK, 354 pp.

Elmegreen, B. and Elfremov, Y.: 1998, "The formation of star clusters", *American Scientist* **86**, 264-273.

Goldreich, P. and Ward, W.R.: 1973, "The formation of planetesimals" *Astrophys. J.* **183**, 1051-1061.

Layzer, D.: 1975, "Galaxy clustering, its description and its interpretation", in *Galaxies and the Universe*, eds., Sandage, A., Sandage, M. and Kristian, J., University of Chicago Press, Chicago, IL, pp. 706-708.

Lin, D.N.C., Woosley, S.E., and Bodenheimer, P.H.: 1991, "Formation of a planet orbiting pulsar 1829-10 from the debris of a supernova explosion", *Nature* **353**, 827-829.

Manuel, O.K.: 1981, "Heterogeneity in meteorite and elemental compositions: Proof of local element synthesis", *Geochemistry International* **18**, 101-125.

Manuel, O.K. and Sabu, D.D.: 1975, "Elemental and isotope inhomogeneities in noble gases: The case for local synthesis of the chemical elements", *Trans. Missouri Acad. Sci.* **9**, 104-122.

Manuel, O.K., Lee, J.T., Ragland, D.E. Macelroy, J.M.D., Bin Li, and Brown, W.K.: 1998, "Origin of the solar system and its elements", *J. Rad. Nucl. Chem.* **238**, 213-225.

Miller, G.E. and Scalo, J.M.: 1979, "The initial mass function and stellar birthrate in the solar neighborhood", *Astrophys. J. Suppl.* **41**, 513-547.

Sandage, A., Tammann, G.A. and Yahil, A.: 1979, "The velocity field of bright, nearby galaxies. I. The variation of mean absolute magnitude with redshift for galaxies in a magnitude – limited sample", *Astrophys. J.* **232**, 352-364.

Schechter, P.L.: 1976, "An analytic expression for the luminosity functions for galaxies", *Astrophys. J.* **203**, 297-306.

Stauffer, H.B.: 1979, "A derivation for the Weibull distribution", *J. Theor. Biol.* **81**, 55-63.

Tammann, G.A., Yahil, A. and Sandage, A.: 1979, "The velocity field of bright, nearby galaxies II. Luminosity functions for various Hubble types and luminosity classes: The peculiar motion of the local group relative to the Virgo Cluster", *Astrophys. J.* **234**, 775-784.

Weaver, H.A. and Danly, L., eds. 1989, *The Formation and Evolution of Planetary Systems*, Cambridge University Press, Cambridge, UK, 344 pp.

Wetherill, G.W.: 1980, "Formation of the terrestrial planets" *Ann. Rev. Astron. Astrophys.* **18**, 77-113.

Wetherill, G.W.: 1989, "The formation of the solar system: consensus, alternatives, and missing factors." In: *The Formation and Evolution of Planetary Systems*, eds., Weaver, H. A. and Danly, L., Cambridge University Press, UK, pp. 1-30.

Wohletz, K.H., Sheridan, M.F., and Brown, W.K.: 1989, "Particle size distributions and the sequential fragmentation/transport theory applied to volcanic ash", *J. Geophys. Res.* **94**, 15703-15721.

Wolszczan, A.: 1994, "Confirmation of earth-mass planets orbiting the millisecond pulsar PSR B1257+12", *Science* **264**, 538-542.

Wolszczan, A. and Frail, D.A.: 1992, "A planetary system around the millisecond pulsar PSR1257+12", *Nature* **355**, 145-147.

# Bipolar Outflows in Stellar Astrophysics

Adam Frank
*Department of Physics* and *Astronomy, University of Rochester, Rochester NY 14625*
*afrank@pas.rochester.edu*

Abstract: Hypersonic bipolar outflows are a ubiquitous phenomena associated with both young and highly evolved stars. Observations of Planetary Nebulae, the nebulae surrounding Luminous Blue Variables such as Eta Carinae, Wolf Rayet bubbles, the circumstellar environment of SN 1987A and Young Stellar Objects all reveal high velocity outflows with a wide range of shapes. In this paper I review the current state of our theoretical understanding of these outflows.

## 1. INTRODUCTION

Bipolar outflows and highly collimated jets are nearly ubiquitous features associated with stellar mass loss. From Young Stellar Objects (YSOs) to Luminous Blue Variables (LBVs) and Planetary Nebulae (PNe) – the stellar cradle to the grave – there exists clear evidence for collimated gaseous flows in the form of narrow high velocity streams or extended bipolar lobes (Figure 1). In YSOs, LBVs and PNe these collimated hypersonic outflows are observed to transport prodigious amounts of energy and momentum from their central stars – enough to constitute a significant fraction of the budgets for the entire system. Thus outflows and jets are likely to play a significant role in the evolution of their parent stars. It is remarkable that such different objects, separated by billions of years of evolution and decades of solar mass, should drive phenomena so similar. The similarity of jets and bipolar outflows across the H-R diagram must tell us something fundamental and quite general about the nature of stellar evolution as well as the interaction of stars with their environments.

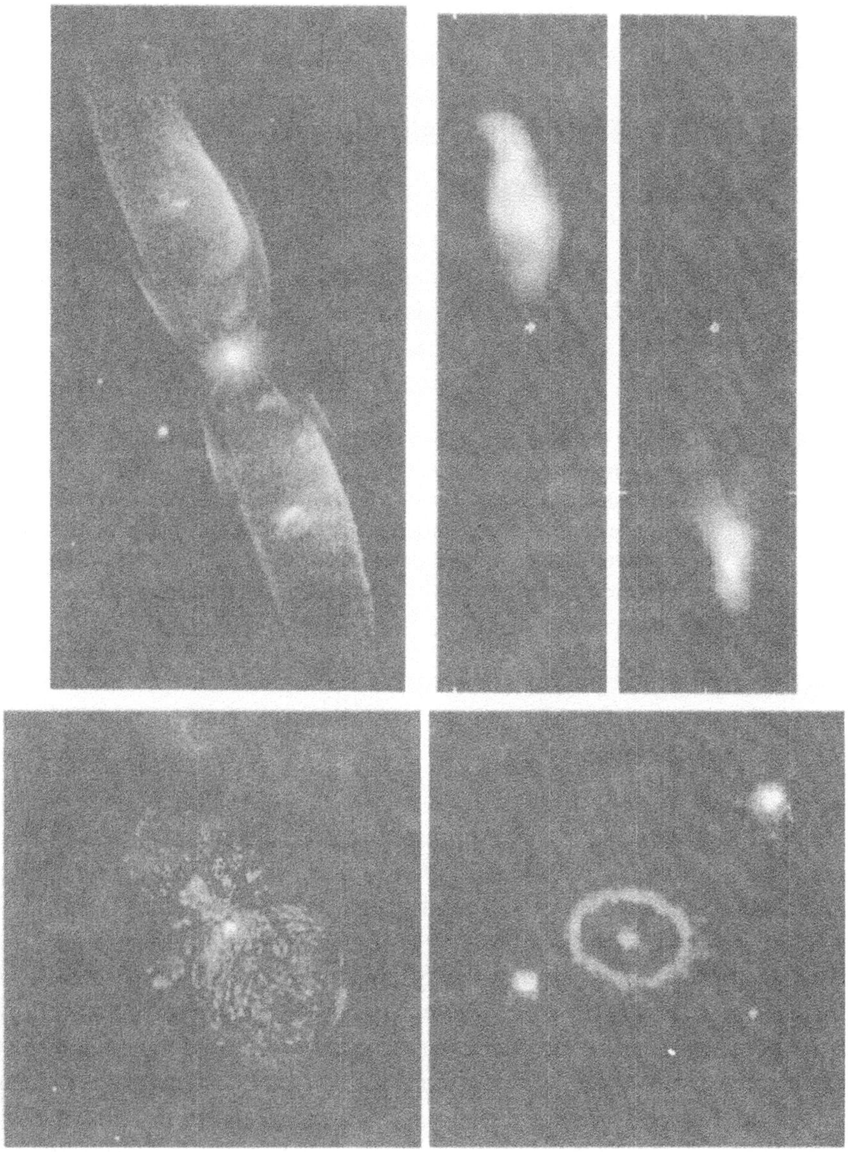

*Figure 1.* Bipolar Outflows across the HR diagram. Four bipolar outflows from different classes of star. LL: the LBV Eta Carinae; UL: PNe M2-9; LR: SN 1987A; UR: YSO 2264G.

In what follows, I review our current understanding of bipolar outflows. There is a unique synergy between theory and observations in this field providing a window into fundamental processes such as shocks, instabilities, ionization dynamics and chemistry. These systems are, however, more than astrophysical laboratories. The tension between an extensive multi-wavelength database and increasingly sophisticated theoretical tools allows bipolar outflows to act as an Archimedean lever yielding insights directly into the birth and death of stars.

## 2. OBSERVATIONAL BACKGROUND

Remarkable progress in understanding bipolar outflows has been achieved in the last two decades. Observationally, the triad of morphologic, kinematic and spectroscopic studies have provided detailed portraits of bipolar outflows both individually and as a class of astrophysical object.

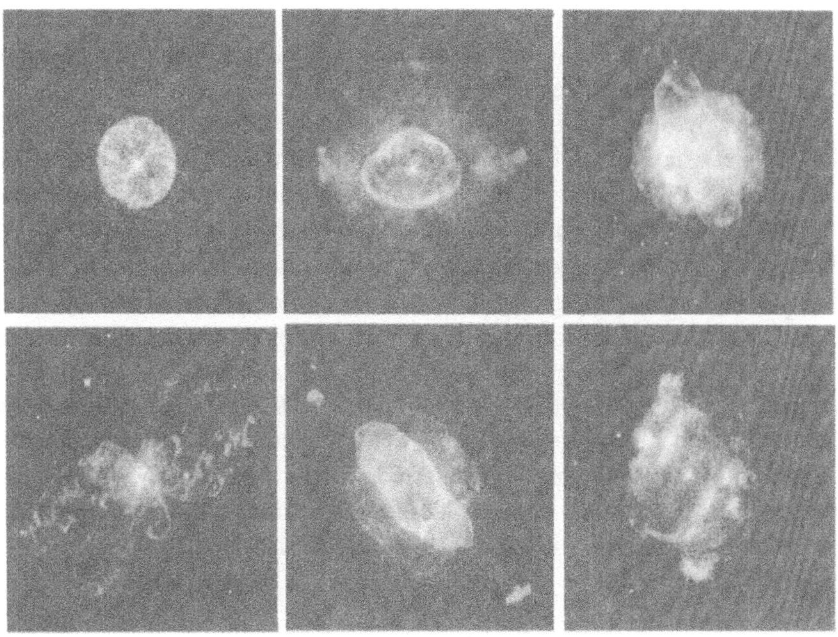

*Figure 2.* Six PNe. From left to right, top to bottom: IC 3568, round; NGC 6826, elliptical with ansae (Balick *et al.*, 1998); NGC 3918, bipolar with jets; Hubble 5, bipolar; NGC 7009, elliptical with jets/ansae; NGC 5307, point-symmetric.

Morphological studies using both the HST and ground based instruments reveal bipolar outflows assuming a wide variety of large scale, L ~ $10^{17}$ cm, *global* configurations Spherical and elliptical outflows are observed in evolved systems such as PNe, Wolf-Rayet nebulae and LBVs (Schwarz *et al.*, 1992; Nota *et al.*, 1995). True bipolar outflows, which appear as two opposing lobes joined at narrow waist (centered on the star) occur in both young and evolved systems. The bipolar lobes exhibit different degrees of collimation ranging from wide figure 8 shapes to long, narrow jets. Figure 2 shows a sample of PNe and underscores the extraordinary diversity of outflow shapes. The appearance of *point symmetry*, where all features are reflected across a central point as in an S (NGC 5307, Figure 2), is a particularly intriguing feature to emerge from recent observations. Morphological studies also reveal an array of small scale features ($L < 10^{16}$ cm) which go by a variety of suggestive labels: ansae, knots, arcs, and cometary globules (Balick *et al.*, 1994; O'Dell and Handron, 1996).

Kinematical studies of bipolar outflows reveal flow patterns that simultaneously possess high degrees of symmetry and complexity (Bryce *et al.*, 1997). The most important point here is the clear presence of globally axisymmetric expansion patterns (Corradi and Schwartz, 1993; Lada and Fich, 1996). The velocities of the bipolar lobes are larger than local sound speeds indicating that the outflow physics will be dominated by processes inherent to hypersonic shock waves.

Spectroscopic studies of line emission provide detailed snapshots of the microphysical state in the outflows. Ionization of the lobes can occur in different ways. Low mass young stars are too cool to produce significant ionizing flux so bipolar outflows associated with these stars can only be collisionally ionized by shocks. Outflows from evolved stars and those associated with high mass young stars are, however, often photoionized by strong stellar UV fluxes. In addition to ions, most classes of bipolar outflow show some evidence for the presence of molecules and molecular chemistry (Latter *et al.*, 1995; Bachiller and Gutierrez, 1997). When spectroscopic studies of ionic and molecular transitions are coupled with shock emission/photoionization models, physical conditions in the outflows (density, temperature, ionization and molecular fractions) can be determined (Hartigan, *et al.*, 1993; Balick *et al.*, 1994; Dopita, 1997).

## 3. THEORY

Rapid progress has also been made in the theory of bipolar outflows. Analytical and numerical studies have recovered many of the observed features of outflows through what we will call the Generalized Wind Blown Bubble (GWBB) paradigm. In this scenario a fast wind from the central source expands into a slowly expanding, or infalling environment. The interaction of the wind and environment produces an expanding bubble bounded by strong shocks.

To understand this paradigm more fully, consider what happens when a stellar wind, initiated at some point in the star's evolution, encounters the circumstellar environment. Initially the wind expands ballistically until enough ambient material is swept up for significant momentum to be exchanged between the two fluids. A triplet of hydrodynamic discontinuities then forms defining an "interaction region" bounded internally (externally) by undisturbed wind (ambient) gas. One can imagine the interaction region as a spherical shock wave layer cake. At the outer boundary is an outward facing shock. It accelerates, compresses and heats ambient material as it propagates. We refer to this feature as the *ambient shock* and denote its position as $R_{as}$. The inner boundary is defined by an inward facing shock which decelerates, compresses and heats the stellar wind. We refer to this feature as the *wind shock*. Its position is $R_{ws}$. A contact discontinuity (CD), $R_{cd}$, separates the shocked wind and shocked ambient material. In the 1-dimensional (1-D) bubble these discontinuities form a sequence in radius: $R_{ws} < R_{cd} < R_{as}$.

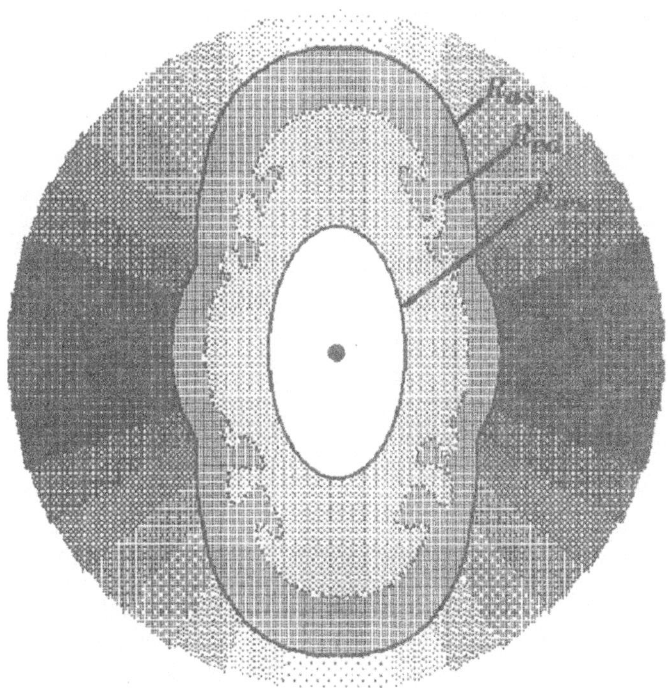


*Figure 3.* Schematic of the Generalized Wind Blown Bubble scenario. A fast wind from the central star expands into a ambient medium with an aspherical density distribution. The shape of the CD indicates the presence of instabilities.

This Wind Blown Bubble (WBB) paradigm can be extended to embrace elliptical and bipolar nebulae by generalizing the model to include an aspherical environmental density distribution with higher density in the equator than at the poles. When an isotropic stellar wind encounters the gaseous torus, *inertial gradients* (as opposed to pressure) allow the ambient shock to expand more rapidly along the poles (Figure 3).

This is the basis of the Generalized Wind Blown Bubble paradigm. The expansion of the fast wind from the central source into a strongly aspherical environment produces shock velocities highest in the direction of lowest density. Thus it is the density gradient in the environment which establishes a preferred axis for the bipolar lobes. This theoretical picture has been remarkably successful at explaining the properties of bipolar outflows in evolved stars. The GWBB paradigm has been applied to almost all forms of bipolar outflows. Models which include both hydrodynamics and microphysics have been able to recover the global morphology, kinematics and ionization patterns in many PNe (Frank and Mellema, 1994) and SN87A (Martin and Arnett, 1995). Strong correspondences also exist between GWBB models and the shapes and kinematics of WR nebulae (Garcia-Seguria and MacLow, 1995), LBVs like Eta Carinia (Frank *et al.*, 1995) and symbiotic stars like R Aquarii (Henney and Dyson, 1992).

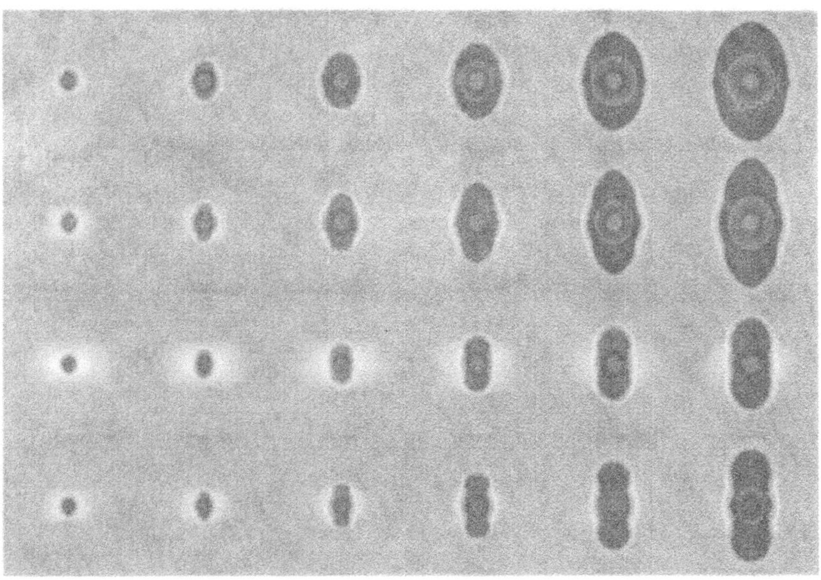

*Figure 4.* Evolution of 4 GWBB planetary nebula models. Shown are grayscale Log density maps at six different times for each model. The pole to equator contrast increases from top to the bottom of the figure. Light scales correspond to high density. The ambient shock and swept-up shell appear as the light grayscales surrounding the bubble. The wind shock appears as the interface between discontinuity in light to dark grayscales. Note the increasing ellipticity of the inner shock as one moves down the figure. These models were computed with a radiation-gasdynamic code and they show that the GWBB model can produce a wide array of bubble shapes. See Frank and Mellema (1994) for details.

Analytical studies of the GWBB model have had limited application due to the complexity of the governing equations (5 coupled non-linear PDEs). Numerical simulations have been necessary in order to explore the full range of behavior possible with the GWBB model. In a series of papers by Frank and Mellema (Frank and Mellema, 1994; Mellema and Frank, 1996 Mellema, 1995) a numerical code was developed which included hydrodynamics, microphysics and radiation transfer from the hot central star. These models tracked cooling and emission behind the ambient shock. Figure 4 shows a series of simulations from Frank and Mellema (1994) which illustrates the dependence of outflow geometry on initial conditions. Figure 4 displays the evolution of density in four PNe simulations. The equator to pole density contrast, $q = \rho(\text{equator})/\rho(\text{pole})$, increases as one moves down the figure. For low values of q, ($q < 2$), the bubble becomes mildly elliptical. Intermediate values of q, ($2 < q < 5$), produce distinct equatorial and polar regions of the ambient shock; *i.e.*, bipolar lobes develop. Larger values of q produce bubbles that are highly collimated. Figure 4 demonstrates the ability of the GWBB model to produce different bubble shapes. The proof of the pudding is in the confrontation with observations and these models are able to recover many of the observed morphological, kinematic and ionization structures seen in real bipolar WBBs.

## 4. CONCLUSIONS

With new images provided by space-based telescopes like the HST and advances in simulation technologies, it is not unreasonable to expect that progress in bipolar outflows studies will someday allow the evolution of fundamental stellar properties to be read off the outflows. The effect of mass loss, rotation, magnetic fields, and binary companions on stellar birth and death are all questions of fundamental importance. The rapid progress being made in bipolar outflow studies will produce insights not only into hydrodynamic and hydromagnetic phenomena in general, but also provide direct links from the physics of these nebulae to the properties of the stellar sources. Taken together these prospects will make the study of bipolar outflows an exciting field for years to come.

## ACKNOWLEDGMENTS

Support for this work was provided at the University of Rochester by NSF grant AST-9702484 and the Laboratory for Laser Energetics.

# REFERENCES


Bachiller, R.: 1996, "Bipolar molecular outflows from young stars and protostars", *Ann. Rev. Astron. Astrophys.*, **34**, 111-154.

Bachiller, R. and Gutierrez, M.P.: 1997, "Shock chemistry in bipolar molecular outflows", in *Herbig-Haro Flows and the Birth of Low Mass Stars*, in *IAU Symposium* No. 182, eds., Reipurth, B. and Bertout, C., pp. 153-162.

Balick, B.: 1987, "The evolution of planetary nebulae. 1. Structures, ionizations, and morphological sequences", *Astron. J.* **94**, 671-678.

Balick, B., Perinotto, M., Maccioni, A., Terzian, Y. and Hajian, A.: 1994, "FLIERs and other microstructures in planetary nebulae, 2", *Ap. J.* **424**, 800-816.

Balick, B., Rugers, M., Terzian, Y. and Chengalur, J.N.: 1993, "Fast, low-ionization emission regions and other microstructures in planetary nebulae", *Ap. J.* **411**, 778-793.

Balick, B., Alexander, J., Hajian, A., Terzian, Y., Perinotto, M. and Patriarchi, P.: 1998, "FLIERs and other microstructures in planetary nebulae. IV. Images of elliptical pns from the Hubble space telescope", *Astron. J.* **116**, 360-371.

Bryce, M., Lopez, J., Holloway, A. and Meaburn, J.: 1997, "A bipolar, knotty outflow with velocities of 500 kilometers per second or above from the engraved hourglass planetary nebula MyCn-18", *Ap. J.* **487**, L161-L164.

Corradi, R.L.M. and Schwarz, H.E.: 1993, "Kinetics of bipolar nebulae", *Astron. Astrophys.* **278**, 247-254.

Dopita, M.: 1997, "What excites FLIERS?", *Ap. J.* **485**, L41-L44.

Frank, A., Balick, B., Icke, V. and Mellema, G.: 1993, "Astrophysical gasdynamics confronts reality – The shaping of planetary nebulae", *Ap. J.* **404**, L25-L27.

Frank, A., Balick, B. and Davidson K.: 1995, "An interacting stellar wind paradigm for Eta carinae", *Ap. J.* **441**, L77-L80.

Frank A., Balick B. and Livio M.: 1996, "A mechanism for the production of jets and ansae in planetary nebulae", *Ap. J.* **471**, L53-L56.

Frank, A. and Mellema, G.: 1994, "From the owl to the eskimo", *Ap. J.* **430**, 800-813.

Frank, A. and Mellema, G.: 1996, "Hydrodynamical models of outflow collimation in young stellar objects", *Ap. J.* **472**, 684-702.

Frank, A., Ryu, D. and Davidson, K.: 1998, "Where is the doughnut? Luminous blue variable bubbles and aspherical fast winds", *Ap. J.* **500**, 291-301.

Garcia-Segura, G. and MacLow, M.: 1995, "Wolf-Rayet bubbles. II. Gasdynamical situations", *Ap. J.* **455**, 160-174.

Hartigan, P., Morse, J.A., Heathcote, S. and Cecil, G.: 1993, "Observations of entrainment and time variability in the HH 47 jet", *Ap. J.* **414**, 121-124.

Henney, W.J. and Dyson, J.E.: 1992, "A dynamical model for the outer nebula of R Aquarii", *Astron. Astrophys.* **261**, 301-313.

Lada, C.J. and Fich, M.: 1996, "The structure and energetics of a highly collimated bipolar outflow: NGC 2264G", *Ap. J.* **459**, 638-652.

Latter, W.B., Kelly, D.M., Hora, J.L. and Deutsch, L.K: 1995, "Investigating the near-infrared properties of planetary nebulae. I. Narrowband images", *Ap. J. S* **100**, 159-167.

Martin, C.L. and Arnett, D.: 1995, "The origin of the rings around SN 1987A: An evaluation of the interacting-winds model", *Ap. J.* **447**, 378-390.

Mellema, G.: 1994, "The gasdynamic evolution of spherical planetary-nebulae-radiation-gasdynamics of Pne-III", *Astron. Astrophys.* **290**, 915-935.

Mellema, G. and Frank, A.: 1995, "Radiation gasdynamics of planetary nebulae – V. Hot bubble and slow wind dynamics", *Mon. Not. Royal Astron. Soc.* **273**, 401-410.

Mellema, G.: 1995, "Radiation gasdynamics of planetary nebulae – VI. The evolution of aspherical planetary nebulae", *Mon. Not. Royal Astron. Soc.* **277**, 173-192.

Mellema, G. and Frank, A.: 1997, "Outflow collimation in young stellar objects", *Mon. Not. Royal Astron. Soc.* **292**, 795-807.

Nota, A., Livio, M., Clampin, M. and Schulte-Ladbeck, R.: 1995, "Nebulae around luminous blue variables: a unified picture", *Ap. J.* **448**, 788-796.

O'Dell, C.R. and Handron, K.D.: 1996, "Cometary knots in the helix nebula", *Astron. J.* **111**, 1630-1645.

O'Dell, C.R., Weiner, L. and Chu, Y.-H.: 1990, "A kinematic determination of the structure of the double ring planetary nebula NGC 2392, the eskimo", *Ap. J.* **362**, 226-234.

Schwarz, H.E., Corradi, R.L.M. and Melnick, J.: 1992, "A catalogue of narrow band images of planetary nebulae", *Astron. Astrophys. Soc.* **96**, 23-113.

PART IV

# NUCLIDES IN THE SUN

# Mini-blackhole at the Solar Center and Isotopic Abundances in the Primitive Solar Nebula

S. Ramadurai
*Astrophysics Group, Tata Institute of Fundamental Research, Homi Bhabha Road, Navy Nagar, Colaba, Mumbai 400 005 INDIA*
*durai@tifr.res.in*

**Abstract:** With the enormous improvement in cosmochemical techniques, it is now possible to obtain accurate estimates of the abundances of even the rarest nuclides in the solar system. While a model of the formation of sun and solar system based on the addition of various nuclear species from diverse sites at various times before the formation of solar system (BSS), is very much favoured by the majority of the cosmochemists, the anomalies in the noble gases are not adequately addressed by this model. This has resulted in a novel suggestion of a single supernova contributing to all the elements in the solar system. It is suggested that despite many difficulties, a modification of the single supernova hypothesis taking into account the exotic suggestion of Clayton *et al.* (1975) of the presence of a mini-blackhole at the solar center may prove beneficial in understanding all the relevant observations.

## 1. INTRODUCTION

It is more than four decades since the basic nucleosynthetic processes have been formulated by $B^2FH$ (Burbidge *et al.*, 1957) and Cameron (1957). Since then, detailed understanding of almost all the processes suggested therein has been achieved (See for a recent summary Arnould and Takahashi, 1999). The cosmochemical techniques have reached such a precision that abundances of even the rarest nuclides have been accurately determined from hydrogen to trans-uranic elements (See papers by Grevesse and Sauval as well as Larimer in this volume). There is good progress in understanding star-formation, both theoretically and observationally (Hartmann, 1998). Several extra-solar system planets have been discovered over the last decade (Marcy and Butler, 1998). Thus we have all the basic ingredients necessary for a proper understanding of formation and evolution of our solar system. In spite of this, it is a pity that we have still not arrived at an universally accepted model of the formation and evolution of the solar system, explaining all the dynamical and physical observations. Hence it becomes necessary to consider some exotic suggestions, even if they are viewed unfavorably by a majority of main-stream workers in the field. The single supernova model of Manuel and Sabu (1975) is a case in point. Lack of a credible explanation of all the correlated noble gas anomalies makes this extreme suggestion appealing. But this suggestion suffers from the failure to explain some of the dynamical aspects of the solar system formation scenario as outlined by Goswami (this volume). Another exotic suggestion is the probable presence of a mini-blackhole at the solar center, investigated by Clayton *et al.* (1975) as a possible explanation of the deficit of the solar neutrinos. Though alternate suggestions based on neutrino physics are favored at present (Bahcall *et al.*, 1998), the possible role of mini-blackholes in the star formation and evolution merits a serious investigation in its own right. The basic ideas about such an investigation are given here.

## 2. SUMMARY OF ABUNDANCES OF NUCLIDES IN THE EARLY SOLAR NEBULA

The available data on the abundances can be classified into two distinct groups: 1. representing the short-lived radio nuclides which are now extinct and 2. concerning the noble gases. While the short-lived species give information about the specific nucleosynthetic process responsible for their production, the noble gases, especially the correlation of the various components observed, give clues to the site of their synthesis and the physico-chemical processes involved in their incorporation into the solar system materials, be it solar wind, planets or meteorites.

The data concerning extinct radioactivities are summarised in Table 1. The mean lives of these nuclides range over nearly three orders of magnitude, from 0.15 Myr for $^{41}$Ca to 149 Myr for $^{146}$Sm. The nuclides include 4 s-process products, 2 r-process products, 2 p-process products, and 1 product each from neutrino process, equilibrium process and hot hydrogen burning process. Thus nucleosynthetic products from the entire range of stellar evolution seems to be present and thus pose a major challenge to the model builders. If one wants to explain all the observations, one has to resort to a somewhat *ad hoc* model in which one brings in the products from diverse sources and mix them to fit the data. Justification of the model rests solely on the plausibility of star formation models supporting such a scenario

(Cameron, 1993). This problem is aggravated by the direct correlation seen for the $^{26}$Al and $^{41}$Ca, which rule out energetic particle irradiation as pointed out by Goswami (this volume).

*Table 2.* Data on extinct radionuclides (Cameron, 1993 modified and updated)

| Nuclide | Meanlife, Myr | Ref. Nuc. | Obs. Ratio | Process |
|---|---|---|---|---|
| $^{41}$Ca | 0.15 | $^{40}$Ca | $2.0 \times 10^{-8}$ | s-process |
| $^{26}$Al | 1.07 | $^{27}$Al | $5.0 \times 10^{-5}$ | hot H burning |
| $^{60}$Fe | 2.2 | $^{56}$Fe | $1.6 \times 10^{-6}$ | s-process |
| $^{135}$Cs | 3.3 | $^{133}$Cs | 3 | s-process |
| $^{53}$Mn | 5.3 | $^{55}$Mn | $4.4 \times 10^{-5}$ | equilibrium |
| $^{107}$Pd | 9.4 | $^{108}$Pd | $2.0 \times 10^{-5}$ | s-process |
| $^{182}$Hf | 5.3 | $^{180}$Hf | $2.0 \times 10^{-4}$ | ν-process |
| $^{129}$I | 23.1 | $^{127}$I | $1.0 \times 10^{-4}$ | r-process |
| $^{92}$Nb | 50 | $^{93}$Nb | $2.0 \times 10^{-5}$ | p-process |
| $^{244}$Pu | 118 | $^{238}$U | $7 \times 10^{-3}$ | r-process |
| $^{146}$Sm | 149 | $^{144}$Sm | $7 \times 10^{-3}$ | p-process |

In the case of noble gases because of the availability of a large body of data involving the several isotopes, measured in different materials, planetary atmospheres, solar wind, gas-rich meteorites, etc., it is possible to classify them into distinct groups (Manuel and Sabu, 1975). Then one can look into the correlations of the different groups. Thus it is seen that planetary He and Ne correlates with the exotic component of Ar, Kr and Xe. Further, from the discussion above on extinct radioactive nuclides, it is clear that products from many stages of the evolution of stars right up to the exploding stage were present in the early solar system. Thus it looks somewhat logical to investigate the possibility that all the nuclides seen in the solar system arise from a single supernova event, whose various zones of the earlier evolutionary stages contribute the diverse short-lived radionuclides as well as the noble gases, in short all the nuclides observed (Manuel and Sabu 1975; Manuel 1998). A snapshot of the current status of this model is presented by Manuel and his coworkers in this meeting.

The single supernova model may qualitatively explain the physicochemical aspects of the nuclear abundances, but a full dynamical explanation of the formation of the sun and solar system has not been forthcoming, in spite of the model being in the field for more than two decades. The basic difficulty is the details of the formation of structures from the material dispersed to enormous distances by the supernova explosion. Additionally the helioseismological data clearly is able to constrain the abundances profile of heavy elements and they are consistent with the standard solar model (Antia and Chitre, 1999), unlike the enrichment of Fe group at the center, as envisaged in the single supernova hypothesis. But detection of planets around pulsar PSR 1257 + 12 (Wolszczan, 1994) makes a strong case for the single supernova hypothesis. Hence an attempt is made to bring in the heretofore neglected suggestion of the presence of a mini-blackhole at the solar center.

## 3. MINI-BLACKHOLE AT THE SOLAR CENTER AND ITS CONSEQUENCES

Hawking (1971) has suggested the possibility of microscopic blackholes remaining from the big bang and one can imagine a protostar forming around one of these. Such primordial blackholes might aid in the star-formation, by providing the necessary gravitational attraction. This possibility is not investigated so far, as the star-formation scenario is already so complex and the theoretical and observational investigation has not reached a stage where such exotic suggestions are given any importance. But as was seen in the last section, the explanation of the origin of elements in the solar system might be the trigger needed for such an investigation. In what follows a summary of the situation when a mini-blackhole contributes to some part of the solar luminosity is first given following the original investigation of Clayton *et al.* (1975). After making sure that the presence of a mini-blackhole is not ruled out by the present observations, a new scenario for the formation of the sun and solar system is qualitatively outlined in the next section.

Let us consider a blackhole present in a dense medium. It will initially accrete mass at the hydrodynamic rate given by

$$\dot{M} \approx 5\,(M_{bh}/M_{\odot})^2\,(\rho/100\ \mathrm{g.cm^{-3}})T_6^{-3/2}\ M_{\odot}s^{-1}$$

where $M_{bh}$ is the blackhole mass and $\dot{M}$ represents the mass accretion rate and $\rho$ is the density of the medium. However this hydrodynamic accretion rate will apply only for a short time initially, as the accreted mass will start radiating with the energy release given by

$$L = \varepsilon \dot{M} c^2$$

where $\varepsilon$ represents the efficiency of matter conversion to energy, which for the gravitational case is usually taken as 0.1. As the blackhole accretes mass, it grows as

$$dM_{bh}/dt = (1 - \varepsilon)\dot{M}$$

As the mass of the blackhole increases so does the accretion rate and hence the luminosity. Eventually a stage is reached where the radiation pressure balances gravity, which is the Eddington Luminosity, $L_{Ed}$. If the Eddington Luminosity is maintained thereafter, the accretion rate is given as

$$\dot{M} c^2 = L_{Ed}/\varepsilon$$

Numerically the value of $L_{Ed}$ is

$$L_{Ed} = (0.13/\kappa)\cdot(M_{bh}/10^{-5}M_{\odot})\cdot L_{\odot}$$

where $\kappa$ is the opacity in cgs units. This accretion rate leads to the expression

$$M_{bh} = M_0 \exp(t/\tau)$$

where

$$\tau = \kappa\{\varepsilon/(1 - \varepsilon)\}\cdot 1.13 \times 10^9 \text{ years.}$$

The initial mass of the blackhole $M_0$ is not the lower limit of the primeval blackhole mass of $10^{-19}M_{\odot}$, but a value which gives the current value of $L_{Ed}$, which is about 3% of the current solar luminosity. This value of 3% is chosen such as not to conflict with the solar seismological observations (Antia and Chitre, 1999). This gives for the current $M_{bh}$ a value of $10^{-6}M_{\odot}$ at the solar center. Thus we can fix one of the parameters in our new scenario for the formation of the solar system. Having obtained a quantitative estimate of the mass of the mini-blackhole at the solar center, one can construct a qualitative scenario of the formation of the sun and solar system.

# 4. MODEL OF THE FORMATION OF THE SUN AND THE SOLAR SYSTEM WITH A MINI-BLACKHOLE AT THE SOLAR CENTER

Keeping in view the current ideas about the solar system formation, the presence of a mini-blackhole is to be examined specifically to explain the different timescales of the extinct radionuclides listed earlier. Thus it is necessary to investigate the role of its presence at each of the stages before formation of the solar system (BSS) mentioned by Cameron (1993):

1. About 140 Myr BSS: What is the role of the mini-blackhole in forming the local 'grandparent' cloud, from the OB Association in the 'great grandparent cloud'?
2. About 120 Myr BSS: Did the mini-blackhole help in accreting the r-process nuclides from the exploding stars in the OB Association ?
3. Beyond this, closer to few tens of million years, did the mini-blackhole help in collecting the different zones of the supernova debris to form the sun and the solar system?

These are some of the questions which are being pursued using numerical investigations and it is hoped that a quantitative and detailed model will emerge in the near future from this investigation.

## ACKNOWLEDGEMENTS

The author wishes to thank the Foundation for Chemical Research, Inc., the ACS Division of Nuclear Chemistry & Technology, the University of Missouri-Rolla and Professors Paul Kuroda and Oliver Manuel for financial assistance which made it possible for the author to attend the meeting.

## REFERENCES

Antia, H.M. and Chitre, S.M.: 1999, "Limits on the p-p reaction cross-section from helioseismology", *Astron. Astrophys.* **347**, 1000-1004

Arnould, M. and Takahshi, K.: 1999, "Nuclear astrophysics", *Rep. Prog. Phys.*, **62**, 395-464.

Bahcall, J.N., Basu, Sarbani, and Pinsonneault, M.H.: 1998, "How uncertain are solar neutrino predictions?", *Phys. Lett.* **433B**, 1-8.

Burbidge, E.M., Burbidge, G.R., Fowler, W.A. and Hoyle, F.: 1957, "Synthesis of elements in stars", *Rev. Mod. Phys.* **29**, 547-650.

Cameron, A.G.W.: 1957, "Stellar evolution, nuclear astrophysics and nucleogenesis", *Chalk River Report,* CRL-41.

Cameron, A.G.W.: 1993, "Nucleosynthesis and star formation", in *Protostars and Planets III*, eds., Levy, E.H. and Lunine, J.I., University Arizona Press, Tucson, AZ, pp. 47-74.

Clayton, D.D., Newman, M.J. and Talbot, Jr., R.J.: 1975, "Solar models of low neutrino-counting rate: the central blackhole", *Ap. J.* **201**, 489-493.

Hartmann, Lee: 1998, *Accretion Processes in Star Formation*, Cambridge University Press, Cambridge, UK, 237 pp.

Hawking, S.W.: 1971, "Gravitationally collapsed objects of very low mass", *Mon. Not. Royal Astron. Soc.* **152**, 75-78.

Manuel, O.: 1998, "Origin of elements in the solar system", preprint, 96 pp.

Manuel, O. and Sabu, D.D.: 1975, "Elemental and isotopic inhomogeneities in noble gases: the case for local synthesis of the chemical elements", *Trans. Missouri. Acad. Sci.* **9**, 104-122.

Marcy, Geoffrey W. and Butler, R. Paul: 1998, "Detection of extrasolar giant planets", *Ann. Rev. Astr. Ap.* **36**, 57-97.

Wolszczan, A.: 1994, "Confirmation of earth-mass planets orbiting the millisecond pulsar", *Science* **264**, 538-542.

# Abundances of the Elements in the Sun

N. Grevesse[1] and A. J. Sauval[2]
[1] *Institut d'Astrophysique et de Géophysique, Université de Liège, B-4000 Liège, Belgium*
[2] *Observatoire Royal de Belgique, B-1180 Bruxelles, Belgium*
*nicolas.grevesse@ulg.ac.be*

**Abstract:** We review the current status of our knowledge of the chemical composition of the sun and its evolution from core to corona. In spite of subtle variations, it is possible to define a standard chemical composition (Table 1), essentially derived from the analysis of the solar photospheric spectrum, together with important contributions from helioseismology as well as from observations of the solar upper atmosphere.

## 1. INTRODUCTION

What is the sun made of? The answer to this fundamental question is of crucial importance not only for modelling the sun itself but more generally for modelling the great variety of stellar objects and the universe as a whole.

Is the solar composition unique or does it vary with time and/or from one solar layer to another? The only obvious expected variation is in the central layers, where, because of thermonuclear reactions, the hydrogen content decreases leading to an increase in the helium content. More subtle and unforeseen variations in the chemical composition have however been observed recently. During the solar lifetime, the convective zone reservoir which fills the outer solar layers, from the photosphere to the corona, is slowly enriched in hydrogen from the radiative zone below it, whereas it slowly looses, through its lower boundary layer, about 10% of all the heavier elements. In the outer atmospheric layers, and essentially in the very heterogeneous solar corona, the observed composition is very variable in different types of solar coronal structures and a fractionation occurs in the low chromospheric layers: elements of low first ionization potential ≤ 10 eV generally show abundances larger than in the photosphere. In the outermost coronal structures, additional gravitational settling acts to further modify the chemical composition.

We shall nevertheless see that in spite of these subtle variations, it is possible to define a standard solar chemical composition, essentially derived from analyses of the solar photospheric spectrum, but with important contributions from other sources.

## 2. HISTORICAL INTRODUCTION

H.N. Russell, whose name is associated with a large number of pioneer researches in astrophysics during the first half of this century (for example the Hertzsprung-Russell diagram) as well as to a series of basic works in atomic spectroscopy (for example the LS or Russell-Saunders coupling), also made the first quantitative analysis of the chemical composition of the solar atmosphere (Russell, 1929). Using eye estimates of solar line intensities measured on the Revised Rowland Atlas of the solar spectrum together with the reversing layer hypothesis, he succeeded to derive the abundances of 56 elements. Russell's mixture was used by almost two generations of astronomers. He also showed that the solar atmosphere and, finally, the universe were essentially made of hydrogen, an observation which took some time to be accepted by the whole astronomical community. Many of the remarkable features correlated to nuclear properties (Figure 1), giving clues to the origin of the different elements, were already present in Russell's results.

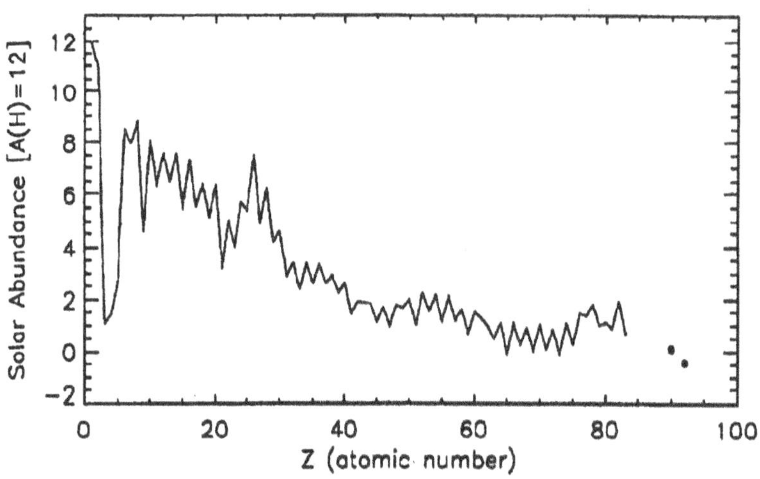


*Figure 1.* Distribution of the abundance of the elements as a function of the atomic number, Z.

Twenty years later, Unsöld (1948) using better observations and better techniques obtained abundance results for 25 elements and noticed that his results were not much different from Russell's values. He concluded that this was not surprising because Russell had an "unvergleichliches spektroskopisches Fingerspitzengefühl" (an incomparable spectroscopic flair).

In the meantime important works of great interest for our understanding of stellar atmospheres have been published, allowing progress in the accuracy of abundance analyses.

Goldberg *et al.* (1960; hereafter GMA) using the curve of growth technique together with a photospheric model, and a careful examination of the observed equivalent widths and, for the first time, of the oscillator strengths of the lines, succeeded in getting the abundances for 42 elements. The results of GMA have been considered as the standard reference work for more than 10 years. In the last 25 years different groups all over the world have been very active in photospheric abundance works: L.H. Aller and his co-workers, D.E. Blackwell and his co-workers, O. Engvold and O. Hauge, H. Holweger and his co-workers, D.L. Lambert, B. Warner, B.J. O'Mara and his co-workers, and our Belgian group including E. Biémont and A. Noels.

Chemical composition data concerning the solar outer layers above the photosphere slowly accumulated during the last 30 years or so. These data, obtained from the ground as well as from space, revealed a great variability from one layer to another, from one event to another. In spite of these variabilities and sometimes large differences with photospheric results, these data play an important role in defining the solar chemical composition and, more importantly, in explaining the physical processes that allow these outermost layers to exist (see different reviews in Fröhlich *et al.*, 1998).

## 3. INTEREST OF SOLAR ABUNDANCES

The chemical composition of the sun is a key data for modelling the sun, the interior as well as the atmosphere. In the photosphere, elements like Mg, Si and Fe, play a particular role because they provide most of the electrons in this layer. The critical role of the opacities and the crucial contributions of the most abundant elements like Fe in the central solar layers and of O and Ne at the bottom of the convective zone has been stressed by many authors (Rogers and Iglesias, 1998; Turck-Chièze, 1998; Turcotte and Christensen-Dalsgaard, 1998; Guzik and Neuforge, this volume).

Progress in the opacity computations allow to say that uncertainties in the opacities due to uncertainties in the abundances are now similar in magnitude to those due to uncertainties in the physics of opacity computations. With the tremendous new achievements of solar seismology, solar interior layers can now be empirically tested (Guzik and Neuforge, this volume). Very slight changes in the abundance profile of the interior layers have profound and observable effects on the sound speed profile obtained with very high accuracy from solar seismic observations.

The sun, being the best known star, has always been considered as the typical star, the reference to which the abundance analyses of other stars are compared. We know that the solar system based standard abundance distribution (SAD) is not universal. Outside of the solar system we obviously observe quite a large number of peculiarities (*i.e.*, differences from the SAD) among different stellar objects. In the solar system itself, we observe the so-called "isotopic anomalies" showing material of very different origins, but these anomalies are confined in a very small mass fraction (See *e.g.*, Ott and many other papers in this volume). These anomalies probably result from incomplete mixing in the primordial solar nebula (See however Manuel, this volume). It has also been suggested that the sun might be somewhat anomalous (See Grevesse *et al.*, 1996 for references), but Gustafsson (1998) has convincingly shown that the sun is like many other stars of the same age in our galaxy and that the claimed slight metal richness of our sun is well within the real cosmic scatter. Abundances are remarkably similar to some extent to the SAD, everywhere we look, and the deviations from the SAD are to be explained by secondary but very important processes.

When studying solar abundances in different solar layers (See section 4), we have also access to important tracers of the structure as well as of the physical processes in the outer solar layers.

The sun is also unique because chemical composition data can be acquired from different objects in the solar system like the earth, moon, planets, comets and meteorites. Few data come from planets; for the terrestrial planets, including the Earth, elements have either evaporated or fractionated. Very few reliable data are available for comets. This is one of the main goals of a future comet rendezvous mission. A very rare class of meteorites, the so-called CI carbonaceous chondrites, is of particular interest (See *e.g.*, Ebihara *et al.*, this volume; Larimer, this volume). These meteorites have preserved the bulk composition of their parent bodies (planetesimals) and have thus retained most of the elements present in the primitive matter of the solar nebula, except for the few most volatile elements. Chemical composition data among these various bodies are important tracers of the origin and evolution of the solar system.

And, last but not least, the standard chemical composition as derived from meteorites is also the basic data that has allowed Burbidge *et al.* (1957) and Cameron (1957) to build the firm basis of the theory of element nucleosynthesis explaining how, where and when all the nuclides are formed. The remarkable pattern seen in Figure 1 is in timely related to nuclear stability.

The different nuclear processes giving rise to this distribution of the chemical elements have been identified: big-bang nucleosynthesis, spallation by galactic cosmic rays, thermonuclear fusion and neutron captures in stars during stationary evolution and supernova explosions (Burbidge, this volume). The standard chemical composition also plays a key role in the chemical evolution of galaxies and the universe (Pagel, 1997).

# 4. SOURCES OF SOLAR ABUNDANCES

Because of its proximity, our sun is unique and, by far, the best known star. Actually, solar abundances can be derived by very different techniques and for very different types of solar matter, from the interior to the outermost coronal layers.

Using spectroscopy in a very large wavelength range, we can derive the chemical composition of the photosphere, chromosphere and corona and also of sunspots. Particle collection techniques from space allow to measure the chemical composition of the solar wind (SW) and of solar energetic particles (SEP). We can eventually get informations on solar flares from gamma-ray spectroscopy. Finally, calibration of solar models and inversion of helioseismic data allow us to derive the solar abundance of helium (See section 5.1). Using similar techniques, we might hope, in the future, to be able to get empirical abundance profiles in the solar interior (Turck-Chièze, 1998; Takada and Shibahashi, 1998; Guzik and Neuforge, this volume) in order to test the selective element migration which takes place in the radiative zone just below the convective zone (See hereafter). Note also that lunar soils also record the past chemical history of the sun (Wieler, 1998).

Below the convective zone, in the radiative zone, selective element settling is at work, thanks to the effects of gravitation, thermal diffusion and radiative acceleration (Vauclair, 1998; Turcotte and Christensen-Dalsgaard, 1998; Guzik and Neuforge, this volume). Although this is a very slow phenomenon, since the birth of the sun, computations show that the reservoir that fills the solar outer layers, could have lost about 10% of He and of all the heavier elements. A varying metallicity profile is thus building up in the radiative zone that has profound effects on the models. Of course, strictly speaking, the chemical composition of the photosphere should be about 10% lower than the composition of the interstellar cloud from which the sun formed 4.6 GY ago.

In the solar outer layers, all the indicators (coronal spectra, SW and SEP measurements, gamma-ray spectroscopy, various measurements in very different coronal structures) show that the observed chemical composition (involving the most abundant elements up to Ge) is very variable in different types of solar matter and that a fractionation occurs, at low chromospheric level, leading to the so-called FIP (First Ionization Potential; or FIT: First Ionization Time) effect: elements of low first ionization potential ($\leq$ 10 eV) show abundances about 4 times larger than in the photosphere whereas elements with higher ionization potential (except for helium; see section 5.1), have the same abundance in the corona and in the photosphere. It has to be mentioned that the FIP effect strongly depends on the observed coronal structure: it can vary from 1, *i.e.*, photospheric abundances, in coronal holes and impulsive flares, up to very large values (15) in long lasting coronal plumes. In the outermost coronal layers, far away from the sun further gravitational settling occurs. The variation in chemical composition among very different coronal structures and the FIP effect are discussed in detail in many recent papers (*e.g.*, Fröhlich *et al.*, 1998; Reames, 1999).

The solar photosphere is without any doubt the layer from which we have the largest number of chemical composition data. Actually, 65 elements, out of 83 stable elements, are present in the photospheric spectrum. The few elements which cannot be measured in the photosphere, the most important being helium, are not absent however. These elements do not show lines in the photospheric spectrum for basic spectroscopic reasons: under the physical conditions in the photosphere, with a typical temperature around 5000 K, no line of these elements (neutral or once-ionized species) falls within the wavelength range covered by the photospheric spectrum. The other sources of solar abundances only concern a limited number of elements: in the SEP, the richest source of elemental abundances, only 21 elements have been measured.

Just above the convection zone, the photosphere is a well mixed region (See, however, Solanki, 1998) whereas the outer solar layers show a very heterogeneous and changing structure. The structure and the physical processes of the photosphere are also rather well known allowing to reach good accuracies. It is also the layer that has been studied quite a long time before the other layers for obvious reasons: the solar photospheric spectrum has been recorded since quite a long time. For all these reasons, photospheric abundances will be adopted as a reference for all the other solar data.

## 5. SOLAR ABUNDANCES

For many reasons given above, the solar chemical composition, to which the results for the other layers will be compared, is the composition derived from the analysis of the solar photospheric spectrum.

Much progress has been made during the last decades. Solar photospheric spectra, the basic data for deriving photospheric abundances, with very high resolution and very high signal over noise ratio, obtained from the ground and from space, are now available for quite a large wavelength range, from the UV to the far IR (See Kurucz, 1995, for a recent review). The strength of an absorption line is directly related to the product of the abundance of the element producing the line and of the transition probability of the line. The exact relation between these two quantities can easily be obtained if the physical conditions and physical processes of the layers where the line is formed are known.

Empirical modelling of the photosphere has now reached a high degree of accuracy [See section 5.5 and Solanki (1998) and Rutten (1998)]. And, last but not least, accurate atomic and molecular data, in particular transition probabilities, have progressively been obtained for transitions of solar interest; these data play a key role in solar spectroscopy.

*Table 1.* Element abundances in the solar photosphere*

| Element | $A_{el}$ | Element | $A_{el}$ | Element | $A_{el}$ |
|---|---|---|---|---|---|
| 01 H | 12.00 | 29 Cu | 4.21 ± 0.04 | 58 Ce | 1.58 ± 0.09 |
| 02 He | [10.93 ± 0.004] | 30 Zn | 4.60 ± 0.08 | 59 Pr | 0.71 ± 0.08 |
| 03 Li | 1.10 ± 0.10 | 31 Ga | 2.88 ± (0.10) | 60 Nd | 1.50 ± 0.06 |
| 04 Be | 1.40 ± 0.09 | 32 Ge | 3.41 ± 0.14 | 62 Sm | 1.01 ± 0.06 |
| 05 B | 2.70 ± 0.16 | 33 As | - | 63 Eu | 0.51 ± 0.08 |
| 06 C | 8.52 ± 0.06 | 34 Se | - | 64 Gd | 1.12 ± 0.04 |
| 07 N | 7.92 ± 0.06 | 35 Br | - | 65 Tb | (-0.1 ± 0.3) |
| 08 O | 8.83 ± 0.06 | 36 Kr | - | 66 Dy | 1.14 ± 0.08 |
| 09 F | [4.56 ± 0.3] | 37 Rb | 2.60 ± (0.15) | 67 Ho | (0.26 ± 0.16) |
| 10 Ne | [8.06 ± 0.10] | 38 Sr | 2.92 ± 0.05 | 68 Er | 0.93 ± 0.06 |
| 11 Na | 6.33 ± 0.03 | 39 Y | 2.24 ± 0.03 | 69 Tm | (0.00 ± 0.15) |
| 12 Mg | 7.58 ± 0.05 | 40 Zr | 2.60 ± 0.02 | 70 Yb | 1.08 ± (0.15) |
| 13 Al | 6.47 ± 0.07 | 41 Nb | 1.42 ± 0.06 | 71 Lu | 0.06 ± 0.10 |
| 14 Si | 7.55 ± 0.05 | 42 Mo | 1.92 ± 0.05 | 72 Hf | 0.88 ± (0.08) |
| 15 P | 5.43 ± 0.05 | 44 Ru | 1.84 ± 0.07 | 73 Ta | - |
| 16 S | 7.33 ± 0.11 | 45 Rh | 1.12 ± 0.12 | 74 W | (1.11 ± 0.15) |
| 17 Cl | [5.5 ± 0.3] | 46 Pd | 1.69 ± 0.04 | 75 Re | - |
| 18 Ar | [6.40 ± 0.10] | 47 Ag | (0.94 ± 0.25) | 76 Os | 1.45 ± 0.10 |
| 19 K | 5.12 ± 0.13 | 48 Cd | 1.77 ± 0.11 | 77 Ir | 1.35 ± (0.10) |
| 20 Ca | 6.36 ± 0.02 | 49 In | (1.66 ± 0.15) | 78 Pt | 1.8 ± 0.3 |
| 21 Sc | 3.17 ± 0.10 | 50 Sn | 2.0 ± (0.3) | 79 Au | (1.01 ± 0.15) |
| 22 Ti | 5.02 ± 0.06 | 51 Sb | 1.0 ± (0.3) | 80 Hg | - |
| 23 V | 4.00 ± 0.02 | 52 Te | - | 81 Tl | (0.9 ± 0.2) |
| 24 Cr | 5.67 ± 0.03 | 53 I | - | 82 Pb | 1.95 ± 0.08 |
| 25 Mn | 5.39 ± 0.03 | 54 Xe | - | 83 Bi | - |
| 26 Fe | 7.50 ± 0.05 | 55 Cs | - | 90 Th | - |
| 27 Co | 4.92 ± 0.04 | 56 Ba | 2.13 ± 0.05 | 92 U | (< -0.47) |
| 28 Ni | 6.25 ± 0.04 | 57 La | 1.17 ± 0.07 | | |

*Elements for which no photospheric value is given are not necessarily absent from the photosphere; see section 4 – Values between square brackets are not derived from the photosphere, but from sun spots (F, Cl), solar corona and solar wind particles (Ne, Ar) – Values between parentheses are less accurate results – For He, see section 5.1; for Th, see Grevesse, *et al.* (1996).

We have recently reviewed in detail this key role and, in particular, the role of transition probabilities in solar spectroscopy, not only in improving the abundance results but also as tracers of the physical conditions and processes in the solar photosphere (Grevesse and Noels, 1993; Grevesse *et al.*, 1995). Most of the progress in our knowledge of the solar photospheric chemical composition during the last decades has been essentially, if not uniquely, due to the use of more accurate transition probabilities as seen in the examples given in the here above mentioned papers. Large discrepancies previously found between the sun and CI meteorites have progressively disappeared as the accuracy of the transition probabilities has been increased. Actually, the dispersion of solar photospheric abundance results reflects the internal accuracy of the transition probabilities used to derive the abundances. The sun is rarely (never?) at fault but unfortunately, older sets of transition probabilities were too often at fault! Hopefully, the techniques now allow to measure transition probabilities with high accuracy, even for rather faint lines. It has to be mentioned that many analyses of solar abundances have resulted from close collaborations between atomic spectroscopists and solar spectroscopists. Too rare groups, however, work to fill the gaps in the many data still needed by the astronomers.

In Table 1, we give the best solar photospheric abundances taken from our latest review (Grevesse and Sauval, 1998). Recent results obtained by Cunha and Smith (1999; B), by ourselves on P (unpublished), by Barklem and O'Mara (1998; Ca), by Barklem and O'Mara (1999; Sr) have been taken into account. Values are given in the logarithmic scale usually adopted by astronomers, $A_{el} = \log N_{el}/N_H + 12.0$, where $N_{el}$ is the abundance by number. Such a scale was originally adopted in order to avoid negative numbers for the least abundant elements. It is still used nowadays even if one value is now negative. We now comment on a few elements.

## 5.1 Helium

In 1868, a new element was discovered in the solar spectrum obtained during an eclipse. The name of the sun was given to the new element, helium. Helium was only discovered on earth in 1895. Nowadays, its primordial abundance is known to a high degree of accuracy (Pagel, 1997).

Despite its name and its very high abundance, He is not present in the photospheric spectrum and is largely lost by the meteorites. Solar wind and solar energetic particles show a very variable and rather low value (*i.e.* low when compared to values observed in hot stars and in the interstellar medium from H II regions around us). Coronal values derived from spectroscopy have large uncertainties: $N_{He}/N_H = 7.9 \pm 1.1\%$ (Gabriel *et al.*, 1995) and $8.5 \pm 1.3\%$ (Feldman, 1998). Giant planets, as observed by the Voyager spacecraft, do not allow to settle the question: Jupiter and Saturn show anomalously low values whereas higher values (9.2 ± 1.7%) are found for Uranus and Neptune (Conrath, *et al.,* 1989). Note that the recent Galileo spacecraft has recently measured an intermediate value on Jupiter, Y = 0.234, or $N_{He}/N_H$ = 7.85% (von Zahn and Hunten, 1996).

Progress in our knowledge of the solar He content has recently come from standard solar models as well as non standard models and the inversion of helioseismic data. While the calibration of the standard models leads to an abundance of He by mass of Y = 0.27 ± 0.01 ($N_{He}/N_H$ = 9.5%) in the protosolar cloud (Christensen-Dalsgaard, 1998), non standard models (*i.e.*, taking element migration, for example, into account) start with an helium abundance of Y = 0.275 (Gabriel, 1997). Inversion of helioseismic data leads to a very accurate, but smaller, value, Y = 0.248 ± 0.002 (*i.e.*, 8.5%) as the value of the present solar abundance of He in the outer convection zone (Dziembowski, 1998). The difference of 10 percent between these two values is now interpreted as due to element migration at the basis of the convection zone during the solar lifetime (Vauclair, 1998; Turcotte and Christensen-Dalsgaard, 1998). The problem of the solar He is also discussed in Guzik and Neuforge (this volume).

In Table 1, we give the present value in the outer layers, Y = 0.248 ± 0.002, or $N_{He}/N_H$ = 8.5%, or $A_{He}$ = 10.93 ± 0.004. The value at birth of the sun is Y = 0.275 ± 0.01, or $N_{He}/N_H$ = 9.8%, or $A_{He}$ = 10.99 ± 0.02.

## 5.2 Lithium, Beryllium, Boron

Recent results obtained for the solar abundances of beryllium (Balachandran and Bell, 1998) and boron (Cunha and Smith, 1999) show that solar data for Be and B now perfectly agree with the meteoritic values.

The Li Be B problem is now reduced to explaining how the sun can deplete Li by a factor 160 (the ratio between the meteoritic and photospheric values) whereas Be and B are not destroyed. Although standard solar models fail to do that, mixing just below the bottom of the convection zone seems to be successful (Blöcker *et al.*, 1998; Vauclair, 1998; Zahn, 1998; Guzik and Neuforge, this volume).

## 5.3 Carbon, Nitrogen, Oxygen

These elements which have largely escaped from meteorites are key elements. Because of their large abundances, they are main contributors to the metallicity (O: 47%, C: 17%, N: 5%); they are also important contributors to the opacity, especially O (Guzik and Neuforge, this volume). Although coronal measurements of the CNO abundances are certainly very helpful, the solar abundances of these three elements will nevertheless heavily rely upon the photospheric values.

The abundances of C, N and O can be derived from a large number of indicators, atoms as well as diatomic molecules made of C, N, O and H. The problems encountered when analyzing all these indicators have been described by Grevesse and Sauval (1994).

Recently, Reetz (1998) has redetermined the solar abundance of oxygen on the basis of the A–X transition of OH, of the O I triplet around 777.3 nm and of vibration-rotation and pure rotation lines of OH in the infrared. He finds $A_O = 8.80 \pm 0.06$.

We have revisited all the lines of all the CNO solar abundance indicators and redone some computations playing with different photospheric models. Unfortunately, our new analysis is not yet finalized because we believe that there are, among other things, many uncleared problems with the atomic as well as molecular data. We can only suggest preliminary values for the revised solar C, N and O abundances: $A_C = 8.52$, $A_N = 7.92$ and $A_O = 8.83$, respectively; we estimate the uncertainties to be of the order of 0.06 dex (Sauval *et al.*, 1999). These values are slightly smaller than the previously recommended values. Even small, these modifications have profound effects on the solar model as shown by Guzik and Neuforge (this volume).

## 5.4 Neon, Argon

These two noble gases do not appear in the solar photospheric spectrum and are largely lost by meteorites. Therefore we have to rely on coronal data as obtained from the coronal spectrum, SW, SEP and gamma-ray spectroscopy.

Ne is an important element because it contributes 10% to the metallicity. Furthermore, it is an important contributor to the opacity at the bottom of the convection zone (Rogers and Iglesias, 1998; Turck-Chièze, 1998).

For Ne, an accurate value has recently been obtained by Widing (1997) who measured the Ne/Mg ratio in photospheric material observed in emerging flux events. His value has been slightly revised by Feldman (1998), $A_{Ne} = 8.12 \pm 0.10$. On the other hand, gradual SEP events lead to a very accurate Ne/O ratio (0.152). If we adopt the hereabove mentioned solar oxygen abundance, we find $A_{Ne} = 8.01 \pm 0.07$, the uncertainty being essentially due to the uncertainty of the photospheric abundance of oxygen. We shall therefore recommend a mean between these two results: $A_{Ne} = 8.06 \pm 0.10$.

The solar abundance of argon has been rediscussed by Young *et al.* (1997) who, mostly on the basis of SW and SEP results, together with B-type stars, as well as lunar soils results, recommended a value $A_{Ar} = 6.47 \pm 0.10$, smaller than the value recently proposed by Feldman (1998), $A_{Ar} = 6.62 \pm 0.12$, from the impulsive flare Ar/Mg ratio. Now recent SEP based Ar abundance results (Reames, 1998) lead to a much smaller value, $A_{Ar} = 6.35 \pm 0.09$, adopting the solar oxygen abundance, $A_O = 8.83$, quoted hereabove. We shall therefore suggest to adopt $A_{Ar} = 6.40 \pm 0.10$, keeping in mind these very different values.

## 5.5 Iron

The longstanding puzzling problem of the possible difference between the photospheric and the meteoritic abundance of iron has been the subject of numerous works by different research groups during the last decade. The debate between the Oxford group (D.E. Blackwell and his co-workers) and the Kiel-Hannover group (H. Holweger and co-workers) as to whether the solar abundance of Fe derived from Fe I lines is high, $A_{Fe} = 7.63$ (Oxford), *i.e.*, larger than the meteoritic value, $A_{Fe} = 7.50$, or low (Kiel-Hannover), *i.e.*, in agreement with the meteorites, is summarized in Grevesse and Sauval (1999) where we give a complete discussion of the problem. Apparently, the cumulative effects on the abundance results of slight differences between the equivalent widths, gf-values absolute scales, microturbulent velocities and empirical enhancement factors of the damping constants, could partly explain the two different abundance results.

We have also rediscussed the problem using most of the lines retained by the two different groups cited hereabove. Our results obtained with the usually adopted Holweger and Müller (1974) photospheric model show that low excitation lines lead to higher abundance values than higher excitation lines. As the low excitation lines are more sensitive to the temperature than the high excitation lines and as they are, on the whole, formed higher in the atmosphere, an easy way to solve for this dependence is to change very slightly the temperature of the photospheric model in the *ad hoc* layers. The new solar model we built has a temperature about 200 K lower at $\log \tau \approx -3$ and has the same T of the Holweger and Müller (1974) model in the deeper layers ($\log \tau \approx -1$). With this new model results from low and high excitation Fe I lines lead to the same abundance value: $A_{Fe} = 7.50 \pm 0.05$, in pretty good agreement with the very accurate meteoritic result (7.50). This value also agrees with results obtained from the analyses of Fe II lines.

# 6. COMPARISON WITH METEORITES

Solar abundance results are generally compared with the very accurate results obtained from a very rare class of meteorites, the CI carbonaceous chondrites (Ebihara *et al.*, this volume; Larimer, this volume). As can be seen from Figure 2, photospheric and meteoritic results agree now perfectly well. It was shown (See section 5, also Anders and Grevesse, 1989) that past discrepancies have gone away as the solar values have become more accurate, especially thanks to transition probabilities of much better accuracy. The few discrepant points can be accounted for by the uncertainty of the photospheric results, much larger than the very small uncertainties of the meteoritic data.

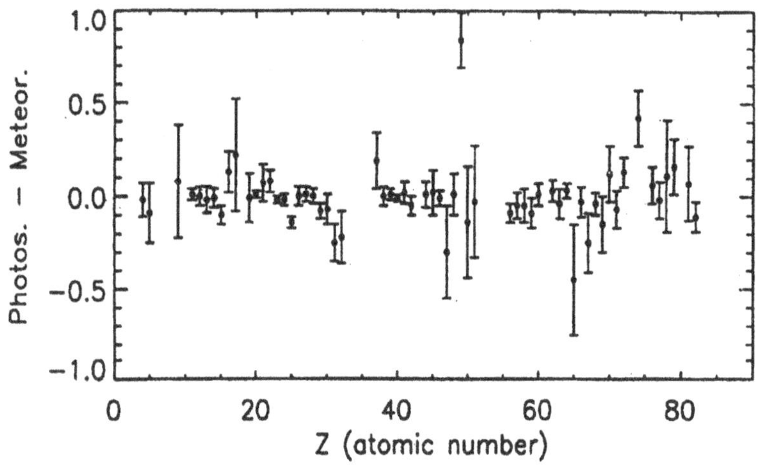


*Figure 2.* Difference between Solar and Meteoritic abundances of elements as a function of Z (Grevesse and Sauval, 1998). Error bars represent the uncertainty of the solar abundance determinations. The point representing Li falls largely outside of this figure (-2.21).

We do believe that this is not just by chance that photospheric and meteoritic abundances agree so well (Figure 2), from refractories to rather highly volatile elements (See Figure 4 of Anders and Grevesse, 1989). This is a strong argument in favor of the CI carbonaceous chondrites as a good representative sample of the well mixed solar nebula.

Meteoritic abundances are generally combined with solar abundances to build the "standard abundance distribution" (SAD; See *e.g.*, Grevesse and Sauval, 1998). For a large number of elements, values from CI's are used rather than solar values because of the higher accuracies of the CI results and because CI and solar data nicely agree. However, for the most abundant elements and the most important ones as far as solar modelling is concerned (H, He, O, C, Ne and N by order of decreasing abundances), the SAD will always rely on solar data.

## 7. SOLAR ISOTOPIC RATIOS

It is very difficult to get any very accurate isotopic ratios from the analysis of the solar photospheric spectrum. Results obtained from the spectroscopic analysis of molecular lines lead to isotopic ratios for C, O and

Mg only: within the rather large error bars, the solar photospheric values agree with the reference terrestrial isotopic abundances (Sauval *et al.*, 1999 for C and O).

Recent solar wind and solar energetic particles measurements of very different coronal structures and events, from various ongoing space experiments (Ulysses, WIND, SOHO, ACE), have allowed to obtain very interesting sets of isotopic data ranging from the $^{3}$He/$^{4}$He ratio up to nickel, including only the most abundant elements. Data also come from lunar surface material which record implanted solar matter since the birth of the Moon. Different review papers concerning these subjects have recently been published by Bochsler (1998), Kallenbach *et al.* (1998), Wieler (1998), Williams *et al.* (1998), Wimmer-Schweingruber *et al.* (1999) and Zurbuchen *et al.* (1998).

It would be out of the scope of the present review to make an exhaustive discussion in a field in very rapid evolution. Helium plays a special role because from the solar $^{3}$He/$^{4}$He ratio, a value of D/H can be obtained which has important consequences for galactic evolution and the big-bang model (Geiss and Gloeckler, 1998). Although this ratio, $^{3}$He/$^{4}$He, is extremely variable, from event to event and from the slow to the fast solar wind, accurate long term averages can be obtained.

The isotopic abundances of refractory elements generally show little variations among different solar system objects including the solar sources. Solar values generally agree with the terrestrial ratios within a few percent, but the slow solar wind could be slightly fractionated (depletion in the heavier isotopes) and some SEP events show large variations probably related to a charge/mass fractionation occuring in SEP.

The situation is quite different and, actually, very complicated for volatile elements like the noble gases Ne, Ar, Kr and Xe (these last two elements have only been observed in lunar soils). For the isotopic ratio $^{20}$Ne/$^{22}$Ne, for example, very different values are observed: 13.7 (SW), 11.0 (SEP, but with large variations down to 6.6 in some events), 9.8 in the terrestrial atmosphere and at least two kinds of "planetary" Ne in meteorites, 8.6 and 10.7, respectively. Although the origin of these rather large differences is unknown, various physical processes (related to acceleration and transport) could lead to these isotopic fractionations in the outer solar layers as well as during the different phases of formation of the planetary bodies (Bochsler, 1998; Williams *et al.*, 1998). Isotopic fractionations in the high speed solar wind do not seem to be very efficient (Wimmer-Schweingruber *et al.*, 1999).

Manuel and Hwaung (1983; Manuel, this volume) took another point of view in order to explain the differences in isotopic composition of noble gases between the solar wind and planetary solids. They suggest that a very efficient fractionation process is operating inside the sun enriching light mass nuclei at the surface, *i.e.*, the photosphere. When the photospheric abundances are corrected for this hypothetical fractionation process, the sun's interior turns out to be essentially made of iron! Hydrogen and helium, the most abundant elements in the universe, become very minor contributors to solar bulk matter. With such a sun, they eventually suggested that the sun formed on the iron core of a precursor supernova which exploded about 5 billion years ago.

We do believe that solar modellers will have tremendous difficulties to compute a shining sun with such a mixture and that helioseismic observations are not in favor of such exotic matter (Guzik and Neuforge, this volume). Furthermore, wherever we look around us, other stellar objects do show, to within important but explainable differences, about the same composition as the solar photospheric one.

In the standard abundance tables, because a large number of isotopic ratios cannot be measured from solar sources, it is generally assumed that the solar isotopic abundances are the same as in terrestrial and meteoritic material, except for very rare cases (Anders and Grevesse, 1989; Ebihara *et al.*, this volume).

## 8. CONCLUSIONS

We have come to a point where photospheric and meteoritic abundances agree pretty well. Does this necessarily mean that there are no real small differences? Probably not because the uncertainties of the photospheric data are still too large. To reduce these uncertainties and make progress in our understanding of the solar photosphere, better and many more atomic data of high precision are required. We have also seen that 1D classical photospheric models have really reached their limits. Progress is now expected if we now turn to more realistic 3D modelling of the photosphere. We could dream to be able to see very slight but meaningful differences between photospheric and meteoritic abundances which would tell us about subtle physical processes at work in the sun.

Progress in our knowledge of the outermost solar layers is obviously going on very rapidly nowadays thanks to very powerful and sensitive instruments already at work in space or to be launched soon. Accumulating data using very different techniques and for extremely different solar structures and dynamic events will allow to better understand these changing solar structures as well as the various fractionations which modify elemental as well as isotopic abundances.

Finally, from progress in helioseismology and solar modelling, we might dream to have direct access to the chemical composition of the interior layers of the sun.

## ACKNOWLEDGEMENTS

We are grateful to P. Bochsler, R.A. Mewaldt and D.V. Reames for helpful discussions. We thank J. Vandekerckhove for his help with the figures. This research has made use of NASA's Astrophysics Data System Abstract Service. N.G. thanks the Belgian Fonds National de la Recherche Scientifique as well as the organizers of the OESS Symposium for financial support.

## REFERENCES

Anders, E. and Grevesse, N.: 1989, "Abundances of the elements: Meteoritic and solar", *Geochim. Cosmochim. Acta* **53**, 197-214.

Balachandran, S.C. and Bell, R.A.: 1998, "Shallow mixing in the solar photosphere inferred from revised beryllium abundances", *Nature* **392**, 791-793.

Barklem, P.S. and O'Mara, B.J.: 1998, "The broadening of strong lines of $Ca^+$, $Mg^+$ and $Ba^+$ by collisions with neutral hydrogen atoms", *Mon. Not. Royal Astron. Soc.* **300**, 863-871.

Barklem, P.S. and O'Mara, B.J.: 1999, "Broadening of lines of Be II, Sr II and Ba II by collisions with hydrogen atoms and the solar abundance of strontium", *Mon. Not. Royal Astron. Soc.*, in press.

Blöcker, T., Holweger, H., Freytag, B., Herwig, F., Ludwig, H.-G. and Steffen, M.: 1998, "Lithium depletion in the sun: A study of mixing based on hydrodynamical simulations", *Space Sci. Rev.* **85**, 105-112.

Bochsler, P.: 1998, "Structure of the solar wind and compositional variations", *Space Sci. Rev* **85**, 291-302.

Burbidge, E.M., Burbidge, G.R., Fowler, W.A. and Hoyle, F.: 1957, "Synthesis of the elements in stars", *Rev. Mod. Phys.* **29**, 547-650.

Cameron, A.G.W.: 1957, "Nuclear reactions in stars and nucleogenesis", *Publ. Astron. Soc. Pac.* **69**, 201-222. See also "Stellar evolution, nuclear astrophysics and nucleosynthesis", Chalk River Report CRL-41, Atomic Energy of Canada, Ltd.

Christensen-Dalsgaard, J.: 1998, "The 'standard' sun", *Space Sci. Rev.* **85**, 19-36.

Conrath, B., Flasar, F.M., Hanel, R., Kunde, V., Maguire, W., Pearl, J., Pirraglia, J., Samuelson, R., Gierasch, P., Weir, A., Bezard, B., Gautier, D., Cruikshank, D., Horn, L., Springer, R. and Schaffer, W.: 1989, "Infrared obervations of the Neptunian system", *Science* **246**, 1454-1459.

Cunha, K. and Smith, V.V.: 1999, "A determination of the solar photospheric boron abundance", *Ap. J.* **512**, 1006-1013.

Dziembowski, W.: 1998, "Shortcomings of the standard solar model", *Space Sci. Rev.* **85**, 37-48.

Feldman, U.: 1998, "FIP effect in the solar upper atmosphere: Spectroscopic results", *Space Sci. Rev.* **85**, 227-240.

Fröhlich, C., Huber, M.C., Solanki, S.K. and von Steiger, R.: 1998, eds., *Solar Composition and its Evolution – From Core to Corona*, Kluwer Academic Publishers, Norwell, MA, (also *Space Sci. Rev.* **85**, R11-R12, 1998).

Gabriel, A.H., Culhane, J.L., Patchett, B.E., Breevelt, E.R., Lang, J., Parkinson, J.H., Payne, J. and Norman, K.: 1995, "Spacelab 2 measurement of the solar coronal helium abundance", *Adv. Space Res.* **15**, 63-67.

Gabriel, M.: 1997, "Influence of heavy element and rotationally induced diffusions on the solar models", *Astron. Astrophys.* **327**, 771-778.

Geiss, J. and Gloeckler, G.: 1998, "Abundance of Deuterium and $^3$He in the protosolar cloud", *Space Sci. Rev.* **84**, 239-250.

Goldberg, L., Müller, E.A. and Aller, L.H.: 1960, "The abundances of the elements in the solar atmosphere", *Ap. J. Suppl. Ser.* **5**, **No 45**, 1-137.

Grevesse, N. and Noels, A.: 1993, "Atomic data and the spectrum of the solar photosphere", *Physica Scripta* **T47**, 133-138.

Grevesse, N. and Sauval, A.J.: 1994, "Molecules in the sun and molecular data", in *Molecular Opacities in the Stellar Environment*, ed., Jørgensen, U.G., Lecture Notes in Physics, Springer-Verlag, New York, NY, **428**, pp.196-209.

Grevesse, N., Noels, A. and Sauval, A.J.: 1995, "Atomic and molecular data in solar photospheric spectroscopy", in *Laboratory and Astronomical High Resolution Spectra*, eds., Sauval, A.J. Blomme, R. and Grevesse, N., ASP Conference Series **81**, pp. 74-87.

Grevesse, N., Noels, A. and Sauval, A.J.: 1996, "Standard abundances", in *Cosmic Abundances*, eds., Holt, S.S. and Sonneborn, G., ASP Conference Series **99**, pp. 117-126.

Grevesse, N. and Sauval, A.J.: 1998, "Standard solar composition", *Space Sci. Rev.* **85**, 161-174.

Grevesse, N. and Sauval, A.J.: 1999, "The solar abundance of iron and the photospheric model", *Astron. Astrophys.* **347**, 348-354.

Gustafsson, B.: 1998, "Is the sun a sun-like star?", *Space Sci. Rev.* **85**, 419-428.

Holweger, H. and Müller, E.A.: 1974, "The photospheric barium spectrum: Solar abundance and collision broadening of Ba II lines by hydrogen", *Solar Physics* **39**, 19-30.

Kallenbach, R., Ipavich, F.M., Kucharek, H., Bochsler, P., Galvin, A.B., Geiss, J., Gliem, F., Gloeckler, G., Grünwaldt, H., Hefti, S., Hovestadt, D. and Hilchenbach, M.: 1998, "Fractionation of Si, Ne and Mg isotopes in the solar wind as measured by SOHO/CELIAS/MTOF", *Space Sci. Rev.* **85**, 357-370.

Kurucz, R.L.: 1995, "The solar spectrum: atlases and line identifications", in *Laboratory and Astronomical High Resolution Spectra*, eds., Sauval, A.J., Blomme, R. and Grevesse, N., ASP Conference Series **81**, pp. 17-31.

Manuel, O.K. and Hwaung, G.: 1983, "Solar abundances of the elements" *Meteoritics* **18**, 209-222.

Pagel, B.E.J.: 1997, *Nucleosynthesis and chemical evolution of galaxies*, Cambridge University Press, Cambridge, UK, 378 pp.

Reames, D.V.: 1998, "Solar energetic particles: sampling coronal abundances", *Space Sci. Rev.* **85**, 327-340.

Reames, D.V.: 1999, "Particle acceleration at the sun and in the heliosphere", *Space Sci. Rev.*, in press.

Reetz, J.: 1998, "Oxygen abundances in cool stars and the chemical evolution of the galaxy", Ph.D. Thesis, Ludwig-Maximilians-Universität München.

Rogers, F.J. and Iglesias, C.A.: 1998, "Opacity of stellar matter", *Space Sci. Rev.* **85**, 61-70.

Russell, H.N.: 1929, "On the composition of the sun's atmosphere", *Ap. J.* **70**, 11-82.

Rutten, R.J.: 1998, "The lower solar atmosphere", *Space Sci. Rev.* **85**, 269-280.

Sauval, A.J., Grevesse, N. and Blomme, R.: 1999, "Revised solar abundances of C, N and O", in preparation.

Solanki, S.: 1998, "Structure of the solar photosphere", *Space Sci. Rev.* **85**, 175-186.

Takada, M. and Shibahashi, H.: 1998, "The refined seismic solar model and the abundance of the heavy elements", in *Structure and Dynamics of the Interior of the Sun and Sunlike Stars*, eds., Korzennik, S.G. and Wilson, A., ESA Special Publications, SP--418, ESTEC, Noordwijk, The Netherlands, pp. 543-548.

Turck-Chièze, S.: 1998, "Composition and opacity in the solar interior", *Space Sci. Rev.* **85**, 125-132.

Turcotte, S. and Christensen-Dalsgaard, J.: 1998, "Solar models with non-standard chemical composition", *Space Sci. Rev.* **85**, 133-140.

Unsöld, A.: 1948, "Quantitative Analyse der Sonnenatmosphäre", *Z. Astrophys.* **24**, 306-326.

Vauclair, S.: 1998, "Element settling in the solar interior", *Space Sci. Rev.* **85**, 71-78.

von Zahn, U. and Hunten, D.M.: 1996, "The helium mass fraction in Jupiter's atmosphere", *Science* **272**, 849-851.

Widing, K.G.: 1997, "Emerging active regions on the sun and the photospheric abundance of neon", *Ap. J.* **480**, 400-405.

Wieler, R.: 1998, "The solar noble gas record in lunar samples and meteorites", *Space Sci. Rev.* **85**, 303-314.

Williams, D.L., Leske, R.A., Mewaldt, R.A. and Stone, E.C.: 1998, "Solar energetic particle isotopic composition", *Space Sci. Rev.* **85**, 379-386.

Wimmer-Schweingruber, R.F., Bochsler, P. and Wurz, P.: 1999, "Isotopes in the solar wind: new results from ACE, SOHO and WIND", in *Solar Wind Nine*, eds., Habbal, S.R., Esser, R., Hollweg, J.V. and Isenberg, P.A., The American Institute of Physics, College Park, MD, pp. 147-152.

Young, P.R., Mason, H.E., Keenan, F.P. and Widing, K.G.: 1997, "The Ar/Ca relative abundance in solar coronal plasma", *Ap. J.* **323**, 243-249.

Zahn, J.-P.: 1998, "Macroscopic transport", *Space Sci. Rev.* **85**, 79-90.

Zurbuchen, T.H., Fisk, L.A., Gloeckler, G. and Schwadron, N.A.: 1998, "Element and isotopic fractionation in closed magnetic structures", *Space Sci. Rev.* **85**, 397-406.

# Isotopic Ratios: The Key to Elemental Abundances and Nuclear Reactions in the Sun

O. Manuel
*Chemistry Department, University of Missouri, Rolla, MO 65401 USA*
*om@umr.edu*

**Abstract:** Differences between the abundances of isotopes of elements in the planetary system, in the solar wind, and in solar flares suggest that a) the most abundant nuclear species in the sun is $^{56}$Fe (the ash of violent thermonuclear reactions at equilibrium) rather than $^{1}$H (the best fuel for thermonuclear reactions), and b) nuclear reactions in the outer regions of the sun produce $^{15}$N by the $^{14}$N($^{1}$H, $\beta^{+}\nu$)$^{15}$N reaction. Diffusion in the sun enriches light-weight elements like H and He at the solar surface, where thermonuclear reactions occur, but the most abundant elements in the bulk sun are Fe, Ni, O, Si, S, Mg, and Ca.

## 1. INTRODUCTION

After cautioning that elemental abundances in the earth's crust and the sun's gaseous envelope may not represent the overall composition of these bodies, William Harkins (1917) used chemical analyses of 443 ordinary meteorites to show that iron is the most abundant element. He found that the seven most abundant elements (Fe, O, Ni, Si, Mg, S, and Ca) all have even atomic numbers, noted a relationship between elemental abundances and nuclear stability, and stated that the abundances of elements are related "*... to the structure of the nuclei of their atoms.*" (Harkins, 1917, p. 856).

Ida and Walter Noddack (1930) used the average density of the moon and the four inner planets (≈ 5.1 g cm-3) to estimate a value for the ratio of iron : troilite : stone = 68 : 9.8 : 100. By assuming that stone meteorites contain 5.5% troilite (FeS), they used the results of analyses on 42 stone meteorites, 19 iron meteorites, and troilite grains from five iron meteorites to show that the seven most abundant elements are O, Fe, Si, Mg, Ni, S, and Ca.

*Origin of Elements in the Solar System: Implications of Post-1957 Observations.*
Edited by O. Manuel, Kluwer Academic/Plenum Publishers, 2000. 

Although the order changed, Noddack and Noddack (1930) and Harkins (1917) concluded that the same seven abundant elements existed in the solar system.

About 99.8% of all the solar system's material is in the sun. Two publications in the late 1930s signaled basic changes in the scientific community's opinion of that body. Goldschmidt (1938) convinced the scientific community that meteorites and the inner planets are rich in elements, like Fe, O, Ni, Si, etc., because they lost volatile elements, like H, He, C, N, etc., and that the sun's atmosphere (Payne, 1925) better represents the overall abundance of the elements. Hans Bethe (1939) suggested that $^{12}C$ might serve as a catalyst for energy production in the sun and other stars by serving as the initial target for a chain of four proton-capture reactions that return the $^{12}C$ nuclide and convert 4 $^{1}H \rightarrow ^{4}He + 2\ \beta^{+} + 2\ \nu$ + energy.

Goldschmidt's 1938 idea remains in vogue even today, but Bethe's CNO cycle lost favor in the scientific community before elements from the sun became available for study. Neutrinos from two intermediate nuclides in Bethe's CNO cycle, $^{13}N$ and $^{15}O$, may exceed the 0.86 Mev threshold of the $^{37}Cl$ detector (Davis, 1955). The embarrassingly low flux of solar neutrinos convinced Davis *et al.* (1972) that Bethe's CNO cycle produces little, if any, of the sun's energy. This is now widely attributed to the proton-proton chain with $E_{\nu} \leq 0.41$ Mev.

## 2. MEASURED ISOTOPIC RATIOS OF THE SUN

About 30 years after the publications by Goldschmidt and Bethe, the Apollo mission returned with lunar soils and breccias containing abundant volatile elements from the solar wind (LSPET, 1969). The solar wind (SW) implanted component masks any indigenous hydrogen, nitrogen, and the noble gases. Subsequent studies also revealed a component of solar flare (SF) elements buried more deeply in the lunar samples (Rao *et al.*, 1991).

Manuel and Hwaung (1983) noted a systematic enrichment of the lighter weight isotopes of hydrogen and the noble gases in the SW-implanted component. The percent enrichment per mass unit decreases in a regular manner as the mass increases across the series: H, He, Ne, Ar, Kr, and Xe. The mass fractionation for the five noble gases is about 200% per amu for SW-He, 27% per amu for SW-Ne, 9% per amu for SW-Ar, 6% per amu for SW-Kr, and 3.5% per amu for SW-Xe. This is shown in Figure 1, where empty rectangles show the relative abundances of noble gas isotopes in planetary material, and filled rectangles show those in the solar wind.

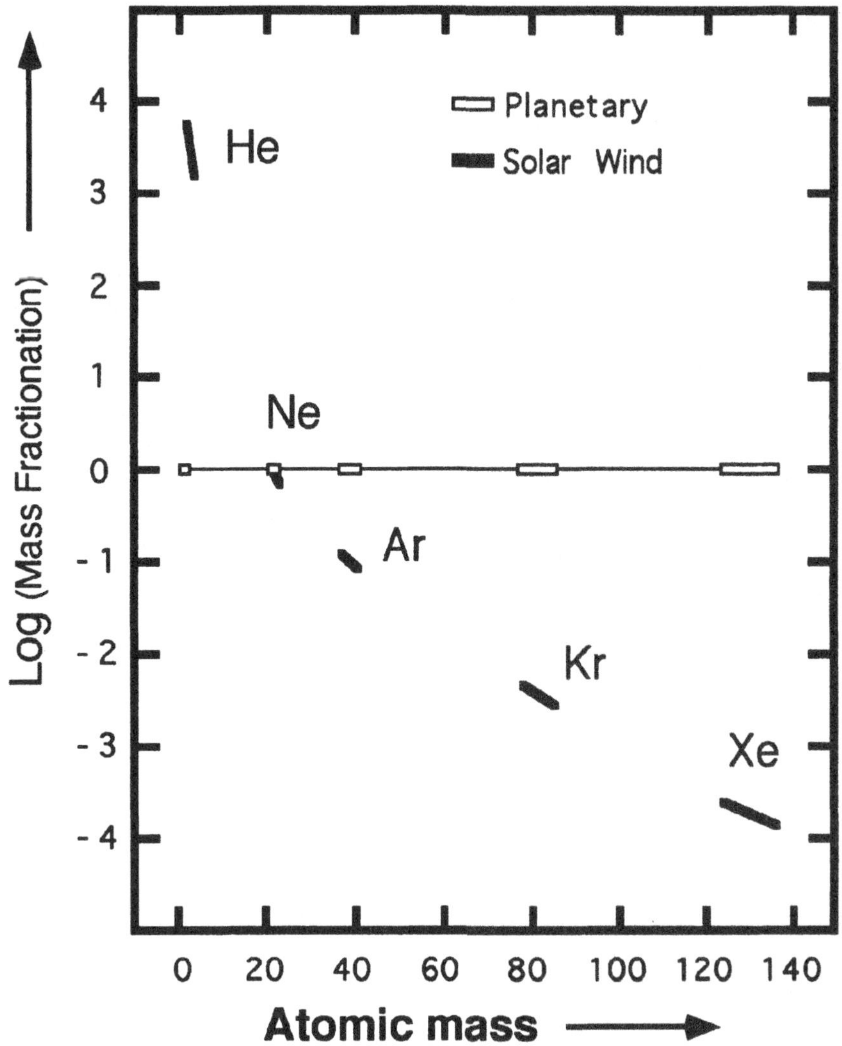


*Figure 1.* Isotope ratios of solar-wind implanted gases show a smooth, mass-dependent fractionation, as expected if intrasolar diffusion enriches lighter nuclides at the solar surface. Mass fractionation is shown relative to $^{20}$Ne.

If the sun and its planetary system formed out of the same batch of elements, then deuterium burning in the sun might have increased the values of the $^{1}$H/$^{2}$H and $^{3}$He/$^{4}$He ratios by converting $^{2}$H into $^{3}$He (Geiss, 1993). This would explain the two largest isotopic anomalies in SW elements, but it does not explain the excess light-weight isotopes for other SW elements. Alternatively, the smooth, mass-dependent fractionation pattern (Figure 1) may mean that ≈ 9-stages of mass fractionation have altered the abundance of any nuclide of mass, m, relative to that of $^{20}Ne$ in the solar wind. Empirically, Manuel and Hwaung (1983) found that:

$$\text{Mass fractionation} = (20/m)^{4.56} \quad (1)$$

where each theoretical stage of mass fractionation alters the isotopic abundance by a factor of $\sqrt{20/m}$.

The light weight isotopes of He, Ne, and Ar are less enriched in solar flare particles than in the solar wind (Rao *et al.*, 1991). Isotopic ratios of Mg in the solar wind (Boschler *et al.*, 1996) also vary systematically with velocity and the heavier mass isotopes become increasingly abundant at higher velocities and most abundant in solar flares (Selesnick *et al.*, 1993). Manuel and Ragland (1997) displayed these isotopic ratios of SW and SF elements in the manner shown in Table 1.

*Table 1.* Isotopic ratios of He, Ne, Mg and Ar in the solar wind and in solar flares.

| Isotopic Ratios | Solar Wind | Solar Flares | SW/SF | $\Delta m/m_{avg.}$ |
|---|---|---|---|---|
| $^{3}He/^{4}He$ | $4.1 \times 10^{-4}$ | $2.6 \times 10^{-4}$ | 1.58 | 0.29 |
| $^{20}Ne/^{22}Ne$ | 13.6 | 11.6 | 1.17 | 0.095 |
| $^{24}Mg/^{26}Mg$ | 7.0 | 6.0 | 1.17 | 0.080 |
| $^{36}Ar/^{38}Ar$ | 5.3 | 4.8 | 1.10 | 0.054 |

In Table 1, column 2 shows isotopic ratios in the solar wind, column 3 shows those in solar flares, and column 4 shows the fractional change in these ratios. If these shifts are caused by partial disruption of the mass fractionation seen in SW-implanted elements (Figure 1), then the fractional changes in the column 4 should scale with the values of _m/m $_{avg.}$ tabulated in column 5. That this is true can be seen by comparing values in column 4 with the sum of one plus twice the value in column 5. Differences between the isotopic ratios of SW and SF elements suggest that SF elements come from deeper within the sun, effectively by-passing ≈ 3.5 stages of mass-dependent fractionation (Ragland and Manuel, 1998).

Isotopic abundances of one element from the sun, N, do not follow the trend shown in Figure 1. Currently, SW-N is enriched in the heavier isotope, $^{15}N$, although the lighter mass isotope, $^{14}N$, was enriched in the early SW-N. Nitrogen is unique in that its isotopic abundances in the sun have changed over geologic time (Kerridge, 1975). Differences in the isotopic abundances of SW-N and SF-N are also unlike the pattern shown in Table 1. The $^{14}N/^{15}N$ ratio in the solar wind is *lower* than that in solar flares (Kerridge, 1993), whereas the light/heavy isotopic ratio for magnesium and the five noble gases in the solar wind is *higher* than that in solar flares (Table 1). This anomalous behavior of nitrogen isotopic ratios in the sun may be the first experimental observation that confirms the existence of a nuclear reaction, and identifies its location, in the sun.

## 2.1 Implications for elemental abundances in the sun

If solids condensed from H, He-rich material of the sun's atmosphere (Payne, 1925), they would likely incorporate only a small fraction of its H and He (Goldschmidt, 1938). The obvious validity of that assertion has led many to accept that planetary solids formed in this way, despite numerous isotopic ratios that directly falsify this view (Manuel, this volume). Nolte and Lietz (this volume) explain why deuterium burning in the sun (Geiss, 1993) is an unlikely explanation for the high $^{3}He/^{4}He$ ratio in the solar wind. Manuel and Hwaung (1983) gave six reasons for rejecting planetary fractionation in favor of solar fractionation as the cause of the mass-dependent fractionation pattern shown in Figure 1. They concluded that diffusion inside the sun is selectively moving lighter elements and the lighter isotopes of individual elements to solar surface. They corrected the photospheric abundances of Ross and Aller (1976) with the empirical mass fractionation effect given by Equation (1) and concluded that the seven most abundant elements in the bulk sun are in decreasing order: Fe, Ni, O, Si, S, Mg, and Ca.

These are the same seven elements that Harkins (1917) and Noddack and Noddack (1930) found as the major constituents of meteorites. However, the paper of Manuel and Hwaung (1983) has been largely ignored, probably because these authors failed a) to mention the earlier work by Harkins (1917) and Noddack and Noddack (1930), and b) to point out how unlikely it is that the empirical mass fractionation pattern of isotopes in the solar wind would select seven trace elements from the solar photosphere that are identical to those which Harkins (1917) and Noddack and Noddack (1930) found to be most abundant in hundreds of ordinary meteorites.

If the observations (Figure 1) defined empirically by Equation (1) are not the result of diffusion in the sun, what are the chances that this misunderstanding would select the same seven elements from the solar photosphere that Harkins (1917) found to comprise about 99% of the material in hundreds of meteorites? The probability that Equation (1) would by chance select Harkins' set of seven elements from the 83 in the sun would be about $7!(83-7)!/83! = 2 \times 10^{-10}$ if each element had an equal chance for selection. Since Fe, Ni, O, Si, S, Mg, and Ca are trace elements in the H,He-rich solar photosphere, with atomic abundances of only $3.0 \times 10^{-5}$, $1.9 \times 10^{-6}$, $6.5 \times 10^{-4}$, $4.2 \times 10^{-5}$, $1.5 \times 10^{-5}$, $3.8 \times 10^{-5}$, and $1.9 \times 10^{-6}$, respectively (Ross and Aller, 1976), the probability for the chance selection of Harkins' set of seven elements would be much less if their probability of selection depended on their atomic abundances in the solar photosphere.

All of the elements in Figure 1 are volatile, and isotopic analysis of a refractory element like Mg was one of three tests that Manuel and Hwaung (1983) advanced for their proposal of diffusion inside the sun. The data in Table 1 confirm correlated shifts in the isotopic ratios of He, Ne, Mg, and Ar, as expected if energetic events at the solar surface disrupt ≈ 3.5 stages of diffusive mass-fractionation (Ragland and Manuel, 1998) possibly by dredging up less mass fractionated material from the solar interior.

## 2.2 Implications for nuclear reactions in the sun

The value of the $^{15}N/^{14}N$ ratio in the SW has increased over geologic time (Kerridge, 1975). This observation suggests that the $^{14}N(^{1}H, \beta^{+}\nu)^{15}N$ part of the CNO cycle proposed by Bethe (1939) may occur in the outer regions of the sun, producing solar neutrinos, violent activity at the solar surface, and increasing the value of the $^{14}N/^{15}N$ ratio over geologic time. Among volatile elements trapped in lunar surface material, Geiss and Bochsler (1982) found that nitrogen is unique in showing isotopic variations ≥ 30%, but they decided against thermonuclear production of $^{15}N$ at the solar surface.

The heavier nitrogen isotope, $^{15}N$, is more abundant in SW than in SF nitrogen (Kerridge, 1993), which is an inversion of the isotopic anomaly patterns observed for SW and SF magnesium and the noble gases. The heavier isotopes of He, Ne, Mg, and Ar are more abundant in SF than in SW elements. These observations suggest that solar flares which dredge up less mass-fractionated He, Ne, Mg, and Ar (Manuel and Ragland, 1997) also bring up nitrogen with less of the $^{14}N(^{1}H, \beta^{+}\nu)^{15}N$ product.

Recently, Kim et al (1995, p. 383) note that "*... the long-term trend in the $^{15}N/^{14}N$ signature, rather than being simply a linear increase with time as originally proposed, has experienced two excursions, to minimum (< -28%) and maximum (> +16%) $\delta^{15}N$ values in a roughly 3.5 Gyr time frame.*" These observations offer a test for the suggestions that the $^{14}N(^{1}H, \beta^{+}\nu)^{15}N$ reaction occurs in the outer regions of the sun and that solar flares carry nitrogen with less of this $^{14}N(^{1}H, \beta^{+}\nu)^{15}N$ product. If so, then fluctuations in values of the $^{15}N/^{14}N$ ratio will be accompanied by excursions in the isotopic ratios of noble gases and magnesium, but in the opposite direction.

# 3. CONCLUSIONS

The Apollo mission returned samples from the moon in 1969. Analyses on the isotopic abundances of elements implanted there from the sun suggest two errors in the mid-20th Century: 1) Acceptance of the proposal that the sun's photosphere (Payne, 1925) represents elemental abundances for the solar system; and 2) Rejection of Bethe's (1939) proposal that the CNO cycle serves as a path for hydrogen fusion in the sun.

Seventeen years into the 20th Century, Harkins (1917) used the wet chemical analyses of hundreds of meteorites to conclude that Fe, O, Ni, Si, Mg, S, and Ca are the seven most abundant elements in the solar system. Seventeen years before the start of the 21st Century, Manuel and Hwaung (1983) combined measurements of isotopic ratios in the solar wind (Eberhardt et al., 1972) with line spectra of elements in the solar photosphere (Ross and Aller, 1976) to conclude that Fe, Ni, O, Si, S, Mg, and Ca are the seven most abundant elements in the sun itself.

The probability that an empirical fit to the abundance pattern of SW isotopes, mass-fractionation = $(20/m)^{4.56}$, would by chance select these seven "trace" elements from the solar photosphere is about one in $5 \times 10^9$.

In terms of nuclear physics, $^{56}$Fe and $^1$H are opposites: One is the ash and the other the best fuel for thermonuclear reactions at equilibrium. Among stable nuclides $^{56}$Fe has the lowest mass per nucleon; $^1$H has the highest. Manuel and Sabu (1975, 1977) explained the origin of such an Fe-rich sun. Additional details are given in this volume by Hwaung and Manuel.

Geiss and Bochsler (1982) used the weak correlation of excess $^{13}$C with excess $^{15}$N in the solar wind (Becker, 1980) to set the following limits on thermonuclear production of $^{15}$N at the solar surface: $2 \times 10^8$K < T < $5 \times 10^8$K; $\rho \approx 0.03$ g cm$^{-3}$; reaction time of $\approx 10^2$ s followed by fast quenching. These nuclear reactions at the sun's surface and the elemental abundances concluded above for its interior are inconsistent with the standard solar model (Dar and Shaviv, 1996).

Harkins (1917) analyzed hundreds of ordinary meteorites to conclude that the most abundant elements are Fe, O, Ni, Si, Mg, S, and Ca. Recent studies have focused instead on elemental abundances in a very rare meteorite group, the CI type carbonaceous chondrites. The results are enigmatic. Some minerals of CI chondrites contain isotopically anomalous elements; others have undergone aqueous alteration (Ebihara *et al.*, this volume). Although they contain only a small fraction of the volatile elements (H, He, C, N, etc.) that are dominant at the sun's surface, nevertheless it is reported that elemental ratios among the remaining ≈ 75 nonvolatile elements are remarkably similar to those in the solar photosphere. This remains as a major puzzle.

## ACKNOWLEDGMENTS

The author acknowledges the kindness of Fate in bringing Professor Paul Kazuo Kuroda, a world-class scholar and an intellectual giant, from the

University of Tokyo to the University of Arkansas to direct the author's graduate research. He is also grateful for the talented students who made his research career possible at the University of Missouri and his secretary, Ms. Phyllis Johnson, who was surrogate mother to his students and a trusted ally since he joined the faculty a few years before the first Apollo mission. The thoughtfulness of good friends, especially Nirmal K. Shastri, the late Dwarka Das Sabu, James E. Johnson, and Marvin W. Rowe, allowed the author to weather the storms of life and complete this project. This manuscript benefited from comments by Jack Koenig, Ramachandran Ganapathy, V. A. Samaranayake, Dan Armstrong, and Margaret Burbidge.

## REFERENCES

Becker, R.H.: 1980, "Evidence for a secular variation in the 13C/12C ratio of carbon implanted in lunar soils", *Earth Planet. Sci. Lett.* **50**, 189-196.

Bethe, Hans A.: 1939, "Energy production in stars", *Phys. Rev.* **55**, 103.

Boschler, P., Gonin, M., Sheldon, R. B., Zurbuchen, Th., Gloeckler, G., Hamilton, D.C., Collier, M. R. and Hovestadt, D.: 1996, "Abundance of solar wind magnesium isotopes determined with wind/mass", in *Solar Wind 8 Proceedings*, 1-4.

Chaussidon, M. and Robert, F.: 1999, "Lithium nucleosynthesis in the sun inferred from the solar wind $^7$Li/$^6$Li ratio", *Nature* **402**, 270-273.

Dar, A. and Shaviv, G.: 1996, "Standard solar neutrinos", *Ap. J.* **468**, 933-946.

Davis, R., Jr.: 1955, "Attempt to detect the antineutrinos from a nuclear reactor by the $^{37}$Cl(n,e-)$^{37}$Ar reaction", *Phys. Rev.* **97**, 766-769.

Davis, R., Jr., Evans, J. C., Radeka, V. and Rogers, L. C.: 1972 "Report on the Brookhaven solar neutrino experiment" in *Neutrino '72, Europhysics Conference Proceedings*, eds., Frenkel, A. and Marx, G., Balatonfüred, Hungary, June 11-17, 1972. Organized by the Hungarian Physical Society, vol. **1**, pp. 5-15.

Eberhardt, P., Geiss, J., Graf, H., Grögler, N., Mendia, M. D., Mörgeli, M., Schwaller, H., Stettler, A., Krähenbühl, U. and von Guten, H. R.: 1972, "Trapped solar wind noble *gases in Apollo 12 lunar fines 12001 and Apollo 11 breccia 10046"*, Proc. 3rd Lunar Sci. Conf. **2**, 1821-1856.

Geiss, J.: 1993, "Primordial abundance of hydrogen and helium isotopes", in *Origin and Evolution of the Elements*, eds., Prantzos, N., Vangioni-Flam, E. and Cassè, M., Cambridge University Press, Cambridge, pp. 89-106.

Geiss, J. and Boschler, P.: 1982, "Nitrogen isotopes in the solar system" *Geochim. Cosmochim. Acta* **46**, 529-548.

Goldschmidt, V. M.: 1938, "Geochemische Verteilungsgestze der Elemente. IX. Die Mengenverhältnisse der Elemente und der Atom-Arten", Skrifter Norske Videnskaps-Akad., Oslo I Math.-Naturv. Klasse. no. 4, 148 pp.

Harkins, W.D.: 1917, "The evolution of the elements and the stability of complex atoms", *J. Am. Chem. Soc.* **39**, 856-879.

Kerridge, J.F.:1975, "Solar nitrogen: Evidence for a secular increase in the ratio of nitrogen-15 to nitrogen-14", *Science* **188**, 162-164.

Kerridge, J.F. :1993, "Long term compositional variations in solar corpuscular radiation: Evidence from nitrogen isotopes in lunar regolith", *Rev. Geophys.* **31**, 423-437.

Kim, J.S., Kim, Y., Marti, K. and Kerridge, J.F.: 1995, "Nitrogen isotope abundances in the recent solar wind", *Nature* **375**, 383-385.

LSPET (The Lunar Sample Preliminary Examination Team): 1969, "Preliminary examination of lunar samples from Apollo 11", *Science* **165**, 1211-1227.

Manuel, O.K. and Hwaung, G.: 1983, "Solar abundances of the elements", *Meteoritics* **18**, 209-222.

Manuel, O.K. and Ragland, D.E.: 1997, Abstract 281, "Diffusive mass fractionation effects across the isotopes of noble gases and magnesium in the solar wind and in solar flares", 1997 Midwest Regional Meeting of the American Chemical Society, Osage Beach, MO, USA, p. 128.

Manuel, O.K. and Sabu, D.D.: 1975, "Elemental and isotopic inhomogeneities in noble gases: The case for local synthesis of chemical elements", *Trans. Missouri Acad. Sci.* **9**, 104-122.

Manuel, O.K. and Sabu, D.D.: 1977, "Strange xenon, extinct superheavy elements and the solar neutrino puzzle", *Science* **195**, 208-209.

Noddack, I. and Noddack, W.: 1930, "Die Häufigkeit der chemischen Elemente", *Naturwissenschaften* **18**, 757-764.

Owen, T., Mahaffy, P., Niemann, H.B., Atreya, S., Donahue, T., Bar-Nun, A. and de Pater, I.: 1999, "A tow temperature origin for the planetesimals that formed Jupiter", *Nature* **402**, 269-270.

Payne, C.H.: 1925, "Stellar atmospheres", in *Harvard Observatory Monographs*, ed., Shapley, H., no. 1, Cambridge, MA, 215 pp.

Ragland, D.E. and Manuel, O.K.: 1998, Abstract Nucl. 029, "The isotope record of diffusive mass fractionation of magnesium and noble gases in the solar wind and in solar flares", 215th National Meeting of the American Chemical Society, Dallas, TX.

Rao, M. N., Garrison, D. H., Bogard, D. D., Badhwar, G. and Murali, A. V.: 1991, "Composition of solar flare noble gases preserved in meteorite parent body regolith", *J. Geophys. Res.* **96**, 19.321-19.330.

Ross, J.E. and Aller, L.H.: 1976, "The chemical composition of the sun", *Science* **191**, 1223-1229.

Selesnick, R.S., Cummings, A.C., Cummings, J.R., Leske, R.A., Mewalt, R.A., Stone, E. C. and von Rosenbinge, T. T.: 1993, "Coronal abundances of neon and magnesium isotopes from solar energetic particles", *Ap. J.* **418**, L45-L48.

*Note added in proof: As this paper went to press, two papers appeared in the 18 November 1999 issue of Nature about elemental abundances and nuclear reactions in the sun.*

- *Owen et al. (1999) report that abundances of heavy noble gases are much greater in Jupiter than first thought. The high abundances of heavy elements may indicate formation of Jupiter at low temperatures, as suggested by the authors, or this may confirm diffusion in the sun that enriches lighter elements at the solar surface.*
- *Chaussidon and Robert (1999) report evidence for the nucleosynthesis of lithium in the outer regions of the sun.*

# Critical Evaluation of CI Chondrites as the Solar System Standard of Elemental Abundances

M. Ebihara[1], K. Shinotsuka[1,2], H. Ozaki[1], and Y. Oura[1]

*[1]Department of Chemistry, Graduate School of Science, Tokyo Metropolitan University, Hachioji Tokyo 192-0397, Japan, [2]National Institute for Fusion Science, Toki, Gifu 509-5292, Japan.*

*ebihara-mitsuru@c.metro-u.ac.jp*

**Abstract:** In compiling the solar system abundance of the elements, analytical data of CI chondrites have been playing an important role for the past several decades. With an increase in the number of high quality data for chondritic meteorites including CI chondrites, it is a good opportunity to critically evaluate the status of CI chondrites as the solar abundance standard. In this article, the abundance data of rare earth elements and refractory siderophile elements in carbonaceous and ordinary chondrites are discussed. Each group of elements are typical refractory elements and are supposed to have behaved similarly in planetary processes in the early solar system. New observations for chemical composition show an apparent fractionation among each group of elements. As we have no criteria for judging which group of meteorites hold the initial composition of the solar nebula from which our solar system formed, we have to return to the composition of the sun. A new measure for such a judgement will be able to be in hand when the solar wind composition is accurately determined for a large number of elements. Fortunately, this will come true in early 21 century by the Genesis space mission project of NASA.

## 1. CI CHONDRITES AS THE SOLAR ABUNDANCE STANDARD?

Solar system abundances of the elements are among the most fundamental of numbers for earth and planetary sciences. These values are often used as normalizing values for analytical data and boundary conditions for theoretical modeling. To compile the solar system abundance of the

*Origin of Elements in the Solar System: Implications of Post-1957 Observations.*
Edited by O. Manuel, Kluwer Academic/Plenum Publishers, 2000. 

elements, two main data sources (meteorites and solar photosphere) have been used since the early compilation of the solar abundances (Goldschmidt, 1938; Suess and Urey, 1956). Meteoritic data are obtained by chemical analysis in the laboratory while solar photospheric data are obtained by spectroscopic observation on the earth. The quality of both data has increased with advances in analytical techniques. Among these two data sets, meteorite data are generally more precise, but comparison of the two data sets has become possible with the great improvement in the photospheric values in recent years (Ebihara, 1996).

Except for highly volatile elements such as hydrogen, carbon, nitrogen and noble gases, the solar system abundances are virtually determined from the chemical analysis of meteorites. Among meteorites, CI chondrites have been the most extensively analyzed for this purpose, because CI chondrites are the richest in volatile components among meteorites. Indeed, the water content (measured as $H_2O$) exceeds 20% (weight %) because the matrix is composed of a large amount of hydrous silicates. When Anders and Ebihara (1982) compiled the abundance table of the elements in the solar system, they surveyed and scrutinized the data available by that time, and deduced the most reliable values for the solar system abundances. This abundance table was widely accepted by the scientific community and was updated in 1989 by Anders and Grevesse. There were few significant changes in solar system abundances of the elements between the two tables, because the analytical data of meteorites reported during that period were limited. In contrast, the solar abundances improved significantly.

As mentioned above, the matrix of CI chondrites consists of fine-grained hydrous silicates for the most part. Although chondrules and aggregates, which are commonly observed in other carbonaceous chondrites, are almost absent in CI chondrites, the presence of high temperature minerals such as olivine and pyroxene have been reported. Indeed, the hydrous silicate materials in matrix are considered to have been produced by the aqueous alteration of these high temperature, anhydrous silicates (Tomeoka and Buseck, 1988). The presence of vein minerals such as sulfates and carbonates (Dufresne and Anders, 1962; Richardson, 1978; Fredriksson and Kerridge, 1988) is another evidence for the aqueous alteration CI chondrites experienced. Thus, CI chondrites undoubtedly experienced some sort of alteration, possibly on their parent body(ies). Now that CI chondrites have been confirmed to have the complex history in their formation, one may question the long-standing status of CI chondrites as the best choice of data source for the solar system abundance of the elements.

In this paper, we critically evaluate CI chondrites based on some data recently obtained in our laboratory for rare earth elements and siderophile elements in CI and other chondrites. As these elements are typical refractory elements, their relative abundances in each group are believed to be invariable in CI chondrites or even in chondrites. After careful determination of these elements, we found a faint but distinct variation in both groups of elements among chondrite groups. In order to evaluate the suitability of meteorites for the solar abundance standard, we need an absolute measure. Finally, we discuss a promising measurement possibility available to us in early 21st century.

## 2. SIDEROPHILE ELEMENT ABUNDANCES IN CHONDRITIC METEORITES

Siderophile elements have a large variety of volatilities. The difference in volatility may have led to a systematic fractionation of elements during condensation and/or accretion processes. As far as refractory siderophile elements such as Re, Os, Ir and Ru are concerned, these elements are thought to be unfractionated from each other in chondritic meteorites, occurring in the same proportion in meteorites. There are, however, several reports indicating that refractory siderophiles were fractionated in chondritic meteorites (Morgan and Lovering,, 1967; Ebihara *et al.*, 1982; Ebihara and Ozaki, 1995). The fractionation of refractory siderophile elements has been a major concern in compiling solar system abundances of the elements. When Anders and Ebihara (1982) compiled their solar abundance table of the elements, they pointed out that Re and Ir were possibly fractionated among chondrites. Anders and Grevesse (1989), however, concluded that the refractory siderophiles were not fractionated in chondritic meteorites and that previously recognized variation in Re/Ir and Re/Os ratios in chondrites were artificial rather than indigenous. When analyzing a large number of Antarctic unequilibrated ordinary chondrites (UOC) for Re, Os and Ir by radiochemical neutron activation analysis, Ebihara and Ozaki (1995) found large fractionations among these elements in Antarctic UOC. Compiling abundance data of these elements for other chondrites groups, they found that Re/Ir and Re/Os ratios in CI chondrites are smaller than those in ordinary and other carbonaceous chondrites. As it became apparent that Re is fractionated among ordinary and carbonaceous chondrites and that the solar abundances of Re (Si-normalized) proposed by Anders and Grevesse (1989) is the lowest value among (Si-normalized) Re abundances in carbon aceous and H chondrites, they suggested a change of solar abundance of Re.

Extending to other groups of chondrites and other siderophile elements, we have further pursued this issue. Figure 1 shows elemental correlation between Ir and other refractory siderophiles (Re, Os and Ru) in various groups of ordinary and carbonaceous chondrites.

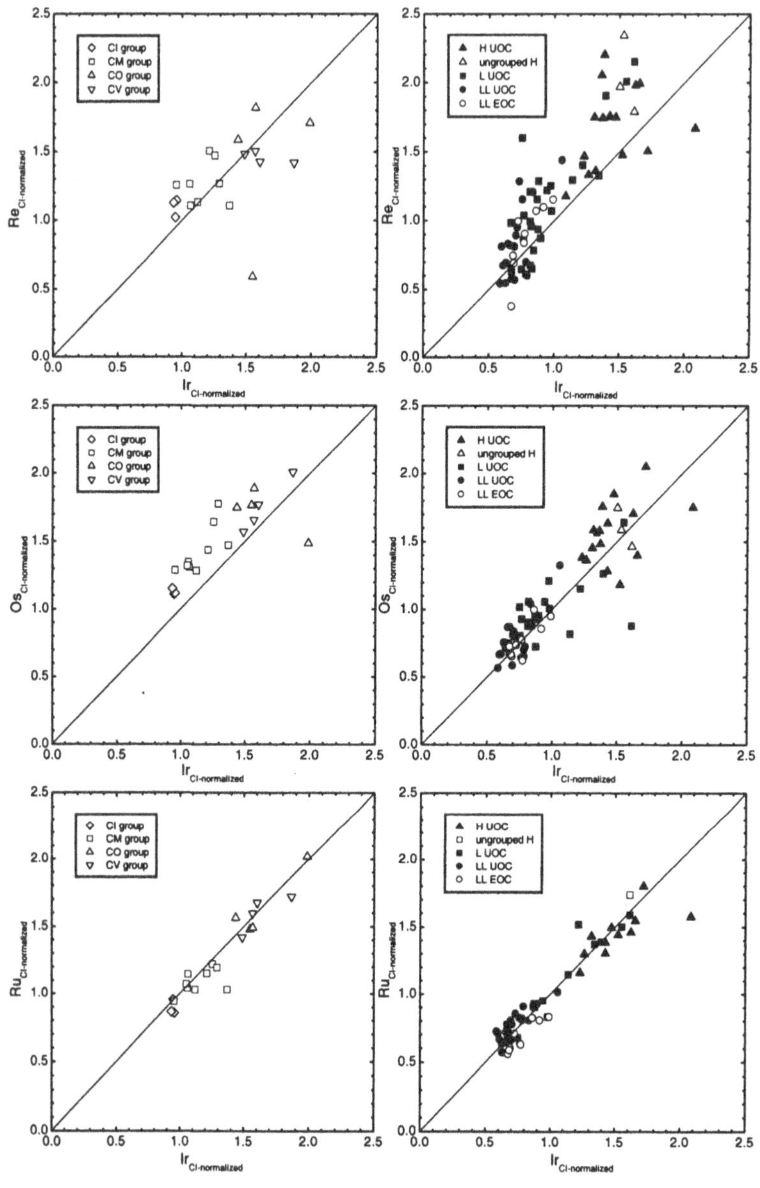


*Figure 1.* Correlations of Re, Os and Ru *vs.*. Ir in chondritic meteorites. These data were obtained by radiochemical neutron activation analysis in our laboratory and individual values will appear elsewhere. A straight line in each correlation plot indicates the CI ratio. Ru/Ir ratios for all groups of chondrites analyzed are in an excellent agreement with the CI ratio. Os and Ir correlate very nicely, but Os/Ir ratios seem to be off the CI line, being slightly shifted upward. This must be due to the small Os value of solar system abundance. The correlation of Re and Ir is poor compared with other pairs. Apparently, Re is fractionated from other refractory siderophile elements in chondritic meteorites.

An excellent correlation can be seen for the pair of Ru and Ir for both carbonaceous and ordinary chondrites. A similarly good correlation can be observed for Os and Ir, but its correlation line does not seem to be superimposed on the CI line. Actually, it is slightly shifted upward. This must be due to the value of Os presented for solar system abundance by Anders and Grevesse (1989). In fact, in revising a solar abundance of Os, Anders and Grevesse (1989) rejected several Os data that Anders and Ebihara (1982) once adopted, leading to a 5.9 % smaller value than the previous value of Anders and Ebihara (1982). We believe that the revision was done in the wrong direction. In contrast to Ru/Ir and Os/Ir ratios, Re/Ir shows a large scattering. Except for several anomalous carbonaceous chondrites, Re/Ir ratios scatter above the CI line for carbonaceous and ordinary chondrites. If literature values for carbonaceous chondrites (not shown in Figure 1) are plotted together, such a scattering and deviation can be clearly recognized. Apparently, Re is highly fractionated from other refractory elements like Os, Ir and Ru. Such a fractionation must have been caused by aqueous alteration on the CI parent body (Ebihara *et al.,* 1982). The question is which group of chondrites represents the original solar system material as far as refractory siderophile abundances are concerned. Because Ir/Os, Re/Os and Re/Ir ratios are individually invariable among Am, CV and CO chondrites, these meteorites could be better candidates than CI.

## 3. RARE EARTH ELEMENT ABUNDANCES IN CHONDRITIC METEORITES

Rare earth elements (hereafter REE; here, Sc and Y are excluded) are typical refractory lithophile elements. Because of their refractoriness and similarity in chemical behavior, REE tend to behave similarly as a group. During the condensation process, REE are predicted to be taken into so-called high temperature oxide minerals, possibly in making a solid solution with these minerals. After being condensed into solid phases, REE must have migrated from one mineral phase to another during the thermal activity on the meteorite parent bodies. The redistribution due to such migration (diffusion) would have occurred on a small scale. As chondrites show no evidence of melting, their parent bodies have not experienced heavy thermal metamorphism caused by intense heating. Thus, even if REE may be heterogeneously distributed in meteorites, such heterogeneity is only in a local scale and their bulk abundance must be invariable if a fairly large amount of chondrites is used for analysis.

Using inductively coupled plasma mass spectrometry (hereafter ICP-MS), we reported very accurate data for all stable REE in several chondritic meteorites (Shinotsuka *et al.*, 1995). In this study, two interesting features in CI-normalized REE abundances (so-called REE abundance pattern) were observed; fractionated REE abundance pattern in Allende and zigzag alteration in all chondrites analyzed. For Allende, we used the Smithsonian reference sample prepared by E. Jarosewich (Jarosewich *et al.*, 1987) and observed a positive anomaly of Tm, a fractionation between light REE and heavy REE. These features are common for Group II inclusions in Allende.

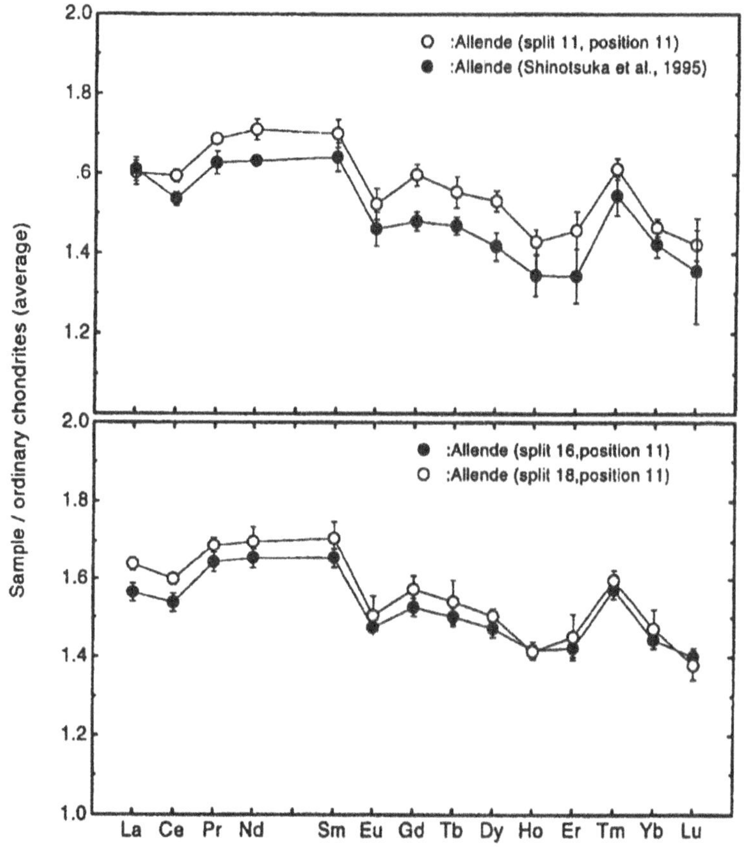


*Figure 2.* Rare earth element abundance patterns for different batches of the Smithsonian Allende reference samples. Average REE content data for ordinary chondrite (see Figure 3) are used for normalization. REE are determined by ICP-MS. As recognized, abundance patterns are not flat, but fractionated among REE. There appear a positive anomaly of Tm and a fractionation between light REE and heavy REE. These features are also noticed for type II CAI. As the Smithsonian reference sample was prepared from a large mass of the meteorite, our observation implies that a whole Allende has the fractionated REE pattern.

Figure 2 shows REE abundance patterns for the Allende reference sample later obtained in our laboratory by ICP-MS. Three different batches of the reference sample were analyzed and all show similar patterns. Considering that the Allende reference sample was prepared by using an amount of 4 kg, such a characteristic pattern is suggested to reflect a typical REE abundance in the whole meteorite. This is the first observation that a whole chondrite has a fractionated REE abundance compared with CI abundance. We are continuing our study of detailed REE abundances for other carbonaceous chondrites including CV.

The second feature pointed above is not limited to Allende, but is commonly observed in all chondrites analyzed. We have analyzed 28 ordinary chondrites, 10 of which are Antarctic meteorites and the rest are non-Antarctic falls. Some of them show apparent fractionated REE patterns, possibly because we used a limited amount of meteorite specimens in which REE are heterogeneously distributed. Using statistics, 28 ordinary chondrites are divided into two groups; one with no fractionated REE abundances (17 chondrites) and another with fractionated REE abundances from a statistical point of view (11 chondrites).

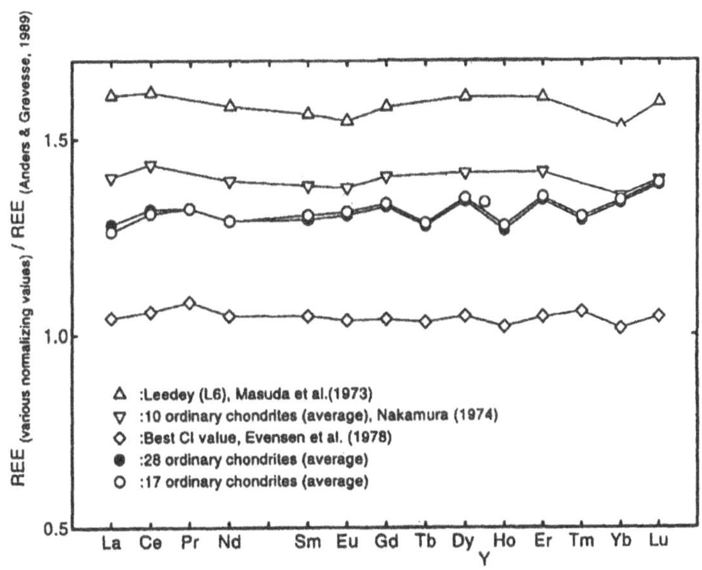


*Figure 3.* Rare earth abundance patterns for ordinary chondrites. Patterns for two sets of ordinary chondrites (averages of 28 and 17 chondrites) show zigzag change from middle to heavy REE span. These data were obtained by ICP-MS in our lab. and their individual values will be reported elsewhere. Apparently, mono-isotopic elements (Tb, Ho and Tm) are plotted with relatively low abundances, possibly caused by relatively larger values of solar system abundances for these elements compiled by Anders and Grevesse (1989).

Figure 3 shows CI-normalized REE abundance patterns for 17 chondrites with unfractionated REE and all 28 ordinary chondrites. There appears to be no significant difference between the two groups. In addition, Figure 3 shows REE patterns for some other chondrite or chondrite group proposed so far for normalizing standards. REE patterns having no data for Pr, Tb, Ho and Tm, which are monoisotopic elements, are based on isotopic dilution analysis. Data for those elements are not determined values for best CI values presented by Evensen *et al.* (1978), but estimated ones. From Figure 3, it is apparent that a zigzag pattern observed for ordinary chondrites by ICP-MS is attributable to erratically high abundances of mono-isotopic REE in the CI values. Note that the REE patterns shown in Figure 2 are normalized to the ordinary chondrite average, so any zigzag change may be offset.

Using REE abundance data for 17 ordinary chondrites having virtually no fractionated REE patterns, REE abundances in the CI chondrites, Orgueil and Ivuna, are normalized and their patterns are shown in Figure 4. We analyzed two Orgueil samples from different museums, one from the American Museum of National History (New York in the figure legend) and another from the National Museum of Natural History, Paris (Paris in the legend). Essentially, the two patterns of Orgueil are identical and unfractionated among REE. Ivuna also shows a similar pattern.

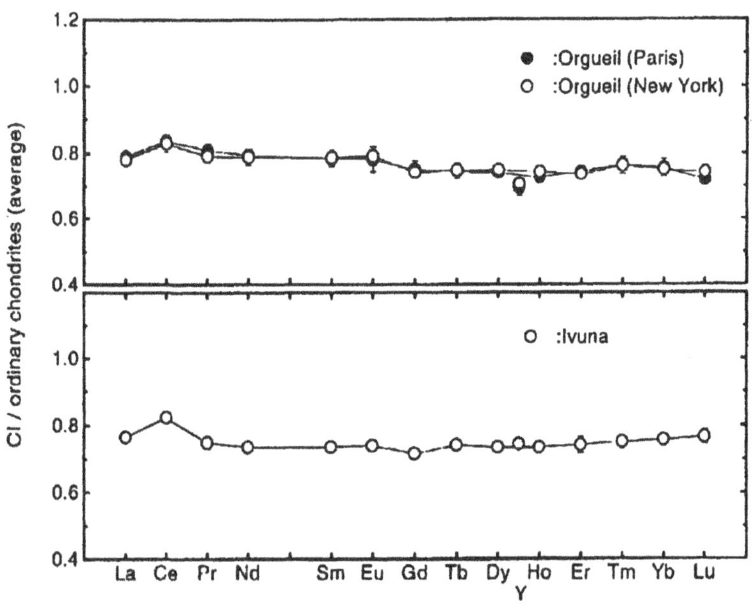


*Figure 4.* REE abundance pattern for CI chondrites obtained by ICP-MS analysis in our laboratory normalized to average values of 17 ordinary chondrite data. Two different specimens of Orgueil were analyzed. As expected, patterns are virtually unfractionated, with a possible positive anomaly of Ce for all specimens and a faint fractionation between light and heavy REE. A positive anomaly of Ce may have been introduced into these specimens before being subjected to analysis.

The absence of a zigzag variation in middle to heavy REE span in Figure 4 indicates that such a variation is not inherent in CI chondrites. Although almost unfractionated patterns in Figure 4, we may notice two faint fractionations; a small positive anomaly of Ce for both Orgueil and Ivuna and a fractionation between light and heavy REE for Orgueil. We are confident with our measurements but cannot insist that our specimens used are representative for CI chondrites. We cannot exclude the possibility that the Ce anomaly reflects contamination that the samples suffered before the analysis.

If REE are not fractionated in chondritic meteorites as long believed, we may use REE abundances in ordinary chondrites rather than those in CI chondrites for the normalizing values, simply because we have a large number of ordinary chondrites. Considering the much higher similarity in chemical characteristics for REE than that for refractory siderophile elements, the adoption of average values of REE from 17 selected ordinary chondrites as solar abundances of REE may be supported. A test of how suitable these newly proposed values are for solar abundance of REE may be tried by drawing a Suess plot of abundances of odd A nuclides (Suess, 1947) like Figure 5.

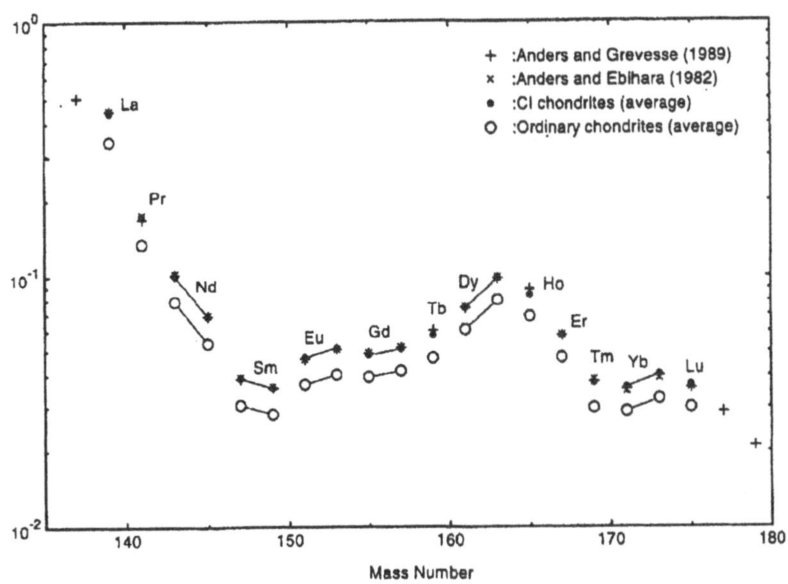


*Figure 5.* Suess plot for odd A nuclides in the region of REE. Plots of ordinary chondrite run parallel to those of CI chondrites as expected. Small changes occur at mono-isotopic elements. The gap in Sm-Eu-Gd region noted for CI chondrites can also be seen for ordinary chondrites, although the gap seems marginally decreased. Whether this gap indicates the limit of the nuclear systematics postulated by Suess must be judged by a new measure for the solar abundance. Such a measurement will be available early in the 21st century.

These plottings are shown for CI values by Anders and Grevesse (1989), CI values (average) by ICP-MS (the same data for Figure 4) and selected 17 ordinary chondrites (average) by ICP-MS. For ordinary chondrites, we used Si values for H, L and LL groups compiled by Wasson and Kallemeyn (1988). As shown, all plottings are very similar, with some minor modification for mono-isotopic elements for ICP-MS-based data. We notice that the discontinuity between Sm and Gd pointed by Anders and Grevesse (1989) still stands for new two sets of data, although the gap between Eu and Gd is marginally decreased for ordinary chondrite average.

## 4. SOLAR WIND - A GOOD CANDIDATE FOR SOLAR ABUNDANCE STANDARD

Solar wind can be a source of data for solar abundances of the elements. Only a few elements (hydrogen, helium, carbon, nitrogen, oxygen, neon, silicon, argon and iron) have so far been determined in the solar wind. Solar wind abundances of noble gas elements having small Z (helium, neon and argon) were determined by the Apollo missions (Geiss *et al.*, 1972), using aluminum and platinum foils that were exposed to the solar wind by spreading on the surface of the moon for several hours to days. These foils were then returned to the earth and analyzed with a noble gas mass spectrometer in the laboratory. Following these experiences, a new space mission project was proposed by Don Burnett of Caltech as a PI. This space mission program was nicknamed Suess-Urey mission at first and later renamed Genesis, because the former name is only recognizable by a limited number of people. The Genesis mission aims at effective collection of solar wind for the study of elemental and isotopic compositions of solar wind. The solar wind is collected by different collection materials in response to purposes of measurements. These materials will be exposed for two years. The Genesis mission project was finally approved as one of the discovery space mission program of NASA. The spacecraft will be launched in early 2001 and the entry capsule in which solar wind-collected materials are shielded will be returned to the earth in mid-2003.

If many elements can be determined in the solar wind with accuracy comparable with that for chemical analysis of meteorites, solar wind can be a good candidate for solar abundance standard. We have examined this possibility in relation to the Genesis mission, in which project a neutron activation method is to be applied for the determination of elements having relatively high Z. For this purpose, the solar wind is collected by a ultra-high purity Si metal (single crystal). In our examination, we aimed at the determination of Se and REE (especially, Eu), because these elements were reported to show some irregularity on a Suess plot like Figure 5 (Burnett *et al.*, 1989; Anders and Grevesse, 1989). The Si crystal was irradiated at the Japan Atomic Energy Research Institute research reactor (JRR-3M) at a thermal neutron flux of 1 x $10^{14}$ $cm^{-2}s^{-1}$ for about 500 h. After stepwise dissolution with chemicals and mechanical polishing of the surface, the interior material was subjected to radiochemical separation of Se and REE. Even with a high efficiency Ge detector and ultra-low background counting facility, no gamma-rays characteristic of radioactive nuclides produced by (n, γ) reactions on Se and REE were detected. Based on these measurements, upper limits in 1 g of Si were 7 x $10^{-14}$ g for Se, 2 x 10'16 g for Eu, 6 x $10^{-14}$g for Gd and 7 x $10^{-15}$g for Tb. In an actual exposure to solar wind, particles (neutral atoms and ions) penetrate the Si collector up to no more than 100 nm. Except for Gd, impurities present in a thin Si layer of 100 nm thickens calculated from the above upper limit values are 2 orders of magnitude lower than the expected amounts with 2 year accumulation, suggesting that the Si material tested (and to be used as a collector for the Genesis mission project) is sufficiently pure in at least Se and some REE. Applying the same experimental conditions for irradiation and chemical separation and the same assumption for accumulation of solar wind, it is concluded that at least 20 $cm^2$ and 80 $cm^2$ of Si collector sheet are required for the quantitative measurement for Se and Eu, respectively. These values are highly realistic for analysis. For Tb, a sheet of 630 $cm^2$ is needed. This value must be a marginal size for actual analysis. The 7600 $cm^2$ required for Gd is unrealistic for analysis.

With these examinations, we are very confident that not a few elements can be determined for the solar wind with high accuracy. It is well known that the chemical composition of the solar wind is not uniform on a basis of relatively short time (say, day). An accumulation for 2 years, however, must diminish time-to-time variation in chemical composition. If we can determine Se and Eu abundances in solar wind with sufficient accuracy, we can judge whether CI chondrites can serve as the solar abundance standard as believed so far. We expect that the Genesis mission project can bring us a new measure to evaluate the status of individual groups of chondrites including CI and further a key to solve the genesis of our solar system.

## ACKNOWLEDGEMENT

This work is supported in part by Grant-in-Aid for Scientific Research of the Ministry of Education, Science and Culture, Japan (No. 08454167 and No. 11440167 to ME).

# REFERENCES

Anders, E. and Ebihara, M.: 1982, "Solar-system abundances of the elements", *Geochim. Cosmochim. Acta.* **46**, 2363-2380.

Anders, E. and Grevesse, N.: 1989, "Abundances of the elements: Meteoritic and solar", *Geochim. Cosmochim. Acta* **53**, 197-214.

Burnett, D.S., Woolum, D.S., Benjamin, T.M., Rogers, P.S.Z., Duffy, C.J. and Maggiore, C.: 1989, "A test of the smoothness of the elemental abundances of carbonaceous chondrites", *Geochim. Cosmochim. Acta* **53**, 471-481.

Dufresne, E.R. and Anders, E.: 1962, "On the chemical evolution of the carbonaceous chondrites", *Geochim. Cosmochim. Acta* **26**, 1085-1114.

Ebihara, M.: 1996, "Solar and solar system abundances of the meteorites", *J. R. Soc. W. Australia* **79**, 51-57.

Ebihara, M. and Ozaki, H.: 1995, "Re, Os and Ir in Antarctic unequilibrated ordinary chondrites and implications for the solar system abundance of Re", *Geophys. Res. Lett.* **22**, 2167-2170.

Ebihara, M., Wolf, R. and Anders, E.: 1982, "Are CI chondrites chemically fractionated? A trace element study", *Geochim. Cosmochim. Acta* **46**, 1849-1861.

Evensen, N.M., Hamilton, P.J. and O'nion, R.K.: 1978, "Rare-earth abundances in chondritic meteorites", *Geochim. Cosmochim. Acta* **42**, 1199-1212.

Fredriksson, K. and Kerridge, J.F.: 1988, "Carbonates and sulfates in CI chondrites: Formation by aqueous activity on the parent body", *Meteoritics* **32**, 35-44.

Geiss, J., Buehler, F., Cerutti, H., Eberhardt, P. and Filleux, C.: 1972 "Solar wind composition experiment", in *Apollo 16 Preliminary Science Report* NASA SP-315, 14-1 - 14-10.

Goldschmidt, V.M.: 1938, "Geochemische verteilungsgezetze der Elemente. IX Die Mengenverhaltnisse der elemente und der atom-arten", *Skrifter utgit av det Norske Videnskaps-Akademi I Oslo. I. Math-naturv*, Klasse, 1937, **4**, 1-148.

Jarosewich, E., Clarke, R.S., Jr. and Barrows, J.N., eds., 1987, "The Allende meteorite reference sample", *Smithsonian Contribution Earth Sci.* **27**, 1-49.

Morgan, J.W. and Lovering, J.F.: 1967, "Rhenium and osmium abundances in chondritic meteorites" *Geochim. Cosmochim. Acta* **31**, 1893-1909.

Richardson, S.M.: 1978, "Vein formation in the C1 carbonaceous chondrites", *Meteoritics* **13**, 141-159.

Shinotsuka, K., Hidaka, H. and Ebihara, M.: 1995, "Detailed abundances of rare earth elements, thorium and uranium in chondritic meteorites: An ICP-MS study", *Meteoritics* **30**, 694-699.

Suess, H.E.: 1947, "Über kosmische Kernhäufigkeiten. I. Einige Häufigkeitsregeln und ihre Anwendung bei der Abschätzung der Häufigigkeitswerte für die mittelschweren Elemente", *Z. Naturforsch* **2a**, 311-321.

Suess, H.E. and Urey, H.C.: 1956, "Abundances of the elements", *Rev. Modern Phys.* **28**, 53-74.

Tomeoka, K. and Busek, P.R.: 1988, "Matrix mineralogy of the Orgueil carbonaceous chondrite", *Geochim. Cosmochim. Acta* **52**, 1627-1640.

Wasson, J.T. and Kallemeyn, G.W.: 1988, "Compositions of chondrites", *Phil. Trans. R. Soc. Lond. A* **325**, 535-544.

# Sensitivity of Solar Oscillation Frequencies to Element Abundances

Joyce A. Guzik and Corinne Neuforge
*Applied Theoretical and Computational Physics Division, Los Alamos National Laboratory, XTA, MS B220, Los Alamos, NM 87545-2345*
*joy@lanl.gov*

Abstract: We are fortunate to live close enough to a pulsating star, the sun, for which we can resolve the disk and thereby measure the frequencies of thousands of its global modes of oscillation. These nonradial p-modes travel to varying depths in the solar interior, and their frequencies have been measured in some cases to a precision of one part in a million; therefore they are extremely sensitive probes of solar interior structure. The process of using solar oscillations to determine the solar interior structure and test the physical input to solar models is called "helioseismology", analogous to the way that seismic waves are used to infer the Earth's interior structure. The modern standard solar model generally has been successful in reproducing observed frequencies and the inferred sound speed of the solar interior. Nevertheless, discrepancies of several tenths of a percent in sound speed persist–these small differences are used as clues to refine the model assumptions and input physics. In this review we discuss the degree to which helioseismology is sensitive to element abundances and composition gradients in the solar interior.

## 1. INTRODUCTION

In this paper we give a general introduction to the assumptions and input physics used in the standard solar evolution model. We also provide a brief introduction to helioseimology, using 5-minute solar oscillation frequencies to infer the interior structure of the sun. Finally we discuss results of the application of helioseismology to infer the sun's interior element abundances

*Origin of Elements in the Solar System: Implications of Post-1957 Observations.*
Edited by O. Manuel, Kluwer Academic/Plenum Publishers, 2000. 

and composition profile, and test models incorporating diffusive settling, radiative levitation, and mixing processes.

## 2. SOLAR MODELING

Solar modelers attempt to derive the interior structure of the present sun by calculating its evolution from the onset of nuclear energy generation in the core to the present day, and comparing with observed quantities. The goal is to match the observed solar luminosity $3.846 \pm 0.005 \times 10^{33}$ erg/sec (Willson *et al.*, 1986), mass $1.9891 \pm 0.0004 \times 10^{33}$ g, radius $6.9599 \times 10^{10}$ cm (Allen, 1973), and surface Z/X abundance ratio $0.0245 \pm 0.0015$ (Grevesse and Noels, 1993) at the present solar age $4.52 \pm 0.04$ billion years (Guenther, 1989). The equations that are solved numerically include those for hydrostatic equilibrium; energy production due to nuclear reactions and conversion of potential energy into thermal energy as the sun's core contracts; and energy transport from the interior to the surface via radiative diffusion or convection. The models are calculated assuming spherical symmetry, negligible mass loss or mass accretion, no rotation or magnetic fields, and that the sun's luminosity is generated by conversion of core hydrogen to helium via nuclear reactions, principally via the proton-proton, but also CNO-cycles. We usually divide the sun into several hundred mass shells from the center to the surface, and divide the evolution into several hundred timesteps. Typically, the initial helium mass fraction ($Y_o$), the initial mass fraction ($Z_o$) of elements heavier than H and He, and a parameter regulating the efficiency of convection (*e.g.*, the mixing length to pressure scale height ratio, $\alpha$), are adjusted to obtain the observed luminosity, radius, and surface Z/X abundance at the present solar age.

Some have also attempted, successfully, to evolve solar models that take into account the additional constraints of the observed solar Li abundance, which is believed to be reduced by a factor of 160 from the primordial value (Grevesse and Sauval, 1998). Solar modelers have so far been unsuccessful in creating models that produce the neutrino fluxes observed by the GALLEX, SAGE, Homstake, Kamiokande, and Super Kamiokande neutrino experiments (Bahcall *et al.*, 1998), while retaining good agreement with observed oscillation frequencies.

Solving the equations requires physical data and other input, including radiative and conductive opacities (as a function of temperature, density, and element composition); equation of state tables giving the pressure, energy, and other thermodynamic quantities as a function of temperature, density and composition; nuclear reaction rates; and a treatment for convective energy transport. Most modern solar models use the OPAL (Iglesias *et al.*, 1992; Iglesias and Rogers, 1996) opacities; the OPAL (Rogers *et al.*, 1996) or MHD (Dappen *et al.*, 1988) equation of state tables; and the nuclear reaction rates of Caughlan and Fowler (1988), Bahcall *et al.* (1995), or those discussed by Brun *et al.* (1998). Since the OPAL opacity tables do not extend to temperatures below 6000 K, these have been supplemented by low-temperature opacity tables, *e.g.*, Kurucz (1992), Alexander and Ferguson (1994, 1995), or Neuforge (1993). While many modelers continue to use the mixing-length treatment of convection (Bohm-Vitense, 1958), some have adopted the treatment of Canuto and Mazzitelli (1991, 1992) which allows for a spectrum of eddy length scales, or an attached envelope calibrated by 2-D or 3-D turbulence models (Rosenthal *et al.*, 1998; Demarque *et al.*, 1997). It has also become standard (and essential) in modern solar models to incorporate some treatment of diffusive settling of helium and heavier elements (See, *e.g.*, Cox *et al.*, 1989; Thoul *et al.*, 1994).

## 3. INTRODUCTION TO HELIOSEISMOLOGY

Solar five-minute oscillations were discovered in 1962 by Leighton *et al.* (1962) from Doppler velocities derived from spectral line profile variations. These velocity variations are manifested as waves on the solar surface with peak velocities of 1000 m $sec^{-1}$. They remain coherent for 6-7 periods over spatial scales of about 0.05 Rs (Figure 1). At first these waves were thought to be superficial phenomena related to convection cells near the surface, but were later interpreted by Ulrich (1970) and independently by Leibacher and Stein (1971) as the interference pattern of over 10 million global acoustic (sound, pressure) modes of oscillation, with frequencies between 1000 and 5000 μHz (periods ~ 5 minutes), and amplitudes less than 15 cm $sec^{-1}$. The sun acts as a resonant acoustic cavity that traps these waves; they are reflected by the rapidly decreasing density as they travel toward the solar surface, and turned around (refracted) by the increasing sound speed as they travel toward the solar center. This discovery launched the new research field of helioseismology. Researchers made rapid progress in modeling solar evolution and pulsations, guided by attempts to match the properties of thousands of solar oscillation modes observed to very high accuracy. There are numerous excellent review articles (*e.g.*, Christensen-Dalsgaard *et al.*, 1996; Harvey, 1995) that describe the properties of solar oscillations and their usefulness for inferring the sun's interior structure.

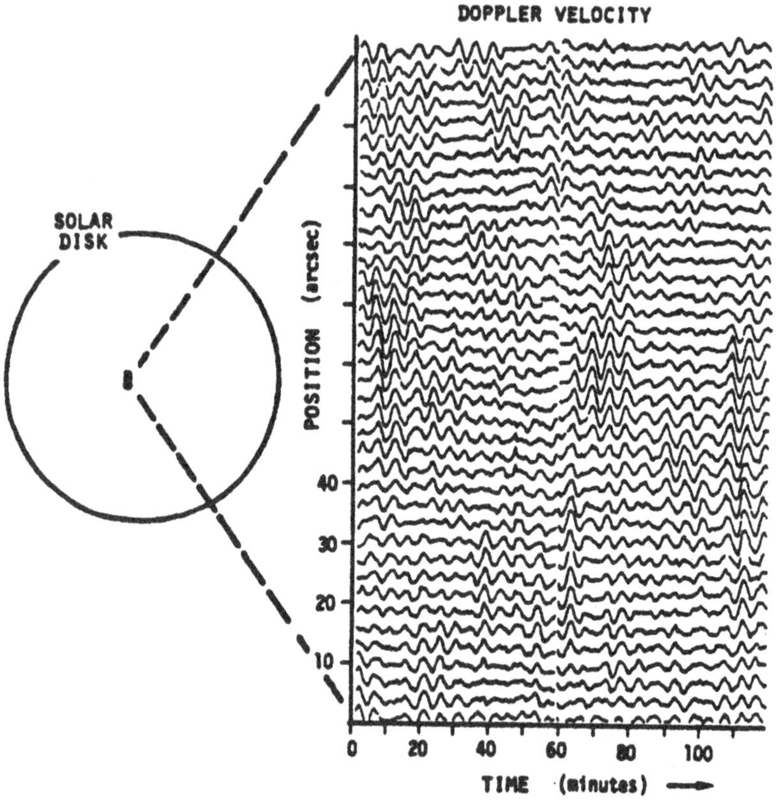


*Figure 1.* Wave packets seen in space- and time-resolved doppler velocity observations of the solar disk. These wave packets are actually the superposition of millions of acoustic resonance modes of oscillation that undergo constructive and destructive interference (from Toomre, 1984).

Mathematically, solar oscillations can be described in terms of a set of normal modes (eigenmodes), consisting of a radial component, with the radial order $n$ equal to the number of nodal points inside the sun (Figure 2), and an angular component represented by spherical harmonics $Y_{l,m}(\theta, \varphi)$ (Figure 3). The angular degree $l$ denotes the total number of nodal lines on the surface. Rotation (and to a lesser extent magnetic fields) breaks the spherical symmetry of the sun. For example, for the sun's slow rotation of about 26 days, each mode with the same degree $l$ is split into $2l+1$ frequencies, separated by approximately the sun's rotational frequency of about 0.45 μHz. The absolute value of the azimuthal order $m$, where $m = -l$, $(-l+1)$, . . ., 0, . . ., $(l-1)$, $l$ , corresponds to the number of nodal lines through the symmetry axis (Figure 3). Note (Figure 2) that modes of low angular degree $l$ penetrate deep within the solar interior, while modes of high degree turn around closer to the solar surface. Thus modes of different $l$ can be used to probe different depths in the solar interior; by considering all of the modes, we can infer the entire interior structure.

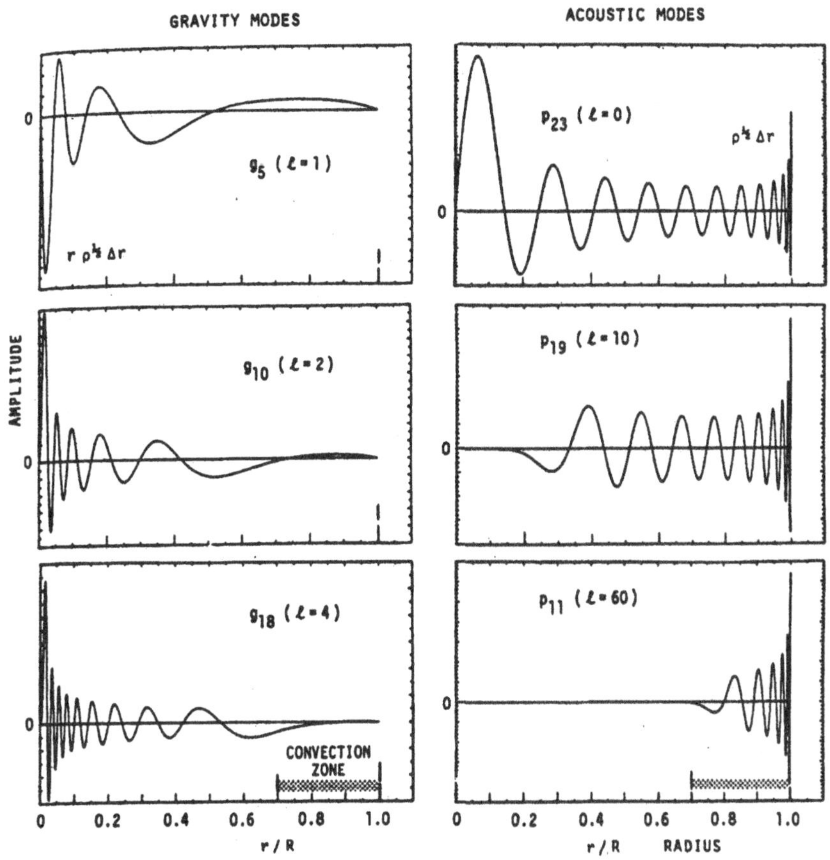


*Figure 2.* Scaled amplitudes versus radius of solar oscillation eigenmodes. The mode order *n* gives the number of nodes in the solar interior. The *g*-modes (left) have large amplitudes near the solar core, but are evanescent in the convection zone. The *p*-modes (right) propagate throughout the entire sun, with modes of low angular degree *l* penetrating farthest into the solar interior (from Toomre, 1984).

Since we can spatially resolve the solar disk, we can obtain the frequencies of modes from degrees $l = 0$ through $\sim 1000$. The discovery of nonradial oscillations in the sun has also prompted searches for nonradial oscillations in sun-like stars, and initiated the field of asteroseismology. For other stars, only frequencies of low-degree modes ($l = 0$, 1, 2, and maybe 3) can be measured from photometric variations, since for these modes amplitude variations averaged over the entire unresolved disk may not fully cancel out.

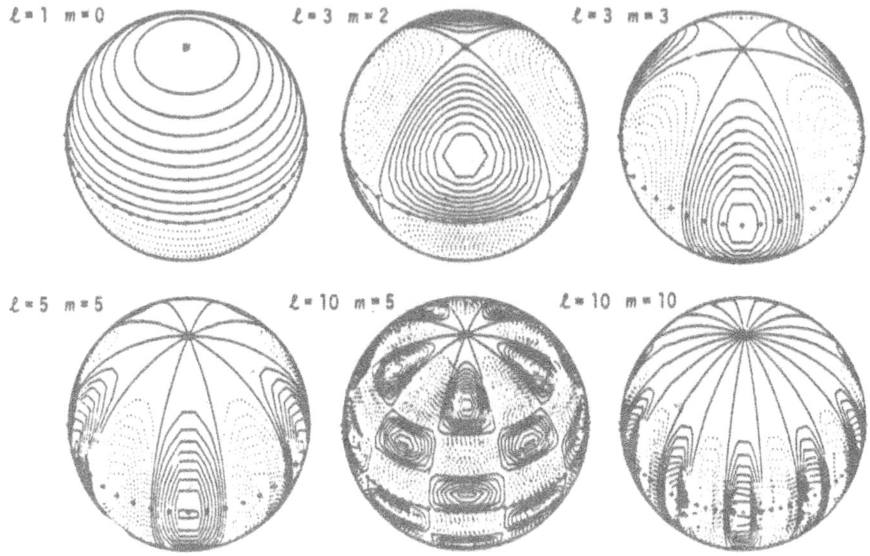


*Figure 3.* Doppler velocity contours for some eigenmodes of solar oscillation. The angular degree *l* gives the number of nodal lines on the solar surface, and the azimuthal order m gives the number of nodal lines through the symmetry axis (from Toomre, 1984).

In addition to the pressure (*p*-) modes, theory also predicts another class of global nonradial modes, the gravity (*g*-) modes, with periods of about 40 minutes. For these waves, the restoring force is gravity instead of pressure. Gravity modes are complementary to *p*-modes, in that they have large amplitudes near the solar core (Figure 2), while only a few of the lowest-degree *p*-modes reach the core. The *g*-mode frequencies would therefore be extremely useful for probing the solar core, and addressing the important questions of the core rotation rate and neutrino flux problem (See Rouse, this volume). However, gravity modes cannot propagate through the solar convection zone, and in spite of many claims, their existence has not been confirmed observationally. The lowest-degree gravity modes may have enough amplitude near the solar surface to eventually be detectable by long time-series space-based observations, for example by the SOHO spacecraft launched in November 1995.

The two approaches to helioseismic comparisons can be categorized as *forward* and *inverse* methods. The objective of the forward method is to test the physical input of solar and stellar models, and not necessarily to derive the absolute structure of the solar interior. In the forward approach, one generally uses the best available physical input to calculate solar evolution as described above, and directly compares the observed and calculated oscillation frequencies. The differences between predictions and observation are used to evaluate and suggest further refinements in physical input. For an excellent overview, see "Testing a Solar Model: The Forward Problem" (Christensen-Dalsgaard, 1995). See Guzik and Swenson (1997) for the results of a recent standard solar model employing the forward method.

The objective of the inverse approach is to derive the internal structure of the sun independent of the physical input of the models. One uses a reference model to derive weighting functions, or *Kernels*, that relate a specific change in model structure to a change in predicted frequency. The differences between observed frequencies and those of the calculated reference model (incorporating as well the observational errors), are used as input to back out differences between the reference model structure and the sun for such quantities as sound speed, density, adiabatic index $\Gamma_1$, or rotation rate as a function of radius. In the case of the solar rotation, the observed rotationally-produced frequency splittings are used to determine the internal rotation profile as a function of latitude as well as of radius. See paper by Rouse (this volume) for additional discussion of forward and inverse methods.

# 4. HELIUM AND HEAVY ELEMENT DIFFUSION

The diffusion of helium and the heavy elements is driven by the gravity and the temperature gradient (gravitational diffusion), the composition gradient (chemical diffusion) and the heat flow (thermal diffusion). A complete derivation of the diffusion and heat flow equations is given in Burgers (1969). In the sun, the convective envelope is completely mixed and homogeneous in composition. Diffusion operates from the base of the envelope and makes the composition of the whole envelope vary with time. This implies that the present surface Z/X is different from the primordial one, $Z_0/X_0$.

## 4.1 Seismological Evidence for Helium Diffusion

Standard solar evolution models require an initial helium mass fraction ($Y_0$) of about 0.274 to generate the current solar luminosity. The models also predict that diffusive settling has reduced the envelope convection zone Y by about 0.026, and the mass fraction of heavier elements by about 5-10% over the sun's 4.5 billion year lifetime. Helioseismology strongly supports this result. For example, Basu (1998) uses the signature in the oscillation frequencies of partial ionization of helium in the convection zone at 50-300,000 K to derive Y = 0.248 ± 0.001. Kosovichev's (1995) direct inversion for the hydrogen abundance profile in the solar interior show a discontinuity in X at the convection zone base, expected from helium draining out of the convection zone and accumulating near the base where the diffusion velocities become slower (Figure 4).

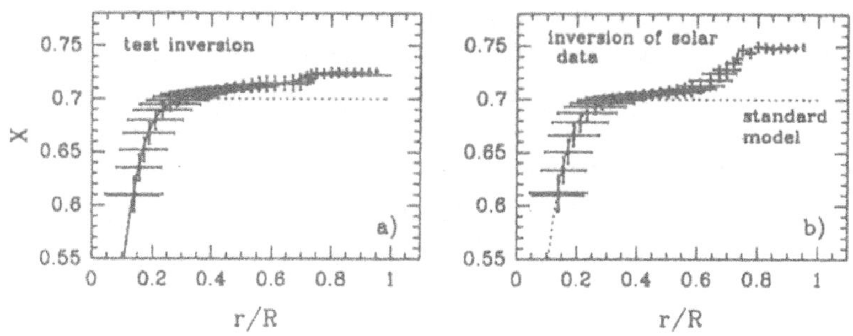


*Figure 4.* Inversion results for the solar interior hydrogen abundance. Panel a) shows how well the profile can be recovered for simulated oscillation data and a test model; panel b) shows the profile using actual solar oscillation data. The inversion shows a higher hydrogen abundance in the solar convection zone (outer 30% of the solar radius), due to diffusion. (Reprinted from *Adv. Space Res.* **15**, Kosovichev, A.G, "Helioseismic measurement of element abundances in the solar interior", 95-99, Copyright 1995, with permission from Elsevier Science.)

Note also that the inversion shows the core hydrogen profile, and the expected depletion due to nuclear processing. The calculated sound speed profile below the solar convection zone also agrees better with the helioseismically-inferred profile when helium diffusion is included, as can be seen in Figure 5 from Gabriel (1997).

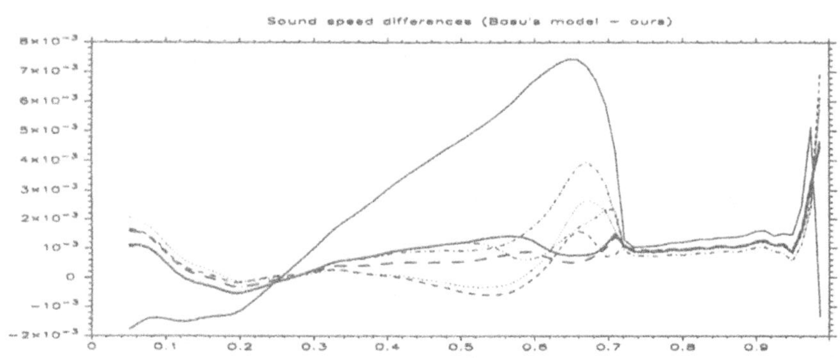


*Figure 5.* Sound speed profile differences between the models of Gabriel (1997) and the seismic reference model of Basu *et al.* (1996), as a function of the fractional radius. A model with helium and heavy element diffusion (dotted line) is in much better agreement with the seismic sun than a model without diffusion (thin solid line).

Diffusion is a difficult problem to tackle and it needs to be treated using various approximations (*e.g.*, Michaud and Proffitt, 1993; Morel *et al.*, 1997; Gabriel, 1997). Unfortunately, the solar models are not sensitive enough to discriminate between the different treatments of diffusion. Moreover, Figure 6 (Turcotte and Christensen-Dalsgaard, 1998a) shows that the differences between the primordial and the present surface abundances resulting from various prescriptions of diffusion are all smaller than the differences between the observed meteoritic and photospheric abundances. Recently, Turcotte *et al.* (1998) calculated solar models including radiative levitation and monochromatic opacities instead of Rosseland mean opacity tables, calculated for fixed relative abundances of the heavy elements. The various chemical species can indeed absorb a part of the net outgoing momentum flux carried by the photons and be pushed upwards, and, because all the elements do not settle at a common rate, their relative abundances vary with time. Figure 7 (Turcotte and Christensen-Dalsgaard, 1998b) shows that these major improvements appear to worsen slightly the calculated and inferred sound speed agreement. This may indicate that other compensating changes to input physics, such as in the opacities, may be needed to restore agreement.

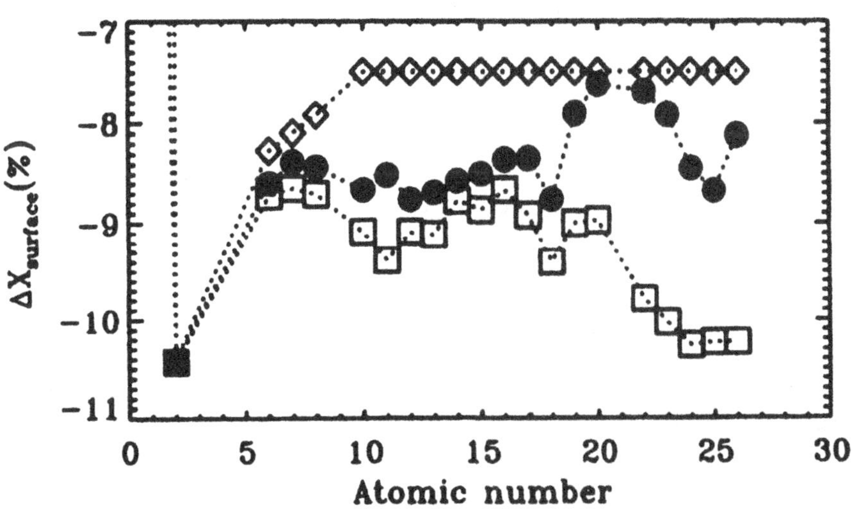


*Figure 6.* Fractional change of the surface abundances of all the elements included in the models of Turcotte and Christensen-Dalsgaard (1998a), at the solar age and as a function of the atomic number. The models are: a solar model assuming a common settling rate for Z > 8 (diamonds), the model including radiative forces and monochromatic opacities (filled circles) and a model including monochromatic opacities but with no radiative forces (squares). This figure from Turcotte and Christensen-Dalsgaard (1998a) is reproduced here with kind permission from Kluwer Academic Publishers.

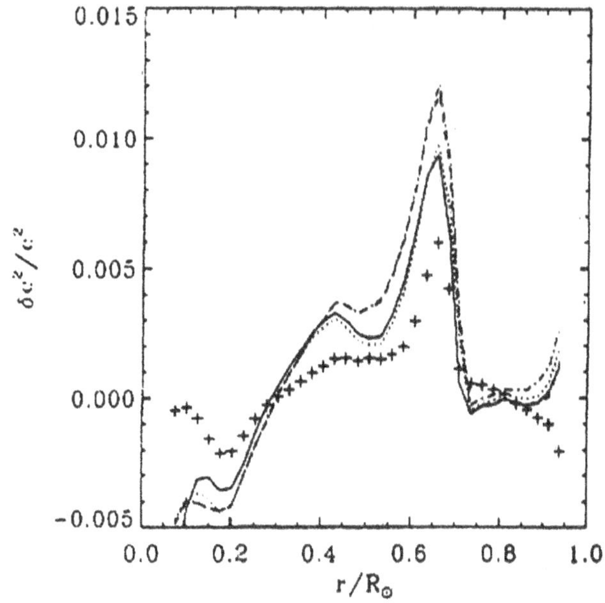


*Figure 7.* Relative differences in squared sound speed between the seismic reference model of Turck-Chièze *et al.* (1998) and theoretical models of Turcotte *et al.* (1998) in the sense (sun)-(model), against fractional radius. The model calculated with no radiative forces and with Rosseland mean opacities (dashed line) appears to be better than the refined model calculated with radiative forces and monochromatic opacities (solid line).

## 5. THE SOLAR MIXTURE

The relative abundances of the heavy elements at the sun's surface (the solar mixture) is used as an input in the computation of Rosseland mean opacity tables needed in model calculations, and Z/X provides a constraint in the modeling of the sun. The most widely used solar abundances are those of Grevesse and Noels (1993).

*Table 1.* Logarithmic C, N, O, S, and Ar abundances in number, relative to hydrogen (=12)

| Element | Grevesse and Sauval (1988) | Grevesse and Noels (1993) |
|---|---|---|
| C | $8.52 \pm 0.6$ | $8.55 \pm 0.05$ |
| N | $7.92 \pm 0.06$ | $7.97 \pm 0.07$ |
| O | $8.83 \pm 0.06$ | $8.87 \pm 0.07$ |
| S | $7.33 \pm 0.11$ | $7.21 \pm 0.06$ |
| Ar | $6.40 \pm 0.06$ | $6.52 \pm 0.10$ |
| Z/X* | 0.023 | 0.0245 |

* Z/X is the present surface mass fraction of heavy elements relative to hydrogen.*

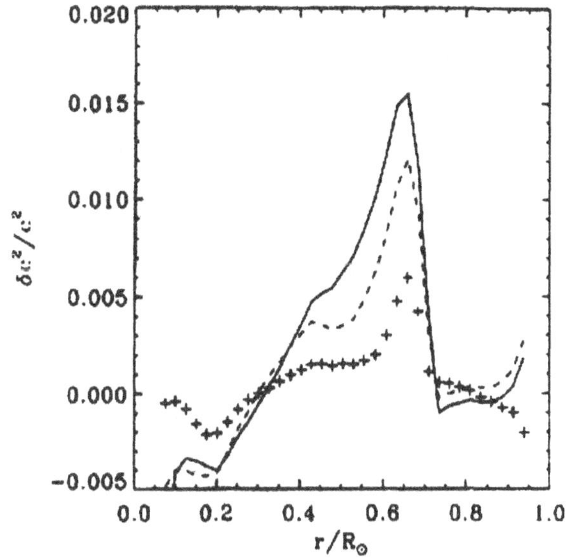


*Figure 8.* Relative differences in squared sound speed between the seismic reference model of Turck-Chièze *et al.* (1998) and theoretical models of Turcotte *et al.* (1998), in the sense (sun)-(model), against fractional radius. The model calculated with the abundances of Grevesse and Noels (1993, dashed line) is in better agreement with the seismic data than the model calculated with the new abundances of Grevesse and Sauval (1998, solid line). Both models include monochromatic opacities but no radiative forces.

Turcotte and Christensen-Dalsgaard (1998b) recalibrated a solar model with the Grevesse and Sauval (1998) revised solar abundances, presented in Table 1. Figure 8 shows that the new data increase the discrepancies between the inferred and calculated sound speed. The most important effects are due to the substantial reductions in Z/X and in the oxygen abundance: oxygen is the most abundant element after H and He and it dominates the opacity just below the convection zone. Nevertheless, the proposed changes for Z/X and oxygen are still within the uncertainties of the Grevesse and Noels (1993) composition.

# 6. ABUNDANCES THROUGH THE SOLAR INTERIOR

Turck-Chièze (1998) suggests that the small bumps superimposed on the general broad deviation in the squared sound speed differences between the model of Brun *et al.* (1998) and the sound speed inversion of Turck-Chièze *et al.* (1997, Figure 9) may result from the signature of the bound-bound processes of different heavy elements in the solar interior. Bound-bound processes give rise to a discontinuous opacity. For a given element, an incorrect abundance at the location where these processes are the most important will produce small discontinuities in the sound speed differences. A one percent effect in the total opacity will lead to a 0.1% effect on the sound speed profile, which should be observable. The seismic data could therefore give us access to the iron abundance near 0.2Rs, the $^3$He abundance at the edge of the nuclear core at 0.27Rs, the silicon around 0.4Rs, the iron again around 0.6Rs, and the oxygen near the base of the convection zone.

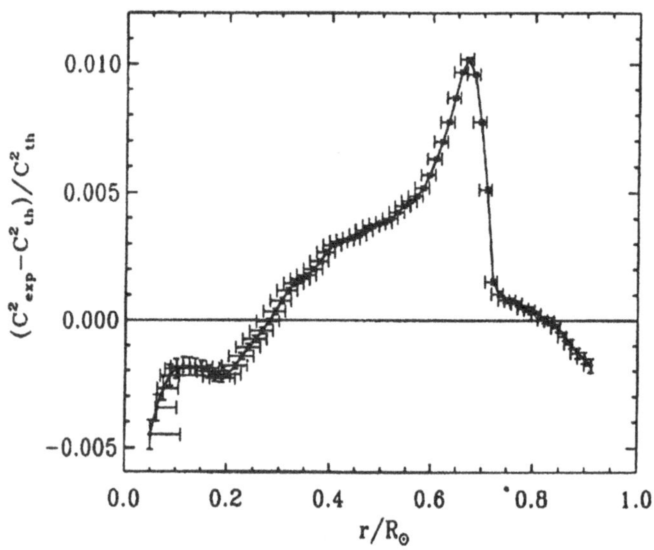


*Figure 9.* Relative differences in squared sound speed between the seismic reference model of Turck-Chièze *et al.* (1997) and the model of Brun *et al.* (1998). One can observe several small bumps superimposed on the general deviation in the squared sound speed differences (from Turck-Chieze, 1998, with kind permission from Kluwer Academic Publishers).

# 7. THE SOLAR LITHIUM PROBLEM

Two explanations have been proposed to account for the depletion of $^7$Li from its presumed protosolar abundance to the present photospheric abundance: Mixing below the solar convection zone, induced by either differential rotation or gravity waves; and mass loss of about 0.1 solar masses early in the sun's main-sequence lifetime. Both mechanisms have the advantage of smearing out the steep composition gradient at the base of the convection zone resulting from helium and element diffusion, and somewhat reducing the small remaining discrepancy between the inferred and calculated sound speed profiles.

Regarding the mass-loss scenario, Guzik and Cox (1995) and Morel *et al.* (1997) conclude that the mass-loss phase must end quite early, no later than about 0.2-0.3 Gyr after the sun arrives on the main sequence, to avoid ruining the good agreement with the sound speed of the models. The diffusion-produced composition gradient at the convection zone base is smoothed by the changing location of the convection zone boundary as the sun loses mass. For example, Anderson *et al.* (1996) show for the Guzik and Cox model that a 0.3% difference in sound speed at the convection zone base is decreased to about 0.2%. However, in the early higher-mass phase, the increased nuclear energy generation rate steepens the resultant composition gradient in the core. Anderson *et al.* (1966) and Morel *et al.* (1997) show that if the mass loss phase does not end early enough, the core sound speed discrepancy is increased from 0-0.2% to about 0.4%.

Mixing induced by rotation or gravity waves, depending on the parametrization of the mixing, can do very well in improving the agreement in sound speed below the convection zone base (Gabriel, 1997; Richard *et al.*, 1996; Guenther and Demarque, 1997). However, the mixing must not extend too deep to avoid ruining the good sound speed agreement in the solar core (Richard and Vauclair, 1997).

# 8. CONCLUSIONS

Recent improvements to the physical input of the standard solar model have resulted in excellent agreement between the calculated and inferred solar structure, or between observed and calculated oscillation frequencies. Calculated low-degree p-mode frequencies agree with observation to within several μHz out of 3000 μHz, or a few tenths of a percent (Guzik and Swenson, 1997). The inferred sound speed profile agrees with the profiles of standard solar evolution models to within 0.2% (Christensen-Dalsgaard *et al.*, 1996). Helioseismology has provided clear evidence of diffusive settling of helium from the solar convection zone during the sun's 4.5 billion year lifetime. Diffusive settling of about 5-10% of elements heavier than helium is expected as well, but the helioseismic signature is smaller due to the much smaller abundance of these elements. However, even small changes in element abundances, mixtures, and interior composition profiles alter the sound speed enough to be detectable by helioseismic methods. The oscillation frequency observations are so accurate that helioseismology promises to be a viable tool for helping to choose between alternative abundance mixtures, diffusion treatments, or proposed mixing processes.

## REFERENCES


Alexander, D.R. and Ferguson, J.W.: 1994, "Low-temperature Rosseland Opacities", *Ap. J.* **437**, 879-891.

Allen, C.W.: 1973, *Astrophysical Quantities*, 3rd Edition, Athlone Press, London, UK, 310 pp.

Anderson, E., Antia, H.M., Basu, S., Chaboyer, B., Chitre, S.M., Christensen-Dalsgaard, J., Eff-Darwich, A., Elliott, J.R., Giles, P.M., Gough, D.O., Guzik, J.A., Harvey, J.W., Hill, F., Leibacher, J.W., Kosovichev, A.G., Monteiro, M.J.P.F.G., Richard, O., Sekii, T., Shibahashi, H., Takata, M. Thompson, M.J., Toomre, J., Vauclair, S. and Vorontsov, S.V.: 1996, "The seismic structure of the sun from GONG", in: *Sounding Solar and Stellar Interiors*, eds., Provost, J. and Schmider, F.-X., IAU, The Netherlands, pp. 151-158.

Bahcall, J.N., Basu, S. and Pinsonneault, M.H.: 1998, "How uncertain are solar neutrino predictions?," *Phys. Lett.* B. **433**, 1-8.

Bahcall, J.N., Pinsonneault, M.H. and Wasserburg, G.J.: 1995, "Solar models with helium and heavy-element diffusion", *Rev. Mod. Phys.* **67**, 781-808.

Basu, S.: 1998, "Effects of errors in the solar radius on helioseismic inferences", *Mon. Not. Royal Astron.* **298**, 719-728.

Basu, S., Christensen-Dalsgaard, J., Schou, J., Thompson, M.J. and Tomczyk, S.: 1996, "The sun's hydrostatic structure from LOWL data", *Ap. J.* **460**, 1064-1070.

Bohm-Vitense, E.: 1958, "Uber die Wasserstoffkonvektionzone in Sternen verschiedener Effektivtemperaturen un Leuchkräfte", *Zeitschrift Astrof.* **46**, 108-143.

Brun, A.S., Turck-Chieze, S. and Morel, P.: 1998, "Standard solar models in the light of new helioseismic constraints. I. The solar core", *Ap. J.* **506**, 913-925.

Burgers, J.M.: 1969, *Flow Equations for Composite Gases*, Acádemic Press, New York, NY, 332 pp.

Canuto, V.M. and Mazzitelli, I.: 1991, "Stellar turbulent convection - A new model and applications", *Ap. J.* **370**, 295-311.

Canuto, V.M. and Mazzitelli, I.: 1992, "Further improvements of a new model for turbulent convection in stars", *Ap. J.* **389**, 724-730.

Caughlan, G.R. and Fowler, W.A.: 1988, "Thermonuclear reactions-rates .5", *Atomic Nuclear Data Tables* **40**, 283-334.

Christensen-Dalsgaard, J.: 1995, "Testing a solar model: The forward problem", in: Proc. VI IAC Winter School, *The Structure of the Sun*, ed., Cortes, T.R., Cambridge Univ. Press, Cambridge, UK, 93 pp.

Christensen-Dalsgaard, J., Dappen, W., Ajukov, S.V., Anderson, E.R., Antia, H.M., Basu, S., Baturin, V.A., Berthomieu, G., Chaboyer, B., Chitre, S.M., Cox, A.N., Demarque, P., Donatowicz, J., Dziembowski, W.A., Gabriel, M., Gough, D.O., Guenther, D.B., Guzik, J.A., Harvey, J.W., Hill, F., Houdek, G., Iglesias, C.A., Kosovichev, A.G., Leibacher, J.W., Morel, P., Proffitt, C.R., Provost, J., Reiter, J., Rhodes, E.J., Rogers, F.J., Roxburgh, I.W., Thompson, M.J. and Ulrich, R.K.: 1996, "The current state of solar modeling", *Science* **272**, 1286-1292.

Cox, A.N., Guzik, J.A. and Kidman, R.B.: 1989, "Oscillations of solar models with internal element diffusion", *Ap. J.* **342**, 1187-1206.

Dappen, W., Mihalas, D., Hummer, D.G. and Mihalas, B.W.: 1988, "The equation of state for stellar envelopes. III - Thermodynamic quantities", *Ap. J.* **332**, 261-270.

Demarque, P., Guenther, D.B. and Kim, Y.-C.: 1997, "The run of superadiabaticity in stellar convection zones. I. The sun", *Ap. J.* **474**, 790-797.

Gabriel, M.: 1997, "Influence of heavy element and rotationally induced diffusions on the solar models", *Astron. Astrophys.* **327**, 771-778.

Grevesse, N. and Noels, A.: 1993, "Cosmic abundances of the elements", in: *Origin and Evolution of the Elements*, eds., Prantzos, N., Vangioni-Flam, E. and Casse, M., Cambridge University Press, New York, NY, pp. 15-23.

Grevesse, N. and Sauval, A.J.: 1998, "Standard solar composition", *Sp. Sci. Rev.* **85**, 161-174.

Guenther, D.B. and Demarque, P.: 1997, "Seismic tests of the sun's interior structure, composition, and age, and implications for solar neutrinos", *Ap. J.* **484**, 937-959.

Guenther, D.B.: 1989, "Age of the sun", *Ap. J.* **339**, 1156-1159.

Guzik, J.A. and Cox, A.N.: 1995, "Early solar mass-loss, element diffusion, and solar oscillation frequencies", *Ap. J.* **448**, 905-914.

Guzik, J.A. and Swenson, F.J.: 1997, "Seismological comparisons of solar models with element diffusion using the MHD, OPAL, and SIREFF equations of state", *Ap. J.* **491**, 967-979.

Harvey, J.: 1995, "Helioseismology", *Physics Today* **48**, 32-38.

Iglesias, C.A. and Rogers, F.J.: 1996, "Updated opal opacities", *Ap. J.* **464**, 943-953.

Iglesias, C.A., Rogers, F.J. and Wilson, B.G.: 1992, "Spin-orbit interaction effects on the Rosseland mean opacity", *Ap. J.* **397**, 717-728.

Kosovichev, A.G.: 1995, "Helioseismic measurement of element abundances in the solar interior", *Adv. Space Res.* **15**, 95-99.

Kurucz, R.L.: 1992, "Finding the missing solar ultraviolet opacity", *Rev. Mexicana Astron. Astrofis.* **23**, 181-186.

Leibacher, J. and Stein, R. F.: 1971, "A new description of the solar five-minute oscillations", *Astrophys. Lett.* **7**, 191-

Leighton, R.B., Noyes, R.W. and Simon, G.W.: 1962, "Velocity fields in the solar atmosphere: I. Preliminary report", *Ap. J.* **135**, 474-499.

Michaud, G. and Proffitt, C.R.: 1993, "Particle transport processes", in: *Inside the Stars*, eds., Baglin, A and Weiss, W.W., ASP, San Fransisco, CA, 246-259.

Morel, P., Provost, J. and Berthomieu, G.: 1997, "Updated solar models", *Astron. Astrophys.* **327**, 349-360.

Neuforge, C.: 1993, "Low temperature Rosseland mean opacities", *Astron. Astrophys.* **274**, 818-820.

Richard, O., and Vauclair, S.: 1997, "Local mixing near the solar core, neutrino fluxes and helioseismology", *Astron. Astrophys.* **322**, 671-673.

Richard, O., Vauclair, S., Charbonnel, C., and Dziembowski, W.A.: 1996, "New solar models including helioseismological constraints and light-element depletion", *Astron. Astrophys.* **312**, 1000-1011.

Rogers, F.J., Swenson, F.J., and Iglesias, C.A.: 1996, "OPAL equation-of-state tables for astrophysical applications", *Ap. J.* **456**, 902-908.

Rosenthal, C.S., Christensen-Dalsgaard, J., Nordlund, A., Stein, R.F. and Trampedach, R.: 1998, "Convective Contributions to the Frequencies of Solar Oscillations," *Astron. Astrophys.*, submitted for publication.

Thoul, A.A., Bahcall, J.N., and Loeb, A.: 1994, "Element diffusion in the solar interior", *Ap. J.* **421**, 828-842.

Toomre, J.: 1984, "Overview of solar seismology: Oscillations as probes of internal structure and dynamics in the sun," in *Solar Seismology from Space*, JPL Publication, pp. 7-39.

Turck-Chièze, S.: 1998, "Composition and opacity in the solar interior", *Sp. Sci. Rev.* **85**, 125-132.

Turck-Chièze, S., Basu, S., Berthomieu, G., Bonanno, A., Brun, A.S., Christensen-Dalsgaard, J., Gabriel, M., Morel, P., Provost, J. and Turcotte, S., the GOLF team: 1998, "Sensitivity of the sound speed to the physical processes included in the standard solar model framework", in: Proceedings of the SOHO6/GONG98 Workshop, *Structure and Dynamics of the Interior of the Sun and Sun-like Stars*, Boston, MA, USA, 1-4 June 1998 (ESA SP-418, 1998), pp. 555-560.

Turck-Chièze, S., Basu, S., Brun, A.S., Christensen-Dalsgaard, J., Eff-Darwich, A., Lopes, I., Hernandez, F.P., Berthomieu, G., Provost, J., Ulrich, R.K., Baudin, F., Boumier, P., Charra, J., Gabriel, A.H., Garcia, R.A., Grec, G., Renaud, C., Robillot, J.M. and Cortes, T.R.: 1997, "First view of the solar core from GOLF acoustic modes", *Sol. Phys.* **175**, 247-265.

Turcotte, S. and Christensen-Dalsgaard, J.: 1998a, "Solar models with non-standard chemical composition", *Sp. Sci. Rev.* **85**, 133-140.

Turcotte, S. and Christensen-Dalsgaard, J.: 1998b, "The effect of differential settling on solar oscillation frequencies", in: Proceedings of the SOHO6/GONG98 Workshop, *Structure and Dynamics of the Interior of the Sun and Sun-like Stars*, Boston, MA, USA, 1-4 June 1998 (ESA SP-418, 1998), pp. 561-566.

Turcotte, S., Richer, J., Michaud, G., Iglesias, C.A., and Rogers, F.J.: 1998, "Consistent solar evolution model including diffusion and radiative acceleration effects", *Ap. J.* **504**, 539-558.

Ulrich, R.: 1970, "The five-minute oscillations on the solar surface", *Ap. J.* **162**, 993-1002.

Ulrich, R.K. and Rhodes, E.J., Jr.: 1983, "Testing solar models with global solar oscillations in the 5-minute band", *Ap. J.* **265**, 551-563.

Willson, R.C., Hudson, H.S., Frolich, C., and Brusa, R.W.: 1986, "Long-term downward trend in total solar irradiance", *Science* **234**, 1114-1117.

# Inverse and Forward Helioseismology: Understanding the Interior Composition and Structure of the Present Sun

Carl A. Rouse

*Rouse Research Incorporated, 627 15th Street, Del Mar, CA 92014*

**Abstract:** Inverse and forward helioseismic approaches are reviewed. Chemical abundance profiles from a solar model generated by Takata and Shibahashi (1998) by inverse helioseismology are used in a forward approach. This forward model is referred to as the author's 'seismic' solar model. This model is compared to a similarly calculated standard evolution solar model (ssm) and the author's reference high-Z core (hzc) solar model. The degree $l = 1$ nonradial oscillation frequencies of the three models are compared, for orders n = 7 to 33. Relative to current observations, there is good agreement between the seismic model predictions and the five-minute-band frequency observations. In addition, there is good agreement between the hzc model frequencies and the seismic model frequencies. Further comparisons are made in terms of fractional differences and/or of ratios of seismic/ssm, hzc/ssm and seismic/hzc quantities such as (1) speed of sound, (2) density and (3) gamma functions, $\Gamma_1$. From prior studies, the effective gamma, $\gamma_e$, defined by the relation, $(\gamma_e - 1) = P/E$, with P, the total pressure and E, the total energy, respectively, is used in the oscillation equations instead of the complete analytic form of the $\Gamma_1$ function. It is shown that between the radius fractions x (= r/R) = 0.05 to 0.95, the fractional differences and ratios seismic/hzc yield good agreements for the speed of sound, density and gamma1 function. Since the hzc model is also consistent with two of the three operating solar neutrino experiments and has the bottom of the convection region at x = 0.6925 with a temperature over $2.4 \times 10^6$ K (hence consistent with the lithium problem without invoking 'overshoot'), it is concluded that the current hzc model is closest to the structure of the real sun. Consequently, an astrophysical solution to the solar neutrino problem is possible as part of an answer to the question of how the real sun formed and evolved to its present state.

# 1. INTRODUCTION

This paper briefly reviews forward and inverse approaches to understanding the structure of the present sun based upon various observed frequencies of solar oscillation and on the various observed solar neutrino counting rates from operating solar neutrino experiments. The current phase of this project with indirect and direct uses of forward and inverse studies is succinctly described by Harvey (1995) as follows:

"Both forward and inverse analyses are used. In a forward analysis a solar model is specified and used to compute some property of the oscillations, such as frequencies. The model is varied slightly to produce a family of predicted values. The model that best matches the observations is then considered to best define the varied parameter. An inverse analysis starts with an observed property of the oscillations and then determines a function of position that is most consistent with the integrals of the function to which the observed properties are sensitive."

Regarding the various inversion techniques, Schou *et al.* (1998) describe seven inversion methods. Their advantages and limitations are also discussed. In particular Schou *et al.* note "*...the intrinsic non uniqueness of all inversion methods, combined with various trade-offs...we are not in a position to advocate which method and resulting solution is the 'best' one.*"

In principle, a forward model based upon the chemical abundance profiles of an inverse model should yield exactly the same frequencies used in the inversion process. However, when the forward and inverse models are calculated with different numerical methods and input physics, exact agreements are not possible. The question to be answered is how close are the frequencies? Further, how close are other aspects of the models such as speed of sound, densities, temperatures, etc., when the forward/inverse model is compared to other models, standard and nonstandard?

This paper is an extension of (a) "Calculation of Stellar Structure IV. Results using a detailed energy generation subroutine," (Rouse, 1995; hereafter Paper I); and (b) "Sensitivity of solar oscillation frequencies in the five-minute-band to different $\Gamma_1$ functions," dated November 25, 1996 (Rouse, 1996; hereafter Paper II). In the present paper an inverse helioseismic model of the sun is added based upon the chemical abundance profiles generated by Takata and Shibahashi (1998; hereafter TS). This is the same approach used in Paper I to generate a standard evolution model of the sun based upon the chemical abundance profiles of Bahcall and Ulrich (1988; hereafter BU).

In Paper I, a review article on "Theories of the Origin of the Solar System 1956-1985" by Brush (1990) is quoted where Brush notes that the origin of the solar system is one of the oldest unsolved problems of modern science. In order to help develop a plausible quantitative explanation of how the sun actually formed–not some statistical average–the author's project seeks an accurate model for the structure of the present sun. When successful, such a model could support one of the existing theories and/or help in deducing an improved theory for how the sun actually formed and evolved.

This study of the relative structure and properties of an inverse helioseismic solar model (seismic), a standard evolution solar model (ssm) and one of the author's high-Z core (HZC or hzc) solar models is an important step in the understanding of the structure of the present sun. I will also indicate to what extent the solution of the solar neutrino problem depends on the actual structure of the present sun and to what extent it may depend on possible new properties of electron neutrinos, if and when proved.

Although the Schou *et al.* (1998) paper is applied to studying internal differential rotation in the sun, the inversion techniques may be applied to other problems of stellar structure. Discussions of various other inversion techniques are also given in Chapter 7 of Unno *et al.* (1989). In this book, the authors divide the various approaches to the inverse problem into two methods, viz., (1) inversion methods based on a reference model and (2) inversion methods not based upon any reference model. The key input for inversions based upon a reference solar model is a kernel (or sensitivity function) that relates the unknown quantity sought to observations. In helioseismology, the observations used are, of course, the observed frequencies of solar oscillation. Since the whole star contributes to the observed frequencies, an integral of the kernel times the unknown function of solar radius is integrated from the center of the sun to the surface, hence, the problem is to solve an integral equation (See Part I of Morse and Feshbach, 1953; Unno *et al.*, 1989; Schou *et al.*, 1998). It is assumed that the standard model of the sun is known accurately enough to derive the kernel of interest.

The basic approach in the second group of inversions is the use of an asymptotic method. As reviewed in Unno *et al.* (1989) through the various references given, "*...the method [is] based upon an asymptotic expansion for eigenfrequencies, from which an Abel type integral equation is derived.*" Unno *et al.* (1989) describes how the asymptotic inversion method is used to infer the solar internal speed of sound distribution for p-mode oscillation data. Under certain conditions, the Abel's integral equation can be analytically inverted to yield r/R equal to an integral expression with the unknown speed of sound and known oscillation frequencies.

However, forward helioseismology has been used by the author to study the sensitivities to various aspects of solar model calculation and the associated oscillation frequency calculations. After deriving an exact solution to the Saha equation for a mixture of any number of elements at a given density and temperature, the application to equation of state calculations was used to study solar structure. Considering that the four, first-order differential equations that describe equilibrium stellar structure are nonlinear and coupled, the need for high precision in the numerical solutions is recognized. Add to that the fact that the equation of state with the real gas properties of ionizing elements is not linear in pressure as the temperature changes at a fixed density, or, as the density changes at a fixed temperature. Along the profile of the sun in the outer half of the model, both the temperature and density change over a wide range with changes in the degrees of ionization of the various elements in the mixture varying nonlinearly in unexpected ways. For example, as the temperature of a solar model increases from the low values at the surface (about 5800 K) where the solar matter is essentially a neutral gas, to higher values in the interior, the ionization of the hydrogen is inhibited by the free electrons produced by the more easily ionized metal atoms. Then after passing a temperature-density threshold, the ionization of the helium is, in turn, inhibited by the free electrons produced by the hydrogen as it begins to ionize 'in force.' In the same way, the abundant CNO (carbon, nitrogen and oxygen) elements also are inhibited from passing to a higher state of ionization by the hydrogen produced free electron density. The same is true for the higher degrees of ionization of the metal atoms. This not only affects the free electron density contributions to the total pressure, but also effects the distribution of ions (more bound electrons than expected) and their effects on the opacities.

This is the first part of the need to solve the stellar structure equations more accurately. Two other parts of the need for increased accuracy are (1) the use of configuration space coordinates rather than mass coordinates in the solution of the stellar structure equations, and (2) use of the equation of state and self-consistent opacity calculations in a subroutine at each temperature-density point along the solar profile. Relative to mass coordinates and space coordinates, they are only equal for matter at a uniform density. In the outer half of a solar model with the rapidly changing densities and temperatures, the use of mass coordinates with nearly equal mass steps will inject an artificial error into the calculations.

The use of an equation of state subroutine, rather than table look-ups, is important because the effects of the interactions of the different elements with different ionization energies on the relative degrees of ionization is more accurately followed with precise equation of state calculations at the temperature-density space points of the model. In addition, the space points must also have small space steps in order to minimize the source of numerical errors inherent in any numerical differencing scheme.

In addition, the author's forward approach includes solving the stellar structure equations in a manner that satisfies the Ince (1926) conditions for the existence of a unique solution to a system of two or more coupled, first-order ordinary differential equations. The main aspect of this theorem is that in the range of the independent variable, values of the dependent variables must be known at one (the same) point in the range of the independent variable. For the sun, the surface is the only point where this is possible. Hence, the only models accepted for evaluation are those that yield, at the surface after continuous calculations from the center, one solar mass, $M_R$ = 1.989 x $10^{33}$ gm; one solar luminosity $L_R$ = 3.846 x $10^{33}$ ergs/sec; one solar radius, R = 6.9599 x $10^{10}$ cm; at R, a temperature of 5780 K and density of 2.778 x $10^{-7}$ g/cc. The pressure corresponding to the surface density and temperature is about 1.036 x $10^5$ dyne/cm$^2$.

In the IEEOS subroutine the solutions of the Saha equation are converged to high precision and used as input to solving the equilibrium stellar structure equations. The stellar structure equations are in turn solved by prediction and correction, with small space steps. And in this system of coupled, nonlinear stellar structure equations, any perturbation of an iterated solution causes deviations from the surface boundary conditions (b.c.). Consequently, to study the effects of perturbations of the structure of the solar core, it is necessary to calculate the outer layers of the model with the same temperature gradient in the convective region and with the same treatment of the superadiabatic layer just below the photosphere. In terms of mixing length theory, this is equivalent to using the same "mixing length" during a given sensitivity study, *i.e.*, not using the mixing length as a free parameter.

In summary, the forward approach used here (1) solves the stellar structure equations in a way that satisfies the Ince (1926) conditions for a unique solution using a given chemical abundance profile throughout the solar interior; (2) solves the Saha equation (with a density-dependent correction) in a subroutine (IEEOS) and avoids table look-ups and extrapolations; (3) uses seven elements in the IEEOS subroutine with their different sets of ionization potentials in order to follow more accurately the variations of the electron pressure and ion distributions along the profile; and (4) uses the numerical method of prediction and correction to solve the equilibrium stellar structure equations with small space steps in order to minimize the numerical errors inherent in all numerical differencing schemes.

The inverse seismic solar model used in this paper was generated by TS based upon a sound speed profile calculated using an asymptotic approach. With this constraint, TS deduced density, pressure, and temperature profiles in the radiative part of the solar interior. The resulting chemical abundance profiles deduced by TS were put into the Rouse forward solar model calculations as the only outside source. The resulting inverse/forward is one of three forward models studied in detail in this report.

Relative to the results in Paper I, significant increases in mathematical and physical precision have been incorporated, mainly in the ionization equilibrium equation of state (IEEOS) subroutine. First, the convergence in the IEEOS is increased to $10^{-7}$ compared to $10^{-6}$ in Paper I. Secondly the number of elements used in the IEEOS is seven, compared to five used in Paper I. Now H, He, C, N, O, Si and Fe are used as singular elements, with Si also given the sum of the number abundances of the remaining Z elements. The Grevesse (1984) solar mixture is used as representative (to be updated in the future). The use of seven elements is important for accurate calculation of the relative abundances of all ionic species at a given temperature and density: the degrees of ionization of individual elements with different ionization energies depend on the free electron density. The opacities depend on the distribution of ions as well as on the free electron density. (Free-free and bound-free opacities, for example, depend on the square of the net core ion charges, hence not a linear response to changes in the ion abundances.) Consequently, the radial variation of the opacities and free electron density in the outer half of the solar radius–particularly just below and in the convective region–are more accurately followed in the present calculations.

All three models are generated with the same numerical and physical input. Only the chemical abundance profiles are different. The number of (configuration) space steps used in each model calculation is about 4300, with about 3800 stored and used in the solar oscillation frequency calculations. The reader is referred to Paper I and the references therein for more detailed discussion of the author's physical input and method of solar model calculation. The 'forward/inverse helioseismic' solar model generated for this report is referred to as 'seismic'; the current reference standard evolution solar model, ssm59b, is referred to as 'ssm'; and the current reference HZC model, hzc26b, is referred to as 'hzc.'

The oscillation frequencies are compared first. Currently this project is mainly concerned with the first-order, one-dimensional structure of the core and interior of the present sun, hence only the degree $l$ = 1 nonradial oscillation frequencies are reported here. The theoretical frequencies are compared with the best available observed frequencies. At this time, the accurate observed oscillation frequencies for orders n = 7 to 33 used are from (1) the BiSON group (Elsworth *et al.*, 1994), n = 7 to 14; (2) the GONG Project (Hill *et al.*, 1996), n = 15 to 25; and (3) Duvall *et al.* (1988), n = 26 to 33. The model properties compared are (1) mass density, $\rho$; (2) mean molecular weight, $\mu$; (3) gamma1 function, $\Gamma_1$; (4) relative mass fraction of hydrogen, $\rho_H$; and (5) speed of sound, c. It is shown that in comparing the seismic and hzc model profiles between x (= r/R) = 0.5 and 0.95, that there are very good agreements with the sound speeds, mass densities and the gamma1 functions. However, in addition, only the hzc model has agreement with other global properties such as consistency with two of the three operating solar neutrino experiments and is consistent with the lithium problem without invoking overshooting at the bottom of the convection zone–the temperature there is over $2.4 \times 10^6$ K. The position of this point in the hzc model is at x = 0.6925 where the radiative temperature gradient equals the convective temperature gradient. (See also Paper I.)

In what follows, Section 2 describes the models, giving values of quantities at points between the center and the surface with four theoretical solar model neutrino counting rates and fluxes; and reviews the author's adiabatic temperature gradient and $\Gamma_1$ function, with applications to the theoretical speed of sound. Section 3 presents eight figures showing variations of the effective gamma function, $\gamma_e$; mean molecular weight, $\mu$; hydrogen mass fraction, $\rho_H$; fractional differences and ratios of seismic, standard and hzc quantities such as c, density, $\rho$, and $\Gamma_1$. Section 4 gives discussion; and Section 5 presents conclusions.

## 2. THE MODELS

Before discussing the models, quantities unique to this project are briefly reviewed. From Paper I and references therein the conventional thermodynamic adiabatic exponent, $\Gamma_1$, is defined by

$$\Gamma_1 = (d\ln P / d\ln\rho)_{adiabat} \tag{1}$$

where $\rho$ is the mass density and P, the pressure, calculated in the IEEOS subroutine. The above gamma function is also referred to as 'gam1'. The conventional gam1 is calculated by an expression derived by Cox and Giuli (1968, p. 418) in their equations 17.69 to 17.72. This expression for gam1 is referred to as $\Gamma_{1S}$, or 'gam1s'. The corresponding expression derived by Rouse (1964, 1968a) is

$$\Gamma_{IR} = gam1r = \Gamma_1 = \gamma_e + \left(\frac{\rho}{\gamma_e - 1}\right)\frac{d\gamma_e}{d\rho}\bigg|_{adiabat} \tag{2}$$

where the effective gamma, $\gamma_e$, is defined by $\gamma_e - 1 = P/E$, with P and E the total pressure and total energy, respectively, calculated in the IEEOS subroutine. Figures 1 and 2 of Paper II give plots of gam1s, gam1r and $\gamma_e$ for a standard evolution solar model and the reference hzc solar model, respectively. For the standard model in Figure 1 of Paper II, $\gamma_e$ (or gamE) differs significantly from gam1r and gam1s only in the outer 30% or so of the solar radius (gam1r and gam1s have no significant differences, from the center to the surface). From Figure 2 of Paper II, this is also true for the hzc model, except that in the inner 2% of the radius, $\gamma_e$ *and* gam1r differ significantly from gam1s.

It is noted in the Paper II study that the theoretical five-minute-band (5MB) solar oscillation frequencies calculated with $\gamma_e$ equal to $\Gamma_1$ give the best fits to the mean observed frequency spacing in the 5MB. From equation (2) above, an adiabatic perturbation at constant $\gamma_e$ gives $\gamma_e \equiv \Gamma_1$. From the IEEOS tables (Rouse, 1968b), it is seen that such a perturbation involves a *simultaneous* increase *or* decrease in density *and* temperature. Such a perturbation can be realized in a radial movement of matter outside the core of a self-gravitating sphere with an homogenous composition. (Inside the hydrogen burning region of a main sequence type star, a constant $\gamma_e$, density-temperature path will not be as simple to describe.)

It follows that the gam1r expression above also modifies the theoretical expression for the speed of sound. Given that the square of the speed of sound is

$$c^2 = \Gamma_1 P/\rho \tag{3}$$

a constant $\gamma_e$ perturbation yields a different theoretical expression for the speed of sound in the outer half of a solar-type star and, in addition, in the inner 2% or so of a hzc-type star.

An important expression in this program that differs is the adiabatic temperature gradient used in the equilibrium stellar structure equations. From Rouse (1964, 1968b),

$$\frac{dT}{dr} = \frac{(\gamma_e - 1)}{\gamma_e}\frac{T}{P}\frac{dP}{dr} + \frac{T}{\gamma_e(\gamma_e - 1)}\frac{d\gamma_e}{dr} + \frac{T}{\mu_e}\frac{d\mu_e}{dr} \tag{4}$$

where $\mu_e \equiv \mu$ is the mean molecular weight defined by $\mu = k\,\rho\,T/m_H\,P$, with k and $m_H$ the Boltzman constant and the mass of the hydrogen atom, respectively, and P the total pressure calculated in the IEEOS subroutine. The models are generated using equation (4) in the stellar structure equations in the convective region, with the boundary at the radius where the radiative temperature gradient equals the adiabatic temperature gradient.

Turning to the models themselves, Table 1 presents various model quantities at four radii from the center to the surface. Theoretical neutrino capture rates and the boron-8 fluxes are also given for each model. Column 2 lists the corresponding values from the TS 'most likely' model. Column 3 lists values for the seismic solar model of this report generated using only the chemical abundance profiles published by TS. Column 4 lists values for the standard solar model for this report generated as described in Paper I using only the chemical abundance profiles from BU. Here, this is viewed as a reference standard evolution solar model that is typical of such models. Finally, Column five gives values for the current reference hzc solar model, also described in Paper I, but calculated for this paper with increased precision, as was the above standard model.

*Table 1.* Quantities at four radii from four solar models, viz., (1) the 'most likely' inverse helioseismic solar model of Takata and Shibahashi (1998) (TS); (2) the forward 'seismic' model for this project generated using only the chemical abundance profiles from TS; (3) the standard evolution solar model for this project (ssm59b) generated using only the chemical abundance profiles of Bahcall and Ulrich (1988) (BU); and (4) the reference high-Z core model, hzc26b (See also Rouse, 1995).

| Model | T S | SEISMIC | SSM59B | HZC26B |
|---|---|---|---|---|
| **At center:** | | | | |
| $T_c\,(10^6\,K)$ | 15.8 | 15.735 | 15.60 | 14.64 |
| $\rho_c$ (g/cc) | 166.0 | 171.58 | 153.28 | 503.8 |
| $P_c\,(10^{17}\,dyn/cm^2)$ | 2.51 | 2.594 | 2.355 | 2.580 |
| $P_e\,(10^{17}\,dyn/cm^2)$ | | 1.528 | 1.374 | 2.508 |
| $\gamma_e$-1 | | 0.66071 | 0.66075 | 0.55694 |
| $\mu$ | | 0.86538 | 0.84388 | 2.377 |
| $X_c$ | 0.316 | 0.3161 | 0.34144 | $1.272\ x10^{-5}$ |
| $Y_c$ | 0.666 | 0.666 | 0.6387 | $5.098\ x10^{-4}$ |
| $Z_c$ | 0.018 | 0.0179 | 0.01996 | 0.9998 |
| **At radius of H burning (99.9% $L_R$):** | | | | |
| $x\ (= r/R)_{core}$ | | 0.270 | 0.272 | 0.334 |
| $X_{core}$ | | 0.7134 | 0.7086 | 0.70954 |
| $Y_{core}$ | | 0.2688 | 0.2716 | 0.27063 |
| $Z_{core}$ | | 0.0178 | 0.0198 | 0.01983 |
| $m/M_R$ | | 0.53435 | 0.5428 | 0.68174 |
| $L/L_R$ | | 0.99896 | 0.99891 | 0.99141 |
| $T_{core}\,(10^6\,K)$ | | 7.461 | 7.453 | 6.233 |
| $\rho_c$ (g/cc) | | 15.20 | 16.88 | 7.827 |
| $P_{core}\ (10^{16}\,dyn/cm^2)$ | | 1.614 | 1.699 | 0.6572 |
| $P_e\,(10^{16}\,dyn/cm^2)$ | | 0.853 | 0.899 | 0.3464 |
| $\gamma_e$-1 | | 0.6593 | 0.6592 | 0.65765 |
| $\mu$ | | 0.61367 | 0.61595 | 0.61721 |
| **At bottom of convection zone:** | | | | |
| $x\ (= r/R)_{conv}$ | 0.709[a] | 0.6965[b] | 0.7720[b] | 0.6925 [b] |
| $X_{conv}$ | 0.749 | 0.7492 | 0.7096 | 0.709575 |
| $Y_{conv}$ | 0.233 | 0.2330 | 0.27059 | 0.27059 |
| $Z_{conv}$ | 0.0183 | 0.0178 | 0.01981 | 0.01980 |
| $m_{conv}/M_R$ [c] | 0.030 | 0.03745 | 0.01581 | 0.0322 |
| $L/L_R$ | | 1.00615 | 1.00648 | 0.992 |
| $T_{conv}\,(10^6\,K)$ | 2.22 | 2.36 | 1.645 | 2.465 |
| $\rho_{conv}$ (g/cc) | 0.194 | 0.241 | 0.080 | 0.244 |
| $P_{conv}\ (10^{13}\,dyn/cm^2)$ | 5.93 | 7.79 | 1.759 | 8.07 |

| Model | T S | SEISMIC | SSM59B | HZC26B |
|---|---|---|---|---|
| **At bottom of convection zone:** | | | | |
| $P_e$ ($10^{13}$ dyn/cm$^2$) | | 4.04 | 0.919 | 4.21 |
| $\gamma_e$ -1 | | 0.64515 | 0.63553 | 0.64488 |
| $\mu$ | | 0.60482 | 0.62147 | 0.62123 |
| **At solar radius R ($6.9599^{10}$ cm)** | | | | |
| $M_R$ (1.989 $10^{33}$g) | | 1.00048 | 1.00008 | 0.9998 |
| $L_R$ (3.846 $10^{33}$ erg/sec) | | 1.00615 | 1.00648 | 0.99237 |
| $T_R$ (K) | | 5774 | 5776 | 5776 |
| $\rho_R$ ($10^{-7}$ g/cc) | | 2.707 | 2.314 | 2.625 |
| $P_R$ (dyn/cm$^2$) | | 1.043 x $10^5$ | 8.589 x $10^4$ | 9.743 x $10^4$ |
| $P_e$ (dyn/cm$^2$) | | 18.10 | 16.81 | 18.32 |
| $\gamma_e$-1 | | 0.66504 | 0.66486 | 0.66494 |
| $\mu$ | | 1.24581 | 1.29382 | 1.29383 |
| Neutrino capture rates and fluxes | | | | |
| Cl (SNU) | 9.39 | 11.2 | 7.14 | 2.43 |
| Ga (SNU)[d] | 137 | 242 | 137 | 102 |
| Ga (SNU)[e] | | 200 | 112 | 88.9 |
| $^8$B ($10^6$/cm$^2$/sec) | 6.74 | 6.00 | 4.84 | 1.30 |

[a]Assumed in Takata and Shibahashi (1998).
[b]At the radius where the radiative and convective temperature gradients are equal.
[c]Mass fraction in the convective region.
[d]Neutrino cross sections from Bahcall and Ulrich (1988).
[e]Neutrino cross sections from Bahcall *et al.* (1982).

As for any relationships between the models, the seismic model takes only the chemical abundance profiles for various isotopes of helium, carbon, nitrogen and oxygen from TS (taken from the preprint of TS) and calculates the hydrogen profile in the IEEOS. This is a check and also guarantees that the sum of the mass fractions of the elements equal 1.0. The seismic model chemical composition is independent of the standard model and the hzc model. The relationship between the standard model and the hzc model is that the hzc model uses the same element abundance profiles used for the standard model, except that a postulated additional high abundance of iron is assumed from the center to a radius of about x = 0.0245, being uniform in the inner region with $x \leq 0.002$, decreasing exponentially to the surface abundance at $x \approx 0.0245$. At this point the solar mass fraction, $q = M_r/M_R = 1.65 \times 10^{-3}$. At x = 0.002, $q = 2.86 \times 10^{-6}$. In the central region the composition is not completely Fe: The Z = 0.9998 in Table 1 includes the other high-Z elements. In this project, the extra Fe is used as a convenient way to control the variation of the mean molecular weight as described in the references of Paper I. The actual Fe or other high-Z element(s) abundances beyond the photospheric abundances must be deduced from a theory of star formation and supported by the accurate measurement and identification of solar g-mode oscillation frequencies. The emphasis on the mean molecular weight variation is important because the various oscillation equations "see" $\mu$ as the nearest explicit information about the chemical composition through the ratio $P/\rho \propto T/\mu$ obtained in the derivations of the various adiabatic oscillation equations (Ledoux and Walraven, 1958; Unno *et al.*, 1989; Boury *et al.*, 1975; Scuflaire, 1974; Cox and Giuli, 1968; Cox, 1980). And note that once ionization begins as the temperature increases and/or the density decreases, $\mu$ also includes the effects of the free electrons.

In a brief outline of this approach to solar model calculation, using only the input chemical abundance profiles, the central temperature and central density are iterated until the model yields one solar mass and one solar luminosity to within one sigma of the measured values. Then the scaled opacity is iterated (perturbed) until the solar temperature and surface density at the solar radius are obtained to within one sigma. Such models are the only models accepted for further analysis: Starting the integration of the stellar structure equations at the center and matching these quantities at the surface in continuous calculations satisfy the Ince (1926) conditions for a unique solution to a system of first order, ordinary differential equations (See also Rouse, 1975; 1983).

In comparison with the most likely TS seismic model, note that the Rouse seismic model central temperature agrees to about 0.4% and the central density agrees to about 3.3%. Relative to the BU standard model, the ssm here agrees with the central temperature to the same three figures given by BU, whereas the central densities differ by about 3.27%. In addition, the central pressures in both cases also differ by about 3%.

All other quantities in Table 1 have their usual meanings, with the definition of $\gamma_e$ discussed above.

In order to show that partial ionization exits well within the inner half of the solar radius, I call attention to the ratios of electron pressure to total pressure, $P_e/P$, from the corresponding values at the center and at the radius where the hydrogen burning is negligible, *i.e.*, where the luminosity is about 99.9% of the surface value. Consider only the seismic model and the ssm. Note that at the center, the $P_e/P$ ratios for the seismic and ssm are about 0.589 and 0.583, respectively. At x = 0.270 and 0.272, the ratios are 0.528 and 0.529, respectively. The approximate 15% difference says that, relative to the ions at the center, significant net recombination has occurred by the distance of a little over a quarter of the solar radius. Consequently, *assuming* all atoms are stripped at temperatures above $10^6$ K (See for example, Basu *et al.*, 1999; hereafter BPB) says that there is no partial ionization well into the convective region of all three models in Table 1, *viz.*, below x = 0.843 in the seismic model; below x = 0.847 in the ssm; and below x = 0.847 in the hzc model, where, respectively, the bottom of the convective zones are at x = 0.6965, 0.773 and 0.6925. Clearly, these radii at

$T = 10^6$ K are over three times the radial distance where partial ionization is evident in these more precise EOS calculations. It follows that the accurate treatment of one- and three-dimensional convective flow must allow for the real gas effects of ionization and recombination above and below the base of the convective zone. In addition, current measurements of solar oscillation p-mode frequencies with accuracies $\leq 0.1$ μHz require these more precise model calculations.

At the surface the ratio of electron pressure to total pressure is less than about $2 \times 10^{-4}$. This means that this solar matter is essentially a neutral gas. This suggests an interesting question about thermal diffusion in this region. According to Chapman and Cowling (1961), top of their page 255, if the molecules are not nearly equal in mass, the heavier molecules tend to diffuse into the cooler regions. This suggests that in the solar surface, helium and the high-Z neutral atoms would tend to diffuse 'upward' toward the minimum temperature in the solar atmosphere. This may possibly explain the relatively high chromosphere helium abundances reported by Unsöld (1964) from spectroscopic observations and by Rouse (1969; 1971) in studies of line and continuum radiation from a solar model photosphere.

Finally for this section, in Table 1, theoretical solar neutrino capture rates and boron-8 neutrino fluxes are given. The Cl(SNU) theoretical rates are to be compared with the Homestake experiment (Lande, 1997) rate of (2.56 ± 0.22) SNU; the Ga(SNU) are to be compared with the two gallium experiment [SAGE (Abazov *et al.*, 1991) and GALLEX (Anselman *et al.*, 1992)] rates of (69.9 ± 9.0) SNU and (76.4 ± 8.0) SNU, respectively (See Svoboda 1997); and the boron-8 values are to be compared to the Kamiokande and SuperKamiokande experiments (Fukuda 1996; Totsuka 1997; see also Fogli *et* al., 1998a,b) results of $(2.80 \pm 0.38) \times 10^6$ and $(2.37 \pm 0.1) \times 10^6$ $cm^{-2}$ $sec^{-1}$, respectively. (One SNU = $10^{-36}$ captures per second.)

## 3. RESULTS

The main results of interest for this report are given in Figures 1 to 8. First, some general statements. In all model oscillation frequency and speed of sound calculations, $\Gamma_1 = \gamma_e$, implying adiabatic radial perturbations at constant $\gamma_e$ (= 1 + P/E). (This does not mean constant P and constant E perturbations, but constant ratio of P/E.) Next, in Figures 3 to 7, the curves present fractional differences and ratios of various quantities, viz., speed of sound, density, and the gamma1 functions. Comparisons are made to the corresponding figures in TS and BPB. However, figures with only fractional changes (FC), where FC = [(f(seismic - f(model))/f(model)], are given in TS and BPB. This quantity is referred to in the present paper as 'fractional differences'. Clearly, f(seismic)/f(model) = FC + 1.0. A node with FC = 0 corresponds to a point where ratio = 1.0. Finally, the abscissas in Figures 1, 4 and 5 have changes of scale. This is done to show the structure in the inner 10% and in the outer 10% of the solar radius in more detail. In all these figures, changes of scale occur at x = 0.006, 0.01, 0.04, 0.1, 0.9, 0.975, 0.9825, 0.990 and 0.998. In a single scale figure with the scale from x = 0.1 to 0.9 as a reference scale, the curves in Figures 1, 4 and 5, from x = 0 to 0.1 would be compressed into the space between 0.1 and the next tick mark to the left between 0.08 and 0.1 on the figure. At the other end of these figures, the curves from x = 0.9 to 1.0 would be compressed into the space between x = 0.9 and the next tick mark to the right between 0.9 and 0.95 on the figure. The structure in the inner 10% and in the outer 10% of the model radius is critical for calculating accurate g-mode and p-mode oscillation frequencies.

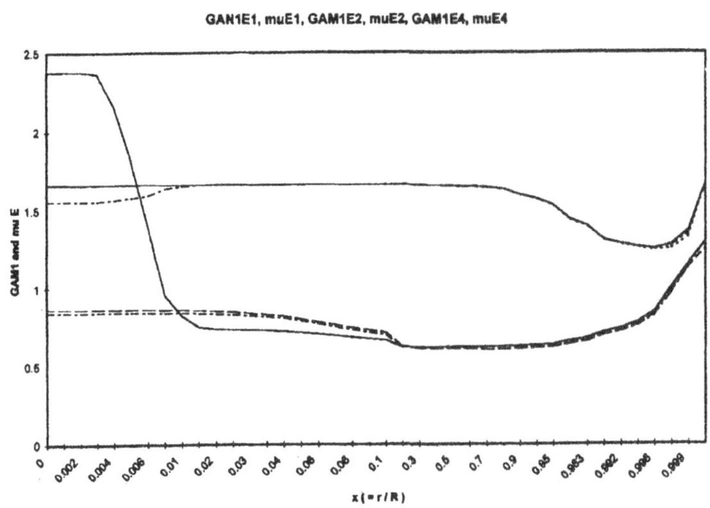


*Figure 1.* Plots, versus x (= r/R), of $\gamma_e$(seismic) (dark solid line); $\gamma_e$(ssm59b) (dotted line); $\gamma_e$(hzc26b) (dash dash dot); $\mu$(seismic) (dash); $\mu$(ssm59b) (dot dash); and $\mu$(hzc26b) (light solid line with a value near 2.5 at x = 0). Along the abscissa, changes of scale occur at x = 0.006, 0.01, 0.04, 0.1, 0.9, 0.975, 0.9825, 0.990 and 0.998.

Figure 1 presents curves of $\gamma_e$ and $\mu$ for the three detailed models, seismic, ssm and hzc as a function of radius fraction x. This figure is like Figures 1 and 2 in Paper II, except that the figures in Paper II include curves of $\Gamma_{1S}$ and $\Gamma_{1R}$ that show the large differences between these complete gamma1 functions and $\gamma_e$ from about x = 0.7 to the surface. In Figure 1 of this report, note that the $\mu$'s and $\gamma_e$'s for the seismic and ssm models are almost equal throughout the interior. Relative to $\gamma_e$, the $\gamma_e$(hzc) matches the other two except below x ≈ 0.2 due to the postulated HZC. Relative to $\mu$(hzc), the large values near the center were expected, but the significantly lower μ(hzc) less than μ(ssm) between x = 0.015 and x = 0.1 was not expected.

It is noted that if all atoms were completely ionized for T greater than $10^6$K, then the values for μ(ssm) and μ(hzc) would be exactly the same between x ≈ 0.02 and the fractional radius where T = $10^6$K because of the same chemical composition.

Figure 2 presents the degree *l* = 1 solar oscillation frequency differences, (Calculated - Observed), in μHz, for orders n = 7 to 33 for the three models.

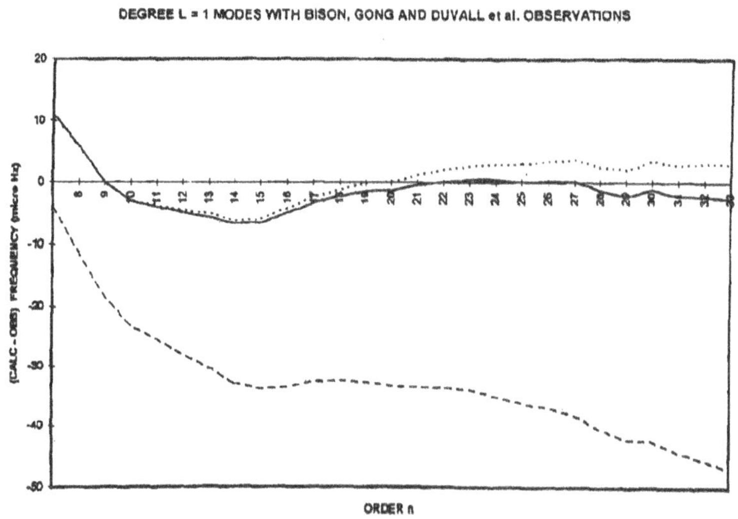


*Figure 2.* Frequency differences , (Calculated - Observed) in μHz, for nonradial degree *l* = 1, orders n = 7 to 33, for the three forward models generated for this project, (1) seismic (solid line); (2) ssm59b (dash line); and (3) hzc26b (dotted line). The sources of the observed frequencies are given in the text.

In all the oscillation frequency calculations, $\Gamma_1 = \gamma_e$ is used in the oscillation equations with the Cowling approximation (Scuflaire, 1974). In Paper II, it is shown that these solutions yield substantially better agreement with observation (*cf.* Figures 3 to 6 in Paper II). (Since for p-modes of oscillation, pressure is the dominant restoring force, this should not be surprising.)

Considering the independence of the TS calculation and the author's calculation, the very good agreements with observation of the frequencies from the author's reproduction of the TS inverse model is support for both calculations. However, the surprise is the relative close agreements between seismic model frequencies and the reference HZC model frequencies! And considering the rather high sensitivity of the 5MB frequencies to the temperature gradient in the convective region (See Rouse, 1990; Vanlommel and Cadez, 1998) and the precision used in the author's model calculations, there may be some global model agreements, with some inputs and/or derived quantities throughout both model computations.

Relative to the ssm's large deviations from the observations, previous sensitivity studies (*cf.* Rouse, 1990) showed that increasing the temperature gradient in the convective region would increase the 5MB frequencies. In (Rouse, 1990) it was demonstrated that with models with different core structures, it is possible to vary the temperature gradients in the convective region and improve agreements with observed 5MB frequencies. Consequently, in order to yield sensitivities to the structure of the core, all recent models of this project use the same real gas adiabatic temperature gradient (equation 4 above) in the present phase of this study.

Figure 3 presents the fractional differences of the sound speeds of the seismic and hzc models, or, [c(seismic) - c(hzc)]/c(hzc), as a function of the fractional radius, x.

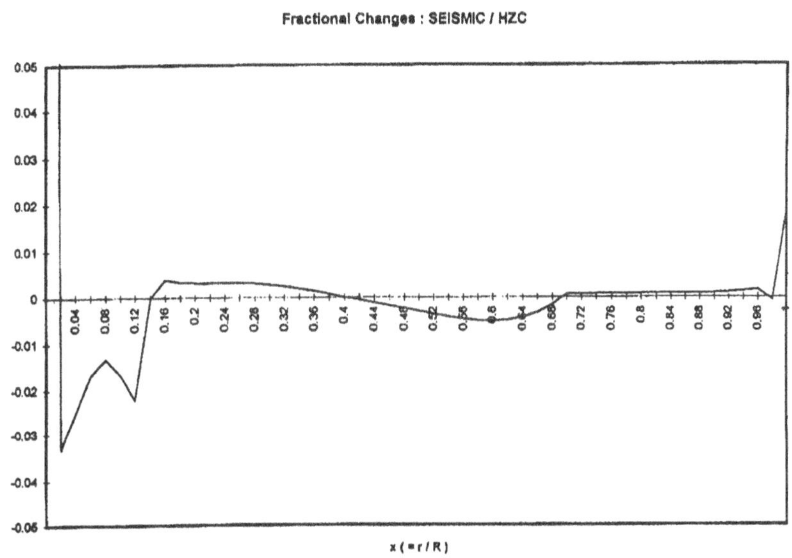


*Figure 3.* Fractional differences, versus x (= r/R), of sound speeds, c, for models seismic and hzc26b, or, (c(seismic) - c(hzc26b))/ c(hzc26b).

The sound speed fractional differences in Figure 3 here are to be compared to (1) the sound speed squared, fractional changes in TS, $[c^2(\text{inverse}) - c^2(\text{model})]/c^2(\text{model})$, in their Figures 2 and 3; and (2) the sound speed fractional changes in BPB, $(\delta c/c)$, in their Figures 2, 6 and 7.

For comparison with TS, only the nodes of the $c^2$ fractional change and the nodes of the c fractional differences should occur at the same value of x. For direct comparison with Figure 2 of BPB, the values in the convective region in Figure 3 here with x ≥ 0.7 have the same magnitude fractional changes as shown in BPB's Figure 2. Between x = 0.3 and x = 0.7, the absolute magnitudes of the fractional changes here are slightly larger at the peaks. Below x = 0.3 going toward zero, the fractional differences here reach a minimum of about -0.033 at x = 0.02 before increasing to values greater than zero, compared with BPB having a minimum of about -0.001 at x = 0.15 (in their Figure 2c), increasing to positive values.

Figure 4 presents the ratios of sound speeds relative to the seismic model and the ssm (dotted curve). Comparing the dotted curve of Figure 4 in this report with the top panel of Figure 2 in TS, the shapes of the two curves are the same. First, the point at x = 0.3 in Figure 4 here where the ratio equals 1.0, occurs at essentially the same point as the node in Figure 2 of TS.

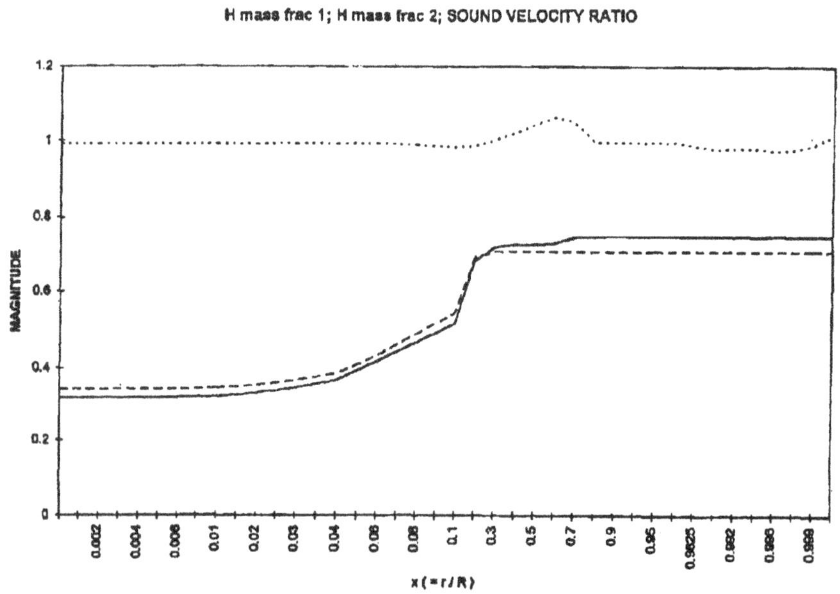


*Figure 4.* Ratios, versus x (= r/R), of sound speeds, c, for models seismic and ssm59b, or, c(seismic)/c(ssm59b) (dotted line). Also shown are the hydrogen mass fractions for the seismic model (solid line) and the standard evolution model, ssm59b (dash line). Scale changes occur along the abscissa at x = 0.006, 0.01, 0.04, 0.1, 0.9, 0.975, 0.9825, 0.990 and 0.998.

The shape of the curve with a dip to the left of x = 0.5 in Figure 4 here to ratio = 0.985 at x = 0.1, and then the rise to ratio = 0.994 at x = 0.05; and then a decrease to ratio = 0.992 at x = 0.01, is the same as the variation in this region of the TS Figure 2. Then to the right of the node in Figure 2 of TS, the peak at about x = 0.65 occurs in Figure 4 here at about x = 0.7. Finally, the decrease of the ratio in Figure 4 here as x increases to 0.998 is the same as the decreases in Figure 2 of TS.

Relative to the agreements with Figure 2 of BPB, the BPB node at x = 0.3 and the peak at about x = 0.65 agree with the ratio = 1.0 at x = 0.3 and the peak at x = 0.65. Beyond x = 0.65, the curve in Figure 4 here decreases. The minimum in the dip in the BPB Figure 4 at x = 0.15 to 0.2 occurs here in Figure 4 at x = 0.1. But, below x = 0.1, instead of increasing to near zero and then decreasing as in the TS curves, the BPB ratio increases for x less than 0.1.

Figure 4 here also presents the hydrogen mass fractions for the seismic model (solid curve) and the ssm (dash curve). Note the increase at x = 0.7 for the seismic curve.

Figure 5 gives the same quantities presented in Figure 4, but for the sound speed ratio, hzc/ssm, and their model hydrogen mass fractions.

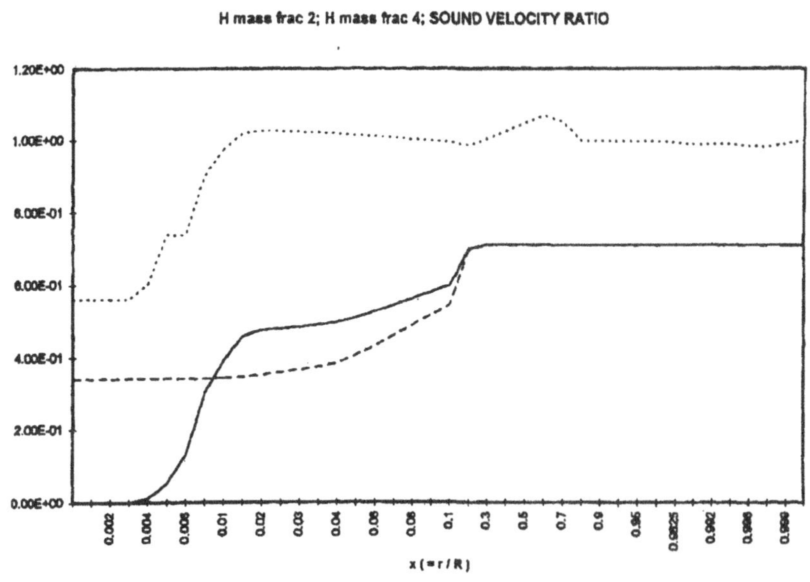


*Figure 5.* Ratios, versus x (= r/R), of sound speeds, c, for models hzc26b and ssm59b, or, c(hzc26b)/c(ssm59b) (dotted line). Also shown are the hydrogen mass fractions for the high-Z core model (solid line) and the standard evolution model, ssm59b (dash line). Scale changes occur along the abscissa at x = 0.006, 0.01, 0.04, 0.1, 0.9, 0.975, 0.9825, 0.990 and 0.998.

For x ≥ 0.1 the sound speed ratio hzc/ssm in Figure 4 and the ratio seismic/ssm here are about equal. Below x = 0.1, the hzc/ssm sound speed ratio, as x decreases, increases to 1.027, before decreasing to values less than 1.0 at x ≈ 0.015, to values of about 0.56 at the center. Hence the speed of sound in the center of the hzc model is considerably less than in the ssm.

Relative to the hydrogen mass fractions, the ssm $\rho_H$ curve in Figure 5 is the same curve as given in Figure 4. However, the hydrogen mass fraction for the hzc model is significantly different from that in the seismic model. Above x = 0.2 the hzc and ssm hydrogen mass fractions are equal as noted above. And from Figure 4, above x = 0.2, the seismic model hydrogen mass fraction is greater than in ssm, with another increase around x = 0.7 as was deduced by TS in their inversion analysis. Hence, in the convective region, the seismic model has less helium ($Y_S = 0.233$) than obtained by BU in their standard evolution model ($Y_S = 0.27059$). Below x = 0.2 down to 0.01, the $\rho_H$ for the hzc is significantly greater than that in both the seismic and ssm models, decreasing to a negligible amount toward the center–as postulated for this hzc model.

Figure 6 is the fractional differences for densities from the seismic and hzc models. These curves are compared directly to the most likely curves in Figure 6(c) of TS and to Figure 3(a) of BPB. In Figure 6 here, the fractional difference [ρ(seismic) - ρ(hzc)]/ ρ(hzc), has nodes at x = 0.1 and x ≈ 0.3. In Figure 6(c) of TS, the most likely curve has nodes at x ≈ 0.15 and x ≈ 0.65.

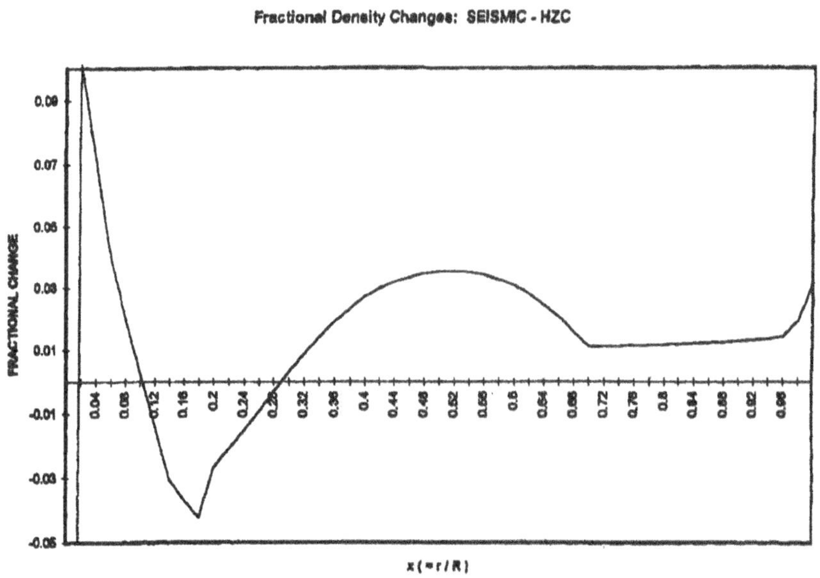


*Figure 6.* Density fractional difference, versus x (= r/R), for models seismic and hzc26b, or [(ρ(seismic) - ρ(hzc26b)]/ρ(hzc26b).

In Figure 3(a) of BPB, the nodes occur at x = 0.15 and x = 0.45. Here in Figure 6, the magnitude of the fractional difference dip at x = 0.18 is about -0.042; in TS the dip is -0.02 at x = 0.3; and in BPB, the magnitude of the dip is about -0.007 at x = 0.3. Relative to Figure 6 here, the positive peak of the curve is about 0.0355 at x = 0.5. TS does not give a value beyond x = 0.7. BPB gives a positive peak equal to about 0.016 at about x = 0.75.

At x < 0.1, in Figure 6 here, the ratio increases to about 1.0 at x = 0.02, then decreases as x decreases. In Figure 6(c) of TS, for x less than 0.15, the fractional difference increases as x decreases toward zero. The same is true for the density fractional difference in BPB. These last two figures do not show a decrease in the density fractional difference below x = 0.02, as is obtained here in Figure 6 for the seismic–hzc density fractional difference.

Figure 7 presents the fractional differences of the equivalent gamma1 functions, $[\gamma_e\text{ (seismic)}) - \gamma_e\text{ (hzc)}]/\gamma_e\text{ (hzc)}$. The fractional difference curve is compared with the fractional change curves in BPB, Figure 3(b). In Figure 7 here, between x = 0.05 and x = 0.95, there are nodes at x = 0.014 and 0.018 (corresponding to a single node at x = 0.016). At x = 0.05, the fractional difference is about -0.00027. At x = 0.95, the fractional difference is about 0.0014, near a peak value of 0.0016 at x = 0.96. A second node occurs at x ≈ 1.0.

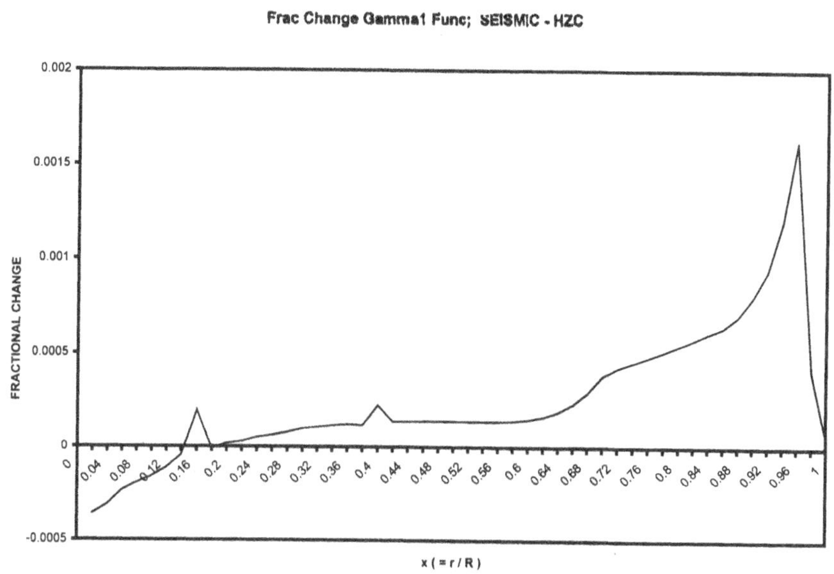


*Figure 7.* Fractional Change Gamma1 Function; SEISMIC - HZC

Finally, Figure 8 presents the same results as given in Figure 2, but with the addition of the degree $l = 1$ oscillation frequency differences from a fourth model. This fourth model was created from the seismic model with a small high-Z core added in the same way a small high-Z core was added to the standard solar model to create the reference high-Z core model, hzc26b. What is shown is how the hzc affected the degree $l = 1$ frequencies relative to the seismic model as compared to how the hzc affected the same frequencies of the standard solar model. In this preliminary result, it is seen that the overall affect of the hzc on the ssm is much greater than the affect on the seismic model. Further sensitivity studies are needed to establish the affect consistently relative to the profile of the extra iron added in the core at the center of the model.

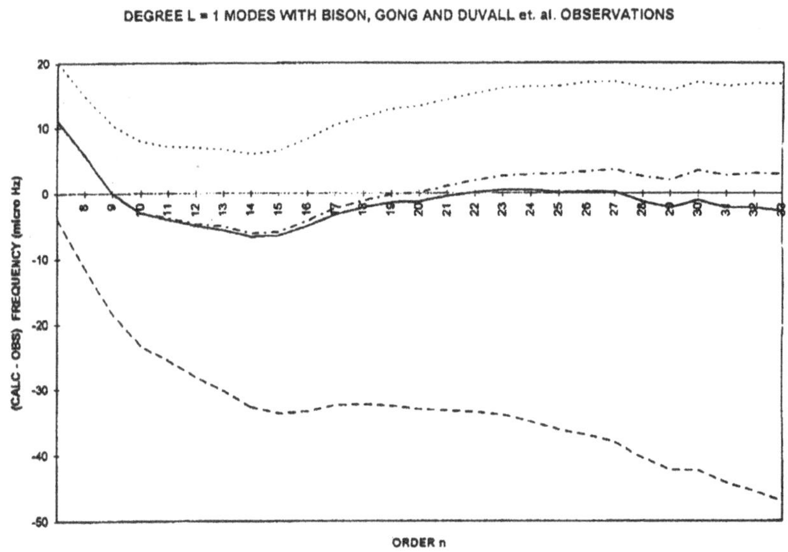


*Figure 8.* Same input as Figure 2, but with the dotted line here being the degree $l = 1$ (Calculated–Observed) frequency differences for a preliminary model generated with a high-z core added to the seismic solar model in the same way that the hzc was added to the standard evolution model to obtain the reference hzc model hzc26b.

For the record, this preliminary hzc variation of the seismic solar model has a central temperature of $15.67 \times 10^6$ K and a central density of 568.8 gm/cc. The theoretical chlorine solar neutrino experiment counting rate is 9.4 SNU and the gallium solar neutrino counting rates are 221 and 184 SNU for the two theoretical neutrino absorption cross sections. The theoretical boron-8 neutrino flux is $4.73 \times 10^6 \text{ cm}^{-2} \text{ s}^{-1}$.

# 4. DISCUSSION

In constructing a first-order, equilibrium model of the present sun, given the equation of state (EOS), opacities, energy generation rates and an expression for the temperature gradient in the convective region, the only remaining quantities needed are the relative abundances of elements (chemical composition) from the center to the surface of the sun. For given element profiles, a unique solution to the four first-order, coupled, nonlinear system of equations that describe an equilibrium star, requires satisfying the Ince (1926) existence theorem for a unique solution. From spectroscopy, the dominant elements that exist in the solar photosphere can be determined–*except helium*. One of the basic outputs of standard evolution theory of the sun is to establish an approximate value for the sun's photospheric helium abundance through iterations through evolution calculations with assumed initial helium abundances until a model of one solar mass, $M_R$, and one solar luminosity, $L_R$, is obtained at a given age for the sun.

Due to hydrogen burning to helium in a varying temperature environment, this yields varying chemical abundance profiles from the center to the outer "edge" of the hydrogen burning central region. Secondary notions are assumed about the distributions of elements not involved in the nuclear energy generation processes. Now this model of the present sun can be used to theoretically study other global solar observations like solar oscillations, solar neutrinos, differential solar rotation, etc. Considering only the first two quantities listed, what if the model is used to theoretically "predict" them–but yield results considerably different from the observations!

Relative to the solar neutrino problem, only forward analysis is possible, *i.e.*, (1) present solar models are varied through alternative star formation and evolution calculations; and (2) different assumptions are made about the structure of the present sun, with implications as to how it formed and evolved.

Relative to solar oscillations, a completely different approach is possible, namely, the various inverse analyses widely used at present. (*cf.* Unno *et al.*, 1989; Harvey, 1995; Gough *et al.*, 1996; Schou *et al.*, 1998.) Here, only the global internal structure determinations of chemical abundance profiles from the center to the surface are of interest. With accurate element profile determinations–if obtained uniquely–it is possible to obtain a unique solution to the equilibrium stellar structure equations as outlined above in Section 2. If unique chemical abundance profiles are not possible, then satisfying the Ince theorem is a *necessary condition* for a unique solution to the stellar structure equations–not a sufficient condition.

With my interest in a unique solution to the solar structure problem, the inversions that yield element abundance profiles in the interior are of fundamental interest. It is at least a quantitative approach to a difficult problem that includes many assumptions. Consequently, when the inverse element abundance profiles of Takata and Shibahashi (1998) were obtained, they were included into this project in a 'forward' analysis of an inverse solution. The model with inverse element abundance profiles incorporated into the Rouse forward sensitivity code is solved as outlined above and in Rouse

(1995) and in the references therein. Here, this solution is referred to as the 'seismic' model even though the element abundance profiles were generated elsewhere by inverse helioseismic methods. Both the TS interior values and the equivalent Rouse model interior values are given in Table 1 as discussed above in Section 2. Of the abundance quantities in the seismic model of Table 1, only the helium-4 profile and the He-3, C, N, O abundance profiles in the TS 'most likely' model are explicitly used here. The Grevesse (1984) Z elements are also used initially for continuity with earlier sensitivity studies, and the hydrogen mass fraction, X, is calculated as a check. Note that the central temperature iterated in the Rouse seismic model is only about 0.4% lower than in the TS model, but the central density is about 3% higher than in the TS model. Relative to the theoretical solar neutrino results, the chlorine experiment rates and the boron-8 fluxes are compatible.

Now, the given chemical profile represents a key input for a unique forward model of the solar interior. We essentially have a model with the chemical abundances generated by inverse helioseismology, but used in a forward helioseismology study. Nevertheless it is still called a 'seismic solar model' here.

Similarly, the reference standard solar model was generated for the Bahcall and Ulrich (1988) standard evolution model as a forward standard solar model (ssm). Lastly, by postulating a small central region of high iron abundance (*cf.*, Rouse 1985, 1987), the hzc model is a forward model relative to both the solar oscillation studies and the solar neutrino studies.

The chemical compositions of the standard model and the hzc model in Table 1 are identical except for the postulated small region of extra iron used to control the variation of the mean molecular weight. This profile will be related to sensitivity studies of solar oscillation–particularly for the g-modes of oscillation and the low degree, low order p-modes of oscillation.

With this exposition of the chemical abundance profiles of the seismic, ssm and hzc solar models of Table 1, all other physical and numerical methods for solving the stellar structure equations are essentially the same. Whereas models ssm and hzc reflect the differences in chemical composition only in the central core region ($x \leq 0.023$), the seismic model and the hzc model, have differences in the chemical abundance profiles from the center to the surface. But all three models have one solar mass, one solar luminosity and one solar radius, all within one sigma of the observed values. The temperatures and densities at one solar radius are in reasonable agreement with the corresponding values of a solar model photosphere profile (*cf.*, Rouse, 1971, 1975). (These values can be iterated to more accurate agreement when warranted.)

Only a few more results in Table 1 will be commented upon. Comparing seismic and hzc models at the center, the significantly different values for $T_c$

and $\rho_c$ yield essentially the same value for the total pressure at the center. Since $P \propto \rho\, T/\mu$, it is the large value of $\mu$ in the hzc at x = 0 that results from the assumed iron that compensates. Next, the radius where the luminosity equals 99.9% of the surface value is about 22% larger in the hzc model compared to the other two models. The fraction of the solar mass is about 26% larger for the hzc model at this point.

Based on the physical condition that the radiative temperature gradient and the convective (adiabatic) temperature gradient are equal at a boundary of a convective zone, the radii at the bottom of the convective zones in both seismic and hzc models are almost equal, with significantly different mass fractions of hydrogen and helium. The total pressures, however, are about equal, again reflecting the $P \propto \rho\, T/\mu$ pressure dependence.

Next, in Figure 3, the near equality of the speeds of sound in the seismic and hzc models between x = 0.1 and the surface–with about ± 1% and less differences–reflects the nearly equal values of density (Table 1 and Figure 6); $\Gamma_1$ (or $\gamma_e$) (Table 1 and Figure 7); and pressure (Table 1).

Finally for this section, relative to the TS and BPB fractional change graphs, it is noted that the reference standard models used for their two papers are different from the reference standard model of this paper. However, the above agreements revealed in this study indicate that the standard models are more or less similar, but not exact. For example, eight of the nine models in Table 1 of BPB have central temperatures and central densities in good agreement with the ssm model in Table 1 of this report, which, incidentally, does not assume diffusion. However, the surface helium mass fractions, $Y_S$, and the depth of the convective zones are significantly different from the ssm in Table 1 here.

## 5. CONCLUSIONS

The chemical abundance profiles of the inverse seismic solar model of TS were used to generate an equivalent forward seismic solar model for this project. For given physical inputs for the equation of state, opacities, energy generation rates and an expression for the adiabatic temperature gradient, the chemical abundance profiles allow for a unique solution to the four nonlinear, coupled, first-order ordinary differential equations. However, since the above inputs may not be precisely accurate throughout the model's interior–particularly the chemical abundance in the region below the convective region, the Ince theorem becomes a *necessary condition* for a unique solution to the stellar structure equations. Comparisons with two other models were made, viz., (1) a reference standard evolution solar model and (2), a representative high-Z core solar model developed by the author. The fractional differences and ratios of several model properties are compared for this report, namely, (1) profiles of mean molecular weight, $\mu$; (2) profiles of $\Gamma_1$ functions (= $\gamma_e$); (3) solar oscillation frequencies for degree $l$ = 1, orders n = 7 to 33; (4) fractional differences of density; (5) fractional differences of the speeds of sound ; and (6) fractional differences of gamma1 functions. The agreements between quantities from the seismic and hzc models are of extreme importance, particularly with respect to the speed of sound. The range of the solar radii with agreement of the speed of sound means that ($\gamma_e$ P/$\rho$) profiles for each model are essentially equal. With $\gamma_e$ and $\rho$ profiles equal means that the P profiles are equal. Finally–and of singular importance–the equality of P/$\rho$ means the equality of the ratio T/$\mu$ in the two models in the radial range. Now, in the hydrogen burning region with essentially completely ionized atoms, the variation of $\mu$ will reflect the hydrogen abundance variation. This, in turn, with the temperature variation, will reflect the neutrino production rates in this region. And with the high sensitivity of the neutrino production rates to the temperature–especially the boron-8 neutrino production–the model with the temperature and hydrogen profiles that yield agreement with the various solar neutrino experiment counting rates must be considered the closest model to the real sun. At present, this is the author's high-Z core solar model.

Further support for any model is to be found in the accurate measurement and identification of the degree $l$ = 1, and lower order p-modes with n < 7, the f-mode and the low degree, low order g-modes. If and when it is established that electron neutrinos emitted from the nuclear reactions in the solar core have mass, then, depending on the magnitude of the mass and on mixing distance, it will be known that the rate of detection of solar electron neutrinos will be less than it would be if electron neutrinos did not change (oscillate) to a form not detected in the present chlorine and gallium solar neutrino detectors. This will mean that either (1) one or more of the neutrino production cross sections inside the sun must be increased to compensate for the decreased detection rate at a given temperature; and/or (2) the temperature profile in the hydrogen burning region must be higher than that in the present hzc model. In the latter case, the mean molecular weight, gamma1 function and density profiles in the hydrogen burning region must be changed in ways to be determined relative to the temperature increase for a given speed of sound. And in the complete model, the T, $\mu$ and $\Gamma_1$ profiles in the outer part of the model must continue to be consistent with the observed p-mode solar oscillation frequencies–particularly in the five-minute-band.

In other words, for the entire solar interior, the unique establishment of the speed of sound throughout the interior may establish the runs of $\Gamma_1$ P/ρ and $\Gamma_1$ T/μ, with the solar neutrino experiments and the solar oscillation frequency observations constraining the model's structure from the center to the surface.

Another valuable study would be to generate an inverse solar model using the physical input of this project. Then the forward model based upon the chemical abundance profiles generated by the appropriate inversion method should yield more accurate agreements with the observed frequencies used in the inversion process.

Further studies in this project will involve sensitivity studies to test for the range of temperature, mean molecular weight and gamma1 function profiles that yield improved agreements with observed and identified solar oscillation frequencies and with solar neutrino counting rates from the various operating solar neutrino experiments.

The very good agreements of the degree $l = 1$ solar oscillation frequencies between the inverse seismic and forward hzc solar models for p-modes orders n = 7 to 33 indicate a nearly equal structure outside the core region. This is supported by small fractional differences in (1) the speed of sound, (2) density and (3) gamma 1 functions. The hzc yields neutrino counting rates consistent with two of the three operating solar neutrino experiments has a lower boundary of the convective zone with a temperature consistent with the lithium problem, and it satisfies the Ince (1926) existence theorem for a necessary condition for a unique solution to the stellar structure equations. This therefore indicates that the forward helioseismic approach yields a more accurate model of the solar interior. Nevertheless, this must still be checked in the future with more accurate observed p-mode oscillation frequencies for low degrees and lower orders, n; the f mode and low degree and low order g-mode frequencies. Perhaps these new more accurate oscillation frequency values will also result in a different inverse helioseismic solar model.

In closing, the current results demonstrate that there is an astrophysical solution to the solar neutrino problem: it is part of the answer to the question of how the real sun actually formed and evolved to its present state. This could take the form of one of the theories reviewed by Brush (1990) or the theory proposed by Manuel and Sabu (1975, 1977).

## ACKNOWLEDGMENTS

I thank F.R. Rouse for discussions, reviewing the manuscript and offering useful comments. I thank F. Hill and J. Pinter for sending the

GONG data and the BISON frequencies used here. This research was supported in part by NSF cooperative agreement ACI-9619020 through computing resources provided by the National Partnership for Advanced Computational Infrastructure at the San Diego Supercomputing Center.

## REFERENCES

Abazov, A.I., Anosov, O.L., Faizov, E.L., Gavrin, V.N., Kalikhov, A.V., Knodel, T.V., Knyshenko, I.I., Kornoukhov, V.N., Mezentseva, S.A., Mirmov, I.N., Ostrinsky, A.V., Pshukov, A.M., Revzin, N.E., Shikhin, A.A., Timofeyev, P.V., Veretenkin, E.P., Wark, D.L., Wilkerson, J.F., Davis, R., Lande, K., Cherry, M.L. and Kouzes, R.T. (SAGE Collab.): 1991, "Search for neutrinos from the sun using the reaction $^{71}Ga(\nu_e, e^-)\,^{71}Ge$", *Phys. Rev. Letters* **67**, 3332-3335.

Anselmann, E., Hampel, W., Heusser, G., Kiko, J., Kirsten, T., Pernicka, E., Plaga, R., Rönn, U., Sann, M., Schlosser, C., Wink, R., Wójcik, M., von Ammon, R., Ebert, K.H., Fritsch, T., Hellriegel, K., Henrich, E., Stieglitz, L., Weyrich, F., Balata, M., Bellotti, E., Ferrari, N., Lalla, H., Stolarczyk, T., Cattadori, C., Cremonesi, O., Fiorini, E., Pezzoni, S., Zanotti, L., von Feilitzsch, F., Mößbauer, R., Schanda, U., Berthomieu, G., Schatzman, E., Carmi, I., Dostrovsky, I., Bacci, C., Belli, P., Bernabei, R., d'Angelo, S., Paoluzi, L., Charbit, S., Cribier, M., Dupont, G., Gosset, L., Rich, J., Spiro, M., Tao, C., Vignaud, D., Hahn, R.L., Hartmann, F.X., Rowley, J.K., Stoenner, R.W. and Weneser, J. (GALLEX Collab.): 1992, "Solar neutrinos observed by GALLEX at Gran Sasso", *Phys. Lett.* **B285**, 376-389.

Bahcall, J. and Ulrich, R.K.: 1988, "Solar models, neutrino experiments, and helioseismology", *Rev. Mod. Phys.* **60**, 297-372.

Bahcall, J., Huebner, W.F., Lubow, S.H., Parker, P.D. and Ulrich, R.K.: 1982, "Standard solar models and the uncertainties in predicted capture rates of solar neutrinos", *Rev. Mod. Phys.* **54**, 767-799.

Basu, S., Pinsonneault, M.H., and Bahcall, J.N.: 1999, "How much do helioseismological inferences depend upon the assumed reference model?" *Ap. J.*, submitted for publication.

Boury, A., Gabriel, M., Noels, A., Scuflaire, R. and Ledoux, P.: 1975, "Vibrational instability of 1 M star toward non-radial oscillation", *Astron. Astrophys.* **41**, 279-285.

Brush, S.G.: 1990, "Theories of the origin of the solar-system 1956-1985", *Rev. Mod. Phys.* **62**, 43-112.

Chapman, S. and Cowling, T.G.: 1961, *The Mathematical Theory of Non-Uniform Gases*, Cambridge at the University Press, Cambridge, UK, p. 255.

Cox, J.P.: 1980, *Theory of Stellar Pulsation*, Princeton Univ Press, Princeton, NJ, 380 pp.

Cox, J. and Giuli, R.T.: 1968, *Principles of Stellar Structure*, Gordon and Breach, New York, NY, p. 418.

Duvall, Jr., T.L., Harvey, J.W., Libbrecht, K.G., Popp, B.D. and Pomerantz, M.A.: 1988, "Frequencies of solar para-mode oscillations", *Ap. J.* **324**, 1158-1171.

Elsworth, Y., Howe, R., Isaak, G.R., McLeod, C.P., Miller, B.A., New, R., Speake, C.C. and Wheeler, S.J.: 1994, "Solar p-mode frequencies and their dependence on solar-activity-recent results from the bison network", *Ap. J.* **434**, 801-806.

Fogli, G.L., Lisi, E. and Montanino, D.: 1998a, "Discriminating MSW solutions to the solar-neutrino problem with flux-independent information at SuperKamiokande and SnO", *Phys. Lett. B* **434**, 333-339.

Fogli, G.L., Lisi, E. and Montanino, D.: 1998b, "The solar neutrino problem after three hundred days of data at SuperKamiokande", *Astropart. Phys.* **9**, 119-130.

Fukuda, Y., Hayakawa, T., Inoue, K., Ishihara, K., Ishino, H., Joukou, S., Kajita, T., Kasuga, S., Koshio, Y., Kumita, T., Matsumoto, K., Nakahata, M., Nakamura, K., Okumura, K., Sakai, A., Shiozawa, M., Suzuki, J., Suzuki, Y., Tomoeda, T., Totsuka, Y., Hirata, K.S., Kihara, K., Oyama, Y., Koshiba, P., Nishijima, K., Horiuchi, T., Fujita, K., Hatakeyama, S., Koga, M., Maruyama, T., Suzuki, A., Mori, M., Kajimura, T., Suda, T., Suzuki, A.T., Ishizuka, T., Miyano, K., Okazawa, H., Hara, T., Nagashima, Y., Takita, M., Yamaguchi, T., Hayato, Y., Kaneyuki, K., Suzuki, T., Takeuzhi, Y., Tanimori, T., Tasaka, S., Ichihara, E., Miyamoto, S. and Nishikawa, K.: 1996, Kamiokande Collaboration, "Solar-neutrino data covering solar-cycle-22", *Phys. Rev. Lett.* **77**, 1683-1686.

Gough, D.O., Leibacher, J.W., Scherrer, P.H. and Toomre J.: 1996, "Perspectives in helioseismology", *Science* **272**, 1281-1283.

Grevesse, N.: 1984, "Abundances of the elements in the sun", in *Frontiers of Astronomy and Astrophysics*, ed., Pallavicini, R., 7th European Regional Meeting, *Firenze: Societa Astronomica Italiana*, Florence, Italy, pp. 71-82.

Harvey, J.: 1995, "Helioseismology", *Physics Today* **48**, 32-38.

Hill, F. Stark, P.B., Stebbins, R.T., Anderson, E.R., Antia, H.M., Brown, T.M., Duvall, T.L., Haber, D.A., Harvey, J.W., Hathaway, D.H., Howe, R., Hubbard, R.P., Jones, H.P., Kenndey, J.R., Korzennik, S. G., Kosovichev, A.G., Leibacher, J.W., Libbrecht, K.G., Pintar, J.A., Rhodes, E.J., Schou, J., Thompson, M.J., Tomczyk, S. Toner, C.G., Toussaint, R. and Williams, W.E.:1996, "The solar acoustic spectrum and eigenmode parameters", *Science* **272**, 1292-1295.

Ince, E.L.: 1926, *Ordinary Differential Equations*, Longmans, Green and Co., Dover, NY, 1956 pp.

Lande, K.: 1997, "Homestake collaboration", in *Proc. 4th Int'l. Solar Neutrino Conf.*, vol. **4**, Heidelberg, Germany, ed., Hampel, W., Max Planck Inst., 1998, p. 85.

Ledoux, P. and Walraven, T.: 1958, "Variable stars," *Handbuch d. Phys.* **51**, 353-604.

Manuel, O.K. and Sabu, D.D.: 1975, "Elemental and isotopic inhomogeneities in noble gases: The case for local synthesis of the chemical elements", *Trans. Missouri Acad. Sci.* **9**, 104-122.

Manuel, O.K. and Sabu, D.D.: 1977, "Strange xenon, extinct superheavy elements and the solar neutrino puzzle", *Science* **195**, 208-209.

Morse, P.M. and Feshbach, H., 1953, *Methods of Theoretical Physics*, McGraw-Hill Book Company, New York, NY, Parts I and II, 1978 pp.

Rouse, C.A.: 1964, "Calculation of stellar structure using an ionization equilibrium equation of state", *University of California, UCRL Report 7820-T*, April 1964 (published as Rouse 1968a), 46 pp.

Rouse, C.A.: 1968a, "Calculation of stellar structure", in *Progress in High Temperature Physics and Chemistry* vol. **2**, ed., C.A. Rouse, Pergamon Press, Oxford, UK, p. 97.

Rouse, C.A. 1968b. "Tables of ionization equilibrium equation of state at stellar temperatures and densities", *NRL Report 6756*, November 22, 1968.

Rouse, C.A. 1969, "Helium abundance determination from solar-model photospheres", *Astron. Astrophys.* **3**, 122-125.

Rouse, C.A.: 1971, "Calculation of stellar structure. II. Determination of the helium abundance of the sun ..." in *Progress in High Temperature Physics and Chemistry*, ed., Rouse, C.A., Pergamon Press, Oxford, UK, vol. **4**, pp. 139-191.

Rouse, C. A.: 1975, "A solar neutrino loophole: Standard solar models", *Astron. Astrophys.* **44**, 237-240.

Rouse, C.A.: 1983, "Calculation of stellar structure. III. Solar models that satisify the necessary conditions for a unique solution to the stellar structure equations", *Astron. Astrophys* **126**, 102-110.

Rouse, C.A.: 1985, "Evidence for a small, high-Z, iron-like solar core", *Astron. Astrophys.* **149**, 65-72.

Rouse, C.A.: 1987, "Evidence for a small, high-Z, iron-like solar core. II. Sensitivity studies of the 5-minute band frequencies to the gravitational perturbation and the 160-minute period of the oscillation to the space mesh", *Solar Phys.* **110**, 211-235.

Rouse, C.A.: 1990, "Sensitivity of the sun's oscillation spectrum and neutrino counting rate to its structure", preprint, revision #5, November 22, 1990.

Rouse, C.A.: 1995, "Calculation of stellar structure. IV. Results using a detailed energy generation subroutine", *Astron. Astrophys* **304**, 431-439.

Rouse, C.A.: 1996, "Sensitivity of solar oscillation frequencies in the five-minute-band to different $\Gamma_1$, functions", preprint, November 25, 1996.

Schou, J., Antia, H.,M., Basu, S., Bogart, R.S., Bush, R.I., Chitre, S.M., Christensen-Dalsgaard, J., Dimauro, M.P., Hoeksema, J.T., Howe, R., Korzennik, S.G., Kosovichev, A.G., Larsen, R.M., Pijpers, F.P., Scherrer, P.H., Sekii, T., Tarbell, T.D., Title, A.H., Thompson, M.J. and Toomre, J.: 1998, "Helioseismic studies of differential rotation in the solar envelope by the solar oscillations investigation using the Michelson doppler imager" *Ap. J.* **505**, 390-417.

Scuflaire, R.: 1974, "The non-radial oscillation of condensed polytropes", *Astron. Astrophys.* **36**, 107-111.

Svoboda, R.: 1997, "Solar neutrinos in SuperKaminokande", in *News about SNU's, International Workshop on Solar Neutrinos*, Inst. of Theoretical Physics, Santa Barbara, CA.

Takata, M. and Shibahashi, H.: 1998, "Solar models based on helioseismology and the solar neutrino problem", *Ap. J.* **504**, 1035-1050.

Totsuka, Y.: 1997, "SuperKamiokande Collaboration", in *Proc. 35th Int'l School of Subnuclear Physics*, Erice, Italy, 1997, in press.

Unno, W., Osaki, Y., Ando, H., Saio, H., Shibahhashi, H.: 1989, *Nonradial Oscillations of Stars*, University of Tokyo Press, Tokyo, Japan, 420 pp.

Unsöld, A.O.J.: 1969, "Stellar abundances in the origin of the elements", *Science* **163**, 1015-1025.

Vanlommel, P. and Cadez, V.M.: 1998, "Influence of temperature profile on solar acoustic modes", *Solar Phys.* **182**, 263-281.

# Heterogeneous Accretion of the Sun and the Inner Planets

Golden Hwaung
*Electrical and Computer Engineering Department, Louisiana State University, Baton Rouge, LA 70803*
*golden@ee.lsu.edu*

Abstract: Some simple arguments are made to question the most popular model of our sun, which assumes that the sun consists mostly of hydrogen and that the fusion of hydrogen at ≅ $10^7$ °K is primarily responsible for its release of energy. An alternative "Fe-Core sun" model is suggested. The correlated chemical and isotopic heterogeneities of elements in meteorites strongly suggest that our solar system was produced from the debris of a single supernova. Cores of the sun and the four inner planets formed in the central Fe-rich region of the supernova debris that made the solar system.

## 1. HYDROGEN SUN MODEL

It is widely assumed that the sun is composed mostly of hydrogen and that the fusion of hydrogen at ≅ $10^7$ °K is primarily responsible for its release of energy. It is also assumed that the sun and its planets were formed separately because of their different elemental composition. Are these three assumptions correct? The "Hydrogen Sun Model" is inconsistent with the results of the neutrino counting experiment (Bahcall and Davis, 1976). Researchers in astrophysics are anxious to explain the neutrino puzzle in order to keep the "Hydrogen Sun Model". Other than the neutrino puzzle, I have three simple arguments that question the validity of the "Hydrogen Sun Model".

[1] The "Hydrogen Sun Model" suggests that the fusion reaction takes place at ≅ $10^7$ °K via tunneling within a distance of $1 \times 10^8$ m from the center

of the sun. Heat can be transferred by black body radiation, conduction and convection. Considering only heat transfer via black body radiation and using the simple black body radiation law and the conservation of energy, the temperature at the surface of the sun is found to be 3.79 x $10^6$ °K. If one considers heat transferred by the black body radiation, conduction and convention and/or considers heat generated by a fusion reaction in the region between 1 x $10^8$ m from the center of the sun and the surface of the sun, then uses complicated equations and calculation, one should get the temperature of more than 3.79 x $10^6$ °K at the surface of the sun. This is because the faster the heat transfers the less steep the temperature gradient. This temperature disagrees with the temperature of ≅ 6000 °K which we observe from the surface of the sun. The "Hydrogen Sun Model" supporters need to explain how a temperature of ≅ $10^7$ °K at 1 x $10^8$ m from the center of the sun is derived using their temperature gradient equation. If this cannot be done, the "Hydrogen Sun Model" must be revised.

[2] The hydrogen fusion reaction is a violent exothermic reaction in that it may cause chain reactions. When the fusion reaction occurs in a region, it changes the density and temperature in and around that region altering the hydrogen fusion reaction rate, since the fusion reaction rate is temperature and density dependent. Thus, according to the "Hydrogen Sun Model", the radiation energy of the sun should fluctuate if fusion reaction is its primary energy source. This is contradictory, however to the observation of constant radiation energy of the sun per minute (or for even shorter time periods) on the surface of the atmosphere of our earth.

[3] The "Hydrogen Sun Model" implies a gradual stellar evolution to the supernova stages with successive stages of burning elements of higher atomic number, starting with a hydrogen burning stage followed by a helium burning stage, then carbon, and silicon burning stages, finally resulting in an iron core. Then Fe core "bounce" occurs, that is the hydrogen rushes into the Fe core resulting in a violent explosion in only a few seconds. The successive burning stages seem reasonable except for the last stage, for the following reasons:

A. What kind of fields can suddenly change to attract huge amounts of hydrogen into Fe core in a few seconds? The kinetic pressure of hydrogen and radiation pressure (Zemansky, 1968) of Fe core should prevent hydrogen from rushing to the Fe core in such a short time. What kind of fields can overcome these pressures?
B. It may be argued that, if Fe core "bounce" does occur, it will generate extremely high gravitational fields that might overcome the hydrogen kinetic pressure and the radiation pressure of an Fe core. General relativity theory (Einstein and Lawson, 1952), however,

implies that the higher the gravitational field, the slower the reaction rate. How can a violent explosion occur in such a short time under such a high gravitational field?

C. Since Fe has the most stable nucleus, any further fusion reaction must be endothermic. This makes the Fe core a fusion chain reaction inhibitor. How can hydrogen rush into an Fe core and cause rapid chain reactions and a violent explosion in seconds? This argument also implies that any other explanation of the violent explosion of a supernova suggested by "gradual stellar evolution model" should fail.

## 2. IRON CORE SUN MODEL

Is there an alternate model of the sun and the origin of our solar system? Manuel, Sabu, Hwaung and co-workers (Manuel and Sabu, 1975, 1977, 1981; Oliver *et al.*, 1981; Hwaung and Manuel, 1982) interpret correlated chemical and isotopic heterogeneities of elements in meteorites as evidence that our solar system was produced from the debris of a single supernova. According to their model, the inner portion of the sun and terrestrial planets formed in the central Fe-rich region of supernova debris. The sun has an Fe-rich core covered with lighter elements and surround by a layer of hydrogen.

The isotopic composition of noble gases in the solar wind shows high enrichment of light isotopes. When corrected for mass fractionation of all five noble gases, they can be resolved in term of the two primitive noble gas components that have been identified in planetary solids (Manuel and Hwaung, 1983a). When the abundance of the elements at the surface of the sun are corrected for this fractionation, it is shown that the atomic abundance for major elements in the bulk sun are (in decreasing order): Fe, Ni, O, Si, S, Mg, and Ca (Manuel and Hwaung, 1983b).

Where does the radiation energy of the sun come from? Hwaung (1982) suggested that most of the radiation energy comes from the black body radiation of residual heat of the Fe-rich core and from the conversion of gravitational energy, with only a small portion of the radiation energy coming from the hydrogen fusion reaction.

## 3. PLANETS

The four inner planets of the solar system were in the Fe-rich region of supernova debris. They originally had Fe-rich cores that, over time, were layered with Si and lighter elements. The most popular model of planetary origins assumes that these four inner planets of the solar system were chemically homogeneous and a differentiation process (*i.e.*, the heavy element Fe sinking to form the present core) resulted in the present form. This popular model ignores three simple facts that argue against the separating process. First, the property of formation of the alloy of iron and silicon disfavors this separating process.

Second, the entropy factor disfavors this process. The higher the temperature, the larger the effect of this entropy factor. If the separating process is not fast enough, the outer layer of chemically homogeneous body will cool down to become solid and the separating process will stop.

Third, when the heavier elements sink to the center, the whole body will expand. This action will increase the gravitational energy so that the net gravitational energy change of this separating process is not necessarily negative. In fact, Hwaung (1982) calculated that the net gravitational energy change of the separating process of a homogeneous earth is positive. A simple question was raised by Hwaung in 1982: Can two immiscible liquids separate into two layers in a zero external gravitational field? The results of an experiment carried out on a NASA space shuttle in 1983 showed that a mixture of oil and water did not separate into two layers in zero external gravitational field. If my simple argument, "The earth cannot result from a chemical homogeneous body by a separating process" is correct, then the four inner planets of our solar system were originally Fe-rich cores. This strongly implies that the sun and four inner planets were in the center of the Fe-rich core region of supernova debris.

## 4. CONCLUSION

The concept of an "Fe-Core Sun Model" is an intèresting although somewhat controversial idea. Its merits, and the positive and negative viewpoints of an "Fe-Core Sun Model" for our solar system, remain to be seen. The important thing is that the model be given consideration in light of the points raised here, leading creative scientists toward still better models of our sun that someday might lead to new types of energy sources in the future.

## REFERENCES

Bahcall, J.N. and Davis, R.: 1976, "Solar neutrinos: A scientific puzzle", *Science* **191**, 264-267.

Einstein, A. and Lawson, R.W.: 1952, (trans), *Relativity*, 15th Edition, Crown Publishers, Inc., New York, NY, 164 pp.

Hwaung, C.-Y. G.: 1982, *The Origin of Solar System*, Thesis, University of Missouri-Rolla, 24 pp.

Hwaung, G. and Manuel, O.K.: 1982, "Terrestrial-type xenon in meteoritic troilite", *Nature* **299**, 807-810.

Manuel, O.K. and Hwaung, G.: 1983a, "Information of astrophysical interest in the isotopes of solar wind implanted noble gases", *Lunar Planet Sci.*, **XIV**, 458-459.

Manuel, O.K. and Hwaung, G.: 1983b, "Solar abundances of the elements", *Meteoritics* **18**, 209-222.

Manuel, O.K. and Sabu, D.D.: 1975, "Elemental and isotopic inhomogeneities in noble gases: The case for local synthesis of the chemical elements", *Trans. Missouri Acad. Sci.* **9**, 104-122.

Manuel, O.K. and Sabu, D.D.: 1977, "Strange xenon, extinct superheavy elements and the solar neutrino puzzle", *Science* **195**, 208-209.

Manuel, O.K. and Sabu, D.D.: 1981, "The noble gas record of the terrestrial planets", *Geochem. J.* **15**, 245-267.

Oliver, L.L., Ballad, R.V., Richardson, J.F. and Manuel, O.K.: 1981, "Isotopically anomalous tellurium in Allende: Another relic of local element synthesis", *J. Inorg. Nucl. Chem.* **43**, 2207-2216.

Zemansky, M.W.: 1968, *Heat and Thermodynamics*, 5th Edition, McGraw-Hill, New York, NY, 431 pp.

# Interstellar Matter, Sun, and the Solar system

M. N. Vahia[1] and D. Lal[2]
*[1]Tata Institute of Fundamental Research, Mumbai 400 005, INDIA*
*[2]Scripps Institution of Oceanography; Geosciences Research Division, 0244, La Jolla, CA 92093-0244 USA*
*vahia@tifr.res.in*

Abstract: We discuss the nature of possible changes in the environment of the solar system considering the presently known information on the physical state of the interstellar matter (ISM) in the vicinity of the sun. We trace the solar motion through the ISM, with particular reference to the consequences of this motion on the composition, flux and energy spectrum of corpuscular radiation in the solar system. We review our present state of knowledge of the ISM, and on the nature of temporal changes in the compositions and fluxes of galactic and anomalous cosmic radiation, due to changes in the solar activity, and the ISM. The plausible changes in the corpuscular radiation fluxes are then estimated, considering the physical state of the ISM at the heliopause-ISM boundary during sun's journey in the past tens of millions of years.

## 1. INTRODUCTION

The solar system is continuously undergoing changes in its physical and electromagnetic structure, being acted upon by the varying sun and the dynamic interstellar matter. These changes produce appreciable variations in the fluxes of corpuscular radiation of galactic and solar origin, and also in the fluxes of interstellar matter accelerated in the vicinity of the heliopause, termed as the ACR (Anomalous Cosmic Radiation). In this presentation, we review our present knowledge of variations in the physical properties of the interstellar matter, namely composition, temperature and density around the sun. This is brought about mainly by the motion of the sun in ISM.

## 2. LONG TERM VARIABILITY OF THE HELIOSPHERE

Recent studies of the starlight absorption from the nearby stars as well as back scattering of light have allowed us to map the local interstellar medium in some detail (see review by Frisch, 1995). These studies reveal that the environment around the sun has been highly variable on a variety of time scales. There are three main reasons for these changes. The first reason is the intrinsic variability of the sun itself (*cf.* Pepin *et al.*, 1980). The second reason for the variations is that the sun is very close to a young star-forming region called the Scorpious Centaurus Association (SCA) (Frisch, 1995). Several supernova explosions are known to have occurred in this region over the last few hundred thousand years. Three major expanding shells from this region are known to have expanded towards the sun. Two of these crossed the present location of the sun while the third one is in the vicinity of the solar system. These supernova bubbles are also known to include cool clouds of very high density and the solar system is currently believed to be in one such cloud. The third reason for the changing environment is that the sun has a velocity of about 16.5 km/s in the local rest frame and the sun is moving in the direction of $l = 53^\circ$, $\beta = +25^\circ$. Bash (1986) has suggested that the sun has passed across the Persius arm to reach the inter-arm region close to the Local Arm. In the last 40 million years or so, the sun was located behind the Persius arm. We estimate the plausible changes in the environment of the sun by tracing backward the path of the sun assuming its present motion.

If we extrapolate back from the current vectors, the sun's path must have crossed the path of the Giant Molecular Cloud of the Orion Nebula. The Orion nebula itself has a very small velocity vector in the local rest frame and hence the sun must have *crossed* the Orion Nebula about 25 million years ago. Prior to that, the sun crossed the Persius Arm and was on the other side of the Orion Nebula (Clube and Napier, 1984; Bash, 1986). It also passed through other changing environments in times that are more recent. Best estimates of the solar movement suggest that a local fluff of high-density cloud currently surrounds the sun. This cloud was probably born out of a disintegrating supernova shell. It entered the cloud about 10,000 years ago. Before, the sun was in a low-density shell of the expanding supernova remnant from a supernova in SCA. This probably occurred about $10^5$ years ago. Prior to that the sun was in the local inter-arm region of low density and low temperature and entered the loose collection of young stars called Goud Belt about 10 million years ago. It also passed close to the Geminga Pulsar but the pulsar itself exploded (about $10^5$ years ago) much after the sun had passed that region. Prior to this, about 25 million years ago it was inside the Orion Giant Molecular Cloud where it resided for about 500,000 years. It entered the Orion from the Persius arm. If we assume that the sun experienced no acceleration during the period then it can be shown that it crossed the Persius arm and the sun was on the other side of the Persius arm about 100 million years ago (Bash, 1986). This charting is done in the two planar dimensions. However, the sun also oscillates perpendicular to the plane of the Galaxy and hence the changes in environment are likely to be more complex than what is suggested here (see also Clube and Napier, 1984). A general summary of this is given in Table 1. For the present study, we concentrate on changing ISM conditions.

*Table 1.* Solar movement in the ISM

| Time(yBP)[(a)] Start | End | DT (y) | Region | Den. ($cm^3$) & (Temp(K)) | Ref | Comments[(b)] |
|---|---|---|---|---|---|---|
| 0 | $1.0\ 10^4$ | $1.0\ 10^4$ | fluff | 0.08 ~ (7000) | 1 | Size 5 pc, From SCA, moving in orthogonal direction to sun @ 20 km/s B= 1.5 μG. Highly Inhomogeneous (1) |
| $1.0\ 10^{5(a)}$ | $3.1\ 10^5$ | $6.0\ 10^3$ | SN(?) | $3(10^4)$ | 2,3 | Geminga?(2,3) |
| $1.0\ 10^4$ | $1.0\ 10^5$ | $(0.1\text{-}1)10^5$ | bubble | $0.04(\sim 10^6)$ | 1,4 | Expanding bubble from SCA. Age 15 My <vel> 4 $10^4$ $km/s^{-1}$ orthogonal to sun, B~ 1.5 μG Crossed the sun 250,000 to 400,000 y ago (1,3,4,5) |
| $1.0\ 10^5$ | $1.0\ 10^7$ | $(0.01\text{-}1)10^7$ | Inter-arm | 0.0005(~ 100) | 1 | (1,6) |
| $1.0\ 10^{7(a)}$ | $2.50\ 10^7$ | $300\ 10^6$ | HI shock | $4(10^4)$ | 6 | From Gould belt (7). Duration is taken as that for the earlier SN.(3,6) |
| $2.5\ 10^{7(a)}$ | $4.0\ 10^7$ | $5.0\ 10^5$ | Orion belt | $10^4(\sim 10)$ | 7 | Size ~ 5pc, sun must have taken 5 $10^5$ yrs to cross at current velocity (7,8) |

| Time(yBP)[(a)] Start | End | DT (y) | Region | Den. ($cm^3$) & (Temp(K)) | Ref | Comments[(b)] |
|---|---|---|---|---|---|---|
| 4.0 $10^7$ | 5.0 $10^7$ | (2.5-4)$10^7$ | Inter-arm | 0.005(10) | 5 | (6) |
| 5.0 $10^7$ | 6.0 $10^7$ | (4-5)$10^7$ | Persius arm | 1(1000) | 6 | Duration uncertain (6) |
| < 6 $10^7$ | | | | | 6 | (9) |

[a]For SN, the start and end time are the interval between which the SN spike crossed the solar system.
[b]Bracket numbers in comments refer to notes.

References: 1 Frisch (1995)
2 Ramadurai (1993)
3 Szabelska *et al.* (1991)
4 Egger *et al.* (1996)
5 Van der Walt and Wolfendale (1988)
6 Bash (1986)
7 Clube and Nupier (1984)

Notes:

1) SCA refers to Scorpius Centaurus association at l = ~ 300-360°. About 35 SNR have been identified in this region (Whiteoak and Green, 1996).
2) While Frisch (1995) claims that the evidence for Geminga is not there, Ramadurai (1993) has claimed the passage of a SN shock around 30,000 yr. ago that arose from a source around the age of Geminga. The sun crossed the vicinity of the parent star of Geminga $10^7$ yr. BP while the SN itself is 3 $10^5$ yrs old (Pavlov *et al.*, 1997). The temperature and density are typical values for SNR (Harwitt, 1973). The shock passage is calculated assuming shock front velocity of 1000 km $s^{-1}$ (1 pc/My) and taking effective thickness of 1 pc (Harwitt, 1973). These parameters are similar to those derived by Szabelska *et al.* (1991). Based on $^{10}$Be data they argue for the passage of a SN shock front near the sun about 30,000 yr. ago.
3) Another SN is just arriving from the SCA association, age ~ 11 Myr BP.
4) The bubble is from the SN shell from the Upper Centaurus Lupus of SCO. Age ~ 14-15 Myr BP.
5) The interarm region is derived from Figure 5 of Breitschwerdt (1998) and Figure 1 of Bash (1986). An approximate plotting of the trajectory is based on the velocity vectors given by Bash (1986).
6) van der Walt and Wolfendale (1988) argue that the gamma ray data suggest that the inter arm cosmic ray are of lower intensity and steeper.
7) More than 130 weak line T Tauri (O type) stars have recently been catalogued in this region (for reference see Wichmann *et al.*, 1997).
8) Based on extrapolation as defined in note 5.
9) Following issues need to be considered in detail:
    a) We are now in the Orion structure in the inter arm region between Persius and Carina arms.
    b) This is uncertain since collision with individual objects is ignored (Bash, 1986). Clube and Nupier, (1984) suggest that in the last 4.5 $10^9$ years, the sun has encountered about 10 molecular clouds. Napier (1985) has done a more detailed analysis to show that the

sun has had close (impact parameter < 20 pc) with 56 GMCs having mass $\geq 3\ 10^3\ M_{sun}$ and 8.2 close encounters with GMCs having $M \geq 10^5\ M_{sun}$.

c) Our galaxy is a highly evolved spiral with about 10% of its mass in diffuse gas; the rest is in stars. The sun is located slightly above the galactic plane, and is moving through space at 16.5 $kms^{-1}$ with respect to the local standard of rest. The sun oscillates in the galactic plane with a period of 33 My (it goes $\pm$ 77 pc, Frisch, 1998).

# 3. DYNAMICS OF VARIABILITY DUE TO CHANGING ISM CONDITIONS

The sun interacts with the ISM in three primary ways (Dyson and Williams, 1997) namely:

By ionization of the ISM gas. This ionization by the solar UV radiation produces a sphere that is referred to as the Stromgren Radius ($R_s$) and extends until the gas opacity for UV reaches unity. $R_s$ for a star with UV flux of $S_*$, in an ISM neutral density of $n_0$ and a recombination rate $\beta_2(T)$ is defined by the equation

$$R_s = (3\ S_* / 4\ \pi\ n_0^2\ \beta_2)^{1/3} \tag{1}$$

Due to photoionization, the gas temperature increases from about $10^2$ K to about $10^4$ K, *i.e.*, by a factor of about a hundred. The ionization process itself increases the number of gas particles and therefore the pressure by a further factor of two. This increased pressure produces an Ionization Front that extends beyond the Stromgren Radius. The functional form for the Ionization Front radius $R_{IF}$ is similar to equation 1 except that the effective limiting density is not the neutral density $n_0$ but the ionized particle density $n_{IF}$ which is typically 0.005 of the neutral density. Hence, $R_{IF}$ is larger than $R_s$ by a factor of about 34 for typical ISM conditions.

The other method of interaction is via the stellar wind (very high-speed continuous mass loss) which produces a mechanical piston that is bound by pressure equilibrium between the stellar wind and ISM. This is conventionally referred to as the Heliopause. The pressure balance between the solar wind and the ISM defines the Heliosphere radius $R_H$. For the densities n and temperature T of the region, this is given by (Longair, 1992).

$$R_H = (n_{sw} T_{sw} / n_{ISM}\ T_{ISM})^{3/4}\ R_{sun} \tag{2}$$

While all three processes occur simultaneously in the interaction of a star with the interstellar medium, in the present study we consider each process independently. In Table 2, we have evaluated the dimensions of each of these interaction regions for solar conditions.

*Table 2.* Parameters of Solar interaction with different ISM Conditions

| | Fluff | Bubble | GMC | Inter-arm | Arm |
|---|---|---|---|---|---|
| *Conditions assumed* | | | | | |
| Electron density ($p/m^3$) | $8\ 10^4$ | $4\ 10^4$ | $1\ 10^{10}$ | $5\ 10^2$ | $1\ 10^4$ |
| Temperature (K) | $7\ 10^3$ | $1\ 10^6$ | $1\ 10^1$ | $1\ 10^2$ | $1\ 10^3$ |
| *Radius (m)* | | | | | |
| Stromgren | $6\ 10^{15}$ | $3\ 10^{16}$ | $5\ 10^{11}$ | $6\ 10^{16}$ | $1\ 10^{16}$ |
| Ionization | $2\ 10^{17}$ | $1\ 10^{18}$ | $2\ 10^{13}$ | $2\ 10^{18}$ | $5\ 10^{17}$ |
| Heliopause | $4\ 10^{12}$ | $2\ 10^{11}$ | $9\ 10^{10}$ | $4\ 10^{15}$ | $8\ 10^{13}$ |
| *Equilibrium time scales (s)* | | | | | |
| Stromgren | $4\ 10^{11}$ | $2\ 10^{11}$ | $8\ 10^8$ | $3\ 10^{13}$ | $1\ 10^{12}$ |
| Ionization | $1\ 10^{14}$ | $6\ 10^{13}$ | $3\ 10^{11}$ | $1.10^{16}$ | $1\ 10^{14}$ |
| Heliopause | $3\ 10^8$ | $1\ 10^6$ | $2\ 10^8$ | $3\ 10^{12}$ | $1\ 10^{10}$ |
| *Energy at the peak flux in modulated CR flux (MeV)* | | | | | |
| Stromgren | $1\ 10^5$ | $1\ 10^7$ | $1\ 10^6$ | $1\ 10^6$ | $1\ 10^6$ |
| Ionization | $1\ 10^5$ | $1\ 10^7$ | $1\ 10^5$ | $1\ 10^3$ | $1\ 10^4$ |
| Heliopause | $1\ 10^3$ | $1\ 10^4$ | $1\ 10^4$ | $1\ 10^2$ | $1\ 10^3$ |

## 4. COSMIC RAY CONNECTION

The largest sample of interstellar matter that reaches the solar system is through the cosmic rays. The cosmic rays consist of several components (Stone *et al.*, 1995). These include, solar wind particles, the solar energetic particles, the cosmic rays produced at the Heliopause and the galactic cosmic rays. Each of these species have their own spectral form. While the solar wind and energetic particles are a direct result of solar activity, other species depend on the solar interaction with the interstaller medium. These species will be severely affected by the changes in the environment of the sun and the interstellar medium. We therefore estimate the cosmic ray modulation parameter $\phi$ (Lockwood and Webber, 1995) to estimate the cosmic ray modulation. We derive the value of $\phi$ from first principals and calculate the turnover in cosmic ray spectrum due to different interaction regions. These results are given in Table 2.

## 5. DISCUSSION AND CONCLUSION

In the present work we have studied the changes in the solar interaction regions for different ISM conditions. We show that the sizes of these regions will vary significantly under different conditions. We then calculate the cosmic ray diffusion parameters for the cosmic rays entering the near sun region and show that the spectral characteristics of the cosmic rays that come to us from the galactic region will change dramatically, even if we assume that the incident particle flux is unchanged.

## REFERENCES

Bash, F.: 1986, "The present, past and future velocity of nearby stars: The path of the sun in $10^8$ years", in *The Galaxy and the Solar System*, eds., Smoluchowski, R., Bahcall, J.N. III, and Shapley, M.M., University of Arizona Press, Tucson, AZ, 483 pp.

Breitschwerdt, D.: 1998, "The local bubble and beyond", *Introductory Lecture: The Local and General Interstellar Medium*, eds., Breitschwerdt, D., Freyberg, M. and Trumper, J., Proceedings of the 166 IAU Symposium, Lecture Notes in Physics, Springer Verlag, New York, NY, pp. 5-18.

Clube, S.V.M. and Napier, W.M.: 1984, "Comet capture from molecular clouds: A dynamical constraint on star and planet formation", *Mon. Not. Royal Astron. Soc.* **208**, 575-588.

Dyson, J.E. and Williams, D.A.: 1997, *The Physics of the Interstellar Medium*, Institute of Physics, UK, 155 pp.

Egger, R.J., Freyberg, M.J. and Morfill, G.E.: 1996, "The local interstellar medium", *Space Sci. Rev.* **75**, 511-536.

Frisch, P.C.: 1995, "Characteristics of nearby interstellar matter", *Space Sci. Rev.* **72**, 499-592.

Frisch, P.C.: 1998, "Interstellar matter and the boundary conditions of the heliosphere", *Space Sci. Rev* **86**, 107-126

Harwitt, M.: 1973, *Astrophysical Concepts*, John Wiley and Sons, New York, NY, 561 pp.

Lockwood, J.A. and Webber, W.R.: 1995, "An estimate of the location of the modulation boundary for E > 70 MeV galactic cosmic rays using voyager and pioneer spacecraft data", *Ap. J.* **442**, 852-860.

Longair, M.S.: 1992, *High Energy Astrophysics*, Second Edition, Cambridge University Press, Cambridge, UK, vol. **1** and **2**, 393 pp.

Nagahama, T., Akira, M., Hideo, O. and Yasuo, F.: 1998, "Spacially complete $^{13}$C J J = 1-0 curve of Orion A cloud", *Astron. J.* **116**, 336-348.

Napier, W.M.: 1985, "Dynamical interactions of the solar system with massive nebula", in *Dissipation of a primordial cloud of comets: Their origin and evolution, IAU Colloquium 83*, eds., Carusi, A. and Valsecchi, G.B., D. Reidel Publishing Co., Dordrecht, Holland, p. 41.

Pavlov, G.G., Welty, A.D. and Cordova, F.A.: 1997, "Hubble space telescope observations of the middle aged pulsar 0650 + 14", *Ap. J.* **489**, L75-L78.

Pepin, R.O., Eddy, J. A. and Merrill, R.B., eds.: 1980, *The Ancient Sun, Geochim. Cosmochim. Acta., Suppl. 13*, Pergamon Press, Oxford, UK, 576 pp.

Poppel, W.: 1997, "The Gould Belt System and the local interstellar medium", *Fundamentals of Physics*, **18**, 1-272.

Ramadurai, S.: 1993, "Geminga as a cosmic ray source", *Bull Astron. Soc. India*, **21**, 391-393.

Stone. E.C., Cummings, A.C. and Webber, W.R.: 1995, "Radial and latitudinal gradients of anomalous cosmic rays in the outer heliosphere", *24th International Cosmic-ray Conference*, Rome, vol. **4**, 796-799.

Szabelska, B., Szabelski, J. and Wolfendale, A.W.: 1991, "Cosmic rays and supernova remnants", *J. Phys.* **G17**, 545-553.

Van der Walt, D.J. and Wolfendale, A.W.: 1988, "Cosmic rays and the galactic spiral arms", *J. Phys.* **G14**, L159-163.

Whiteoak, J.B.Z. and Green, A.J.: 1996, "The MOST supernova remnant catalogue (MSC)", *Astron. Astrophys.* **118**, 329-380.

Wichmann, R., Sterzik, M., Krautter, J., Metanomski, A. and Voges, W.T.: 1997, "T Tauri Stars and the Gould Belt near Lupus", *Astron. Astrophys.* **326**, 211-217.

PART V

# NUCLIDES IN THE SUN'S PLANETARY SYSTEM

# Isotope Anomalies in Tellurium in Interstellar Diamonds

R. Maas[1,5], R. D. Loss[1], K. J. R. Rosman[1], J. R. De Laeter[1], U. Ott[2], R.S. Lewis[3], G. R. Huss[3,6], E. Anders[3] and G. W. Lugmair[2,4]
*[1]Curtin University, GPO Box U1987 Perth 6845, Western Australia; [2]Max - Planck - Institut für Chemie, Mainz F.R.G.; [3]University of Chicago, Chicago, IL 60637, USA; [4]University of California, San Diego, CA 92093-0212, USA; [5]Now at La Trobe University, Victoria, Australia.; [6]Now at the California Institute of Technology, Pasadena, CA 91125, USA*
*rdelaeter@cc.curtin.edu.au*

**Abstract:** The isotopic composition of Te has been measured in a purified sample of interstellar microdiamonds from the Allende meteorite. Small positive anomalies were only found in $^{128}$Te (4.0 ± 0.3‰) and $^{130}$Te (9.3 ± 2.0‰) from three analyses of the Allende microdiamond sample EB#2. The magnitude of the anomalies are smaller than in the heavy noble gases, although the absolute amount of anomalous Te is greater than that of anomalous Xe in the same sample. The anomalies in Te can be interpreted in terms of standard r-process nucleosynthesis followed by rapid separation of the stable isotopes from their radioactive precursors.

## 1. INTRODUCTION

Nanometer-sized diamonds in primitive chondritic meteorites are the host of Xe-HL, an enigmatic Xe component that has defied explanation by familiar scenarios, *e.g.*, transuranic fission, normal r-process nucleosynthesis (Anders and Zinner, 1993). The Xe-H part, which is strongly enriched in $^{134,136}$Xe, has been explained as a product of a distinct nucleosynthetic environment termed the "mini r-process" (Heymann and Dziczkaniec, 1979) or the "neutron burst" (Clayton, 1989; Howard *et al.*, 1992), which is thought to operate in type II supernovae. Recently, Ott (1996) has questioned the necessity of invoking a special process such as the neutron burst to account for the appreciable quantities of Xe-H in primitive meteorites. Ott (1996) suggested that Xe-H could be derived from the average (and therefore common and widespread) r-process if the irradiation is followed by an early separation of the stable Xe daughters, already produced from the precursor isotopes that have not yet decayed. This "time-scale" model can reproduce Xe-H with unprecedented accuracy. Three possible separation processes were discussed: early condensation of non-volatiles, rapid implantation followed by recoil loss from the minute diamond grains (this process is dependent only on the precursor lifetimes and is therefore the purest form of the timescale model), and charge-state separation (Ott, 1996).

An obvious method of distinguishing between the various mechanisms which have been postulated to explain Xe-H, is to analyse presolar diamonds for other elements in the near vicinity of Xe in order to determine if anomalies are present which can be correlated with those in Xe. Tellurium is a logical candidate in that it has Z = 52 (as compared to Z = 54 for Xe) and possesses two r-only process nuclides ($^{128}$Te, $^{130}$Te) which are situated on an r-process peak. From a mass spectrometric point of view, Te possesses eight isotopes spanning a mass range from A = 120 to 130, but has a high ionization potential of 9.01 eV. Tellurium is unique in that it possesses three s-only isotopes $^{122,123,124}$Te. In addition it has one p-only isotope $^{120}$Te, two isotopes ($^{125,126}$Te) nucleosynthesized by a combination of s- and r-processes, and the two r-only nuclides, $^{128,130}$Te. The small relative abundance of $^{120}$Te of 0.09% makes this isotope difficult to measure mass spectrometrically in small samples, and therefore extremely difficult to test for p-process effects in prestellar diamond samples.

Numerous authors have argued that Te should show r-process anomalies associated with the Xe-H anomalies (*e.g.* Heymann and Dziczkaniec, 1979). Ballad *et al.* (1979) reported isotopically anomalous Te extracted from acid-treated samples of the carbonaceous chondrite Allende using neutron activation analysis. However, Loss *et al.* (1990) found no evidence for anomalous Te extracted from acid-resistant residues from Allende using thermal ionization mass spectrometry. Heymann and Dziczkaniec (1981) proposed that isotopically anomalous Te would be produced in zones of supernovae which synthesize Xe-HL, and quantified the magnitude of the anomalous $^{130}$Te/$^{128}$Te as being approximately four. They also argued that the long-lived isotope, $^{126}$Sn, which decays to $^{126}$Te with a half life of approximately $10^5$ yr, would be produced under the same set of nucleosynthetic conditions as Xe-HL and $^{128,130}$Te. Clayton (1989) argued that a neutrino burst from the collapsed core in Type II supernovae acted as a neutron source which generated Xe-H in the helium shell followed by implantation of some of the Xe-H in diamonds.

Clayton (1989) predicted that the anomalous $^{130}$Te/$^{128}$Te would be approximately five in presolar diamonds extracted from primitive meteorites.

Howard *et al.* (1992) subsequently re-examined this neutron burst model and deduced a neutron irradiation environment which not only reproduced the isotopic pattern in Xe-H, but enabled other heavy element abundances to be predicted. In the case of Te the amount of $^{125,126,128,130}$Te generated would be 0.43, 0.59, 0.38 and 1.51 respectively, to give a $^{130}$Te/$^{128}$Te ratio of approximately four. However if the products of the neutron burst model were mixed with Te of solar composition, the anomalous $^{130}$Te/$^{128}$Te ratio would be diluted.

The time-scale model of Ott (1996) explains the presence of Xe-H in interstellar diamonds by r-process nucleosynthesis coupled with a separation of Xe from I and Te precursors within a short time after termination of the r-process. This model (Figure 1) predicted an over-abundance of $^{130}$Te, but with negligible enhancements of $^{128}$Te and $^{126}$Te.

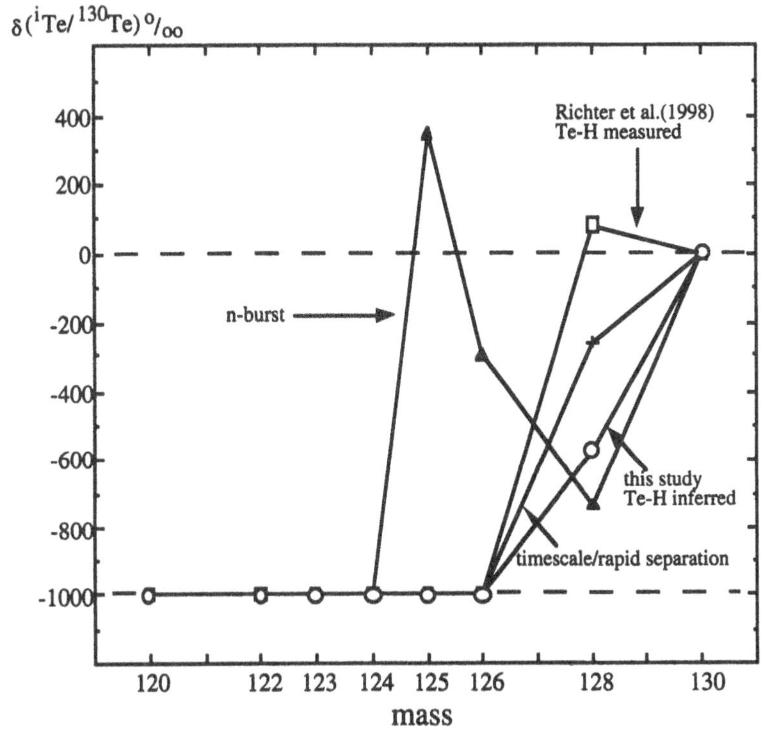


*Figure 1.* Comparison of Te isotopic compositions in Te-H (normalized to $^{130}$Te =1). The neutron burst (n-burst) pattern is taken from Howard *et al.* (1992), assuming all $^{124}$Te is contamination.

A more detailed examination of the precursor half-lives for the Te isotopes indicates enhancements at both $^{130}$Te and $^{128}$Te, with the relative enhancement at $^{128}$Te about half that of $^{130}$Te, without any enhancement in any of the other isotopes. Richter *et al.* (1998) measured the Te isotope ratios from Allende diamonds by NTIMS multi-ion counting of Te ion beams produced from directly loaded 0.01 mg diamond fractions. They found essentially pure Te-H composed entirely of $^{128}$Te and $^{130}$Te, with negligible $^{125\text{-}126}$Te and no detectable $^{120\text{-}124}$Te. The absence of effects at $^{125}$Te and $^{126}$Te is significant because the neutron burst model predicts significant production of both isotopes (Figure 1). By contrast, a pure time-scale model (the recoil model of Ott, 1996) predicts both the presence of $^{128}$Te and $^{130}$Te, and the absence of $^{125\text{-}126}$Te which are held up by long-lived radioactive precursors $^{125}$Sb (2.77 yr) and $^{126}$Sn ($\approx 10^5$ yr), respectively (Richter *et al.*, 1998).

While the results of Richter *et al.* (1998) support the pure time-scale model, their measured $^{128}$Te/$^{130}$Te ratio in Te-H is a problem. Aggregate precursor lifetimes are such that final $^{130}$Te should be established faster than $^{128}$Te, implying that the value of ($^{128}$Te /$^{130}$Te)$_H$, should always be equal to or less than the solar (or r-process) ratio. The diamond data (Richter *et al.*, 1998), however, show that ($^{128}$Te/$^{130}$Te)$_H$ is some 8% higher than the solar ratio. This can only be reconciled with the time-scale model if the r-process is allowed to have structure similar to that found in the s-process (Richter *et al.*, 1998).

## 2. RESULTS AND DISCUSSION

This paper presents isotopic and elemental abundance data from diamonds extracted from the Allende meteorite. The diamond sample (Allende EB#2) was prepared at the University of Chicago following the procedures of Huss and Lewis (1994a) and an aliquot of EB#2 gave typical diamond Xe-HL and Kr-H patterns (Huss and Lewis, 1994b). The remaining 35.8 mg of EB#2 was combusted in an oxygen plasma excited by R-F induction at Scripps Institute of Oceanography in San Diego. The sample was processed by ion exchange chemistry to yield two fractions containing Ba-Sr-Rb and Te-Pd-Cd-Zn. Details of the extraction procedure are given by Lewis *et al.* (1991a) who also reported small isotopic anomalies in Sr and Ba from the first fraction. The Te analyses were carried out on the second fraction of EB#2 at Curtin University. Te was measured by conventional PTIMS methods using a VG354 mass spectrometer equipped with a Daly collector. Aliquots representing *ca.* 5 mg of combusted diamond were loaded directly onto single rhenium filaments without further ion exchange, electroplated, and analysed using a Si gel/boric acid emitter. Data were averaged from three runs typically yielding $\approx$ (1 - 4) x $10^{-14}$ A total Te ion current. Total analytical blanks were on the order of 1 pg Te and can be ignored.

The Te abundance (640 ppb) in EB#2 was measured by the isotope dilution technique. This value is high relative to most other trace components in the diamonds and is most likely related to trace impurities rich in Platinum Group Elements (Lewis *et al.*, 1991b). Isotopic ratios (relative to s-only $^{124}$Te as reference mass) are close to solar but show a greater degree of linear mass fractionation (with light mass enhancement) than the Te laboratory standard. Superimposed on this trend (Figure 2) are resolvable non-linear excesses at $^{128\text{-}130}$Te; whereas $^{125\text{-}126}$Te do not appear to show any anomalous Te. This pattern is preserved if a correction for an average 1.3‰ amu$^{-1}$ mass fractionation (as observed in standard loads of comparable size) is applied. It appears therefore that EB#2 contains a small component of non-solar Te.

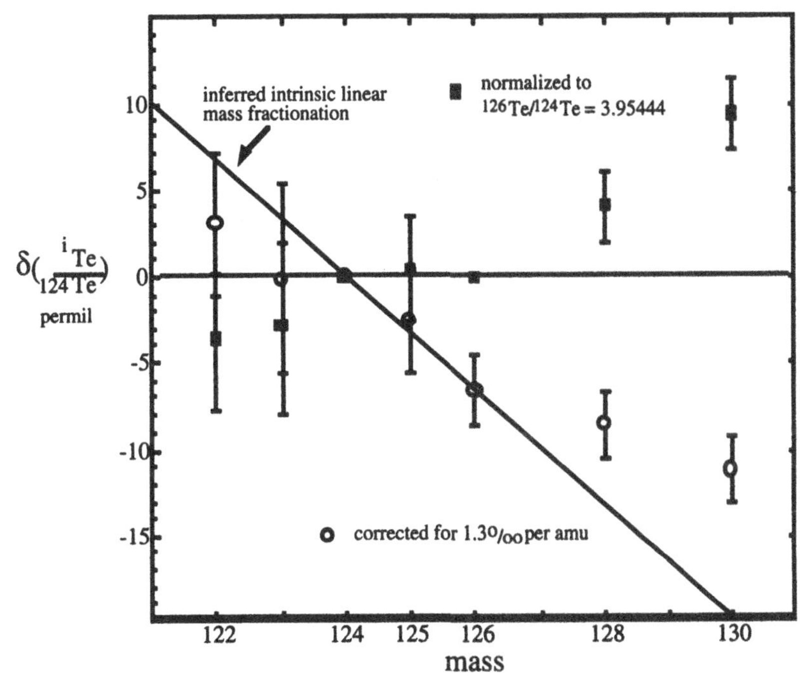


*Figure 2.* Measured isotope ratios of Te extracted from combusted Allende diamond sample EB#2. Ratios as deviations (in ‰) from laboratory standard (= solar Te), using $^{124}$Te as the reference mass, corrected for a 1.3‰ amu$^{-1}$ mass fractionation (average observed in standard Te loads of similar size); and after normalization to a constant $^{126}$Te/$^{124}$Te, using a power law.

We note that no reasonable choice of fractionation correction can generate the pattern of enrichment suggested in mixtures of solar and neutron burst-type Te-H (Howard *et al.*, 1992). By contrast, the time-scale model and the Te data of Richter *et al.* (1998) indicate essentially zero enhancement at $^{126}$Te which justifies the use of a constant (and solar) $^{126}$Te/$^{124}$Te (3.9544, our laboratory standard) to normalize the data for EB#2.

This normalization yields $^{120\text{-}125}Te/^{124}Te$ ratios that are normal within uncertainty, but well-resolved excesses at $^{128}Te$ ($\delta^{128}Te/^{124}Te = 4.0 \pm 0.3$), and $^{130}Te$ ($\delta^{130}Te/^{124}Te = 9.3 \pm 2.0$), as shown below.

*Table 1.* Te isotopic results for Allende diamond separate EB#2 (2 sigma errors)

| $^{i}Te/^{124}Te$* | $^{122}Te$ | $^{123}Te$ | $^{125}Te$ | $^{126}Te$ | $^{128}Te$ | $^{130}Te$ |
|---|---|---|---|---|---|---|
| diamond | 0.53911 | 0.18806 | 1.48938 | ≡ 3.9544 | 6.65134 | 7.14121 |
| $\delta(^{i}Te/^{124}Te)$ ‰ | -3.7 ± 4.0 | -3.0 ± 5.0 | 0.4 ± 3.0 | ≡ 0.0 | 4.0 ± 0.3 | 9.3 ± 2.0 |

Te 640 ppb, Xe 6.5 ppb, $\delta^{136}Xe$ 1098 ± 10‰ (Lewis *et al.*, 1991a)

*Solar Te isotopic ratios determined using $^{124}Te$ as the reference: 122/123/124/125/126/128/130 = 0.54109/0.18861/1.0000/1.4887/3.9544/6.6249/7.0753

Comparison of these data with those by Richter *et al.* (1998) and with the patterns expected for the neutron burst and time-scale models (Figure 1) illustrates several important points. Our results show no trace of the prominent $^{125}Te$ effect predicted by the neutron burst model. Enhancement of $^{128}Te$ and $^{130}Te$ only, on the other hand, is exactly what the time-scale model in its pure form predicts. Both isotopes are r-only produced and their normal ratio of $^{128}Te/^{130}Te = 0.9363$ (our laboratory standard) is the production (or solar) ratio for the average r-process. The $(^{128}Te/^{130}Te)_H$ ratio inferred for EB#2 (0.43 ± 0.22) is lower than both the ratio measured in Te-H of 1.008 ± 0.018, (Richter *et al.*, 1998) and the production ratio, as it should be in a time-scale scenario. For a production ratio of 0.9363, the measured $(^{128}Te/^{130}Te)_{excess}$ (or $\delta(^{128}Te/^{130}Te)_H = -574 \pm 235$ ‰) is reached approximately 3800 s after termination of the r-process, broadly consistent with the timescale inferred from $^{134}Xe/^{136}Xe$ (Ott, 1996). A 9.3‰ excess of $^{130}Te$, together with a Te concentration of 640 ppb, corresponds to a $^{130}$Te-H concentration of $9.3 \times 10^{12}$ atoms $g^{-1}$, compared to a $^{136}$Xe-H concentration of $2.4 \times 10^{12}$ atoms $g^{-1}$ (Lewis *et al.*, 1991a; Huss and Lewis, 1994b). The resulting atomic ratio, $^{130}$Te-H/$^{136}$Xe-H = 3.9, determined in this way is tantalizingly close to the solar system abundance ratio of these two r-only isotopes of 4.4 (Anders and Grevesse, 1989), which is also the ratio predicted from the pure time-scale model as opposed to a ratio of 0.6 for the neutron burst scenario (Howard *et al.*, 1992). This agreement indicates little, if any, elemental fractionation in the H component.

## 3. CONCLUSIONS

The present experiment has revealed positive anomalies in $^{128}Te$ (4.0 ± 0.3‰) and $^{130}Te$ (9.3 ± 2.0 ‰) from microdiamonds extracted from Allende and combusted in an oxygen plasma. Despite the experimental difficulties, three separate analyses of EB#2 gave consistent results for the Te isotopes. In contrast to the neutron burst model of Clayton (1989) and Howard *et al.* (1992), no anomalies were found in $^{125,126}$Te. Because of its extremely small abundance, $^{120}$Te could not provide any definitive information on p-process systematics. The anomalies in $^{128,130}$Te are consistent with the time-scale model of Ott (1996), taking into account new data on the half-lives of the Te precursors. The concentration of 640 ppm of Te in EB#2, together with the magnitude of anomalous $^{130}$Te, gives a $^{130}$Te-H concentration of 9.3 x $10^{12}$ atoms $g^{-1}$ in contrast to that of 2.4 x $10^{12}$ atoms $g^{-1}$ for $^{136}$Xe-H. The apparent disagreement between the ($^{128}$Te/$^{130}$Te) ratios measured here and by Richter *et al.* (1998) remains unexplained. However, the present data fit the pure time-scale model substantially better than do the results of Richter *et al.* (1998).

## REFERENCES

Anders, E. and Grevesse, N.: 1989, "Abundances of the elements: Meteoritic and solar", *Geochim. Cosmochim. Acta* **53**, 197-214.

Anders, E. and Zinner, E.: 1993, "Interstellar grains in primitive meteorites: diamond, silicon carbide, and graphite", *Meteoritics* **28**, 490-514.

Ballad, R.V., Oliver L.L., Downing, R.G. and Manuel O.K.: 1979, "Isotopes of Te, Xe and Kr in Allende meteorite retain record of nucleosynthesis", *Nature* **227**, 615-620.

Clayton, D.D.: 1989, "Origin of heavy xenon in meteoritic diamonds", *Ap. J.* **340**, 613-619.

Heymann, D. and Dziczkaniec M.: 1979, "Xenon from intermediate zones of supernovae", *Lunar Planet Sci.* **10**, 1943-1959.

Heymann, D. and Dziczkaniec M.: 1981, "Tellurium, should it be isotopically anomalous in the Allende Meteorite?" *Geochim. Cosmochim. Acta* **45**, 1829-1834.

Howard, W.M, Meyer, B.S. and Clayton, D.D.: 1992, "Heavy-element abundances from a neutron burst that produces Xe-H", *Meteoritics* **27**, 404-412.

Huss, G.R. and Lewis, R.S.: 1994a, "Noble gases in presolar diamonds I: Three distinct components and their implications for diamond origins", *Meteoritics* **29**, 791-810.

Huss, G.R. and Lewis R.S.: 1994b, "Noble gases in presolar diamonds II: Component abundances reflect thermal processing", *Meteoritics* **29**, 811-829.

Lewis, R.S., Huss, G.R. and Lugmair, G.: 1991a, "Finally, Ba and Sr accompanying Xe-HL in diamonds from Allende", *Lunar Planet. Sci.* **22**, 807-808.

Lewis, R.S., Huss, G.R., Anders, E., Liu, Y.G. and Schmitt, R.A.: 1991b, "Elemental abundance patterns in presolar diamonds", *Meteoritics* **26**, 363-364.

Loss, R.D., Rosman, K.J.R. and De Laeter, J.R.: 1990, "The isotopic composition of Zn, Pd, Ag, Cd and Te in the Allende meteorite", *Geochim. Cosmochim. Acta* **54**, 3525-3536.

Ott, U.: 1996, "Interstellar diamond xenon and timescales of supernova ejecta", *Ap. J.* **463**, 344-348.

Richter, S., Ott, U. and Begemann, F.: 1998, "Tellurium in pre-solar diamonds as an indicator, for rapid separation of supernova ejecta", *Nature* **391**, 261-26

# Isotope Abundance Anomalies in Meteorites: Clues to Yields of Individual Nucleosynthesis Processes

Ulrich Ott
*Max-Planck-Institut für Chemie, Becherweg 27, D-55128 Mainz, Germany*
*ott@mpch-mainz.mpg.de*

**Abstract:** Isotopic abundance anomalies are found in a variety of meteoritic phases. They are most pronounced in grains of presolar origin that survived intact the formation of the Solar system and for which an overview is given. While also large anomalies are observed in the light elements, interpretation with regard to the nucleosynthetic processes is more straightforward in the case of the heavy elements. Specifically discussed are the large overabundances of s-process isotopes in trace elements in presolar silicon carbide. These data allow a direct determination, without correction for r- or p-process contributions, of the complete s-process pattern. Several key elements for which such data have been obtained, allow one to put constraints on physical conditions during the process. Some hints regarding r-process nucleosynthesis are touched upon.

## 1. ROLE OF METEORITES AND ISOTOPE ABUNDANCE ANOMALIES FOR THE STUDY OF THE ORIGIN OF THE ELEMENTS

The study of meteorites has played and is continuing to play an important role in our attempts at understanding the origin of the chemical elements. This is because–like any other theory–our ideas about the processes by which the elements were made must be compared to a "ground truth": the sum of these processes (in the appropriate proportions) must reproduce the "solar abundances" as they are observed in the sun and in the most primitive meteorites (Anders and Grevesse, 1989). In fact, the first "modern" table of solar abundances by Suess and Urey (1956) was essential in enabling Burbidge *et al.* (1957) and Cameron

*Origin of Elements in the Solar System: Implications of Post-1957 Observations.*
Edited by O. Manuel, Kluwer Academic/Plenum Publishers, 2000. 

(1957) to identify a variety of processes that still today form the basis of how we believe that the elements from carbon on upwards were made in stars.

It is probably fair to say that, of all the observations by which the correctness of the ideas outlined in the ground-breaking work of $B^2FH$ (Burbidge *et al.*, 1957) can be tested, the observation of isotope abundance anomalies in meteorites is the one that even these enormously prescient workers are most unlikely to have foreseen. This is in spite of the fact that in 1957 the first observations of isotopic anomalies, those of excess $^{129}$Xe from the decay of now-extinct $^{129}$I (meanlife 23 Ma) in 1960 (Reynolds, 1960) and in 1964 of the strange Xenon nowadays called Xe-HL (Reynolds and Turner, 1964; see Section 5 below) were only a few years away. The abundance anomalies occur in matter that we can hold in our hands and analyze in the laboratory, and in favorable cases we can infer from isotopically peculiar phases the isotopic composition of a given element as it was made in an individual nucleosynthesis process identified by $B^2FH$.

## 2. ISOTOPE ABUNDANCE ANOMALIES IN METEORITES: THE CASE OF PRESOLAR DUST GRAINS

Isotope abundance anomalies in meteorites can be divided into four different types (Table 1): On the upper left are those that are due to the decay of now extinct radionuclides that had been present in the early solar system (such as $^{129}$I) and on the lower left are those that have a non-radiogenic origin and are hosted by phases that formed within the solar system. On the right are these same two sets of anomalies in grains that predate the solar system and survived the processes during its formation. Anomalies due to the former presence of now-extinct radionuclides in solar system phases in principle bear chronological information on timescales for the early solar system (Goswami, this volume), while in presolar grains such anomalies are "fossil" attesting to the fact that the grains must have formed from stellar outflows or explosions fast enough so that the radioisotopes were still alive at the time of grain formation. They do not put constraints on when that happened on an absolute scale.

*Table 1.* Types of isotope abundance anomalies in meteorites

| | |
|---|---|
| Solar system phase, extinct nuclide | Presolar grain, extinct nuclide |
| Solar system phase, non-radiogenic | Presolar grain, non-radiogenic |

In this contribution I will mainly focus on a discussion of isotopic patterns in grains of presolar origin (and there on anomalies in the heavy elements) because in this case observed abundance anomalies are much larger (*cf.* Section 5 below). The reason is that these grains formed directly in the outflows (winds) of stars or from supernova debris, from where they found their way largely unaltered into the solar system and into the meteorites in which we find and can study them now (Figure 1).

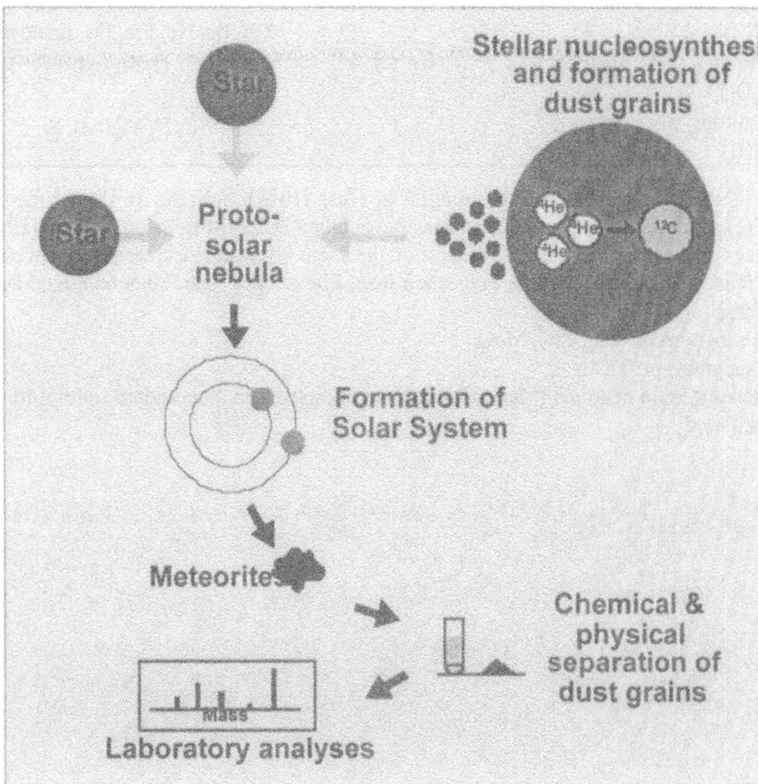


*Figure 1.* The path of circumstellar grains from stars to laboratory. Courtesy P. Hoppe.

An overview of the types of presolar grains identified in primitive meteorites so far, their properties, and the elements for which isotope abundance anomalies have been observed in them is given in Table 2. For a more complete overview see the review by Zinner (1998) and the contributions in Bernatowicz and Zinner (1997). Pictures of the most abundant grain types – diamond and silicon carbide - obtained by electron microscopy are shown in Figures 2 and 3.

The most abundant presolar phase is nanometer-sized diamond (Figure 2) which occurs in the most primitive meteorites on a level of nominally more than one permil.

*Table 2.* Presolar grains identified in primitive meteorites

| Phase | Abundance [ppm] | Typical Size [μm] | Isotopically Anomalous Elements |
|---|---|---|---|
| Graphite[a] | ~ 10 | ~ 1-20 | C, N, O, Mg(Al), Si, K, Ca, Ti, noble gases |
| Diamond | ~ 1400 | ~ 0.002 | N, Sr, Te, Ba, noble gases |
| Silicon carbide[b] (SiC) | ~ 14 | ~ 0.3-20 | C, N, Mg(Al), Si, Ca, Ti, Sr, Zr, Mo, Ba, Nd, Sm, Dy, noble gases |
| Corundum[c] ($Al_2O_3$) | ~ 0.01 | ~ 0.5-3 | O, Mg(Al) |
| Silicon nitride ($Si_3N_4$) | ≥ 0.002 | ~ 1 | C, N, Mg(Al), Si |

Listed abundances are from the compilation of Huss (1997) and refer to the CI meteorite Orgueil (exception: silicon nitride, where the number is for Murchison). Generally the abundances scale with the matrix fraction of the meteorites and decrease with increasing metamorphism (Huss, 1997). Size ranges are from Zinner (1997) and may be biased against small grains.

[a]contains subgrains of: TiC, ZrC, MoC.

[b]contain subgrains of: TiC.

[c]in addition one grain of spinel ($MgAl_2O_4$) and two of mixed spinel-corundum compositions have been found.

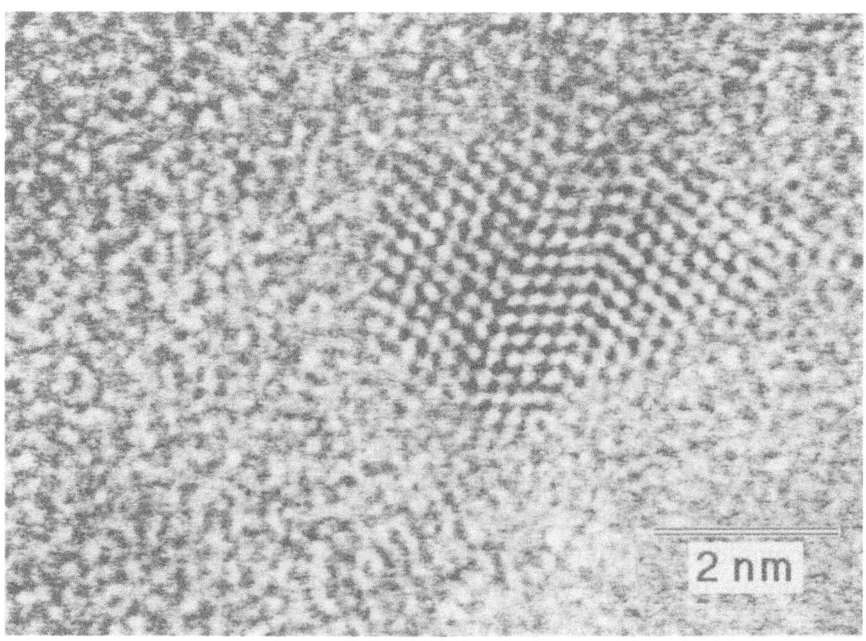


*Figure 2.* TEM photograph of a presolar diamond grain. Courtesy F. Banhart.

We cannot rule, however, that most of these nanodiamonds are actually of local, not presolar, origin. This is, because the structural element, carbon has an overall isotope ratio $^{12}C/^{13}C$ (~ 92) that is within the range of normal compositions and – at least with current techniques – it is not possible to analyze individual diamond grains, which on the average consist of only about 1000 atoms of $^{12}C$ (and hence 11 atoms of $^{13}C$) for their isotopic composition. The presolar origin for at least a fraction of the diamond grains is indicated by the occurrence within them of isotopically strange trace elements such as Xe-HL (Lewis, *et al.*, 1987) and Te-H (Richter *et al.*, 1998; Maas *et al.*, this volume), but these trace elements occur on such a low level, that there is only one atom of them per about a million diamond grains.

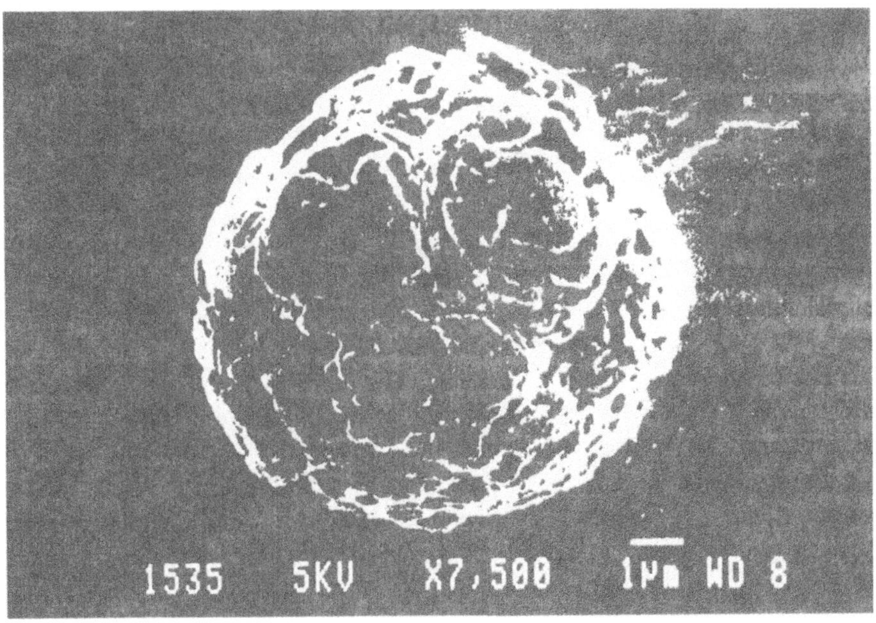


*Figure 3.* SEM photograph of one (of the larger) presolar silicon carbide grains. Courtesy P. Hoppe.

The situation is different for the other grain types in Table 2, the size of which is on the order of a fraction of a μm up to several μm (Figure 3). Grains larger than ~ 0.5 μm have been individually analyzed for the isotopic composition of the structural and several of the more abundant trace elements and show isotope abundance anomalies in many elements (Table 2). The next generation of ion microprobes, which should come into operation within the near future, will extend the range where individual analysis is possible, to even smaller sizes (Zinner, 1997).

# 3. LIGHT *VS.* HEAVY ELEMENT ISOTOPE ANOMALIES

Interpretation of the isotopic structures in the light elements such C, N, O and Si is complicated. Figure 4 shows – as an example – carbon and nitrogen isotopic compositions measured in a number of single grains of silicon carbide. The isotopic ratios show enormous variations: Ratios for $^{12}C/^{13}C$ between 2 and 7000 have been measured (solar system: 89) and $^{14}N/^{15}N$ ratios between 7 and 19000 (solar: 272).

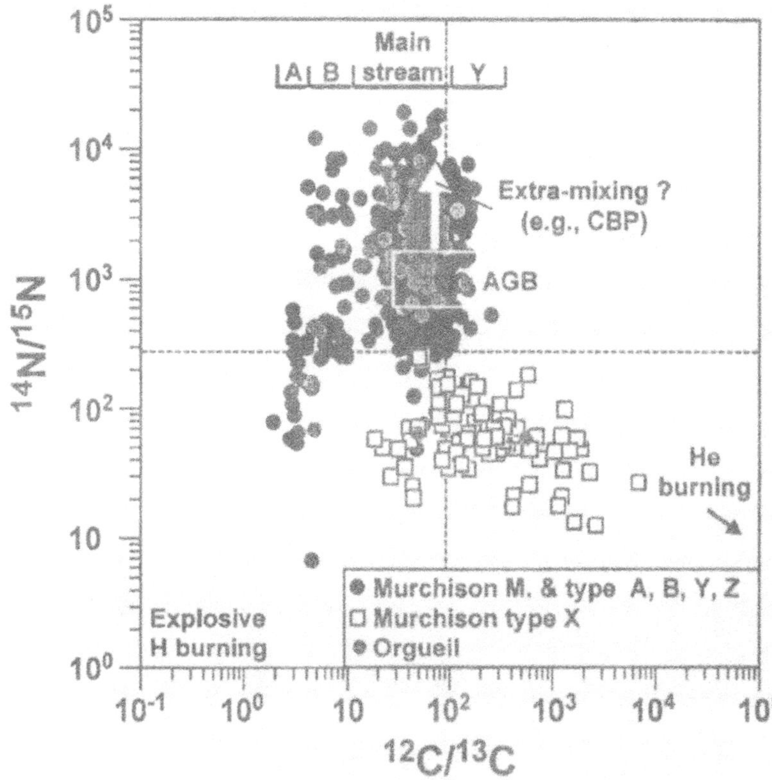


*Figure 4.* C and N isotopic composition of a number of individual silicon carbide grains from the Murchison and Orgueil carbonaceous chondrites as measured with the ion microprobe (courtesy P. Hoppe). The likely source for the majoritiy of grains ("mainstream") are Red Giant stars in the asymptotic giant branch (AGB) phase. Also indicated are the effects on the isotopic compositions of several nucleosynthesis and mixing processes. For details see Hoppe and Ott (1997).

The data can be used to infer the type of stars from which the grains formed and to draw conclusions regarding the extent of mixing in stars and galactic chemical evolution (See, *e.g.*, the reviews by Zinner, 1998; Hoppe and Ott, 1997; and other contributions in Bernatowicz and Zinner, 1997). However, with often only 2 isotopes, a variety of nucleosynthesis processes contributing, plus the effects from mixing in stars and galactic chemical evolution, it is not possible to derive from this type of data for the light elements the detailed isotopic composition produced in an individual process.

# 4. HEAVY ELEMENT ISOTOPIC STRUCTURES IN PRESOLAR SIC GRAINS AND INFERENCES FOR S-PROCESS OF NUCLEOSYNTHESIS

The situation is different for the heavy elements which are primarily produced by two types of neutron capture processes: slow (s-process) and rapid (r-process), and which are only barely influenced by the charged-particle reactions responsible for the synthesis of the lighter elements.

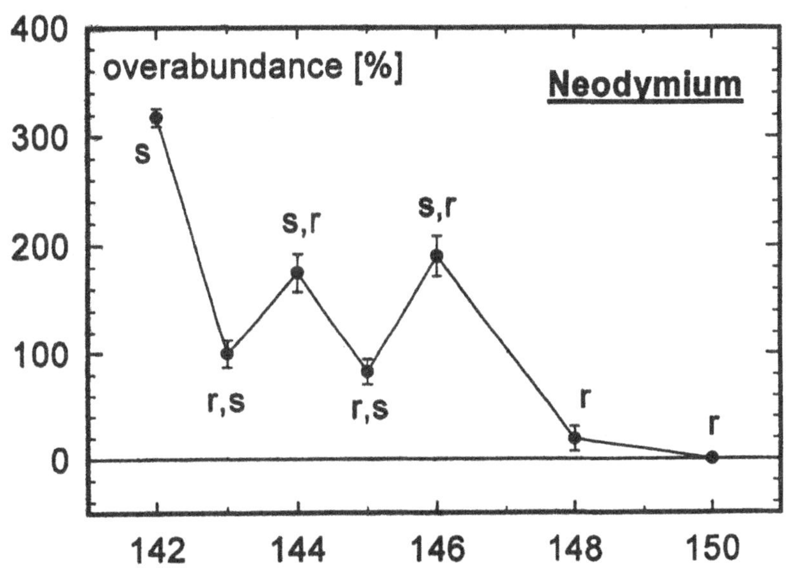


*Figure 5.* Isotopic composition of neodymium as measured by thermal ionization mass spectrometry on an ensemble of SiC grains extracted from the Murchison meteorite. Shown is the deviation in percent of ($^{i}Nd/^{150}Nd$) ratios from the composition of isotopically normal neodymium. Also indicated are the processes responsible for the synthesis of the various isotopes. Relative to $^{150}Nd$, which is not produced in the s-process, those with s-process contributions show overabundances which range up to a factor of ~ 4 (for s-only $^{142}Nd$).

In fact, the large anomalies – overabundances of isotopes made by the s-process – in heavy trace elements contained in presolar SiC, such as, *e.g.*

neodymium (Figure 5) allow one to derive very precisely – without having to make corrections for contributions from r- or p-process and essentially without having to make any assumptions – the isotopic composition of these elements as they are produced by the s-process. This in turn allows one to draw conclusions regarding the physical conditions of the process. As the compositions show the s-process recorded in the presolar SiC grains corresponds to the s-process "main component" (Käppeler *et al.*, 1989) identified in the structure of solar system abundances. The stellar sources – in all likelihood – are Red Giant stars during the AGB (asymptotic giant branch) phase (*e.g.* Gallino *et al.*, 1993, 1998).

While the eventual aim must be the comparison with detailed astrophysical model calculations (*e.g.* Gallino *et al.*, 1998), comparison with the so-called "classical model" is useful for a basic understanding of the nuclear processes involved. In the classical model (*e.g.*, Käppeler *et al.*, 1989) the essential physical parameters are considered as constant and feeding the observed s-process compositions into the formalism of the classical model allows one to derive "effective values" of the various parameters by which the process is described. So far s-process isotopic compositions have been obtained for Kr, Sr, Zr, Mo, Xe, Ba, Nd, Sm and Dy in presolar SiC and for a few elements in presolar graphite as well.

*Neutron exposure.* Over a wide mass range nuclide abundances produced in the main component of the s-process can be described by the "local approximation": the product $\sigma$ x N of neutron capture cross section, $\sigma$, and abundance, N, is approximately constant for neighboring nuclides. The simple relation breaks down where the neutron cross section is small, because in this case the "propagator" $(1+(\sigma x \tau_0)^{-1})^{-1}$, which is introduced in the full description of process in the classical model–with $\tau_0$ the mean neutron exposure–noticeably deviates from 1. Neutron capture cross sections are small at magic neutron numbers where shell closure occurs, hence among the nuclei that have been measured the ratios of the abundances of $^{88}$Sr and $^{138}$Ba relative to the neighboring Sr and Ba isotopes are sensitive to neutron exposure. With the interpretation of the Sr data complicated by the branching (See below) at $^{85}$Kr, data from barium provide the most clear-cut answer. Based on the ratio of $^{138}$Ba to s-only $^{136}$Ba effective neutron exposures between 0.14 and 0.17 mb$^{-1}$ have been derived, half of the value of ~ 0.30 mb$^{-1}$ characteristic for the solar system abundance distribution (Ott and Begemann, 1990; Gallino *et al.*, 1993). This may indicate that, on the average, the sources of the presolar SiC grains may have been of higher metallicity then the sources that supplied the solar system inventory of s-process nuclides in the mass range above A ~ 90 (Gallino *et al.*, 1993, 1998).

*Branchings.* Branching points in the s-process occur at nuclides where both neutron capture and $\beta$-decay occur in non-negligible proportions. One of the branch points among the elements for which data have been obtained for presolar grains in SiC is $^{147}$Nd. Most of it decays to $^{147}$Pm (unstable and another branch point) with a halflife of 11 days, but a small fraction captures another neutron leading to the synthesis via the s-process of a small amount of $^{148}$Nd. This is evidenced by the fact that in presolar SiC $^{148}$Nd is slightly overabundant relative to $^{150}$Nd (Figure 5), which, since its formation would require bridging also the gap at unstable $^{149}$Nd, is a "real" r-process only isotope. It is possible to deduce from thc observed branching ratio $f_n = \lambda_n / (\lambda_n + \lambda_\beta)$ the neutron capture rate $\lambda_n$ and from that the effective neutron density $n_n$ as being on the order of $3 \times 10^8$ cm$^{-3}$ (Hoppe and Ott, 1997).

Some of the branchings for which data are available, involve nuclides with β-decay half-lives that are sensitive to the physical conditions at the place and time where the s-process took place. These provide constraints on temperature (branchings at $^{79}$Se and $^{151}$Sm) as well as on electron and mass density (via the abundance of $^{163}$Dy which decays via bound state β-decay). For details see Hoppe and Ott (1997) and references therein.

*Astrophysical model and single grain analyses.* As pointed out above, the results from the isotopic analyses and the "effective values" derived finally must be prepared with realistic astrophysical models which take into account stellar evolution and the variations of neutron density and temperature during the period of s-process nucleosynthesis. Current models assume the s-process to take place in the He-burning shell during the TP-AGB (thermally pulsing asymptotic giant branch) phase of Red Giant stars. According to these models (Gallino *et al.*, 1998) most of the exposure is provided by the $^{13}$C(α,n)$^{16}$O neutron source and is characterized by low neutron densities of ~ $10^7$cm$^{-3}$ and a low temperature of ~ 8 keV on long time scales (several $10^4$ yr). What essentially is recorded in the branchings, however, is the action of the $^{22}$Ne(α,n)$^{25}$Mg neutron source, which is marginally activated at the end of each pulse for a few years and during which higher neutron densities in the range $10^{10}$ cm$^{-3}$ and temperatures in the range 30 keV are reached.

Of special value for the comparison are the results that have been obtained for Sr, Zr and Mo in single grains of presolar graphite and silicon carbide (Nicolussi *et al.*, 1997, 1998a, 1998b), since they may have recorded the compositions established at various stages of TP-AGB evolution. The data for Sr (Nicolussi *et al.*, 1998b) have large errors and essentially confirm the observations of Podosek *et al.* (1999) on grain aggregates for s-process $^{87}$Sr/$^{86}$Sr and $^{88}$Sr/$^{86}$Sr to differ comparably little from normal ratios. Compared to this, the Zr data appear remarkable. This is because how much (if any) of $^{96}$Zr produced in the process depends on the branching at $^{95}$Zr (half-life = 64 days). Several of the grains analyzed seem to be devoid of $^{96}$Zr, an observation not predicted by current models (Nicolussi *et al.*, 1997, 1998a).

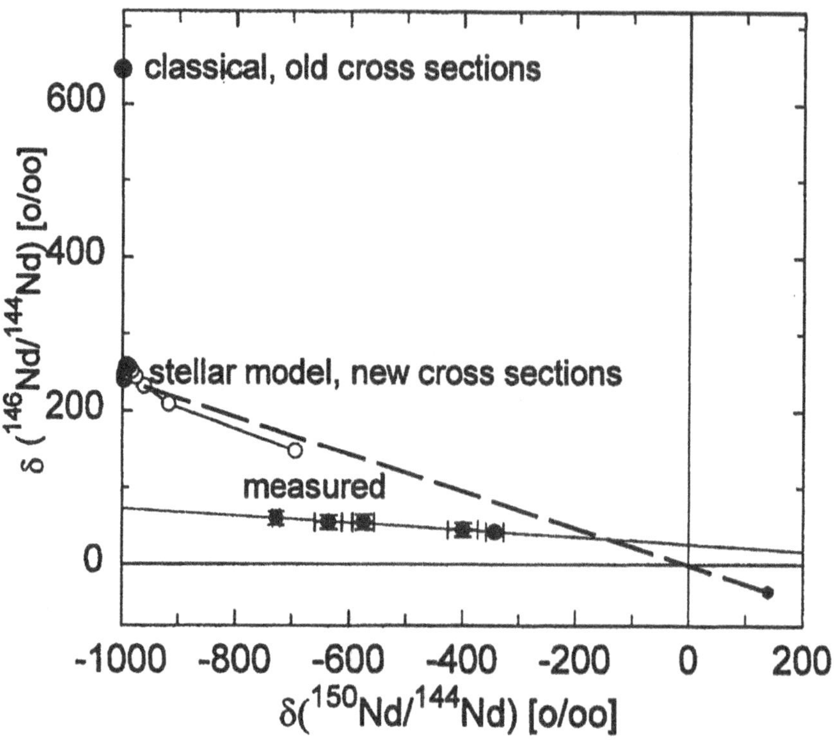


*Figure 6.* Three-isotope diagram of $^{146}$Nd/$^{144}$Nd *vs.* $^{150}$Nd/$^{144}$Nd. Plotted are δ-values, *i.e.*, deviations from the normal composition in permil. A comparison is made between the ratios measured in presolar SiC, the extrapolated s-process compositions and model predictions (See text for discussion).

While this may indicate some problems of even the currently most-sophisticated stellar models with reproducing the observations, a closer look at the Nd (and also the Ba) data discussed above and a comparison with the models indicates that there may be problems with our understanding of heavy element nucleosynthesis in general. This is illustrated in Figure 6, which is a 3-isotope diagram of δ($^{146}$Nd/$^{144}$Nd) *vs.* δ($^{150}$Nd/$^{144}$Nd), where δ($^{i}$Nd/$^{144}$Nd) is the deviation of the $^{i}$Nd/$^{144}$Nd ratio from the normal terrestrial composition. Earlier predictions for the composition of s-process Nd differed significantly (by more than 50%) from the observation-derived composition, which is given by the ordinate intercept of the line fitted to the data points at vanishing abundance of r-only $^{150}$Nd. While the discrepancy has been mostly due to errors in the neutron capture cross sections (*cf.* Gallino *et al.*, 1993) and has largely disappeared with the new cross sections (Toukan *et al.*, 1995), it has not completely disappeared. Even with the new cross sections and the astrophysical model (Gallino, 1998) there exist differences (~ 15% in the case of $^{146}Nd/^{144}Nd$) that are not covered by the combined nominal errors of grain and cross section measurements, which are both reported to be at the %-level. Nor can the discrepancy be easily explained by choosing different stellar conditions: neutron capture cross sections of the Nd isotopes are large (Toukan *et al.*, 1995) so that the ratio is not sensitive to neutron exposure, and neither is there a branching involved.

If, as our current understanding goes, r- and s-process are the essential sources of $^{144}Nd$, $^{146}Nd$ and $^{150}Nd$, with only small contributions by the p-process, and all these processes always produce the same well-defined composition, any possible composition must lie on a mixing line which joins the s-process composition on the left with the r-process composition further to the right and which must pass through the normal Nd isotopic composition (which represents a specific mixture of s-and r-process contributions). The fact that the line defined by the grain data points does not extrapolate to the theoretical s-process composition and also misses the normal composition must be seen as evidence that there is still something wrong with either the grain or capture cross section measurements or with our basic understanding of heavy element nucleosynthesis.

## 5. R-PROCESS ABUNDANCE ANOMALIES

Our current database does not contain clear evidence for large, unaltered, overabundances caused by the addition of matter produced in the r-process. There is convincing evidence for a r-process signature in Nd and Sm of the FUN Ca-Al rich inclusion EK1-4-1 (Lee, 1988), but the effects are only on the order of permil, much smaller than, *e.g.*, the s-process overabundances in silicon carbide (Figure 7). These small effects cannot be uniquely assigned to (a) specific isotope(s), because mass-dependent fractionation, which occurs during isotopic analysis in a mass spectrometer as well as fractionation effects that may have occurred in nature are of similar size. It is necessary to correct for this mass fractionation by assuming a fixed value for one "normalizing ratio", generally a ratio considered likely to be normal. In the case of Nd in EK1-4-1 the correction was chosen in such a way that resulting overabundances of $^{148}Nd$ and $^{150}Nd$ which are predominantly of r-process origin were about equal (Figure 7). Similarly, in the case of Sm in EK1-4-1, mass fractionation correction was performed by assuming the ratio of the two s-only isotopes $^{148}Sm$ and $^{150}Sm$ to be normal resulting in an r-process overabundance pattern (Lee, 1988).

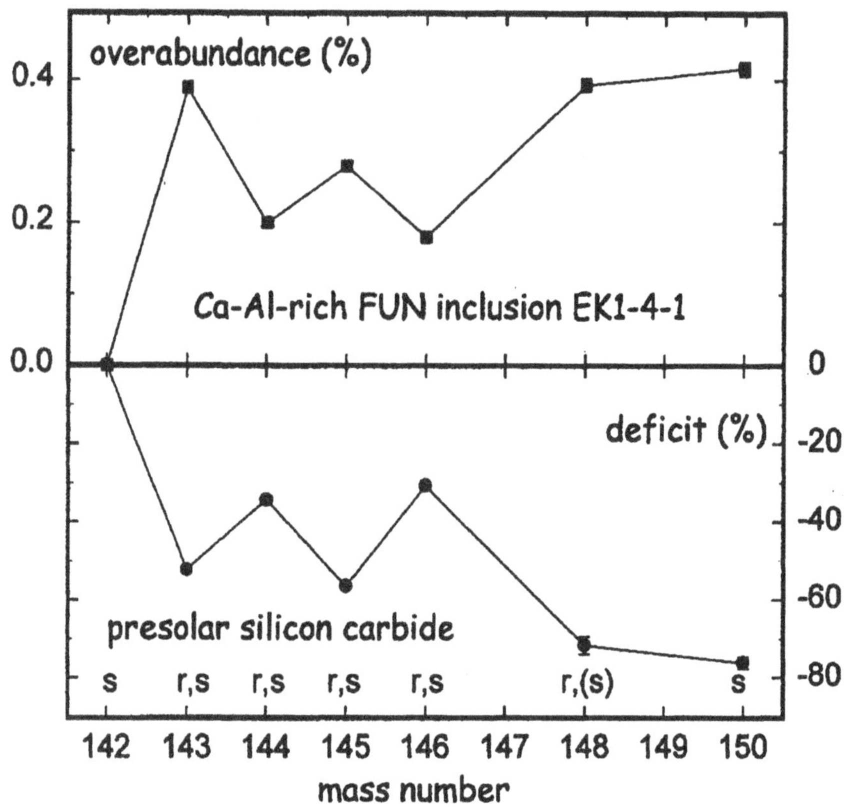


*Figure 7.* Comparison of isotopic anomalies of neodymium in Ca-Al-rich FUN inclusion EK1-4-1 and a sample of presolar silicon carbide grains. The former shows an overabundance of r-process products of a few permil, the latter a deficit of tens of percent (corresponding to an overabundance of s-process products by up to a factor 4; *cf.* Figures 5, 6).

Mass fractionation corrections are negligible in the case of the large anomalies found in grains of presolar origin such as silicon carbide (Figure 7). Similarly large anomalies associated with r-process overabundances appear to be present in xenon and tellurium within the presolar diamonds, but here the situation is not straightforward for a different reason: while the diamonds contain xenon with large overabundances of the two r-only isotopes $^{134}$Xe and $^{136}$Xe (Xe-H; as well as the two p-only isotopes $^{124}$Xe and $^{126}$Xe: Xe-L), the r-only isotopes do not occur in "average r-process" proportions (nor do the p-only isotopes occur in the "average p-process ratio"), *i.e.*, the relative overabundances at the r-only isotopes are different (Figure 8).

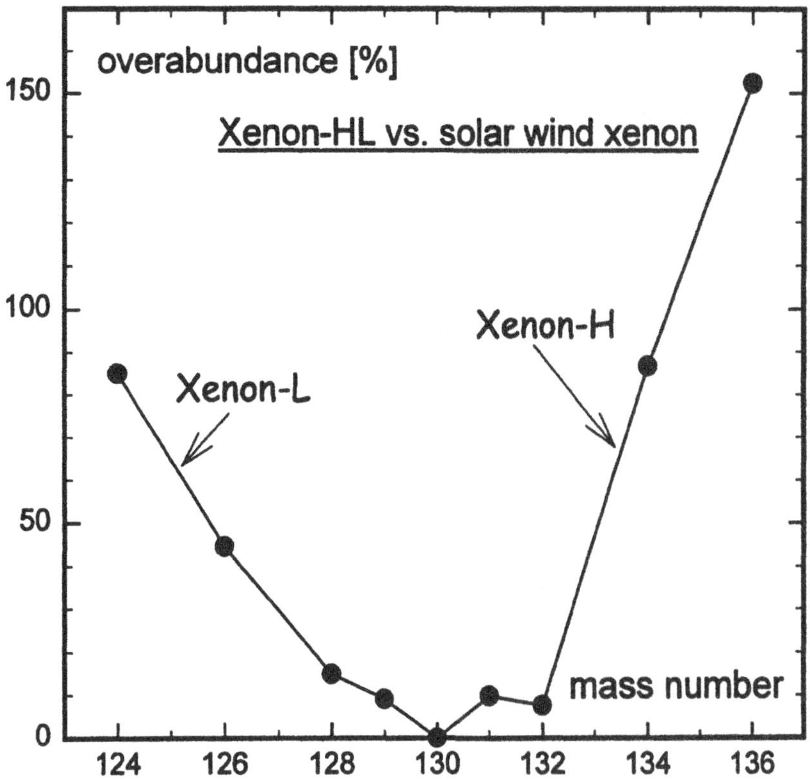


*Figure 8.* Normalized to s-only $^{130}$Xe, xenon-HL shows large overabundances relative to solar wind xenon of the p- and r-process isotopes. However, relative overabundances at the two r-only isotopes $^{134}$Xe and $^{136}$Xe are not the same, nor are the relative overabundances of the two p-only isotopes $^{124}$Xe and $^{126}$Xe

A possible way out of the dilemma has been suggested by Ott (1996) who showed that the observed pattern could be obtained from "average r-process matter" in a scenario in which an early separation occurred between precursors, in the decay chain leading from the very neutron-rich nuclei produced in the r-process to the stable end products, and those end products. While it is unclear how such a separation may have occurred on the inferred time scale (hours) after the process (generally thought to be associated with the explosion of a supernova), the validity of the general approach seems to be confirmed by recent data on tellurium within the diamonds (Richter *et al.*, 1998; Maas *et al.*, this volume). The data of Maas *et al.* (this volume) within their larger uncertainties show perfect agreement with the predictions from the Ott (1996) model. Richter *et al.* (1998) found much larger effects with correspondingly smaller errors on the derived Te-H composition, but used the direct-loading method which is more susceptible to isobaric interferences. Taken at face value, their data may require to give up the concept of an average r-process source within the early separation model. If true, this would not be the only evidence, however, from isotope abundance anomalies for a non-unique r-process pattern. Also the problems with the interpretation of the s-process Nd and Ba data (Section 4; Figure 6) could be caused by a non-solar r-process, and further arguments have been advanced based on the relative abundances in the early solar system of the radionuclides $^{129}I$ and $^{182}Hf$ of r-process origin (Wasserburg *et al.*, 1996; Qian *et al.*, 1998).

## 6. SUMMARY AND OTHER ISOTOPE ABUNDANCE ANOMALIES

In this overview, I have concentrated on the large anomalies found in grains of presolar origin, and there on the anomalies in the heavy elements, the reason being that for this type of anomaly inferences regarding the processes of nucleosynthesis are the most direct. There are, however, isotope abundance anomalies in meteorites besides those in the presolar grains which are important in their own right. A few, like the r-process anomalies in EK1-4-1 have been touched upon above. Those due to extinct radionuclides are covered in the contribution by Goswami (this volume). Among the others are:

- the widespread $^{16}O$ anomaly, which may be of nucleosynthetic origin (*cf.* Clayton, 1993).
- not well understood anomalies in nitrogen, primarily enhancements of $^{15}N$ (*e.g.*, Prombo and Clayton, 1985; Sugiura and Hashizume, 1992).
- overabundances of the most neutron-rich nuclei of the Fe-peak elements in Ca-Al-rich inclusions (FUN and otherwise) due to a kind of neutron-rich nuclear statistical equilibrium (Lee, 1988) or r-process (Kratz *et al.*, 1995).
- anomalies not obtained or confirmed by mass spectrometric analysis, and hence generally regarded with scepticism, of osmium (Kumar and Goel, 1991) and mercury (Thakur, 1997).

Among the presolar grains, interpretation of the light element isotopic compositions in terms of nucleosynthetic processes is complicated by the fact that there are often only two isotopes and several processes contributing as well as the effects due to mixing in stars and galactic chemical evolution. The large anomalies in xenon and tellurium within presolar diamonds are indicative of r-process contributions, but appear not to be unadulterated r-process material. By and large, the most direct constraints on the physics of nucleosynthesis and its astrophysical context exists in the case of the s-process patterns observed in presolar graphite and silicon carbide grains.

# REFERENCES


Anders, E. and Grevesse, N.: 1989, "Abundances of the elements: meteoritic and solar", *Geochim. Cosmochim. Acta* **53**, 197-214.

Bernatowicz, T.J. and Zinner, E., eds.: 1997, *Astrophysical Implications of the Laboratory Study of Presolar Materials*, AIP Conf. Proc 402, 750 pp.

Burbidge, E.M., Burbidge, G.R., Fowler, W.A. and Hoyle, F.: 1957, "Synthesis of elements in stars", *Rev. Mod. Phys.* **29**, 547-650.

Cameron, A.G.W.: 1957, "Stellar evolution, nuclear astrophysics and nucleogenesis", *Chalk River Report, AECL*, Atomic Energy of Canada Ltd, CRL-41.

Clayton, R.N.: 1993, "Oxygen isotopes in meteorites", *Annu. Rev. Earth Planet. Sci.* **21**, 115-149.

Gallino, R., Raiteri, C.M. and Busso, M.: 1993, "Carbon stars and isotopic Ba anomalies in meteoritic SiC grains", *Ap. J.* **410**, 400-411.

Gallino R.: 1998, private communication.

Gallino, R., Arlandini, C., Busso, M., Lugaro, M., Travaglio, C., Straniero, O., Chieffi, A. and Limongi, M.: 1998, "Evolution and nucleosynthesis in low-mass asymptotic giant branch stars. II. Neutron captures and the s-process", *Ap. J.* **497**, 388-403.

Hoppe, P. and Ott, U.: 1997, "Mainstream silicon carbide grains from meteorites", in *Astrophysical Implications of the Laboratory Study of Presolar Materials*, eds., Bernatowicz, T.J. and Zinner, E., AIP Conf. Proc 402, pp. 27-58.

Huss, G.R.: 1997, "The survival of presolar grains in solar system bodies", in *Astrophysical Implications of the Laboratory Study of Presolar Materials*, eds., Bernatowicz, T.J. and Zinner, E., AIP Conf. Proc. 402, pp. 721-748.

Käppeler, F., Beer, H. and Wisshak, K.: 1989, "s-Process nucleosynthesis - nuclear physics and the classical model", *Rep. Prog. Phys.* **52**, 945-1013.

Kratz, K.-L., Mueller, A.C. and Thielemann, F.-K.: 1995, "'FUN' durch Supernovaexplosionen und Kernphysik mit LISE", *Phys. Bl.* **51**, 183-186.

Kumar, P. and Goel, P.S.: 1991, "Variations in the $^{190}Os/^{184}Os$ ratio in some stone meteorites and acid residues of Sikhote Alin iron meteorite", *Geochem. J.* **25**, 399-409.

Lee, T.: 1988, "Implications of isotopic anomalies for nucleosynthesis" in *Meteorites and the Early Solar system* eds., Kerridge, J.F. and Matthews, M.S., University of Arizona Press, Tucson, AZ, pp. 1063-1089.

Lewis, R.S., Tang, M., Wacker, J.F., Anders, E. and Steel, E.: 1987, "Interstellar diamonds in meteorites", *Nature* **326**, 160-162.

Nicolussi, G.K., Davis, A.M., Pellin, M.J., Lewis, R.S., Clayton, R.N. and Amari, S.: 1997, "s-Process zirconium in presolar silicon carbide grains", *Science* **277**, 1281-1283.

Nicolussi, G.K., Pellin, M.J., Lewis, R.S., Davis, A.M., Amari, S. and Clayton, R.N.: 1998a, "Molybdenum isotopic composition of individual presolar silicon carbide grains from the Murchison meteorite", *Geochim. Cosmochim. Acta* **62**, 1093-1104.

Nicolussi, G.K., Pellin, M.J., Lewis, R.S., Davis, A.M., Clayton, R.N. and Amari, S.: 1998b, "Strontium isotopic composition in individual circumstellar silicon carbide grains: a record of s-process nucleosynthesis", *Phys. Rev. Lett.* **81**, 3583-3586.

Ott, U. and Begemann, F.: 1990, "Discovery of s-process barium in the Murchison meteorite", *Ap. J.* **353**, L57-L60.

Ott, U.: 1996, "Interstellar diamond xenon and time scales of supernova ejecta", *Ap. J.* **463**, 344-348.

Podosek, F.A., Prombo, C.A., Amari, S. and Lewis, R.S.: 1999, "s-Process Sr isotopic compositions in presolar SiC from the Murchison meteorite", *Ap. J.*, in press.

Prombo, C.A. and Clayton, R.N.: 1985, "A striking nitrogen isotope anomaly in the Bencubbin and Weatherford meteorites", *Science* **230**, 935-937.

Qian, Y.-Z., Vogel, P. and Wasserburg, G.J.: 1998, "Diverse supernova sources for the r-process", *Ap. J.* **494**, 285-296.

Reynolds, J.H.:1960, "Determination of the age of the elements", *Phys. Rev. Lett.* **4**, 8-10.

Reynolds, J.H. and Turner, G.: 1964, "Rare gases in the chondrite Renazzo", *J. Geophys. Res.* **69**, 3263-3281.

Richter, S., Ott, U. and Begemann, F.: 1998, "Tellurium in pre-solar diamonds as an indicator for rapid separation of supernova ejecta", *Nature* **391**, 261-263.

Suess, H.E. and Urey, H.C.: 1956, "Abundances of the elements", *Rev. Mod. Phys.* **28**, 53-74.

Sugiura, N. and Hashizume, K.: 1992, "Nitrogen isotope anomalies in primitive ordinary chondrites", *Earth Planet. Sci. Lett.* **111**, 441-454.

Thakur, A.N.: 1997, "Measurement of mercury isotopic ratio in stone meteorites by neutron activation analysis", *J. Radioan. Nucl. Chem.* **216**, 151-159.

Toukan, K.A, Debus, K., Käppeler, F. and Reffo, G.: 1995, "Stellar neutron capture cross sections of Nd, Pm, and Sm isotopes", *Phys. Rev.* **C51**, 1540-1550.

Wasserburg, G.J., Busso, M. and Gallino, R.: 1996, "Abundances of actinides and short-lived nonactinides in the interstellar medium: diverse supernova sources for the r-process", *Ap. J.* **466**, L109-L113.

Zinner, E.: 1997, "Presolar material in meteorites: an overview", in *Astrophysical Implications of the Laboratory Study of Presolar Materials*, eds., Bernatowicz, T.J. and Zinner, E., AIP Conf. Proc 402, pp. 3-26.

Zinner, E.: 1998, "Stellar nucleosynthesis and the isotopic composition of presolar grains from primitive meteorites", *Annu. Rev. Earth Planet. Sci.* **26**, 147-188.

# Variation of Molybdenum Isotopic Composition in Iron Meteorites

Qi Lu and Akimasa Masuda
*Department of Chemistry, The University of Tokyo. Correspondence to: A. Masuda at COLGA, 449 S-Yasmats, Tokorozawa, Saitama, 359-0024, Japan*
*mansfield@green.ocn.ne.jp*

Abstract: The isotopic composition of molybdenum was determined with high precision for eight iron meteorites and terrestrial samples. The s,r nuclides, $^{95}$Mo and $^{98}$Mo, were chosen for normalization in order to cancel effects of mass-dependent isotope fractionation. After this normalization, another s,r nuclide $^{97}$Mo indicates little isotope anomaly. The only s-process nuclide, $^{96}$Mo, shows an anomaly ($\varepsilon_{96}$) ranging from -2.33 to + 1.94. There is a clear positive correlation between $\varepsilon_{96}$ and $\varepsilon_{94}$, while there is a clear negative correlation between $^{96}$Mo and the only pure r-process nuclide $^{100}$Mo. The relationship between $^{96}$Mo and the p-process nuclide, $^{92}$Mo, is grossly positive, but the relevant points are rather scattered. The production process of $^{92}$Mo is judged to be partially related with that of $^{94}$Mo and partially independent of it. These results are very significant for nucleosynthesis and the evolution of the solar system. Great chemical caution was taken against the isobaric effects coming from Zr and Ru, and a special technique was devised to secure a stable, intense ion beam of Mo$^{+}$ in the mass spectrometer.

## 1. INTRODUCTION

Molybdenum consists of seven stable isotopes, $^{92}$Mo, $^{94}$Mo, $^{95}$Mo, $^{96}$Mo, $^{97}$Mo, $^{98}$Mo and $^{100}$Mo, which are from various processes of nucleosynthesis (Table 1). Although a considerable number of elements have seven or more than seven stable isotopes, the number "seven" is large. In addition, the stable isotopes of molybdenum are uniquely advantageous for investigation of isotopic composition by precise measurement of isotopic abundance ratios because all of the molybdenum isotopes have comparable abundances (Table

1). The isotope with the highest abundance is $^{98}Mo$ (24.2%), while that with the lowest abundance is $^{94}Mo$ (9.2%). Moreover, it merits attention that the abundance of the p-nuclide, $^{92}Mo$, is as high as 14.7%. This exceptionally high abundance of $^{92}Mo$ as a p-nuclide is due to the fact that the neutron number for this nuclide is 50, an outstanding magic number. In spite of the advantages that molybdenum has in nuclear physics and isotopic abundances, much caution should be taken against interference from the isobaric effects, derived from the impurities of $_{40}Zr$ and $_{44}Ru$ (Table 1), in measurement of the isotopic abundance ratios of $_{42}Mo$. Another technical difficulty has been to secure a stable and intense ion current for Mo in the mass spectrometer.

*Table 1.* Nucleosynthetic processes of Mo isotopes, with relevant notes.

| Nuclide | Formation Process[1] | Abundance (%) | Isobars[2] |
|---|---|---|---|
| $^{92}Mo$ | p | 14.73 | $^{92}Zr$ (17.15%) |
| $^{94}Mo$ | p(s) | 9.21 | $^{94}Zr$ (17.39%) |
| $^{95}Mo$ | s,r(p) | 15.89 | |
| $^{96}Mo$ | s | 16.67 | $^{96}Zr$ (2.80%), $^{96}Ru$ (5.54%) |
| $^{97}Mo$ | s,r | 9.57 | |
| $^{98}Mo$ | s,r | 24.22 | $^{98}Ru$ (1.87%) |
| $^{100}Mo$ | r | 9.70 | $^{100}Ru$ (12.60%) |

[1]After Allen *et al.* (1971).

[2]Percentage in parentheses refers to the percentage abundance of the nuclide as an isotope. Figures on Zr are from Sahoo and Masuda (1997), and those on Ru are from Poths *et al.* (1987) and Huang *et al.* (1996). Abundance data on Mo are from the present work.

This is the first successful study of variations in isotope abundance ratios of a metallic element in iron meteorites. One of the initial motivations was to detect the decay product of extinct $^{97}Tc$ in terrestrial molybdenites and/or meteoritic samples, but it was difficult to attain this goal definitely.

## 2. EXPERIMENTAL

The essential points required of the experimental method are a high degree of completeness in the chemical separation of Mo from Zr, Ru and matrix elements (mainly Fe) and delicate procedures for pretreatment of the filament, for loading the activator (boric acid) plus sample, and for stepwise increases in the filament current during mass spectrometry.

The flow chart of the total analytical procedures used for separation and purification of microgram quantities of Mo from gram quantities of iron meteorites is shown in Figure 1. More descriptive explanations of the chemical treatment and mass spectrometry are given as an APPENDIX at the end of the paper.

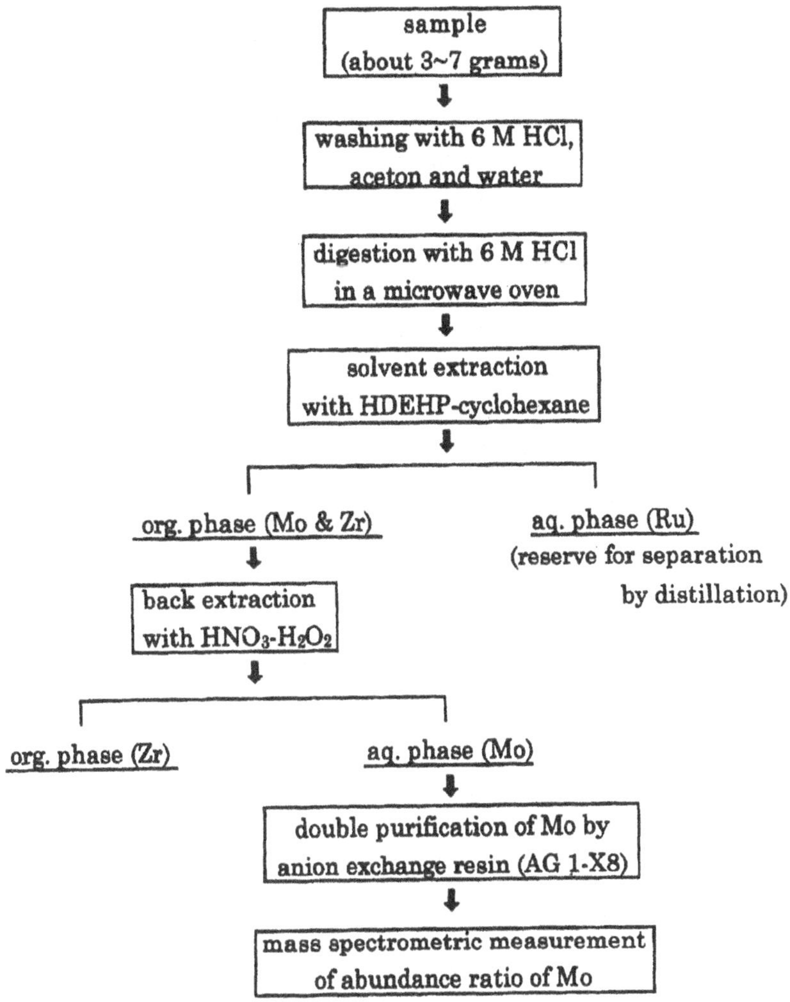


*Figure 1.* Flow chart for the chemical separation and purification of Mo from iron meteorites and the isotopic abundance measurement of Mo.

# 3. RESULTS AND DISCUSSION

## 3.1 Normalization

We found that the isotopic abundance ratios of molybdenum in iron meteorites are variable, even after corrected for the strictly mass-dependent isotope fractionation that takes place during the mass spectrometric measurement. One cannot rule out the possibility that similar mass-dependent fractionation may have occurred in the solar nebula, too. Therefore, in order to disclose a non-mass-dependent variation or isotope anomaly, it is necessary to cancel the mass-dependent fractionation effect by selecting the appropriate pair of isotopes for normalization. This choice poses an essential and serious problem. Table 1 shows the isotopes of molybdenum classified by their nucleosynthesis origin.

It would be reasonable to assume that the isotopes produced by similar processes of nucleosynthesis will display a nearly constant abundance ratio and can be taken as the adequate isotopic pair for normalization. The effect of magnification of mass spectrometric error should also be taken into account. In view of both of these points, we chose $^{98}$Mo and $^{95}$Mo as the normalizing pair, because both of these are s,r combined nuclides. It should be borne in mind that the pattern of the resultant aberration ("isotope anomaly") is seriously affected by the choice of the normalizing pair.

The abundance ratio (0.655964, Table 2) obtained for the standard (Aldrich molybdenum) is taken as a normalization value for the $^{95}$Mo/$^{98}$Mo ratio. The value of the average for 17 terrestrial molybdenites from wide locations of the earth plus Aldrich Mo (Table 2) will be also considered together with the values for Aldrich Mo alone.

*Table 2.* The abundance ratios[1] for terrestrial molybdenum

| Isotope ratios | Standard (Aldrich, $MoO_3$)[2] | Earth (average)[3] |
|---|---|---|
| $^{92}$Mo/$^{98}$Mo | 0.607926 | 0.607888 |
| $^{94}$Mo/$^{98}$Mo | 0.380200 | 0.380191 |
| $^{95}$Mo/$^{98}$Mo | 0.655964 | 0.655964 |
| $^{96}$Mo/$^{98}$Mo | 0.688146 | 0.688138 |
| $^{97}$Mo/$^{98}$Mo | 0.394947 | 0.394944 |
| $^{100}$Mo/$^{98}$Mo | 0.400129 | 0.400147 |

[1]Normalized against $^{95}$Mo/$^{98}$Mo = 0.655964.
[2]Average for seven sequences of measurements made on different dates.
[3]Average for 18 terrestrial samples (17 molybdenites plus "standard").

Thus, the "degrees of anomalies" or "isotope anomaly" for molybdenum isotopes in samples investigated by us is based on the correcting the obtained values by assuming that $^{95}$Mo/$^{98}$Mo should be 0.655964. The resultant apparent isotope anomalies are shown in Figure 2. The anomaly of $^{96}$Mo intrigues us, because it is the only pure s-nuclide in contrast with s,r nuclides. It turns out that the apparent anomaly for $^{96}$Mo ranges between - 2.33 and + 1.94 in $\varepsilon$. It may be significant that the isotope anomaly of $^{97}$Mo which is also s,r nuclide like $^{95}$Mo and $^{98}$Mo falls close to zero, exhibiting very small variation in resultant normalized values.

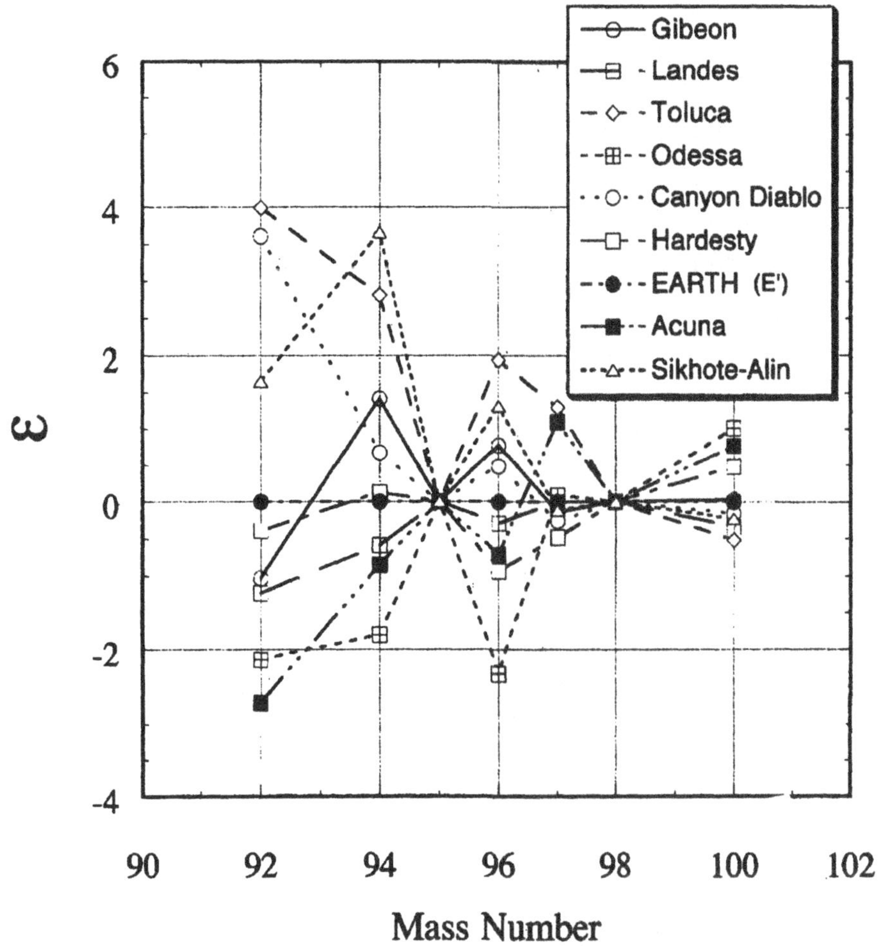


*Figure 2.* Isotope anomaly patterns for Mo in iron meteorites, taking $^{95}Mo$ and $^{98}Mo$ as an isotopic pair for normalization.

## 3.2 Relationship between the isotope anomalies

In Figure 3, the values of $\varepsilon_{97}$ are plotted against those of $\varepsilon_{96}$. As stated above, the relation line is almost horizontal with a nearly zero level for $\varepsilon_{97}$. Conceivably the positive aberrations for A (Acuna) and T (Toluca) may be due to the decay of extinct $^{97}Tc$ (Howard *et al.*, 1991) whose half-life is 2.6 x $10^6$ yr, although one cannot rule out other fortuitous factors. Figure 4 shows the good positive correlation between the values of $\varepsilon_{94}$ and $\varepsilon_{96}$. As a whole, the slope for $\varepsilon_{94}/\varepsilon_{96}$ is about unity but *appears slightly concave upwards*, perhaps reflecting the involvement of complicated secondary effects. According to Allen *et al.* (1971), $^{94}$Mo is a p(s) nuclide unlike $^{96}$Mo, a pure s-nuclide. However, it merits attention that both of $^{94}$Mo and $^{96}$Mo are shielded by $^{94}$Zr and $^{96}$Zr against beta-decay series originating from very rapid neutron-capture processes (starting with seed elements of atomic numbers smaller than 40) and from nuclear fission of transuranium elements. The lightest molybdenum isotope, $^{92}$Mo, is also shielded from similar effects by $^{92}$Zr (and $^{92}$Nb).

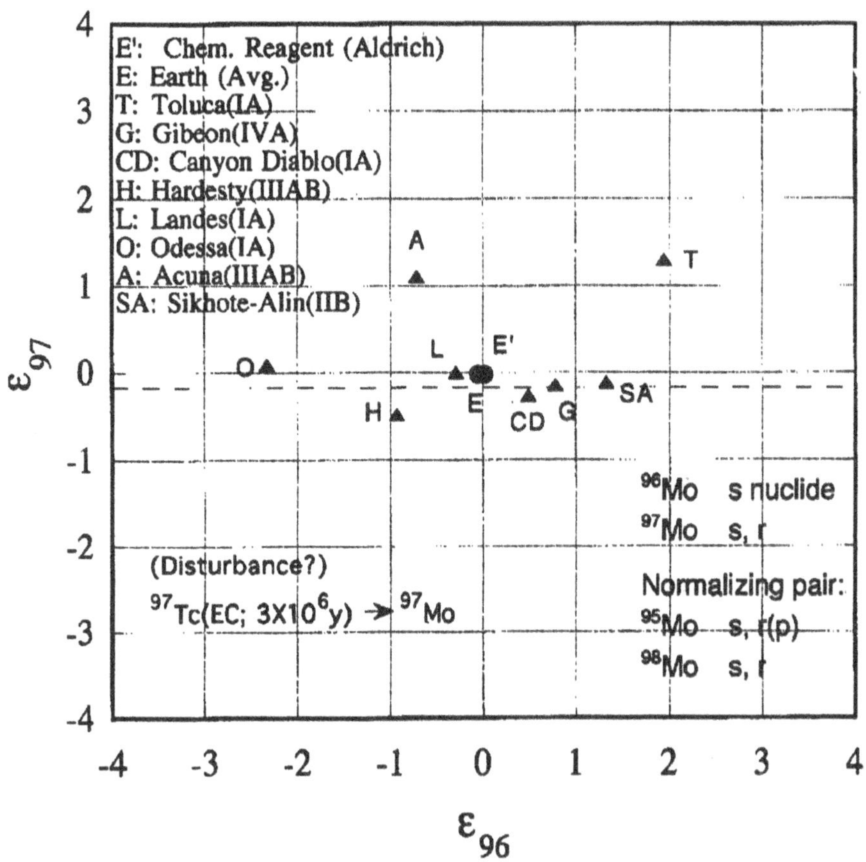


*Figure 3.* The relationship between the values of $\varepsilon_{97}$ and $\varepsilon_{96}$ for the iron meteorites analyzed in this study.

It is noteworthy that $^{94}$Mo is bypassed by the s-process in the major nucleosynthesis route "if the time scale for neutron capture on $^{93}$Zr is much less than 1.5 x 10$^{6}$ yr, the half-life of $^{93}$Zr" (Woosley and Howard, 1978). That is, if the neutron capture in question is comparable with 1.5 x 10$^{6}$ yr, the contribution of s-process can become great. The potential heterogeneity in the site and/or stage for s-nuclide production should be also considered.

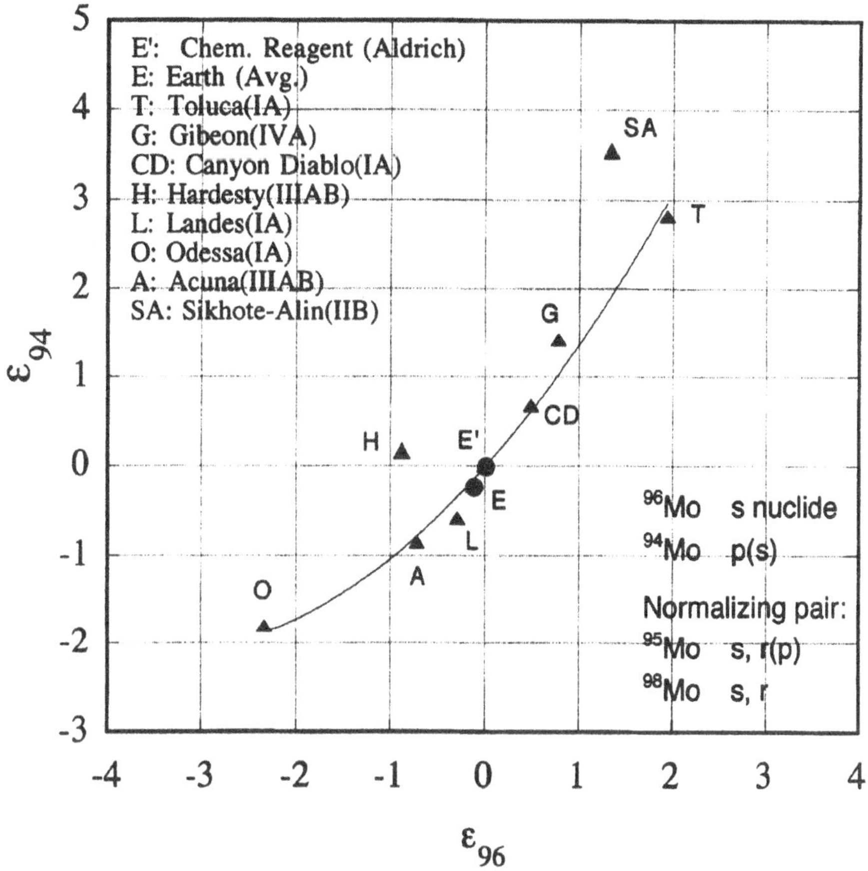


*Figure 4.* The relationship between the values of $\varepsilon_{94}$ and $\varepsilon_{96}$ for the iron meteorites analyzed in this study.

On the other hand, the $\varepsilon_{100}$ versus $\varepsilon_{96}$ diagram (Figure 5) shows a negative correlation between the relative abundance aberrations of $^{100}$Mo and $^{96}$Mo, normalized to $^{95}$Mo/$^{98}$Mo. The slope of $\varepsilon_{100}$ /$\varepsilon_{96}$ is about -0.42 ~ -0.47.

Qualitatively the negative relationship between ε values in Figure 5 are reasonable, because $^{96}$Mo comes only from the s-process, a slow process in massive or AGB stars, while $^{100}$Mo comes only from the r-process, which occurs in Type II supernova, and the normalization is based on r,s nuclides. Our observation endorses a separation of the s-process and the r-process. This is the first finding of this separation for a metallic element, though it has been observed in meteorites for the noble-gas elements, Kr and Xe.

The fact that the absolute value of the slope of $\varepsilon_{100}/\varepsilon_{96}$ is about 0.45 can be a clue to the contribution of r and s nuclides for the s,r isotopes of Mo and to the rate of homogenization between r- and s-process nuclides. Simply, the latter factor (a slower rate of homogenization of $^{96}$Mo than $^{100}$Mo) may be dominant over the former one.

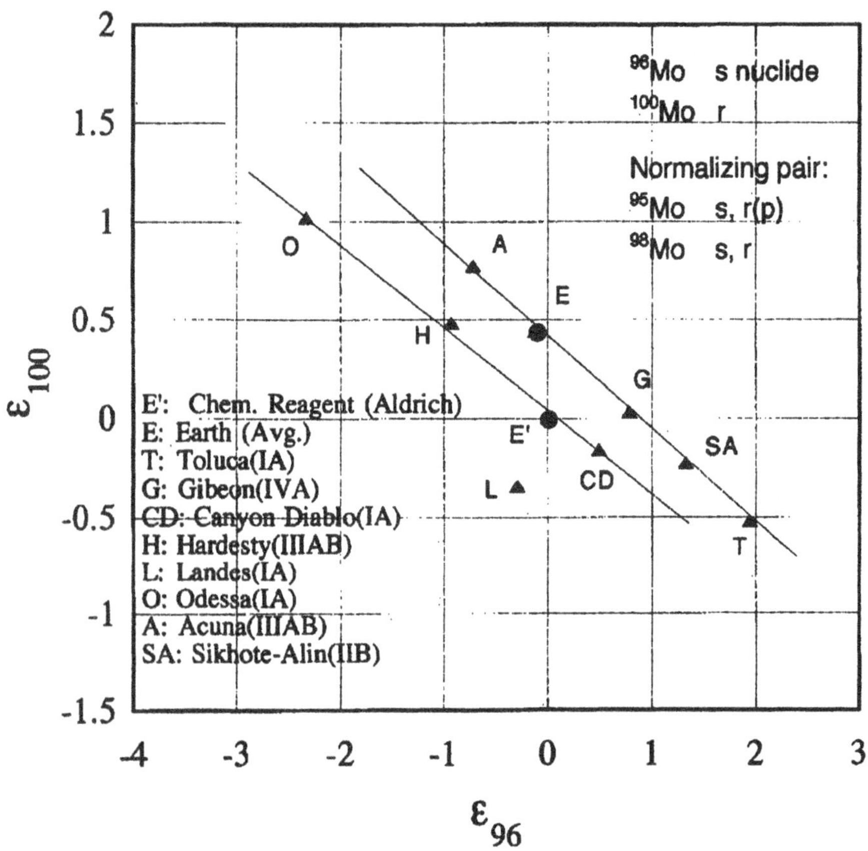


*Figure 5.* The relationship between the values of $\varepsilon_{100}$ and $\varepsilon_{96}$ for the iron meteorites analyzed in this study.

In Figure 6 is shown the relationship between the values of $\varepsilon_{92}$ versus $\varepsilon_{96}$. The points are rather scattered in this diagram, although one can recognize a general positive relation between the aberrations in the abundances of these two Mo isotopes.

It should be noted that $^{92}$Mo is a p-nuclide and that the neutron number of this nuclide is 50. It may merit some attention that the points (E and

E') for the earth are plotted in the central part of the diagram but are somewhat away from most of the points for iron meteorites. The p-process (or gamma-process) is an interesting problem in nucleosynthesis and has been discussed by Woosley and Howard (1978), Howard *et al.* (1991) and Hoffman *et al.* (1996). Meyer (1994) reviewed the r-, s- and p-processes in nucleosynthesis.

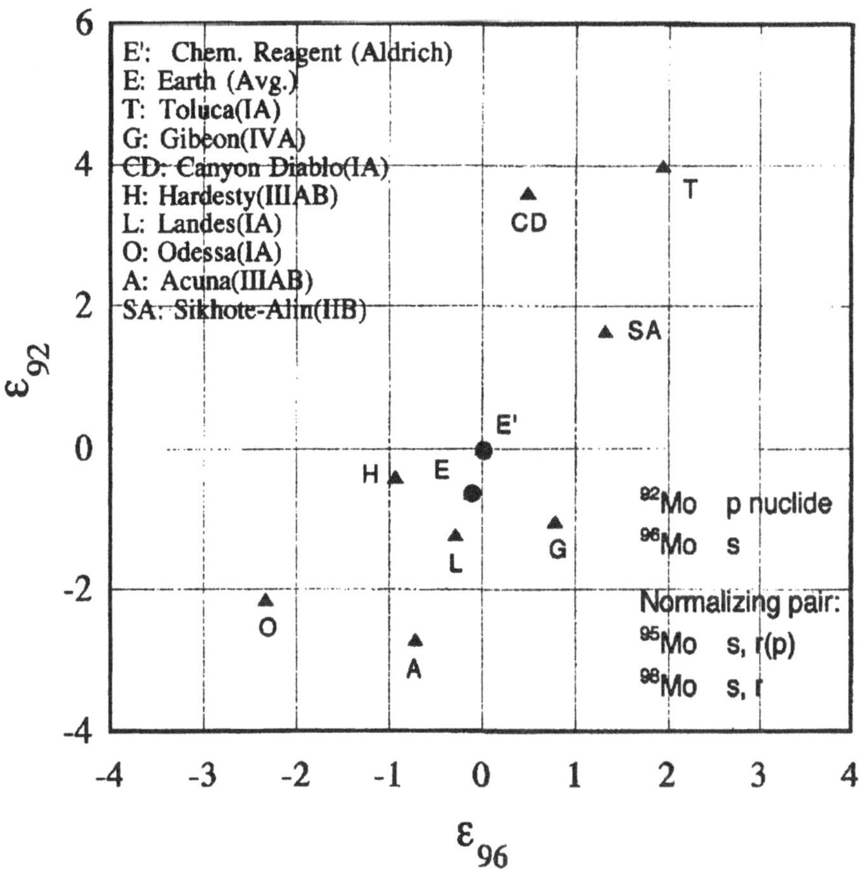


*Figure 6.* The relationship between the values of ε92 and ε96 for the iron meteorites analyzed in this study.

Since the genesis of $^{92}$Mo can be more or less related with that of $^{94}$Mo, values of $\varepsilon_{92}$ are plotted against $\varepsilon_{94}$ (Figure 7). Howard (1998) has suggested that "the best models would predict that $^{92}$Mo is made from a different process than $^{94}$Mo, although some $^{92}$Mo production (perhaps equal amounts) should be related to $^{94}$Mo." Taking Howard's suggestion into account, the lines with a variation rate of unity are drawn in Figure 7.

One can recognize two subgroups falling along the solid lines, except for T and CD.

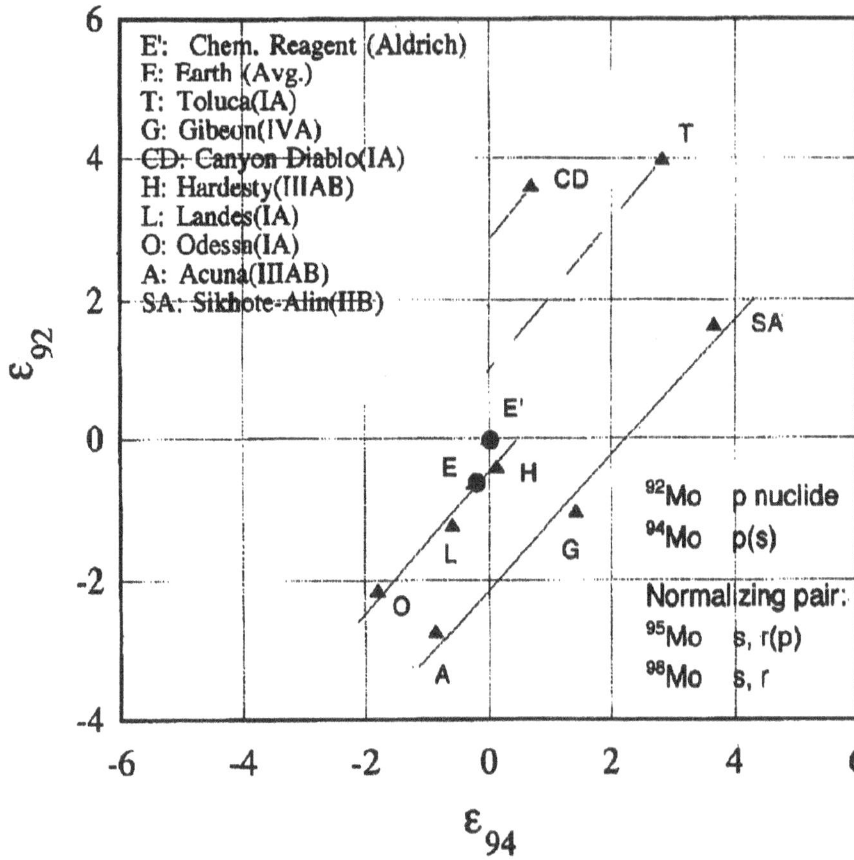


*Figure 7.* The relationship between the values of $\varepsilon_{92}$ and $\varepsilon_{94}$ for the iron meteorites analyzed in this study.

## 3.3 Potential bearings on groups of iron meteorites

Wasson (1967) and Wasson and Kimberlin (1967) classified iron meteorites into genetic groups (*cf.* Wasson, 1974). It intrigues us whether one can recognize any relationship between the iron meteorite groups and the isotopic aberrations for molybdenum isotopes. In Figure 8 ($\varepsilon_{92}$ versus $\varepsilon_{94}$ diagram), solid squares designate group IA, while open triangles designate groups other than IA. Three (SA, G and A) of four iron meteorites, except one (H), with open triangles fall close to the lowermost line.

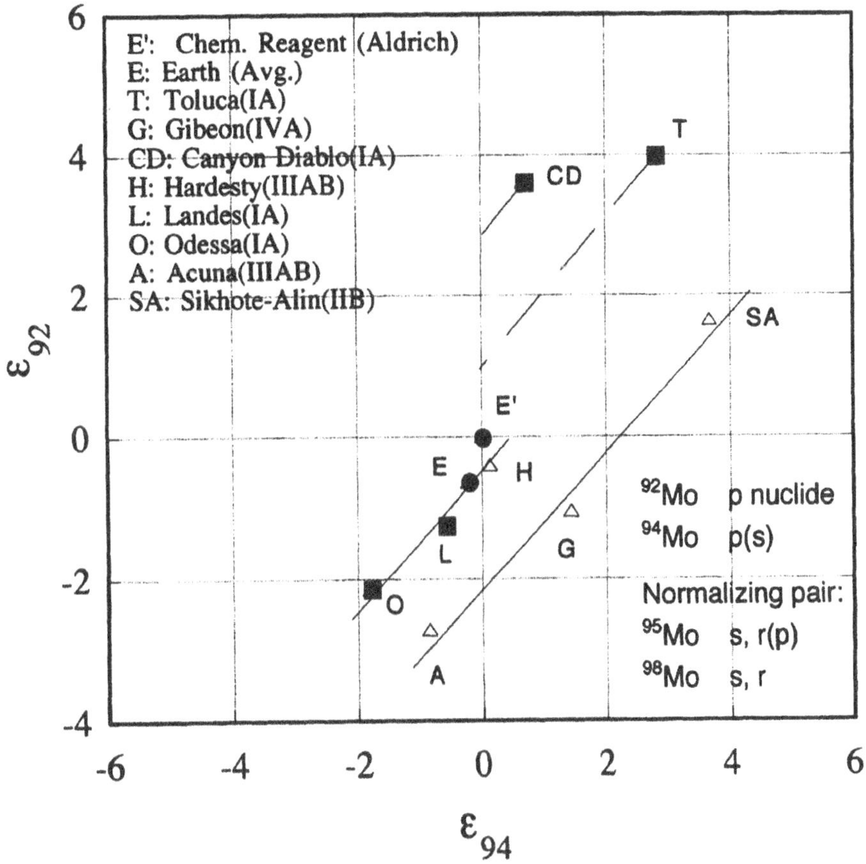


*Figure 8.* Classified relationship between the values of $\varepsilon_{92}$ and $\varepsilon_{94}$. The solid squares refer to the IA group and the open triangles to other groups.

The same symbols are used in Figure 9 ($\varepsilon_{100}$ versus $\varepsilon_{96}$ plot). It merits attention that also three points (SA, G and A) fall again on the same upper line, except H. The same grouping (alignment) in the isotopic anomaly diagram conceivably may indicate a similar origin. If such a viewpoint is followed, more than two different splitted alignments would suggest that the production sites in question are not singular. The features observed in Figure 8 may be considered to be consistent with Howard's (1998) suggestion. However, since the number of samples is limited, it may be premature to conclude decisively that there is a genetic relationship between the Mo isotopic composition and the grouping of iron meteorites. Hoffman *et al.* (1996) predicted that the r-process nuclei and part of some light p-process nuclei may be coproduced at probably similar sites. It is difficult to reconcile our observation with their prediction.

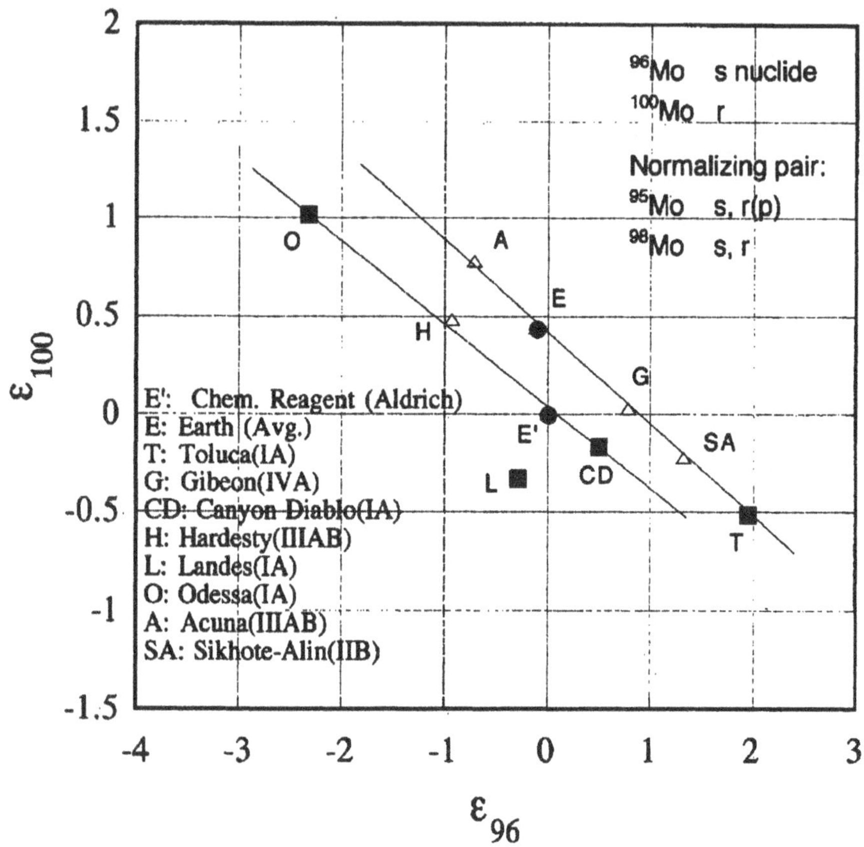


*Figure 9.* Classified relationship between the values of $\varepsilon_{100}$ and $\varepsilon_{96}$. The solid squares refer to the IA group and the open triangles to other groups, as in Figure 8.

## 3.4 Additional comments

Molybdenum is a chalcophile and/or siderophile element, while technetium is considered to be rather lithophile (or less chalcophile and/or siderophile than molybdenum). If so, one can expect to detect notable effects derived from extinct $^{97}$Tc in terrestrial molybdenite occurring in the earth's crust. However, we could not detect any sizable isotope anomaly for $^{97}$Mo in terrestrial molybdenites.

It deserves a comment that in nucleogenetic diagrams (Figures 4-7), meteorites A and T are always plotted in opposite sides relative to the zero point. But in Figure 3 only where radiogenic effects are plotted, A and T show the positive aberrations in common. These facts strengthen the argument that the excess $^{97}$Mo may be the decay product of $^{97}$Tc.

As shown in the above diagrams for isotope anomalies emerging after normalization based on $^{95}Mo/^{98}Mo$ ratio, the isotopic composition of Mo for the earth (Table 2) is almost an average of Mo in iron meteorites, although the isotopic composition of Mo in iron meteorites is variable and reflects nucleosynthetic effects. This may be significant for the origin and evolution of the solar system, as well as for isotopic mixing processes.

It is enigmatic whether the Mo isotope anomalies in iron meteorites observed here reflect the same effects as those found in pre-solar SiC grains of carbonaceous chondrites (Zinner *et al.*, 1987, 1991). Very likely, the isotope anomalies in question correspond to different stages and/or different sites of nucleosynthesis.

The investigation of the isotopic heterogeneity of Mo within a comparatively big mass or a great fall (*e.g.*, Sikhote-Alin) of an identical iron meteorite has been abandoned on the way, before gaining the confident results. Although this study is laborious work, it deserves a further attempt; heterogeneity in a fragment may be conceivable. In general, our measurements were carried out with precision of $\pm$ 0.1 ~ 0.2ε.

## ACKNOWLEDGEMENTS

We are much obliged to Dr. W. Michael Howard, Lawrence Livermore National University, CA, for lasting interest and helpful suggestions. Our thanks are extended to Professor Oliver K. Manuel, University of Missouri, for his kind arrangements.

## APPENDIX (EXPERIMENTAL)

**Chemical Treatment:**

The chemical reagents and waters to be used have been prepared with much caution to minimize contamination by impurities. Also the experimental vessels and tools have been carefully cleaned by the high-purity acids and waters.

*Mo in iron meteorites*

An accurately weighed, 3-7 gram quantity of iron meteorite is put into a 120 mL teflon vessel. Approximately 10 mL of 6 M HCl is added to the vessel for acid digestion to dissolve the sample. (The further detailed conditions and cautions required to complete the dissolution of the sample are omitted here. Generally, it takes more than 6 hours.) The solution obtained is mixed with 25 mL of 0.75 M bis(2-ethylhexyl) hydrogen-phosphate (HDEHP) in cyclohexane in a 50 mL separatory funnel. After shaking for 5 minutes, the aqueous phase is removed, and the organic phase is washed three times with 25 mL of 5 M $HClO_4$, 2 times with 25 mL of 10 M $HNO_3$. (Ru, partially partitioned into the organic phase at the initial solvent extraction stage, can be removed by washing the organic phase with $HClO_4$ or $HNO_3$.)

For back-extraction of Mo, 5 mL of 10 M $HNO_3$ -3% $H_2O_2$ is added to the organic phase. After shaking 5 minutes, the aqueous phase is removed to a teflon beaker. This back-extraction process is repeated again.

The aqueous phase is evaporated to dryness at low temperature. The residue is dissolved in 1 M HF-0.01 M HCl for further purification by anion exchange.

The resultant solution containing Mo is poured into a 15 mL teflon column of anion exchange resin (AG 1 X-8 100-200 mesh) chemically pretreated to remove the impurity trace elements and equilibrated with 0.01 M HCl-1.0 M HF. The Mo is separated from Fe, Ni, Co, Cu, Ru and Zr by elution of different acid solutions. (The elution is conducted by 1M HF-0.01 M HCl (for Fe, Ni, Co and Cu), 6M HCl (for Zr), 1M HCl or 7M $HNO_3$ (for Mo) and 14M $HNO_3$ (for Ru); further details of chemical conditions are omitted.) The Mo eluate is evaporated to dryness. The residue is dissolved by $NH_4OH$ and/or acids. However, since the solution still appears somewhat yellowish due to iron, the solution is poured into a small column (0.5 mL) of anion exchange resin, and the differential elution is repeated in the same way as for a bigger column. The colorless solution of Mo eluate in 2mL of 7M $HNO_3$ is condensed by evaporation and loaded on the side filament for mass spectrometry.

*Molybdenite*

About 0.3 to 0.5 g of homogenized molybdenite ($MoS_2$) powder (>150 mesh) is placed to a 120 mL teflon digestion vessel. Approximately 5 g of distilled water, 5 g of 14 M $HNO_3$ and 3 g of concentrated $H_2SO_4$ are added to the vessel. After sealing, the acid digestion is made according to a carefully designed program. Mo and Zr in perchloric or nitric solution is extracted into HDEHP-cyclohexane phase. By back-extraction using 3% $H_2O_2$ in 10 M $HNO_3$, Mo and Zr can be separated (See Figure 1). ($H_2O_2$ reduces Mo to lower valencies, facilitating the extraction into the aqueous phase.) The aqueous phase is transferred to a teflon beaker and evaporated to dryness at 5% power in the microwave oven (Model MDS-81D, CEM Corp.). Then the residue is dissolved in 3% $NH_4OH$ and evaporated to dryness at 5% power in the oven again. The salt (ammonium paramolybdate) is taken up in water and diluted to ~10 μg/μL, a small droplet of which is loaded on the filament and dried for mass spectrometric measurement. This procedure was used for 17 molybdenites analyzed in this study.

*Standard*

Molybdenum oxide ($MoO_3$, 99.999%) manufactured by Aldrich Co., USA, is used repeatedly as a standard, in parallel with the measurements for terrestrial and meteoritic samples. The Mo solution is purified by the solvent extraction as described above.

**Mass spectrometry:**

Employing a thermal ionization mass spectrometer, VG Sector 54-30, we examined the static multi-collector system, the dynamic multi-collector system and a single collector system (peak jumping mode) for accurate measurement of isotopic abundance ratio of molybdenum. Obviously, the single collector system, based on the peak jump mode, has disadvantages in that a set of ion intensity measurements of seven isotopes of Mo (92, 94, 95, 96, 97, 98 and 100) needs a long measuring time and may be subject to fluctuations of the ion beam intensity during the measurement. Nevertheless, it has a merit in that it is simply and completely free from the effect of the sensitivity bias among detectors which should be corrected when using the multi-collector system. According to our examination, the single collector measurement proved to be dependable and sufficiently rewarding after we succeeded in developing a stable, intense ion beam of $Mo^+$ that permits the longer measurement time.

*Procedures for mass spectrometry*

Molybdenum is loaded as ammonium paramolybdate (*cf.* Chemical Treatment) together with 2 μL of a saturated aqueous solution of boric acid and 1 μL of 1 M $HNO_3$ on an outgassed side filement of 5-pass zone refined Re. Normally, 20 μg of molybdenum with 1 μg B is sufficient to yield fairly stable $^{96}Mo$ ion intensity of $2 \times 10^{-12}$ A at about 3.6 amps of central filament current (~ 1750 °C) for about 5 hours. A special triple filament technique has been used for the molybdenum isotopic analysis, where no electric current is applied to the side filament loaded with the sample. It is believed that Mo on the side filament is volatilized by radiation heat from the central filament and this effect could improve much the stability of beam of Mo ion which is produced by vaporization from side filament of relatively lower temperatures and ionization at central filament of higher temperatures. Further, the possibility of production of cluster ions of alkali atoms or alkali halides can be eliminated. For obtaining the intense and stable $Mo^+$ ion current, one should follow carefully the way to increase the central filament current, which has been devised by us.

Special caution is paid to the emission of $Zr^+$ (and $Ru^+$) ions from filament material. Formation of $K_2F^+$ ion is also checked.

In our measurements of relative isotopic abundance, $^{98}Mo$ is taken as the denominator. Further detailed descriptions of techniques for mass spectrometry and chemical purification are given by Qi-Lu (1991).

# REFERENCES


Allen, B.J., Gibbons, J.H. and Macklin, R.L.: 1971, "Nucleosynthesis and neutron-capture cross sections", *Adv. Nucl. Phys.* **4**, 205-259.

Hoffman, R.D., Woosley, S.E., Fuller, G.M. and Meyer, B.S.: 1996, "Production of the light p-process nuclei in neutrino-driven winds", *Ap. J.* **460**, 478-488.

Howard, W.M.: 1998, personal communication.

Howard, W.M., Meyer, B.S. and Woosley, S.E.: 1991, "A new site for the astrophysical gamma-process", *Ap. J.* **373**, L5-L8.

Huang, M., Liu, Y.-Z and Masuda, A.: 1996, "Accurate measurement of ruthenium isotopes by negative thermal ionization mass spectrometry", *Anal. Chem.* **68**, 841-844.

Meyer, B.S.: 1994, "The r-, s-, and p-processes in nucleosynthesis", *Ann. Rev. Astron. Astrophys.* **32**, 153-190.

Poths, H., Schmitt-Strecker, S. and Begemann, F.: 1987, "On the isotopic composition of ruthenium in the Allende and Leoville carbonaceous chondrites", *Geochim. Cosmochim. Acta* **51**, 1143-1149.

Qi-Lu: 1991, "High accuracy measurement of isotopic abundance ratios of molybdenum in iron meteorites and molybdenites and the fundamental studies for chemical separation and mass spectrometry", *Doctoral Thesis*, The University of Tokyo, Japan.

Sahoo, S.K. and Masuda, A.: 1997, "Precise measurement of zirconium isotopes by thermal ionization mass spectrometry", *Chem. Geol.* **141**, 117-126.

Wasson, J.T.: 1967, "The chemical classification of iron meteorites: I. A study of iron meteorites with low concentrations of gallium and germanium", *Geochim. Cosmochim. Acta* **31**, 161-180.

Wasson, J.T.: 1974, *Meteorites: Classification and Properties*, Springer-Verlag, New York, NY 360 pp.

Wasson, J.T. and Kimberlin, J.: 1967, "The chemical classification of iron meteorites - II. Irons and pallasites with germanium concentrations between 8 and 100 ppm", *Geochim. Cosmochim. Acta* **31**, 2065-2093.

Woosley, S.E. and Howard, W.M.: 1978, "The p-process in supernovae", *Ap. J. Suppl.* **36**, 285-304.

Zinner, E., Tang, M. and Anders, E.: 1987, "Large isotopic anomalies of Si, C, N and noble gases in interstellar silicon carbide from the Murray meteorite", *Nature* **330**, 730-732.

Zinner, E., Amari, S. and Lewis, R.S.: 1991, "s-Process Ba, Nd, and Sm in presolar SiC from the Murchison meteorite", *Ap. J. Lett.* **382**, L47-L50.

# Iron Meteorites and Paradigm Shifts

E. Calvin Alexander, Jr.
Geology and Geophysics Department, University of Minnesota, Minneapolis, MN 55455
alexa001@ tc.umn.edu

Abstract: The assumption that iron meteorites were formed by planetary differentiation of asteroid-scaled parent bodies is deeply imbedded in the standard solar system formation model. This assumption is directly falsified by the trace element distributions in one significant group of iron meteorites and is not supported by chronologic studies of primordial or extinct radioisotopes. There is growing evidence that iron meteorites may be individually and, as a class, isotopically heterogeneous. The paradigm of iron meteorite formation is undergoing a fundamental shift.

## 1. INTRODUCTION

Most research on origin of the solar system assumes that iron meteorites are produced by planetary-scale differentiation of asteroid-scale parent bodies. The standard model of iron meteorite formation postulates a sequence of events:

- Condensation of solids from an isotopically homogenized solar nebula. Ca, Al-rich inclusions (CAIs) are one of the early phases. Iron condenses as a separate phase at about the same temperature as the major silicate phases.
- Separation of the condensed solids from the remaining gas phase.
- Aggregation of the condensed solids into progressively larger objects up to asteroidal-scale parent bodies. Chondritic meteorites are the undifferentiated products of this stage.
- Differentiation of at least some of those parent bodies to form iron cores and silicate mantles.

*Origin of Elements in the Solar System: Implications of Post-1957 Observations.*
Edited by O. Manuel, Kluwer Academic/Plenum Publishers, 2000. 

- Slow cooling to allow the crystallization of the Widmanstätten pattern.
- Continued evolution of the asteroidal parent bodies. Some accrete to form planets. Many are ejected from the inner solar system. Some are disrupted by collision to expose their iron cores, stored in stable orbits for 4.5 billion years (Gyr), and then perturbed into earth-crossing orbits.

In this model iron meteorites should be younger than the undifferentiated meteorites from which they formed.

The standard model may be presented as a simple assertion, for example: "*Iron meteorites preserve a record of planetary differentiation, metal-silicate segregation, and core formation*" (Stewart *et al.*, 1996, p. 1); or it may be implicit in the research. The standard model is deeply embedded in current planetary science. This raises two related questions: *1) Are iron meteorites older or younger than CAIs, chondrites or achondrites, and 2) Are iron meteorites produced by the planetary-scale differentiation of asteroid-scale parent bodies?* The distribution of siderophile trace elements between and within individual meteorites provides important clues to origin of the irons. Data from long-lived radioisotopes or from the decay of shorter-lived, now-extinct radioisotopes can in principle falsify or support the relative age sequence inherent in the standard model.

## 2. SIDEROPHILE ELEMENT DISTRIBUTIONS

The distribution of siderophile elements in iron meteorites forms the basis of their classification (Choi *et al.*, 1995; Wasson *et al.*, 1998). Wasson and co-workers have identified three groups of iron meteorites: a magmatic group, which includes the bulk of classified iron meteorites, a nonmagmatic group (chemical groups IAB, IIE, and IIICD), and "unclassified" irons (about 15% of the total) which do not fit into the established chemical groups. Magmatic iron meteorites are interpreted to be pieces of the cores of asteroidal-scaled parent bodies that have melted, differentiated, fractionally crystallized and cooled slowly. Very little work has been done to see if other models can explain the observed geochemical trends as well or better than a fractionated asteroid core model. The composition trends in nonmagmatic irons do not fit those predicted by fractional crystallization. The nonmagmatic IAB irons contain ubiquitous trapped melt and chondritic (or subchondritic) silicate inclusions. Choi *et al.'s* (1995, p. 593) preferred model for nonmagmatic irons is "*individual IAB irons are interpreted as melt pools produced by impacts into a chondritic megaregolith*", but they also discuss other possible models.

## 3. PRIMORDIAL RADIOISOTOPES

The five long-lived primordial parent isotopes of the standard isotope chronometers, U,Th/Pb, Rb/Sr, K/Ar, and Sm/Nd are all (apparently by chance) lithophile elements and have negligibly small concentrations in iron meteorites. One primordial long-lived radioisotope, $^{187}$Re ($t_{1/2}$ ~ 50 Gyr), is a siderophile and is present in useful quantities in iron meteorites. Re/Os isochrons have been constructed for groups of iron meteorites by several research groups (Shen *et al.*, 1996; Smoliar *et al.*, 1996; Ding *et al.*, 1997; Horan *et al.*, 1998). Under the standard assumptions of geochronology, the slopes of those isochrons correspond to ages around 4.5 Gyr.

There are two fundamental limitations on the interpretation of these data. First, the physically measured half-life of $^{187}$Re is currently different from the geochronologically adjusted value by about 17% (Renne *et al.*, 1998). This discrepancy introduces a formal uncertainty into a Re/Os age at 4.5 Gyr of ± 800 million years (Myr). Until the half-life uncertainty is resolved the Re/Os ages cannot be meaningfully compared with those obtained by other chronometers.

The half-life uncertainty does not enter into the comparison of the two Re/Os ages. Smoliar *et al.* (1996) report an age difference between the IIIA and IVA irons of 102 ± 25 Myr. The same authors report an apparent spread of about 500 Myr in iron meteorite ages based on the evolution of the initial $^{187}$Os/$^{188}$Os ratios.

Such comparison are constrained, however, by the second fundamental limitation on the data. No evidence has been presented that iron meteorites, individually or collectively, have ever been parts of an isotopically homogeneous object. Ding *et al.* (1997) have presented evidence that individual iron meteorites may never have been isotopically homogeneous. Attempts to construct Re/Os ages for chondrites have not yielded well defined isochrons (Walker *et al.*, 1999). The Re and Os data for chondrites scatter about the iron meteorite isochrons. Such results are consistent with chondrites' well documented isotopic heterogeneity at the grain scale (Choi *et al.*, 1998; Dougherty *et al.*, 1999).

Although the parent isotopes of the standard isotope chronometers are not present in usable quantities in irons, they do occur in the inclusions that are commonly found in the nonmagmatic irons. Some of these inclusions appear to be very primitive. For example, Stewart *et al.* (1996) report that a silicate inclusion in the Caddo IAB iron meteorite contains the lowest $^{142}$Nd/$^{143}$Nd age measured to that point in solar system material. The fact that the usable inclusions are in the nonmagmatic iron meteorites, however, renders these measurements moot on the question of irons as the products of asteroidal-scale planetary differentiation.

## 4. EXTINCT RADIOISOTOPES

Measurements of the stable decay products of now extinct radioisotopes in meteorites have offered the hope of a more precise chronology of solar system formation since John Reynolds (1960) discovered $^{129}$Xe excesses from the decay of extinct $^{129}$I. The study of isotope anomalies in meteoritic materials has grown explosively in the past three decades, and a review of that vast literature is beyond the scope of this note. Suffice to say that as the analytical improvements have allowed high precision isotope measurements on individual meteorite mineral grains, nucleosynthetic and extinct isotope anomalies have proven to be ubiquitous in meteorites. It is clear that primitive and ordinary chondrites contain individual, separable mineral grains that predate the formation of the solar system (See for example, Choi *et al.*, 1998). Such grains contain evidence of a range of stellar nucleosynthetic environments but provide no information about the chronology of solar system formation.

In the list of isotopes with half-lives between 0.7 Myr and 700 Myr, there are several parent-daughter pairs in which one or both of the isotopes have siderophile affinities. The best documented extinct isotope in iron meteorites is $^{107}$Pd which decays to $^{107}$Ag with a $t_{1/2}$ of 6.5 Myr. Chen and Wasserburg (1990, p. 1729) demonstrated that excess $^{107}$Ag from the decay of $^{107}$Pd was present "*in almost all iron meteorites studied with a ratio of $^{108}$Pd/$^{109}$Ag greater than ~ 400*". They and many other workers argue that the correlation of excess $^{107}$Ag with $^{108}$Pd demonstrates that the $^{107}$Pd parent was present when the iron meteorites crystallized. The range of inferred $^{107}$Pd/$^{108}$Pd ratios corresponds to an age difference of ~ 40 Myr "*if this variation is interpreted to represent an age difference*" (Chen and Wasserburg, 1990, p. 1729). However, Ding *et al.* (1997, p. 414) presented initial evidence for the preservation of "*presolar grains at sub-μm scales in iron meteorites*". If Ding *et al.'s* (1997) results are verified, the chronologic implications of the $^{107}$Ag anomalies will disappear.

The extinct isotope $^{182}$Hf, which decays to $^{182}$W with a $t_{1/2}$ of 9 Myr, is an interesting case (Harper and Jacobsen, 1996; Podosek, 1999). Hf is strongly lithophilic whereas W is moderately siderophilic. In a core forming event, Hf should fractionate into the silicate phase, while W should fractionate into the iron phase. If this separation occurs before $^{182}$Hf has decayed, W in the metal phase will be depleted in $^{182}$W, but that in the silicate phase will be enriched. Terrestrial rocks and undifferentiated meteorites appear to have the same W composition. Iron meteorites are depleted by 2 to 4 parts in $10^4$, but lunar samples are enriched by up to 30 parts in $10^4$. Nevertheless, as with $^{107}$Ag excesses, models in which the anomalies are inherited from presolar grains, which contain the excesses and depletions preformed, need to be investigated.

The recent report by Dougherty *et al.* (1999) in which the apparently isotopically normal Cr in the ordinary chondrite Semarkona is composed of chemically separable components, one of which is enriched in $^{53}$Cr while the other is depleted in $^{53}$Cr, is a clear wakeup call. The measurement of a "normal" isotopic composition in a macroscopic sample of a meteorite is not evidence that the element in question has a homogenous isotopic composition at some arbitrarily smaller scale.

## 5. DISCUSSION

*Are iron meteorites older or younger than CAIs, chondrites or achondrites?* We do not know. The uncertainty in our knowledge of the half-lives of the long-lived isotopes limits any meaningful comparison of the results of different chronometers, yet the geochemistry of these objects necessitates such comparisons. More fundamentally, an isotopic "age" determination must assume an isotopically homogeneous initial state. It is clear that CAIs and chondrites have never been isotopically homogeneous. There is growing evidence that some, perhaps all, iron meteorites are also isotopically heterogeneous. Isotopic techniques have not answered, and may never be able to answer. this question.

*Are iron meteorites produced by planetary-scale differentiation of asteroid-scale parent bodies?* In some cases no. The trace element distributions in the nonmagmatic iron meteorites do not fit the trends predicted by fractional crystallization of the iron cores of asteroids. For this group of iron meteorite, the standard model has been falsified. The clear scientific consensus is that nonmagmatic iron meteorites are not the product of asteroidal-scale planetary differentiation.

For the magmatic iron meteorites, the scientific jury is still out. The trace element distributions in these meteorites can be explained by the standard model, but it remains to be seen if other models do a better job of explaining the observations. If the initial indications that magmatic iron meteorites are internally isotopically heterogeneous is verified, planetary differentiation models will be untenable.

The iron meteorite formation paradigm is shifting in response to new data and new interpretations of those data.

## ACKNOWLEDGMENTS

I thank Oliver Manuel and the organizers of the OESS Symposium for the invitation to present this brief review.

## REFERENCES


Chen, J.H. and Wasserburg, G.J.: 1990, "The isotopic composition of Ag in meteorites and the presence of $^{107}$Pd in protoplanets", *Geochim. Cosmochim. Acta* **54**, 1729-1743.

Choi, Byeon-Gak, Ouyang, Xinwei and Wasson, John T.: 1995, "Classification and origin of IAB and IIICD iron meteorites", *Geochim. Cosmochim. Acta* **59**, 593-612.

Choi, Byeon-Gak, Huss, Gary R., Wasserburg, G.J. and Gallino, Roberto: 1998, "Presolar corundum and spinel in ordinary chondrites: Origins from AGB stars and supernova", *Science* **282**, 1284-1289.

Ding, G.-J., Kilius, L.R., Wilson, G.C., Zhao, X.-L. and Rucklidge, J.C.: 1997, "Evidence for anomalous $^{107}$Ag and $^{109}$Ag compositions in iron meteorites", *Nuclear Instruments and Methods in Physics Research B* **123**, 414-423.

Dougherty, J.R., Brannon, J.C., Nichols, Jr., R.H. and Podosek, F.A.: 1999, "Thoroughly anomalous Cr in the ordinary chondrite Semarkona", in *Lunar Planet. Sci. XXX*, abs. #1451, Lunar Planetary Institute, Houston, TX (CDROM).

Harper, Charles L. and Jacobsen, Stein B.: 1996, "Evidence for $^{182}$Hf in the early Solar System and constraints on the timescale of terrestrial accretion and core formation", *Geochim. Cosmochim. Acta* **60**, 1131-1153.

Horan, M.F., Smoliar, M.I. and Walker, R.J.: 1998, "$^{182}$W and $^{187}$Re-$^{187}$Os systematics of iron meteorites: Chronology for melting , differentiation, and crystallization in asteroids", *Geochim. Cosmochim. Acta* **62**, 545-554.

Podosek, Frank A.: 1999, "A couple of uncertain age", *Science* **283**, 1863-1864.

Renne, Paul R., Karner, Daniel B. and Ludwig, Kenneth R.: 1998, "Absolute ages aren't exactly", *Science* **282**, 1840-1841.

Reynolds, J.H.: 1960, "Determination of the age of the elements", *Phys. Rev. Letters* **4**, 8-10 .

Shen, J.J., Papanastassiou, D. A. and Wasserburg, G. J.: 1996, "Precise Re-Os determinations and systematics of iron meteorites", *Geochim. Cosmochim. Acta* **60**, 2887-2900.

Smoliar, Michael I., Walker, Richard J. and Morgan, John W.: 1996, "Re-Os ages of group IIA, IIIA, IVA, and IVB iron meteorites", *Science* **207**, 1099-1102.

Stewart, Brian, Papanastassiou, D.A. and Wasserburg, G.J.: 1996, "Sm-Nd systematics of a silicate inclusion in the Caddo IAB iron meteorite", *Earth Planet. Sci. Lett.* **143**, 1-12.

Walker, R.J., Becker, H. and Morgan, J.W.: 1999, "Comparative Re-Os isotope systematics of chondrites: Implications regarding early solar system processes", in *Lunar Planet. Sci. XXX*, abs. #1208, Lunar Planetary Institute, Houston, TX (CDROM).

Wasson, John T., Choi, Byeon-Gak, Jerde, Eric A. and Ulff-Møller, Finn: 1998, "Chemical classification of iron meteorites: XII. New member of the magmatic groups", *Geochim. Cosmochim. Acta* **62**, 715-724.

# Chronology of Early Solar System Events: Dating with Short-lived Nuclides

Jitendra Nath Goswami
*Physical Research Laboratory, Ahmedabad 380009, India*
*goswami@prl.ernet.in*

Abstract: Isotopic records in meteorites provide evidence for the presence of several short-lived nuclides in the early solar system with half-lives varying from 100,000 years to 82 million years. These nuclides serve as excellent chronometers for dating early solar system events. A stellar origin for these now-extinct nuclides is generally favored, even though there are suggestions that some of them could be products of energetic particle interactions in the early solar system. A stellar origin for the short-lived nuclides, and $^{41}Ca$ in particular, constrains the time scale for the collapse of the protosolar molecular cloud leading to the formation of the sun and some of the first solar system solids to less than a million years. This short time scale argues for a triggered collapse of the protosolar cloud. Records of short-lived nuclides in meteoritic objects that have formed in the solar nebula suggest nebular processes to be operative for a time scale of a few million years. Discovery of fossil records of the short-lived nuclide $^{26}Al$ in a differentiated meteorite confirms this radionuclide as the heat source for early melting of planetesimals in the solar system. This observation and results obtained from studies of the short-lived nuclides $^{53}Mn$ and $^{182}Hf$ suggest that accretion, heating, melting and differentiation of planetesimals took place within the first five million years of the origin of the solar system. Fossil records of $^{107}Pd$ and $^{182}Hf$ in iron and stony-iron meteorites show that silicate-metal fractionation and the formation of metallic core in planetesimals were also complete within the first ten millions years of the formation of the solar system.

## 1. INTRODUCTION

Our present understanding of the origin and the early evolution of the solar system is based on what we know about star formation processes and also a study of the solar system objects. About 4.6 Gyr back some external perturbation triggered the collapse of a solar mass molecular cloud fragment within a molecular cloud complex. This led to infall of material towards the center of this collapsing cloud leading to the formation of a massive object at its center. This object became self-luminous due to the continuous input of gravitational potential energy of the infalling material and the proto-sun was born. A rotating gas and dust disk, referred to as the solar nebula, formed around the new-born sun. Most of the presolar grains initially present in the nebula were perhaps destroyed in the high-energy environment around the active proto-sun, and as the nebula cooled condensation of some of the first solar system grains took place. These grains must be highly refractory in composition to survive in the ambient high-temperature environment. Other less refractory grains started forming as the temperature in the nebula went down with time. The newly formed grains as well as any surviving presolar grains will slowly settle down to the mid-plane of the nebula under the influence of the gravitational attraction of the sun. These grains can grow in size considerably during this settling process via collisional sticking and aggregation with other grains that they will encounter as they move towards the mid-plane of the nebula. In the denser environment within the central plane of the nebula the growth process is accelerated and formation of meter-sized objects and finally ten to hundred kilometer-sized objects took place perhaps within a time scale of less than a million years. Further growth of these large objects, termed as planetesimals, was governed by collisional aggregation. If any one of the planetesimals acquired sufficient mass to gravitationally perturb and deflect other smaller objects towards it, a run away growth occurred leading to the formation of planetary-sized objects.

The above scenario for the origin and early evolution of the solar system is mainly based on analytical studies of molecular cloud collapse and on formation and growth of objects in the solar nebula. However, recent advances in observational astronomy and in laboratory studies of solar system objects now allow us to further refine this scenario by adding additional details. Infra-red observations made in the eighties and recent observations made with the Hubble Space Telescope provide spectacular evidence for the formation of a rotating gas and dust disk around new born stars. Even though the formation and the growth of grains to planetesimals and planets remain primarily a domain of analytical investigations, results obtained from studies of meteorites provide some very important time constraints for the various events taking place during this early evolutionary phase of the solar system. This has become feasible following the identification of refractory grains in primitive meteorites, whose compositions match those predicted for the first grains to form in the solar nebula, and the subsequent discovery of fossil records of several now-extinct, short-lived nuclides in these grains. Studies of these records allow us to address questions like how long it could have taken for the collapse of the protosolar cloud to form the sun and some of the first solar system solids. Fossil records of now-extinct nuclides found in different types of meteorites also allow us to deduce the time scales of nebular processes and the time taken for the formation of planetesimals and their subsequent differentiation. This paper will address these aspects of meteorite research, as well as the question of the plausible source(s) of the now-extinct short-lived nuclides present in the early solar system, and consider the implications for the formation and early evolution of the solar system.

## 2. SHORT-LIVED NUCLIDES: THE METEORITIC RECORDS

Isotopic studies of meteorites have provided evidence for the presence of several short-lived nuclides in the early solar system. The half-lives of these nuclides vary from ~ 0.1 Myr ($^{41}$Ca) to ~ 82 Myr ($^{244}$Pu). These short-lived nuclides may be considered as "now-extinct" for all practical purposes, and their presence in the early solar system is inferred from an observed excess of their daughter nuclide abundance in meteorite samples. The very first such observation was made in 1960 with the detection of excess abundance of $^{129}$Xe, a decay product of the short-lived nuclide $^{129}$I (half-life = 15.7 Myr), in samples of the Richardton meteorite (Reynolds, 1960; Jeffreys and Reynolds, 1961). This was soon followed by the discovery of "excess xenon" from the fission of $^{244}$Pu in meteorites (Rowe and Kuroda, 1965). Although the potential use of these nuclides as chronometers of early solar system processes was emphasized, their half-lives are rather long to delineate processes taking place within time scales of a few million years. The real breakthrough came in 1976 when excess $^{26}$Mg from the decay of the short-lived nuclide $^{26}$Al (half-life = 0.74 Myr) was identified unambiguously in some rare refractory phases isolated from the primitive carbonaceous chondrite, Allende (Lee *et al.*, 1976). As already mentioned, the mineralogy and chemistry of these refractory phases match those expected for the first solids to form in the solar nebula (See, *e.g.*, Grossman, 1980). Since then, the presence of more than half a dozen short-lived nuclides in the early solar system have been inferred from isotopic records found in meteorite samples. A list of these nuclides is given in Table 1. In the case of a few nuclides, the experimental data only provide a hint for their presence in the early solar system. Further data from more refined experiments will be needed to confirm these observations.

*Table 1.* Short-lived nuclides in the early solar system

| Nuclide | Half-life* (Myr) | Daughter nuclide | Initial abundance** | Ref. | Stellar production Sites*** |
|---|---|---|---|---|---|
| $^{41}$Ca | 0.1 | $^{41}$K | $1.5 \times 10^{-8}$ ($^{41}$Ca/$^{40}$Ca) | [1] | SN, AGB, W-R |
| $^{26}$Al | 0.74 | $^{26}$Mg | $5 \times 10^{-5}$ ($^{26}$Al/$^{27}$Al) | [2] | SN, N, AGB, W-R |
| $^{60}$Fe | 1.5 | $^{60}$Ni | $4 \times 10^{-9}$ ($^{60}$Fe/$^{56}$Fe) | [3] | SN, AGB |
| | | | $2 \times 10^{-9\,\$}$ | | |
| $^{53}$Mn | 3.74 | $^{53}$Cr | $4 \times 10^{-5}$ ($^{53}$Mn/$^{55}$Mn) | [4] | SN |
| | | | $4.7 \times 10^{-6}$, $\sim 10^{-5\,\$}$ | [5] | |
| $^{107}$Pd | 6.5 | $^{107}$Ag | $2 \times 10^{-5}$ ($^{107}$Pd/$^{108}$Pd) | [6] | SN, AGB, W-R |
| | | | $4.5 \times 10^{-5\,\$}$ | | |
| $^{182}$Hf | 9 | $^{182}$W | $2 \times 10^{-4\,\$}$ ($^{182}$Hf/$^{180}$Hf) | [7] | SN |
| $^{129}$I | 15.7 | $^{129}$Xe | $10^{-4}$ ($^{129}$I/$^{127}$I) | [8] | SN |
| $^{244}$Pu | 82 | α, SF ● | $7 \times 10^{-3}$ ($^{244}$Pu/$^{238}$U) | [9] | SN |
| $^{99}$Tc # | 0.21 | $^{99}$Ru | $\sim 10^{-4}$ ($^{99}$Tc/$^{99}$Ru) | [10] | AGB |
| $^{36}$Cl # | 0.3 | $^{36}$Ar | $1.4 \times 10^{-6}$ ($^{36}$Cl/$^{35}$Cl) | [11] | SN, AGB, W-R |
| $^{205}$Pb # | 15.3 | $^{205}$Tl | $< 3 \times 10^{-4}$ ($^{205}$Pb/$^{204}$Pb) | [12] | AGB, W-R |
| $^{92}$Nb # | 34.7 | $^{92}$Zr | $2 \times 10^{-5}$ ($^{92}$Nb/$^{93}$Nb) | [13] | SN |

* Values taken from Table of Isotopes (1996 Edition).

** At the time of formation of Ca, Al-rich inclusions (CAIs). See references for precise values and associated errors.

*** SN = supernova, AGB = asymptotic giant branch star, N = nova, W-R = Wolf-Rayet star.

$ Extrapolated value to the time of CAI formation.

● alpha particle (α) and spontaneous fission (SF) products.

# Suggestive evidence only, needs confirmation.

[1] Srinivasan *et al.* (1994,1996); [2] Lee *et al.* (1976); see also MacPherson *et al.* (1995); [3] Shukolyukov and Lugmair (1993 a,b) and Lugmair *et al.* (1995); [4] Birck and Allègre (1985); [5] Lugmair and Shukolyukov (1998); [6] Kelly and Wasserburg (1978); see also Chen and Wasserburg (1996); [7] Lee and Halliday (1995, 1996); Harper and Jacobsen (1996); [8] Jeffery and Reynolds (1961); see also Hohenberg *et al.* (1998); [9] Rowe and Kuroda (1965); see also Hudson *et al.* (1988); [10] Yin *et al.* (1992); [11] Murty *et al.* (1997); [12] Chen and Wasserburg (1987); [13] Harper (1996).

It is important to establish two features to confirm the presence of a now-extinct, short-lived nuclide at the time of formation of specific meteorite samples (*e.g.*, the refractory phases). First, a demonstrable excess in the abundance of the daughter nuclide and second a proof that the excess is indeed due to the decay of the short-lived nuclide. A correlation of the excess in the daughter nuclide abundance (*e.g.*, $^{26}$Mg in the case of $^{26}$Al and $^{41}$K in the case of $^{41}$Ca) with the abundance of the parent element (*e.g.*, aluminum and calcium in these cases) constitutes this proof. Such a correlation is clearly evident in Figure 1, where we show the data for Ca-K isotopic systematics in refractory samples from several meteorites obtained from ion microprobe studies (Srinivasan *et al.*, 1994, 1996; Sahijpal *et al.*, 1998, 1999). The measured $^{41}$K/$^{39}$K ratios in samples with high Ca/K are distinctly above the normal value of 0.072, and the excess correlates with the $^{40}$Ca/$^{39}$K ratio, providing evidence that the excess $^{41}K$ is related to Ca and resulted from *in-situ* decay of $^{41}Ca$ in the analyzed samples. One can write a simple linear relation for this correlation in terms of the measured parameters as:

$$[^{41}K/^{39}K]_{meas.} = [^{41}K/^{39}K]_{init.} + [^{41}Ca/^{39}K]$$

$$= [^{41}K/^{39}K]_{init.} + [^{41}Ca/^{40}Ca]_{init.} \times [^{40}Ca/^{39}K]_{meas.}$$

Thus, in Figure 1, the slope of the correlation line gives the initial value of $^{41}Ca/^{40}Ca$, and the intercept the initial value of $^{41}K/^{39}K$ at the time of formation of the analyzed refractory samples from the different meteorites. Note also that the measured values for terrestrial analog samples yield the normal $^{41}K/^{39}K$ ratio, as expected.

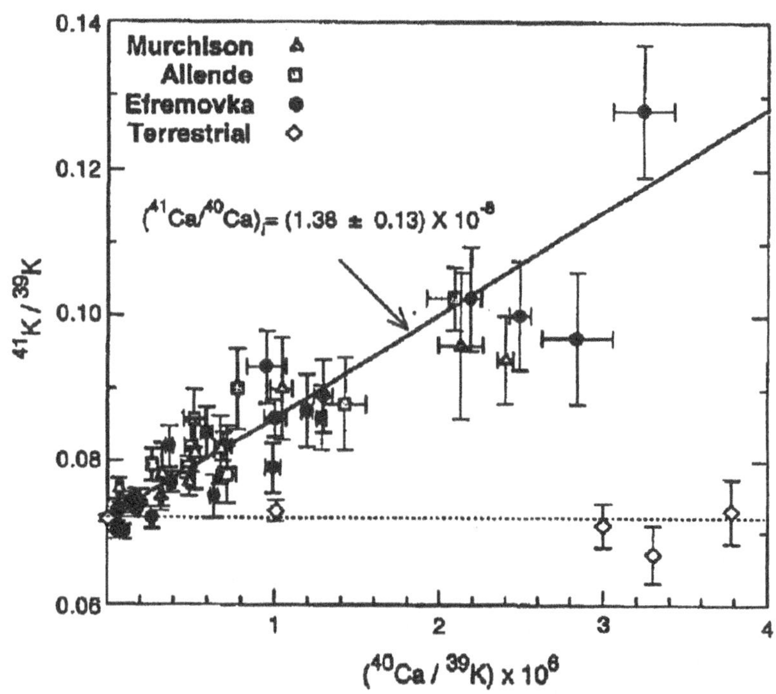


*Figure 1.* The potassium isotopic ratio, $^{41}K/^{39}K$, in refractory oxide and silicate phases from three primitive meteorites, Murchison, Allende and Efremovka, plotted as a function of their $^{40}Ca/^{39}K$ ratio. The dotted horizontal line represents normal $^{41}K/^{39}K$ ratio of 0.072. The linear correlation of the excess $^{41}K/^{39}K$ with the $^{40}Ca/^{39}K$ ratio indicates that the excess in $^{41}K$ is related to Ca and resulted from *in-situ* decay of "live" $^{41}Ca$ (half-life ~ $10^5$ years) within the refractory phases (data from Srinivasan *et al.*, 1996 and Sahijpal *et al.*, 1999). The measured values for the terrestrial analog phases are consistent with normal $^{41}K/^{39}K$ ratio.

## 2.1 Short-lived Nuclides as Chronometers

It is important to realize that the data for the short-lived nuclides in meteorites provide information *only* on the relative time scales of events and do not provide absolute chronology. However, if the absolute age of one of the analyzed samples can be determined by using independent chronometers that rely on very long-lived radionuclides (*e.g.*, $^{238,235}U$-$^{206,207}Pb$, $^{147}Sm$-$^{143}Nd$, $^{187}Re$-$^{187}Os$, $^{87}Rb$-$^{87}Sr$ systems), it is possible to tie up the relative chronology to absolute time scale. This is indeed possible as the formation age of the refractory Ca-Al-Inclusions (CAIs) found in primitive meteorites and considered to be the first solids to form in the solar nebula has been determined by using different chronometers. The U-Pb chronometer provides the best precision and the currently accepted formation age of CAIs is 4566 ± 2 Myr (Manhès *et al.*, 1988; see also Allègre *et al.*, 1995). However, the focus in the present paper is primarily on relative chronology and we shall not dwell much on the absolute ages of events taking place in the early solar system.

An important assumption made in establishing relative chronology for events in the early solar system from the short-lived nuclide records in meteorites is that of a homogeneous distribution of these nuclides in the solar nebula. The time difference between any two events in the early solar system can then be simply obtained from the difference in the initial abundance of a short-lived nuclide between samples that formed during these specific events. Obviously, if the distribution was not homogeneous, such difference need not necessarily translate to a difference in time between the events. The question of the distribution of the short-lived nuclides in the solar nebula has been extensively discussed by MacPherson *et al.* (1995) for $^{26}Al$, and by Lugmair and Shukolyukov (1998) for $^{53}Mn$, and they concluded that a homogeneous distribution may be considered to be a valid proposition. A similar conclusion has also been drawn by Sahijpal *et al.* (1999) based on an analysis of the presently available data for $^{41}Ca$. We shall, therefore, assume a homogeneous distribution of these and other short-lived nuclides while discussing the meteorite data. In the following, we provide a brief summary of the various short-lived nuclides that may be used to date early solar system events; they are arranged in order of increasing half-life.

*$^{41}Ca$-$^{41}K$ ($t_{1/2}$ = 0.1 Myr)*: This radionuclide has the shortest half-life among all the now-extinct nuclides whose presence in the early solar system has been established form isotopic studies of meteorites. The presence of $^{41}Ca$ in the early solar system was demonstrated by Srinivasan *et al.* (1994, 1996) from a study of CAIs from the Efremovka carbonaceous chondrite. A hint for its possible presence was suggested earlier by Hutcheon *et al.* (1984), who studied a couple of CAIs from another carbonaceous chondrite, Allende. The

presence of $^{41}$Ca in the early solar system has been also confirmed by Ireland *et al.* (1999).

*$^{26}$Al-$^{26}$Mg ($t_{1/2}$ = 0.74 Myr)*: The presence of this radionuclide in the early solar system was conclusively established by Lee *et al.* (1976) from a study of Allende CAIs following initial hints provided by the work of Gray and Compston (1974) and Lee and Papnastassiou (1974). This is the most widely studied short-lived nuclide in meteorites. In addition to the CAIs, records of $^{26}$Al have also been found in meteoritic chondrules (See, *e.g.*, Russell *et al.*, 1996) and in a differentiated meteorite (Srinivasan *et al.*, 1999).

*$^{60}$Fe-$^{60}$Ni ($t_{1/2}$ = 1.5 Myr)*: The record of this short-lived nuclide has so far been found in two differentiated meteorites (Shukolyukov and Lugmair, 1993a, b). The isotopic systematics in mineral separates appeared to be disturbed and the linear correlation expected between $^{60}$Ni/$^{58}$Ni and $^{56}$Fe/$^{58}$Ni is seen only for the data for bulk samples. More data are needed to establish a widespread distribution of this nuclide in the early solar system.

*$^{53}$Mn-$^{53}$Cr ($t_{1/2}$ =3.7 Myr)*: The presence of this short-lived nuclide in the early solar system was first established from a study of CAIs and bulk samples of the Allende carbonaceous chondrite (Birck and Allègre, 1985). Since then a large data base of $^{53}$Mn abundances in different meteorites has become available (See, *e.g.*, Lugmair and Shukolyukov, 1998).

*$^{107}$Pd-$^{107}$Ag ($t_{1/2}$ = 6.5 Myr)*: Studies of samples of iron and stony-iron meteorites yielded records for the presence of the short-lived nuclide $^{107}$Pd in the early solar system (Kelly and Wasserburg, 1978; also see Chen and Wasserburg, 1996). These meteorites are fragments of planetesimals that have undergone differentiation leading to separation of silicate and metal and formation of a metallic core. The abundance of this nuclide along with that of $^{182}$Hf (See below) constrain the time scale for the differentiation and subsequent formation of metallic core in planetesimals in the early solar system.

*$^{182}$Hf-$^{182}$W ($t_{1/2}$ = 9 Myr)*: A very important feature of the Hf-W system is the distinctly different chemical affinities of the two elements. Hf is a lithophile element with strong affinity towards silicates, while W, a moderately siderophile element, prefers to go with metals. Thus, a study of Hf-W isotopic systematics in meteorites allows us to infer the time scale of segregation of metal due to silicate-metal fractionation. If this happens rather quickly, before significant decay of $^{182}$Hf can take place, the metal will be depleted in $^{182}$W compared to the silicates that will incorporate most of the Hf. In fact, the discovery of $^{182}$Hf in the early solar system is based on the observed depletion in the $^{182}$W/$^{184}$W ratio in iron meteorites compared to that in carbonaceous chondrites (Lee and Halliday, 1995; Harper and

Jacobsen, 1996). Hf-W isotopic systematics in differentiated meteorites have also been reported (Lee and Halliday, 1996, 1997).

The two other short-lived nuclides $^{129}$I and $^{244}$Pu, whose presence in the early solar system was established from studies of xenon isotopes in meteorites in the sixties (Reynolds 1960; Jeffreys and Reynolds, 1961; Rowe and Kuroda, 1965), have relatively longer half-lives and are not discussed here. It may be noted that attempts to use the I-Xe chronometer to infer time scales of processes in the early solar system have not been successful due to a lack of knowledge of the carrier(s) of iodine in meteorite samples (See, *e.g.*, Swindle and Podosek, 1988). It is now believed that this chronometer can be best utilized to date secondary events taking place in planetesimals during their early evolution (Hohenberg *et al.*, 1998; Brazzle *et al.*, 1999). We shall also not discuss the records of the four other short-lived nuclides, $^{99}$Tc (Yin *et al.*, 1992), $^{36}$Cl (Murty *et al.*, 1997), $^{205}$Pb (Chen and Wasserburg, 1987) and $^{92}$Nb (Harper, 1996), whose inferred presence in early solar system needs to be substantiated by further experimental work.

The records of all the now-extinct short-lived nuclides cannot be seen in a single suite of meteorite samples such as the refractory CAIs or the differentiated meteorites. This has got to do with both the time scales of processes and chemistry. The short lived nuclides $^{41}$Ca and $^{26}$Al belong to the refractory group of elements and they get preferentially incorporated during the formation of the refractory CAIs, while other nuclides such as $^{60}$Fe, $^{53}$Mn and $^{107}$Pd are inhibited. Even though trace amounts of these nuclides may be present in the CAIs, identification of their decay products becomes a formidable experimental challenge. On the other hand by the time of formation of the differentiated meteorites, $^{41}$Ca could have become extinct, the abundance of $^{26}$Al will become very low and the records of the longer-lived nuclides will be more prominent in these meteorites. Thus, we have to synthesize the data obtained from analyses of different meteorite samples to build up a self-consistent scenario for the formation and the early evolution of the solar system. Several review papers dealing with the meteoritic records of the short-lived nuclides have appeared recently (Podosek and Nichols, 1997; Goswami 1998; Goswami and Vanhala, 1999; Wadhwa and Russell, 1999).

## 2.2 Source(s) of the Short-lived Nuclides in the Early Solar System

The short-lived nuclides present in the early solar system are generally considered to be of stellar origin (See, *e.g.*, Cameron 1993; also Table 1). However, the possibility that some of the relatively shorter-lived nuclides such as $^{41}$Ca, $^{26}$Al and $^{53}$Mn are products of energetic particle interactions has also been suggested. There could be two settings in which this can happen. These nuclides could be produced by interactions of low energy particles with gas and dust in the protosolar cloud (Clayton, 1994) or they could be produced by energetic particles from the early sun interacting with gas and dust in the solar nebula (Heymann and Dziczkaniec, 1976; Wasserburg and Arnould, 1987; Clayton and Jin, 1995; Shu *et al.*, 1997; Goswami *et al.*, 1997). It is important to identify the most plausible source among these alternatives. If the short-lived nuclides were present in thc protosolar cloud at the time of its collapse, their "live" presence at the time of formation of the first solar system objects (CAIs) provides a very strong constraint on the time scale of cloud collapse. However, if these nuclides were produced later by interactions of solar energetic particles with material in the solar nebula they do not provide any information on presolar processes and they may be used as time markers only for early solar system processes. Results obtained from several analytical studies of energetic particle production of these nuclides show that co-production of $^{41}$Ca and $^{26}$Al to match their initial abundances in the early solar system (See Table 1) is virtually impossible (Ramaty *et al.*, 1996; Shu *et al.*, 1997; Goswami *et al.*, 1997; Lee *et al.*, 1998). The production of $^{26}$Al in the required amount leads to more than an order of magnitude higher abundance of $^{41}$Ca than observed. One may therefore propose additional contributions from a stellar source that provided most of the $^{26}$Al, while $^{41}$Ca was primarily produced by interactions of solar energetic particles. However, any stellar object that synthesizes $^{26}$Al will also be a significant source of $^{41}$Ca as well. Further, a recent experiment (Sahijpal *et al.*, 1998) clearly demonstrates that the presence as well as the abundance of both $^{41}$Ca and $^{26}$Al are tightly correlated even at a very small spatial scale of about ten microns within individual early solar system grains (See Figure 2). This observation can be best explained if these two nuclides were cogenetic and were introduced into the solar nebula together and followed the same pathways in the nebula before getting incorporated "live" into the early solar system grains. Thus, a single stellar source for these two nuclides appears to be more plausible than their production by both a stellar source as well as solar energetic particles. Analytical calculations of production of short-lived nuclides in different stellar sources suggest a thermally pulsing asymptotic giant branch (TP-AGB) star or a supernova as plausible sources of the short-lived nuclides $^{41}$Ca, $^{26}$Al, $^{60}$Fe, $^{53}$Mn and $^{107}$Pd present in the early solar system (Wasserburg *et al.*, 1994, 1995; Cameron *et al.*, 1995). It is of course difficult to completely rule out any contribution from energetic particle production and in fact there is a possibility that a significant amount of $^{53}$Mn was produced by solar energetic particles (Goswami *et al.*, 1997, 1999). However, as far as $^{41}$Ca and $^{26}$Al are concerned, the contribution from energetic particle interactions can be neglected and this is also true for $^{60}Fe$ and $^{107}Pd$ (See, Goswami and Vanhala, 1999, for a detail discussion). In the following, we consider a stellar source for the now-extinct nuclides and consider the constraint they put on the time scale of different events in the early solar system.

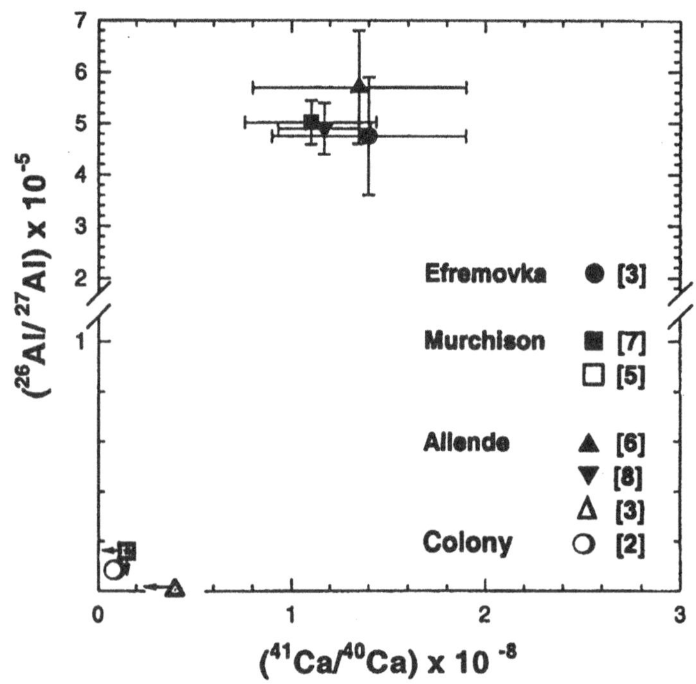


*Figure 2.* A plot of initial $^{26}Al/^{27}Al$ ratio *vs.* initial $^{41}Ca/^{40}Ca$ ratio in several sets of refractory oxide and silicate phases from four carbonaceous chondrites. The number of analyses performed in each case is given within parenthesis. Data from Sahijpal *et al.* (1998, 1999) and Marhas *et al.* (1999).

## 3. DATING OF EARLY SOLAR SYSTEM EVENTS

The various stages of the early evolution of the solar system, from a meteoritic perspective, may be depicted schematically as in Figure 3. As illustrated in this figure the records of the short-lived nuclides in different meteoritic objects allow us to obtain a chronology of the following early solar system events.

1. Formation of the sun
2. Formation of early solar system objects (such as the CAIs and chondrules) that are considered to be nebular products.

3. Formation and differentiation of planetesimals representing parent bodies of differentiated meteorites such as the achondrites, the irons and the stony-irons.

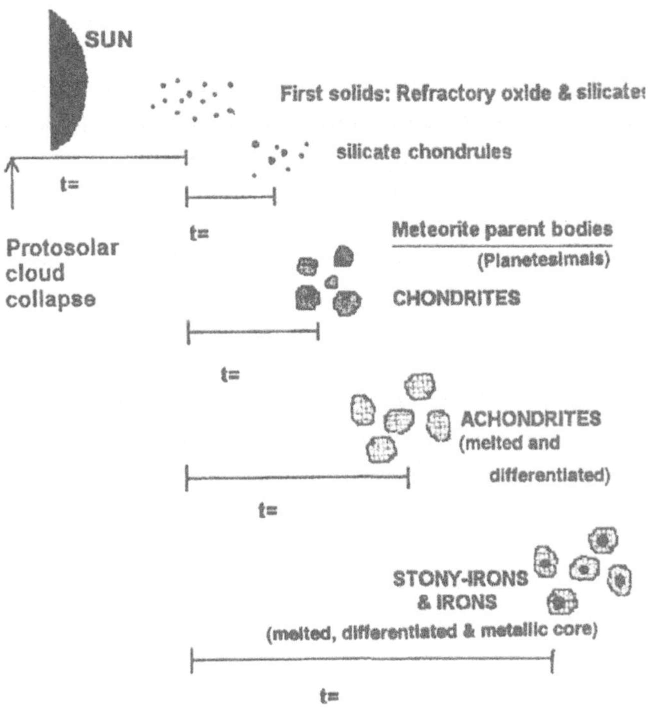


*Figure 3.* A schematic representation of the formation of the sun and early evolution of the solar system objects from a meteoritic perspective. Records of now-extinct, short-lived nuclides in refractory phases and chondrules of chondrites, as well as in bulk samples of chondrites, achondrites, stony-iron and iron meteorites, allow us to determine the time scales indicated by the horizontal bars.

## 3.1 Formation of the Sun

A stellar origin for the short-lived nuclides provide a very strong constraint on the time scale of collapse of the protosolar cloud leading to the formation of the sun and some of the first solar system objects (such as the CAIs). The most stringent constraint comes from the nuclide with the shortest half-life, $^{41}$Ca ($t_{1/2}$ = 0.1 Myr), whose presence at the time of formation of the CAIs indicates this time scale to be less than a million years. A self consistent explanation for the presence of the short-lived nuclides $^{41}$Ca, $^{26}$Al, $^{60}$Fe and $^{107}$Pd with their observed abundances in the early solar system (See

Table 1) would require this time scale to be 0.6 Myr if a TP-AGB star was the source of these nuclides (Wasserburg *et al.*, 1995). The time scale for cloud collapse depends on the density of the cloud and for the simplest case (free-fall collapse) this time scale is given as, $t_{ff} \sim 10^8 \rho^{1/2}$ years, where ρ is the number density of molecules (basically $H_2$) in the cloud. Thus, the protosolar molecular cloud must have had a density, $\rho \geq 10^4(H_2)\ cm^{-3}$, which is higher than the typical density observed in dense cores within molecular cloud complexes. Furthermore, analytical calculations of cloud collapse suggest a time scale of ~ 5-10 Myr for the formation of a sun-like star through unassisted collapse of a molecular cloud fragment (Mouschovias, 1989; Shu, 1995). The much shorter time scale for cloud collapse inferred from the meteorite data thus led to the revival of the suggestion for a triggered origin of the solar system (Cameron and Truran, 1977) with the stellar source (a TP-AGB star or a supernova) of the short-lived nuclides itself acting as the trigger. Recent numerical simulation studies of the impact of stellar outflows/ejecta with dense cloud cores show that a triggered collapse of the protosolar cloud is indeed a viable proposition (Foster and Boss, 1996, 1997; Vanhala and Cameron, 1998; see also Goswami and Vanhala, 1999). These studies also demonstrated that the freshly synthesized nuclides carried by the stellar outflows can be injected into the collapsing protosolar cloud core via Rayleigh-Taylor instabilities that will develop as the stellar material gathers on the top of the collapsing cloud. The short-lived nuclide data in meteorites, therefore, suggest formation of the sun within a million years of protosolar cloud collapse and favor a triggered collapse of the protosolar cloud.

## 3.2 Formation of the CAIs and the age of the solar system

The refractory Ca-Al-Inclusions (CAIs) found in primitive meteorites (See Figure 4) represent some of the first solar system objects that formed in the nebula soon after the formation of the sun. The presence of fossil records of the short-lived nuclides $^{41}$Ca and $^{26}$Al in CAIs suggests that they formed within the first million year of the origin of the solar system (Srinivasan *et al.*, 1994, 1996). A large data-base exists for $^{26}$Al abundance in the solar nebula at the time of formation of the CAIs (See, *e.g.*, compilation by MacPherson *et al.*, 1995). The data for the initial abundance of $^{41}$Ca are also available for a few dozen CAIs from several primitive meteorites (Srinivasan *et al.*, 1994,1996; Sahijpal *et al.*, 1998, 1999). The inferred initial abundances of $^{26}$Al and $^{41}$Ca in the nebula, relative to their stable counterparts, $^{27}$Al and $^{40}$Ca, at the time of formation of the CAIs, are ~ $5\times10^{-5}$ ($^{26}$Al/$^{27}$Al) and ~ $1.5 \times 10^{-8}$ ($^{41}$Ca/$^{40}$Ca), respectively (See, Figures 1 and 2).

Experimental data also reveal that the initial $^{26}Al/^{27}Al$ ratios in some of the CAIs are lower than the above value, while others are devoid of both $^{26}Al$ and $^{41}Ca$ (See Figure 2). Formation of some of the CAIs within the collapsing protosolar cloud, prior to the arrival of the short-lived nuclides injected by a stellar source, and, secondary processing leading to late disturbances in the Mg and K isotopic systematics in other CAIs may explain these features (See, *e.g.*, MacPherson *et al.*, 1995; Sahijpal and Goswami, 1998). We have noted earlier that the absolute age of the CAIs is determined to be 4566 ± 2 Myr. Since the formation of the CAIs took place within the first million year of the solar system history, this age can be taken as the age of the solar system.

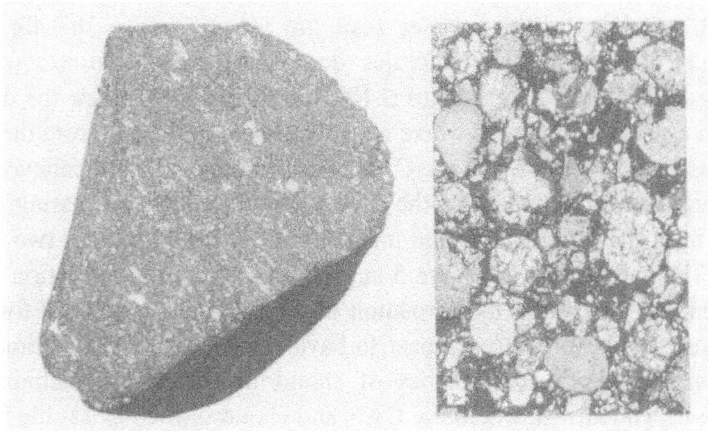

*Figure 4.* A cut slice of the Allende carbonaceous chondrite, a few cm across, with several large whitish refractory Ca-Al-Inclusions (CAIs) in it (Photograph on the left). The photograph on the right shows examples of millimeter-sized chondrules, spheroidal silicate objects found in chondrites, as seen in a thin section of the chondrule-rich meteorite Tieschitz. A close scrutiny will also reveal a large number of chondrules in the Allende slice.

## 3.3 Short-lived nuclides in chondrules: Time scale of nebular processing

The chondrites, that are further sub-divided into several groups, are the most abundant type of meteorites. A ubiquitous constituent of chondritic meteorite is the "chondrule", a spheroidal silicate object that varies in size from a few tens of microns to a few millimeters. A representative example of chondrules is shown in Figure 4. In carbonaceous chondrites, both CAIs and chondrules co-exist side by side along with other fine-grained silicate material indicating that the accretion of the meteoritic objects took place following the formation of the refractory CAIs and the silicate chondrules. The morphology and petrography of the chondrules suggest that they formed via rapid cooling of molten silicate droplet during some high temperature events in the solar nebula (See, *e.g.*, Grossman, 1988). Following the discovery of $^{26}$Al in CAIs, attempts have been made to look for the possible presence of $^{26}$Al in chondrules. The presence of excess $^{26}$Mg due to decay of $^{26}$Al in a chondrule-like object was first reported by Hutcheon and Hutchison (1989), who estimated an initial $^{26}$Al/$^{27}$Al ratio of $\sim 8 \times 10^{-6}$ at the time of formation of this object. However, there exists some ambiguity about the identification of this object as a genuine chondrule and the first unambiguous results that showed the presence of $^{26}$Al at the time of chondrule formation were reported by Russell *et al.* (1996). The highest value for initial $^{26}$Al/$^{27}$Al abundance in chondrules was found to be $\sim 2.5 \times 10^{-5}$, a factor of two lower than the value of $5 \times 10^{-5}$ for CAIs. Furthermore, many of the chondrules do not have resolved excess $^{26}$Mg indicating initial $^{26}$Al/$^{27}$Al values of $\leq 10^{-6}$. In Figure 5, we show the data for more than a dozen chondrules where a resolved excess of $^{26}$Mg from the decay of $^{26}$Al has been observed. As the CAIs and the chondrules are believed to be nebular products, an estimate of the time scale of nebular processing can be obtained from the difference in the initial $^{26}$Al/$^{27}$Al between these two sets of objects. The data shown in Figure 5 suggest that chondrule formation started about a million years after the formation of the CAIs and continued for $\sim$ 4-5 Myr. Thus, nebular processes appear to have been active over this time scale and the accretion of parent bodies of chondritic meteorites, including the carbonaceous chondrites, where the CAIs and chondrules co-exist side by side (See Figure 4), took place during or after this time period. Unfortunately, this estimate of the time scale of nebular processing appears to be rather long for two specific reasons. First, the problem of storage of the earlier formed CAIs in the solar nebula for a million years or more until the time of chondrule formation. Simple analytical calculations show that the sub-millimeter to millimeter-sized CAIs in the solar nebula will drift inward and fall onto the sun within a time scale of $< 10^5$ years due to gas drag (Weidenschilling, 1977). Thus, storage of the CAIs in the nebula for a million years is a difficult proposition. Second, data for $^{26}$Al, $^{53}$Mn and $^{182}$Hf in differentiated meteorites argue for an early accretion and differentiation of large-sized (> Km) objects in the solar nebula, perhaps within a few million years (See next two sections). Several solutions have been proposed to overcome these problems. These include possible late disturbances in the Al-Mg isotopic systematics in the chondrules resulting in the lower initial $^{26}$Al/$^{27}$Al ratios, a non-nebular origin for the chondrules, and presence of turbulence in the solar nebula that may retard the drift of the early forming CAIs and chondrules towards the sun. The exact time scale of nebular processing remains an open question at present.

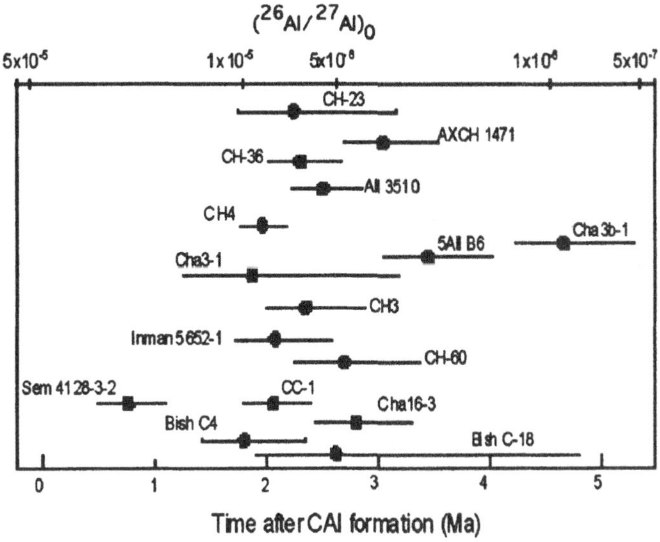


*Figure 5.* Initial $^{26}Al/^{27}Al$ ratio in chondrules from several chondritic meteorites [Chainpur (CH, Cha), Semarkona (Sem, CC), Bishunpur (Bish), Allende (All), Inman and Axtell]. All the values are lower than the CAI value of $5\times10^{-5}$. The elapsed time after CAI formation, assuming that the lower values are due to decay of $^{26}Al$, is also shown (data from compilation by Wadhwa and Russell, 1999; Mostefaoui *et al.*, 1999; Kita, 1999).

## 3.4 Formation and differentiation of planetesimals

Chemical and petrographic signatures present in many of the meteorites suggest that they have undergone various degrees of thermal metamorphism and differentiation during their evolutionary history. Mild to moderate degrees of thermal metamorphism is evident in the different types of chondrites (See, *e.g.*, McSween *et al.*, 1988). On the other hand, most of the achondrites and stony iron and iron meteorites are considered to have been derived from planetesimals that were completely or partially molten and differentiated. Records of several short-lived nuclides have been found in the differentiated meteorites. In the following, we discuss these results in terms of formation and differentiation of asteroidal-sized planetesimals representing the parent bodies of these meteorites.

### 3.4.1 Formation of achondrites and early melting of planetesimals

The meteorites belonging to the achondrite group were derived from planetesimals that have experienced melting and differentiation. The absolute ages of achondrites determined by long-lived radionuclide chronometers (*e.g.*, U-Pb, Sm-Nd, Rb-Sr systems) suggest their formation very early in the history of the solar system, perhaps within the first few tens of million years, indicating an early melting of planetesimals in the solar system. Several suggestions have been made for the heat source responsible for this early melting. The most prominent among these are the suggestions for radioactive heating due to $^{26}$Al decay (Urey, 1955) and solar wind induction heating (Sonett *et al.*, 1968). The search for fossil records of $^{26}$Al in basaltic achondrites by Schramm *et al.* (1970) yielded negative results. Since then several groups have tried to detect $^{26}$Al in differentiated meteorites without success. The discovery of excess $^{26}$Mg due to decay of $^{26}$Al in the differentiated meteorite Piplia Kalan, belonging to the eucrite group, was reported recently by Srinivasan *et al.* (1999). A clear excess of $^{26}$Mg was found in plagioclase grains from this meteorite with extremely high Al/Mg ratio (See Figure 6).

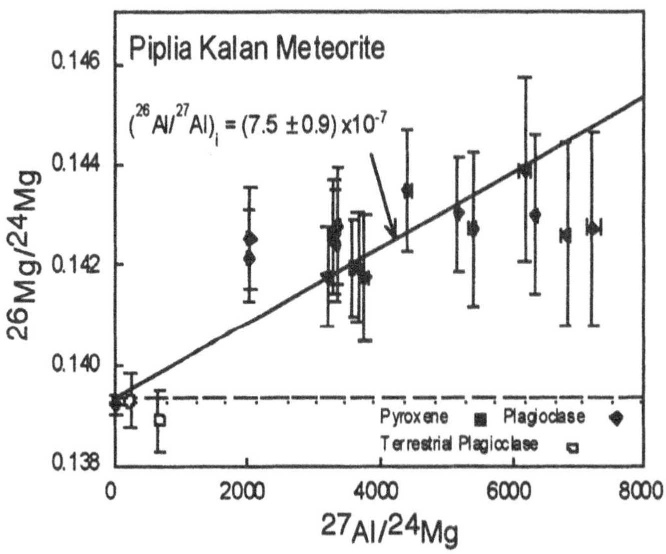


*Figure 6.* Measured $^{26}$Mg/$^{24}$Mg ratios in plagioclase grains with very high $^{27}$Al/$^{24}$Mg ratio in the Piplia Kalan eucrite reveal presence of excess $^{26}$Mg due to the decay of $^{26}$Al. Co-existing pyroxenes with low Al/Mg ratio and terrestrial plagioclase have normal magnesium isotopic composition ($^{26}$Mg/$^{24}$Mg = 0.13932; the dashed horizontal line). The inferred initial $^{26}$Al/$^{27}$Al at the time of formation of the Piplia Kalan eucrite is also shown. Data from Srinivasan *et al.* (1999).

The eucrites are members of a particular group of basaltic achondrites, known as the HED group (for Howardites, Eucrites and Diogenite), and the asteroid 4 Vesta is considered to be the probable parent body of this group of meteorites (See, *e.g.*, Drake, 1979; Binzel and Xu, 1993). The initial $^{26}$Al/$^{27}$Al at the time of formation of Piplia Kalan, inferred from the isotopic data shown in Figure 6, is ~ $7 \times 10^{-7}$, almost two orders of magnitude below the value of $5 \times 10^{-5}$ for the CAIs. This difference indicates a time interval of ~ 5 Myr between the formation of the CAIs and the Piplia Kalan meteorite. This would imply that accretion, heating, melting, differentiation and formation of basaltic crust on the parent body of Piplia Kalan was complete within the first five million years of the solar system history. The role of $^{26}$Al as the heat source for early melting of planetesimals is also established from the results obtained by Srinivasan *et al.* (1999).

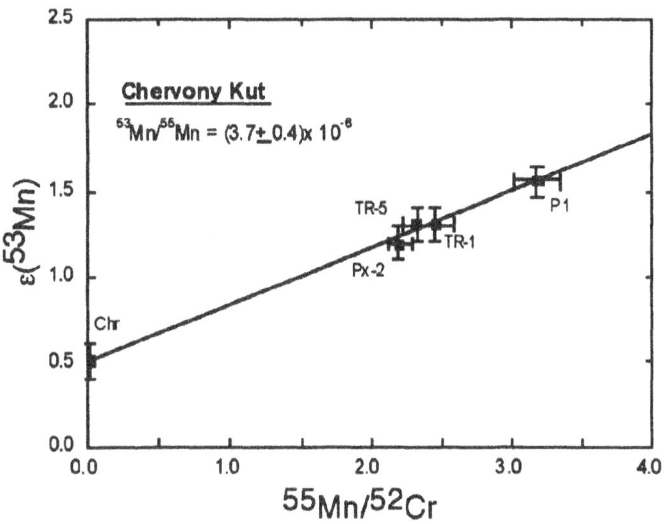


*Figure 7.* Deviation in the measured $^{53}$Cr/$^{52}$Cr ratio from the normal value, expressed in parts per ten thousand, ε($^{53}$Mn), plotted as a function of the $^{55}$Mn/$^{52}$Cr ratio for different samples of the eucrite, Chervony Kut. The inferred initial $^{53}$Mn/$^{55}$Mn ratio at the time of its formation is also shown. Data from Lugmair and Shukolyukov (1998).

The short time scale for melting and differentiation of planetesimals such as the asteroid 4 Vesta (the most probable parent body of the HED group of meteorites) is also supported by the results obtained from the study of the short-lived nuclides $^{53}$Mn ($t_{1/2}$ = 3.7 Myr) in HED meteorites (Lugmair and Shukolyukov, 1998). The $^{53}$Mn data suggest that the initial $^{53}$Mn/$^{55}$Mn ratio at the time of completion of differentiation in the parent body of the HED meteorites was ~ $4.7 \times 10^{-6}$ (Lugmair and Shukolyukov, 1998). In particular, they also found that at the time of formation of the two eucrites, Chervony Kut and Juvinas, that are very similar to Piplia Kalan, the initial $^{53}$Mn/$^{55}$Mn values were ~ $3.7\times10^{-6}$ and ~ $3\times10^{-6}$, respectively (See Figure 7). A straight forward estimate for the time interval between the formation of the CAIs and the eucrites is difficult in these cases as the initial $^{53}$Mn/$^{55}$Mn ratio in the CAIs is not well established (Birck and Allègre, 1985; Lugmair and Shukolyukov, 1998;

Nyquist *et al.*, 1999; see Table 1). However, Lugmair and Shukolyukov (1998) combined data for absolute ages of the CAIs and certain achondrites and the data for $^{53}$Mn in these achondrites and the eucrites, Chervony Kut and Juvinas, to infer a time interval of ~ 3-6 million years between the formation of the CAIs and these eucrites. This value is consistent with the time scale inferred from the $^{26}$Al data in Piplia Kalan eucrite. A recent analytical study of $^{26}$Al heating of the asteroid 4 Vesta suggest an accretion time of 2.85 Myr, core formation by 4.58 Myr and formation of basaltic crust by 6.58 Myr for this asteroid (Ghosh and McSween, 1998). The data for $^{182}$Hf in eucrites (See Figure 8) also suggest that silicate-metal segregation in the HED parent body took place within the first 5-15 million years of the solar system history (Lee and Halliday, 1997).

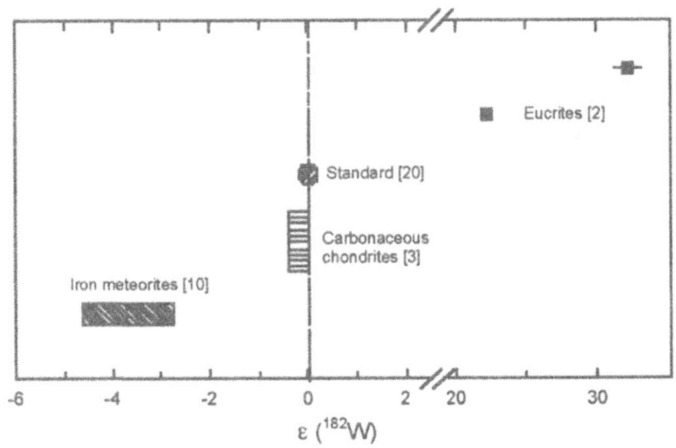


*Figure 8.* Deviation in the measured $^{182}$W/$^{184}$W ratios in differentiated meteorites relative to that in carbonaceous chondrites, expressed in parts per ten thousand ε($^{182}$W). The depletion in iron meteorites and enrichment in eucrites suggest early differentiation and silicate-metal segregation in the parent bodies of these meteorites. Data from Lee and Halliday (1995, 1996, 1997); the number of analyses performed in each case is given within brackets.

### 3.4.2 Time scale of silicate/metal fractionation and formation of metallic core in planetesimals

The stony-iron and iron meteorites represent fragments of asteroids that have experienced silicate-metal fractionation leading to their segregation and the formation of a metallic core. Excess of $^{107}$Ag, due to decay of $^{107}$Pd, ($t_{1/2}$ = 6.5 Myr) has been found in a large number of iron and stony-iron meteorites (Chen and Wasserburg, 1996). These records allow us to constrain the time scale of magmatism, differentiation, and metallic core formation in planetesimals representing parent bodies of these meteorites. The spread in the measured initial $^{107}Pd/^{108}Pd$ in these meteorites suggest a time scale of ~ 12 Myr within which large-scale magmatism, silicate-metal fractionation, and formation of Fe-Ni core in the parent bodies of these meteorites took place. Since the planetesimal accretion time scale is much smaller (perhaps a few million years), the $^{107}Pd$ data hints at a very early differentiation and formation of metallic core in planetesimals. This inference is further corroborated by the data for the short-lived nuclide, $^{182}Hf$, in chondrites and iron meteorites.

The short-lived nuclide $^{182}Hf$ ($t_{1/2}$ = 9 Myr) decays to $^{182}W$, and the Hf-W isotopic system is an ideal tracer to monitor time of silicate-metal fractionation in planetesimals representing parent bodies of the achondrites, iron, and stony-iron meteorites. The initial abundances of Hf and W in the planetesimals are expected to be close to the solar (chondritic) values. If silicate-metal fractionation took place in such a planetesimal prior to significant decay of $^{182}Hf$, one would expect to find a depletion in the $^{182}W/^{184}W$ ratio in the metallic fraction and an enrichment in the silicate fraction, relative to the chondritic value. The data obtained by Lee and Halliday (1995, 1996, 1997) and Harper and Jacobsen (1996) indeed reveal depletion in the measured $^{182}W/^{184}W$ ratios for several iron meteorites (metal) and enrichment for a couple of eucrites (silicate) relative to the measured $^{182}W/^{184}W$ ratio in carbonaceous chondrites (See Figure 8). These results hint at an early silicate-metal segregation in the parent bodies of these meteorites. As already noted, the $^{182}Hf$ data for the eucrites suggest silicate-metal fractionation and core formation in these objects within a time scale of ~ 5-15 million years. In the case of the iron meteorites, the data indicate formation of a core in their parent bodies within a few million years. However, fractional crystallization within the metallic core and formation of different types of iron meteorites could have continued for a much longer time.

## 4. SUMMARY AND FUTURE PERSPECTIVES

The presence of several now-extinct, short-lived nuclides in the early solar system has been established from studies of isotopic records in meteorites. The relatively shorter-lived nuclides, $^{41}Ca$, $^{26}Al$, $^{60}Fe$, $^{53}Mn$ and $^{107}Pd$, with mean life < 10 Myr, were most probably synthesized and injected into the protosolar cloud from a single stellar source such as a thermally pulsing asymptotic giant branch star or a supernova. Production of some of the nuclides such as $^{41}Ca$, $^{26}Al$ and $^{53}Mn$ is also feasible by energetic particle interactions within the protosolar cloud or later in the solar nebula.

However, the contribution from energetic particle production appears to be negligible compared to the stellar input, except perhaps for $^{53}$Mn. A stellar origin for the short-lived nuclides constrains the time scale of protosolar cloud collapse and the formation of the sun to less than a million years. This short time scale argues for a triggered collapse of the protosolar cloud. Analytical studies of cloud collapse indeed suggest the viability of a triggered origin of the solar system.

The records of the now-extinct, short-lived nuclides in meteorites also provide a chronology of the events during the early evolution of the solar system. The refractory Ca-Al-Inclusions (CAIs) found in primitive meteorites, that are considered to be some of the first objects to form in the solar nebula, contain records of the now-extinct, short-lived nuclide $^{41}$Ca indicating that they formed within the first million years of the evolution of the solar system. The absolute age of 4566 ± 2 Myr for the CAIs thus closely approximate the age of the solar system. The time scale of nebular processing, inferred from the records of the short-lived nuclide, $^{26}$Al, in CAIs and chondrules, that are considered to be nebular products, is a few million years. This appears to be rather long and is in conflict with our present understanding of the dynamical evolution of small objects in the solar nebula. Observation of fossil records of $^{26}$Al in a differentiated meteorite confirms this short-lived nuclide as the heat source responsible for early melting of planetesimals. Data for the short-lived nuclides, $^{26}$Al, $^{53}$Mn, $^{107}$Pd and $^{182}$Hf, in meteorites that are fragments from differentiated planetesimals suggest a short time scale of about five to ten million years within which accretion, heating, melting, silicate-metal fractionation and core formation in these plenetesimals took place.

Even though significant advances have been made in our understanding of the early evolution of the solar system from the study of now-extinct, short-lived nuclide records in meteorites, several aspects need further attention. It is important to confirm the presence of the short-lived nuclides $^{99}$Tc, $^{36}$Cl, $^{205}$Pb and $^{92}$Nb in the early solar system by more refined experiments. This will allow us to pinpoint the exact stellar source of these and other short-lived nuclides. The initial abundance of $^{53}$Mn needs to be established without ambiguity; production of this nuclide by solar energetic particles is a distinct possibility if the proposed lower initial abundance is indeed correct. The stellar nucleosynthesis models, both for a supernova and a TP-AGB star, need further refinement. The dynamics of the interactions of the ejecta/outflows from these stellar sources with the stellar neighborhood and with the interstellar medium are not well understood at present. The spatial resolutions of the numerical simulation studies of cloud collapse and of injection of stellar material into the collapsing protosolar cloud need to be improved so that predictions can be made at scale lengths relevant to meteorite records. Finally, astronomical observations of triggered star formation will bolster the proposal for a triggered origin for the solar system.

## Acknowledgment

This paper was written during a short visit to the Lunar and Planetary Institute (LPI), Houston and I thankfully acknowledge the support provided by LPI.

## References

Allègre, C.J., Manhès, G. and Göpel, C.: 1995, "The age of the earth", *Geochim. Cosmochim. Acta* **59**, 1445-1456.

Binzel, R.P. and Xu, S.: 1993, "Chips off of asteroid 4 Vesta: Evidence for the parent body of basaltic achondrite meteorites", *Science* **260**, 186-191.

Birck, J.L. and Allègre, C.J.: 1985, "Evidence for the presence of $^{53}$Mn in the early solar system", *Geophys. Res. Lett.* **12**, 745-748.

Brazzle, R.H., Pravdivtseva, O.V., Meshik, A.P. and Hohenberg, C.M.: 1999, "Verification and interpretation of the I-Xe chronometer", *Geochim. Cosmochim. Acta* **63**, 739-760.

Cameron, A.G.W.: 1993, "Nucleosynthesis and star formation", in *Protostars and Planets III,* eds., Levy, E.H. and Lunine, J.L., Univ. Arizona Press, Tucson, AZ, pp. 47-73.

Cameron, A.G.W. and Truran, J.W.: 1977, "The supernova trigger for the formation of the solar system", *Icarus* **30**, 447-461.

Cameron, A.G.W., Höflich, P., Myers, P.C. and Clayton, D.D.: 1995, "Massive supernova, Orion gamma rays and the formation of the solar system", *Ap. J. Lett.* **447**, L53-L57.

Chen, J.H. and Wasserburg, G.J.: 1987, "A search for evidence of extinct lead 205 in iron meteorites", *Lunar Planet. Sci.* **XVIII**, 165–166.

Chen, J.H. and Wasserburg, G.J.: 1996, "Live $^{107}$Pb in the early solar system and implications for planetary evolution", in *Earth Processes: Reading the Isotopic Code*, eds., Basu, A. and Hart, S., *AGU Geophys. Mon. Ser.* 95, pp. 1-20.

Clayton, D.D.: 1994, "Production of $^{26}$Al and other extinct radionuclides by low-energy heavy cosmic rays in molecular clouds", *Nature* **368**, 222–224.

Clayton, D.D. and Jin, L.: 1995, "A new interpretation of $^{26}$Al in meteoritic inclusions" *Ap. J. Lett.* **451**, L87–L91.

Drake, M.J.: 1979, "Geochemical evolution of the eucrite parent body: Possible nature and evolution of asteroid 4 Vesta?", in *Asteroids*, eds., Gehrels, T. and Matthews, M.S., Univ. Arizona Press, Tucson, AZ, pp. 765-782.

Foster, P.N. and Boss, A.P.: 1996, "Triggering star formation with stellar ejecta", *Ap. J.* **468**, 784-796.

Foster, P.N. and Boss, A.P.: 1997, "Injection of radioactive nuclides from the stellar source that triggered the collapse of the presolar nebula", *Ap. J.* **489**, 346–357.

Ghosh, A. and McSween H.Y., Jr.: 1998, "A thermal model for the differentiation of asteroid 4 Vesta, based on radiogenic heating", *Icarus* **134**, 187-206.

Goswami, J.N.: 1998, "Short-lived nuclides in the early solar system", Proc. *Indian Acad. Sci., (Earth Planet. Sci.)* **107**, 401-411.

Goswami, J.N. and Vanhala, H.A.T.: 1999, "Extinct radionuclides and the origin of the solar system", in *Protostars and Planets IV*, eds., Mannings, V., Russell, S. and Boss, A., Univ. Arizona Press, Tucson, AZ, in press.

Goswami, J.N., Marhas, K.K. and Sahijpal, S.: 1997, "Production of short-lived nuclides by solar energetic particles in the early solar system", Abs., *Lunar Planet. Sci. XXVIII*, 439–440.

Goswami, J.N., Marhas, K.K. and Sahijpal, S.: 1999, "Did solar energetic particles produce some of the short-lived nuclides in the early solar system?", in preparation.

Gray, C.M. and Compston, W.: 1974, "Excess $^{26}$Mg in Allende meteorite", *Nature* **251**, 495-497.

Grossman, J.N.: 1988, "Formation of chondrules", in *Meteorites and the early solar system*, eds., Kerridge, J.F. and Matthews, M.S., Univ. Arizona Press, Tucson, AZ, pp. 680-696.

Grossman, L.: 1980, "Refractory inclusions in the Allende meteorite", *Ann. Rev. Earth Planet. Sci.* **8**, 559–608.

Harper, C.L., Jr.: 1996, "Evidence for $^{92g}$Nb in the early solar system and evaluation of a new p-process cosmochronometer from $^{92g}$Nb/$^{92}$Mo", *Ap. J.* **466**, 437-456.

Harper, C.L., Jr., and Jacobsen, S.B.: 1996, "Evidence for $^{182}$Hf in the early solar system and constraints on the timescale of terrestrial accretion and core formation", *Geochim. Cosmochim., Acta.* **60**, 1131-1153.

Heymann, D. and Dziczkaniec, M.: 1976, "Early irradiation of matter in the solar system: Magnesium (proton, neutron) scheme", *Science* **191**, 79-81.

Hohenberg, C.M., Brazzle, R.H., Pravdivtseva, O.V. and Meshik, A.P.: 1998, "The I-Xe chronometer", *Proc. Indian Acad. Sci. (Earth Planet. Sci.)* **107**, 413-427.

Hudson, G.B., Kennedy, B.M., Podosek, F.A. and Hohenberg C.M.: 1988, "The early solar system abundance of Pu as inferred from the St. Severin chondrite", *Proc. Lunar Planet. Conf.* **19**, 547-557.

Hutcheon, I.D. and Hutchison, R.H.: 1989, "Evidence from the Semarkona ordinary chondrite for the $^{26}$Al heating of small planets", *Nature* **337**, 238-241.

Hutcheon, I.D., Armstrong, J.T. and Wasserburg, G.J.: 1984, "Excess $^{41}$K in Allende CAI: A hint reexamined". *Meteoritics* **19**, 243–244.

Ireland, T.R., Zinner, E., Sahijpal, S. and McKeegan, K.D.: 1999, "Confirmation of excess $^{41}$K from $^{41}$Ca decay in refractory inclusion", *Meteoritics Planet. Sci.* **34**, A57.

Jeffery, P.M. and Reynolds, J.H.: 1961, "Origin of excess $^{129}$Xe in stone meteorites", *J. Geophys. Res.* **66**, 3582-3583.

Kelly, W.R. and Wasserburg, G.J.: 1978, "Evidence for the existence of $^{107}$Pd in the early solar system" *Geophys. Res. Lett.* **5**, 1079-1082.

Kita, Noriko: 1999, personal communication.

Lee, D.-C. and Halliday, A.N.: 1995, "Hafnium-tungsten chronometry and the timing of terrestrial core formation", *Nature* **378**, 771-774.

Lee, D.-C. and Halliday, A.N.: 1996, "Hf-W isotopic evidence for rapid accretion and differentiation in the early solar system" *Science* **274**, 1876-1879.

Lee, D.-C. and Halliday, A.N.: 1997, "Core formation on Mars and differentiated asteroids" *Nature* **388**, 854-857.

Lee, T. and Papanastassiou, D.A.: 1974, "Mg isotopic anomalies in the Allende meteorite and correlation with O and Sr effects", *Geophys. Res. Lett.* **1**, 225-228.

Lee, T., Papanastassiou, D.A. and Wasserburg, G.J.: 1976, "Demonstration of $^{26}$Mg excess in Allende and evidence for $^{26}$Al", *Geophys. Res. Lett.* **3**, 109-112.

Lee, T., Shu, F.H., Shang, H., Glassgold, A.E. and Rehm, K.E.: 1998, "Protostellar cosmic rays and extinct radionuclides in meteorites", *Ap. J.* **506**, 898–912.

Lugmair, G.W. and Shukolyukov, A.: 1998, "Early solar system timescales according to $^{53}$Mn-$^{53}$Cr systematics", *Geochim. Cosmochim. Acta* **62**, 2863–2886.

Lugmair, G.W., Shukolyukov, A. and MacIsaac Ch.: 1995, "The abundance of Fe in the early solar system", in *Nuclei in the cosmoc III*, eds., Busso, M., Gallino, R. and Raiteri, C.M., AIP, pp. 591-594.

MacPherson, G.J., Davis, A.M. and Zinner, E.K.: 1995, "The distribution of aluminum-26 in the early solar system - A reappraisal", *Meteoritics* **30**, 365–386.

Manhès, G., Göpel, C. and Allègre, C.J.: 1988, "Systematique U-Pb dans les inclusions refractaires d' Allende: le plus vieux materiau solaire", *Comptes Rendus de l'ATP Planetologie* (In French) 323-327.

Marhas, K.K., Goswami, J.N., Davis, A.M. and Russell, S.: 1999, "Radiogenic and stable isotopic anomalies in CM and CO meteorites", *Meteoritics Planet Sci.* **34**, A71.

McSween, H.Y., Jr., Sears, D.W.G. and Dodd, R.T.: 1988, "Thermal metamorphism" in *Meteorites and the early solar system*, eds., Kerridge, J.F. and Matthews, M.S., Univ. Arizona Press, Tucson, AZ, pp.102-113.

Mostefaoui, S., Kita, N.K., Nagahara, S., Togashi, S. and Morishita, Y.: 1999, "Aluminium-26 in two ferromagnesian chondrules from a highly unequilibrated ordinary chondrite: Evidence of a short period of chondrule formation", *Meteoritics Planet. Sci.* **34**, A84.

Mouschovias, T.Ch.: 1989, "Magnetic fields in molecular clouds: Regulators of star formation", in *The Physics and Chemistry of Interstellar Molecular Clouds*, eds., Winnewasser, G. and Armstrong, J.T, Springer-Verlag, Berlin, Germany, pp. 297-312.

Murty, S.V.S., Goswami, J.N. and Shukolyukov, Yu. A.: 1997, "Excess $^{36}$Ar in Efremovka meteorite : A strong hint for the presence of $^{36}$Cl in the early solar system", *Ap. J. Lett.* **475**, L65-L68.

Nyquist, L., Shih, C.-Y., Wiesmann, H., Reese, Y., Ulyanov, A.A. and Takeda, H.: 1999, "Towards a Mn-Cr timescale for the early solar system" *Lunar Plan. Sci. XXX*, #1604.

Podosek, F.A. and Nichols, R.H., Jr.: 1997, "Short-lived radionuclides in the solar nebula", in *Astrophysical implications of the laboratory study of presolar materials*, eds., Bernatowicz, T.J. and Zinner, E., Woodbury, AIP, 617-647.

Ramaty, R., Kozlovsky, B. and Lingenfelter, R.E.: 1996, "Light isotopes, extinct radioisotopes and gamma ray lines from low energy cosmic ray interactions", *Ap. J.* **456**, 525–540.

Reynolds, J.H.: 1960, "Determination of the age of the elements", *Phys. Rev. Lett.* **4**, 8-10.

Rowe, M.W. and Kuroda, P.K.: 1965, "Fissiogenic xenon from the Pasamonte meteorite", *J. Geophys. Res.* **70**, 709-714.

Russell, S.S., Srinivasan, G., Huss, G.R., Wasserburg, G.J. and MacPherson G.J.: 1996, "Evidence for widespread $^{26}$Al in the solar nebula and constraints for nebula time scales", *Science* **273**, 757-762.

Sahijpal, S. and Goswami, J.N.: 1998, "Refractory phases in primitive meteorites devoid of $^{26}$Al and $^{41}$Ca: Representative samples of first solar system solids?", *Ap. J. Lett.* **509**, L137–L140.

Sahijpal, S., Goswami, J.N., Davis, A.M., Grossman, L. and Lewis, R.S.: 1998, "A stellar origin for the short-lived nuclides in the early solar system", *Nature* **391**, 559–561.

Sahijpal, S., Goswami, J.N. and Davis, A.M.: 1999, "K, Mg, Ti and Ca isotopic compositions and refractory trace element abundances in hibonites from CM and CV meteorites: Implications for early solar system processes", *Geochim. Cosmochim. Acta*, submitted for publication.

Schramm, D.N., Tera, F., and Wasserburg, G.J.: 1970, "The isotopic abundance of $^{26}$Mg and limits on $^{26}$Al in the early solar system", *Earth Planet. Sci. Lett.* **10**, 44-59.

Shu, F.H.: 1995, "The birth of sunlike stars" *in Molecular clouds and star formation*, eds., Yuan, Chi and You, Junhan, World Scientific, Singapore, pp. 97-148.

Shu, F.H., Shang, H., Glassgold, E. and Lee, T.: 1997, "X-rays and fluctuating X-wind from protostars", *Science* **277**, 1475–1479.

Shukolyukov, A. and Lugmair, G.W.: 1993a, "Live iron-60 in the early solar system", *Science* **259**, 1138–1142.

Shukolyukov, A. and Lugmair, G.W.: 1993b, "$^{60}$Fe in eucrites", *Earth Planet. Sci. Lett.* **119**, 159–166.

Sonett, C.P., Colburn, D.S. and Swartz, K.: 1968, "Electrical heating of meteorite parent bodies and planets by dynamo induction from a pre-main sequence T-Tauri solar wind", *Nature* **219**, 924-926.

Srinivasan, G., Ulyanov, A.A. and Goswami, J.N.: 1994, "$^{41}$Ca in the early solar system" *Ap. J. Lett.* **431**, L67–L70.

Srinivasan, G., Sahijpal, S., Ulyanov, A.A. and Goswami, J.N.: 1996, "Ion microprobe studies of Efremovka CAIs: II. Potassium isotope composition and $^{41}$Ca in the early solar system", *Geochim. Cosmochim. Acta* **60**, 1823–1835.

Srinivasan, G., Goswami, J.N. and Bhandari, N.: 1999, "$^{26}$Al in eucrite Piplia Kalan: Plausible heat source and chronology of formation of eucrite parent bodies", *Science* **284**, 1348-1350.

Swindle, T.D. and Podosek, F.A.: 1988, "Iodine-xenon dating", in *Meteorites and the early solar system*, eds., Kerridge, J.F. and Matthews, M.S., Univ. Arizona Press, Tucson, AZ, pp. 1127–1146.

Urey, H.C.: 1955, "The cosmic abundances of potassium, uranium, and thorium and the heat balance of the earth, the moon , and Mars", *Proc. Natl. Acad. Sci. U.S.* **41**, 127-144.

Vanhala, H.A.T. and Cameron, A.G.W.: 1998, "Numerical simulations of triggered star formation. I. Collapse of dense molecular cloud cores", *Ap. J.* **508**, 291–307.

Wadhwa, M. and Russell, S.S.: 1999, "Timescales of accretion and differentiation in the early solar system: The meteoritic evidence", in *Protostars and Planets IV*, eds., Mannings, V., Russell, S. and Boss, A., Univ. Arizona Press, Tucson, AZ, in press.

Wasserburg, G.J. and Arnould, M.: 1987, "A possible relationship between extinct $^{26}$Al and $^{53}$Mn in meteorites and early solar system", in *Lecture Notes in Physics 287, 4th workshop on Nuclear Astrophysics*, eds., Hillebrandt, W. et al., Springer-Verlag, NY, pp. 262–276,.

Wasserburg, G.J., Busso, M., Gallino, R. and Raiteri, C.M.: 1994, "Asymptotic giant branch stars as a source of short-lived radioactive nuclei in the solar nebula", *Ap. J.* **424**, 412–428.

Wasserburg, G.J., Gallino, R., Busso, M., Goswami, J.N. and Raiteri, C.M.: 1995, "Injection of freshly synthesized $^{41}$Ca in the early solar nebula by an asymptotic giant branch star" *Ap. J. Lett.* **440**, L101–L104.

Weidenschilling, S.J.: 1977, "Aerodynamics of solid bodies in the solar nebula", *Mon. Not. Roy. Astron. Soc.* **180**, 57-70.

Yin, Q., Jagoutz, E. and Wanke, H.: 1992, "Re-Search for extinct $^{99}$Tc and $^{98}$Tc in the early solar system", *Meteoritics* **27**, 310.

# Xenology, FUN Anomalies and the Plutonium-244 Story

P.K. Kuroda[1] and W.A. Myers[2]
*[1]4191 Del Rosa Court, Las Vegas, Nevada 89121 USA, [2]Department of Chemical Engineering, University of Arkansas, Fayetteville, Arkansas 72701 USA*
*wam@engr.uark.edu*

Abstract: It has long been generally accepted that $^{244}$Pu fission xenon is absent in lunar fines, while carbonaceous chondrites contain the so-called xenon-HL, which is a strange fission-like xenon component, whose isotopic composition is different from that of the $^{244}$Pu fission xenon. Re-examination of a vast amount of existing xenon isotope data reveals, however, that lunar fines and carbonaceous chondrites both contain appreciable amounts of $^{244}$Pu fission xenon and they began to retain their xenon almost 5.1 billion years ago, when the $^{244}$Pu to $^{238}$U ratio in the solar system was roughly 1 to 2 (atom/atom). In the present paper, re-examination of the experimental data is extended to the lead/lead and potassium/argon dating methods.

## 1. INTRODUCTION

In 1957, the Soviet Union launched Sputnik and Burbidge *et al.* (1957) published an important paper concerning the synthesis of the elements in stars. Three years later, Reynolds (1960) published his discovery of the occurrence of $^{129}$I in the early solar system, and Kuroda (1960) put forward the hypothesis that $^{244}$Pu should also have been present in the early solar system and that experimental evidence for its existence could be secured by searching for the presence in meteorites of excess heavy xenon isotopes $^{131}$Xe, $^{132}$Xe, $^{134}$Xe and $^{136}$Xe which are produced by the spontaneous fission of $^{244}$Pu with a half-life of 6.55 x $10^{10}$ years. Then, newly elected President

John F. Kennedy declared that the United States would send a Man to the Moon before the decade was over.

A search for $^{244}$Pu fission xenon in meteorites thus began in 1960 and it is worthy of note that three years later Reynolds (1963) was aware of the fact that the carbonaceous chondrite Murray contained at least 20 x $10^{-12}$ $cm^3$STP (standard temperature and pressure)/g of *excess* $^{136}$Xe and in the following year Reynolds and Turner (1964) reported that the carbonaceous chondrite Renazzo contained as much as 50 x $10^{-12}$ $cm^3$STP/g of *excess* $^{136}$Xe.

Many years have elapsed since the discovery of $^{129}$I and $^{244}$Pu in the early solar system, but investigators in this field of study encountered numerous difficulties in explaining the origin of these extinct radionuclides in the early solar system. A brief summary of the extremely complicated sequence of events which took place in this field of study since the early 1960's will be presented in this report.

## 2. XENOLOGY, FUN-ANOMALIES AND THE $^{244}$PU STORY

In his classic paper entitled *Xenology,* Reynolds (1963) stated that: "*Xenology means to us the detailed study of the abundance of Xe isotopes evolved from meteorites by heating or other means and the inferences that can be drawn from these studies about the early history of the meteorites and the solar system. To the classicists Xenology means study of a strange substance, which is also appropriate.*"

In this paper, Reynolds (1963) discussed xenology in the context of theories of the origin of the heavy elements by Burbidge *et al.* (1957) and by Cameron (1958) and a theory of the origin of Xe isotope anomalies in meteorites by Kuroda (1960) and Cameron (1962). He called large variations of relative abundances of $^{129}$Xe the *special anomalies*, and the less spectacular variations observed at all mass numbers except 129, the *general anomalies*. Then he went on to state that the general anomalies are explained by two processes: (a) relative abundances at mass numbers 131, 132, 134 and 136 will be enhanced by the addition of the spontaneous fission product of $^{244}$Pu, as pointed out by Kuroda (1960) and (b) xenon in the sun has been exposed to heavy neutron irradiation during the deuterium-burning phase of the evolution of the sun and hence the transfer of solar xenon to earth would have the effect of enhancing the relative abundances at mass numbers 128, 129 and 132, as pointed out by Cameron (1962).

The presence of excess $^{244}$Pu fission products $^{131\text{-}136}$Xe was first detected by Rowe and Kuroda (1965). Several supporting evidences for the presence of excess $^{244}$Pu fission xenon in achondrites and ordinary chondrites were reported (Fleischer *et al.*, 1965; Rowe and Bogard, 1966; Pepin, 1966; Hohenberg *et al.*, 1967a; Wasserburg *et al.*, 1969) during the second half of the 1960's. It thus appeared to be quite certain at that time that the samples of the Moon, which were soon to become available to man, would clearly demonstrate the presence of the decay products of the extinct radionuclides $^{129}$I and $^{244}$Pu.

In the first report on the examination of the lunar samples brought to earth from the 20 July 1969 Apollo 11 landing on the Moon, members of the Lunar Sample Preliminary Examination Team (LSPET, 1969) wrote, however, that the isotopic compositions of xenon found in the fines and in a breccia resembled those of trapped xenon found in carbonaceous chondrites and the decay products of $^{129}$I and $^{244}$Pu both appeared to be absent in lunar samples.

An extremely important report entitled "Isotopic analysis of rare gases from stepwise heating of lunar fines and rocks" then appeared, in which Reynolds *et al.* (1970) concluded that (a) excess $^{244}$Pu fission xenon was absent in the Apollo 11 lunar fines 10084,59 and (b) a small excess of $^{238}$U fission xenon was detected in the Apollo 11 fine grained crystalline rock 10057,20.

It appeared that the conclusions reached by LSPET (1969) and Reynolds *et al.* (1970) concerning the isotopic compositions of xenon found in the Apollo 11 lunar samples were consistent with each other and these results could only be interpreted to mean that the Apollo 11 lunar samples must have begun to retain their xenon much later than the carbonaceous chondrites. Thus, at the beginning of 1970, most investigators in the field of xenology seemed to have accepted the conclusion that the Moon must be younger than the meteorites.

Meanwhile, in a short article entitled "Temperature of the sun in the early history of the solar system", Kuroda (1971) pointed out that the only way to explain the differences in the isotopic compositions of xenon found in carbonaceous chondrites, in lunar samples and in the earth's atmosphere is to consider the alterations of the isotopic composition of xenon by a combined effect of (a) mass-fractionation, (b) spallation and (c) *stellar-temperature* neutron-capture reactions. Kuroda (1971) thus felt that the conclusion reached by Reynolds *et al.* (1970) concerning the age of the moon must be in error, because of the fact that these investigators have always been interpreting the xenon isotope data without taking into account the above-mentioned processes (a), (b) and (c).

In an important paper entitled "Plutonium-244: Confirmation as an Extinct Radioactivity", Alexander *et al.* (1971, p. 837) reported: "*The mass spectrum of xenon from spontaneous fission in a laboratory sample of plutonium-244 is precisely what meteoriticists predicted it would be; this discovery completes a web of proof that this nuclide is a bona fide extinct radioactivity of galactic origin, that r-process nucleosynthesis was ongoing in the galaxy at the time of the birth of the sun, and that the early meteoritic abundances of plutonium-244, heretofore tentative, can be utilized with confidence in models for the chronology of galactic nucleosynthesis. The search for an explanation for anomalous fission-like xenon in carbonaceous chondrites can now be narrowed.*"

Hoffman *et al.* (1971, p. 132) published a paper entitled "Detection of Plutonium-244 in Nature", in which they reported "*Mass spectrometric measurements of plutonium isolated from Precambrian bastnasite confirm the presence of $^{244}$Pu in nature. Although the existence of $^{244}$Pu as an extinct radioactivity has been postulated to explain the xenon isotope ratios observed in meteorites, this is the first indication of its present existence in nature*".

It is interesting to recall that thirteen years earlier Seaborg (1958, p. 75) remarked: "*The isotope $^{244}$Pu (half-life, 7.6 x $10^7$ years) is especially interesting, because of its long half life, and offers intriguing possibilities for the more distant future*". Although he did not clearly state what the intriguing possibilities were, it seems to be quite likely that what he had in his mind in 1958 was the possibility of detection of $^{244}$Pu in nature, either in meteorites or in terrestrial samples of minerals or ores.

It is worthy of note here, however, that the report of Hoffman *et al.* (1971) was based solely on a single measurement, which was never repeated or reproduced. Moreover, the following year Fleischer and Naeser (1972, p. 465) reported: "*We found that bastnaesite, a rare earth fluorocarbonate, from the Precambrian Mountain Pass deposit has an apparent Cretaceous fission track age, and hence does not reveal any anomalous fission tracks due to $^{244}$Pu.*"

Manuel *et al.* (1972a) reported that carbonaceous chondrites contain two isotopically distinct components of trapped xenon, which can not be explained by the occurrence of nuclear or fractionation processes within these meteorites. This meant that the seemingly complicated theory proposed by Kuroda (1971) did not explain all the existing experimental data. According to Manuel *et al.* (1972a) the heaviest xenon isotopes, $^{134}$Xe and $^{136}$Xe, and the lightest xenon isotopes, $^{124}$Xe and $^{126}$Xe, were simultaneously enriched in carbonaceous chondrites.

It is to be noted here that the method of treatment of the xenon isotopes used by Manuel *et al.* (1972a) was essentially the same as the one commonly used during the 1960's, but their arguments seemed to lead us to a new concept, that the r- and the p-process nucleosynthesis products may not have been initially well mixed with the solar nebula. Moreover, the fact that another strange xenon component was added to the list of strange xenon components six years later seemed to strengthen the case for Manuel *et al.*(1972a) arguments. They called the new trapped xenon component, in which the r- and the p-process products were simultaneously enriched, xenon-X and this became widely known as xenon-HL (Pepin and Phinney, 1978) during the 1970's and the 1980's.

Srinivasan and Anders (1978) reported that the Murchison carbonaceous chondrite contained a new type of xenon component, enriched by up to 50 percent in five of the nine stable xenon isotopes, mass numbers 128 to 132. They pointed out that their results were strongly suggestive of processes believed to take place in red giants, the s-process nucleosynthesis, and if this interpretation were correct, then primitive meteorites contain yet another kind of alien, presolar material, dust grains ejected from the red giants.

Sakamoto (1974) published a noteworthy article entitled "*Possible cosmic dust origin of terrestrial plutonium-244*". He reinvestigated the implications of its present existence in nature in terms of three possibilities: (1) survival of the primeval $^{244}$Pu, (2) influx as a heavy cosmic-ray component, and (3) inflow as a cosmic-dust component from supernova remnants. Results from his detailed calculations seemed to favor the possibility (3) and he concluded (p. 131): "*An experimental search for $^{244}$Pu in deep sea sediment would be feasible if the present hypothesis is valid. Further studies along this line are under way at our laboratory.*"

Then Wasserburg and Papanastassiou (1982) published a review article entitled "Some short-lived nuclides in the early solar system: A connection with the placental ISM", in which they reviewed, in detail, the steps that led to the discovery of the third *bona fide* extinct radionuclide $^{26}$Al with a half-life of 0.72 million years. They also pointed out that virtually every element analyzed in two Allende inclusions C-1 and EK 1-4-1 had an anomalous isotopic composition and these have been called "FUN" samples to denote mass-fractionation (F) and unknown nuclear anomalies (UN).

That same year Kuroda (1982) presented a new idea that the origin of the isotopic anomalies observed in many elements (the so-called FUN anomalies) may be attributable to the x-process nucleosynthesis, which must have occurred at the same time the r-process produced heavy elements, such as $^{244}$Pu, during the explosion of the last supernova.

In his 1983 Nobel Lecture, Fowler (1984) attached a great significance to the occurrence of $^{129}$I and $^{26}$Al, but not $^{244}$Pu, in the early solar system, and he expressed his opinion that the fact that anomalies produced by short-lived isotopes to normal abundances all were of the order of $10^{-4}$ (atom/atom) despite the wide range in their mean lifetime, indicated that this anomaly range must be the result of inhomogeneous mixing of *exotic* materials with much larger quantities of *normal* solar system materials.

Three years later in his 1986 Crafoord Lecture, Wasserberg (1987), presented a schematic diagram showing rates of nucleosynthesis determined from relatively short-lived nuclei, according to which the most recent *blip* with a concentration of approximately $10^{-4}$ solar masses, took place within a few million years before formation of the solar system and the last significant "r" process contributions (several percent) took place a few hundred million years before that and contributed $^{244}$Pu and some U and Th and other associated transuranic nuclei.

The opinions expressed by Fowler (1984) and Wasserburg (1987) had a profound influence on the way of thinking of most investigators during the 1980's. Podosek and Swindle (1988a,b) and Swindle and Podosek (1988) thus reported that, among the three extinct radionuclides known to man at that time, $^{26}$Al with the shortest half-life of only 0.72 million years was likely to be the most useful in the study of the events which took place in the early solar system, while despite the availability of data from more than 75 meteorites, the question of whether the I-Xe system was a chronometer and, if so, what it was dating had not been conclusively answered.

Then, in the following year two important discoveries were announced by the University of Chicago group. One was the discovery by Lewis *et al.* (1989) that the strange xenon component HL found in carbonaceous chondrites is carried by interstellar grains of tiny diamonds (less than 3 nm in diameter), while Zinner *et al.* (1989) reported that the so-called s-type xenon is carried by interstellar grains of silicon carbide.

In a review article entitled "Interstellar grains in meteorites", Ott (1993) reported that primitive meteorites contain grains of silicon carbide, graphite and diamond formed outside of the solar system and probably before its birth. In another review article which appeared in the same year Anders and Zinner (1993) reported that these microdiamonds contain xenon-HL, which shows the signature of the p- and r-process and thus is apparently derived from supernovae, while silicon carbide inclusions show the signature of the s-process and apparently come from red giant carbon stars of 1 - 3 solar masses. It is worthy of note here, however, that they concluded (p. 506): "*The most pristine, unaltered interstellar grains provide little information on the early solar system, bearing no memory of their gentle arrival ...*"

It so happened, however, that this puzzling conclusion was the direct consequence of the fact that they *assumed* the questionable conclusions reached by a majority of previous investigators to have been correct: that the strange xenon components are isotopically *pure* substances. If that were indeed the case, there would have been no way of dating the time of arrival of the interstellar grains sometime during the early history of the solar system.

We therefore reported (Kuroda and Myers, 1998a) recently that, although it became increasingly popular to attribute the isotopic anomalies observed in meteorites to the result of inhomogeneous mixing of *exotic* substances from outside of the solar system over a short time, existing experimental data can be explained in a straightforward manner as due to the result of free decay of the extinct radionuclides created in the nucleosynthesis processes which took place in a supernova more than 5 billion years ago.

Meanwhile, in a series of papers (Kuroda 1989, 1993; Myers and Kuroda 1989, 1991a, 1991b, 1992, 1995; Kuroda and Myers 1991a, 1991b, 1991c, 1992a, 1992b, 1993a, 1993b, 1994a, 1995, 1996, 1997, 1998a, 1998b, 1999) published since 1989, we have re-examined a vast amount of experimental data which have been accumulated in the literature since the 1960's and reported that some of the carbonaceous chondrites contain more than 70 x $10^{-12}$ ($cm^3$STP/g) of excess $^{136}$Xe ($^{136f}$Xe) from the spontaneous fission of $^{244}$Pu, while many of the lunar fines contain from 100 to almost 1,000 x $10^{-12}$ ($cm^3$STP/g) of the $^{244}$Pu fission xenon and the most primitive carbonaceous chondrites and lunar fines appear to have started to retain their xenon more than 5 billion years ago, when the ratio of $^{244}$Pu to $^{238}$U in the solar system was about 2 to 1 (atom/atom).

## 3. $^{244}$PU DATING OF THE EARLY SOLAR SYSTEM

Basford *et al.* (1973) published a 40-page article entitled "*Krypton and xenon in lunar fines*". In this important paper, they stated (p. 1915): "*The composition of surface correlated Xe is significantly variable from sample to sample; the systematics of the variations provide evidence for the presence of specific components tentatively attributed to neutron capture in iodine, decay of extinct $^{129}$I, and spontaneous fission of $^{244}$Pu. ... Assumption that the second volume components is fissiogenic leads to derivation of a spallation spectrum for lunar fines which is in generally excellent agreement with measurements in lunar rocks, and to an isotopic composition for the fissiogenic Xe which approaches that from $^{244}$Pu but does not quite agree with it within error ...*"

Basford *et al.* (1973) were aware of the important roles played by the process of neutron-capture reactions in altering the isotopic composition of xenon found in lunar fines but were unable to identify the presence of excess $^{244}$Pu xenon in these samples. The reason was that they did not take into account the conclusion reached by Kuroda (1971) that the only way to explain the differences in the isotopic compositions of xenon found in carbonaceous chondrites, in lunar fines and in the earth's atmosphere is to consider the alterations of the isotopic composition of xenon by a combined effect of (a) mass-fractionation, (b) spallation and (c) stellar temperature neutron-capture reactions. Let us, therefore, proceed to re-examine the xenon isotope data reported by Basford *et al.* (1973) by the use of the method, which we reported in a series of papers since 1993 (Kuroda 1993, Kuroda and Myers 1994a, 1994b, 1995, 1997).

Table 1 shows the relationship between the isotopic compositions of xenon found in the bulk samples of the fines 10084,48 ($< 1\mu$) and in the earth's atmosphere. Note that the alterations of the abundances of xenon isotopes caused by the effects of (a) mass-fractionation and (b) spallation reactions are considered here.

*Table 1.* Comparison of the isotopic compositions of xenon found in the Apollo 11 fines 10084,48 (< 1 μ) and in the atmosphere

| Mass No | (I) 10084,48(< 1μ), Basford *et al.* (1973) | (II) Corrected For mass-fractionation | (III) Atmosphere | (IV) Spallation Xe removed From II | (V) Difference ($\delta_i$) =(IV)-(III) |
|---|---|---|---|---|---|
| 124 | 0.01691 | 0.01202 | 0.0108 | 0.01034 | -0.00046 |
| | | | | -0.00168 | |
| 126 | 0101710 | 0.01290 | 0.0101 | 0.0101 | ≡0.0000 |
| | | | | -0.0028 | |
| 128 | 0.28720 | 0.2296 | 0.216 | 0.2256 | +0.0096 |
| | | | | -0.0040 | |
| 129 | 3.4870 | 2.86885 | 2.981 | ≡(2.8689) | (-0.1122) |
| 130 | 0.5548 | 0.46965 | 0.460 | 0.46705 | +0.00705 |
| | | | | -0.0026 | |
| 131 | 2.7488 | 2.39367 | 2.388 | 2.3839 | -0.00409 |
| | | | | -0.0098 | |
| 132 | 3.3400 | 2.99132 | 3.032 | 2.9897 | -0.04230 |
| | | | | -0.00162 | |
| 134 | 1.2432 | =1.1770 | 1.177 | ≡0.0000 | |
| 136 | =1.0000 | =1.0000 | =1.0000 | ≡0.0000 | |

$$\frac{\delta_{132}}{\delta_{130}} = \frac{-0.04230}{+0.00705} = -6.000(atom/atom)$$

The equation derived by Aston (1923) was used here to correct for the effect of mass fractionation:

$$r = \sqrt[\frac{m_2+m_1}{m_2-m_1}]{\frac{\text{Initial Volume}}{\text{Final Volume}}} \quad (1)$$

Where r is the enrichment by diffusion of the residue as regard the heavier constituent, and $m_1$ and $m_2$ are the masses of the lighter and heavier isotopes, respectively.

The cosmic-ray production ratios of xenon isotopes reported by Clark *et al.* (1967) were used here to correct for the effect of excess spallation xenon present in the meteorite:

$$^{124}Xe:^{126}Xe:^{128}Xe:^{130}Xe:^{131}Xe:^{132}Xe =$$

$$0.60:1.00:1.43:0.93:3.49:0.58 \text{ (atom/atom)} \quad (2)$$

As shown in Table 1, differences in the isotopic compositions of xenon found in the fines 10084,48 (< 1μ) and in the earth's atmosphere are generally not too large, but noteworthy differences do exist among them.

The facts that (a) the relative abundances of $^{124}$Xe and $^{126}$Xe are much greater in the former relative to the latter and (b) $^{126}$Xe to $^{124}$Xe ratio is greater than one in the former, while it is smaller than one in the latter tell us that the lunar fines 10084,48 (< 1$\mu$) contains a considerable excess of spallation xenon.

It is also highly significant that the value of $\delta_{129}$ for the sample 10084,48 (< 1$\mu$) is negative relative to the earth's atmosphere. This means that the atmospheric xenon contains an excess of $^{129}$Xe relative to the xenon from the moon, although Basford *et al.* (1973) reported that the reverse was the case.

One more important fact which emerges from the results of calculations shown in Table 1 is that the values of $\delta_{128}$ and $\delta_{130}$ are positive, while the values of $\delta_{131}$ and $\delta_{132}$ are negative for the xenon found in the moon. This means that the xenon from the moon is likely to contain a significant excess of $^{244}$Pu fission xenon.

Kuroda (1993) drew attention to the important fact that $^{130}$Xe and $^{132}$Xe are the only isotopes of xenon whose abundances are altered solely by the neutron-capture reactions on the neighboring isotopes of xenon: $^{129}$Xe and $^{131}$Xe, and he reported that, if the atmospheric xenon is irradiated with 10 keV (stellar temperature) neutrons, the abundances of $^{130}$Xe and $^{132}$Xe are both enhanced and the following relationship holds:

$$\Delta_{132}/\Delta_{130} = 0.767(\text{atom/atom}) \tag{3}$$

and hence this ratio can be used as an "*indicator*" for the purpose of finding the end point value (f*) or the exact amount of $^{244}$Pu fission xenon in the sample.

The $^{244}$Pu spontaneous fission yields for the mass range 131 to 136 have been accurately measured by Alexander *et al.* (1971):

$$^{131}\text{Xe}:^{132}\text{Xe}:^{134}\text{Xe}:^{136}\text{Xe} = 0.246:0.870:0.921:100(\text{atom/atom}) \tag{4}$$

Table 2 shows the changes of the values of $\delta_i$ and $\delta_{132}/\delta_{130}$ when small amounts of $^{244}$Pu fission xenon ($^{136f}$Xe) are removed little by little from the sample. Note that the amount of $^{244}$Pu fission to be removed here is expressed in terms of

$$f = {}^{136f}\text{Xe}/\Sigma^{136}\text{Xe} \tag{5}$$

It is interesting to note in Table 2 that the initial value of $\delta_{132}/\delta_{130}$, which is -6.000(atom/atom), increases steadily as the amount (f) of $^{244}$Pu fission xenon to be removed is increased little by little and reaches the end-point value of f* = $\Delta 132/\Delta 130$ = 0.767 (atom/atom) is reached when f = f* = 0.08215 (atom/atom).

*Table 2.* Amount of $^{244}Pu$ fission xenon ($^{136f}Xe$) in the Apollo 11 fines 10084,48 (< 1$\mu$) calculated from the xenon isotope data reported by Basford *et al.*

| $^{244}Pu$ fission Xe removed, $f = {}^{136f}Xe/\Sigma^{136}Xe$ | $\delta_{124}$ | $\delta_{126}$ | $\delta_{128}$ | $\delta_{129}$* | $\delta_{130}$ | $\delta_{131}$ | $\delta_{132}$ | $\delta_{132}/\delta_{130}$ |
|---|---|---|---|---|---|---|---|---|
| 0.0000 | -0.00046 | ≡0.0000 | 0.0096 | (-0.1122) | 0.00705 | -0.00409 | -0.04230 | -6.000 |
| 0.0100 | -0.00050 | ≡0.0000 | 0.0095 | (-0.1101) | 0.00807 | 0.00212 | -0.03582 | -4.439 |
| 0.0200 | -0.00055 | ≡0.0000 | 0.0094 | (-0.1082) | 0.00908 | 0.00831 | -0.02931 | -3.230 |
| 0.0300 | -0.00060 | ≡0.0000 | 0.0093 | (-0.1064) | 0.01007 | 0.01448 | -0.02279 | -2.264 |
| 0.0400 | -0600065 | ≡0.0000 | 0.0092 | (-0.1049) | 0.01104 | 0.02063 | -0.01625 | -1.472 |
| 0.0500 | -0100071 | ≡0.0000 | 0.0090 | (-0.1036) | 0.01200 | 0.02676 | -0.00970 | -0.8081 |
| 0.0600 | -0.00076 | ≡0.0000 | 0.0088 | (-0.1025) | 0.01294 | 0.03285 | -0.00313 | -0.2418 |
| 0.0700 | -0.00081 | ≡0.0000 | 0.0086 | (-0.1016) | 0.01386 | 0.03892 | 0.00345 | 0.2491 |
| 0.0800 | -0.00087 | ≡0.0000 | 0.0084 | (-0.1009) | 0.01476 | 0.04494 | 0.01005 | 0.6806 |
| 0.08100 | -0.00087 | ≡0.0000 | 0.0084 | (-0.1009) | 0.01485 | 0.04554 | 0.01071 | 0.7209 |
| 0.08200 | -0.00088 | ≡0.0000 | 0.0083 | (-0.1008) | 0.01494 | 0.04614 | 0.01137 | 0.7608 |
| 0.08210 | -0.00088 | ≡0.0000 | 0.0083 | (-0.1008) | 0.01495 | 0.04620 | 0.01143 | 0.7649 |
| 0.08215(=f*) | -0.00088 | ≡0.0000 | +0.00834 | (-0.1008) | -0.01495 | +0.04613 | +0.01147 | 0.7668 |

Table 3 shows the relationship between the isotopic compositions of xenon found in the bulk samples of the fines 10084,48 (< 1$\mu$) and in the earth's atmosphere, in which the alterations of the abundances of xenon isotopes caused by the effects of (a) mass-fractionation, (b) spallation and (c) stellar temperature (10 keV) neutron-capture reactions, as well as the presence of excess $^{244}Pu$ fission xenon, are considered. The total $^{136}Xe$ content ($\Sigma^{136}Xe$) of the fines 10084,48 (< 1$\mu$) is 4.47 x $10^{-8}$ ($cm^3$STP/g) and hence the content of $^{136f}Xe$ is (4.47 x $10^{-8}$) (0.08215) = 3,672 x $10^{-12}$ ($cm^3$STP/g).

*Table 3.* $^{244}Pu$ fission xenon in the Apollo 11 fines 1008448 (< 1$\mu$) calculated from the xenon isotope data reported by Basford *et al.* (1973).

| Mass No. | (I) 10084,48 (< 1$\mu$), Basford *et al.*, (1973) | (II) Fission xenon removed | (III) Re-normalized To $^{136}Xe$=1.0000 | (IV) Corrected for M.F. | (V) Spallation xenon removed | (VI) Atmosphere | $\Delta_i$ +(V)-(VI) |
|---|---|---|---|---|---|---|---|
| 124 | 0.01691 | | 0.01842 | 0.01135 | 0.00992<br>-0.00143 | 0.0108 | -0.00088 |
| 126 | 0.01710 | | 0.01863 | 0.01249 | 0.0101<br>-0.00239 | 0.0101 | =0.0000 |
| 128 | 0.2872 | | 0.31291 | 0.22775 | 0.22434<br>-0.00341 | 0.216 | +0.00834 |
| 129 | 3.4870 | | 3.79910 | 2.88020 | | 2.981 | (-0.1008) |
| 130 | 0.5548 | | 0.60446 | 0.47717 | 0.47495<br>-0.00222 | 0.460 | +0.01495 |
| 131 | 2.7488 | 2.72818<br>-0.02062 | 2.97236 | 2.44256 | 2.43413<br>-0.00843 | 2.388 | +0.04613 |
| 132 | 3.3400 | 3.26804<br>-0.07196 | 3.56053 | 3.04485 | 3.04396<br>-0.00139 | 3.032 | +0.01147 |

| Mass No. | (I) 10084,48 (< 1μ), Basford *et al.*, (1973) | (II) Fission xenon removed | (III) Re-normalized To $^{136}$Xe=1.0000 | (IV) Corrected for M.F. | (V) Spallation xenon removed | (VI) Atmosphere | $\Delta_I$ +(V)-(VI) |
|---|---|---|---|---|---|---|---|
| 134 | 1.2432 | 1.16754<br>-0.07566 | 1.27204 | =1.177 | =1.177 | 1.177 | =0.000 |
| 136 | =1.0000 | 0.91785<br>-0.08215 | =1.0000 | =1.0000 | =1.0000 | =1.000 | =0.000 |

$$\frac{D_{132}}{D_{130}} = \frac{0.01147}{0.01495} = 0.767$$

It is also interesting to note that this sample contains not only a large quantity of $^{244}$Pu fission xenon, but also the isotopic composition of the trapped xenon shows a striking resemblance to those of the primitive xenon reported by Takaoka (1972) and *U*-xenon reported by Pepin and Phinney (1978) which, in turn, are quite similar to those of the mass-fractionated atmospheric xenon.

Table 4 shows a comparison of the isotopic composition of the trapped xenon found in the Apollo 11 fines 10084,48 (< 1μ) and those of the primordial (or primitive) xenon reported by Takaoka (1972) and Pepin and Phinney (1978). They show a striking resemblance with each other and small differences which they display are likely to be attributed to the effects of spallation reactions and the presence of $^{244}$Pu fission xenon in the samples.

Reynolds *et al.* (1970) concluded that (a) excess $^{244}$Pu fission xenon was absent in the Apollo 11 lunar fines 10084,59 and (b) a small excess of $^{238}$U fission xenon was present in the Apollo 11 fine-grained crystalline rock 10057,20. The results of calculations which we have carried out in this study demonstrate, however, that (a) lunar fines contain not only appreciable amounts of $^{244}$Pu fission xenon, but also (b) a trapped xenon, whose isotopic composition is least affected by the mass-fractionation, spallation and stellar-temperature neutron-capture reactions as shown in Table 4.

*Table 4.* A 0 of the isotopic composition of the trapped xenon found in the Apollo 11 fines 10084,48 (< 1μ) and those of the primordial xenon.

| Mass No. | (I) Trapped xenon found in fines 10084,48 (< 1μ) (See Table 3, Column III) | (II) U-type xenon (Pepin and Phinney, 1978) | (III) Atmospheric Xenon (Nier, 1950) | (IV) Mass-fractionated atmospheric Xe ($^{134}$Xe/$^{136}$Xe = 1.280) | (V)Mass-fractionated atmospheric Xe ($^{134}$Xe/$^{136}$Xe = 1.290) | (VI) Primitive xenon (Takaoka's Xe, 1972) |
|---|---|---|---|---|---|---|
| 124 | 0.01842 | 0.01772 | 0.0108 | 0.0182 | 0.0191 | (0.018) |
| 126 | 0.01863 | 0.01528 | 0.0101 | 0.0156 | 0.0162 | (0.016) |

| Mass No. | (I) Trapped xenon found in fines 10084,48 (< 1μ) (See Table 3, Column III) | (II) U-type xenon (Pepin and Phinney, 1978) | (III) Atmospheric Xenon (Nier, 1950) | (IV) Mass-fractionated atmospheric Xe ($^{134}Xe/^{136}Xe$ ≡ 1.280) | (V)Mass-fractionated atmospheric Xe ($^{134}Xe/^{136}Xe$ ≡ 1.290) | (VI) Primitive xenon (Takaoka's Xe, 1972) |
|---|---|---|---|---|---|---|
| 128 | 0.31291 | 0.3058 | 0.216 | 0.3045 | 0.34143 | 0.310 |
| 129 | 3.79910 | ——— | 2.981 | 4.0214 | 4.2062 | ——— |
| 130 | 0.60446 | 0.6012 | 0.460 | 0.5940 | 0.6081 | 0.610 |
| 131 | 2.97236 | 3.003 | 2.388 | 2.9525 | 3.0100 | 2.99 |
| 132 | 3.56053 | 3.636 | 3.032 | 3.5905 | 3.6393 | 3.63 |
| 134 | 1.27204 | 1.28 | 1.177 | ≡1.280 | ≡1.290 | 1.29 |
| 136 | ≡1.0000 | ≡1.000 | ≡1.000 | ≡1.000 | ≡1.000 | ≡1.00 |

Table 5 shows a comparison of the values of $\Delta_{130}$ and $\Delta_{132}$ calculated from the reported xenon isotope data for the trapped xenon found in various meteorites and lunar samples. Note that the trapped xenon found in the Apollo 11 fines 10084,48 (< 1$\mu$) has the smallest values of $\Delta_{130}$ and $\Delta_{132}$, indicating that it has been least affected by the process of stellar-temperature neutron capture reactions.

*Table 5.* A comparison of the values of $\Delta_{130}$ and $\Delta_{132}$ calculated from the xenon isotope data for various meteorites and lunar samples.

| No. Sample (References) | $\Delta_{130}$ | $\Delta_{132}$ | $\Delta_{130}/\Delta_{132}$ (atom/atom) |
|---|---|---|---|
| (1) Apollo fines 10084,48 (1 <μ) (Basford *et al.*, 1973) | 0.01495 | 0.01147 | 0.767 |
| (2) Luna 16 fines 16-g-7 (Kaiser, 1972 | 0.01596 | 0.01479 | 0.767 |
| (3) Apollo 11 fines 10084,59 (Reynolds *et al.*, 1970) | 0.01648 | 0.01264 | 0.767 |
| (4) Orgueil (C1) (Eugster *et al.*, 1967) | 0.02300 | 0.01765 | 0.767 |
| (5) Orgueil (C1) (Krummenacher *et al.*, 1962) | 0.02348 | 0.01829 | 0.767 |
| (6) Orgueil (C1) (Pepin and Phinney, 1978) | 0.02662 | 0.02042 | 0.767 |
| (7) Murray (C2) (Kuroda *et al.*, 1974) | 0.02754 | 0.02113 | 0.767 |
| (8) Murchison (C2) (Kuroda *et al.*, 1975) | 0.02925 | 0.02243 | 0.767 |
| (9) Allende (CV3) (Manuel *et al.*, 1972b) | 0.4100 | 0.03148 | 0.767 |
| (10) Atmospheric xenon irradiated with a total flux of $\phi{\cdot}\tau = 10^{22}$(n/cm$^2$) of stellar-temperature (10 keV) neutrons (Kuroda and Myers, 1998b) | 0.0335 | 0.0257 | 0.767 |

Table 6 shows the $^{244}$Pu/$^{136}$Xe ages of the size separated Apollo 11 fines 10084 calculated from the xenon isotope data reported by Basford *et al.* (1973). The value of 544 ppb for the uranium content reported by Tatsumoto and Rosholt (1970) was used here for the calculation. Note that sample No. 1 (< 1$\mu$) has by far the highest content of ($^{136f}$Xe)Pu and seems to have started to retain its xenon 5,071 million years ago when the $^{244}$Pu to $^{238}$U ratio in the solar system was as high as 0.444 (atom/atom).

*Table 6.* $^{244}$Pu/$^{136}$Xe Ages of the size-separated Apollo 11 fines 10084,48 studied by Basford *et al.* (1973).

| No. Size-separated sample ($\mu$) | U (ppb) | $\Sigma^{136f}$Xe | ($^{136f}$Xe)$_U$ | ($^{136f}$Xe)$_{Pu}$ (10$^{-12}$cm$^3$STP/g) | ($^{136f}$Xe)$_{Pu}$/U (10$^{-4}$cm$^3$STP/g) | $^{244}$Pu/$^{238}$U (atom/atom) | $^{244}$Pu/$^{136}$Xe Age (10$^6$ year) |
|---|---|---|---|---|---|---|---|
| (1) < 1 | a | 3,672 | 2 | 3,670 | 67.5 | 0.444 | 5,071 |
| (2) 1-4 | a | 973 | 2 | 971 | 17.9 | 0.118 | 4,911 |
| (3) 4-8 | a | 399 | 2 | 397 | 7.30 | 0.0480 | 4,803 |
| (4) 4-8 (repeat) | a | 77 | 2 | 75 | 1.38 | 0.00907 | 4,602 |
| (5) 16-25 | a | (-32) | --- | ----- | ------ | ----- | ----- |
| (6) 25-37 | a | 269 | 2 | 267 | 4.91 | 0.0323 | 4,755 |
| (7) 37-74 | a | (-35) | --- | ----- | ----- | ----- | ----- |
| (8) 74-105 | a | 74 | 2 | 72 | 1.32 | 0.00867 | 4,597 |
| (9) 105-147 | a | 81 | 2 | 79 | 1.45 | 0.00953 | 4,608 |
| (10) 147-250 | a | 40 | 2 | 38 | 0.70 | 0.00460 | 4,520 |
| (11) 250-590 | a | 213 | 2 | 211 | 3.88 | 0.0255 | 4,727 |
| (12) 590-1000 | a | 168 | 2 | 166 | 3.05 | 0.0200 | 4,698 |
| (13) Unsieved | a | 337 | 2 | 335 | 6.16 | 00.0405 | 4,783 |

a) A value of 544 ppb reported by Tatsumoto and Rosholt (1970) was used for the calculation.

Table 7 shows the $^{244}$Pu/$^{136}$Xe ages of several other Apollo 11 fines 10084 calculated from the xenon isotope data reported by various investigators and the uranium content of 544 ppb reported by Tatsumoto and Rosholt (1970). Samples No. 1 and No. 2 studied by the Lunar Sample Preliminary Examination Team (LSPET, 1969) and by Reynolds *et al.* (1970) seem to have started to retain their xenon about 4,700 million years ago when the $^{244}$Pu to $^{238}$U ratio in the solar system was 0.025 to 0.027 (atom/atom), and their ages are roughly the same as that of the 25-37 $\mu$ fraction of the fines 10084,48, studied by Basford *et al.* (1973). Sample No. 4, 10084,12 (< 37$\mu$) studied by Podosek *et al.* (1971), on the other hand, seems to have started to retain its xenon 4,819 million years ago when the $^{244}$Pu to $^{238}$U ratio in the solar system was 0.0549 (atom/atom) somewhat earlier than 4,800 million years ago, when the sample No. 3 (10084,48, 4 - 8$\mu$), shown in Table 6, began to do so.

*Table 7.* $^{244}$Pu/$^{136}$Xe ages of the Apollo 11 fines 10084.

| No. Sample (Reference) | U (ppb) | $\Sigma^{136f}$Xe | ($^{136f}$Xe)$_U$ | ($^{136f}$Xe)$_{Pu}$ (10$^{-12}$cm$^3$STP/g) | ($^{136f}$Xe)$_{Pu}$/U (10$^{-4}$cm$^3$STP/g) | $^{244}$Pu/$^{238}$U (atom/atom) | $^{244}$Pu/$^{136}$Xe Age (10$^6$ year) |
|---|---|---|---|---|---|---|---|
| (1) 10084 (LSPET, 1969) | 544 | 212 | 2 | 3.86 | 0.0254 | 0.0254 | 4,726 |

| No. Sample (Reference) | U (ppb) | $\Sigma^{136f}Xe$ | $(^{136f}Xe)_U$ | $(^{136f}Xe)_{Pu}$ (10$^{-12}$cm$^3$STP/g) | $(^{136f}Xe)_{Pu}/U$ (10$^{-4}$cm$^3$STP/g) | $^{244}Pu/^{238}U$ (atom/atom) | $^{244}Pu/^{136}Xe$ Age (10$^6$ year) |
|---|---|---|---|---|---|---|---|
| (2) 10084,59 (Reynolds *et al.*, 1970) | 544 | 226 | 2 | 4.12 | 0.0271 | 0.0271 | 4,734 |
| (3) 10084,59 (Hohenberg *et al.*, 1970) | 544 | (-122) | --- | --- | --- | ----- | ----- |
| (4) 10084,12 (Podosek *et al.*, 1971) | 544 | 458 | 2 | 8.35 | 0.0549 | 0.0549 | 4,819 |
| (5) 10084,29 (Marti *et al.*, 1970) | 544 | 118 | 2 | 2.13 | 0.0140 | 0.0140 | 4,655 |

Table 8 shows the $^{244}Pu/^{136}Xe$ ages of the Luna 16 and 20 samples studied by Kaiser (1972) and Shukolyukov *et al.* (1989). It is interesting to note that the Luna 16 regolith 1608-3,26 (Shukolyukov *et al.* in 1989) contains as much as 940 x $10^{-12}$ (cm$^3$STP/g) of $^{244}Pu$ fission xenon and it seems to have started to retain their xenon more than 5,000 million years ago, when the $^{244}Pu$ to $^{238}U$ ratio in the solar system was as high as 0.247 (atom/atom). As we recently pointed out (Myers and Kuroda, 1991b), the isotopic compositions of the trapped xenon found in this sample shows a striking resemblance to those of the primordial xenon, such as Pepin and Phinney's (1978) U-xenon and Takaoka's (1972) primitive xenon.

*Table 8.* $^{244}Pu/^{136}Xe$ ages of the Luna 16 and 20 samples of fines and regoliths.

| No. Sample (Reference) | U (ppb) | $\Sigma^{136f}Xe$ | $(^{136f}Xe)_U$ | $(^{136f}Xe)_{Pu}$ (10$^{-12}$cm$^3$STP/g) | $(^{136f}Xe)_{Pu}/U$ (10$^{-4}$cm$^3$STP/g) | $^{244}Pu/^{238}U$ (atom/atom) | $^{244}Pu/^{136}Xe$ Age (10$^6$ year) |
|---|---|---|---|---|---|---|---|
| (1) Luna 16 fines 16-G-7 (Kaiser, 1972) | 375 | 442 | 1 | 441 | 11.8 | 0.0774 | 4,861 |
| (2) Luna 16 regolith 1608-3, 26 (Shukolyukov *et al.*, 1989) | 250 | 940 | 1 | 940 | 37.6 | 0.247 | 5,001 |
| (3) Luna 20 regolith 2004,37 (Skukolyukov *et al.*, 1989) | 500 | 117 | 2 | 115 | 2.30 | 0.0151 | 4,664 |

Table 9 shows that the unseparated sample of the Apollo 12 soil 12001, ba-2 studied by Eberhardt *et al.* (1972), contains as much as 973 x $10^{-12}$ (cm$^3$STP/g) of $^{244}Pu$ fission xenon and the isotopic composition of the trapped xenon found in this soil sample also resembles that of the primordial xenon, as reported by us (Myers and Kuroda, 1991b) recently.

*Table 9.* $^{244}Pu/^{136}Xe$ ages of the Apollo 12 soils studied by Eberhardt (1972)

| No. Sample (Reference) | U (ppb) | $\Sigma^{136f}Xe$ | $(^{136f}Xe)_U$ | $(^{136f}Xe)_{Pu}$ | $(^{136f}Xe)_{Pu}/U$ ($10^{-4}cm^3STP/g$) | $^{244}Pu/^{238}U$ (atom/atom) | $^{244}Pu/^{136}Xe$ Age ($10^6$ year) |
|---|---|---|---|---|---|---|---|
| | | ($10^{-12}cm^3STP/g$) | | | | | |
| (1) Apollo 12 Soil 12001 ba-2 (unseparated) | (1,500) | 973 | (5) | (968) | (6.45) | (0.0424) | (4,788) |
| (2) BB-12(151μ) | ____ | -123 | ____ | ____ | ____ | ____ | ____ |
| (3) BB-13 (82μ) | ____ | -66 | ____ | ____ | ____ | ____ | ____ |
| (4) BB-14 (44μ) | ____ | -25 | ____ | ____ | ____ | ____ | ____ |
| (5) BB-15 (20μ) | ____ | -52 | ____ | ____ | ____ | ____ | ____ |
| (6) bb-16 (10.4μ) | (1,500)[a] | 499 | (5) | (494) | (3.29) | (0.0217) | (4,707) |
| (7) BB-17 (2μ) | (1,500)[a] | 275 | (5) | (271) | (1.81) | (0.0119) | (4,635) |
| (8) BB-18 (1.3μ) | ____ | -656 | ____ | ____ | ____ | ____ | ____ |

a) According to LSPET (1970), the uranium content of Apollo 12 fines 12070 is 1,500 (±200) ppb.

Table 10 shows the $^{244}Pu/^{136}Xe$ ages of a few samples of Apollo 14, 15 and 17 fines and soils. It is again worthy of note that the Apollo 17 soil 72701 studied by Bogard *et al.* (1974) contains as much as 942 x $10^{-12}$ ($cm^3STP/g$) of $^{244}Pu$ fission xenon and it seems to have began to retain its xenon just about the same time as the Luna 16 fines 16-G-7 studied by Kaiser (1972).

*Table 10.* $^{244}Pu/^{136}Xe$ ages of the Apollo 14, 15 and 17 fines or soils.

| No. Sample (Reference) | U (ppb) | $\Sigma^{136f}Xe$ | $(^{136f}Xe)_U$ | $(^{136f}Xe)_{Pu}$ | $(^{136f}Xe)_{Pu}/U$ ($10^{-4}cm^3STP/g$) | $^{244}Pu/^{238}U$ (atom/atom) | $^{244}Pu/^{136}Xe$ Age ($10^6$ year) |
|---|---|---|---|---|---|---|---|
| | | ($10^{-12}cm^3STP/g$) | | | | | |
| (1) Trench Soil 14149,52 (Drozd *et al.*, 1975) | 3900± 600 | 149 | 13 | 136 | (0.034)[b] | ____ | ____ |
| (2) Fines 15601 (Srinivasan *et al.*, 1972) | ____ | 9[b] | ____ | ____ | ____ | ____ | ____ |
| (3) Soil 72701 Bogard *et al.*, 1974) | 808[a] | 942 | 3 | 939 | 11.6 | 0.0764 | 4,859 |
| (4) Soil 76501, 12 (Bogard *et al.*, 1974) | 405[a] | 196 | 2 | 194 | 4.79 | 0.0315 | 4,752 |

a) Nunes *et al.* (1974); b) These values are abnormally low and unreliable.

Table 11 shows a comparison of the $^{244}Pu/^{136}Xe$ ages calculated for the Apollo 14 lunar breccia studied by Drozd *et al.* (1972, 1975) and Behrmann *et al.* (1973). It appears that these samples of breccia began to retain their xenon approximately 4,700 to 4,500 million years ago.

*Table 11.* $^{244}Pu/^{136}Xe$ ages of the Apollo 14 lunar breccias.

| No. Sample (Reference) | U (ppb) | $\Sigma^{136f}Xe$ ($10^{-12}cm^3STP/g$) | $(^{136f}Xe)_U$ | $(^{136f}Xe)_{Pu}$ | $(^{136f}Xe)_{Pu}/U$ ($10^{-4}cm^3STP/g$) | $^{244}Pu/^{238}U$ (atom/atom) | $^{244}Pu/^{136}Xe$ Age ($10^6$ year) |
|---|---|---|---|---|---|---|---|
| (1) 14313 (Behrmann *et al.*, 1973) | 3,210 | 1,654 | 11 | 1,643 | 5.12 | 0.0336 | 4,760 |
| (2) 14055,3 (Drozd *et al.*, 1975) | 3,680 | 515 | 12 | 503 | 1.37 | 0.00900 | 4,601 |
| (3) 14301 (Drozd *et al.*, 1972) | 3,630 | 375 | 12 | 363 | 1.00 | 0.00657 | 4,564 |
| (4) 14318 (Behrmann *et al.*, 1973) | 3,800 | 273 | 13 | 260 | 0.68 | 0.00448 | 4,516 |

Table 12 shows the $^{244}Pu/^{136}Xe$ ages of the Apollo 11 and 12 lunar rocks. As was mentioned earlier, Reynolds *et al.* (1970) reported that a small excess of $^{238}U$ fission xenon was present in the Apollo 11 fine-grained crystalline rock 10057,20. We found, however, in the present work that this sample contained about 4 times more $^{244}Pu$ fission xenon than $^{238}U$ fission xenon and it seems that it began to retain its xenon when the $^{244}Pu$ to $^{238}U$ ratio in the solar system was 0.00083 (atom/atom) 4,314 million years ago.

*Table 12.* $^{244}Pu/^{136}Xe$ ages of the Apollo 11 and 12 lunar rocks.

| No. Sample (Reference) | U (ppb) | $\Sigma^{136f}Xe$ ($10^{-12}cm^3STP/g$) | $(^{136f}Xe)_U$ | $(^{136f}Xe)_{Pu}$ | $(^{136f}Xe)_{Pu}/U$ ($10^{-4}cm^3STP/g$) | $^{244}Pu/^{238}U$ (atom/atom) | $^{244}Pu/^{136}Xe$ Age ($10^6$ year) |
|---|---|---|---|---|---|---|---|
| (1) 10057,20 (Reynolds *et al.*, 1970) | 780 | 12.3 | 2.5 | 9.8 | 0.126 | 0.00083 | 4,314 |
| (2) 10044,20 (Hohenberg *et al.*, 1970) | 280 | 0.3 | 0.9 | < 0 | ___ | ___ | ___ |
| (3) 12013,10 (Alexander, 1970) | 9,700 | 25 | 30 | < 0 | ___ | ___ | ___ |

According to Sears (1978), the C1 chondrites such as the carbonaceous chondrite Orgueil, are usually considered to be the most primitive – in the sense of least altered – sample of the solar system known to man. It is therefore surprising that only few reports exist on the isotopic composition of the xenon found in this group of meteorites.

Table 13 shows that we were able to obtain positive values for the ratio of $(^{136}Xe)_{Pu}/U$ in only two out of four sets of the xenon isotope data reported in the literature: the samples Orgueil studied by Eugster *et al.* (1967) and Pepin and Phinney (1978). The former was the isotopic composition of xenon measured in a single total melting experiment more than three decades ago, however, while the latter was a set of xenon isotope data obtained in stepwise heating experiments.

*Table 13.* $^{244}Pu/^{136}Xe$ Ages of the C1 Carbonaceous Chondrite Orgueil.

| No. Sample (Reference) | U (ppb) | $\Sigma^{136f}Xe$ ($10^{-12}cm^3STP/g$) | $(^{136f}Xe)_U$ | $(^{136f}Xe)_{Pu}$ | $(^{136f}Xe)_{Pu}/U$ ($10^{-4}cm^3STP/g$) | $^{244}Pu/^{238}U$ (atom/atom) | $^{244}Pu/^{136}Xe$ Age ($10^6$ year) |
|---|---|---|---|---|---|---|---|
| (1) Euster *et al.*, (1967) | 8.2[a] | 73.9 | < 0.1 | 73.9 | 90.1 | 0.593 | 5,106 |
| (2) Pepin and Phinney (1978) | 8.2[a] | $\geq$ 59.7 | < 0.1 | $\geq$ 59.7 | $\geq$ 72.8 | $\geq$ 0.478 | $\geq$ 5,080 |

a) Tatsumoto *et al.* (1976)

Unfortunately, however, the relative abundance of $^{134}Xe$ was not reported for the 1,100°C fraction of xenon, which made it impossible to calculate the amount of $^{244}Pu$ fission released at this temperature. Consequently, the ratio of $(^{136f}Xe)_{Pu}/U$ is shown in Table 14 as greater than or equal to 72.8 x $10^{-4}$ ($cm^3STP/g$) and the $^{244}Pu/^{136}Xe$ age of this meteorite is at least 5,080 million years; which is not far from the $^{244}Pu/^{136}Xe$ age of 5,071 million years calculated for the less than 1$\mu$ fraction of the Apollo 11 fines 10084,48 (See Table 6). These results clearly demonstrate the need for further studies on the isotopic composition of xenon found in this important meteorite.

*Table14.* $^{244}Pu/^{136}Xe$ Ages of C2 and CV3 Carbonaceous Chondrites.

| No. Sample (Reference) | U (ppb) | $\Sigma^{136f}Xe$ ($10^{-12}cm^3STP/g$) | $(^{136f}Xe)_U$ | $(^{136f}Xe)_{Pu}$ | $(^{136f}Xe)_{Pu}/U$ ($10^{-4}cm^3STP/g$) | $^{244}Pu/^{238}U$ (atom/atom) | $^{244}Pu/^{136}Xe$ Age ($10^6$ year) |
|---|---|---|---|---|---|---|---|
| (1) Cold Bokkeveld (C2) (Pepin and Phinny, 1978) | 11.0[b] | 46.9 | <0.1 | 46.9 | 42.3 | 0.278 | 5,015 |
| (2) Murray (C2) (Kuroda *et al.*, 1975) | 10.8[a] | 28.2 | <0.1 | 28.2 | 26.1 | 0.171 | 4,957 |
| (3) Murchison (C2) (Kuroda *et al.*, 1975) | 11.0[a] | 26.8 | <0.1 | 26.8 | 24.4 | 0.160 | 4,948 |
| (4) Renazzo (C2) (Reynolds and Turner, 1964) | 11.5[b] | 21.6 | <0.1 | 21.6 | 18.8 | 0.123 | 4,917 |
| (5) Mokoia (CV3) (Manuel *et al.*, 1972b) | 14.8[a] | 24.6 | <0.1 | 24.6 | 16.6 | 0.109 | 4,902 |
| (6) Allende (CV3) (Srinivasan *et al.*, 1978) | 14.8[a] | 22.4 | <0.1 | 22.4 | 15.1 | 0.0992 | 4,891 |

| No. Sample (Reference) | U (ppb) | $\Sigma^{136f}Xe$ | $(^{136f}Xe)_U$ | $(^{136f}Xe)_{Pu}$ (10$^{-12}$cm$^3$STP/g) | $(^{136f}Xe)_{Pu}/U$ (10$^{-4}$cm$^3$STP/g) | $^{244}Pu/^{238}U$ (atom/atom) | $^{244}Pu/^{136}Xe$ Age (10$^6$ year) |
|---|---|---|---|---|---|---|---|
| (7) Allende (CV3) (Manuel *et al.*, 1972b) | 14.8[a] | 21.7 | <0.1 | 21.7 | 14.7 | 0.0965 | 4,887 |

a) Tatsumoto *et al.* (1976); b) Morgan and Lovering (1968)

Table 14 shows a comparison of the $^{244}Pu/^{136}Xe$ ages of C2 and CV3 carbonaceous chondrites. Isotopic compositions of various C2 and CV3 carbonaceous chondrites have been reported by a large number of investigators, but most of the xenon isotope data were obtained in total melting experiments and hence many of the results were more or less unreliable. We have therefore listed in this table only those samples whose isotopic compositions of xenon were measured in stepwise heating experiments. It is to be noted here that the C2 carbonaceous chondrites seem to have started to retain their xenon 5,000 to 4,900 million years ago, when the $^{244}Pu$ to $^{238}U$ ratio in the solar system was decreasing from 0.278 to 0.123 (atom/atom).

Table 15 shows a comparison of the $^{244}Pu/^{136}Xe$ ages of ordinary chondrites. It is interesting to note that these meteorites seem to have begun retaining their xenon about 4,640 to 4,439 million years ago, when the $^{244}Pu$ to $^{238}U$ ratio in the solar system was decreasing from 0.0124 to 0.00219 (atom/atom).

*Table 15.* $^{244}Pu/^{136}Xe$ Ages of Ordinary Chondrites.

| No. Sample (Reference) | U (ppb) | $\Sigma^{136f}Xe$ | $(^{136f}Xe)_U$ | $(^{136f}Xe)_{Pu}$ (10$^{-12}$cm$^3$STP/g) | $(^{136f}Xe)_{Pu}/U$ (10$^{-4}$cm$^3$STP/g) | $^{244}Pu/^{238}U$ (atom/atom) | $^{244}Pu/^{136}Xe$ Age (10$^6$ year) |
|---|---|---|---|---|---|---|---|
| (1) Guarena (H6) (Hagee *et al.*, 1990) | 9.1[a] | 1.74 | 0.03 | 1.71 | 1.88 | 0.0124 | 4,640 |
| (2) Pultusk (H5) (Hagee *et al.*, 1990) | 10[a] | 1.50 | 0.03 | 1.47 | 1.47 | 0.00966 | 4,610 |
| (3) St. Séverin, Light (LL6) (Wasserburg *et al.*, 1969) | 8.7[a] | 1.31 | 0.03 | 1.28 | 1.47 | 0.00966 | 4,610 |
| (4) Bruderheim (L6) (Merrihue, 1966) | 15[b] | 1.96 | 0.05 | 1.91 | 1.27 | 0.00834 | 4,592 |
| (5) Olivenza (LL5) (Hagee *et al.*, 1990) | 10.9[a] | 1.03 | 0.04 | 0.99 | 0.91 | 0.00598 | 4,552 |
| (6) Marion (L6) (Hagee *et al.*, 1990) | 10.9[a] | 0.77 | 0.04 | 0.73 | 0.68 | 0.00447 | 4,517 |
| (7) St. Séverin, Dark (LL6) (Wasserburg *et al.*, 1969) | 14[a] | 0.60 | 0.05 | 0.55 | 0.357 | 0.00219 | 4,439 |

a) Hagee *et al.* (1990); b) Tilton (1973)

*Table 16.* $^{244}Pu/^{136}Xe$ Ages of Achondrites (1966 to 1970).

| No. Sample (Reference) | U (ppb) | $\Sigma^{136f}Xe$ ($10^{-12}cm^3STP/g$) | $(^{136f}Xe)_U$ | $(^{136f}Xe)_{Pu}$ | $(^{136f}Xe)_{Pu}/U$ ($10^{-4}cm^3STP/g$) | $^{244}Pu/^{238}U$ (atom/atom) | $^{244}Pu/^{136}Xe$ Age ($10^6$ year) |
|---|---|---|---|---|---|---|---|
| (1) Angra dos Reis (Hohenberg, 1970) | 206[a] | 19.95 | 0.70 | 19.25 | 0.935 | 0.00615 | ≡4,555 |
| (2) Angra dos Reis (Munk, 1967) | 206[a] | 19.90 | 0.70 | 19.20 | 0.932 | 0.00613 | ≡4,555 |
| (3) Nuevo Laredo (Munk, 1967) | 127[a] | 9.7 | 0.43 | 9.3 | 0.732 | 0.00481 | 4,526 |
| (4) Pasamonte I (Hohenberg *et al.*, 1967a) | 100[b] | 7.5 | 0.34 | 7.2 | 0.720 | 0.00473 | 4,524 |
| (5) Pasamonte II (Hohebnerg *et al.*, 1967a) | 100[b] | 6.9 | 0.34 | 6.6 | 0.660 | 0.00433 | 4,513 |
| (6) Juvinas (Kuroda *et al.*, 1966) | 210[c] | 11.9 | 0.71 | 11.2 | 0.533 | 0.00350 | 4,488 |
| (7) Sioux County (Kuroda *et al.*, 1966) | 107[c] | 6.0 | 0.36 | 5.6 | 0.523 | 0.00343 | 4,485 |
| (8) Lafayette (Rowe *et al.*, 1966) | 40 | 0.86 | 0.13 | 0.73 | 0.180 | 0.00118 | 4,362 |
| (9) Stannern (Kuroda *et al.*, 1966) | 176[c] | 2.7 | 0.59 | 2.1 | 0.119 | 0.00078 | 4,310 |
| (10) Nakhla (Rowe *et al.*, 1966) | 40 | (-0.5) | — | — | — | — | — |

a) Tatsumoto *et al.* (1973); b) Unruh *et al.* (1977); c) Kuroda *et al.* (1966); d) Clark *et al.* (1967)

*Table 17.* $^{244}Pu/^{136}Xe$ ages of achondrites. These ages were calculated from recently reported data by Miura *et al.* (1998).

| No. Sample (Reference) | | U (ppb) | $\Sigma^{136f}Xe$ ($10^{-12}cm^3STP/g$) | $(^{136f}Xe)_U$ | $(^{136f}Xe)_{Pu}$ | $(^{136f}Xe)_{Pu}/U$ ($10^{-4}cm^3STP/g$) | $^{244}Pu/^{238}U$ (atom/atom) | $^{244}Pu/^{136}Xe$ Age ($10^6$ year) |
|---|---|---|---|---|---|---|---|---|
| (1) Béréba | | 94 | 4.25 | 0.32 | 3.93 | 0.418 | 0.00295 | 4,458 |
| (2) Binda | | 26 | 1.52 | 0.09 | 1.43 | 0.550 | 0.00388 | 4,491 |
| (3) Camel Donga | #1 | 112 | 6.36 | 0.38 | 5.98 | 0.533 | 0.00376 | 4,487 |
| (4) | #2 | 112 | 7.00 | 0.38 | 6.62 | 0.590 | 0.00416 | 4,500 |
| (5) | #3 | 112 | 6.94 | 0.38 | 6.56 | 0.585 | 0.00412 | 4,499 |
| (6) | #4 | 112 | 2.86 | 0.38 | 2.48 | 0.221 | 0.00156 | 4,381 |
| (7) | #5 | 112 | 5.21 | 0.38 | 4.83 | 0.430 | 0.00303 | 4,461 |
| (8) | #6 | 112 | 4.92 | 0.38 | 4.54 | 0.405 | 0.00286 | 4,454 |
| (9) | #7 | 112 | 5.05 | 0.38 | 4.67 | 0.416 | 0.00293 | 4,457 |
| (10) Juvinas | #1 | 86 | 5.30 | 0.29 | 5.01 | 0.583 | 0.00411 | 4,498 |
| (11) | #2 | 86 | 6.84 | 0.29 | 6.55 | 0.762 | 0.00537 | 4,530 |
| (12) Millbillillie | #1 | 310 | 6.96 | 1.04 | 5.92 | 0.191 | 0.00135 | 4,364 |
| (13) | #2 | 190 | 4.65 | 0.64 | 4.01 | 0.211 | 0.00149 | 4,376 |
| (14) | #3 | 310 | 9.26 | 1.04 | 8.22 | 0.265 | 0.00187 | 4,403 |
| (15) | #4 | 190 | 4.52 | 0.64 | 3.88 | 0.204 | 0.00144 | 4,372 |
| (16) | #5 | 190 | 4.80 | 0.64 | 4.16 | 0.218 | 0.00154 | 4,380 |
| (17) Millbillillie (cont.) | #6 | 190 | 5.06 | 0.64 | 4.42 | 0.232 | 0.00164 | 4,387 |

| No. Sample (Reference) | | U (ppb) | $\Sigma^{136f}Xe$ | $(^{136f}Xe)_U$ | $(^{136f}Xe)_{Pu}$ | $(^{136f}Xe)_{Pu}/U$ ($10^{-4}cm^3STP/g$) | $^{244}Pu/^{238}U$ (atom/atom) | $^{244}Pu/^{136}Xe$ Age ($10^6$ year) |
|---|---|---|---|---|---|---|---|---|
| | | | ($10^{-12}cm^3STP/g$) | | | | | |
| (1) Béréba | | 94 | 4.25 | 0.32 | 3.93 | 0.418 | 0.00295 | 4,458 |
| (2) Binda | | 26 | 1.52 | 0.09 | 1.43 | 0.550 | 0.00388 | 4,491 |
| (3) Camel Donga | #1 | 112 | 6.36 | 0.38 | 5.98 | 0.533 | 0.00376 | 4,487 |
| (4) | #2 | 112 | 7.00 | 0.38 | 6.62 | 0.590 | 0.00416 | 4,500 |
| (5) | #3 | 112 | 6.94 | 0.38 | 6.56 | 0.585 | 0.00412 | 4,499 |
| (6) | #4 | 112 | 2.86 | 0.38 | 2.48 | 0.221 | 0.00156 | 4,381 |
| (7) | #5 | 112 | 5.21 | 0.38 | 4.83 | 0.430 | 0.00303 | 4,461 |
| (8) | #6 | 112 | 4.92 | 0.38 | 4.54 | 0.405 | 0.00286 | 4,454 |
| (9) | #7 | 112 | 5.05 | 0.38 | 4.67 | 0.416 | 0.00293 | 4,457 |
| (10) Juvinas | #1 | 86 | 5.30 | 0.29 | 5.01 | 0.583 | 0.00411 | 4,498 |
| (11) | #2 | 86 | 6.84 | 0.29 | 6.55 | 0.762 | 0.00537 | 4,530 |
| (12) Millbillillie | #1 | 310 | 6.96 | 1.04 | 5.92 | 0.191 | 0.00135 | 4,364 |
| (18) | #7 | 190 | 6.00 | 0.64 | 5.36 | 0.282 | 0.00199 | 4,411 |
| (19) | #8 | 310 | 5.66 | 1.04 | 4.62 | 0.149 | 0.00105 | 4,334 |
| (20) Stannern | #1 | 185 | 8.27 | 0.62 | 7.65 | 0.413 | 0.00291 | 4,457 |
| (21) | #2 | 185 | 9.06 | 0.62 | 8.44 | 0.456 | 0.00322 | 4,469 |

Table 18 shows the $^{244}Pu/^{136}Xe$ ages of a special class achondrites, the so-called SNC meteorites calculated from the xenon isotopes data reported by Miura *et al.* (1998), Swindle *et al.* (1995), and Murthy and Mohapatra (1997). All except one of the ages calculated by the use of our latest method of the treatment of data are smaller than the 4,555 million years age of the Angra dos Reis meteorite and lie within the range of about 4,300 to 4,500 million years and these results indicate that the process of resetting the nuclear clock based on the ratio of $^{244}Pu$ to $^{136}Xe$ took place many times during the early history of the solar system.

*Table 18.* $^{244}Pu/^{136}Xe$ Ages of the so-called SNC meteorites.

| No. Sample (Reference) | U (ppb) | $\Sigma^{136f}Xe$ | $(^{136f}Xe)_U$ | $(^{136f}Xe)_{Pu}$ | $(^{136f}Xe)_{Pu}/U$ ($10^{-4}cm^3STP/g$) | $^{244}Pu/^{238}U$ (atom/atom) | $^{244}Pu/^{136}Xe$ Age ($10^6$year) |
|---|---|---|---|---|---|---|---|
| | | ($10^{-12}cm^3STP/g$) | | | | | |
| (1) ALH84001,29 (Swindle *et al.*, 1995) | 10[a] | 0.68 | 0.03 | 0.65 | 0.65 | 0.00428 | 4,511 |
| (2) ALH84001#1 (Miura *et al.*, 1995) | 10[a] | 0.42 | 0.03 | 0.39 | 0.39 | 0.00257 | 4,450 |
| (3) ALH84001#2 (Miura *et al.*, 1995) | 10[a] | 0.50 | 0.03 | 0.47 | 0.47 | 0.00331 | 4,472 |
| (4) ALH84001 (Murty & Mohapatra, 1997) | 10[a] | 1.00 | 0.03 | 0.97 | 0.97 | 0.00639 | 4,560 |

a) Dreibus *et al.* (1994)

The results of calculations of the $^{244}Pu/^{136}Xe$ ages of lunar samples and various meteorites, which have been summarized in Tables 6 through 18 indicate that the formation of the solar system began soon after the last supernova exploded very close to 5.1 billion years ago. These results also indicate that the process of re-setting of the nuclear clock must have occurred numerous times throughout the early history of the solar system.

## 4. RADIOACTIVE FALLOUT IN ASTRONOMICAL SETTINGS

In his famous monograph entitled "The planets: Their origin and development", Urey (1952, p. ix) stated: "*Dr. Edward Teller remarked recently that the origin of the earth was somewhat like the explosion of the atomic bomb: the physical effects are often temporary, but the chemical effects, such as radioactive and non radioactive elements, remain. It is possible by a study of these substances to learn much about the bomb, and also about the origin of the earth*".

On October 31st of the same year 1952, the U.S. Government conducted the first thermonuclear explosion, a 10.4-megaton device named "MIKE", at the Enewetak Atoll in the Pacific Ocean. It was a remarkable coincidence that the year 1952 was the $100^{th}$ anniversary of the birth of Becquerel (1896), who discovered radioactivity.

Eight years later Öpik (1960) described in detail the fate of fresh nuclear debris created by a supernova in his classic monograph entitled "The Oscillating Universe".

Figure 1 shows a scenario which Öpik (1960) had in his mind, according to which fresh debris such as the CRAB nebula, which is a remnant of the supernova observed in A.D. 1054, will attain a diameter of 70 light years in 23,000 years; 140 light years in 260,000 years; 210 light years in 1.3 million years; and 280 light years in 4 million years, which is the maximum attainable diameter. Thus, fresh debris formed in a supernova explosion will eventually become mixed with old debris from earlier supernova explosions, just as in the case of the fallout particles created by a thermonuclear weapon's test.

According to Öpik (1960), the Orion Halo is near the end of its expansion, judging by its size. Its age is estimated to be from one to 10 million years, according to the mass of the original supernova. Perhaps 5 million years ago a star exploded somewhere around the present center of the Halo, in the middle of the constellation, near the star Eta Orionis and west of the Belt. The explosion swept up and compressed into an expanding shell a mass of interstellar gas of the order of 15,000 solar masses, 1,000 times greater than the original mass of the supernova.

When and where the birth of the solar system occurred in relation to the supernova explosion is unknown, but it is important to note that Öpik (1960, p. 99) stated: "*Compressed gas and dust are believed to offer favorable conditions for the condensation of diffuse matter into stars. Thus, the expanding shell of a supernova is a place where, if anywhere in space, new stars may be born."*

This means that it will be much more likely that the place where the solar system was born was very close to the place of a supernova explosion, as shown in Figure 1.

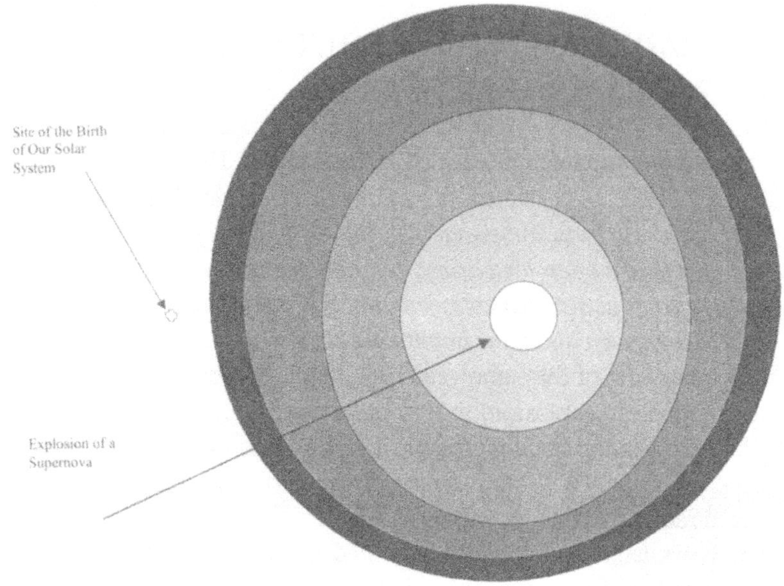


*Figure 1.* The birth of our solar system in the close vicinity of an exploding supernova.

In a paper entitled "The time interval between nucleosynthesis and formation of the earth", Kuroda (1961), therefore, pointed out that (a) in the case of the single event model, in which it is assumed that the solar system formed from a single supernova explosion,

$$({}^{244}Pu/{}^{238}U)_0 = \rho_{244}/\rho_{238} = 1\ (atom/atom) \tag{6}$$

where $\rho_{244}$ and $\rho_{238}$ are the production rates of $^{244}$Pu and $^{238}$U in the supernova explosion, respectively, and (b) in the case of the continuous-synthesis model, in which it is assumed that the solar system material was synthesized by a large number of supernova explosions during a period of billions of years preceding the formation of the solar system,

$$({}^{244}Pu/{}^{238}U)_0 = \frac{\rho_{244}}{\rho_{238}} \bullet \frac{\lambda_{238}-\lambda_*}{\lambda_{244}-\lambda_*} \bullet \frac{\exp(-\lambda_*\Delta)-\exp(-\lambda_{244}\Delta)}{\exp(-\lambda_*\Delta)-\exp(-\lambda_{238}\Delta)} \tag{7}$$

where $\lambda_*$ is the rate constant for the nucleosynthesis (the rate of nucleosynthesis is assumed to decay exponentially since the mean time of beginning with the decay constant $\lambda_*$), and $\Delta$ is the duration of nucleosynthesis, and $\lambda_{244}$ and $\lambda_{238}$ are the decay constants of $^{244}$Pu, respectively. Equation (7) yields a value of:

$$(^{244}Pu/^{238}U)_0 = 0.0215\ (atom/atom) \tag{8}$$

if it is assumed that $\Delta = 6 \times 10^9$ years and $\lambda_*$ is $0.1 \times 10^{-9}$ year$^{-1}$.

It is to be noted here that the above calculations represent two extreme cases and the initial ratio of $^{244}$Pu to $^{238}$U in the solar system at the time of the cessation of nucleosynthesis should lie intermediate between the two extreme cases of approximately 1 (atom/atom) and about 0.02 (atom/atom), depending upon the amount of contribution from the last supernova. Thus, the following relationship should hold:

$$(^{244}\mathrm{Pu}/^{238}\mathrm{U})_0 = 0.0215(\mathrm{atom/atom}) \tag{9}$$

where $\alpha$ is the fraction of the contribution from the last supernova and (1 - $\alpha$) is that from many supernova exploded earlier.

The value of $\alpha$ is unknown, but if $\alpha$ = 0.5, which means that the contribution from the last supernova amounted to 50 percent of the total, we have a value of ($^{244}$Pu/$^{238}$U)o $\cong$ 0.51 (atom/atom). If, on the other hand, the contribution from the last supernova amounted to 10 percent of the total, or $\alpha$ = 0.10, we obtain a value of ($^{244}$Pu/$^{238}$U)o = 0.12 (atom/atom).

It is interesting to note in Table 15 that the ratio of $^{244}$Pu to $^{238}$U in the solar system was 0.123 and 0.109 (atom/atom) for the C2 carbonaceous chondrites Renazzo and Mokoia, respectively. It is to be noted here, however, that one must be careful not to interpret these results as due to the fact that these two meteorites formed suddenly in space when the ratio of $^{244}$Pu to $^{238}$U in the solar system was about 0.12 (atom/atom) since the process of re-setting the nuclear clock based on the ratio of $^{136}$Xe/$^{244}$Pu must have occurred many times during the early history of the solar system.

It is worthy of note here that Fowler (1972), reported that the initial ratio of $^{244}$Pu to $^{238}$U in the solar system at the time the last supernova explosion took place had to be

$$\left(\frac{^{244}Pu}{^{238}U}\right)_\Delta = \frac{\lambda_{244}\tau_{244}}{\lambda_{238}\tau_{238}} \frac{[1-\exp(-\Delta/\tau_{244})]}{[1-\exp(-\Delta/\tau_{238})]}$$

$$= (.09 \pm 0.1)\frac{0.118\,[1-\exp(-7.7/0.118)]}{6.51\,[1-\exp(-7.7/6.51)]}$$

$$= 0.024 \pm 0.004\ (atom/atom) \tag{10}$$

This means that he did not take into account of the extremely important fact that the expanding shell of a supernova is a place where, if anywhere in space, new stars may be born.

Thermonuclear weapons tests, such as the one conducted by the U.S. Government on 31 October 1952 in the South Pacific Ocean, may be regarded as an attempt by man to reproduce the supernova explosion on our planet earth. In his classic monograph entitled "Nuclear Astrophysics", Fowler (1967) showed the spontaneous fission decay curve of the californium fraction isolated from the fresh nuclear debris collected at the test site.

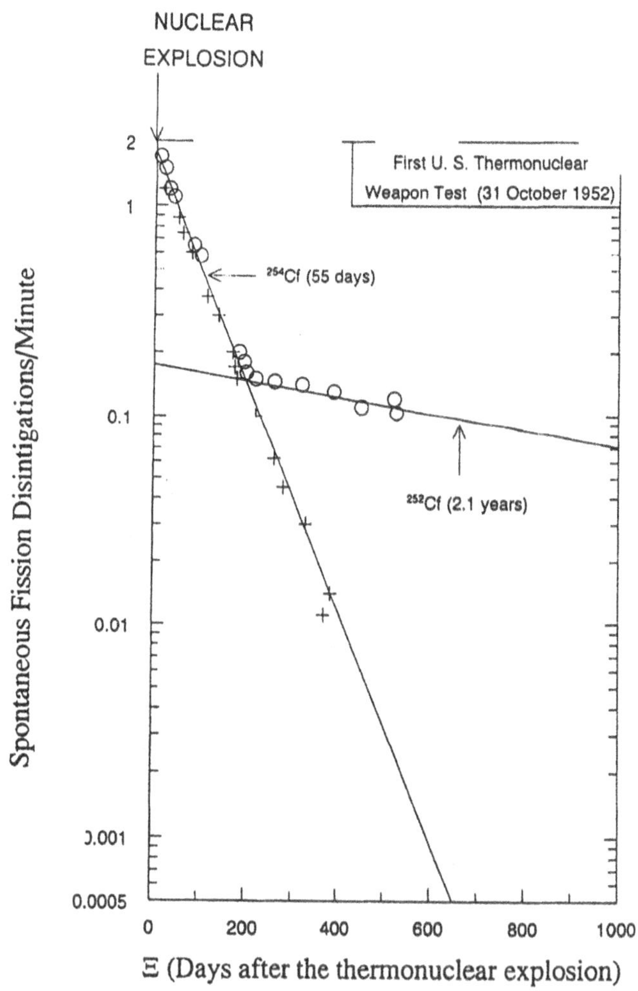


*Figure 2.* Formation of californium isotopes in a thermonuclear weapon's test, according to Fowler (1967).

Figure 2 shows the spontaneous fission decay curve of the californium fraction from the 1952 thermonuclear test. According to Fowler (1967), simple analysis showed that fifty times as much $^{252}$Cf relative to $^{254}$Cf was produced in the Enewetak test, and one needed only to assume equal production of these isotopes in supernova events to obtain an energy input into supernova debris which would match the light output for at least 400 days.

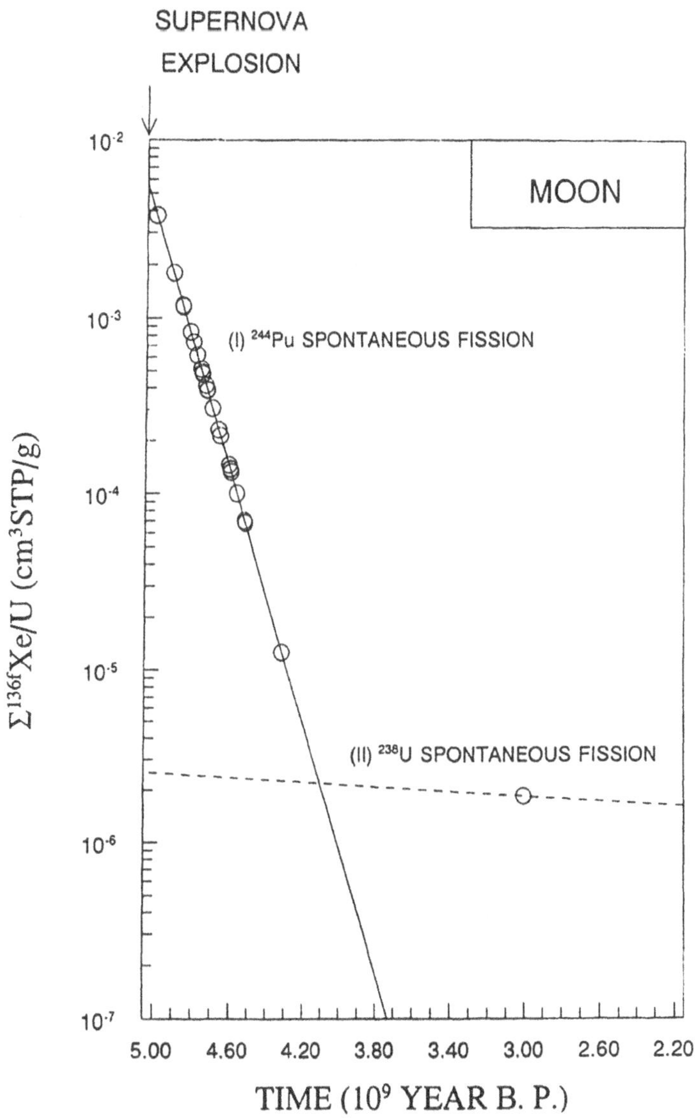


*Figure 3.* Spontaneous fission decay of $^{244}$Pu and $^{238}$U in lunar samples from the fresh nuclear debris created by a supernova explosion. (A 3.0 billion year-old sample is the old terrestrial granite studied by Butler *et al.*, 1963).

Figure 3 and Figure 4 show plots of the ratios of $^{136f}$Xe/U *versus* the $^{244}$Pu ages of the lunar samples and the meteorites, respectively, which are shown in Tables 6 through 18. It is to be noted here that the $^{244}$Pu fission xenon dominates over the $^{238}$U fission xenon for a period of almost ten half-lives of $^{244}$Pu after the explosion of a supernova in both cases.

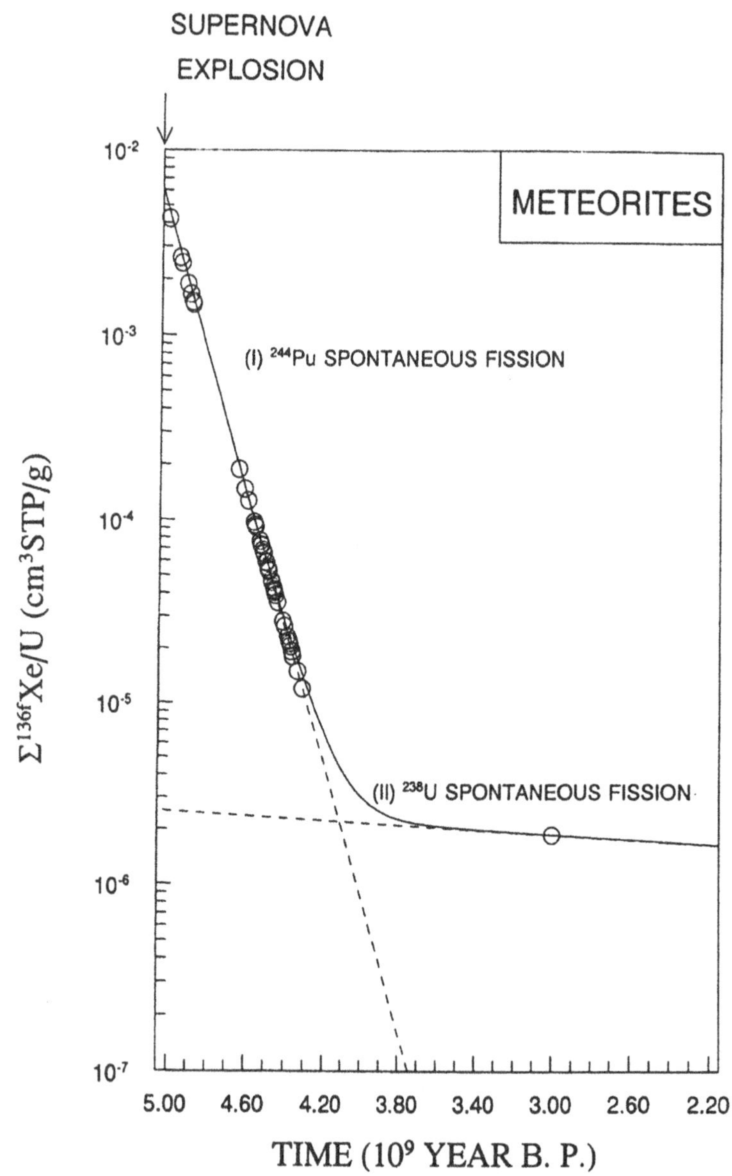


*Figure 4.* Spontaneous fission decay of $^{244}$Pu and $^{238}$U in meteorites from the fresh nuclear debris created by a supernova explosion. (A 3.0 billion year-old sample is an old terrestrial granite studied by Butler *et al.*, 1963).

In 1967, two important papers were published by the Berkeley group. The first was an article entitled "Xenon-iodine dating: sharp isochronism in chondrites" by Hohenberg *et al.* (1967b) in which they reported that measurements of the accumulation of $^{129}$Xe from radioactive decay of extinct $^{129}$I in meteorites showed that the $^{129}$I/$^{127}$I ratio in high-temperature minerals of diverse chondrites was 10-4 (atom/atom) at the time of cooling and the uniformity in the ratio indicated that minerals cooled simultaneously within 1 or 2 million years.

The second was a review article entitled "Fissiogenic xenon in the Renazzo and Murray meteorites" by Funk *et al.* (1967). These investigators reported: "We reinvestigated the xenon data from a stepwise heating of the Murray and Renazzo carbonaceous chondrites by Reynolds and coworkers and confirm the exceptionally large amounts of fission-produced xenon postulated by Reynolds' group. Our interpretation supports the existence of three components of xenon in Murray and Renazzo, probably best explained as (i) a primitive trapped component, (ii) a fission component, and (iii) atmospheric-like component. The excessive amounts of fissiogenic xenon cannot be accounted for by atmospheric contamination, cosmic-ray xenon or diffusion effects."

Funk *et al.* (1967) went on to re-calculate the initial ratio of $^{244}$Pu to $^{238}$U in the solar system at the time these meteorites began to retain their xenon to be ≥ 0.45 (± 0.17) (atom/atom) in the case of the Murray meteorite and 0.08 (± 0.02) (atom/atom) in the case of the Renazzo meteorites.

It is interesting to compare these ratios with those which we have obtained in the present work more than three decades later. The value of ≥ 0.45 (± 0.17) (atom/atom) which Funk *et al.* (1967) obtained is about three times the value of 0.171 (atom/atom) which we obtained for the Murray meteorite, (see Table 15 of this report), while the value of 0.08 (± 0.02) (atom/atom) which Funk *et al.* (1967) obtained for the Renazzo meteorite is in approximate agreement with the value of 0.123 (atom/atom) which we obtained in the present work (see Table 15).

Funk *et al.* (1967) went on to conclude, however, that the fission-like xenon component, which they found in carbonaceous chondrites was not the product of in situ decay of the extinct radionuclide $^{244}$Pu in the early solar system on the ground that the theory of element synthesis in stars predicted that the ratio of $^{244}$Pu to $^{238}$U had to be ≤0.05 (atom/atom), according to Burbidge *et al.* (1957) and Fowler (1967).

It is worthy of note here that the paper by Burbidge *et al.* (1957) had been published three years before the existence of $^{244}$Pu in the early solar system was predicted by Kuroda (1960), and hence the fact that the initial ratio of $^{244}$Pu to 2$^{38}$U had to be less than 0.05 (atom/atom) was not mentioned in the 103-page article by Burbidge *et al.* (1957). In the paper by Funk *et al.* (1967), on the other hand, Fowler is quoted simply as personal communication (1966). One of us (PKK) therefore felt that it was most likely a personal opinion held by Professor Fowler.

Investigators in the field of xenology thus found themselves in a peculiar situation in which the experimentally obtained ratios of $^{129}I$ to $^{127}I$ in meteorites were abnormally low, whereas the ratios of $^{244}Pu$ to $^{238}U$ were abnormally high.

Six years later, Clayton *et al.* (1973) made the important discovery that the high temperature phases in carbonaceous chondrites contain oxygen with anomalous isotopic composition: the proportion of the most abundant isotope was enriched by up to 5 percent, while the ratio of $^{17}O/^{18}O$ remained essentially constant in these mineral phases. Nuclear reactions within the solar nebula of solar system appeared to have been inadequate to cause so large an isotopic perturbation in such an abundant element and hence these investigators have attributed the $^{16}O$ excess to galactic nucleosynthesis processes which occurred in stars.

Variations in the isotopic composition of magnesium in the Allende meteorite which were not attributable to isotopic fractionation were discovered by Lee and Papanastassiou (1974) and Gray and Compston (1974). Lee *et al.* (1977) found large excesses of up to 10 percent $^{26}Mg$ in different phases of a Ca-Al rich inclusion in the Allende meteorite. They concluded that the data provided definitive evidence for the presence of $^{26}Al$ (half-life $7.3 \times 10^5$ years) in the early solar system. These data indicated that the initial $^{26}Al/^{27}Al$ ratio was $(5.1 \pm 0.6) \times 10^{-5}$ (atom/atom).

Lee *et al.* (1978) then found isotopic anomalies in calcium, while McCulloch and Wasserburg (1978) found similar anomalies in barium and neodymium, and McCulloch *et al.* (1978) also found anomalies in samarium. Furthermore, Heydegger *et al.* (1979) reported that some inclusion materials from the Allende meteorite had a statistically significant enhancement, of the order of one per mil, in the $^{50}Ti/^{49}Ti$ ratio, probably due to a nucleogenetic anomaly in $^{50}Ti$ abundance.

In a series of articles published during the 1970's (Kuroda 1975, 1976, 1979a,b,c,d,e,f,g), one of us (PKK) expressed his opinion that these so-called FUN anomalies, as well as the $^{129}Xe$ anomalies observed in various meteorites were likely to be attributed to the x-process nucleosynthesis. In Chapter 7 of his monograph entitled "The origin of the chemical elements", Kuroda (1982) concluded that the origin of the isotopic anomalies observed in meteorites appears to be attributable to the x-process nucleosynthesis, which was responsible for the production of the deficient elements lithium, beryllium and boron during the earliest stage of the history of the solar system.

Neither Reynolds (1960, 1963) nor Wasserburg and Papanastassiou (1982) or Wasserburg (1987) have yet succeeded in establishing the iodine/xenon and $^{26}Al/^{27}Al$ dating method for meteorites in a straightforward manner. The reason for this is that these investigators have always assumed the age of the solar system to be close to 4.5 to 4.6 billion years and that meteorites and the earth suddenly appeared in space roughly 4.5 to 4.6 billion years ago. We thus feel that their difficulties can be eliminated if the experimental data are interpreted in such a manner that $^{26}Al$ and $^{129}I$ both must have started to retain their daughters at about the same time, 5.1 billion years ago, and the generally accepted 4.55 billion year age of the solar system refers, instead, to the time of breakup of the meteorite parent body (Kuroda and Myers 1998a, Myers and Kuroda 1989).

In our opinion, both meteorites and lunar samples began to retain their xenon about 5.1 billion years ago, soon after the explosion of the last supernova, and the re-setting process of the nuclear clocks based on the extinct radionuclides $^{26}Al$ and $^{129}I$ must have occurred at this time. The half-lives of $^{26}Al$ and $^{129}I$ are much shorter than that of $^{244}Pu$. Both $^{26}Al$ and $^{129}I$, therefore, decayed at a much faster rate than $^{244}Pu$ and must have disappeared from the solar system long before the breakup of the parent body occurred some 4.55 billion years ago.

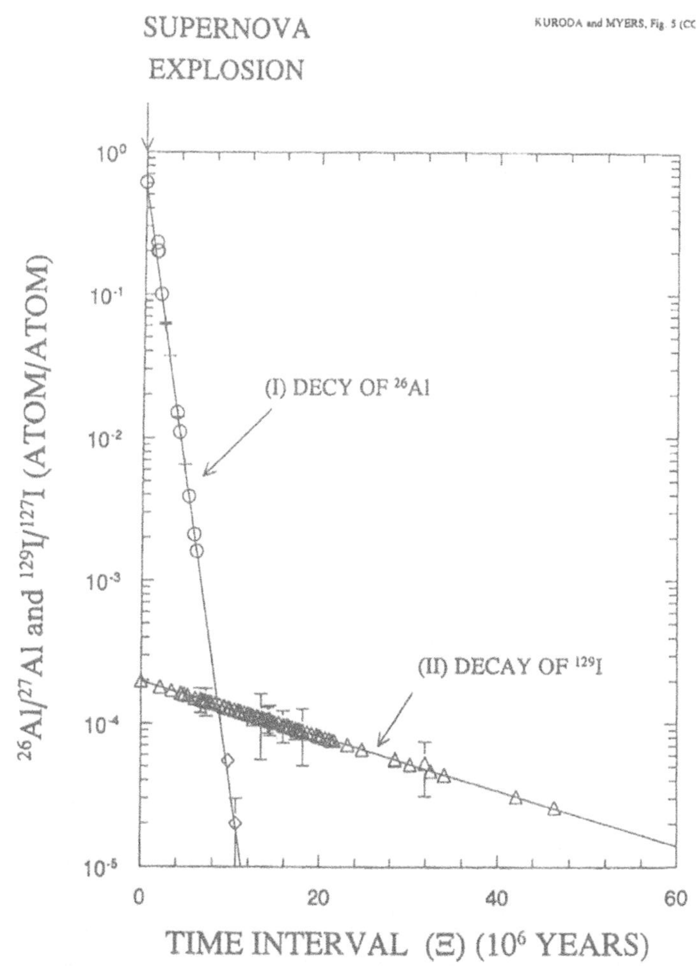


*Figure 5.* Decay of $^{26}Al$ and $^{129}I$ in the early solar system, according to Kuroda and Myers (1998a).

Figure 5 is a plot of the ratios of $^{26}Al$ to $^{27}Al$ and $^{129}I$ to $^{127}I$ *versus* time shortly after the explosion of the last supernova took place; which we have recently reported (Kuroda and Myers, 1988b). Note that the nuclear clock based on the extinct radionuclide $^{26}Al$ ceased to be operative about 10 million years after the supernova explosion, whereas the nuclear clock based on the extinct radionuclide $^{129}I$ did so, about 50 million years after the last supernova explosion.

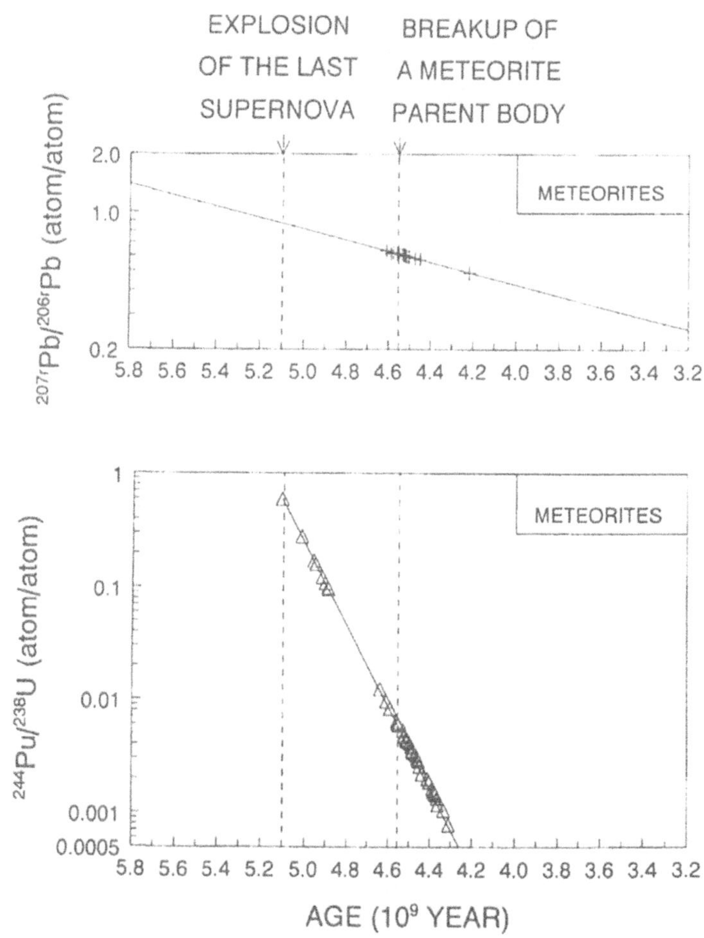


*Figure 6.* Comparison of the $^{207}Pb/^{206}Pb$ ages and the $^{244}Pu/^{238}U$ ages of meteorites.

Figure 6 shows a comparison of the $^{244}Pu/^{238}U$ ratios in the solar system at the time the meteorites began to retain their xenon (See Table 1) and the $^{207}Pb/^{206}Pb$ ages of meteorites studied by Tatsumoto (1970a, 1970b, 1973), by Tatsumoto and Rosholt (1970), by Nunes *et al.* (1973, 1974), by Tatsumoto and Unruh (1976), and by Tatsumoto *et al.* (1970, 1972, 1976).

Note that most meteorites began to retain their lead about 4.5 to 4.6 billion years ago, whereas the $^{244}Pu/^{136}Xe$ ages of the meteorites cover a period of about 800 million years between 5.1 to 4.3 billion years ago.

Figure 7 shows a comparison of the $^{244}Pu/^{238}U$ ratios in the solar system at the time the lunar samples began to retain their xenon (see Table 1) and the $^{207}Pb/^{206}Pb$ ages of lunar samples studied by Tatsumoto and his coworkers (Tatsumoto 1970a, 1970b, 1973; Nunes *et al.*, 1973; Tatsumoto and Unruh, 1976; Tatsumoto *et al.*, 1970, 1972, 1976). Note that the $^{244}Pu/^{136}Xe$ ages of meteorites cover a period of about 800 million years between 5.1 and 4.3 billion years, but the $^{207}Pb/^{206}Pb$ ages of lunar samples cover period of 3.2 billion years from 5.7 billion years ago to 3.5 billion years ago.

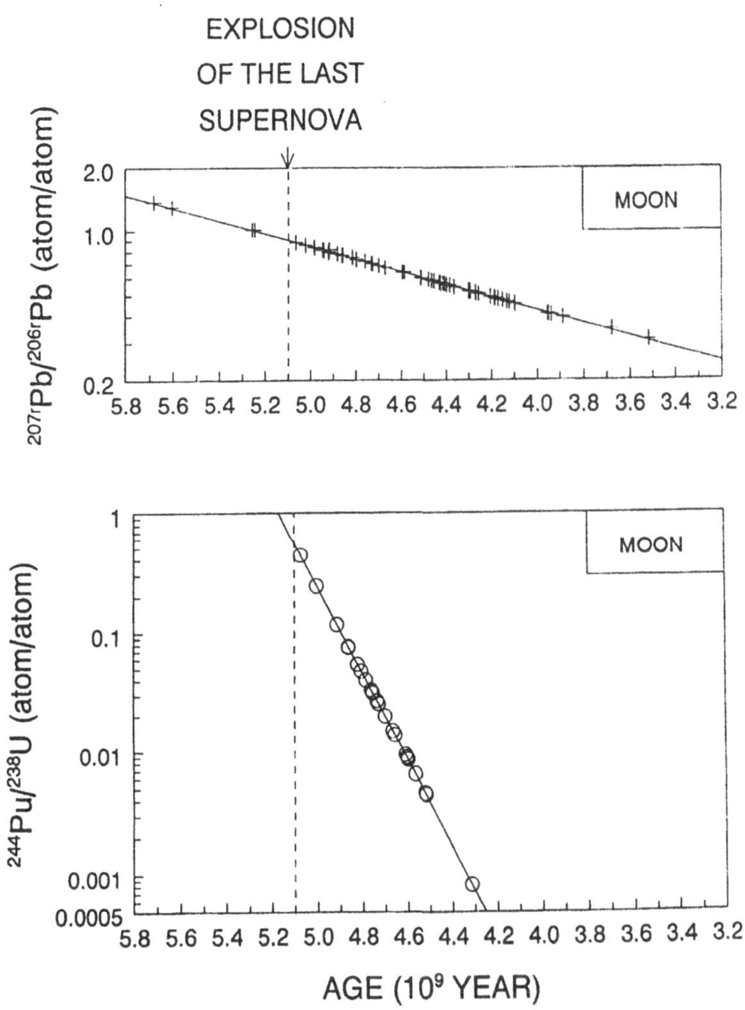


*Figure 7.* Comparison of the $^{207}Pb/^{206}Pb$ ages and the $^{244}Pu/^{238}U$ ages of lunar samples.

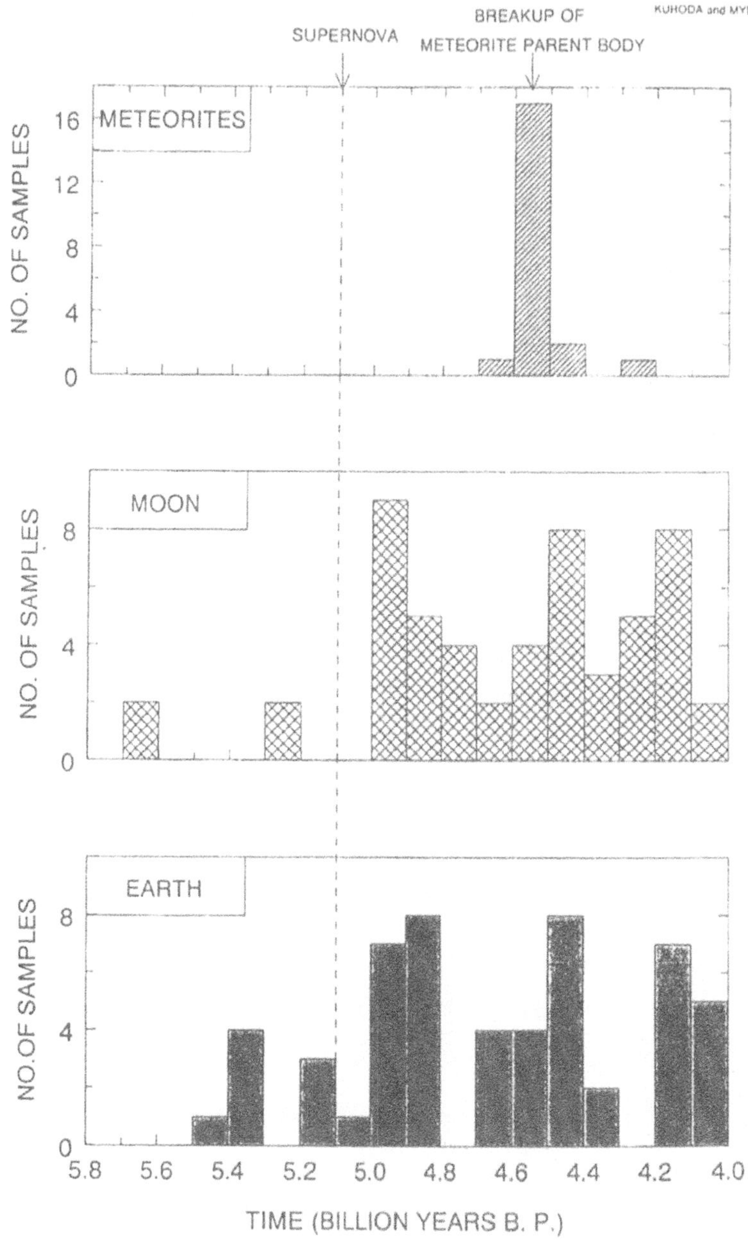


*Figure 8.* Comparison of the $^{207}Pb/^{206}Pb$ ages and the $^{244}Pu/^{238}U$ ages of meteorites, the moon and the earth.

Figure 8 shows the histograms of the lead/lead ages, which we have calculated from the lead isotope data reported by various investigators since the 1940's. The lead/lead ages for the extra-terrestrial samples were calculated from the data reported since 1970 by Tatsumoto and his coworkers, while the lead/lead ages for the terrestrial samples were calculated from the data reported by Nier (1938), Nier *et al.* (1941), Collins *et al.* (1952, 1953), Russell *et al.* (1954), and reviewed by Rankama (1954).

It is to be noted here that in these calculations of the lead/lead ages from the lead isotope ratios reported by many investigators since the 1940's, we used the latest decay constants for the uranium isotopes and the uranium isotopic ratios recommended by Tatsumoto *et al.* (1976).

$$\lambda^{238} = 0.155125\,(10^{-9}/yr)\ and\ \lambda_{235} = 0.98485\,(10^{-9}/yr) \qquad (11)$$

$$\text{and, } (^{238}U/^{235}U)_{today} = 137.88\,(atom/atom) \qquad (12)$$

These results indicate that the birth of planets took place most likely in the following manner: a) the explosion of a supernova occurred about 5.1 billion years ago and a dense, spinning cloud of gas and dust formed, which consisted of a mixture of fresh nuclear debris from the supernova explosion and older nuclear debris from earlier supernova explosions; b) a new star formed within 0 to 1 million years after the supernova explosion and the planets began to form soon thereafter; c) approximately 600 million years later, the breakup of a meteorite parent body occurred and the process of re-setting of the nuclear clocks based on the $^{207}Pb/^{206}Pb$ ratio took place in the meteorites; and d) this means that the generally accepted 4.55 billion year age of the earth (Patterson, 1956) refers to the breakup of the meteorite parent body.

## 5. ISOTOPIC COMPOSITION OF THE SO-CALLED XENON-HL

Reynolds (1960) and Reynolds and Turner (1964) reported the existence of a strange xenon component in carbonaceous chondrites and this component later became widely known as xenon-HL (Pepin and Phinney, 1978). It displays a V shaped anomaly, in which the light xenon isotopes (L) and the heavy isotopes (H) are simultaneously enriched.

Figure 9 shows the isotopic composition of the xenon fraction released at 800°C from the CV3 carbonaceous chondrite Allende, as reported by Manuel *et al.* (1972b). Note that the light xenon isotopes and the heavy xenon isotopes are simultaneously enriched in this sample.

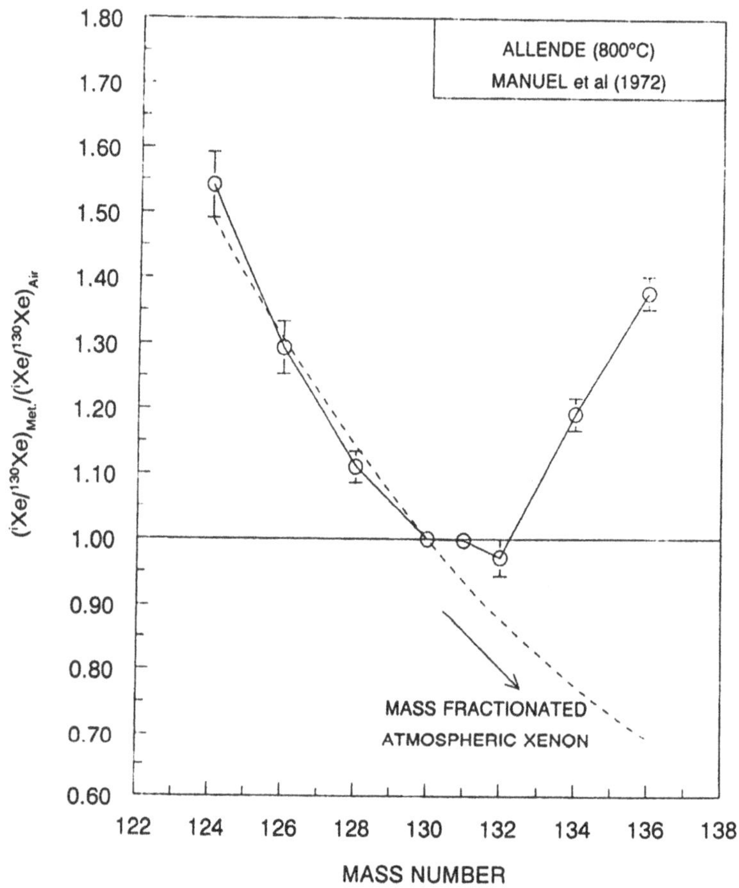


*Figure 9.* The puzzle of the xenon isotopes. I. Allende 800$^0$C fraction.

Figure 10 shows a comparison of the isotopic composition of the so-called average carbonaceous chondrite xenon (AVCC) and of the atmospheric xenon, according to Reynolds (1978). Note that the carbonaceous chondrites appear to contain an excess of fission xenon (*f*) relative to the atmospheric xenon.

It is worthy of note here that both Manuel *et al.* (1972b) and Reynolds (1978) have normalized the xenon isotope data to the fission-shielded $^{130}$Xe in Figure 9 and Figure 10, respectively. Kuroda (1971) pointed out, however, that the only way to correctly explain the differences in the isotopic compositions of xenon found in carbonaceous chondrites, in lunar samples, and in the earth's atmosphere is to normalize the xenon isotope data to the heaviest isotope $^{136}$Xe and to consider the alterations of the isotopic composition of xenon by a combined effect of (a) mass fractionation, (b) spallation and (c) *stellar-temperature* neutron capture reactions. Let us therefore proceed to re-examine the existing xenon isotope data in terms of the values of $D_i$ after normalizing to the heaviest isotope $^{136}Xe$.

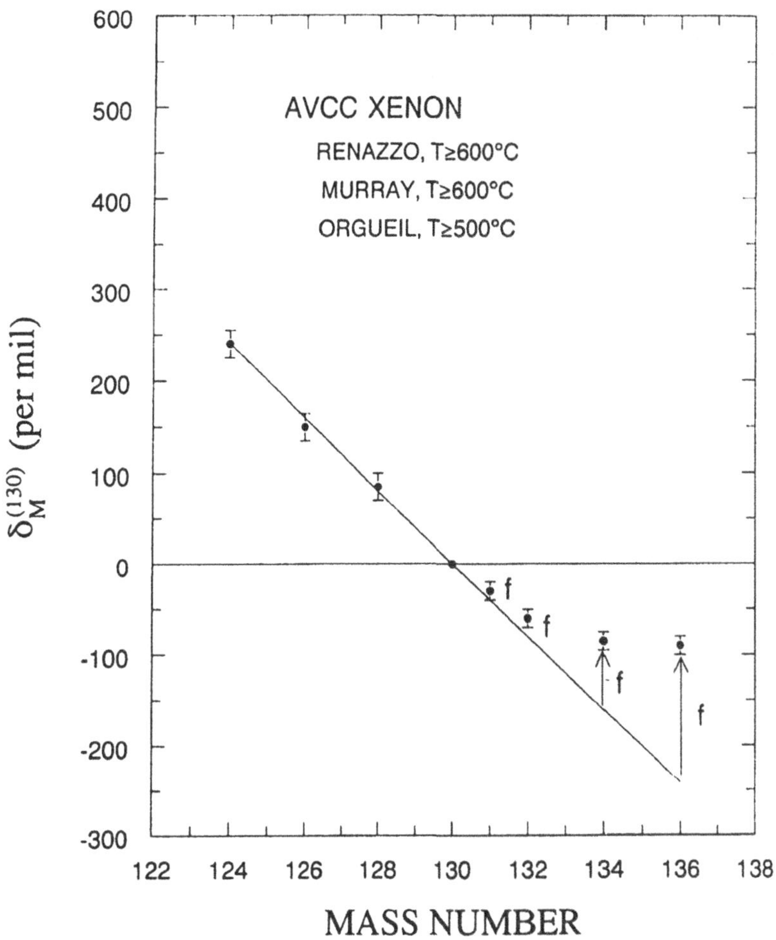


*Figure 10.* The puzzle of the xenon isotopes. II. The so-called AVCC xenon, according to Reynolds (1978).

$$D_i = \left\{ \frac{({}^{i}Xe/{}^{136}Xe)sample}{({}^{i}Xe/{}^{136}Xe)atmosphere} - 1 \right\} \bullet 100\,(\text{percent}) \qquad (13)$$

Manuel *et al.* (1972b) assumed that the abundance of $^{130}Xe$ is not affected by the processes (b) spallation and (c) stellar temperature neutron-capture reactions. Moreover, they made an additional assumption that the effect of the process (a) mass fractionation on the xenon isotope data is always in such a manner that abundances of the lighter isotopes of xenon are systematically enhanced relative to the heavier isotopes of xenon.

Figure 11 shows a comparison of the isotopic compositions of xenon found in the C2 carbonaceous chondrite Murchison (Kuroda *et al.*, 1975) and in the earth's atmosphere. Note that the values of $D_i$ steadily decrease as the mass number (i) increases, except for the fact that the value of $D_{126}$ for the 400°C fraction is much greater than the value of $D_{124}$.

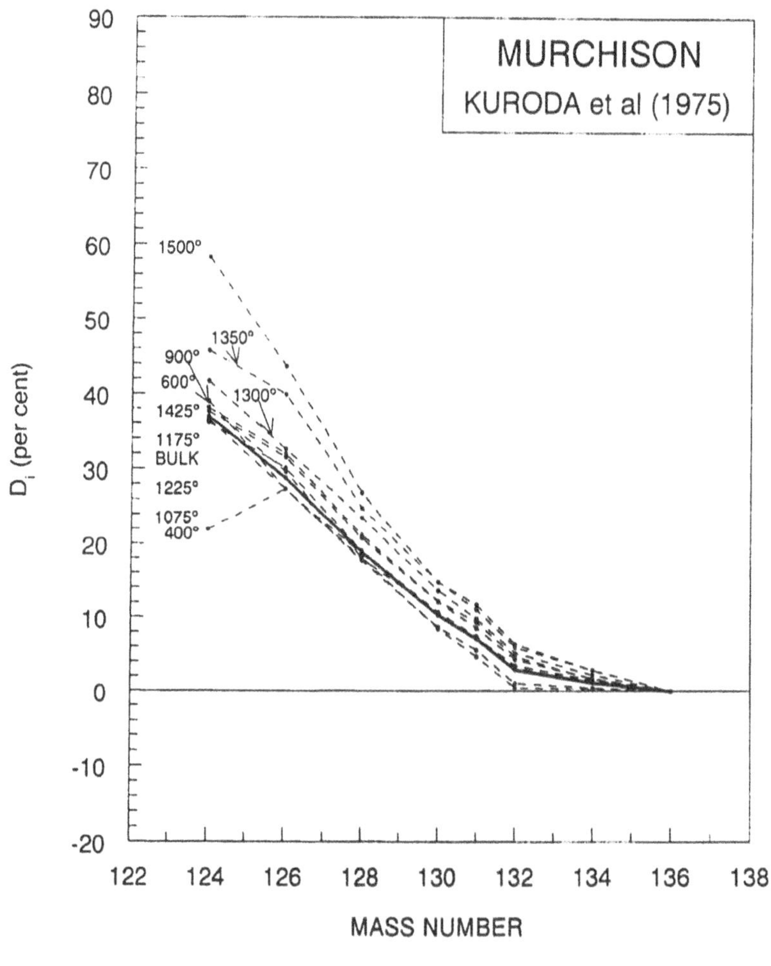


*Figure 11.* Isotopic compositions of xenon released from the Murchison meteorite.

Reynolds (1978) has therefore assumed that the pattern of variations of the xenon isotope data for carbonaceous chondrites such as shown here could be attributed solely to the effect of mass-dependent fractionation.

Figure 12 shows a comparison of the isotopic compositions of trapped xenon released from the Murchison meteorite at different temperatures (See Kuroda and Myers 1994b). Note that the plots shown here as Figure 11 and Figure 12 are hardly distinguishable from each other and the reason for this is that the content of $^{244}$Pu fission xenon ($^{136f}$Xe) of this meteorite is 26.9 x $10^{-12}$ (cm$^3$STP/g), while the content of ($\Sigma^{136}$Xe) is 3,971 x $10^{-12}$ (cm$^3$STP/g) and hence the ratio, ($^{136f}$Xe/$\Sigma^{136}$Xe) = 26.8/3971 = 0.0067 or only 0.67 percent.

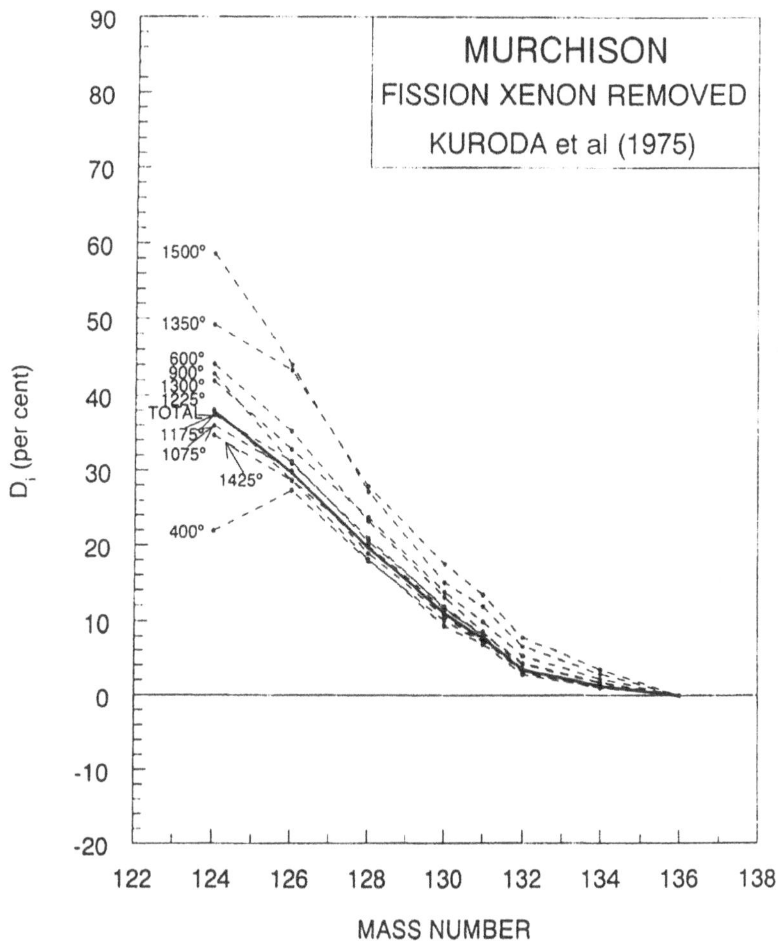


*Figure 12.* Isotopic compositions of xenon released from the Murchison meteorite (fission xenon removed).

Figure 13 shows a comparison of the isotopic compositions of xenon found in the C2 carbonaceous chondrite Renazzo studied by Reynolds and Turner (1964). It is to be noted here that the concentration of $^{136}$Xe in this meteorite ($\Sigma^{136}$Xe) is 1,110 x $10^{-12}$ cm$^3$STP/g and the content of fission xenon ($^{136f}$Xe) is 21.6 x $10^{-12}$ (cm$^3$STP/g) as shown in Table 14 of this paper. Hence the ratio of ($^{136f}$Xe/$\Sigma^{136}$Xe) = 21.6/1,110 = 1.95 (percent). It is therefore to be expected that the effect of the presence of $^{244}$Pu fission xenon should become visible in the case of this meteorite. Indeed, the values of $D_{132}$ and $D_{134}$ for the 700°, 800° and 900°C fractions of the Renazzo meteorite are all negative indicating that appreciable quantities of $^{244}$Pu fission xenon are being released at these temperatures.

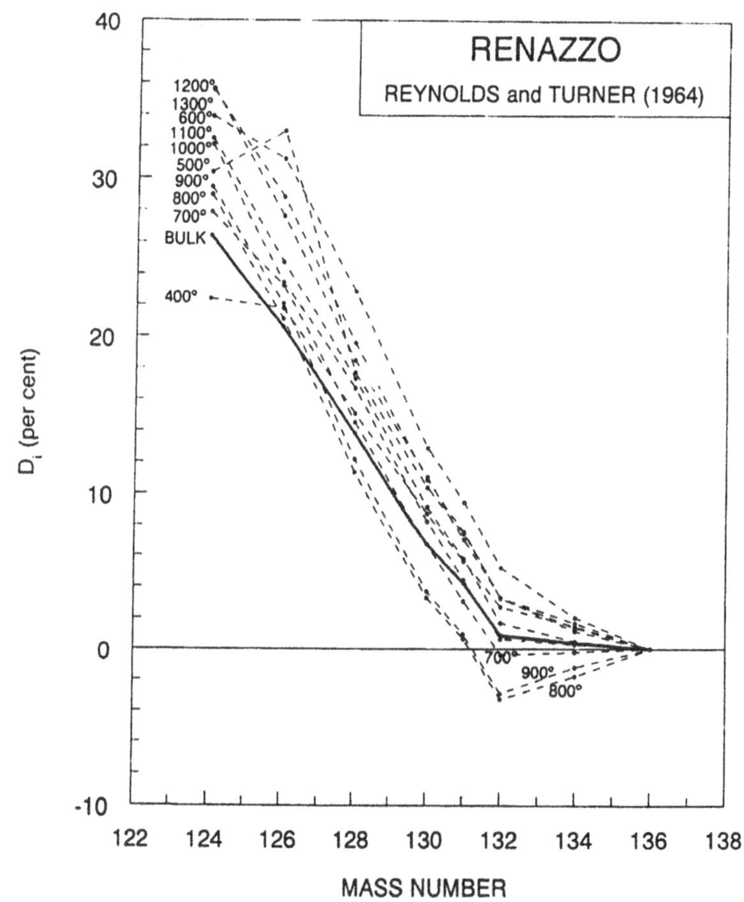


*Figure 13.* Isotopic compositions of xenon released from the Renazzo meteorite.

Figure 14 shows a comparison of the isotopic compositions of *trapped* xenon released from the Renazzo meteorite at different temperatures (Kuroda and Myers 1994a). The most interesting feature of the plot shown here is that the trapped xenon released from this meteorite at 800°C has negative values of $D_{132}$ and $D_{134}$.

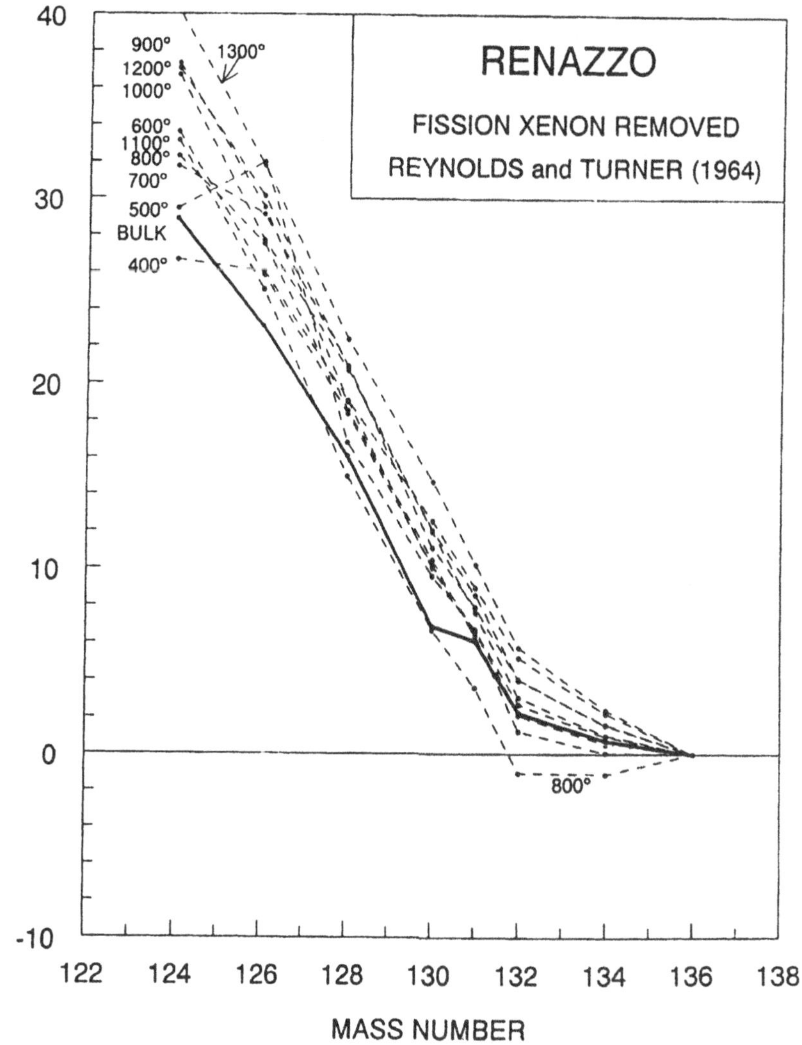


*Figure 14.* Isotopic compositions of xenon released from the Renazzo meteorite (fission xenon removed).

Figure 15 shows a comparison of the isotopic compositions of xenon found in the Allende meteorite and in the earth's atmosphere. The isotopic compositions of xenon released from this meteorite in stepwise heating experiments were reported by Manuel *et al.* (1972b). The concentration of $^{136}$Xe in this meteorite ($\Sigma^{136}$Xe) is 560 x $10^{-12}$ ($cm^3$STP/g) and the content of fission xenon ($^{136f}$Xe) is 21.7 x $10^{-12}$ ($cm^3$STP/g), as shown in Table 14 of this paper. Hence the ratio of ($^{136f}$Xe/$^{136}$Xe) = 21.7/560 = 0.03875 or 3.875 (percent), which is twice the value of 1.95 (percent), which we obtained earlier for the Renazzo meteorite, and hence the effect of the presence of

$^{244}$Pu fission xenon is far more clearly visible in the case of the Allende meteorite than in the case of the Renazzo meteorite.

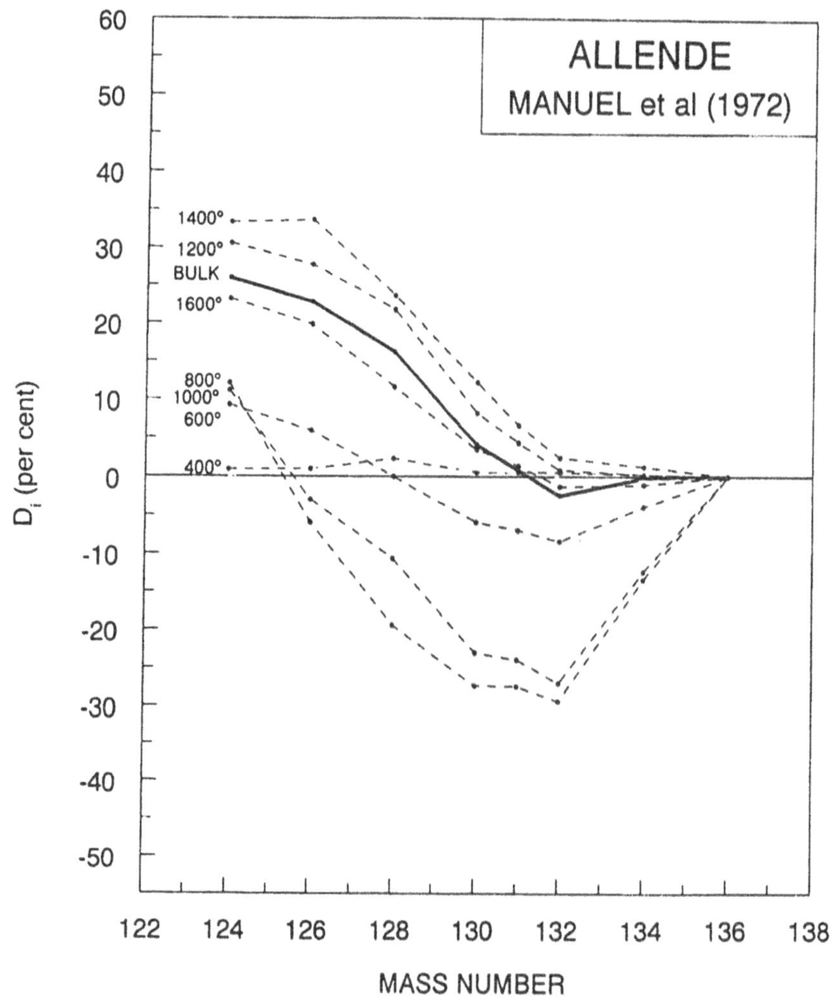


*Figure 15.* Isotopic compositions of xenon released from the Allende meteorite.

Figure 16 shows a comparison of the isotopic compositions of *trapped* xenon released from the Allende meteorite at different temperatures (Kuroda and Myers, 1994a) and of the atmospheric xenon. The most interesting feature of the plot shown here is that the values of $D_{130}$, $D_{131}$, $D_{132}$ and $D_{134}$ for the xenon fractions released at 600°, 800° and 1,000°C remain negative, in spite of the fact that the $^{244}$Pu fission xenon was removed from these xenon fractions. The reason for this is that the $^{136}$Xe/$^{136}$Xe ratio is altered not solely by the presence of $^{244}$Pu fission xenon in the sample, but the effect of mass-fractionation plays an important role here.

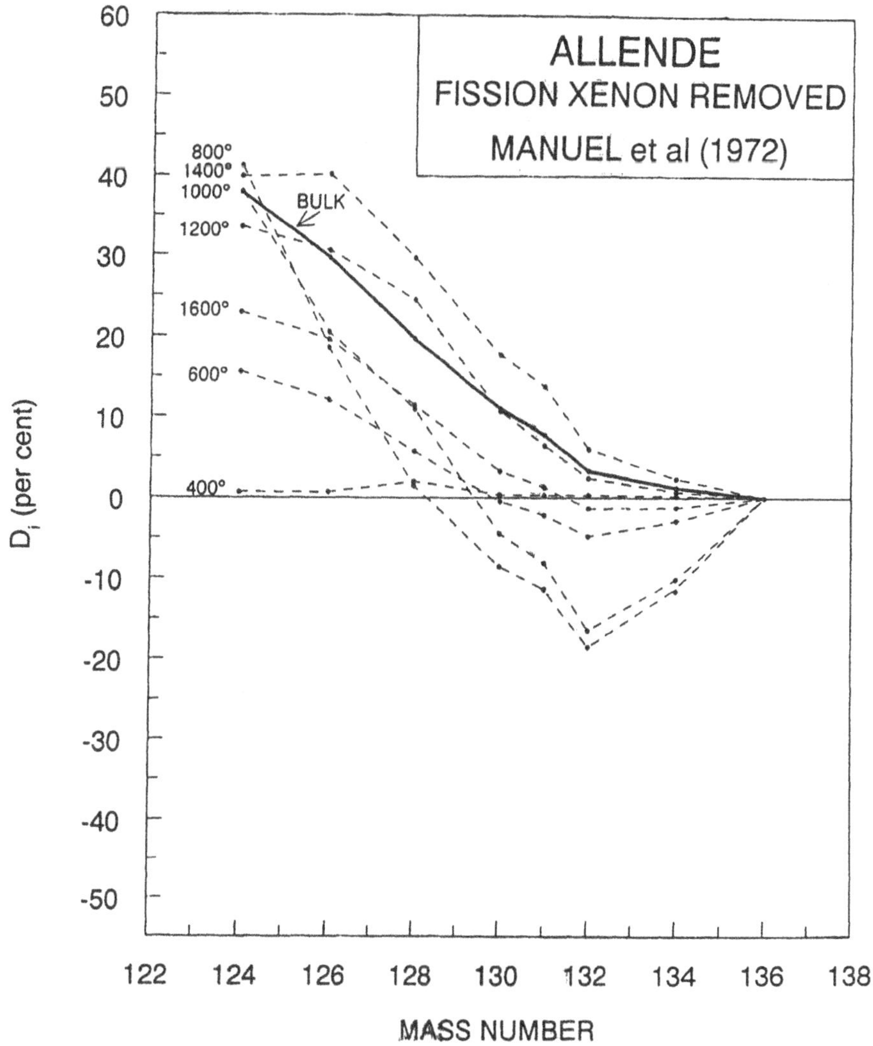


*Figure 16.* Isotopic compositions of xenon released from the Allende meteorite (fission xenon removed).

Figure 17 shows a comparison of the values of $D_i$ for the 800°C fraction of the Allende meteorite and for the *trapped* (fission xenon removed) xenon released at this temperature. Note that the 800°C xenon fraction of the Allende meteorite contains a large excess of $^{244}$Pu fission xenon, but the *trapped* xenon released from the Allende meteorite is severely mass-fractionated in such a manner that the abundances of the heavier xenon isotopes are systematically enhanced relative to the lighter xenon isotopes.

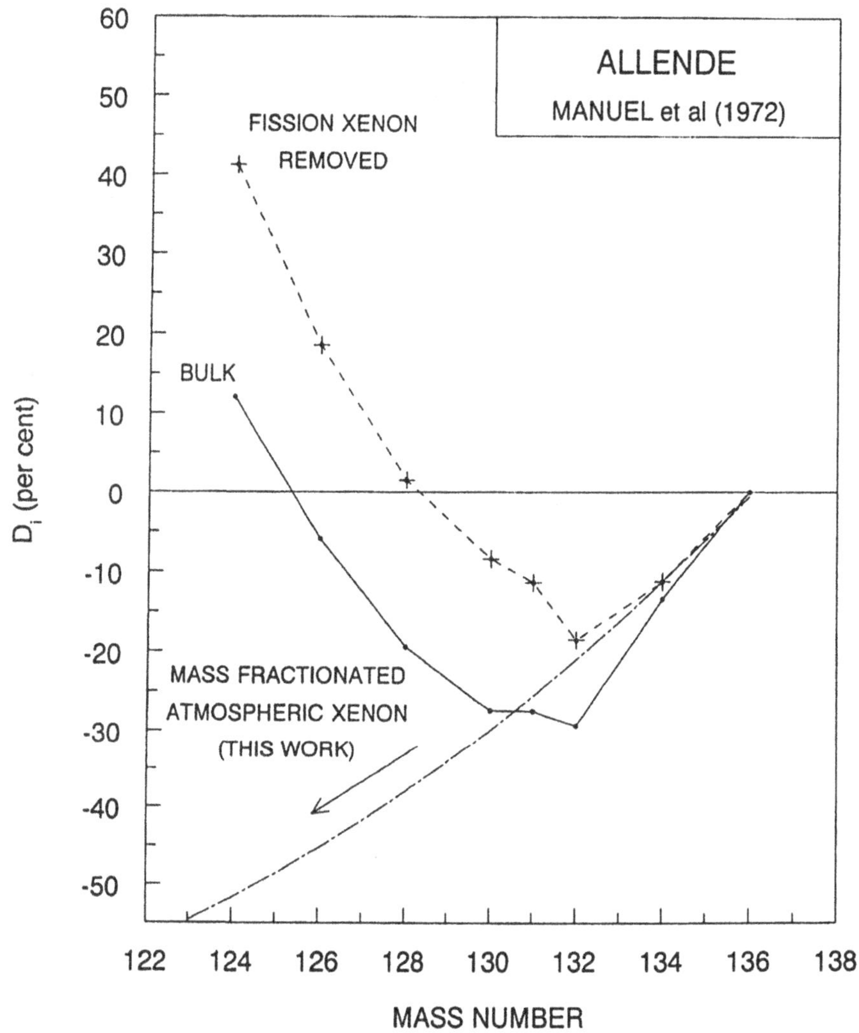


*Figure 17.* Comparison of the isotopic compositions of xenon released from the Allende meteorite at 800°C and of the atmospheric xenon, according to the present work.

Figure 18 shows a comparison of the values of $D_i$ for the 1,200°C fraction of the Allende meteorite and for the *trapped* (fission xenon removed) xenon released at this temperature. Note that the *trapped* xenon released from the Allende meteorite is mass fractionated in such a manner that the abundances of the lighter xenon isotopes are systematically enhanced relative to the heavier xenon isotopes.

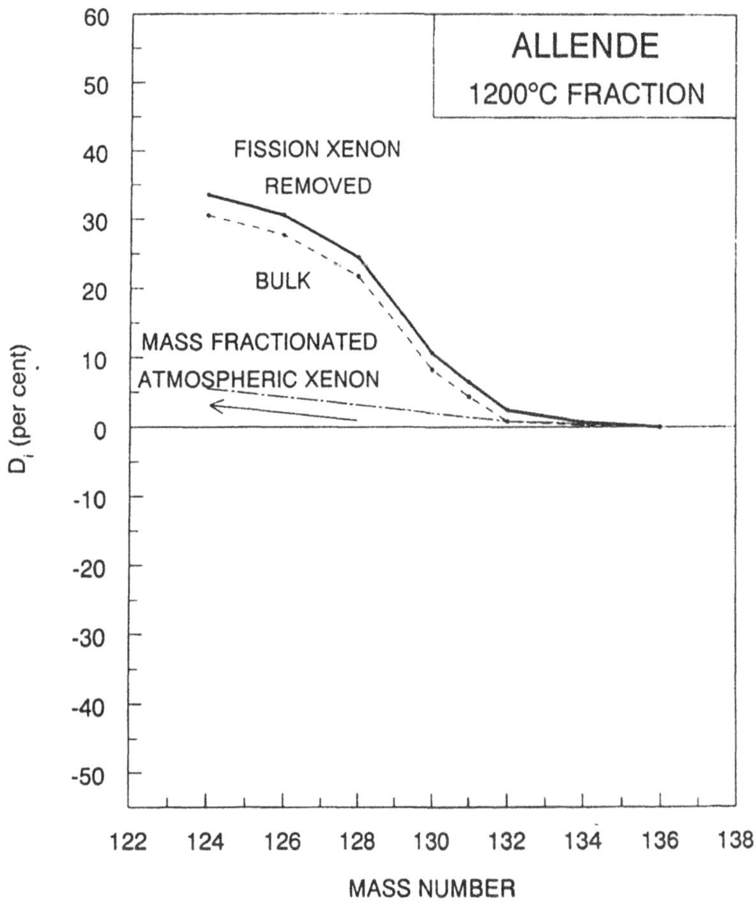


*Figure 18.* Comparison of the isotopic compositions of xenon released from the Allende meteorite at 1,200°C and of the atmospheric xenon, according to the present work.

As shown in Figure 9, Manuel *et al.* (1972b) interpreted the xenon isotope data for the 800°C fraction of the Allende meteorite *assuming* that the abundances of lighter xenon isotopes are systematically enhanced relative to the heavier xenon isotopes. Figure 17 shows, however, that the reverse is the case and the abundances of the heavier xenon isotopes happen to be systematically enhanced relative to the lighter xenon isotopes in the trapped xenon released from the Allende meteorite at 800°C. It was indeed unfortunate that Manuel *et al.* (1972b), as well as Reynolds (1978), were convinced during the 1970's that the isotopic compositions of *trapped* xenon released from carbonaceous chondrites at different temperatures were all severely mass fractionated relative to the atmospheric xenon in such a manner that the lighter xenon isotopes are systematically enriched relative to the heavier xenon isotopes.

Figure 19 shows a comparison of the isotopic compositions of xenon found in the Mokoia meteorite and in the earth's atmosphere. The isotopic compositions of xenon released from this meteorite in stepwise heating experiments were reported by Manuel *et al.* (1972b). The concentration of $^{136}$Xe in this meteorite ($\Sigma^{136}$Xe) is 899 x $10^{-12}$ (cm$^3$STP/g) and the content of $^{244}$Pu fission xenon ($^{136f}$Xe) is 24.6 x $10^{-12}$ (cm$^3$STP/g) and hence the ratio of ($^{136f}$Xe/$\Sigma^{136}$Xe) = 24.6/899 = 0.0274 or 2.74 (percent), which is intermediate between the corresponding values calculated for the Renazzo meteorite and the Allende meteorite.

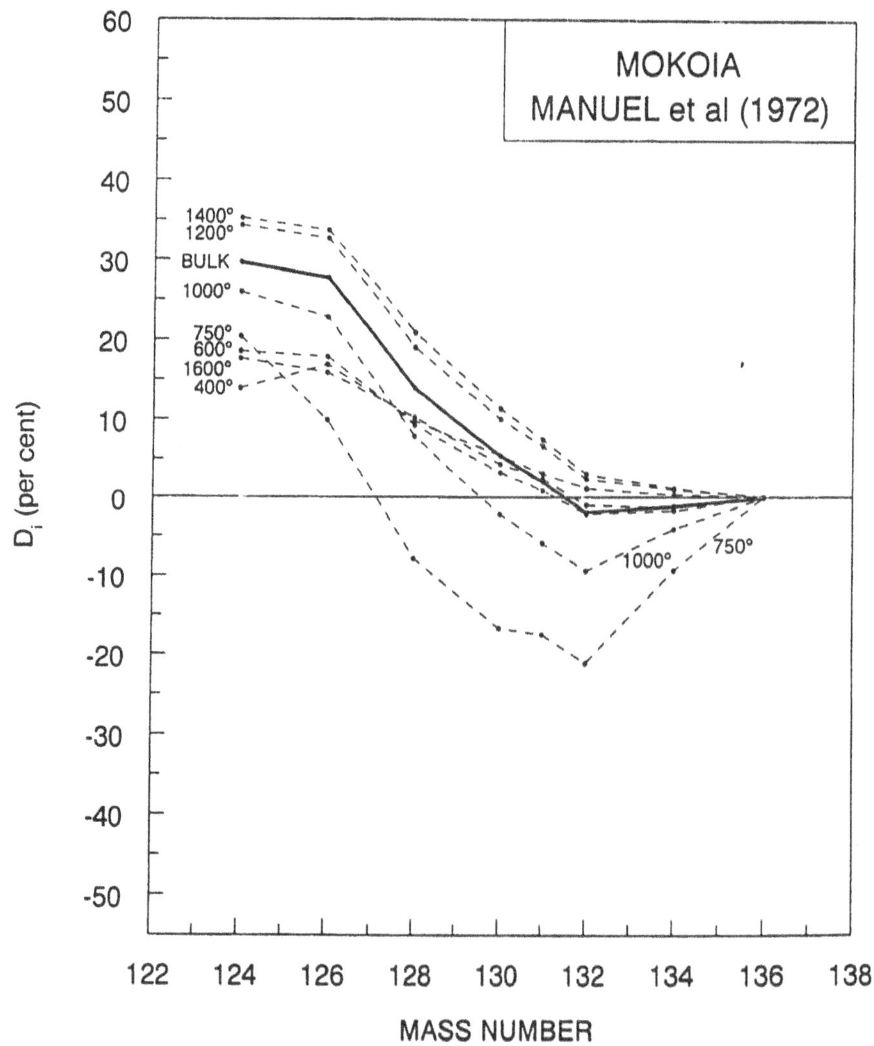


*Figure 19.* Isotopic compositions of xenon released from the Mokoia meteorite.

Figure 20 shows a comparison of the isotopic compositions of xenon found in the trapped component of xenon in the Mokoia meteorite and in the earth's atmosphere. It is interesting to note here that there exists a certain resemblance between the plots shown here and those shown in Figure 14 for the *trapped* xenon in the Renazzo meteorite.

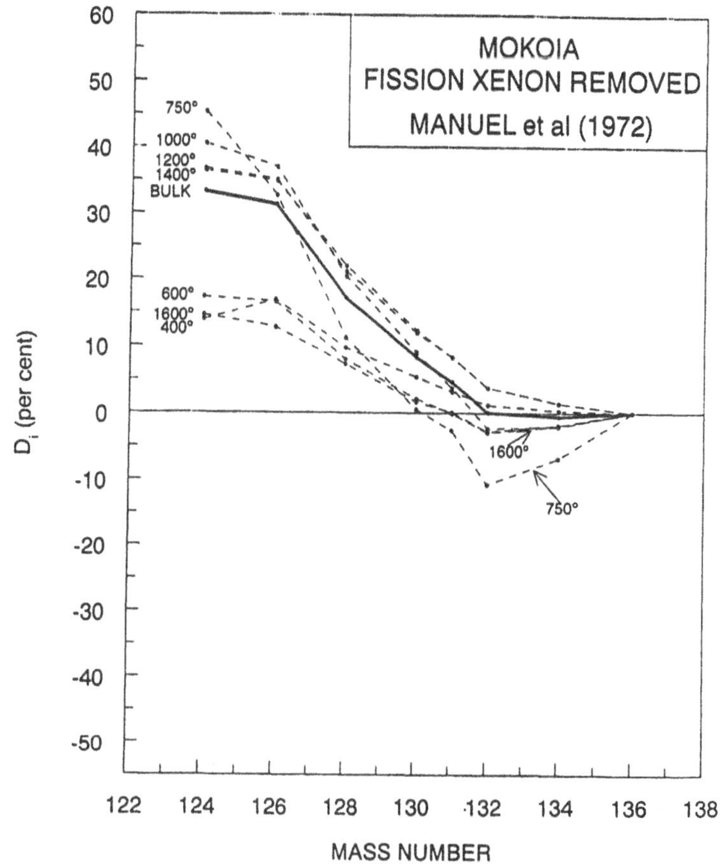


*Figure 20.* Isotopic compostitions of xenon released from the Mokoia meteorite (fission xenon removed).

Huss and Lewis (1994) reported the results of measurements of the isotopic compositions of xenon in high-purity separates of presolar diamond, which had been prepared from 14 primitive chondrites, including the carbonaceous chondrite Allende. Although they interpreted the rare gas isotope data in such a manner that they are complex mixtures of several strange components, it seems that the xenon isotope data which they obtained can be explained in a more straightforward manner as in the case of the 800°C fraction of the Allende meteorite shown in Figure 15.

Figure 21 shows a comparison of the pattern of variations of the values of $D_i$ calculated for the bulk xenon and temperature fractions released from (33.3 ± 0.2) microgram sample of Allende diamond 'EB'. Note that the isotopic compositions of xenon released from the diamond separates from the Allende meteorite, which are shown here, are remarkably similar to that of the xenon released from this meteorite at 800°C (see Figure 17). The reason for this is that the xenon found in the Allende diamond contains an appreciable quantity of $^{244}$Pu fission xenon.

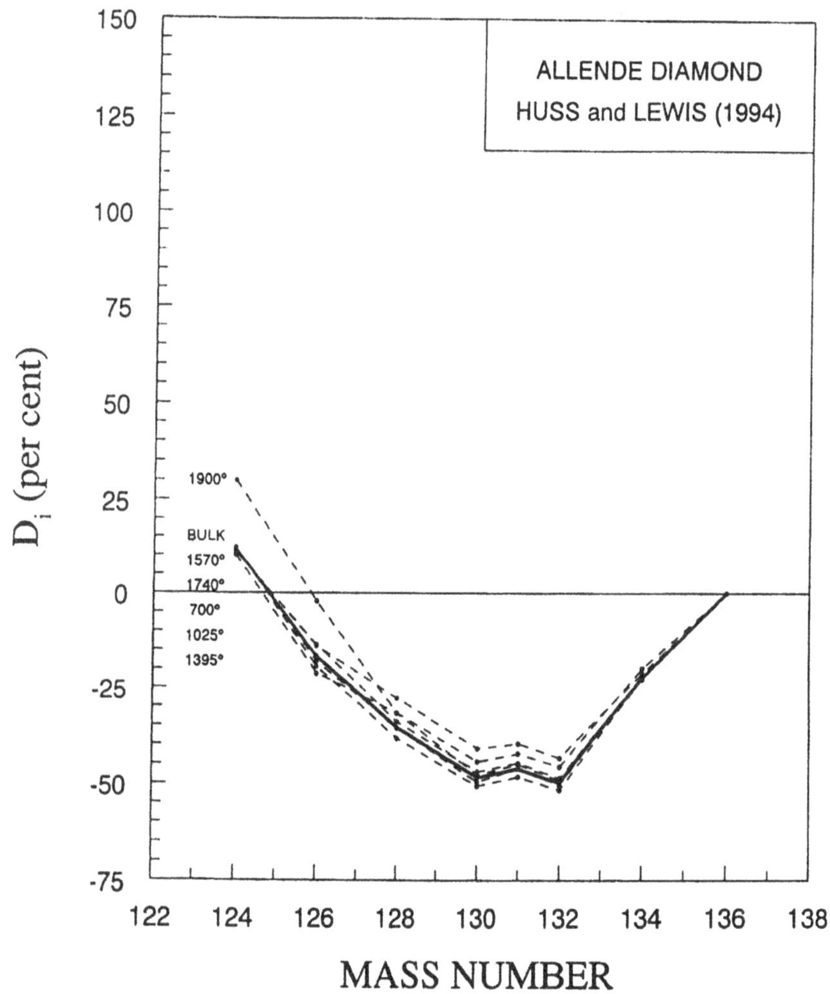


*Figure 21.* Isotopic compositions of xenon released from the Allende diamond 'EB'.

Figure 22 shows a comparison of the isotopic compositions of trapped xenon released form the Allende diamond 'EB' and of the atmospheric xenon. Note here that the values of $D_{130}$, $D_{131}$, $D_{132}$ and $D_{134}$ remain negative for all the xenon fractions, in spite of the fact that the $^{244}$Pu fission xenon was removed from these xenon fractions.

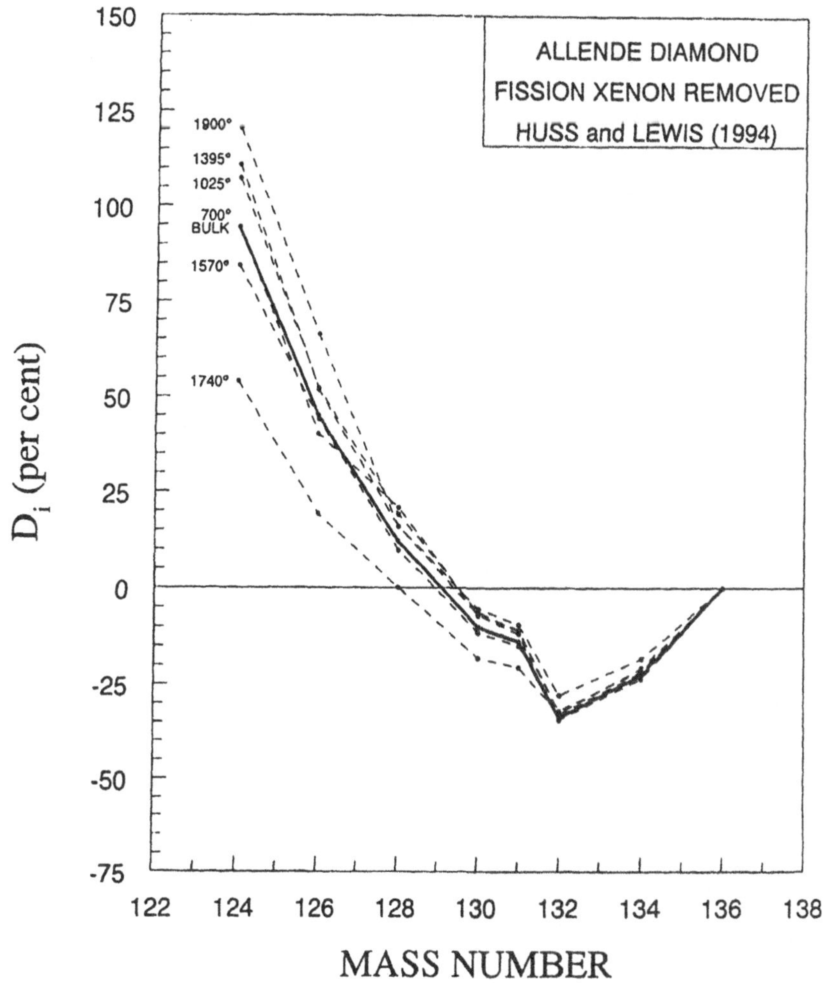


*Figure 22.* Isotopic compositions of xenon released from the Allende diamond "EB" (fission xenon removed).

It is also worthy of note in Figure 22 that the relative abundances of the light isotopes $^{124}$Xe, $^{126}$Xe and $^{128}$Xe are markedly enhanced in the trapped component of xenon found in the Allende diamond 'EB'. The reason for these strange patterns of variation of the isotopic compositions cannot be explained at the present time, but these results indicate that we are now entering the area of xenology, where the effects of processes taking place inside an exploding star are being revealed to us for the first time.

# 6. ISOTOPIC COMPOSITION OF THE SO-CALLED S-TYPE XENON

Srinivasan and Anders (1978) reported that the Murchison carbonaceous chondrites contains a new type of xenon component, the so-called s-type xenon and they pointed out that the results which they obtained were strongly suggestive of processes believed to take place in red giant stars. Re-examination of the xenon isotope data obtained by other investigators during the 1980s reveals, however, that the origin and nature of the strange xenon found in primitive meteorites can be explained in straightforward manner by considering the alterations of the isotopic compositions of xenon caused by a combined effect of (a) mass-fractionation, (b) spallation and (c) stellar-temperature neutron-capture reactions.

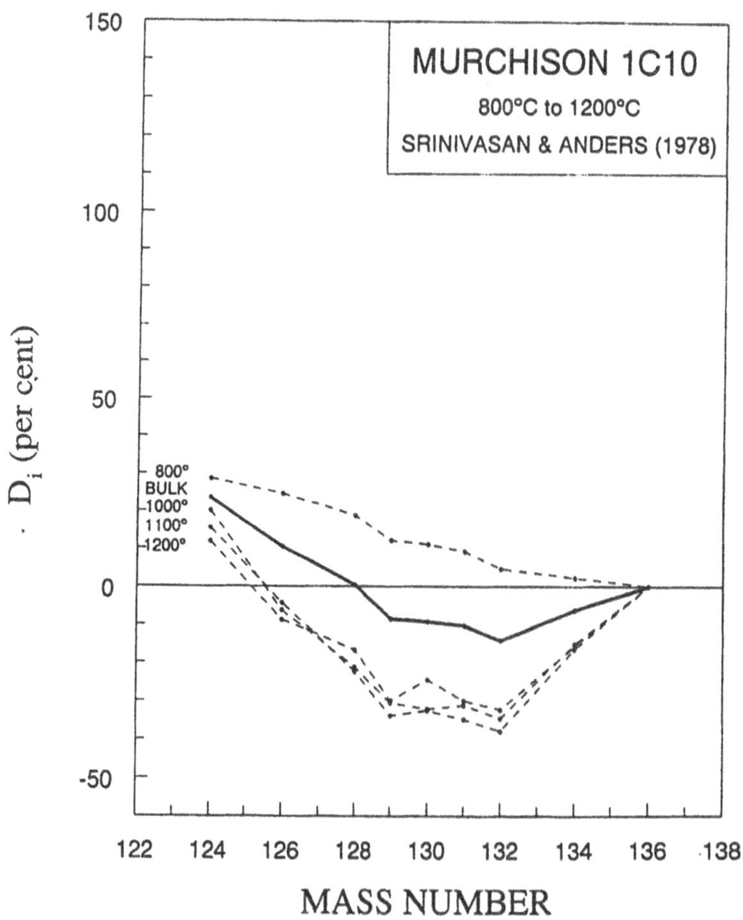


*Figure 23.* Isotopic compositions of xenon released from the acid residue Murchison 1C10 at temperatures 800°C to 1,200°C.

Figure 23 shows a comparison of the isotopic compositions of the bulk xenon and xenon fractions released from acid residue 1C10 Murchison (Srinivasan and Anders, 1978) at temperatures between 800°C and 1,200°C. The isotopic composition of the xenon found in the bulk sample of the Murchison residue 1C10 resembles that of the xenon released from the Allende meteorite at temperatures 600°, 800° and 1,000°C, indicating that this sample contains an appreciable quantity of the so-called xenon-HL. The isotopic composition of the xenon released from this sample at 800°C, on the other hand, resembles that of the xenon found in the bulk sample of the Murchison meteorite (see Figure 11), and the isotopic compositions of xenon released from Murchison 1C10 at 1,000°, 1,100° and 1,200°C resemble those of the xenon found in the 600°, 800° and 1,000°C fractions of the Allende meteorite.

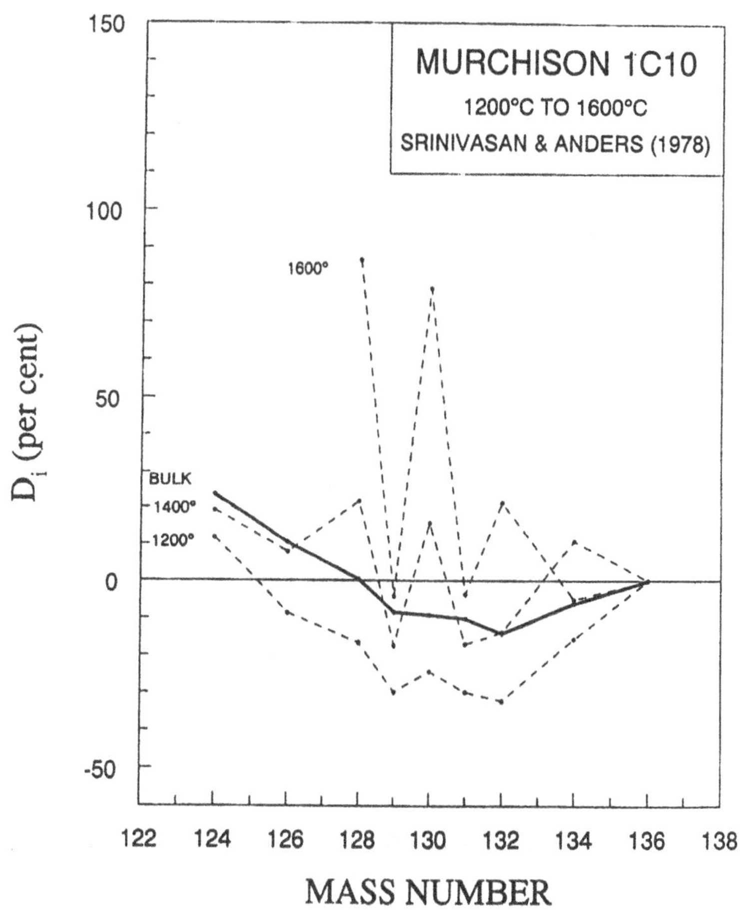


*Figure 24.* Isotopic compositions of xenon released from the acid residue Murchison 1C10 at temperatures 1,200°C to 1,600°C.

Figure 24 shows a comparison of the isotopic compositions of the 1,200°, 1,400° and 1,600° fractions and of the bulk sample of the Murchison 1C10 (Srinivasan and Anders, 1978). Note that the values of $D_{128}$, $D_{130}$ and $D_{132}$ show a spectacular increase as the release temperature is raised from 1,400° to 1,600°C. These results indicate that we are dealing here with the xenon fractions which had been exposed to an extremely high flux of stellar-temperature neutrons, most likely in the immediate vicinity of the surface or the interior of an exploding supernova.

Figure 25 shows a comparison of the isotopic compositions of xenon released from the Murchison residue 2C10f (< 1$\mu$) studied by Alaerts *et al.* (1980). The results obtained by these investigators are similar to those reported by Srinivasan and Anders (1978) two years earlier (see Figures 19 and 20), but less complete.

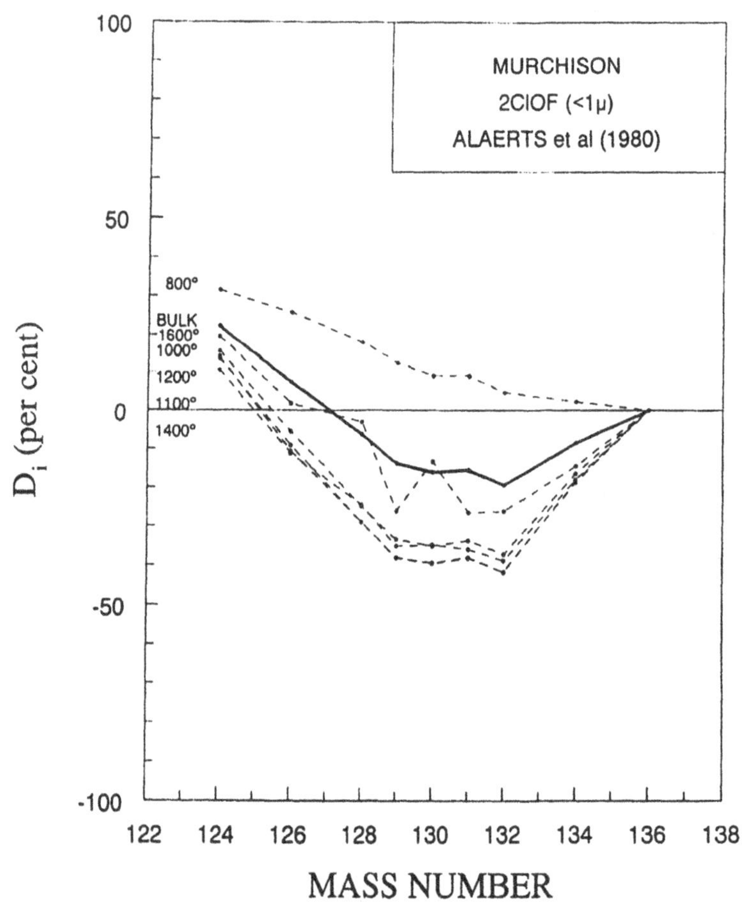


*Figure 25.* Isotopic compositions of xenon released from the acid residue Murchison 2C10f (< 1$\mu$).

Figure 26 shows a comparison of the isotopic compositions of xenon released from the Murchison residue 2C10m (1 - 3$\mu$) studied by Alaerts *et al.* (1980), which resembles the plots shown as in Figure 20. Sharp increases in the values of $D_{124}$, $D_{126}$, $D_{128}$, $D_{130}$ and a small increase in the value of $D_{132}$ displayed by the 1,600°C xenon fraction are telling us that appreciable amounts of spallation-produced xenon and the xenon exposed to an extremely high flux of stellar-temperature neutrons are being released at this temperature. Unfortunately, however, the isotopic compositions of the xenon released from this sample at temperatures 1,200° and 1,400°C were not measured by these investigators.

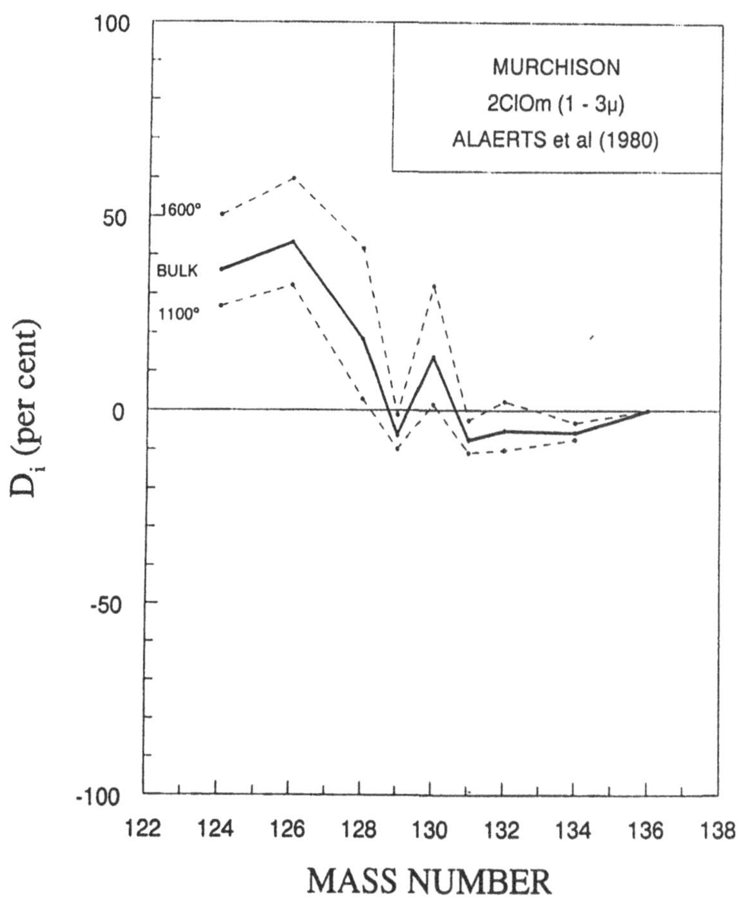


*Figure 26.* Isotopic compositions of xenon released from the acid residue Murchison 2C10m(1-3$\mu$).

Figure 27 shows a comparison of the isotopic compositions of the bulk xenon and xenon fractions released from the Murchison SiC, HX, studied by

Zinner *et al.* (1989). The plot shown here is similar to those shown in Figure 19 and in Figure 23 for the Murchison acid residues 1C10 and 2C10f, respectively, and shows that the Murchison, SiC, *HX*, contains appreciable quantities of the so called xenon-*HL*.

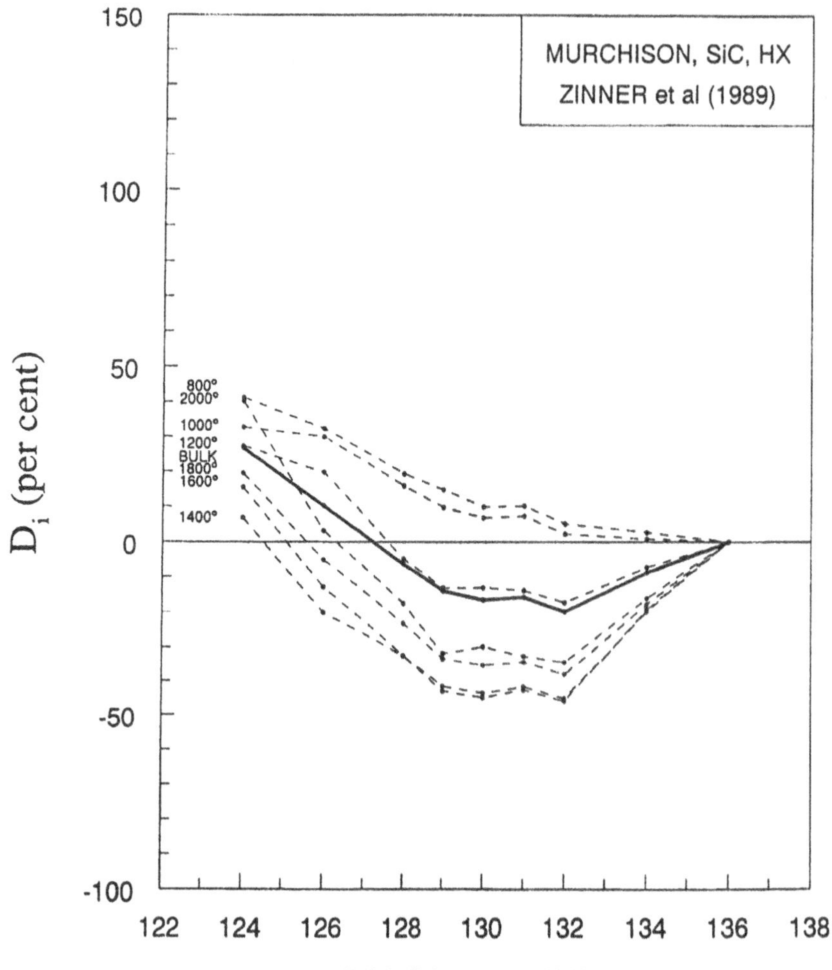


*Figure 27.* Isotopic compositions of xenon released from Murchison SiC, HX.

Figure 28 shows a comparison of the isotopic compositions of the bulk xenon and xenon fractions released from the Murchison SiC, *HM*, studied by Zinner *et al.* (1989). Note that the relative abundances of $^{128}$Xe, $^{130}$Xe and $^{132}$Xe are markedly enhanced in this sample. It is also worthy of note that appreciable quantities of spallation-produced xenon isotopes, $^{124}$Xe and

$^{126}$Xe, were removed from this sample when the released temperature was raised above 1,200°C.

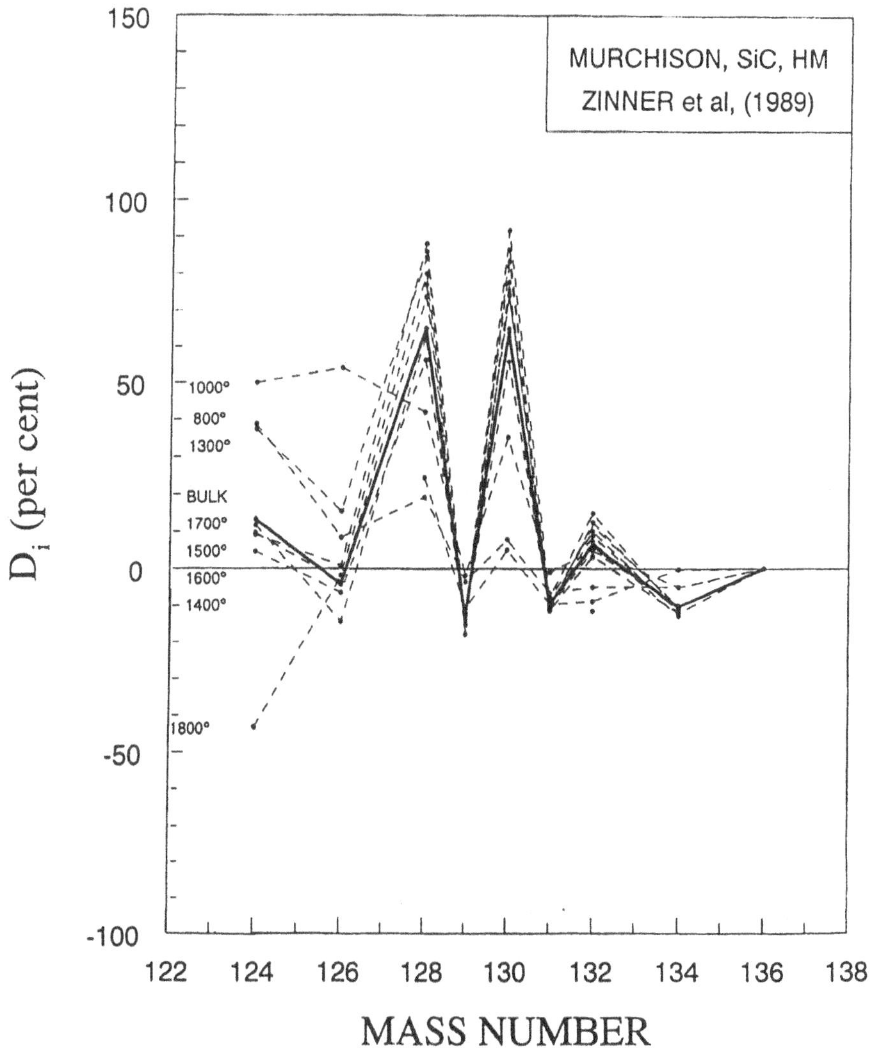


*Figure 28.* Isotopic compositions of xenon released from Murchison SiC, HM.

Figure 29 shows a comparison of the isotopic compositions of the bulk xenon and xenon fractions released from the Murchison SiC, *HN*, studied by Zinner *et al.* (1989). Note that the relative abundances of $^{128}$Xe, $^{130}$Xe and $^{132}$Xe are markedly enhanced and also that appreciable quantities of the spallation-produced xenon isotopes ($^{124}$Xe and $^{126}$Xe) were removed from this sample when the release temperature was raised to 1,500°C.

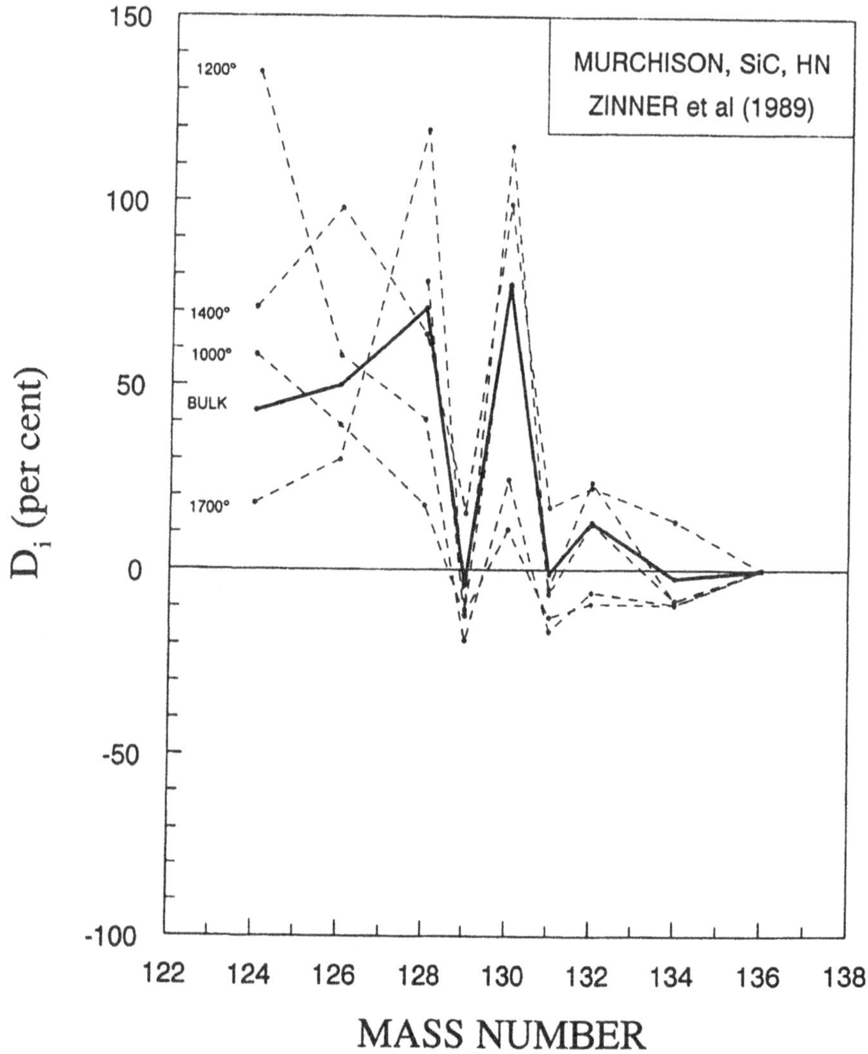


*Figure 29.* Isotopic compositions of xenon released from Murchison SiC, HN.

Figure 30 shows a comparison of the isotopic compositions of the xenon found in the bulk samples of the Murchison, SiC, *HM*, *HN*, *HX* and *HO*. Note that the relative abundances of $^{128}$Xe, $^{130}$Xe and $^{132}$Xe are markedly enhanced in *HM* and *HN*, indicating that the so-called *s*-type xenon is enriched in these samples. In the cases of *HX* and *HO*, on the other hand, the values of $D_{129}$, $D_{130}$, $D_{131}$, $D_{132}$ and $D_{134}$ are all negative indicating that the so-called xenon-*HL* is enriched in these samples.

Lewis *et al.* (1994) reported that they analyzed He, Ne, Ar, Kr, and Xe in fourteen size fractions of interstellar SiC, isolated from the Murchison C2 chondrite and found that all are mixtures of a highly anomalous component bearing the isotopic signature of the astrophysical s-process and a more normal component, generally solar-like but with anomalies of up to 30 percent in the heavy isotopes. These investigators therefore concluded that as these two components strikingly resembled predictions for the He-burning shells and envelopes of red giant carbon stars, it appeared that the SiC grains were pristine circumstellar condensates from such stars.

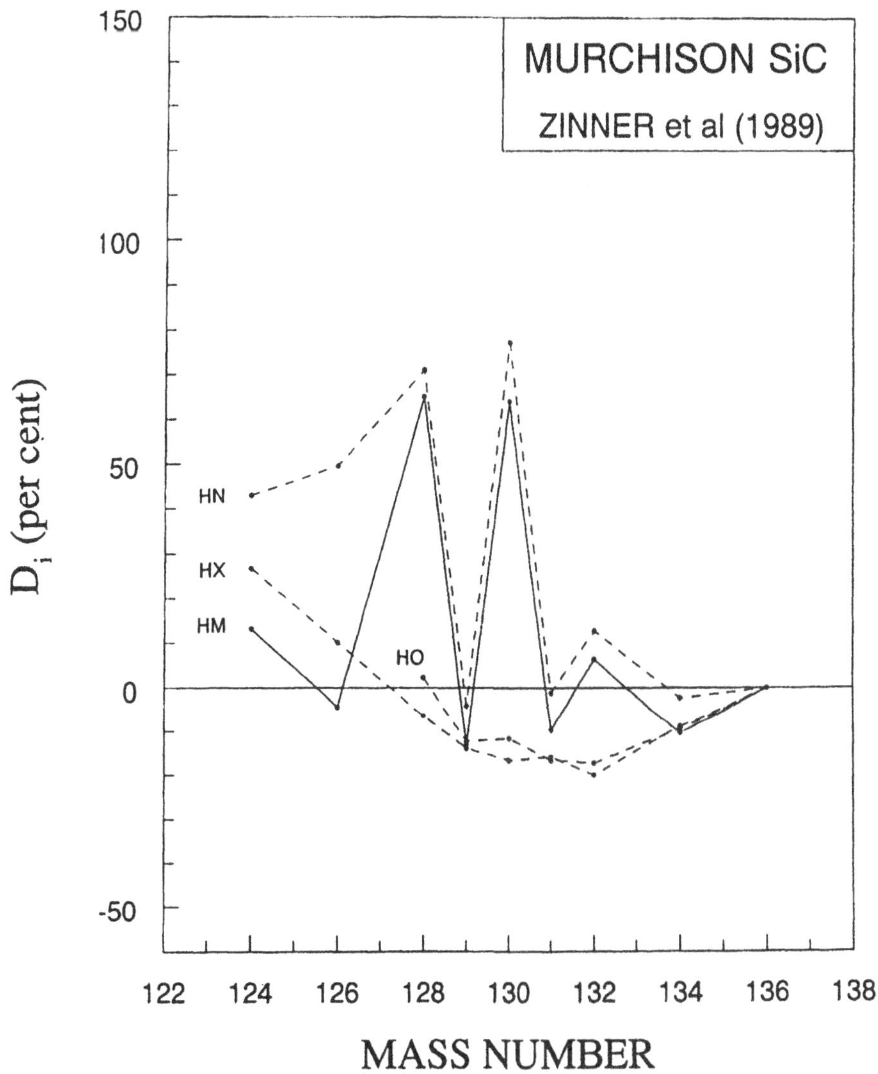


*Figure 30.* Comparision of the isotopic compostions of xenon found in various SiC samples.

Figure 31 shows a comparison of the isotopic compositions of the xenon found in the bulk samples and in several temperature fractions of the

Murchison SiC, *KJA*, studied by Lewis *et al.* (1994). Note that the relative abundances of $^{128}$Xe, $^{130}$Xe and $^{132}$Xe are markedly enhanced in this sample, indicating that we are dealing here with a sample of xenon, whose isotopic compositions are drastically altered by the effect of *stellar-temperature* neutron-capture reactions. It is also interesting to note in Figure 29 that much of the spallation-produce xenon isotopes ($^{124}$Xe and $^{126}$Xe) are being removed as the release temperature is raised above 1,000°C.

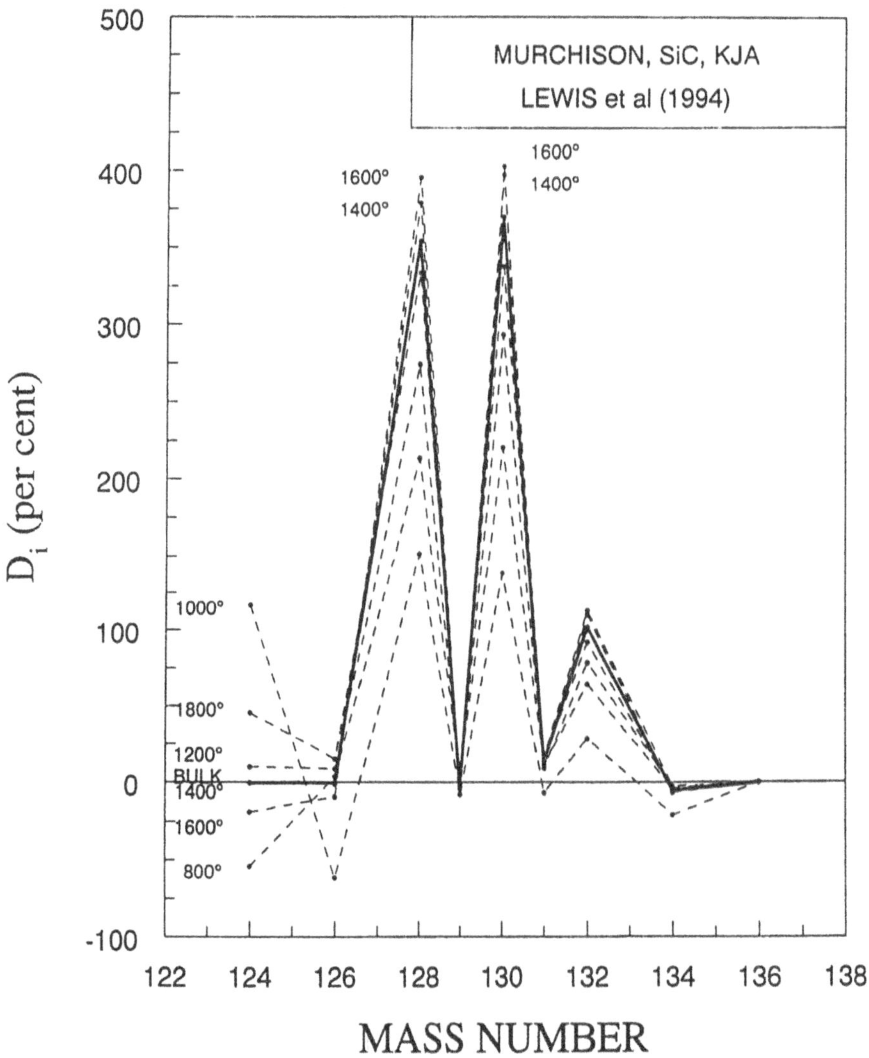


*Figure 31.* Isotopic compositions of xenon released from Murchison SiC, KJA.

Figure 32 shows a similar plot for the Murchison SiC, *KJB*, studied by Lewis *et al.* (1994). Note that the spallation-produced xenon isotopes ($^{124}$Xe and $^{126}Xe$) are almost completely removed from this sample and the isotopic composition of the trapped xenon found in this sample is quite similar to that of the atmospheric xenon.

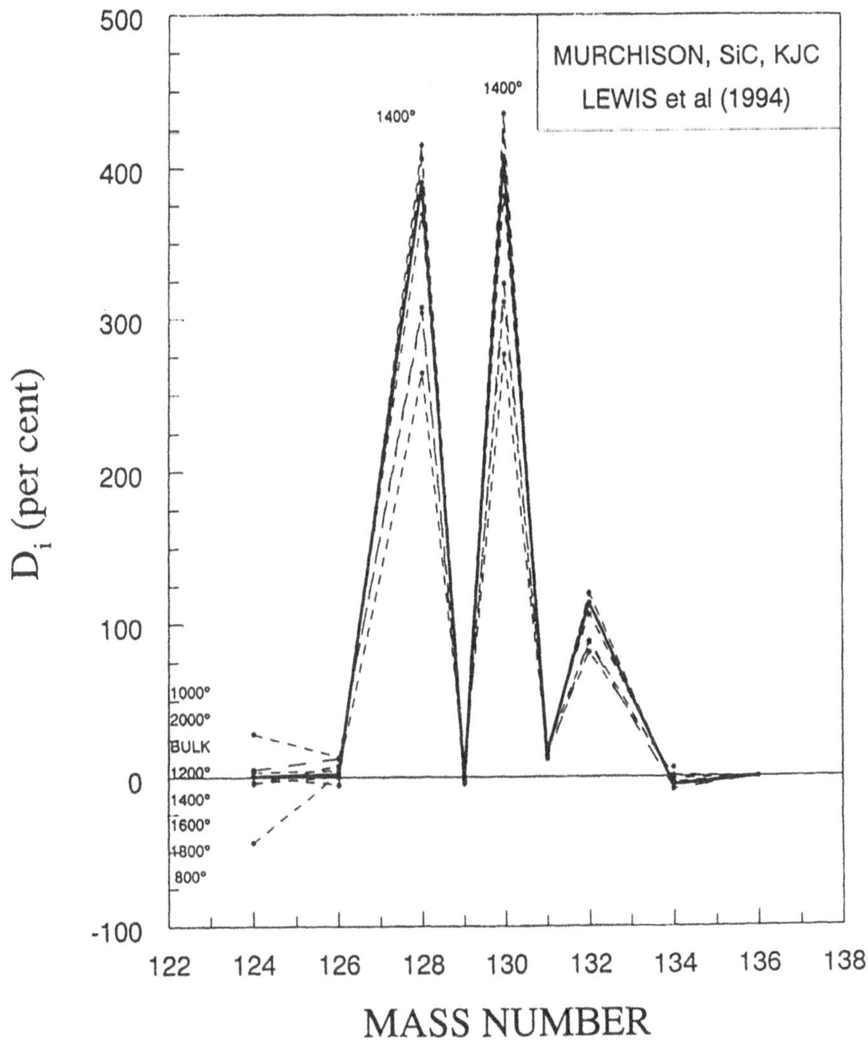


*Figure 32.* Isotopic compositions of xenon released from Murchison SiC, KJB.

Figure 33 shows a similar plot for the Murchison SiC, *KJC*, studied by Lewis *et al.* (1994). Note that the spallation-produced xenon isotopes ($^{124}Xe$ and $^{126}Xe$) are almost completely removed also from this sample and the isotopic composition of the trapped xenon found in this sample is quite similar to that of the atmospheric xenon.

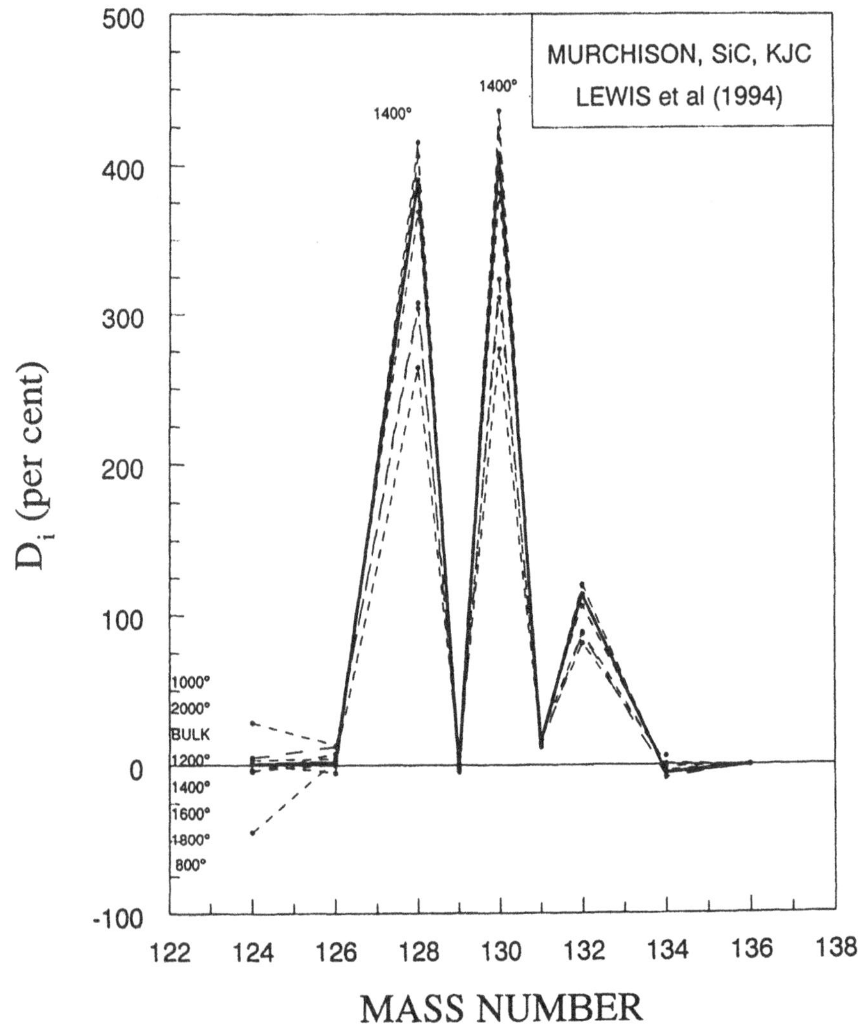


*Figure 33.* Isotopic compositions of xenon released from Murchison SiC, KJC.

These results indicate that we are dealing here with a sample of xenon, whose isotopic compositions are drastically altered by the effect of *stellar-temperature* neutron-capture reactions. This means that data can be explained in a straightforward manner as due to the alteration of the isotopic composition of xenon caused by a combined effect of (a) mass-fractionation, (b) spallation and (c) *stellar-temperature* neutron-capture reactions, as pointed out by Kuroda (1971).

Amari *et al.* (1994) reported that they studied the isotopic compositions of noble gases released from interstellar graphite grains in stepwise heating experiments and the results indicated that the experimental data showed the signature of the s-process and hence these grains must have come from AGB (Asymptotic Giant Branch) star of 1 to 3 solar masses.

Figure 34 shows a comparison of the isotopic compositions of xenon released from the graphite grains of the Murchison meteorite studied by Amari *et al.* (1994) and of the atmospheric xenon. The pattern of variations of the values of $D_i$ shown here shows a remarkable resemblance to the case of Murchison SiC grains studied by Zinner *et al.* (1989) (See Figure 28), and these results indicate that we are dealing here with mixtures of the so-called xenon-*HL* and *s*-type xenon.

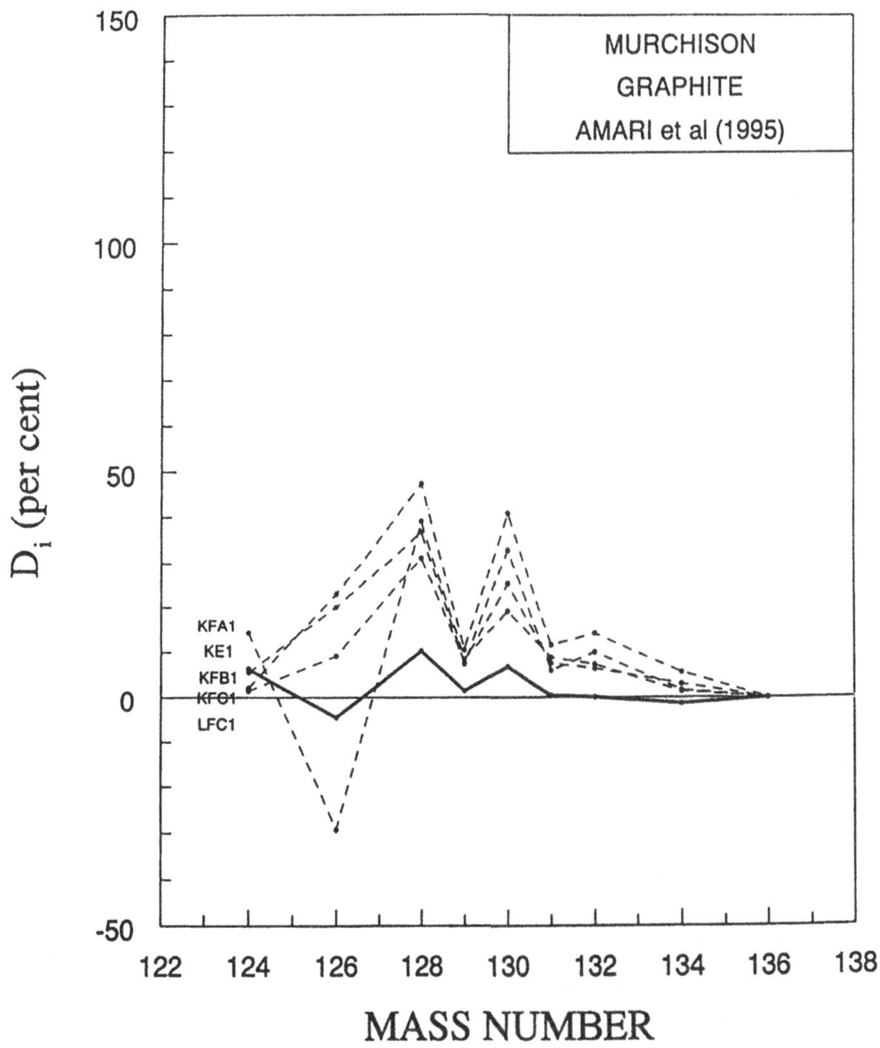


*Figure 34.* Isotopic compositions of xenon released from Murchison graphite samples.

# 7. K/AR AGES OF SOME TERRESTRIAL AND EXTRATERRESTRIAL SAMPLES

As shown in Figure 8, there exists a striking resemblance between the $^{207}Pb/^{206}Pb$ ages of the samples from the earth and the moon. In view of the fact that we found these totally unexpected results from the studies concerning the extinct radionuclide $^{244}Pu$ and the near-extinct radionuclide $^{235}U$ in the early solar system, we felt that it would be interesting to investigate the case of another near-extinct radionuclide $^{40}K$ in the early solar system.

Turner (1970) reported that seven crystalline rock samples returned by Apollo 11 have been analyzed in detail by means of the $^{40}Ar/^{39}Ar$ dating technique. The extent of radiogenic argon loss in these samples ranged from 7 percent to ~ 48 percent. Potassium/argon ages, corrected for the effects of this loss, clustered relatively closely around the value of 3.7 billion years, while in the same year, Marti *et al.* (1970) reported that less than 30 percent of the $^{40}Ar$ in the Apollo 11 fines could be accounted for on the basis of the measured K and a K/Ar age of 4.5 billion years.

Kaiser (1972) then reported on a detailed study on the release of rare gases from the Luna 16 fines 16-G-7 by the use of stepwise heating technique, and he stated that the release of large excesses of $^{40}Ar$ occurred from this sample. The "excess" $^{40}Ar$ had a maximum release in the temperature range of 320°C to 400°C and also at 1,000°C. He reported that the K/Ar age based on the "excess" $^{40}Ar$ and the K content of 880 ppm led to an unreasonably high age of 6.73 billion years and a K/Ar age based on only the high temperature "excess" $^{40}Ar$ was also unreasonable 5.03 billion years.

It is interesting to recall here that Tatsumoto (1973) reported the age of a sample of Luna 20 soil calculated from the ratios of $^{206}Pb/^{238}U$, $^{207}Pb/^{235}U$, $^{207}Pb/^{206}Pb$ and $^{208}Pb/^{232}Th$ was 5.13, 4.98 and 4.95 billion years, respectively, and these results indicate that the K/Ar ages of some of the lunar samples may be indeed greater than 5 billion years.

Jessberger and Dominik (1979) reported that K/Ar and/or $^{40}Ar/^{39}Ar$ ages of the Allende samples (whole rock, matrix, chondrules, white inclusions) ranged from 3.8 billion years for matrix to some 5.0 billion years for some white inclusions, but that they clustered strongly near 4.53 billion years, indicating that the dominant K/Ar resetting process for Allende materials occurred at this time. These investigators reported in 1980 that possible explanations for the apparent presolar ages (greater than 4.6 billion years) included: ~20 percent loss of $^{39}Ar$; ~40 percent loss of $^{40}K$ about 3.8 billion years ago with no loss of $^{40}Ar$; trapped argon of unique $^{40}Ar/^{36}Ar$ isotopic composition; and a mixture of "very old" presolar grains (see also Jessberger *et al.*, 1980).

In an article entitled "K-Ar isochron dating of Zaire cubic diamonds", Zashu *et al.* (1986) reported that the $^{40}Ar/K$ and $^{40}Ar/^{39}Ar$ plots, which they obtained for 10 samples of Zaire diamonds, both showed good linear correlations, which are most easily understood as isochrons; that is, the observed linear arrays reflected the correlation between K and K-driven $^{40}Ar$ in the diamonds. They concluded, however, that: "*The formal age calculated from the slope of these plots is (6.0 ± 0.3) billion years: however, this is older than the Earth and is therefore unacceptable* ..." (page 71) (Emphases added.)

Ozima *et al.* (1989) then reported that cubic diamonds from Zaire showed excellent correlation between potassium content and $^{40}Ar/^{36}Ar$ ratio, and between $^{40}Ar/^{36}Ar$ and $^{39}Ar/^{36}Ar$, which could be interpreted to yield an isochron age of about 6 billion years; but they added the following concluding remark: "*Now the discovery of a correlation between chlorine content and $^{40}Ar$, and together with recent mineralogical and geochemical work concerning the origin of cubic diamonds, strongly suggests that the $^{40}Ar$ and its associated potassium are contained in sub-micrometer inclusions of mantle-derived fluid*".

The primary objective of the studies on the occurrence of extinct radionuclides in the early solar system should be to determine the age of the solar system as precisely as possible. We feel, however, that most of investigators in the field of cosmochemistry have been attempting to interpret the experimental data *assuming* that the age of the solar system to be 4.55 billion years (see, for example, Dalrymple 1991 and Allérge *et al.*, 1995).

## 8. SUMMARY AND CONCLUDING REMARKS

In 1661 Robert Boyle in his classic essays entitled "The Sceptical Chymist", taught us that the most important lesson to be learned in the field of science is always try to interpret the experimental data without any pre-conceived ideas. It is also worthy of note here that the origin of the concept of chemical elements can itself be traced back to his early ideas which he put forward more than three and one third centuries ago.

In his first proposition, Boyle (1661, p. 51) stated: "*It seems not absurd to conceive that at the first production of mixt bodles, the universal matter whereof they among other parts of the universe consisted, was actually divided into little particles of several sizes and shapes variously moved.*"

He then proceeded to his PROPOSITIONS II and III and finally to his concept of elements in his PROPOSITION IV, in which he stated: "*It may likewise be granted, that those distinct substances, which concretes generally either afford or made up of, may without very much inconvenience be called the elements or principles of them.*" (p. 51).

Three centuries after the publication of "The Sceptical Chymist" by Boyle, in 1661, the Soviet Union launched Sputnik in 1957 and Burbidge *et al.* (1957) published an important paper concerning the synthesis of the elements in stars. Four years later in 1961, newly elected President John F. Kennedy declared that the United States would send Man to the Moon before the decade was over.

In his classic paper entitled Xenology, Reynolds (1963, p. 2939) stated that: "*Xenology means to us the detailed study of the abundance of Xe isotopes evolved from meteorites by heating or other means and the inferences that can be drawn from these studies about the early history of the meteorites and the solar system.*"

The field of Xenology had a turbulent history since the 1960's as it was reviewed in this paper. The primary cause for this was that both Reynolds (1960) and Kuroda (1960) were attempting to interpret the xenon isotope data with a pre-conceived idea that the age of the solar system to be very close to 4.5 to 4.6 billion years.

Re-examination of a vast amount of existing xenon isotope data has revealed, however, that lunar fines and carbonaceous chondrites both contain appreciable quantities of $^{244}$Pu fission xenon and that they began to retain their xenon almost 5.1 billion years ago, when the $^{244}$Pu to $^{238}$U ratio in the solar system was roughly 1 to 2 (atom/atom).

We have extended the re-examination of the experimental data to the field of studies on the lead/lead and potassium/argon dating methods and the results indicated that some of the extra terrestrial and terrestrial samples also have the very old ages determined by these methods, which are greater than 5.1 billion years.

## ACKNOWLEDGEMENT

We wish to express our deep gratitude to Professor Oliver K. Manuel of the University of Missouri-Rolla, who was kind enough to invite us to this important symposium.

## REFERENCES

Alaerts, L., Lewis, R.S., Matsuda, J. and Anders, E.: 1980, "Isotopic anomalies of noble gases in meteorites and their origins", VI. Presolar components in the Murchison C2 chondrite, *Geochim. Cosmochim. Acta* **44**, 189-209.

Alexander, E.C. Jr.: 1970, "Rare gases from stepwise heating of lunar rock 12013", *Earth Planet. Sci. Lett.* **9**, 201-207.

Alexander, E.C. Jr., Lewis, R.S., Reynolds, J.H. and Michel, M.C.: 1971, "Plutonium-244: Confirmation as an extinct radioactivity", *Science* **172**, 837-840.

Amari, S., Lewis, R.S. and Anders, E.: 1994, "Interstellar grains in meteorites: I. Isolation of SiC, graphite, and diamond; size distributions of SiC and graphite", *Geochim. Cosmochim. Acta* **58**, 459-470.

Anders, E. and Zinner, E.: 1993, "Interstellar grains in primitive meteorites: Diamond, silicon carbide, and graphite", *Meteoritics* **28**, 490-514.

Aston, F.W.: 1933, *Mass-spectra and Isotopes*, Edward Arnold and Company, London, p. 220.

Basford, J.R., Dragon, J.C., Pepin, R.O., Coscio, M. and Murthy, R.: 1973, "Krypton and xenon in lunar fines", *Proc. 4th Lun. Sci. Conf.* (Suppl. **4**, *Geochim. Cosmochim. Acta*) **2**, 1915-1955.

Becquerel, H.: 1896, "On the invisible radiations emitted by phosphorescent substances", *Compt. Rend. Acad. Sci. Paris* **122**, 501 (2 March) 501-503.

Behrmann, C.J., Drozd, R.J. and Hohenberg, C.M.: 1973, "Extinct lunar radioactivities: Xenon from $^{244}$Pu and $^{129}$I in Apollo 14 breccias", *Earth Planet. Sci. Lett.* **17**, 446-455.

Bogard, D.D., Hirsch, W.C. and Nyquist, L.E.: 1974, "Noble gases in Apollo 17 fines: Mass fractionation effects in trapped Xe and Kr", *Proc. 5th Lunar Sci. Conf.* **2**, 1975-2003.

Boyle, R.: 1661, *The Sceptical Chymist*, J. M. Dent & Sons LTD, London, E. P. Dutton & Co., Inc., New York, No. 559 of Everyman's Library, 1910, pp. 29-36.

Burbidge, E.M., Burbidge, G.R., Fowler, W.A. and Hoyle, F.: 1957, "Synthesis of the elements in stars", *Rev. Mod. Phys.* **29**, 547-650.

Butler, W.A., Jeffery, P.M., Reynolds, J.H. and Wasserburg, G.J.: 1963, "Isotopic variations in terrestrial xenon", *J. Geophys. Res.* **68**, 3283-3291.

Cameron, A.G.W.: 1958, "Nuclear astrophysics", *Ann. Rev. Nuc. Sci.* **8**, 299-326.

Cameron, A.G.W.: 1962, "The formation of the sun and planets", *Icarus* **1**, 13-69.

Clark, R.S., Rao, M.N. and Kuroda, P.K.: 1967, "Fission and spallation xenon in meteorites", *J. Geophys. Res.* **72**, 5143-5148.

Clayton, R.N., Grossman, L. and Mayeda, T.K.: 1973, "A component of primitive nuclear composition in carbonaceous meteorites", *Science* **182**, 485-488.

Collins, C.B., Farguhar, R.M. and Russell, R.D.: 1952, "Variations in the relative abundances of the isotopes of common lead", *Phys. Rev.* **88**, 1275-1276.

Collins, C.B., Russell, R.D. and Farguhar, R.M.: 1953, "The maximum age of the elements and the age of the earth's crust", *Can. J. Phys.* **31**, 402-418.

Dreibus, G., Burghele, A., Jochum, K.P., Spettel, B., Wlotzka, F. and Wänke, H.: 1994, "Chemical and mineral composition of ALH84001: A martian orthopyroxenite", *Meteoritics* **29**, 461.

Drozd, R., Hohenberg, C.M. and Ragan, D.: 1972, "Fission xenon from extinct $^{244}$Pu in 14301", *Earth Planet. Sci. Lett.* **15**, 338-346.

Drozd, R.J., Hohenberg, C. and Morgan, C.: 1975, "Krypton and xenon Apollo 14 samples: Fission and neutron capture in gas-rich samples", *Proc. 6th Lunar Sci. Conf.*, 1857-1877.

Eberhardt, P., Geiss, J., Graf, H., Grögler, N., Mendia, M. D., Mörgeli, N., Schwaller, H., Stettler, A., Krähenbühl, U. and Von Gunten, H.R.: 1972, "Trapped solar wind noble gases in Apollo 12 lunar fines 12001 and Apollo 11 breccia 10046", *Proc. 3rd Lunar Sci. Conf.* **2**, 1821-1856.

Eugster, O., Eberhardt, P. and Geiss, J.: 1967, "Krypton and xenon isotopic composition in three carbonaceous chondrites", *Earth Planet Sci. Lett.* **3**, 249-257.

Fleischer, R.L. and Naeser, C.W.: 1972, "Search for plutonium-244 tracks in Mountain Pass bastnaesite", *Nature* **240**, 465.

Fleischer, R.L., Price, P.B. and Walker, R.M.: 1965, "Spontaneous fission tracks from extinct $^{244}$Pu in meteorites and the early history of the solar system", *J. Geophys. Res.* **70**, 2703-2707.

Fowler, W.A.: 1967, "Nuclear Astrophysics" *Jayne Lectures for 1965*, American Philosophical Society, pp. 63-65.

Fowler, W.A.: 1972, in *Cosmology, Fusion & Other Matters*, Ed., Reines, F., Colorado Associated University Press, Boulder, CO, pp. 67-123.

Fowler, W.A.: 1984, "The Quest for the Origin of the Elements", *Nobel Lecture*, 8 December 1983, Science **226**, 922-935.

Funk, H., Podosek, F. and Rowe, M.W.: 1967, "Fissiogenic xenon in the Renazzo and Murray meteorites", *Geochim. Cosmochim. Acta* **31**, 1721-1732.

Gray, C.M. and Compston, W.: 1974, "Excess $^{26}$Mg in the Allende meteorite", *Nature* **251**, 495-497.

Hagee, B., Bernatowicz, T.J., Podosek, F.A., Johnson, M.L., Burnett, D.S. and Tatsumoto, M.: 1990, "Actinide abundances in ordinary chondrites", *Geochim. Cosmochim. Acta* **54**, 2847-2858.

Heydegger, H.R., Foster, J.J. and Compston, W.: 1979, "Evidence of a new isotopic anomaly from titanium isotopic ratios in meteoritic materials", *Nature* **278**, 704-707.

Hoffman, D.C., Lawrence, F.O., Mewherter, J.L. and Rourke, F.M.: 1971, "Detection of plutonium-244 in nature", *Nature* **234**, 132-134.

Hohenberg, C.M.: 1970, "Xenon from the Angra dos Reis meteorite", *Geochim. Cosmochim. Acta* **34**, 185-191.

Hohenberg, C.M., Podosek, F.A. and Reynolds, J.H.: 1967a, "Xenon-iodine dating: Sharp isochronism in chondrites", *Science* **156**, 233-236.

Hohenberg, C.M., Munk, M.N. and Reynolds, J.H.: 1967b, "Spallation and fissiogenic xenon and krypton from stepwise heating of the Pasamonte achondrite; the case for extinct plutonium-244 in meteorites; relative ages of chondrites and achondrites", *J. Geophys. Res.* **72**, 3139-3177.

Hohenberg, C.M., Davis, P.K., Kaiser, W.A., Lewis, R.S. and Reynolds, J.H.: 1970, "Trapped and cosmogenic rare gases from stepwise heating of Apollo 11 samples", *Proc. Apollo 11 Lun. Sci. Conf.* **2**, 1283-1309.

Huss, G.R. and Lewis, R.S.: 1994, "Noble gases in presolar diamonds I: Three distinct components and their implications for diamond origins", *Meteoritics* **29**, 791-810.

Jessberger, E. and Dominik, B.: 1979, "Gerontology of the Allende meteorite", *Nature* **277**, 544-556.

Jessberger, E.K., Dominik, B., Staudacher, T. and Herzog, G.F.: 1980, "$^{40}$Ar-$^{39}$Ar ages of Allende", *Icarus* **42**, 380-405.

Kaiser, W.A.: 1972, "Rare gas studies in Luna-16-G-7 fines by stepwise heating technique: A low fission solar wind Xe", *Earth Planet. Sci. Lett.* **13**, 387-399.

Krummenacher, D., Merrihue, C.M., Pepin, R. O. and Reynolds, J.H.: 1962, "Meteoritic krypton and barium versus the general isotopic anomalies in meteoritic xenon", *Geochim. Cosmochim. Acta* **26**, 231-249.

Kuroda, P.K.: 1960, "Nuclear fission in the early history of the earth", *Nature* **187**, 36-38.

Kuroda, P.K.: 1961, "The time interval between nucleosynthesis and formation of the earth", *Geochim. Cosmochim. Acta* **24**, 40-47.

Kuroda, P.K.: 1971, "Temperature of the sun in the early history of the solar system", *Nature Phys. Sci.* **230**, 40-42.

Kuroda, P.K.: 1975, "The puzzle of the rare gases and of oxygen in materials within the solar system", *Geochem. J.* **9**, 51-62.

Kuroda, P.K.: 1976, "Xenology: the enigma of xenon in carbonaceous chondrites", *Geochem. J.* **10**, 121-136.

Kuroda, P.K.: 1979a, "Isotopic anomalies in the early solar system", *Geochem. J.* **13**, 83-90.

Kuroda, P.K.: 1979b, "Palladium-107 in the early solar system", *Geochem. J.* **13**, 135-136.

Kuroda, P.K.: 1979c, "Meteoritic barium and cerium versus the general isotopic anomalies in meteoritic xenon", *Geochem. J.* **13**, 137-140.

Kuroda, P.K.: 1979d, "Isotopic composition of gadolinium in meteorites", *Geochem. J.* **13**, 281-285.

Kuroda, P.K.: 1979e, "More mysteries from Pandora's box: Samarium isotopic anomalies in the Allende inclusions EK1-4-1 and C1", *Geochem. J.* **13**, 287-298.

Kuroda, P.K.: 1979f, "Time scales inferred from the cosmochronometers iodine-129, plutonium-244 and aluminum-26", *Geochem. J.* **13**, 291-296.

Kuroda, P.K.: 1979g, "Correlated calcium and titanium isotopic anomalies in Allende inclusions", *Geochem. J.* **13**, 297-300.

Kuroda, P.K.: 1982, *The Origin of the Chemical Elements and the Oklo Phenomenon*, Springer-Verlag, Berlin, Germany, Heidelberg, Germany, pp. 115-146.

Kuroda, P.K.: 1989, "Plutonium-244 dating of the early solar system. 50 years with nuclear fission", eds. Behrens, J.W. and Carlson, A.D., American Nuclear Society, Inc., pp. 901-908.

Kuroda, P.K.: 1993, "Plutonium-244, supernova and hot atom chemistry", *Radiochim. Acta* **62**, 27-33.

Kuroda, P.K. and Myers, W.A.: 1991a, "Plutonium-244 dating I. Initial ratio of plutonium to uranium in the Allende meteorite", *J. Radioanalyt. Nucl. Chem.* **150**, 35-51.

Kuroda, P.K. and Myers, W.A.: 1991b, "Plutonium-244 dating II. Initial ratios of plutonium to uranium in the Murray and Murchison meteorites", *J. Radioanalyt. Nucl. Chem.* **150**, 53-69.

Kuroda, P.K. and Myers, W.A.: 1991c, "Plutonium-244 dating III. Initial ratios of plutonium to uranium in lunar samples", *J. Radioanalyt. Nucl. Chem.* **150**, 71-87.

Kuroda, P.K. and Myers, W.A.: 1992a, "Plutonium-244 dating VII. Initial abundances of uranium-235 and plutonium-244 in lunar samples", *J. Radioanalyt. Nucl. Chem.* **158**, 437-453.

Kuroda, P.K. and Myers, W.A.: 1992b, "Plutonium-244 dating VIII. A note on the ages of lunar rocks 10057 and 12013", *J. Radioanalyt. Nucl. Chem.* **159**, 281-284.

Kuroda, P.K. and Myers, W.A.: 1993a, "Plutonium-244 dating IX. Fission xenon in carbonaceous chondrites", *J. Radioanalyt. Nucl. Chem.* **173**, 219-227.

Kuroda, P.K. and Myers, W.A.: 1993b, "Plutonium-244 dating X. Initial abundance of iodine-129 in carbonaceous chondrites", *J. Radioanalyt. Nucl. Chem.* **173**, 229-237.

Kuroda, P.K. and Myers, W.A., 1994a, "Plutonium-244 in the early solar system", *Radiochim. Acta* **64**, 155-165.

Kuroda, P.K. and Myers, W.A.: 1994b, "Plutonium-244 fission xenon in the most primitive meteorites", *Radiochim. Acta* **64**, 167-174.

Kuroda, P.K. and Myers, W.A.: 1995, "Plutonium-244 in the early solar system and xenology", *Radiochim. Acta* **68**, 81-90.

Kuroda, P.K. and Myers, W.A.: 1996, "Aluminum-26 in the early solar system", *J. Radioanalyt. Nucl. Chem.* **211**, 539-555.

Kuroda, P.K. and Myers, W.A.: 1997, "Iodine-129 and plutonium-244 in the early solar system", *Radiochim. Acta* **77**, 15-20.

Kuroda, P.K. and Myers, W.A.: 1998a, "Plutonium-244 fission xenon and primordial xenon in lunar samples and meteorites", *J. Radioanalyt. Nucl. Chem.* **230**, 175-195.

Kuroda, P.K. and Myers, W.A.: 1998b, "Plutonium-244 fission xenon and primordial xenon in the Allende meteorite", *J. Radioanalyt. Nucl. Chem.* **230**, 197-211.

Kuroda, P.K. and Myers, W.A.: 1998c, "Extinct Radionuclides $^{26}$Al and $^{129}$I in the early solar system", *Naturwissenschaften* **85**, 180-182.

Kuroda, P.K. and Myers, W.A.: 1999, "Age of the earth and the moon", *J. Radioanalyt. Nucl. Chem.* **241**, 655-658.

Kuroda, P.K., Rowe, M.W., Clark, R.S. and Ganapathy, R.: 1966, "Galactic and solar nucleosynthesis", *Nature* **212**, 241-243.

Kuroda, P.K., Beck, J.N., Efurd, D.W. and Miller, D.K.: 1974, "Xenon isotopic anomalies in the carbonaceous chondrite Murray", *J. Geophys. Res.* **79**, 3981-3992.

Kuroda, P.K., Sherrill, R.D., Efurd, D.W. and Beck, J.N.: 1975, "Xenon isotopic anomalies in the carbonaceous chondrite Murchison", *J. Geophys. Res.* **80** 1558-1570.

Lee, T., Papanastassiou, D.A. and Wasserburg, G.J.: 1977, "Aluminum-26 in the early solar system: Fossile or fuel", *Astrophys. J.* **211**, L107-L110.

Lee, T., Papanastassiou, D.A. and Wasserburg, G.J.: 1978, "Calcium isotopic anomalies in the Allende meteorite", *Astrophys. J.* **220**, L21-L25.

Lee, T. and Papanastassiou, D.A.: 1974, "Mg isotopic anomalies in the Allende meteorite and correlation with O and Sr effect", *Geophys. Lett.* **1**, 225-228.

Lewis, R.S., Anders, E. and Draine, T.B.: 1989, "Properties, detectability and origin of interstellar diamonds in meteorites", *Nature* **339**, 117-121.

Lewis, R.S., Amari, S. and Anders, E.: 1994, "Interstellar grains in meteorites: II. SiC and noble gases", *Geochim. Cosmochim. Acta* **58**, 471-494.

LSPET (The Lunar Sample Preliminary Examination Team): 1969, "Preliminary examination of lunar samples from Apollo 11", *Science* **165**, 1211-1227.

LSPET (The Lunar Sample Preliminary Examination Team): 1970, "Preliminary examination of lunar samples from Apollo 12", *Science* **167**, 1325-1339.

Manuel, O.K., Hennecke, E.W. and Sabu, D.D.: 1972a, "Xenon in carbonaceous chondrites", *Nature Phys. Sci.* **240**, 99-101.

Manuel, O.K., Wright, R.J., Miller, D.K. and Kuroda, P.K.: 1972b, "Isotopic composition of rare gases in the carbonaceous chondrites Mokoia and Allende", *Geochim. Cosmochim. Acta* **36**, 961-983.

Marti, K., Lugmair, G.W. and Urey, H.C.: 1970, "Solar wind gases, cosmic ray spallation products, and the irradiation history", *Science* **167**, 548-550.

McCulloch, M.T. and Wasserburg, G.J.: 1978, "Barium and neodynimum isotopic anomalies in the Allende meteorite", *Astrophys. J.* **220**, L15-L19.

McCulloch, M.T., Wasserburg, G.J. and Papanastassiou, D.A.: 1978, "More mysteries from Pandora's box", *Geological Survey Open-File Report (U.S.)* **78-702**, 282-285.

Merrihue, C.: 1966, "Xenon and krypton in the Bruderheim meteorite", *J. Geophys. Res.* **71**, 263-313.

Miura, Y.N., Nagao, K., Sugiura, N., Sagawa, H. and Matsubara, K.: 1995, "Orthopyroxenite ALH84001 and shergottite ALH77005: Additional evidence for a martian origin from noble gases", *Geochim Cosmochim. Acta* **59**, 2105-2113.

Miura, Y.N., Nagao, K., Sugiura, N., Fujitani, T. and Warren, P.H.: 1998, "Noble gases, $^{81}$Kr-Kr exposure ages and $^{244}$Pu-Xe ages of six eucrites, Béréba, Binda, Camel Donga, Juvinas, Millbillillie, and Stannern", *Geochim. Cosmochim. Acta* **62**, 2369-2387.

Morgan, J.W. and Lovering, J.F.: 1968, "Uranium and thorium abundances in chondritic meteorites", *Talanta* **15**, 1079-1095.

Munk, M.N.: 1967, "Argon, krypton, and xenon in Angra dos Reis, Nuevo Laredo, and Norton County achondrites: The case for two types of fission xenon in achondrites", *Earth Planet Sci. Lett.* **3**, 457-465.

Murty, S.V.S. and Mohapatra, R.K.: 1997, "Nitrogen and heavy noble gases in ALH 84001: Signatures of ancient Martian atmosphere", *Geochim. Cosmochim. Acta* **61**, 5417-5428.

Myers, W.A. and Kuroda, P.K.: 1989, "Plutonium-244 fission xenon in the solar system. 50 years with nuclear fission", eds. Behrens, J.W. and Carlson, A.D., American Nuclear Society, Inc., pp. 909-915.

Myers, W.A. and Kuroda, P.K.: 1991a, "Plutonium-244 dating IV. Initial ratios of plutonium to uranium in the Renazzo, Mokoia, and Groznaya meteorites", *J. Radioanalyt. Nucl. Chem.* **152**, 99-116.

Myers, W.A. and Kuroda, P.K.: 1991b, "Plutonium-244 dating V. Initial ratios of plutonium to uranium in some ordinary chondrites", *J. Radioanalyt. Nucl. Chem.* **152**, 409-434.

Myers, W.A. and Kuroda, P.K.: 1992, "Plutonium-244 dating VI. Initial ratios of plutonium to uranium in achondrites", *J. Radioanalyt. Nucl. Chem.* **157**, 217-238.

Myers, W.A. and Kuroda, P.K.: 1995, "Plutonium-244 and strange xenon components in the solar system", *J. Radioanalyt. Nucl. Chem.* **195**, 335-342.

Nier, A.O.: 1938, "Variations in the relative abundances of the isotopes of common lead from various sources", *J. Am. Chem. Soc.* **60**, 1571-1576.

Nier, A.O.: 1950, "A redetermination of the relative abundances of the isotopes of neon, krypton, rubidium, xenon, and mercury", *Phys. Rev.* **79**, 450-454.

Nier, A.O., Thompson, R.W. and Murphy, B.F.: 1941, "The isotopic constitution of lead and the measurement of geological time III", *Phys. Rev.* **60**, 112-116.

Nunes, P.D., Tatsumoto, M., Knight, R.J., Unruh, D.M. and Doe, B.R.: 1973, "U-Th-Pb systematics of some Apollo 16 lunar samples", *Proc. Fourth Lunar Sci. Conf.* (Supplement 4, *Geochim. Cosmochim. Acta*) **2**, 1797-1822.

Nunes, P.D., Tatsumoto, M. and Unruh, D.M.: 1974, "U-Th-Pb systematics of some Apollo 17 lunar samples and implications for a lunar basin excavation chronology", *Proc. Fifth Lunar Sci. Conf.* (Supplement 5, *Geochim. Cosmochim. Acta*) **2**, 1487-1514.

Öpik, E.J.: 1960, *Oscillating Universe*, Mentor Book, The New American Library, New York, NY, pp. 99-100.

Ott, U.: 1993, "Interstellar grains in meteorites", *Nature* **364**, 25-32.

Ozima, M., Zashu, S., Takigami, Y. and Turner, G.: 1989, "Origin of the anomalous $^{40}$Ar-$^{39}$Ar age of Zaire cubic diamonds: Excess $^{40}$Ar in pristine mantle fluids", *Nature* **337**, 226-229.

Patterson, C.: 1956, "Age of meteorites and the earth", *Geochim. Cosmochim. Acta* **10**, 230-237.

Pepin, R.O.: 1966, "Heavy rare gases in silicates from the Estherville mesosidrite", *J. Geophys. Res.* **71**, 2815-2829.

Pepin, R.O. and Phinney, D.: 1978, "Components of xenon in the solar system," University of Minnesota, Space Science Center, 164 pp.

Podosek, F.A. and Swindle, T.D.: 1988a, "Extinct radionuclides", in *Meteorites and the Early Solar System*, eds. Kerridge, J.F. and Mathews, M.S., The University of Arizona Press, Tucson, AZ, pp. 1093-1113.

Podosek, F.A. and Swindle, T.D.: 1988b, "Nuclear cosmochemistry", in *Meteorite, and the Early Solar System*, eds., Kerridge, J.E. and Mathews, M.S., The University of Arizona Press, Tucson, pp. 1114-1126.

Podosek, F.A., Huneke, J.C., Burnett, D.S. and Wasserburg, G.J.: 1971, "Isotopic composition of xenon and krypton in the lunar soil and in the solar wind", *Earth Planet. Sci. Lett.* **10**, 199-216.

Rankama, K.: 1954, *Isotope Geology*, McGraw-Hill Book Co., Inc., New York, Pergamon Press Ltd., London, UK, 535 pp.

Reynolds, J.H.: 1960, "Determination of the age of the elements", *Phys. Rev. Lett.* **4**, 8-10.

Reynolds, J.H.: 1963, "Xenology", *J. Geophys. Res.* **68**, 2939-2956.

Reynolds, J.H.: 1978, *Proc. Robert A. Welch Foundation, Conf. on Chemical Res. XXI. Cosmochemistry*, pp. 201-244.

Reynolds, J.H. and Turner, G.J.: 1964, "Rare gases in the chondrite Renazzo", *J. Geophys. Res.* **69**, 3263-3281.

Reynolds, J.H., Hohenberg, C.M., Lewis, R.S., Davis, P.K. and Kaiser, W.A.: 1970, "Isotopic analysis of rare gases from stepwise heating of lunar fines and rocks", *Science* **167**, 545-548.

Rowe, M.W. and Bogard, D.D.: 1966, "Xenon anomalies in the Pasamonte meteorite", *J. Geophys. Res.* **71**, 686-687.

Rowe, M.W. and Kuroda, P.K.: 1965, "Fissiogenic xenon from the Pasamonte meteorite", *J. Geophys. Res.* **70**, 709-714.

Rowe, M.W., Bogard, D.D. and Kuroda, P.K.: 1966, "Mass yield spectrum of cosmic-ray produced xenon", *J. Geophys. Res.* **71**, 4679-4684.

Russell, R.D., Farquhar, R.M., Cumming, G.L. and Wilson, J.T.: 1954, "Dating galenas by isotopic constitutions", *Transactions Am. Geophys. Union* **35**, 301-309.

Sakamoto, K.: 1974, "Possible cosmic dust origin of terrestrial plutonium-244", *Nature* **248**, 130-132.

Seaborg, G.T.: 1958, *The Transuranium Elements*, Yale University Press, New Haven, CT, p. 75.

Sears, D.W.: 1978, *The Nature and Origin of Meteorites*, Oxford University Press, New York, p. 65.

Shukolyukov, Y.A. and Minh, D.V.: 1985, "Anomalous xenon in certain carbonaceous chondrites", *Geochem. International* **22**, 125-136.

Shukolyukov, Y.A., Minh, D.V. and Tarasov, L.S.: 1989, "Xe compositions and concentrations in Luna 16, 20, and 24 samples", *Geochem. International* **26**, 64-71.

Srinivasan, B. and Anders, E.: 1978, "Noble gases in the Murchison meteorite: Possible relics of s-process nucleosynthesis", *Science* **201**, 51-56.

Srinivasan, B., Hennecke, E.W., Sinclair, D.E. and Manuel, O.K.: 1972, "A comparison of noble gases released from lunar fines (#15601.64) with noble gases in meteorites and in the earth", *Proc. 3rd Lunar Sci. Conf.* **2**, 1927-1945.

Srinivasan, B., Lewis, R.S. and Anders, E.: 1978, "Noble gases in the Allende and Abee meteorites and a gas rich mineral fraction: investigation by stepwise heating", *Geochim. Cosmochim. Acta* **42**, 183-198.

Swindle, T.D. and Podosek, F.A.: 1988, "Iodine-xenon dating", in *Meteorites and the Early Solar System*, eds. Kerridge, J.F. and Mathews, M.S., The University of Arizona Press, Tucson, pp. 1127-1146.

Swindle, T.D., Grier, J.A. and Burkland, M.K.: 1995, "Noble gases in orthopyroxenite ALH84001: A different kind of martian meteorite with an atmospheric signature", *Geochim. Cosmochim. Acta* **59**, 793-801.

Takaoka, N.: 1972, "An interpretation of general anomalies of xenon and the isotopic composition of primitive xenon", *Mass Spectrometry* **20**, 287-302.

Tatsumoto, M.: 1970a, "Age of the moon: An isotopic study of U-Th-Pb systematics of Apollo 11 lunar samples – II", *Proc. Apollo 11 Lunar Sci. Conf.* **2**, 1595-1612.

Tatsumoto, M.: 1970b, "U-Th-Pb age of Apollo 12 rock 12013", *Earth Planet. Sci. Lett.* **9**, 193-200.

Tatsumoto, M.: 1973, "U-Th-Pb measurements of Luna 20 soil", *Geochim. Cosmochim. Acta* **37**, 1079-1086.

Tatsumoto, M. and Rosholt, J.H.: 1970, "Age of the moon: An isotopic study of uranium-thorium lead systematics of lunar samples", *Science* **167**, 461-463.

Tatsumoto, M. and Unruh, D.M.: 1976, "KREEP basalt age: Grain by grain U-Th-Pb systematics study of the quartz monzodiorite clast 15405,88", *Proc. 7th Lunar Sci. Conf.*, 2107-2129.

Tatsumoto, M., Knight, R.J. and Doe, B.R.: 1970, "U-Th-Pb systematics of Apollo 12 lunar samples", *Proc. Second Lunar Sci. Conf.* **2**, 1521-1546.

Tatsumoto, M., Hedge, C.E., Doe, B.R. and Unruh, D.M.: 1972, "U-Th-Pb and Rb-Sr measurements on some Apollo 14 lunar samples", *Proc. Third Lunar Sci. Conf.* (Supplement 3, *Geochim. Cosmochim. Acta*) **2**, 1531-1555.

Tatsumoto, M., Knight, R.J. and Allégre, C.J.: 1973, "Time differences in the formation of meteorites as determined by $^{207}Pb/^{206}Pb$", *Science* **180**, 1279-1283.

Tatsumoto, M., Unruh, D.M. and Desborough, G.A.: 1976, "U-Th-Pb and Rb-Sr systematics of Allende and U-Th systematics of Orgueil", *Geochim. Cosmochim. Acta* **40**, 617-634.

Tilton, G.R.: 1973, "Isotopic lead ages of chondritic meteorites", *Earth Planet. Sci. Lett.* **19**, 321-329.

Turner, G.: 1970, "Argon-40/argon-39 dating of lunar rock samples", *Science* **167**, 466-468.

Unruh, D.M., Nakamura, N. and Tatsumoto, M.: 1977, "History of the Pasamonte achondrite: Relative susceptibility of the Sm-Nd, Rb-Sr, and U-Pb systems to metamorphic events", *Earth Planet. Sci. Lett.* **37**, 1-12.

Urey, H.C.: 1952, *The Planets: Their Origin and Development*, Yale University Press, New Haven, preface, p. ix.

Wasserburg, G.J.: 1987, "Isotopic abundances: Inferences on the solar system and planetary evolution", Crafoord Lecture, 24 September 1986, *Earth Planet. Sci. Lett.* **86**, 129-173.

Wasserburg, G.J. and Papanastassiou, D.A.: 1982, "Some short-lived nuclides in the early solar system: A connection with the placental ISM", in *Essays in Nuclear Astrophysics*, eds. Barnes, C.A., Clayton, D.D. and Schramm, D.N., Chapter 6, pp. 77-140.

Wasserburg, G.J., Huneke, J.C. and Burnett, D.S.: 1969, "Correlation between fission tracks and fission type xenon in meteoritic whitlockite", *J. Geophys. Res.* **74**, 4221-4232.

Zashu, S., Ozima, M. and Nitoh, O.: 1986, "K-Ar isochron dating of Zaire cubic diamonds", *Nature* **323**, 710-712.

Zinner, E., Tang, M. and Anders, E.: 1989, "Interstellar SiC in the Murchison and Murray meteorites: Isotopic composition of Ne, Xe, Si, C, and N", *Geochim. Cosmochim. Acta* **53**, 3273-3290.

# Extinct $^{244}$Pu: Chronology of Early Solar System Formation

Marvin W. Rowe
*Department of Chemistry, P. O. Box 30012, Texas A & M University, College Station, TX 77842-3012*
*mwrowe@tamu.edu*

Abstract: Early and recent work on Pu-Xe dating of achondrites (eucrites) is discussed here; the latter has focused on establishing the time of closure for Xe, not on the estimation of the time interval between the cessation of nucleosynthesis and the crystallization of solid bodies in our solar system. By using various Angra dos Reis dates as standard, the Xe closure of some eucrites were found to be early in the solar system, about 4.56 Gyr ago, synchronous within error with chondrites. The light cosmic ray produced Xe isotopes were found to be a good surrogate for the Nd content. Using the measured cosmic ray exposure age allows accurate closure times to be established. Combining this new information with data from CAIs and chondrules and chondrites suggests that the achondritic differentiation was accomplished within about 10 Myr or less. Pu-Xe dating in chondrites is not covered in this review.

## 1. INTRODUCTION

Consider the following project. You are to measure the isotopic composition and abundance of Xe in a solid sample (meteorite, earth, moon, etc.) that evolved from the origin of our solar system. In a simplified scenario, that Xe will contain a number of components: initially, an inherent trapped component from the combination of several nuclear processes in nucleosynthesis - and of unknown composition, but probably *grossly* similar to either atmospheric Xe (Nier, 1950) or Xe from carbonaceous chondrites (Reynolds, 1960b). Nucleosynthesis had presumably occurred in the solar system in a hot state where

***Origin of Elements in the Solar System: Implications of Post-1957 Observations.***
**Edited by O. Manuel, Kluwer Academic/Plenum Publishers, 2000.** 

almost all material was in a gaseous state. As things began to condense into solids (Larimer, this volume), some of the trapped gas was incorporated into the solids and, the Xe began to be altered by the addition of two nuclear decay processes: 17 million year (Myr) $^{129}$I (Reynolds, 1960a) and 82 Myr $^{244}$Pu (Rowe and Kuroda, 1965). The solids then went through an ill-defined accumulation process until asteroidal sized bodies were formed. This was followed by geological re-working of the solid body, including bombardment of the surface (constantly changing) with smaller sized solids that might introduce new solid material with a distinct Xe composition into the mix and new Xe from the solar wind. After billions ($10^9$, Gyr) of years, the larger bodies collided in such a way meter-sized pieces of the original body broke loose, becoming exposed to cosmic rays in the vacuum of inter-planetary solar system space for millions of years. The cosmic-ray bombardment resulted in the formation of $^{124,126,128,129,130,131,132}$Xe (Rowe *et al.*, 1965) that is intimately mixed with the pre-existing Xe mixture. The exact composition of the spallation Xe depends on the elemental make-up of the solids being bombarded, the length of time the exposure occurs, the amount of shielding, and the energy of the irradiating particles (including secondary particles produced by nuclear interactions with the solid material, etc.). Then, the sample will eventually be attracted by the earth's gravitational pull, and enter the upper atmosphere at speeds of tens of km/sec and impact violently, without benefit of thermal shields or a braking mechanism, with the earth's surface. The sample, which had been under vacuum for millions of years, will now be exposed for an unknown length of time, but short on cosmic scales, to the strongly oxidizing atmosphere and the weathering processes of the earth. The sample, in this case a meteorite, is finally collected and placed under curation under generally unknown conditions until a sample is brought into your mass spectrometry laboratory for analysis.

You must take this sample, extract the Xe and deconvolute it into its different components. An awesome feat, this has been done with more or less success for the past forty years, spurred by the introduction of high vacuum, static operation, noble gas mass spectrometers (Reynolds, 1956). During the 1960s, many discoveries were made that illuminated (to a greater or lesser extent) the different components enumerated above, especially in Ca-rich achondrites. For example, a good *estimate* of the primordial trapped component was found (Reynolds, 1960b) in Xe-rich carbonaceous chondrites. Reynolds (1960a) first discovered $^{129}$Xe from the decay of extinct $^{129}$I when a very large excess of $^{129}$Xe was seen in a chondritic meteorite. Use of that couple as a chronometer has been limited primarily by the fact that the nuclear processes producing the $^{129}$I are not well enough known yet. The demonstration of cosmic ray produced Xe and the first estimation of its isotopic signature (known to be some-what variable) came in 1965 and later (Rowe *et al.*, 1965, 1966; Marti *et al.*, 1966; Pepin, 1966; Hohenberg *et al.*, 1967; Wasserburg *et al.*, 1969; Rowe, 1970), and has been further refined as additional experiments have been done.

The discovery of Xe from extinct $^{244}$Pu decay (Rowe and Kuroda, 1965) and confirmation soon after (Marti *et al.*, 1966; Pepin, 1966; Rowe and Bogard, 1966; Hohenberg *et al.*, 1967; Wasserburg *et al.*, 1969; Rowe, 1970) were all found in Ca-rich achondritic meteoritic material. There are several attributes that make Ca-rich achondrites likely samples for analyses of both cosmogenic and fissiogenic Xe: they have a low concentration of the trapped Xe component, coupled with relatively large amounts of cosmogenic and fissiogenic Xe, so that the exact composition of the trapped component has little effect on the separation of these components and their interpretation. The mass yield of $^{244}$Pu-produced Xe was determined in meteorite samples before it was determined by extraction and measurement of Xe isotopes from $^{244}$Pu produced in a hydrogen bomb (Alexander *et al.*, 1971). Table 1 shows these early determinations of the isotopic composition of $^{244}$Pu-produced $^{131,132,134,136}$Xe. Amounts of fissiogenic Xe can also be reasonably accurately analyzed in Ca-rich achondritic meteorites.

*Table 1.* Relative Xe isotopic yields of spontaneous fission of $^{244}$Pu derived from meteorite studies compared with that measured from bomb-made $^{244}$Pu.

| Investigators | Yields | | | |
|---|---|---|---|---|
| | $^{131}$Xe | $^{132}$Xe | $^{134}$Xe | $^{136}$Xe |
| Eberhardt and Geiss (*1966*) with Rowe and Bogard (*1966*) data | 33 ± 3 | 91 ± 3 | 91 ± 3 | 100 |
| Hohenberg, Munk, and Reynolds (*1967*) | 25 ± 3 | 88.5 ± 3 | 94 ± 5 | 100 |
| Wasserburg, Huneke and Burnett (*1969*); recalculation | 24 ± 8 | 86.6 ± 1.3 | 91.8 ± 1.9 | 100 |
| Rowe (*1970*) | 26 ± 3 | 88 ± 4 | 91 ± 5 | 100 |
| Alexander *et al.* (*1971*) $^{244}$Pu | 24.6 ± 2.2 | 87.0 ± 3.1 | 92.1 ± 2.7 | 100 |

## 2. EXTINCT $^{244}$PU

This paper reviews the $^{244}$Pu-$^{136}$Xe decay couple as a tool for early solar system chronology. This area of research was reviewed about a dozen years ago (Podosek and Swindle, 1988), so we will focus here on new studies of the past decade. $^{244}$Pu ($t_{1/2}$ = 82 x $10^6$ years) decays primarily by $\alpha$-emission to $^{232}$Th. Since $^{232}$Th is monoisotopic, there is no real chance to detect $^{244}$Pu via $^{232}$Th excesses. But an alternative decay by spontaneous fission was recognized soon after the discovery of $^{244}$Pu. This mode of decay was detected in meteorites through two means in 1965: (i) Excesses of the heavy isotopes of Xe produced as fission fragments (Rowe and Kuroda, 1965). This was experimentally feasible because the daughter Xe isotopes are highly depleted so that small additions of nuclear components are identifiable in the isotopic anomalies. (ii) Microscopic "tracks", lattice disorder in mineral grains caused by fission fragments passing through the meteoritic material (Fleischer *et al.*, 1965). These were revealed by selective dissolution in laboratory etching.

Fissiogenic Xe from $^{244}$Pu-decay is potentially detectable in bulk chondritic meteorites, but it is far more readily seen and measured in samples that are enriched in Pu and depleted in normal trapped Xe such as igneous, *i.e.*, Ca-rich, achondrites, specific phases of chondrites such as refractory rich inclusions in carbonaceous chondrites, or phosphates in ordinary chondrites. The present review will focus on studies in achondritic meteorites. Since 1988, the bulk of the work on $^{244}$Pu in bulk chondrites has been published by and reviewed here by Kuroda and Myers (this volume). Nothing more will be said here about chondritic meteorite studies.

## 3. $^{244}$PU-$^{136}$XE DATING OF XE CLOSURE IN ACHONDRITES: THE 1990s

$^{244}$Pu must have been produced in the r-process of stellar nucleosynthesis. Thus it is a good candidate for chronology of the early solar system. For that use, however, we need the relative cosmic abundance for $^{244}$Pu compared to some other relatively stable isotope, *i.e.*, normalization to some other element, there being no stable or long-lived isotope of Pu. Nd appears to be a suitable normalization element for Pu (Lugmair and Marti, 1977). No value for this cosmic abundance ratio ($^{244}$Pu/Nd) has been accepted as definitive yet. Until that is accomplished with accuracy, it is difficult to calculate the time interval between the cessation of nucleosynthesis and the formation of solid solar system bodies *(i.e.*, the time of closure for Xe loss from the solid bodies).

In the past decade, emphasis has shifted in $^{244}$Pu-$^{136}$Xe dating. Earlier on, the major thrust seemed to be to estimate the time interval between the end of nucleosynthesis until the formation of the solid solar system bodies. Recent work on $^{244}$Pu in achondrites has related the time of Xe closure to formation times measured by other dating systems, *e.g.*, Pb-Pb, Sm-Nd and K-Ar systems.

Work in the 1990s is based on several observations of eucritic and angritic Xe inventories: (1) The fraction of $^{136}$Xe due to $^{244}$Pu is relatively large; (2) The fraction of cosmic ray spallation $^{124}$Xe and $^{126}$Xe is relatively large; (3) Amounts of spallation $^{124}$Xe and $^{126}$Xe formed are proportional to: (i) concentrations of Ba and light rare earth elements serving as targets for irradiation that produces the cosmic ray produced isotopes; (ii) the cosmic ray exposure time; and (iii) to a lesser extent, the degree of shielding of the sample during cosmic ray bombardment.

Hohenberg *et al.* (1991) suggested an improvement in analysis that overcame a major problem in previous attempts to use $^{244}$Pu-$^{136}$Xe dating, namely that Nd concentrations had been measured on different samples than Xe. Instead of requiring Nd analysis, $^{124}Xe$ and $^{126}Xe$ formed by cosmic ray bombardment were substituted. Since amounts of $^{124}Xe$ or $^{126}Xe$ are proportional to Nd concentration, as well as Ba and light rare earth elements (LREE), both $^{136}Xe$ and Nd concentrations (through $^{124}Xe$, *e.g.*) can be done with a single determination of Xe isotopic composition. Since eucrites usually have similar LREE patterns, replacement of $^{126}Xe_{Nd}$ by $^{126}Xe_{REE}$ does not seem to introduce significant error. The cosmic ray exposure age can be determined by measurement of $^{81}Kr$ in the same sample used for Xe measurement or on a separate sample, as cosmic ray age is expected to be constant for a given meteorite.

Several papers exploiting traditional Pu dating systematics have been published in the 1990s (Swindle *et al.*, 1990; Eugster *et al.*, 1991; Michel and Eugster, 1994; Weigel *et al.*, 1997), but I focus this discussion on recent studies using new parameters for calculating the $^{244}Pu/^{136}Xe$ closure ages (Shukolyukov and Begemann, 1996a). No significant differences in conclusion are found over the earlier approaches, but higher precision is obtained.

Both Shukolyukov and Begemann (1996a) and Miura *et al.* (1998) used three isotope plots (either $^{134}Xe/^{132}Xe$ *vs.* $^{136}Xe/^{132}Xe$ or $^{134}Xe/^{130}Xe$ *vs.* $^{136}Xe/^{130}Xe$) to establish that fissiogenic Xe is dominated by $^{244}Pu$ decay, with only 6-8% from $^{238}U$ decay, as has been done since the initial studies on achondritic Xe. Data points for $^{244}Pu$-Xe falls on the best-fit line for eucritic bulk samples. Xe of atmospheric composition represents the trapped component, although whether indigenously or by adsorption of atmospheric Xe is unknown.

Similarly, three isotope plots (either $^{124}Xe/^{126}Xe$ *vs.* $^{130}Xe/^{126}Xe$ or $^{126}Xe/^{130}Xe$ *vs.* $^{124}Xe/^{130}Xe$) were used to establish the character of the spallation Xe, again as had been done in earlier studies. Shukolyukov and Begemann (1996a) found that target element composition of eucrites is similar to chondrites. Ba and LREE contributions had only minor scatter, probably from different shielding conditions. Although different isotope ratios are observed in shielding-sensitive pairs, production rate of $^{126}Xe$ was found to be almost constant.

# 4. METHOD FOR DETERMINATION OF PU-XE AGES

Shukolyukov and Begemann (1996a) assumed that at the time of onset of solid body formation, all eucrites had the same Pu/Nd ratio. The difference between the onset of Xe retention in any two of them is given by decay in the solid bodies. They derived the necessary equations to estimate the time of closure compared to Angra dos Reis, by calculating fractions of $^{126}Xe_{REE}$ via [Ba]/[REE] ratios rather than from the isotopic composition of spallation Xe. Compared to the traditional method of Pu-Xe dating, the absolute concentration of Nd, or of the REEs, is not needed; rather only element ratios of Xe. Hence, this approach is an improvement over the traditional method because in individual eucrites, [Ba]/[LREE] ratios are more constant than are the concentrations.

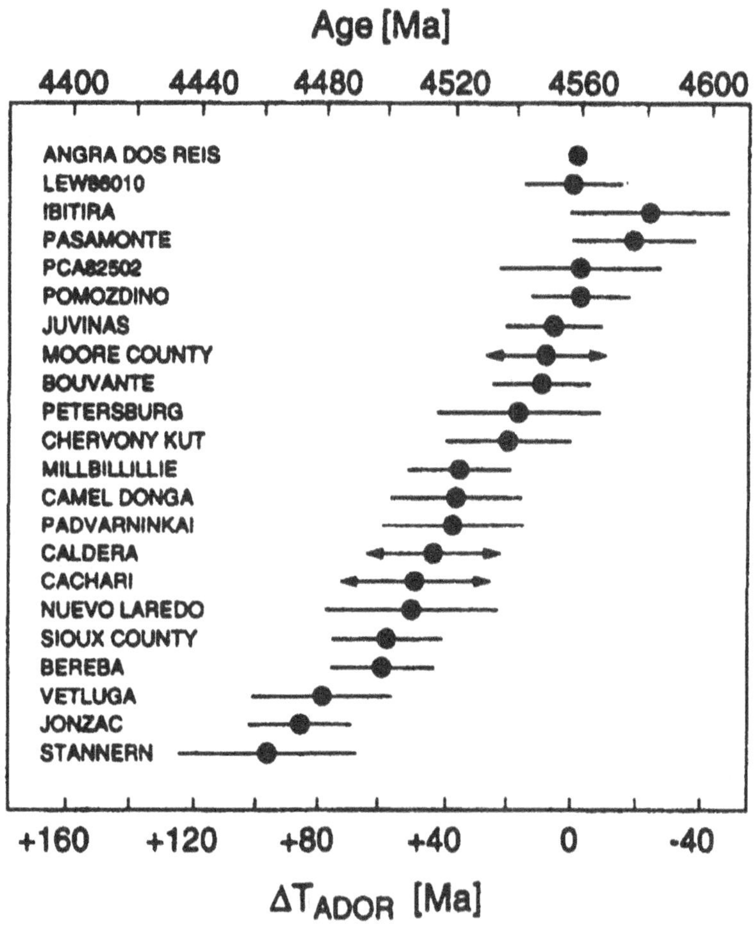


*Figure 1.* Distribution of the average relative (lower abscissa) and absolute Pu-Xe ages. For Cachari, Caldera, and Moore County the arrows indicate minimal possible errors. From Shukolyukov and Begemann (1996a).

Angra dos Reis was chosen as the standard for comparison because of extensive data available on Angra dos Reis, not only Xe isotopic measurements and Ba and REE compositions, but for accurate cosmic ray exposure age, and a precise Pb-Pb age of 4.5578 ± 0.0004 Gyr (Lugmair and Marti, 1977; Lugmair and Galer, 1992). Figure 1 in Shukolyukov and Begemann (1996a) shows a comparison of the calculated Pu-Xe ages for multiple samples of different meteorites; reproducibility is good.

Pu-Xe closure ages of 21 eucrites are reproduced here in Figure 1 (Shukolyukov and Begemann, 1996a). Clearly many eucrites formed as solid bodies early on. In Figure 1, a positive age ($\Delta T_{ADOR}$) indicates formation at a later stage than Angra dos Reis (ADOR). Positive $\Delta T_{ADOR}$ present an enigma with regard to cause. However, correlation between $^{40}$Ar retention and Pu-Xe ages ($^{136}Xe_{Pu}$ retention), suggests that young Pu-Xe ages result from partial Xe loss from samples at some later time, associated with the event dated by the K-Ar ages.

Seven of the eight eucrites studied by Shukolyukov and Begemann (1996a) for which both Pu-Xe and Sm-Nd ages were available agree very closely. Pb-Pb ages were often younger than either the Pu-Xe or Sm-Nd ages, probably due to later metamorphic events that affected the Pb-Pb systematics while leaving the Sm-Nd and Pu-Xe ages unaffected. Later work of Miura *et al.* (1998) on six eucrites confirmed the Shukolyukov and Begemann (1996a) conclusions that follow below.

Goswami (this volume) summarizes the formation times of CAIs, chondrules and chondrites. All of those appear to have formed within less than 500 kyr of one another. Shukolyukov and Begemann (1996a) Pu-Xe closure ages indicate that eucrites formed within 10-15 Myr or so (approximate error in Pu-Xe closure ages) of those less differentiated meteoritic materials, possibly contemporaneously with them. Differentiation of achondritic material occurred rapidly on a cosmic time scale.

## 5. CONCLUSIONS

(1) Although elemental concentrations between samples of the same meteorite vary widely, Pu-Xe ages calculated with an assumption of constant Ba/REE generally agree.
(2) Agreement in Pu-Xe ages suggest the Pu/Nd ratio was constant to ± 10-15%.
(3) Old Pu-Xe ages of Bouvante, Chervony Kut, Ibitira, Juvinas, LEW86010, Millbillillie fines, Moore County, Pasamonte, PCA-82502, Petersburg and Pomozdino are close to 4.56 Gyr and probably represent the crystallization age of the meteorites.
(4) Younger Pu-Xe ages of Caldera, Nuevo Laredo, Sioux County and Béréba, that agree with the Pb-Pb ages, are consistent with either (i) time

of crystallization; (ii) time of metamorphism, or (iii) partial Xe loss during later events that caused Ar loss in the meteorite parent body.

(5) Pu-Xe isotope systems provide an accurate chonometer for dating of eucrites when coupled with other chronometric systems.

(6) The paucity of radiogenic $^{129}$Xe from in situ decay of 17 Myr $^{129}$I remains puzzling; less than 1% of the expected amount is seen. This may reflect differences in location of fissiogenic Xe (retentively sited in silicate crystals) and radiogenic $^{129}$Xe that may be located in iodine-bearing phases with very low retention of Xe. Or, the iodine content of the eucrites may be highly overestimated, being too high due to terrestrial contamination. Or, there may have been a heterogeneous distribution of extinct $^{129}$I in the solar system.

Since the Pu-Xe chronometer agrees with Sm-Nd and Pb-Pb systematics in establishing an early crystallization age and presumably short differentiation age for many of the eucrites, calculation of the time interval between cessation of nucleosysnthesis and crystallization of solid solar system bodies awaits only an accurate determination of the Pu/Nd at the end of nucleosynthesis.

## ACKNOWLEDGEMENTS

I am grateful to Ann Miller for assisting with figures and to Elmo J. Mawk for assisting with figures and literature search.

## REFERENCES

Alexander, E.C., Jr., Lewis, R.S., Reynolds, J.H. and Michels, M.C.: 1971, "Plutonium-244: confirmation as an extinct radioactivity", *Science* **172**, 837-840.

Eberhardt, P. and Geiss, J.: 1966, "On the mass spectrum of fission xenon in the Pasamonte meteorite", *Earth Planet. Sci. Lett.* **40**, 63-70.

Eugster, O., Michel, T. and Niedermann, S.: 1991, "$^{244}$Pu-Xe formation and gas retention age, exposure history, and terrestrial age of angrites LEW86010 and LEW87051; comparison with Angra dos Reis", *Geochim. Cosmochim. Acta* **55**, 2957-2964.

Fleischer, R.L., Price, P.B. and Walker, R.M.:1965, "Spontaneous fission tracks from $Pu^{244}$ in meteorites and the early history of the solar system", *J. Geophys. Res.* **70**, 2703-2707.

Hohenberg, C.M., Bernatowitz, T.J. and Podosek, F.A.: 1991, "Comparative xenology of two angrites", *Earth Planet. Sci. Lett.* **102**, 167-177.

Hohenberg, C.M., Hudson, B., Kennedy, B.M. and Podosek, F.A.: 1981, "Xenon spallation systematics in Angra dos Reis", *Geochim. Cosmochim. Acta* **45**, 1909-1915.

Hohenberg, C.M., Munk, M.N. and Reynolds, J.H.: 1967, "Spallation and fissiogenic xenon and krypton from stepwise heating of the Pasamonte achondrite: The case for extinct plutonium-244 in meteorites; relative ages of chondrites and achondrites", *J. Geophys. Res.* **72**, 3139-3177.

Lugmair, G.W. and Galer, S.J.G.: 1992, "Age and isotopic relationships among the angrites Lewis Cliff 86010 and Angra dos Reis", *Geochim. Cosmochim. Acta* **56**, 1673-1694.

Lugmair, G.W. and Marti, K.: 1977, "Sm-Nd-Pu timepieces in the Angra dos Reis meteorite", *Earth Planet. Sci. Lett.* **35**, 273-284.

Marti, K., Eberhardt, P. and Geiss, J.: 1966, "Spallation, fission, and neutron capture anomalies in meteoritic krypton and xenon", *Z. Natursforschg.* **21a**, 398-413.

Michel, T. and Eugster, O.: 1994, "Primitive xenon in diogenites and plutonium-244-fission xenon ages of a diogenite, a howardite, and eucrites", *Meteoritics* **29**, 593-606.

Miura, Y.N., Nagao, K., Sugiura, N., Fujitani, T. and Warren, P.H.: 1998, "Noble gases, $^{81}$Kr-Kr exposure ages and $^{244}$Pu-Xe ages of six eucrites, Béréba, Binda, Camel Donga, Juvinas, Millbillillie, and Stannern", *Geochim. Cosmochim. Acta* **62**, 2369-2387.

Nier, A.O.: 1950, "A redetermination of the relative abundances of the isotopes of neon, krypton, rubidium, xenon, and mercury", *Phys. Rev.* **79**, 450-454.

Pepin, R.O.: 1966, "Heavy rare gases in silicates from the Estherville mesosiderite", *J. Geophys. Res.* **71**, 2815-2829.

Podosek, F. and Swindle, T.D.: 1988, "Extinct radioactivities", in *Meteorites and the Early Solar System*, eds., Kerridge, J.F.and Matthews, The University of Arizona Press: Tucson, AZ, pp. 1093-1126.

Reynolds, J.H.: 1956, "High sensitivity mass spectrometer for noble gas analysis", *Rev. Sci. Instr.* **27**, 928-934.

Reynolds, J.H.: 1960a, "Determination of the age of the elements", *Phys. Rev. Lett.* **4**, 351-354.

Reynolds, J.H.: 1960b, "Isotopic composition of primordial xenon", *Phys. Rev. Lett.* **4**, 8-10.

Rowe, M.W.: 1970, "Evidence for decay of extinct $Pu^{244}$ and $I^{129}$ in the Kapoeta meteorite", *Geochim. Cosmochem. Acta* **34**, 1019-1028.

Rowe, M.W. and Bogard, D.D.: 1966, "Xenon anomalies in the Pasamonte meteorite", *J. Geophys. Res.* **71**, 686-687.

Rowe, M.W., Bogard, D.D., Brothers, C.E. and Kuroda, P.K.: 1965, "Cosmic-ray-produced xenon in meteorites", *Phys. Rev. Lett.* **15**, 843-845.

Rowe, M.W., Bogard, D.D. and Kuroda, P.K.: 1966, "Mass yield spectrum of cosmic-ray-produced xenon", *J. Geophys. Res.* **15**, 4679-4684.

Rowe, M.W. and Kuroda, P.K.: 1965, "Fissiogenic xenon from the Pasamonte meteorite", *J. Geophys. Res.* **70**, 709-714.

Shukolyukov, A. and Begemann, F.: 1996a, "Cosmogenic and fissiogenic noble gases and $^{81}$Kr-Kr exposure age clusters of eucrites", *Geochim. Cosmochim. Acta* **60**, 2453-2471.

Shukolyukov, A. and Begemann, F.: 1996b, "Pu-Xe dating of eucrites", *Meteoritics Planet. Sci.* **31**, 60-72.

Swindle, T.D., Garrison, D.H., Goswami, J.N., Hohenberg, C.M., Nichols, R.H.and Olinger, C.T.: 1990, "Noble gases in the howardites Bholgati and Kapoeta", *Geochim. Cosmochim. Acta* **54**, 2183-2194.

Wasserburg, G.J., Huneke, J.C. and Burnett, D.S.: 1969, "Correlation between fission tracks and fission-type xenon from an extinct radioactivity", *Phys. Rev. Lett.* **22**, 1198-1201.

Weigel, A., Eugster, O., Koeberl, C. and Krahenbuhl, U.: 1997, "Differentiated achondrites Asuka 881371, an angrite, and Divnoe: Noble gases, ages, chemical composition, and relation to other meteorites", *Geochim. Cosmochim Acta* **61**, 239-248.

# A Search for Natural Pu-244 in Deep-Sea Sediment: Progress Report

K. Sakamoto, Y. Hashimoto and T. Nakanishi
*Department of Chemistry, Faculty of Science, Kanazawa University, Kakuma-machi, Kanazawa 920-1192 Japan*
*sakamoto@cacheibm.s.kanazawa-u.ac.jp*

Abstract: "Now-extant" Pu-244 discovered in a Precambrian bastnaesite by D.C. Hoffman *et al.* in 1971 was explained by one of the authors (K.S.) in 1974 as due to possible inflow to the Earth as a cosmic dust component from supernova remnants, not as due to survival of the primeval Pu-244 nor to cosmic ray component. A search for the "live" Pu-244 has recently been started with a 80 kg of wet sediment from 5800m in depth off Hawaii to see the inflow of the extra-Solar System material, and a preliminary result on alpha-counting for Pu-244 separated from a small portion corresponding to 400g of the sample indicates an upper limit that is comparable to the expected amount of Pu-244 in deep-sea sediment. The implications of this result and our future program will be discussed.

## 1. INTRODUCTION

The existence of $^{244}$Pu ($T_{1/2}$ = 8.08 x $10^7$ yr) in nature offers direct proof of the r-process described by $B_2FH$ (Burbidge *et al.*, 1957). The agreement of the xenon isotopic spectrum from spontaneous fission of $^{244}$Pu (Alexander *et al.*, 1971) with that found earlier in meteorites (Rowe and Kuroda, 1965) further supports the existence of $^{244}$Pu in the early solar system. However, the present existence of $^{244}$Pu in terrestrial materials raises other concerns: If the solar system has been closed against the influx of $^{244}$Pu since its formation 4.6 x $10^9$ years ago, then the present average terrestrial abundance of $^{244}$Pu would be $10^{-26}$ g/g. The steady state influx of cosmic rays and/or interstellar dusts could, however, be possible sources for terrestrial $^{244}Pu$ now.

In 1971, D.C. Hoffman (Hoffman *et al.*, 1971) reported the discovery of (2.0 + 0.3-0.7) x $10^7$ atoms of $^{244}Pu$ in a 85 kg of ore containing bastnaesite ($10^{-18}$ g-$^{244}$Pu/g-mineral), which was not inconsistent with an upper limit of 3x$10^{-17}$ g-$^{244}$Pu/g-ore obtained from a 1.6 kg gadolinite (Be Fe $Y_2$ $Si_2$ $O_{10}$) by Fields (Fields *et al.*, 1966). Hoffman *et al.* (1971) discussed the implications of their results in establishing the last time of formation of the heavy elements prior to the formation of the solar system, but left further studies to decide whether the $^{244}Pu$ is primordial or originated from a cosmic ray flux.

One of the present authors (K.S.) reinvestigated the implications of the present existence of $^{244}Pu$ in nature in terms of three possibilities: (1) Survival of primordial $^{244}Pu$; (2) Influx as a heavy component of cosmic rays; and (3) Inflow as a cosmic dust component from supernova remanants (Sakamoto, 1974). The results showed that there are some difficulties with the first two possibilities, both of which were thought to be plausible at the time of the discovery. Assuming that 10% of the cosmic dust influx of $10^5$ tons per year onto the entire Earth's surface (Barker and Andrew, 1968; Tanaka et al., 1972; Crocket and Kuo, 1979; LaViolette, 1985) originates in frequent supernova explosions in our galaxy and that the dust material is chondritic with $^{244}Pu/^{238}U$ = 0.013 and $^{238}U$ = 20 ppb, the steady state influx is expected to be 2 x $10^{-26}$g (or 4 x $10^{-5}$ atoms) of $^{244}Pu$ $cm^{-2}sec^{-1}$. Using the further assumption that the probability of survival against nuclear disintegration during passage through the atmosphere is 0.1, the amount accumulated during the mean life of $^{244}Pu$ (1.2 x $10^8$ yr) causes an average concentration of 1.6 x $10^{-22}$ g-$^{244}$Pu/g in the 10 km of the Earth's crust. A similar calculation resulted in an influx of 7 x $10^{-28}$ g-$^{129}$I $cm^{-2}sec^{-1}$ and an average concentration of 2 x $10^{-19}$ g-$^{129}$I/g in the crust. This value is not inconsistent with the reported $^{129}Xe$ excess of 2.2 x $10^{-12}$ ccSTP/g of natural iodyrite (Srinivasan *et al.*, 1971).

On the other hand G.A. Cowan (1972) estimated an influx of $10^{-10}$ atoms of $^{244}Pu$ $cm^{-2}sec^{-1}$ on the Earth's surface, when the solar system is in the empty region, $N_H$ = 0.1 $cm^{-3}$ of the galaxy. If the solar system spends 7% of time in a dense region of $N_H$ =10 $cm^{-3}$, the time-averaged flux is $10^{-9}$ atoms of $^{244}Pu$ $cm^2sec^{-1}$, which is lower by a factor of $10^{4-5}$ than the value deduced above.

As noted by D.C. Hoffman (1972) and the present author (Sakamoto, 1974), the best collector of cosmic dust would be the ocean. An accumulated $^{244}Pu$ in the top surface (1 cm) of deep sea sediment with a sedimentation rate of 1 mm/$10^3$ y would be (1.7 x $10^{-26}$)(3.2x$10^7$)x($10^4$)/0.5 = 1.1 x $10^{-14}$ g-$^{244}$Pu/g-sediment = 3 x $10^7$ atoms/g-sediment. Here the specific gravity of the dry sediment is assumed to be 0.5 g/$cm^3$, and there is negligible decay of $^{244}Pu$ during $10^4$ y; the decay is only 0.8% even at 1 m from the top surface of sediment. The amount in 5 kg of dry sediment corresponds to 40 μBq, which is detectable in a Si charged-particle detector. If the estimate by G.A. Cowan is correct, the same amount of $^{244}$Pu is expected in about 250 kg of the sediment.

The amount of $^{244}$Pu observed by Hoffman *et al.* (1971) could possibly be explained by the bastnaesite formation from a magma source contaminated with subducted sea sediment containing 3 x $10^7$ atoms/g-surface sediment. A dilution of the surface $^{244}$Pu by a factor of 4 x $10^4$ might have occurred in the process of sediment incorporation into the magma source, which is not inconsistent with $^{10}$Be observed in fresh volcanic arc lavas (Morris and Tera, 1989).

A further calculation for the amounts of $^{129}$I ($T_{1/2}$ = 1.57 x $10^7$ yr), $^{247}$Cm ($T_{1/2}$ = 1.56 x $10^7$ yr) and $^{146}$Sm ($T_{1/2}$ = 1.03 x $10^8$ yr) expected in deep-sea sediment leads to about 2 x $10^{-16}$ g (5 x $10^5$ atoms)/g-sediment for $^{129}$I and $^{247}$Cm and to almost the same concentration for $^{146}$Sm as for $^{244}$Pu.

In this paper a preliminary result is presented on the search for $^{244}$Pu in deep-sea sediment in which an upper limit corresponding to the expected content of $10^7$ atoms per g of sediment has now been attained with a sample of 1 kg dry sediment.

## 2. EXPERIMENTAL

### 2.1 Sample

Some 20 years after the above mentioned proposal, about 80 kg of wet red clay sediment (Sample No. 92 SAD 01) was provided by the late Professor K. Yamakoshi, who had received it from the Dai-ni Hakurei-maru of the Japan Mineral Mining Corporation, one ton dredged from a water depth of 5,800 m in the Pacific Ocean off Hawaii (9°30'N, 174°18'W) in 1992. The sample was washed with 6 ℓ of water and passed through 0.117 mm pores. A 12.85 kg sample of dry sediment was then obtained, after filtering off the nodules.

### 2.2 Chemical Isolation of Plutonium

The outline of the chemical procedure is shown in Figure 1. A 1,054 g aliquot was divided into 34 aliquots of 31 g each. Each 31 g portion of the 1,054 g aliquot was well mixed and 30 g was fused with 120 g of NaOH for 15 hrs at 400°C. The 34 HCl-leachates from 1,020 g of dry sediment were subjected to plutonium separation by anion exchange, after removing iron by isopropyl-ether extraction. The Pu was electrodeposited on a 25 mmϕ stainless-steel disc and then subjected to α–spectrometry for 18 months.

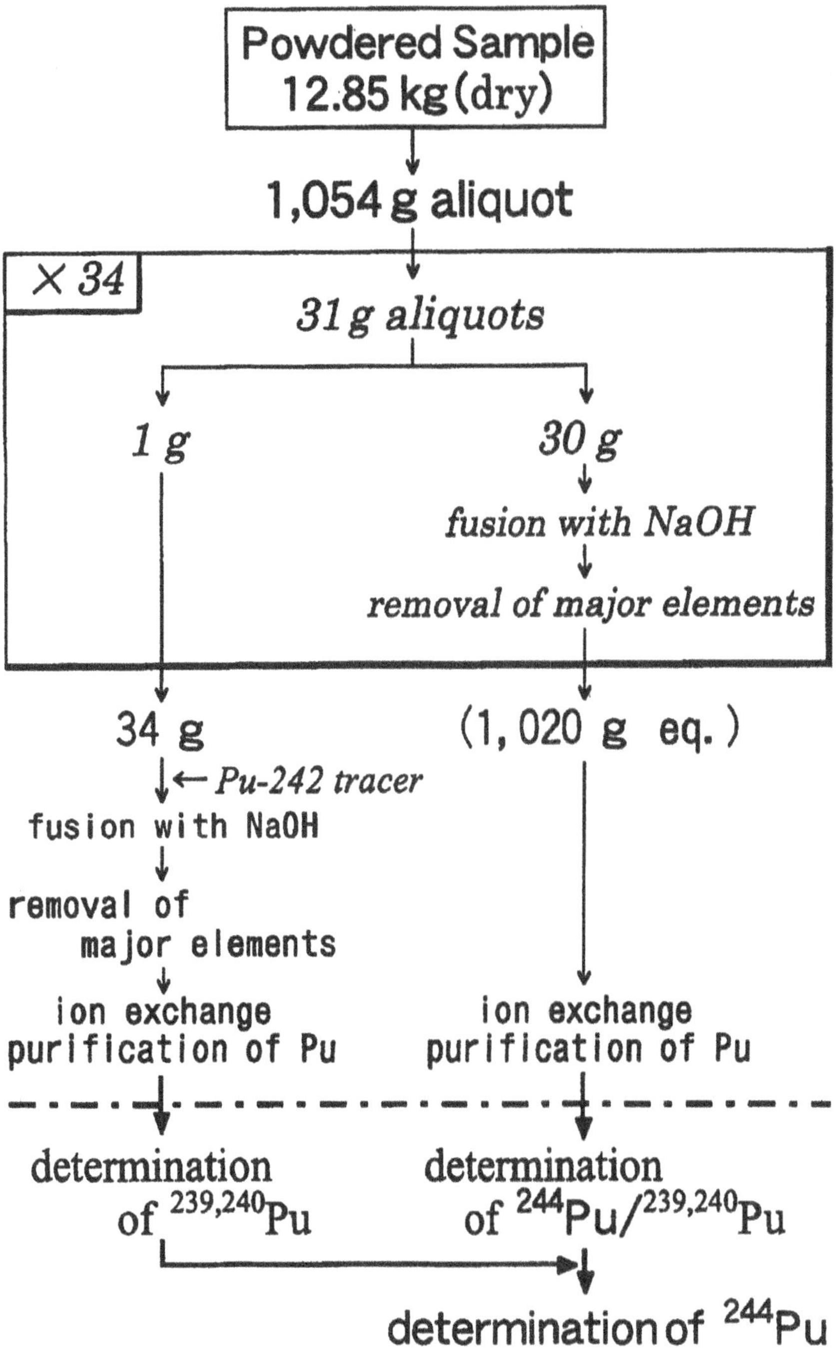


*Figure 1.* The outline of the chemical procedure to isolate plutonium from the 92 SAD 01 deep-sea sediment.

Alpha-spectrometry was carried out using two silicon surface barrier detectors (450 $mm^2$) connected to an EG & G Ortec Octete™ PC alpha-spectrometer. Measurements of the counting efficiency (~ 25%) and energy calibration (~10 keV/channel) of the spectrometer were performed using a $^{209}Po+^{210}Po$ reference source. All of the 1 g fractions from the 34 portions were combined and fused with NaOH, after addition of $^{242}Pu$ tracer, to determine $^{239,240}Pu$ of fall-out origin. This serves as chemical yield tracer for $^{244}Pu$.

# 3. RESULTS

## 3.1 Pu-239,240 in sediment sample

For the 34 g sample, the concentration of $^{239,240}Pu$ was found to be (35 ± 5) μBq/g. When combined with two other determinations of (42 ± 11) μBq/g and (33 ± 3) μBq/g on samples of 4 g and 62 g, respectively, the concentration of $^{239,240}Pu$ in the dried sediment sample was determined to be (34±3) μBq/g.

## 3.2 Pu-244 in sediment sample

The α–counting which started on December 25, 1997, has continued until August 7, 1999 (350 d + 196 d), with 19 breaks for data saving and background counting. The resultant α-spectrum for the 350 d counting is shown in Figure 2. From the $^{239,240}Pu$ peak, the chemical yield was determined to be 40%. The gross count in the expected region for 4.59 and 4.55 MeV-α from $^{244}Pu$ was 158 per 47,142,025 sec, which includes the detector background and the tailings from $^{239,240}Pu$ and $^{234}U$. Therefore, one can deduce the upper limit on $^{244}Pu$ to be 13 μBq/kg.

The value that we obtained for the upper limit is only a factor of 1.6 higher than the expected one of 8 μBq/kg, which may have an ambiguity of orders of magnitude although it is reasonably well consistent with the finding of Hoffman *et al.* (1971). The present result encourages further measurements to lower the limit and better constrain the steady state inflow of $^{244}Pu$ from the intersteller medium to the Solar System.

It is obvious that a purification of the sample from U is required to reduce the tailing, as well as an increase in the amount of the sediment sample for treatment.

A small peak ascribable to $^{238}Pu$ is always seen in the Pu spectra for deep-sea sediment in the Pacific Ocean.

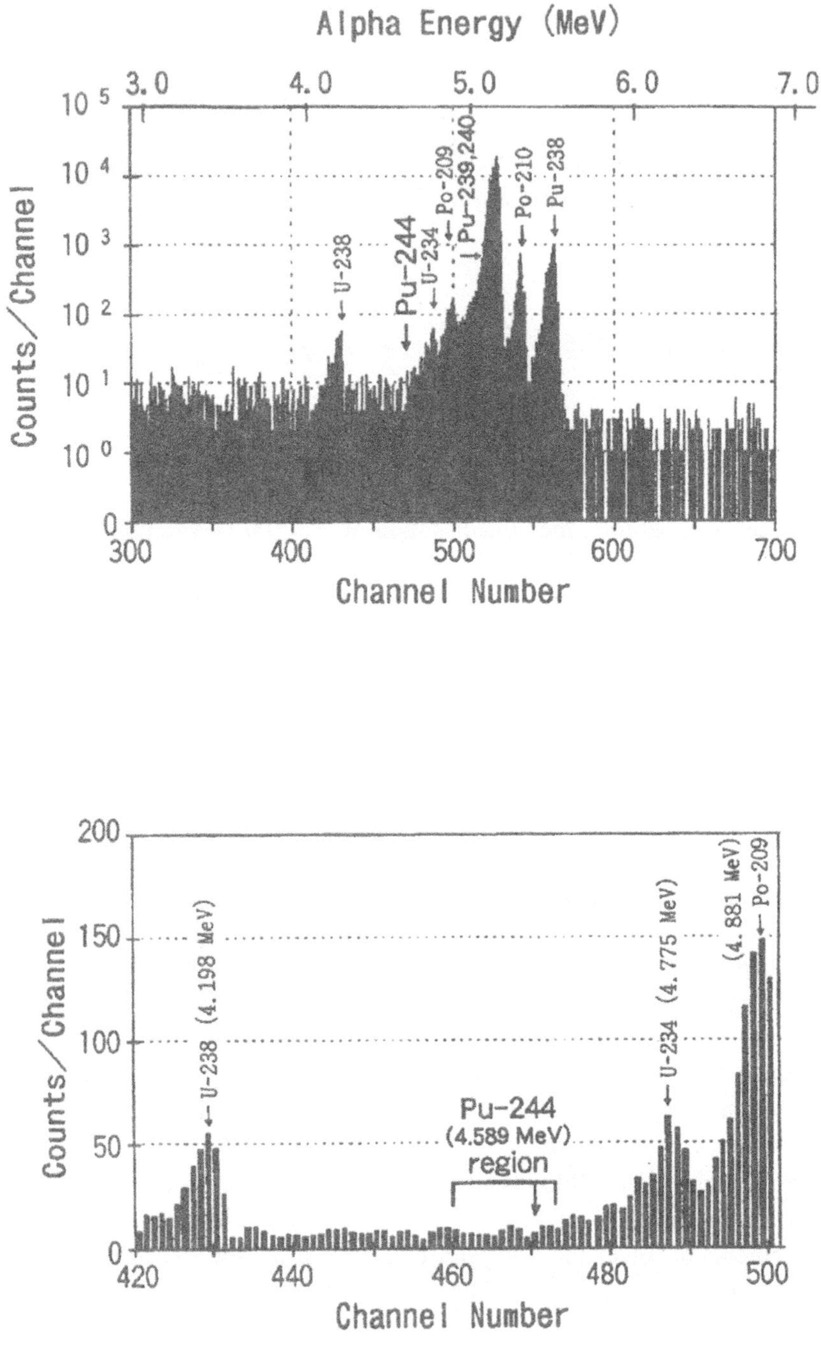


*Figure 2.* The α-spectra of plutonium fraction isolated from 1,020g of 92 SAD 01 deep sea sediment. Upper: Raw spectrum of 350 d counting. Lower: Smoothed and magnified region of upper spectrum.

## 3.3 Future Study

Since a large amount of the fall-out $^{239,240}$Pu in the Pu fraction interferes with the deduction of α–counts from $^{244}$Pu, the distribution of $^{239,240}$Pu was studied by a stepwise dissolution of a 53 g sample. It was observed that 21, 27, 11 and 40% of the $^{239,240}$Pu were in carbonates (leacheate with 1M $CH_3COONa$ adjusted to pH5 with $CH_3COOH$ at room temperature), hydrous Fe-Mn oxides (leacheate with 0.04M $NH_2OH{\bullet}HCl$ in 25% $CH_3COOH$ adjusted to pH2 with NaOH at 100°C), organic matter (leacheate with 30% $H_2O_2$ adjusted to pH2 with $HNO_3$ at 40°C) and silicates (the final residue), respectively. Also the magnetic fraction of 57.8 mg separated from 5.47 kg dry-sediment is also under investigation for $^{26}$Al and $^{239,240}$Pu. The candidate fractions such as the silicate and magnetic separates are now to be subjected for Pu separation, with an improvement for the procedure to increase chemical yields.

Also the isolations of Sm will be performed from the reserved wastes to look for $^{146}$Sm. All the Pu and Sm samples will then be subjected to mass spectrometry, and the assumptions involved in estimating the $^{244}$Pu inflow will be reinvestigated by the use of these results.

## REFERENCES

Alexander, E.C., Jr., Lewis, R.S., Reynolds, J.H. and Michel, M.C.: 1971, "Plutonium-244: Confirmation as an extinct radioactivity", *Science* **172**, 837-840.

Barker, J.L. and Anders, E.: 1968, "Accretion rate of cosmic matter from iridium and osmium contents of deep-sea sediments", *Geochim. Cosmochim. Acta* **32**, 627-645

Burbidge, E.M., Burbidge, G.R., Fowler, W.A. and Hoyle, F.: 1957, "Synthesis of the elements in stars", *Rev. Mod. Phys.* **29**, 547-650.

Cowan, G.A.: 1972, "$^{244}$Pu as a possible indicator of interstellar dust within the solar system", in *Symposium on Cosmochemistry*, Cambridge, MA, U.S.A., 12 pp.

Crocket, J.H. and Kuo, H.Y.: 1979, "Sources for gold, palladium and iridium in deep-sea sediments", *Geochim. Cosmochim. Acta* **43**, 831-842.

Fields, P.R., Friedman, A.M., Milsted, J., Lerner, J., Stevens, C.M., Metta, D. and Sabine, W.K.: 1966, "Decay properties of plutonoim-244 and comments on its existence in nature", *Nature* **212**, 132-134.

Hoffman, D.C., Lawrence, F.O., Mewherter, J.L. and Rourke, F.M.: 1971, "Detection of $^{244}$Pu in nature", *Nature* **234**, 132-134.

Hoffman, D.C.: 1972, "The search for plutonium-244 and other heavy element isotopes in nature", Lectures at Los Alamos Sci. Lab., Los Alamos, NM. U.S.A., Inst. for Nucl. Res., Karlsruhe, Germany; Nucl. Chem. Inst., Univ. Oslo, Oslo, Norway. Also see *Univ. Calif., Irvine, Calif., U.S.A., Rept. LA-UR-73-251, LASL*, Univ. Calif., NM (1973).

LaViolette, P.A.: 1985, "Evidence of high cosmic dust concentrations in late pleistocene polar ice (20,000-14,000 years Bp)", *Meteoritics* **20**, 545-558.

Morris, J. and Tera, F.: 1989, "Be-10 and Be-9 in mineral separates and whole rocks from volcanic arcs – Implications for sediment subduction", *Geochim. Cosmochim. Acta* **53**, 3197-3206.

Rowe, M.W. and Kuroda, P.K.: 1965, "Fissionogenic xenon from the Pasamonte meteorite", *J. Geophys. Res.* **70**, 709-714.

Sakamoto, K.: 1974, "Possible cosmic dust origin of terrestrial plutonium-244", *Nature* **248**, 130-132.

Srinivasan, B., Alexander, E.C., Jr., and Manuel, O.K.: 1971, "Iodine-129 in terrestrial ores", *Science* **173**, 327-328.

Tanaka, S., Sakamoto, K. and Komura, K.: 1972, "Aluminum-26 and manganese-53 produced by solar flare particles in lunar rock and cosmic dust", *J. Geophys. Res.* **23**, 4281-4288.

# Strange Xenon Isotope Ratios in Jupiter

K. Windler
*Chemistry Department, University of Illinois, Urbana-Champaign, IL 61801 USA*
*windler@uiuc.edu*

Abstract: In January 1998, the public release of the data collected by the Galileo space probe's mass spectrometer on its descent into Jupiter's atmosphere allowed scientists a glance at the composition of the giant planet (Goldin, 1998). Most importantly, the spectrometer took a direct measurement of the isotopic composition of a planet previously untouched since the birth of the solar system. One of the most studied and debated elements, xenon, provided a good point of focus both because of the controversy surrounding it and because of its place in a high-mass, low-contamination region of the spectrum. A full analysis of the isotopes of xenon and contamination showed the presence of isotopically "strange" xenon like that found in carbonaceous chondrites. Interestingly, the nebular model (Wood, 1999) for making the solar system does not explain these anomalies.

## 1. INTRODUCTION

Three years after its descent into the Jovian atmosphere, the Galileo Space Probe again came to public attention. The director of NASA, Dr. Daniel S. Goldin, released the probe's mass spectrometer data on national television (Goldin, 1998). This data told much about the composition of the planet and gave clues about our solar system. Most importantly, it measured isotope abundance ratios of noble gasses.

Trying to analyze the entire range of data would take years at best, so the mass ranges corresponding to each of the noble gasses were separated and studied first. Of these, the high-mass region of xenon proved to be the least contaminated. Because of this, and the fact that controversy has surrounded xenon since the discovery of Xe-X, xenon gas with an unusually high

*Origin of Elements in the Solar System: Implications of Post-1957 Observations.*
Edited by O. Manuel, Kluwer Academic/Plenum Publishers, 2000. 

$^{136}Xe/^{134}Xe$ ratio created from rapid neutron capture in a supernova (Burbidge *et al.*, 1957; Manuel *et al.*, 1972; 1977), analysis of xenon began first.

Two main problems hindered the analysis at the outset. Many of the abbreviations and symbols used on the raw data made no sense, and the sheer amount of data present required a computer to rapidly and accurately process.

Fortunately, the first problem had an easy solution: the members of the Galileo Probe Mass Spectrometer (GPMS) team were contacted through e-mail. They had published isotopic ratios for some of Jupiter's lighter elements (Niemann *et al.*, 1996) and they answered all the questions related to the abbreviations and the symbols used in the data.

The rapid calculation of the data with the computer proved a more time-consuming process. A computer program to perform calculations went through several versions, but finally resulted in a quick and easy way to transform the raw data into a usable format. It would correct for the background noise seen in the spectrum and would average several scans of the same region together to decrease error. Many calculations by hand checked the work of the computer, ensuring the accuracy of the results.

With the computer program working, the calculations began in earnest. During the flight into the atmosphere, the mass spectrometer had taken measurements at different mass numbers every half of a second. By cycling through mass numbers this quickly, the instrument was able to complete several replicate analyses at each mass number. Unfortunately, not all of these replicates could be distinguished from the background.

The amount of xenon gas in our atmosphere is quite low. The same is true of the Jovian atmosphere. Because of this, the probe used two devices to contain and concentrate the xenon gas (and other noble gasses) called enrichment cells. Without them, all the xenon measurements would have been indistinguishable from the background. These cells contained a hydrocarbon adsorbent material called Carbosieve (Niemann *et al.*, 1992), which was designed to remove some of the contaminants from the noble gasses. Carbosieve removes larger molecular weight hydrocarbons more efficiently, meaning that xenon would have the least contaminants of all the noble gasses. This was especially true for gasses from the first cell, which were analyzed while the mass spectrometer was closed to the atmosphere. While the gasses from the second enrichment cell were analyzed, a direct leak of atmospheric gasses continued, flooding the Carbosieve with contaminants.

## 2. EXPERIMENTAL DATA

A schematic diagram of the GPMS system shown is below. All heaters are labeled as H, valves are labeled with a V, enrichment cells with C, and getters with G (Niemann *et al.*, 1992).

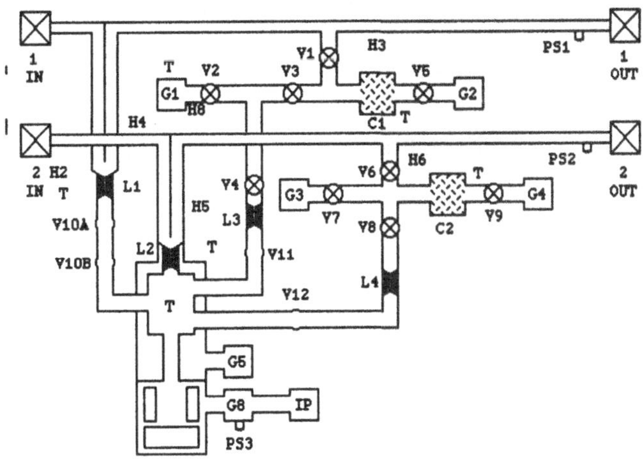


*Figure 1*

The only scans distinguishable from the background were those taken during the analysis of the enrichment cells. During the analysis of the first cell, gasses flowed as shown in the diagram below:

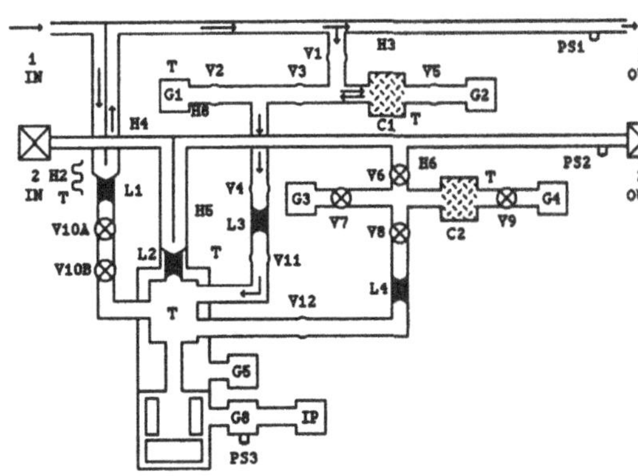

*Figure 2.*

Six scans above background level were taken during this time. Five of the six had an ionizing voltage of 75eV, while the sixth had an ionizing voltage of 15eV. The 15eV scan, which is only slightly above the first ionization potential of xenon, 12ev, was dropped from the statistical analysis to reduce the error. The remaining five scans were averaged together statistically for xenon isotopes at $m/z$ = 128, 129, 130, 131, 132, 134 and 136, giving the most weight to that number which had the least error. Xenon isotopes at $m/z$ = 124 and 126 are too low in abundance to distinguish from background. Raw data for the other seven xenon isotopes from the first enrichment cell (C1) are shown in Table 1.

Values of ($m/z$ = 136)/($m/z$ = 134) for the raw Jupiter data are shown on the right and compared with the $^{136}Xe/^{134}Xe$ ratio in the solar wind (SW) and in Xe-X from the 3CS4 mineral separate of the Allende meteorite (Kaiser, 1972; Lewis *et al.*, 1975).

*Table 1.* Raw data from EC1 of the GPMS at 75ev for xenon isotopes at $m/z$ = 128 - 136.

| Step number | Mass/charge, m/z | Counts | (m/z = 136)/(m/z = 134) |
|---|---|---|---|
| 2590 | 128 | 43 | |
| 2591 | 129 | 529 | |
| 2592 | 130 | 73 | |
| 2593 | 130 | 67 | |
| 2594 | 131 | 352 | |
| 2595 | 132 | 468 | |
| 2597 | 134 | 183 | |
| 2599 | 136 | 188 | 1.03 |
| | | | |
| 2871 | 128 | 55 | |
| 2872 | 129 | 726 | |
| 2873 | 130 | 111 | |
| 2874 | 131 | 537 | |
| 2875 | 132 | 667 | |
| 2877 | 134 | 244 | |
| 2879 | 136 | 287 | 1.18 |
| | | | |
| 2898 | 128 | 66 | |
| 2899 | 128 | 70 | |
| 2900 | 129 | 810 | |
| 2901 | 130 | 101 | |
| 2902 | 131 | 589 | |
| 2903 | 132 | 696 | |
| 2905 | 134 | 267 | |
| 2907 | 136 | 255 | 0.96 |
| | | | |
| 2926 | 128 | 67 | |
| 2927 | 129 | 722 | |
| 2928 | 130 | 99 | |
| 2929 | 131 | 568 | |
| 2930 | 132 | 752 | |
| 2932 | 134 | 222 | |
| 2933 | 134 | 233 | |

| Step number | Mass/charge, m/z | Counts | (m/z = 136)/(m/z = 134) |
|---|---|---|---|
| 2935 | 136 | 194 | 0.85 |
| | | | |
| 3072 | 128 | 63 | |
| 3073 | 129 | 714 | |
| 3074 | 130 | 109 | |
| 3075 | 131 | 636 | |
| 3076 | 132 | 752 | |
| 3080 | 134 | 259 | |
| | 136 | 230 | 0.89 |
| | | | |
| **Net Jupiter EC1** | | | **1.01** |
| | | | |
| **SW Xenon** | | | **0.80** |
| | | | |
| **Xe-X in Allende's 3SC4** | | | **1.04** |

The analysis of the second cell had slightly more contamination, due to the direct leak from the atmosphere being open, and the fact that the GPMS was lower in the atmosphere, which meant contaminants were more plentiful. The diagram is shown below:

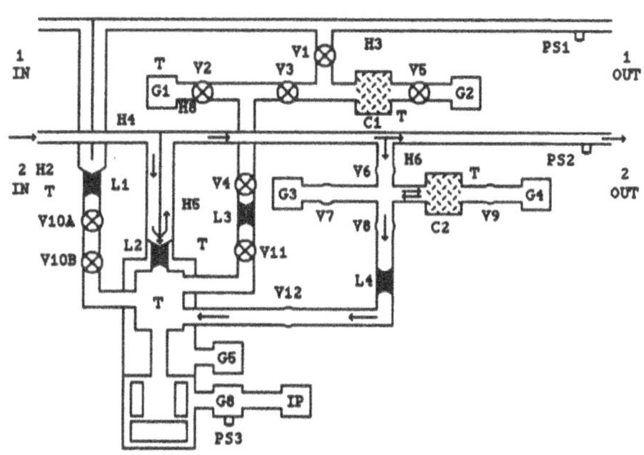


*Figure 3*

Despite this, eight reliable scans were taken, six of which had ionizing voltages of 75eV. The seventh had an ionizing voltage of 15eV, while the eighth one had 25eV. Both the seventh and eighth were discounted to minimize the error. The remaining six at 75eV were statistically averaged together in the same way the scans of the first enrichment cell were to reduce error.

Raw data for these six sweeps of $^{128-136}$Xe isotopes from the second enrichment cell (C2) are shown in Table 2. As in Table 1, the ($m/z$ = 136)/($m/z$ = 134) ratio for raw Jupiter data are shown on the right and compared with the $^{136}$Xe/$^{134}$Xe ratio in the solar wind (SW) and in Xe-X from the 3CS4 mineral separate of the Allende meteorite (Kaiser, 1972; Lewis *et al.*, 1975).

*Table 2.* Raw data from EC2 of the GPMS at 75ev for xenon isotopes at $m/z$ = 128 - 136.

| Step number | Mass/charge, $m/z$ | Counts | ($m/z$ = 136)/($m/z$ = 134) |
|---|---|---|---|
| 4584 | 128 | 121 | |
| 4585 | 129 | 1709 | |
| 4586 | 130 | 203 | |
| 4587 | 131 | 938 | |
| 4588 | 132 | 1619 | |
| 4590 | 134 | 492 | |
| 4592 | 136 | 507 | 1.03 |
| | | | |
| 4903 | 128 | 141 | |
| 4904 | 129 | 1865 | |
| 4905 | 130 | 247 | |
| 4906 | 131 | 1157 | |
| 4907 | 132 | 1791 | |
| 4909 | 134 | 506 | |
| 4911 | 136 | 421 | 0.83 |
| | | | |
| 5222 | 128 | 119 | |
| 5223 | 129 | 1533 | |
| 5224 | 130 | 216 | |
| 5225 | 131 | 1099 | |
| 5226 | 132 | 1629 | |
| 5227 | 132 | 1647 | |
| 5229 | 134 | 464 | |
| 5231 | 136 | 422 | 0.91 |
| | | | |
| 5287 | 128 | 149 | |
| 5288 | 129 | 1567 | |
| 5289 | 130 | 217 | |
| 5290 | 131 | 982 | |
| 5291 | 132 | 1225 | |
| 5293 | 134 | 424 | |
| 5295 | 136 | 384 | 0.91 |
| | | | |
| 5303 | 128 | 157 | |
| 5304 | 129 | 1613 | |
| 5305 | 130 | 204 | |
| 5306 | 131 | 971 | |
| 5307 | 132 | 1509 | |
| 5310 | 134 | 434 | |

| Step number | Mass/charge, *m/z* | Counts | (*m/z* = 136)/(*m/z* = 134) |
|---|---|---|---|
| 5312 | 136 | 354 | 0.82 |
| | | | |
| 5319 | 128 | 114 | |
| 5320 | 129 | 1667 | |
| 5321 | 130 | 216 | |
| 5322 | 131 | 977 | |
| 5323 | 132 | 1493 | |
| 5325 | 134 | 404 | |
| 5327 | 136 | 358 | 0.89 |
| **Net Jupiter C2** | | | **0.90** |
| **SW Xenon** | | | **0.80** |
| **Xe-X in Allende's 3CS4** | | | **1.04** |

## 3. DISCUSSION

It has been shown in meteorites that isotopically "strange" Xe-X is always closely associated with primordial He, but there is little or no He trapped with isotopically "normal" Xe (Manuel *et al.*, 1977; Sabu and Manuel, 1980). Primordial helium abounds in the Jovian atmosphere, providing a basis for comparison to meteorites. Our main work focused on answering the question: Are the xenon isotope ratios in Jupiter like the Xe-X found in the Allende meteorite, or are they like those ratios found in the solar wind? A way to distinguish between these two ratios had to be found.

Data had to first be refined; the raw data shown above in Tables 1 and 2 had not been corrected for background counts. Once this correction was made, the net values for the $^{136}Xe/^{134}Xe$ ratio in the first and second enrichment cells, C1 and C2, shifted slightly to 0.99 ± 0.12 and 0.90 ± 0.08, respectively (Manuel *et al.*, 1998). Calculated isotope ratios revealed that some degree of contamination remained; at the 95% confidence interval for the data, the error was fairly large. In order to reduce error from contamination, a suitable spot on the spectrum was chosen to serve as a marker for hydrocarbon contamination. This marker, at *m/z* = 77, was plotted against the averages of the different isotope ratios from each cell to give a line of best fit (Manuel *et al.*, 1998) and then extrapolated back to give the point of zero contamination. Each point on the line represents an average of the points contained in the cells themselves and is calculated for a

95% confidence interval. Each of the following ratios were calculated in this manner: $^{128}Xe/^{130}Xe$, $^{129}Xe/^{132}Xe$, $^{131}Xe/^{132}Xe$, and $^{136}Xe/^{134}Xe$.

Table 3 shows a comparison of these ratios from Jupiter's atmosphere with those found in the solar wind (Kaiser, 1972) and those found in mineral separate 3CS4 of the Allende meteorite (Lewis *et al.*, 1975). 3CS4 contains xenon-X (alias Xe-HL). Most of the solar wind ratios and the Xe-X ratios are close in number, which makes them virtually indiscernible when dealing with data with errors as large as those in the Jupiter data. In particular, values of the $^{128}Xe/^{130}Xe$ ratio are 0.51 in the solar wind and 0.55 in Xe-X. Unfortunately, the Jupiter data is 0.56 ± 0.10, which completely overlaps both values. The $^{129}Xe/^{132}Xe$ ratios for the solar wind and Xe-X are indistinguishable even under laboratory conditions, being 1.06 in both. The $^{131}Xe/^{132}Xe$ ratios give a hint at what Jupiter contains, being 0.82 in the solar wind, 0.84 in Xe-X, and 0.89 ± 0.06 in Jupiter, which does not quite overlap the solar wind value. Were this the only evidence on which to base a conclusion, it would be rather questionable. Coupled with the $^{136}Xe/^{134}Xe$ data, though, it becomes compelling. These numbers are shown in the table below:

*Table 3.* Isotopic ratios of Xe in Jupiter, the solar wind and Xe-X from Allende's 3CS4

| Isotopic Ratio | GPMS EC1 | GPMS EC2 | Net Jupiter | Solar Wind | Allende's 3CS4 (Xe-X) |
|---|---|---|---|---|---|
| $^{128}Xe/^{130}Xe$ | 0.57 ± 0.07 | 0.57 ± 0.10 | 0.56 ± 0.10 | 0.51 | 0.55 |
| $^{129}Xe/^{132}Xe$ | 1.05 ± 0.90 | 1.08 ± 0.10 | 1.04 ± 0.12 | 1.06 | 1.06 |
| $^{131}Xe/^{132}Xe$ | 0.80 ± 0.04 | 0.66 ± 0.07 | 0.89 ± 0.06 | 0.82 | 0.84 |
| $^{136}Xe/^{134}Xe$ | 0.99 ± 0.12 | 0.90 ± 0.08 | 1.03 ± 0.14 | 0.80 | 1.04 |

Numbers on the right half of Table 3 show that only values of the $^{136}Xe/^{134}Xe$ ratio definitively distinguish the Jupiter data between the solar wind and Xe-X. This is because the difference between the solar wind ratio and that of Xe-X is much larger than the others; the solar wind ratio is 0.80, while that of Xe-X is 1.04.

An interesting comparison can be made by plotting the $^{136}Xe/^{132}Xe$ ratio against the contamination marker, which gives a deceiving value close to that of the solar wind. Unfortunately, this test does not give any information because it incorrectly compares the products of two different nuclear processes. The $^{136}Xe/^{134}Xe$ ratio is comprised of pure r-process products while the $^{136}Xe/^{132}Xe$ ratio compares an r-process product ($^{136}Xe$) to an s-process product ($^{132}Xe$), which is not diagnostic. To indicate the presence of Xe-X, products from a pure nucleosynthetic process must be compared.

Surprisingly, the Jupiter $^{136}Xe/^{134}Xe$ ratio falls at 1.03 ± 0.14. While this is a large error, even for the 95% confidence interval for which it was calculated, it overlaps the ratio of 1.04 very well and does not come close to that of the solar wind. Along with the ratios from $^{131}Xe/^{132}Xe$, it constitutes strong evidence that Xe-X accompanies the primordial He that is present in Jupiter's atmosphere.

The nebular model of forming the solar system (Wood, 1999) predicts solar-type Xe rather that Xe-X in Jupiter. What does the presence of Xe-X imply? It certainly raises more questions than it answers. The discovery of Xe-X in meteorites sparked heated debate over its origin (*e.g.*, Manuel *et al.*, 1977). While current theories hold that Xe-X is either pre-solar matter injected from a nearby supernova or is dust that blew in during our solar system's formation, they do not explain why Xe-X is associated with primordial He in meteorite minerals, as well as in Jupiter.

Perhaps these theories need to be re-examined, or perhaps there is another theory which explains the presence of Xe-X. Whatever the case, the great judge of history, time, will reveal this mystery to us.

## REFERENCES

Burbidge, E.M., Burbidge, G.R., Fowler, W.A. and Hoyle, F.: 1957, "Synthesis of the elements in stars", *Rev. Mod. Phys.* **29**, 547-650.

Goldin, D.S.: 1998, "Future of Space Science", Speech at the 191st Meeting of the American Astronomical Society, Washington, D.C.; *C-SPAN tape 98-01-07-22-1*, Purdue University Public Affairs Video Archives, Item 98526.

Kaiser, W.A.: 1972, "Rare gas studies in Luna-16-G-7 fines by stepwise heating technique. A low fission solar wind xenon", *Earth Planet. Sci. Lett.* **13**, 387-399.

Lewis, R.S., Srinivasan, B. and Anders, E.: 1975, "Host phase of strange xenon component in Allende", *Science* **190**, 1251-1262.

Manuel, O.K., Hennecke, E.H. and Sabu, D.D.: 1972, "Xenon in carbonaceous chondrites", *Nature* **240**, 99-101.

Manuel, O.K., Sabu, D.D., Lewis, R.S., Srinivasan, B. and Anders, E.: 1977, "Strange xenon, extinct superheavy elements and the solar neutrino puzzle", *Science* **195**, 208-210.

Manuel, O., Windler, K., Nolte, A., Johannes, L., Zirbel, J. and Ragland, D.: 1998, "Strange xenon in Jupiter", *J. Radioanal. Nucl. Chem.* **238**, 119-121.

Niemann, H.B., Harpold, D.N., Atreya, S.K., Carignan, G.R., Hunten, D.M. and Owen, T.C.: 1992, "Galileo probe mass spectrometer experiment", *Space Sci. Rev.* **60**, 111-142.

Niemann, H.B., Atreya, S.K., Carignan, G.R., Donahue, T.M., Haberman, J.A., Harpold, D.N., Hartle, R.E., Hunten, D.M., Kasprzak, W.T., Mahaffy, P.R., Owen, T.C., Spencer, N.W. and Way, S.H.: 1996, "The Galileo probe mass spectrometer: Composition of Jupiter's atmosphere", *Science* **272**, 846-849

Sabu, D.D. and Manuel, O.K.: 1980, "Noble gas anomalies and synthesis of the chemical elements", *Meteoritics* **15**, 117-138.

Wood, J.A.: 1999, "Forging the planets: The origin of our Solar System", *Sky & Telescope* **97**, 36-48.

# Abundances of Hydrogen and Helium Isotopes in Jupiter

Adam Nolte and Cara Lietz
*Chemical Engineering Department, University of Missouri, Rolla, MO 65401, USA*
*nolte@umr.edu*

Abstract: Abundances of hydrogen and helium isotopes in Jupiter and other giant planets can answer important questions on the origin of elements in the solar system and the nature of processes in the sun. The Galileo Probe entered Jupiter in late 1995. In January of 1998, raw data from the Galileo Probe Mass Spectrometer (GPMS) were placed on the internet at the website that is given below: http://webserver.gsfc.nasa.gov/code915/gpms/datasets/gpmsdata.html. From the raw data we estimate values of $^3He/^4He = (2.17 \pm 0.03) \times 10^{-4}$ and $^2H/^1H \approx 1.0 \times 10^{-4}$. These are higher than expected if the solar system formed from a homogeneous nebula (Wood, 1999) with subsequent production of excess $^3He$ by deuterium burning in the sun (Geiss, 1993). It appears that Jupiter formed instead from elements with some of the same chemical and isotopic irregularities observed in meteorites.

## 1. INTRODUCTION

Niemann *et al.* (1996) reported values of $^4He/H_2 = 0.156 \pm 0.006$, $^3He/^4He = (1.1 \pm 0.1) \times 10^{-4}$, and $^2H/^1H = (5 \pm 2) \times 10^{-5}$ for the abundances of these four lightest nuclides in Jupiter. Values for these isotope ratios in Jupiter were later changed to $^3He/^4He = (1.66 \pm 0.05) \times 10^{-4}$ and $^2H/^1H = (2.6 \pm 0.7) \times 10^{-5}$ (Mahaffy *et al.*, 1998). Because of the importance of these measurements to our understanding of the origin of elements in the solar system, this note re-examines key experimental data that form the basis for these reports.

In the solar wind the $^3He/^4He$ ratio is higher and the $^2H/^1H$ ratio is lower than the values reported in Jupiter. Niemann *et al.* (1996) and Mahaffy *et al.*

(1998) suggest that differences in these isotopic ratios are consistent with the nebular model for forming the solar system (Wood, 1999) and with the production of $^3$He by deuterium burning in the sun (Geiss, 1993). According to this model, Jupiter and the sun formed out of the same nebular material, but lighter mass isotopes of hydrogen and helium are now enriched in the solar wind because deuterium burning in the sun converted $^2$H into $^3$He.

Deuterium burning in the Sun seems likely, but another process may have produced the high $^3$He/$^4$He ratio in the solar wind. Manuel and Hwaung (1983) pointed out that the lighter isotopes of all five noble gases, He, Ne, Ar, Kr, and Xe, are systematically enriched in the solar wind. The enrichment is only ≈ 4% per amu for Xe isotopes (Kaiser, 1972; Bernatowicz and Podosek, 1978) but steadily increases for lighter elements, becoming ≈ 27% per amu for Ne isotopes and ≈ 200% per amu for He isotopes. Such a mass-dependent fractionation pattern is expected from thermal diffusion in an ionized gas. Chapman and Cowling (1952, Section 14.71, p. 255) note that, "*This must happen in the sun and the stars, where thermal diffusion will assist pressure diffusion in concentrating the heavier nuclei towards the hot central regions.*"

According to this model, diffusion enriches lighter isotopes of each element and lighter weight elements like H and He at the solar surface. Thus, the H, He-rich solar skin may hide an interior of Fe, Ni, O, Si, S and Mg (Manuel and Hwaung, 1983). These abundant elements of the inner planets may also be major components of the solar interior, which contains ≈ 99.8% of all solar system material (Wiens *et al.*, 1999).

In the following sections we re-examine the experimental basis for reported abundances of $^1$H, $^2$H, $^3$He and $^4$He in Jupiter and compare them with solar abundances to evaluate the merits of these two models. As noted above, abundances of these four lightest stable nuclides may contain a record of the origin and abundance of elements in the solar system and the nature of processes in the sun.

It should be noted that the isotope ratios reported here are based solely on the raw GPMS data reported on the website. We did not have access to instrumental mass discrimination for the GPMS when the measurements shown in Tables 1 and 2 were made. We also were unable to obtain information on the source of the GPMS signal at 4 amu when the instrument was operated at 15 eV (Table 2) to prevent ionization of He.

## 2. EXPERIMENTAL DATA

Both Niemann *et al.* (1996) and Mahaffy *et al.* (1998) note that counts at 3 amu are a mixture of $HD^+$, $^3He^+$ and $H_3^+$ ions. Niemann *et al.* (1996)

report that the value of $^3He/^4He$ = (1.1 ± 0.1) x $10^{-4}$ in Jupiter was determined from 3 amu and 4 amu data obtained with the noble gas cell (NGC) sample. They note that hydrogen was effectively absent from this gas sample. Mahaffy *et al.* (1998) report that the value of $^3He/^4He$ = (1.66 ± 0.04) x $10^{-4}$ in Jupiter was determined from 3 amu, 4 amu and 16 amu (or 12 amu) data from NGC, from the first enrichment cell (EC1), and from calibration data of the Flight Unit (FU). They note that methane and noble gases are major components of the NGC sample.

The value of the $^3He/^4He$ ratio in Jupiter is best defined by the NGC sample. These NGC gases were collected through the first direct leak (DL1) into the first enrichment cell (EC1), where getter pumps were used to eliminate hydrogen and other chemically active species from the NGC sample. The mass spectrometer was isolated from atmospheric gases leaking directly into instrument during the analysis of NGC and EC1 gases. Data from the second enrichment cell (EC2) are less useful in defining the $^3He/^4He$ ratio because a direct leak of atmospheric gases into the mass spectrometer continued during the analysis of those gases.

Table 1 shows spectra over the mass range of 2-4 amu from the Galileo Probe Mass Spectrometer (GPMS) for gases from EC1. The first column on Table 1 gives the step number. The second column gives the mass/charge ratio, m/z. The third column gives the number of counts obtained. The ionizing potential was 75 eV for all of the data shown in Table 1. The first six scans (steps 2165-2232) in Table 1 show the mass spectra of the "gettered" gases from NGC. There was very little hydrogen, but abundant helium, in this NGC sample. Thus, column 5 of Table 1 shows the lowest observed values for the ratio of counts at (m/z = 2) relative to those at (m/z = 4). The last two scans (steps 2775-2940) in Table 1 include gases released when the cell was heated to drive off some of the hydrogen trapped there.

Jupiter consists mostly of hydrogen and helium. Thus, the $H_2^+$ ion is responsible for essentially all counts at m/z = 2. Likewise, the $^4He^+$ ion is responsible for essentially all counts at m/z = 4. However, the counts at m/q = 3 are expected to be a mixture of $^3He^+$, $HD^+$, and perhaps $H_3^+$ (Niemann *et al.*, 1996; Mahaffy *et al.*, 1998).

The last two columns in Table 1 show values of count ratios for (m/z = 3)/(m/z = 4) and (m/z = 2)/(m/z = 4). It can be seen that the (m/z = 3)/(m/z = 4) count ratio increased from a value of ≈ 2.5 x $10^{-4}$ to ≈ 2.7 x $10^{-4}$ to ≈ 16 x $10^{-4}$ as the value of the (m/z = 2)/(m/z = 4) count ratio increased from a value of ≈ 0.0005 to ≈ 0.0010 to ≈ 0.035. A time span of ≈ 6.45 minutes is represented by steps 2165-2940 tabulated in Table 1.

The data in Table 1 are valuable for determining the $^3He/^4He$ ratio because hydrogen is essentially absent. Likewise, other data are most useful for determining the $HD/H_2$ ratio when helium is essentially absent. This was accomplished by reducing the ionizing voltage potential below 24.48 electron volts, the first ionization potential of He.

*Table 1.* Data from NGC and ECl of the GPMS at 75 eV for species at $m/z = 2, 3$ and 4

| Step Number | Mass/charge, $m/z$ | Counts | $(m/z = 3)/(m/z = 4)$ x $10^4$ | $(m/z = 2)/(m/z = 4)$ x $10^4$ |
|---|---|---|---|---|
| 2165 | 2 | 3102 | | |
| 2166 | 3 | 1521 | 2.545 | 5.191 |
| 2167 | 4 | 5976064 | | |
| 2253 | 2 | 8028 | | |
| 2254 | 3 | 2254 | 2.767 | 9.854 |
| 2255 | 4 | 8146944 | | |
| 2278 | 2 | 8084 | | |
| 2279 | 3 | 2222 | 2.744 | 9.983 |
| 2280 | 4 | 8097792 | | |
| 2288 | 2 | 7964 | | |
| 2289 | 3 | 2218 | 2.714 | 9.746 |
| 2290 | 4 | 8171520 | | |
| 2291 | 2 | 9352 | | |
| 2292 | 3 | 2198 | 2.731 | 11.619 |
| 2293 | 4 | 8048640 | | |
| 2319 | 2 | 7996 | | |
| 2320-2321 | 3 | 2228 | 2.765 | 9.925 |
| 2322 | 4 | 8056832 | | |
| 2775 | 2 | 349952 | | |
| 2776 | 3 | 15864 | 16.205 | 357.480 |
| 2777 | 4 | 9789440 | | |
| 2938 | 2 | 341248 | | |
| 2939 | 3 | 15800 | 15.953 | 344.550 |
| 2940 | 4 | 9904128 | | |

Niemann *et al.* (1996) give an upper limit of $HD/H_2 = (1.1 \pm 0.3) \times 10^{-4}$, corresponding to an atomic ratio of $D/H = 5.5 \times 10^{-5}$, for gases entering the mass spectrometer via inlet DL2 during descent at atmospheric pressures of 8.21 to 21 bars. At that time, the ionizing potential was reduced to 15 eV for steps 3960-3967, 4104-4119, and 4211-4222. Under those conditions, there are essentially no He ions (Niemann *et al.*, 1996).

The results are shown in Table 2 for the 2-4 amu mass range. The first three columns of Table 2 are the same as in Table 1. The last two columns of Table 2 are the count ratios normalized to the number of counts at 2 amu. All of the data in Table 2 were taken at an ionization potential of 15 eV. Because of this low ionization potential, column 5 of Table 2 shows the lowest observed values for the ratio of counts at ($m/z = 4$) relative to those at ($m/z = 2$).

*Table 2.* Data at $m/z$ = 2, 3 and 4 for gases directly leaked *via* DL2 into the GPMS at 15 eV

| Step Number | Mass/charge, $m/z$ | Counts | $(m/z = 3)/(m/z = 2)$ x $10^4$ | $(m/z = 4)/(m/z = 2)$ x $10^4$ |
|---|---|---|---|---|
| 3960 | 2 | 558592 | | |
| 3961 | 3 | 103 | 1.84 | 1.61 |
| 3962 | 4 | 90 | | |
| 3964 | 2 | 527872 | | |
| 3965 | 3 | 113 | 2.14 | 1.46 |
| 3966 | 4 | 77 | | |
| 4104 | 2 | 609792 | | |
| 4105 | 2 | 591360 | | |
| 4106 | 3 | 123 | 2.01 | 1.79 |
| 4107 | 3 | 119 | | |
| 4108 | 4 | 119 | | |
| 4109 | 4 | 96 | | |
| 4112 | 2 | 589312 | | |
| 4113 | 3 | 144 | 2.44 | 1.60 |
| 4114 | 4 | 94 | | |
| 4116 | 2 | 581120 | | |
| 4117 | 3 | 109 | 1.88 | 1.70 |
| 4118 | 4 | 99 | | |
| 4211 | 2 | 652800 | | |
| 4212 | 3 | 124 | 1.90 | 1.47 |
| 4213 | 4 | 96 | | |
| 4215 | 2 | 621056 | | |
| 4216 | 3 | 135 | 2.17 | 1.72 |
| 4217 | 4 | 107 | | |
| 4219 | 2 | 624128 | | |
| 4220 | 3 | 130 | 2.08 | 1.73 |
| 4221 | 4 | 108 | | |

## 3. INTERPRETATION

We believe that the data in Table 1 and Table 2 are the most important sets of observations for determining values of the $^3He/^4He$ and $HD/H_2$ ratios in Jupiter. Table 1 shows the counts at 2, 3 and 4 amu when the mass spectrometer had the highest ratio of $^4He^+/H_2^+$ ions. Table 2 shows the counts at 2, 3 and 4 amu when the mass spectrometer had the lowest ratio of $^4He^+/H_2^+$ ions.

### 3.1 The $^3He/^4He$ Ratio in Jupiter

Data from Table 1 are plotted as (m/z = 3)/(m/z = 4) vs (m/z = 2)/(m/z = 4) in Figure 1. As noted earlier, the (m/z = 3)/(m/z = 4) count ratio increased from a value of ≈ 2.5 x $10^{-4}$ to ≈ 16 x $10^{-4}$ as the value of the (m/z

= 2)/(m/z = 4) count ratio increased from ≈ 0.0005 to ≈ 0.035. Such a correlation is expected if the signals at 2 amu, 4 amu, and 3 amu consist of $H_2^+$, $^4He^+$, and ($^3He^+$ plus $HD^+$) ions, respectively. In that case, the y intercept at $H_2/^4He = 0$ would correspond to the value of the (m/z = 3)/(m/z = 4) ratio for pure helium, *i.e.*, it would be the value of the $^3He/^4He$ ratio if the instrument sensitivity is identical for $^3He$ and $^4He$. Likewise, the slope of the line would correspond to the value of the (m/z = 3)/(m/z = 2) ratio for pure hydrogen, *i.e.*, it would be the value of the $DH/H_2$ ratio if there were no reaction producing $H_3^+$ ions from $H_2$ and the instrument sensitivity was identical for HD and $H_2$. The intercept in Figure 1 is $(2.34 \pm 0.03) \times 10^{-4}$ and the slope is $(3.91 \pm 0.02) \times 10^{-2}$.

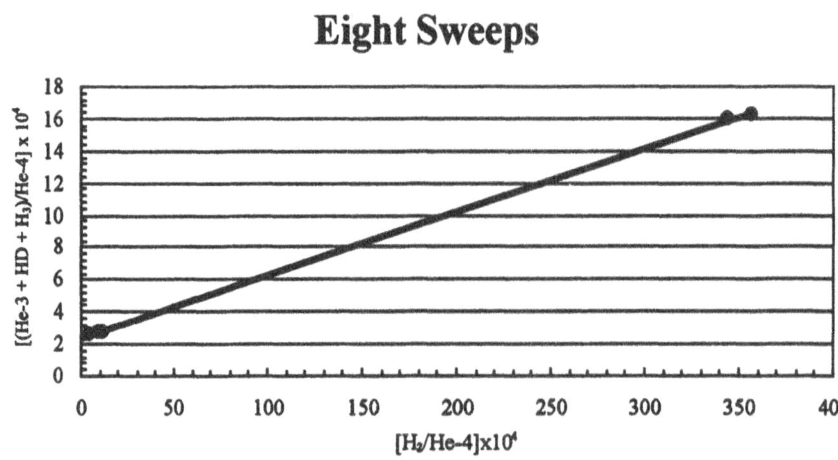


*Figure 1.* This graph of the raw data from Table 1 illustrates the corrleation between the signals for the 3 amu/4 amu ratio and the signals corresponding to the hydrogen/helium ratio.

Niemann *et al.* (1996) and Mahaffy *et al.* (1998) stress the presence of $H_3^+$ ions in the total counts at 3 amu. Thus, the vertical axis in Figure 1 is labeled as $[(He\text{-}3 + HD + H_3)/He\text{-}4] \times 10^4$. They note that $H_3^+$ ions are produced by the dissociative ionization of $CH_4$ and by ion-molecule reactions on $H_2$. $H_3^+$ ions from reactions on $H_2$ should disappear when the $H_2/^4He$ ratio goes to zero. To correct for $H_3^+$ from $CH_4^+$, we use a value of $H_3^+/ CH_4^+ = 1.85 \times 10^{-5}$ as the fraction of $CH_4$ counts that contribute to the counts at 3 amu for the NGC and EC1 data shown in Table 1 (Mahaffy *et al.*, 1998). The signal at 16 amu, as measured at steps 2180, 2381, 2790 and 2952, increased by about 16% during the time span represented by the data in Table 1. From the value of $H_3^+/CH_4^+ = 1.85 \times 10^{-5}$, we estimate the counts of $H_3^+$ ions from $CH_4^+$ to be 124, 132, 134, 135, 135, 138, 141, and 143 counts when the signal at 3 amu was counted in step numbers 2166, 2254, 2279, 2289, 2292, 2320-2321, 2776, and 2939, respectively.

Figure 2 shows the correlation of (m/z = 3)/(m/z = 4) with (m/z = 2)/(m/z = 4) after the above correction for the contribution of $CH_4$ to $H_3^+$ ions. The vertical axis in Figure 2 is still labeled as [(He-3 + HD + $H_3$)/He-4] x $10^4$. Although the contribution of $H_3^+$ ions from $CH_4$ was subtracted from the total count at 3 amu, there may still be a contribution of $H_3^+$ ions from the ion-molecule reaction on $H_2$. If so, that will tend to increase the slope of the line in Figure 2 and make it higher than the value of the HD/$H_2$ ratio. The vertical intercept in Figure 2 suggests that pure Jovian helium, free of interference from HD and $H_3$, has $^3He/^4He = (2.17 \pm 0.03) \times 10^{-4}$.

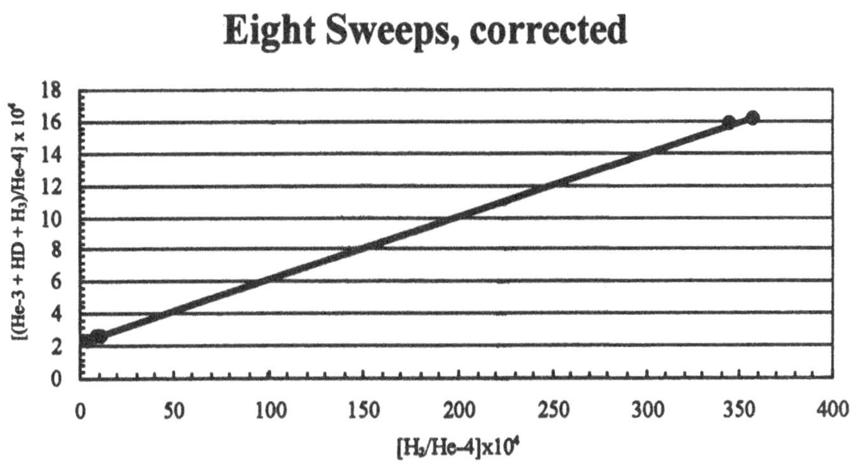


*Figure 2.* This graph represents Figure 1 after correction for $H_3^+$ interference.

The slope of the line in Figure 2 is $(3.92 \pm 0.02) \times 10^{-2}$. This is ≈ 3 orders-of-magnitude steeper than the slope expected from HD interference if $^2H/^1H = (2.6 \pm 0.7) \times 10^{-5}$ (Mahaffy *et al.*, 1998) and ≈ 2 orders-of-magnitude steeper than the slope expected from HD if $^2H/^1H \approx 1.0 \times 10^{-4}$. Large isotopic fractionation effects are expected in chemical and physical reactions involving hydrogen because it has the largest isotopic mass ratio of any element (Geiss and Reeves, 1981). Thus, the processed hydrogen gas from NGC and EC1 may be enriched in HD so that the vertical component of the line slopes in Figures 1 and 2 represents a mixture of counts from $HD^+$ ions at 3 amu with the product of ion molecule reactions on $H_2$ that also produce counts of $H_3^+$ ions at 3 amu.

Interference from $H_3^+$ ions produced by the ion-molecule reaction on $H_2$ might be reduced by using only the gas from NGC, before the cell was heated to drive off trapped $H_2$. Niemann *et al.* (1996, p. 847) state that, "*During probe descent, the $^3He/^4He$ ratio was determined from 3 amu and 4 amu data obtained with the noble gas cell (NGC) sample (6). Hydrogen was* *effectively absent from this gas sample.*" This NGC sample is represented by first six scans (steps 2165-2232) in Table 1.

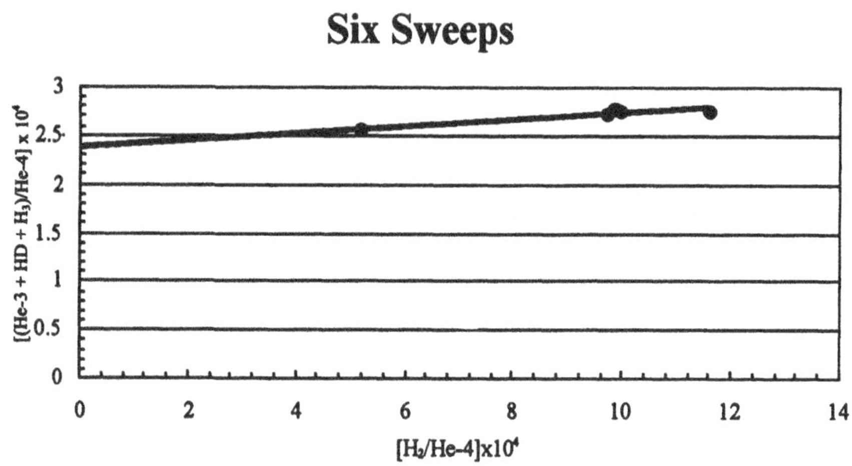


*Figure 3.* This graph of raw data for only the first six sweeps from Table 1 shows the same correlation and intercept, within experimental error, as in Figure 1.

Figure 3 shows the correlation of (m/z = 3)/(m/z = 4) with (m/z = 2)/(m/z = 4) for these first six scans of the NGC sample, with no correction for the contribution of $CH_4$ to $H_3^+$ ions. The vertical intercept in Figure 3 suggests an upper limit of $^3He/^4He = (2.39 \pm 0.08) \times 10^{-4}$. The slope of the line in Figure 3, $(3.5 \pm 0.8) \times 10^{-2}$, is poorly defined because of the very limited range of values in the first six sweeps.

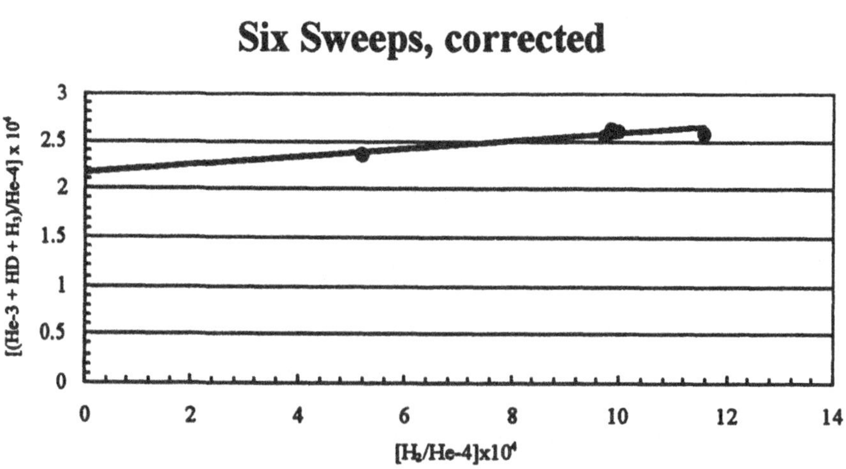


*Figure 4.* This graph represents Figure 3 after correction for $H_3^+$ interference.

Figure 4 shows the correlation of (m/z = 3)/(m/z = 4) with (m/z = 2)/(m/z = 4) for the first six scans in Table 1 (the NGC sample) after a value of $H_3^+$/ $CH_4$ = 1.85 x $10^{-5}$ (Mahaffy *et al.*, 1998) was used to correct for the contribution of $CH_4$ to $H_3^+$ ions. This correction did not improve the uncertainty on the intercept or the slope. The intercept in Figure 4 yields a value of $^3He/^4He = (2.15 \pm 0.09) \times 10^{-4}$. The slope of the line in Figure 4 is $(4 \pm 1) \times 10^{-2}$.

As noted, values of the intercepts and the slopes in Figures 3 and 4 are indistinguishable from those in Figures 1 and 2, respectively. This means that interference at 3 amu from $HD^+$ and $H_3^+$ ions, produced by the ion-molecule reaction on $H_2$, was not significantly reduced by using only the gas from NGC.

From the data in Table 1 and Figures 1-4 we conclude that the best value for the $^3He/^4He$ ratio in Jupiter is $(2.17 \pm 0.03) \times 10^{-4}$ (See Figure 2). Eliminating data from the last two sweeps increases the uncertainty but does not significantly change the value of the $^3He/^4He$ ratio (See Figures 3 and 4). $^3He/^4He = (2.17 \pm 0.03) \times 10^{-4}$ is about a factor of 2 higher than the value reported by Niemann *et al.* (1996) and 31% higher than the value concluded by Mahaffy *et al.* (1998).

In computing the above values for the $^3He/^4He$ ratio, we assumed that the mass spectrometer sensitivity is identical for $^3He$ and $^4He$. This seems to be a reasonable assumption. Mahaffy *et al.* (1998) used a refurbished Engineering Unit (EU) to duplicate the performance of the Flight Unit (FU). They report (Mahaffy *et al.*, 1998, p. 257) that "Additional EU experiments show that the instrument response to the two helium isotopes introduced to the ion source in the molecular flow regime is flat for $^3He$ and $^4He$ so this count ratio represents a measure of the Jovian $^3He/^4He$ ratio."

## 3.2 The $^2H/^1H$ Ratio in Jupiter

As noted above, slopes of the lines in Figures 1-4 suggest that counts at 3 amu from $HD^+$ ions and $H_3^+$ ions amount to ≈ 4% of the counts from $H_2$ at 2 amu for the operational conditions used to obtain the data shown in Table 1. This instrumental barrier, with (m/z = 3)/(m/z = 2) = 4 x $10^{-2}$, prevents use of the data in Table 1 to determine if the $HD/H_2$ ratio in Jupiter is ≈ $10^{-4}$, as seen for hydrogen here on Earth and in most phases of meteorites, or ≈ $10^{-5}$ as seen in some meteorite phases and as proposed for the protosolar nebula, for atmospheres of the Jovian planets, and for primordial galactic hydrogen (Geiss and Reeves, 1981; Geiss, 1993).

In addition to interference at 3 amu from $H_3^+$ ions, measurement of the $HD/H_2$ ratio is also hindered by the presence of $^3He^+$ ions. Fortunately, interference from $^3He^+$ and $H_3^+$ ions are both diminished by reducing the electron energy in the ion source. Niemann *et al.* (1996) used gases entering the mass spectrometer via inlet DL2, when analyzed with a reduced electron energy in the ion source, to obtain an upper limit of $HD/H_2$ = (1.1 ± 0.3) x $10^{-4}$. Table 2 shows the data obtained when the ionizing potential was reduced to 15 eV. Under those conditions, no He ions were produced and Niemann *et al.* (1996) assumed that there was also "*no $H_3^+$ production*" (p. 847).

If there is no interference from $H_3^+$ for the data tabulated in Table 2, then it appears that $HD/H_2$ = (2.1 ± 0.2) x $10^{-4}$ if the mass spectrometer sensitivity is identical for HD and $H_2$. This is the average value for the ratio of signals, (m/z = 3)/(m/z = 2), for the data shown in Table 2. This value of the $HD/H_2$ molecular ratio is equivalent to a value of $^2H/^1H \approx 1.0 \times 10^{-4}$ for the atomic ratio.

The values obtained here for the $HD/H_2$ and $^2H/^1H$ ratios are about a factor of 2 higher than those reported by Niemann *et al.* (1996). This might be attributed to a factor of 2 difference in the sensitivity of the mass spectrometer for HD and $H_2$, although we found no mention of this instrumental discrimination in the report by Niemann *et al.* (1996).

Mahaffy *et al.* (1998, p. 257) state that, "*Final corrections to the [3]/[2] ratio to reflect the $HD/H_2$ abundance arise from instrumental discrimination between HD and $H_2$. This effect is approximately 10 in L1 as determined from introduction of a known mixture of HD and $H_2$ into the EU L1. It is even higher in L2...*" The gases used here to estimate the $HD/H_2$ (Table 2) entered the mass spectrometer via L2 and might therefore exhibit an instrumental discrimination factor between HD and $H_2$ that is > 10.

However, application of the instrumental discrimination factor reported by Mahaffy *et al.* (1998) to the data in Table 2 does not yield the value of $^2H/^1H$ reported by Mahaffy *et al.* (1998). Use of an instrumental discrimination factor > 10 on the data in Table 2 would indicate a value of $^2H/^1H < 1.0 \times 10^{-5}$. This is significantly less than the value of $^2H/^1H$ = (2.6 ± 0.7) x $10^{-5}$ reported by Mahaffy *et al.* (1998).

## 4. DISCUSSION AND CONCLUSION

The value concluded here for the $^3He/^4He$ ratio in Jupiter, 2.2 x $10^{-4}$, is higher than that of the $^3He/^4He$ ratio in primitive meteorites, 1.4-1.5 x $10^{-4}$ (Manuel and Hwaung, 1983; Geiss, 1993). The difference between the $^3He/^4He$ ratio in primitive meteorites and in the Jovian atmosphere may indicate: a) selective enrichment of lighter mass helium isotope in the upper atmosphere of Jupiter, or b) selective loss of the lighter mass isotope from meteorite minerals. Thus, a value of $^3He/^4He$ = 2.2 x $10^{-4}$ in Jupiter is credible, although it is a factor of 2 higher than the value reported by Niemann *et al.* (1996).

The value concluded here for the $^2H/^1H$ ratio in Jupiter, ≈ $1.0 \times 10^{-4}$, lies at the lower end of the range of values reported for the $^2H/^1H$ ratio in meteorites, $(0.9 - 11) \times 10^{-4}$ (Robert *et al.*, 1987a,b). This value for the $^2H/^1H$ ratio in Jupiter is almost a factor of 4 higher than the value that Geiss (1993) concludes for the protosolar nebula, but it is within the range of values reported for the primordial $^2H/^1H$ ratio in intergalactic gas clouds at high redshifts (Songaila, *et al.*, 1997; Webb *et al.*, 1997).

The value of $^2H/^1H \approx 1.0 \times 10^{-4}$ in Jupiter is a factor of 2 higher than that reported by Niemann *et al.* (1996), $^2H/^1H = (5 \pm 2) \times 10^{-5}$, and almost a factor of 4 higher than the value reported by Mahaffy *et al.* (1998), $^2H/^1H = (2.6 \pm 0.7) \times 10^{-5}$.

It should be stressed that values reported for the $^2H/^1H$ ratio depend on the count ratio at (m/z = 3)/(m/z = 2) arising from the ionic $HD^+/H_2^+$ ratio. Table 2 shows the lowest measured count ratios at (m/z = 3)/(m/z = 2) in the raw GPMS data on the internet at http://webserver.gsfc.nasa.gov/code915/gpms/datasets/gpmsdata.html. So far as we know, the GPMS team observed no sample in Jupiter with $(m/z = 3)/(m/z = 2) < 10^{-4}$.

Niemann *et al.* (1996) and Mahaffy *et al.* (1998) have much more detailed information on the operation of the GPMS than we and should therefore have more reliable values for isotopic ratios. However, the lowest values observed for the (m/z = 3)/(m/z = 2) count ratio are included in Table 2, and the instrumental discrimination factor reported by Mahaffy *et al.* (1998) does not appear to make those data agree with the value that Mahaffy *et al.* (1998) report for Jupiter's $^2H/^1H$ ratio.

The average terrestrial $^2H/^1H$ ratio is $1.6 \times 10^{-4}$. Relative to terrestrial hydrogen, that in Jupiter displays a value of δD ≈ -375 ‰. This is similar to the low-δD hydrogen that Robert *et al.* (1987a) reported in the amorphous matrix surrounding chondrules of the Chainpur chondritic meteorite.

Finally, abundances of H and He isotopes in Jupiter can be compared with predictions of the supernova and the nebular models for the formation of the solar system. The first imagines that the solar system inherited most of its chemical and isotopic irregularities directly from heterogeneous supernova (SN) debris (Manuel and Sabu, 1975, 1977). The inner planets consist mostly of Fe and other elements from the SN interior; the outer planets consist mostly of H, He and other elements that remained in the cooler, outer SN layers.

The nebular model suggests that the entire solar system formed from a homogeneous cloud and secondary processes made the present chemical and isotopic heterogeneities (Wood, 1999). According to the nebular model, the earth and other rocky planets with iron-rich cores were produced in the inner part of the solar system by chemical differentiation and loss of volatile elements like H, He and C. Nuclear reactions in the sun converted D into $^3$He (Geiss, 1993), but the composition of the protosolar nebula has been preserved in giant, gaseous planets like Jupiter and Saturn that reside outside the asteroid belt.

Niemann *et al.* (1996) and Mahaffy *et al.* (1998) claim that differences in abundances of H and He isotopes in Jupiter and in the solar wind (SW) agree with predictions of the nebular model for forming the solar system. For example, the first report of results from the Galileo Probe Mass Spectrometer measurement of Jupiter states that, "*Together, the D/H and $^3He/^4He$ ratios are consistent with conversion in the sun of protosolar deuterium to present-day $^3He$.*" (Niemann *et al.*, 1996, p. 846, last sentence of abstract). The values of both isotope ratios were later changed, but the conclusion remained the same: "*We have established the anticipated result that Jupiter indeed represents a repository of solar nebula material unmodified by nuclear reactions for the last 4.55 Gy.*" (Mahaffy *et al.*, 1998, pp. 261-262).

If Jupiter and the sun initially formed out of the same material and only deuterium burning altered the abundances of nuclides at the solar surface, then the value of the $^3$He/$^4$He ratio there today would be the same as that of the ($^2$H + $^3$He)/ $^4$He ratio in the starting material. Table 3a shows values expected for the $^3$He/$^4$He ratio in the sun today if deuterium burning acted on Jupiter-like starting material with the $^4$He/$H_2$ ratio reported by Niemann *et al.* (1996) and isotopic compositions reported in Niemann *et al.* (1996), in Mahaffy *et al.* (1998), and in this paper.

*Table 3a.* Solar values expected for $^3$He/$^4$He after deuterium burning of Jupiter-like material

| Reference | Initial Composition | After D-burning | $^3$He/$^4$He |
|---|---|---|---|
| Niemann *et al.* (1996) | $^1$H = 10000; $^2$H = 0.50<br>$^4$He = 780; $^3$He = 0.086 | $^1$H = 10000; $^2$H = 0.00<br>$^4$He = 780; $^3$He = 0.59 | 7.5 x $10^{-4}$ |
| Mahaffy *et al.* (1998) | $^1$H = 10000; $^2$H = 0.26<br>$^4$He = 780; $^3$He = 0.13 | $^1$H = 10000; $^2$H = 0.00<br>$^4$He = 780; $^3$He = 0.39 | 5.0 x $10^{-4}$ |
| This work | $^1$H = 10000; $^2$H ≈ 1.0<br>$^4$He = 780; $^3$He = 0.17 | $^1$H = 10000; $^2$H = 0.00<br>$^4$He = 780; $^3$He ≈ 1.17 | ≈ 15 x $10^{-4}$ |

The solar value of the $^3$He/$^4$He ratio is usually assumed to be that in the solar wind (SW), where $^3$He/$^4$He ≈ 4 x $10^{-4}$ (Geiss and Bochsler, 1991; Manuel and Hwaung, 1983). Deuterium burning of Jupiter-like material as described in the first GPMS report would produce $^3$He/$^4$He = 7.5 x $10^{-4}$ (Upper right section of Table 3a). Because of large errors on isotopic ratios in the first GPMS report (Niemann *et al.*, 1996), this is within error limits of the SW $^3$He/$^4$He ratio. There were smaller errors in the revised GPMS report (Mahaffy *et al.*, 1998). Deuterium burning of this material would produce $^3$He/$^4$He = 5.0 x $10^{-4}$ (Middle right section of Table 3a), and might indicate better agreement of the nebular model with the SW $^3$He/$^4$He ratio. However, deuterium burning of material with the $^3He/^4He$ and the $^2H/^1H$ ratios concluded here would not yield the $^3He/^4He$ ratio observed in the solar wind (Lower right section of Table 3a). Deuterium burning of this material would produce ≈ 4 times the value of the $^3He/^4He$ ratio in the solar wind.

Saturn might be used instead of Jupiter to represent the composition of the solar nebula. There have been no direct measurements on isotopic ratios of H and He in Saturn. However, Voyager measurements indicate a much lower value of the $^4He/H_2$ ratio in Saturn than in Jupiter (Conrath *et al.*, 1984). Table 3b shows values expected for the $^3He/^4He$ ratio in the sun today if deuterium burning acted on Saturn-like starting material with the $^4He/H_2$ ratio reported by Conrath *et al.* (1984) and the isotopic compositions reported in Jupiter by Niemann *et al.* (1996), Mahaffy *et al.* (1998), and this paper.

*Table3b.* Solar values expected for $^3He/^4He$ after deuterium burning of Saturn-like material

| Reference | Initial Composition | After D-burning | $^3He/^4He$ |
|---|---|---|---|
| Niemann *et al.* (1996)<br>Conrath *et al.* (1984) | $^1H$ = 10000; $^2H$ = 0.50<br>$^4He$ = 172; $^3He$ = 0.019 | $^1H$ = 10000; $^2H$ = 0.00<br>$^4He$ = 172; $^3He$ = 0.52 | 30 x $10^{-4}$ |
| Mahaffy *et al.* (1998)<br>Conrath *et al.* (1984) | $^1H$ = 10000; $^2H$ = 0.26<br>$^4He$ = 172; $^3He$ = 0.028 | $^1H$ = 10000; $^2H$ = 0.00<br>$^4He$ = 172; $^3He$ = 0.29 | 17 x $10^{-4}$ |
| This work<br>Conrath *et al.* (1984) | $^1H$ = 10000; $^2H$ ≈ 1.0<br>$^4He$ = 172; $^3He$ = 0.037 | $^1H$ = 10000; $^2H$ = 0.00<br>$^4He$ = 172; $^3He$ = 1.037 | ≈ 60 x $10^{-4}$ |

It can be seen from the right column of Table 3b that deuterium burning of Saturn-like material would, for all three estimates of the initial H and He ratios, produce $^3He/^4He$ ratios that are higher than that observed in the solar wind.

From Tables 3a and 3b, we conclude that the Galileo Probe Mass Spectrometer provides no evidence in support of the nebular model for forming the solar system. The $^3He/^4He$ and $^2H/^1H$ ratios in Jupiter are so high that deuterium burning of this material would produce more $^3He$ than is observed in the solar wind. Jupiter formed from elements with some of the same chemical and isotopic heterogeneities seen in meteorites, including hydrogen and strange xenon (Manuel *et al.*, 1998). The discrepancy between observations and the nebular model increases if Saturn-like material formed the sun.

Conrath *et al.* (1984) and other proponents of the nebular model claim that the lower value of the $^4He/H_2$ ratio in Saturn may indicate that He may have migrated to the core of that planet in the same fashion they imagine Fe may have migrated to the core of the earth and other rocky planets. According to this view Jupiter has been spared the fate suffered by other planets and remains as proof of the nebular model: "*Thus, helium differentiation appears to have not yet begun in Jupiter, and the atmospheric helium abundance should be representative of the bulk composition of the planet.*" (Conrath *et al.*, 1984, p. 807).

The nebular model was a useful assumption when the classical papers on stellar nucleosynthesis were published (Burbidge *et al.*, 1957; Cameron, 1957). However, it has outlived its usefulness and cannot be sustained by such *ad hoc* explanations for the multitude of discordant observations since 1957, including the results presented in many other papers at this symposium and those from the Galileo Probe Mass Spectrometer shown here in Tables 1-3 and in Figures 1-4.

## REFERENCES

Bernatowicz, T.J. and Podosek, F.A.: 1978, "Nuclear components in the atmosphere", in *Terrestrial Rare Gases*, eds., Alexander, Jr., E.C. and Ozima, M., Center for Academic Publications Press, Tokyo, pp. 99-135.

Burbidge, E.M., Burbidge, G.R., Fowler, W.A. and Hoyle, F.: 1957, "Synthesis of the elements in stars", *Rev. Mod. Phys.* **29**, 547-650.

Cameron, A.G. W.: 1957, "Nuclear reactions in stars and nucleosynthesis", *Publ. Astron. Soc. Pac.* **69**, 201-222.

Chapman, S. and Cowling, T.G.: 1952, *The Mathematical Theory of Non-Uniform Gases*, Cambridge University Press, Cambridge, 431 pp.

Conrath, B.J., Gautier, D., Hanel, R.A. and Hornstein, J.S.: 1984, "The helium abundance of Saturn from Voyager measurements", *Ap. J.* **282**, 807-815.

Geiss, J.: 1993, "Primordial abundance of hydrogen and helium isotopes", in *Origin and Evolution of the Elements*, eds., Prantzos, N., Vangioni-Flam, E. and Cassè, M., Cambridge University Press, Cambridge, pp. 89-106.

Geiss, J. and Bochsler, P.: 1991, "Long-term variations in solar wind properties. Possible causes versus observations", in *The Sun in Time*, eds. Sonnett, C.P., Giampapa, M.S., and Matthews, M.S., Univ. Arizona Press, pp. 98-117.

Geiss, J. and Reeves, H.: 1981, "Deuterium in the solar system", *Astron. Astrophys. J.* **93**, 189-199.

Kaiser, W.A.: 1972, "Rare gas studies in Luna-16-G-7 fines by stepwise heating technique. A low fission solar wind xenon", *Earth Planet. Sci. Lett.* **13**, 387-399.

Mahaffy, P.R., Donahue, T.M., Atreya, S.K., Owen, T.C. and Niemann, H.B.: 1998, "Galileo probe measurements of D/H and $^3$He/$^4$He in Jupiter's atmosphere", *Space Science Reviews*, **84**, 251-263.

Manuel, O.K. and Hwaung, G.: 1983, "Solar abundances of the elements", *Meteoritics* **18**, 209-222.

Manuel, O.K. and Sabu, D.D.: 1975, "Elemental and isotopic inhomogeneities in noble gases: The case for local synthesis of the chemical elements", *Trans. Missouri Acad. Sci.* **9**, 104-122.

Manuel, O.K. and Sabu, D.D.: 1977, "Strange xenon, extinct superheavy elements and the solar neutrino puzzle", *Science* **195**, 208-209.

Manuel, O., Windler, K., Nolte, A., Johannes, L., Zirbel, J. and Ragland, D.: 1998, "Strange xenon in Jupiter", *J. Radioanal. Nucl. Chem.* **238**, 119-121.

Niemann, H.B., Atreya, S.K., Carignan, G.R., Donahue, T.M., Haberman, J.A., Harpold, D.N., Hartle, R.E., Hunten, D.M., Kasprzak, W.T., Mahaffy, P.R., Owen, T.C., Spencer,

N.W. and Way, S.H.: 1996, "The Galileo probe mass spectrometer: Composition of Jupiter's atmosphere", *Science* **272**, 846-848.

Robert, F., Javoy, M., Halbout, J., Dimon, B. and Merlivat, L.: 1987a, "Hydrogen isotope abundances in the solar system. Part I. Unequilibrated chondrites", *Geochim. Cosmochim. Acta* **51**, 1787-1805.

Robert, F., Javoy, M., Halbout, J., Dimon, B. and Merlivat, L.: 1987b, "Hydrogen isotope abundances in the solar system. Part II. Meteorites with terrestrial-like D/H ratio", *Geochim. Cosmochim. Acta*, **51**, 1807-1822.

Songaila, A., Wampler, E.J. and Cowie, L.L.: 1997, "A high deuterium abundance in the early universe", *Nature* **385**, 137-139.

Webb, J.K., Carswell, R.F., Lanzetta, K.M., Ferlet, R., Lemoine, M., Vidal-Madjar, A. and Bowen, D.V.: 1997, "A high deuterium abundance at redshift z = 0.7", *Nature* **388**, 250-252.

Wiens, R.C., Huss, G.R. and Burnett, D.S.: 1999, "The solar system oxygen-isotopic composition: Predication and implications for solar nebula processes", *Meteoritics. Planet. Sci.* **34**, 99-107.

Wood, J.A.: 1999, "Forging the planets: The origin of our solar system", *Sky and Telescope*, **97**, 36-48.

# The Possible Role of PeP Weak Interactions in the Early History of the Earth

Thomas E. Ward
*U.S. Department of Energy, Washington, DC 20585*
*Thomas.Ward@ns.doe.gov*

Abstract: The $p(pe^-, \nu)d$ reaction (Q = 1.442 MeV) is examined in view of the possible Gamow Factor Cancellation (GFC) theorized by Kim and Zubarev (1995). The lifetime for the pep reaction in a metal hydride environment, such as FeH in the core of the earth, was calculated to be 2.54E(+12) yrs. The duration of the earth's melting and degassing prior to the formation of the oceans can be calculated using the present ratio D/H = 1.5E(-4) for the oceanic value, the pep lifetime in Fe and the primordial ratio of D/H = 2.6E(-5). The time interval was calculated to be 630 ± 160 million years, yielding an oceanic formation age of 3.92 billion years ago. The $d(pe^-, \nu)t$ reaction (Q = 5.474 MeV) was also examined in view of the subsequent decay of tritium to $^3He$ and its possible role as a radiochronometer.

## 1. INTRODUCTION

The pep reaction lifetime in the sun is approximately 4.34E(+12) years because (a) the core temperature (energy < 1.35 keV), pressure and particle densities are sufficiently high to allow a substantial reaction rate between the three free particles, and (b) the effect of the quantum tunneling through the coulomb barrier of the two charged protons, the Gamow penetrability factor (Bahcall, 1989; Clayton, 1968). It is not at all obvious that this process should proceed in low temperature geologic settings. However, recent theoretical optical model investigations into ultra-low energy two- and three-body nuclear interactions by Kim and Zubarev (1995, 1996, 1997) indicate that a combination of Thomas-Fermi electronic shielding (Lindhard *et al.*, 1968; Zeigler *et al.*, 1985) coupled with a very weak long range attractive imaginary (elastic + fusion) scattering potential allows for the Gamow Factor Cancellation, the GFC effect. Additionally, mineral physics of the lattice vibrations thermally excited above the Debye temperature allow for particle-hole excitations within the electronic conduction band (Poirier, 1991; Kittel, 1986) uniquely producing highly mobile protons which share the electronic environment of the lattice. The added benefit of a long range attractive interaction (LRAI) in the imaginary scattering potential (Kim and Zubarev, 1995, 1996, 1997) is the formation of an internal electric field ($E_{int}$) in the lattice given by

$$E_{int} = -\nabla U_{sc} - \frac{\partial A_c}{\partial t} \qquad (1)$$

with $U_{sc}$ the electronic screening potential and $A_c$ the Fermi contact term or potential for the electron-proton at $r = 0$ (Yosida, 1996; Low, 1997), the LRAI. This form of the induced internal electric field is commonly known as a displacement potential.

In this paper I will apply the Kim-Zubarev GFC mechanism with a LRAI to investigate the pep and ped weak interaction decay lifetimes in the earth's early geological environment when melting, degassing and differentiation were occurring to create the first oceans and atmosphere. The premise is that the earth's primordial hydrogen, *chemically bound as hydrides* in either the Fe core or the Fe-silicate mantel, decayed into deuterium via pep weak interaction fusion during the time interval between the accretion of the earth 4.55 billion years ago and the degassing and formation of the early oceans and atmosphere. The basic assumption is that the majority of pep reactions occur in either the earth's Fe core or Fe-silicate mantel at thermal energies above the Debye temperature (470-750 K), and that the rate is largely determined by the Thomas-Fermi electron screening energy of the protons in Fe, the GFC effect and the proton density.

The $d(pe^-, \nu)t$ weak interaction fusion reaction was also examined because the subsequent tritium beta decay to $^3He$ has important consequences with regard to a possible radiogenic component. The question of whether radiogenic $^3He$ has contributed appreciably to the $^3He/^4He$ ratio anomalies observed between the atmospheric, oceanic seawater, Mid-Oceanic Ridge Basalts (MORB), Oceanic Island Basalts (OIB) and deep mantel "hot-spot" plumes is an interesting and timely one (see commentary Ladbury, 1999). The standard interpretation of helium isotopic anomalies reported over the past three decades has been based on a "primordial or protosolar" $^3He$ content of ocean basalts and deep mantle plumes (Clarke *et al.*, 1969; Craig *et al.*, 1975; Lupton and Craig, 1975; Kurz *et al.*, 1982; McKenzie and O'Nions, 1983; Kaneoka, 1983; Allegre *et al.*, 1983; O'Nions and Oxbugh,

1983; Jochum *et al.*, 1983; Porcelli and Wasserburg, 1995; Rocholl *et al.*, 1996; Patterson *et al.*, 1997; Niedermann *et al.*, 1997; Hoffmann, 1997; Kamijo *et al.*, 1998; Eiler *et al.*, 1998; Zhang, 1998; Hanyu *et al.*, 1999; Pedroni *et al.*, 1999). The interpretation of a possible radiogenic source of $^3He$ could help in resolving major differences between the atmospheric, MORB and plume helium anomalies and in better defining the D/H, U and Th abundances in the upper and lower crust (Wedepohl, 1995) and in the differentiated mantle (See references 1-13 in Jochum *et al.*, 1983).

## 2. RESULTS AND DISCUSSIONS

*Mineral Physics and Lattice Vibrations.* At temperatures above the Debye temperature ($\theta_D$) of a metal or ceramic hydride, hydrogen becomes mobile as a proton hole state with the electrons occupying the conduction band. The ceramic perovskites with oxygen vacancies created at elevated temperatures readily absorb water, followed by the hydroxyl groups filling the vacancies thus creating mobile protons at temperatures above $\theta_D \approx 750\ K$ (Navrotsky, 1999; Kreuer, 1997, 1996; Norby, 1990; Nowick and Du, 1995; Bose and Navrotsky, 1998). The high density and short O-O distance in MgSiOxides favors hydration and proton migration. Thermally excited proton-electron particle-hole continuum states with hopping frequencies > 1E(+13) per sec are effectively developed by electronic screening potentials of the order of a keV in the vibrational lattice of Fe materials commonly found in geological settings, such as the earth's Fe core or in the mantle perovskites, Mg(Fe)SiOxides. The screening potential yields an "effective energy or temperature" for the pep or ped reactions in the metal or metal-silicate. The electronic screening potential for protons in Fe was calculated to be 2.496 keV (See equation 5 below), a value greater than the equivalent central core temperature of the sun (< 1.35 keV).

*Optical Model Formulation of Ultra-Low Energy Fusion Reactions.* For non-resonance reactions which are relevant to primordial and stellar nucleosynthesis at low energies (a few keV), as in the weak interaction fusion reactions, the cross-section for production is given by the following very well known formula (Kim and Zubarev, 1995, 1996, 1997; Bahcall, 1989; Clayton, 1968),

$$\sigma_G(E) = \frac{S(E)}{E} e^{-2\pi\eta(E)} \tag{2}$$

where $\eta(E) = Z_a Z_b e^2/\hbar v$ and $e^{-2\pi\eta(E)}$ is the Gamow factor representing the probability of bringing two charged nuclei to zero separation distance, and

S(E) is expected to be a slowly varying function of E. The GFC effect results from the appearance of a very weak long range attractive interaction (LRAI) in the imaginary part of the T-matrix of the Optical Model. The proof of separability of the imaginary part of the T-matrix, $T_{Im}$ = $\mathrm{Im}\langle r| T| r'\rangle = -U_l(r,r')$, for the S-wave ($l=0$) elastic and fusion channels is given in the theory by Kim and Zubarev (1995, 1996, 1997) with further formulation given below. $U_0(r,r')$ can be further parameterized by λ (strength/length) and $\beta^{-1}$ (range) in a separable form for estimating the S-wave cross-section, $\sigma_0(E)$, for the two-channel case, as in $U_0(r,r') = \lambda g(r)g(r')$. Thus, using $g(r) = e^{-\beta r}/r$, the low energy zero energy cross-section, Equation (2), becomes

$$\sigma_{\wedge}(E)=\frac{4\pi\lambda}{kE}\left[\int_0^\infty dr\psi_0^c(r)e^{-\beta r}/r\right]^2=\frac{4\pi^2\lambda}{E}R_b\frac{(e^{-2\phi\eta}-1)^2}{(e^{2\pi\eta}-1)}e^{4\phi\eta} \tag{3}$$

where $e^{4\phi\eta}=exp\left[4\alpha\frac{\mu c^2}{\hbar c}(Z_aZ_b/k)tan^{-1}\left(\frac{k}{\beta}\right)\right]$, $\phi$ = $\tan^{-1}(k/\beta)$, and the

Bohr radius of the system is $R_B=\hbar^2/(2\mu Z_aZ_be^2)$. The energy dependence of λ is expected to be weak and several general forms of g(r) also lead to the same enhancement factor $e^{4\pi\eta}$, with the following result for $\sigma_0(E)$,

$$\sigma_0(E)=\frac{S_0(E)}{E}e^{4fh}e^{-2ph} \tag{4}$$

The enhancement factor $e^{4\phi\eta}$ is $e^{2/R_B\beta}$ at zero energy and decreases, as E increases, to $e^{\pi/kR_B}$ = 1 for large E. The important result is that in the limit $\beta\rightarrow 0,\phi=tan^{-1}(k/\beta)=\pi/2$, and $e^{4\phi\eta}=e^{2\pi\eta}$ which just cancels the Gamow factor, $e^{-2\pi\eta}$, for zero energy reactions.

Above the Debye temperature of the material the mobile protons share the electronic environment of the host lattice and the S-factor becomes modified accordingly, $S_G(E)\approx f_U(E)S(E)$ where an enhancement factor $f_U\approx$ $\exp(\pi\eta U_{sc}/E)$ is due to the electron screening effect on the protons in the host metal hydride. $U_{sc}$ is the Thomas-Fermi (Lindhard *et al.*, 1968; Zeigler *et al.*, 1985) screening energy which is given by

$$U_{sc}=30.7Z_aZ_b(Z_a^{2/3}+Z_b^{2/3})^{1/2}eV \tag{5}$$

with $Z_a$ = 1 and $Z_b$ = 26, yielding a screening potential energy of 2.496 keV for the three body system, $ppe^-$, in Fe or Fe-Silicate lattice. Note the two body reactions, $pp \rightarrow dve^+$ and $pd \rightarrow {}^3He\gamma$, are not allowed since the screening potential between the two protons without the assistance of the electronic Fermi contact potential in Fe, is only 43.4 eV. This "effective energy or temperature" is insufficient to produce a noticeable GFC effect, $e^{-2\pi\eta} \approx$ 1.4E (-21) compared with 1.3E(-4) with shielded protons in Fe.

The pep reaction rate at "effective energies or temperatures" comparable to the solar plasma, 3-15(E+6) $K$ (260-1300 eV), is given by:

$$R_{pep} \cong 1.102E(-4)\left(\rho/\mu_e\right)T_6^{-1/2}(1+0.02T_6)R_{pp} \tag{6a}$$

or reduced to,

$$R_{pep} = \chi_{pep} R_{pp} \tag{6b}$$

where $\mu_e$ is the mean molecular weight per electron, $\rho$ the local density (Bahcall, 1989) and $\chi_{pep}$ is understood to represent the quantities in front of $R_{pp}$ in equation(6a). The proton-proton rate, $R_{pp}$, is given by:

$$R_{pp} = \frac{n_a n_b}{1+\delta ab}\langle\sigma v\rangle \tag{7}$$

where $n_{a,b}$ are the particle densities, $1 + \delta_{ab}$ is the Kronecker delta which prevents double counting and the average cross section times velocity is approximated by $\langle\sigma v\rangle = \sigma v$ with $\sigma(E)$ given by equation (4). Equation (4) is modified appropriately to include the electron screening enhancement factor, $f_u \approx e^{2\pi\eta U_{sc}/E}$, for this comparison of the rates in the earth ($E_e = U_{sc}$) and the sun ($E_{sun}$).

The comparison of the rates, $R_{pep\,(earth)}/R_{pep\,(sun)}$, reduces to the following equation:

$$R_{pep}^{earth} = \frac{1}{1.01E(+10)yrs}\left(\frac{\chi_{earth}}{\chi_{sun}}\right)\left(\frac{n_{a,b}^e}{n_{a,b}^s}\right)\left(\frac{E_{sun}}{U_{sc}}e^{2\pi\eta U_{sc}/E_{sun}}\right)\left(\frac{v_e}{v_s}\right) \tag{8}$$

The energies are converted to $T_6$ temperatures, $U_{sc} = k_b T_6$, the mean molecular weight per electron (earth) = 2, average mantle density ranging between $\rho$ = 3.56 to 4.11 $g/cm^3$ and the mantle proton particle density $n_{a,b}^e$ = 1.44 to 1.65E(+22), equivalent to one conduction electron or particle-hole state per perovskite molecule, a 0.6% proton fraction. The solar parameters were taken from Bahcall (1989). See his tables 4.4 and 4.5. The final result is a calculated pep weak interaction fusion lifetime in the mantle of the earth of 2.54(38)E(+12) years which is comparable to the lifetime in the sun, 4.34E(+12) years. The uncertainty, in parentheses is approximately ± 15% based on a 10% uncertainty in the overall pep solar rate (Bahcall, 1989) and the range of mantle densities (Poirier, 1991).

*The Formation Age of the Oceans.* The formation interval (Ξ) between the accretion of the earth 4.55 Gyr ago and the degassing and formation of the early oceans and atmosphere was calculated assuming the difference between seawater, D/H = 1.5(1)E(-4), and the best estimate (Geiss and Reeves, 1972) for protosolar gas, D/H = 2.5E(-5), or the measured Jovian atmospheric isotope ratio (Nieman *et al.*, 1996), D/H = 2.6(7)E(-5), resulted from the pep reaction in the earth's early history. The formation interval equation is given as:

$$\delta(D/2H) = (1-e^{-\lambda\Xi}) \tag{9}$$

where $\lambda = \tau_{pep}^{-1}$ and $\tau_{pep}$ = 2.54(38)E(+12) years and δ(D/2H) = 2.48(32) E(-4). Using the Jovian measurement value (Nieman *et al.*, 1996) yields, Ξ = 630 ± 160 Myr. Half of the estimated uncertainty is from the measured uncertainties in the D/H ratios and the remainder from the pep lifetime calculation (Bahcall, 1989). The formation age of oceans can be estimated using the accretion age of the solar system of 4556 ± 2.0 Myr (Lugmair and Shukolyukov, 1998) and the formation interval calculated above, yielding 3.92 Gyr before the present. This age agrees well with the appearance of the first ancient continental crust or cratons and other geologic evidence such as the water-lain sediments dated approximately to 3.8 Gyr ago (Titayeva, 1994).

*The d(pe⁻,ν)t Lifetime and the Helium Anomalies.* Following the formation of the oceans and the development of a differentiated core, mantle and crust the pep and ped weak interactions would continue to produce radiogenic deuterium and tritium (and subsequently $^3$He). The ped reaction lifetime is much shorter than the pep lifetime due to the greater beta decay energy. The nuclear beta matrix elements are constant and the decay constant follows the well known power law, $\tau^{-1} \propto E_\beta^{4.697}$. The ped lifetime compared with the pep reaction is estimated to be $[(E_{pep}/E_{ped})^{4.697} = 9.5E(-4)]$ 1/2 $\tau_{pep}$ = 2.43E(+9) years and is expected to be in secular equilibrium with the pep reaction.

Comparing the $^3He/^4He$ ratios for the atmosphere ($R_{air}$ = 1) to the average MORB ($R/R_{air}$ = 8.18 ± 0.73) (*cf* compilation by Kerr, 1999) one could infer that the upper mantle has been degassed and the MORB source represents a depleted upper/lower mantle. The increased He ratio would therefore represent a decreased average U and Th abundance relative to D/H. Using the averaged crustal values (Wedepohl, 1995) of 2.7 ppm U and 8.5 ppm Th the MORB ratio would decrease these values to 333 ppb and 1039 ppb, respectively. Alternatively Jochum *et al.* (1983) have measured N-type MORB concentration averages of 75 ppb U and 189 ppb Th. This approach warrants further investigation.

The deep mantle plumes with $R_{plume}/R_{air} \approx 30$-43 would seem to indicate that either "primordial" $^3$He locked within the undegassed deep mantle (the accepted theory) or a further depletion of U and Th abundances (less likely) was occurring to increase the ratio substantially to what is now observed for MORB. Recent models (Kerr, 1999; van der Hilst and Karason, 1999; Kellogg *et al.*, 1999; Kaneshima and Helffrich, 1999) indicate that the deep mantle may be isolated from the upper portions over periods of several billion years and is not depleted relative to the upper portions of the mantle and crust. If that is the case, then the weak interaction fusion reactions considered here could provide a new radiogenic point of view. Assuming the difference in $R/R_{air}$ of the plumes and MORB is not due to U, Th and D/H concentration changes but rather that MORB is degassed whereas the deep mantle plumes are not, then one can calculate the increased concentration of radiogenic $^3He$ over the past 4.55 Gyr in the deep mantle. The plume to MORB ratio is calculated to be $R_{plume}/R_{MORB} = 5.40 \pm 1.35$ using the pep(dep) lifetime. Correcting for the decay of U and Th over the last 4.55 Gyr yields a plume ratio of $R_{plume}/R_{air} \approx 44 \pm 11$, a value in good agreement with what is observed experimentally.

## 3. SUMMARY

The pep and ped hydride lifetimes in the earths interior were calculated using the Kim-Zubarev optical model approach. A LRAI in the imaginary potential results in the GFC effect and a screening potential enhancement factor for protons in Fe. The pep lifetime was calculated to be 2.54E(+12) years which together with the net D/H ratio in seawater was used to calculate the time interval between the earth's accretion 4.55 Gyr ago and the formation of the first oceans and atmosphere. The time interval was calculated to be $\Xi = 630$ Myr, yielding a formation age of 3.92 Gyr ago. The He anomalies associated with the MORB, OIB and deep mantle plumes can be accounted for by assuming a radiogenic $^3$He component and a combination of mantle depletion of U and Th and degassing of the upper/lower mantle. A well separated deep mantle which has not been degassed since the first differentiation into deep and upper/lower mantle components has a much larger He isotopic anomaly than the MORB values.

## ACKNOWLEDGMENTS

I would like to thank Prof. Yeong Kim and Dr. Alex Zubarev for valuable discussions and suggestions. This work was supported in part by the U.S. Department of Energy.


## REFERENCES


Allegre, J., Staudacher, T., Sarda, P. and Kurz, M.: 1983, "Constraints on evolution of Earth's mantle from rare gas systematics", *Nature* **303**, 762-766.

Bahcall, J.: 1989, *Neutrino Astrophysics*, Cambridge University Press, New York, NY, 567 pp. (See Chapter 3).

Bose, K. and Navrotsky, A.: 1998, "Thermochemistry and phase equilibria of hydrous phases in the system MgO-$SiO_2$-$H_2O$: Implications for volatile transport to the mantle", *J. Geophys. Res.* **103**, 9713-9719.

Clarke, W., Beg, M. and Craig, H.: 1969, "Excess He-3 in the sea: Evidence for terrestrial primordial helium", *Earth Planet Sci. Lett.* **6**, 213-220.

Clayton, D.: 1968, *Principles of Stellar Evolution and Nucleosynthesis*, McGraw Hill, New York, NY, 612 pp. (See Chapters 4 and 5).

Craig, H., Clarke, W. and Beg, M.: 1975, "Excess $He^3$ in deep water on the East Pacific Rise", *Earth Planet Sci. Lett.* **26**, 125-132.

Eiler, J., Farley, K. and Stolpher, E.: 1998, "Correlated helium and lead isotope variations in Hawaiian lavas", *Geochim. Cosmochim. Acta* **62**, 1977-1984.

Geiss, J. and Reeves, H.: 1972, "Cosmic and solar system abundances of deuterium and helium-3", *Astron. Astrophys.* **18**, 126-132.

Hanyu, T., Kaneoka, I. and Nagao, K.: 1999, "Noble gas study of HIMU and EM ocean island basalts in the Polynesian region", *Geochim. Cosmochim. Acta* **63**, 1181-1201.

Hoffmann, A.: 1997, "Mantle geochemistry: The message from oceanic volcanism", *Nature* **385**, 219-229.

Jochum, K., Hoffmann, A., Ito, E., Seufert, H. and White, W.: 1983, "K, U and Th in mid-ocean ridge basalt glasses and heat production, K/U and K/Rb in the mantle", *Nature* **306**, 431-436.

Kamijo, K., Hashizume, K. and Matsuda, J.-I.: 1998, "Noble gas constraints on the evolution of the atmosphere-mantle system", *Geochim. Cosmochim. Acta* **62**, 2311-2321.

Kaneoka, I.: 1983, "Noble gas constraints on the layered structure of the mantle", *Nature* **302**, 698-700.

Kaneshima, S. and Helffrich, G.: 1999, "Dipping low-velocity layer in the mid-lower mantle: Evidence for geochemical heterogeneity", *Science* **283**, 1888-1892.

Kellogg, L., Hager, B. and van der Hilst, R.: 1999, "Compositional stratification in the deep mantle", *Science* **283**, 1881-1886.

Kerr, R.: 1999, "A lava lamp model for the deep Earth", *Science* **283**, 1826-1827.

Kim, Y. and Zubarev, A.: 1995, "Optical theorem and finite-range effect for nuclear reactions in astrophysics", *Few-Body Systems Suppl.* **8**, 334-336.

Kim, Y. and Zubarev, A.: 1996, "Optical theorem and effective finite-range nuclear interaction for low-energy nuclear-fusion reactions", *Nuovo Cimento* **A108**, 1009-1018.

Kim, Y. and Zubarev, A.: 1997, *Neutrino '96, Proc. 17th International Conf. On Neutrino Physics and Astrophysics*, eds., Enqvist, K., Huitu, K. and Maalanpi, J., World Scientific, Singapore, pp. 120-126.

Kittel, C.: 1986, *Introduction to Solid State Physics*, 6th Edition, John Wiley and Sons, New York, NY, pp. 108-143.

Kreuer, K.D.: 1996, "Proton conductivity: Materials and applications", *Chem. Mater.* **8**, 610-641.

Kreuer, K.D.: 1997, "On the development of proton conducting materials for technological applications", *Solid State Ionics* **97**, 1-15.

Kurz, M., Jenkins, W. and Hart, S.: 1982, "Helium isotopic systematics of ocean islands and mantle heterogeneity", *Nature* **297**, 43-47.

Ladbury, R.: 1999, "Model suggests deep-mantel topography goes with the flow", *Physics Today* **52**, no. 8, 21-24.

Lindhard, J., Nielsen, V. and Scharff, M.: 1968, "Thomas-Fermi model", *Mat. Phys. Medd. Dan. Vid. Selsk* **36**, no. 10, 47 pp.

Low, F.: 1997, *Classical Field Theory*, John Wiley and Sons, New York, NY, pp. 24-66.

Lugmair, G. and Shukolyukov, A.: 1998, "Early solar system time scales according to $^{53}$Mn-$^{53}$Cr systematics", *Geochim. Cosmochim. Acta* **62**, 2863-2886.

Lupton, J. and Craig, H.: 1975, "Excess $^{3}$He in oceanic basalts: Evidence for terrestrial primordial helium", *Earth Planet Sci. Lett.* **26**, 133-139.

McKenzie, D. and O'Nions, R.: 1983, "Mantle reservoirs and ocean island basalts", *Nature* **301**, 229-231.

Navrotsky, A.: 1999, "A lesson from ceramics", *Science* **284**, 1788-1789.

Niedermann, S., Bach, W. and Erzinger, J.: 1997, "Noble gas evidence for a lower mantle component in MORBs from the southern East Pacific Rise: Decoupling of helium and neon isotope systematics", *Geochim. Cosmochim. Acta* **61**, 2677-2715.

Niemann, H.B., Atreya, S.K., Carignan, G.R., Donahue, T.M., Haberman, J.A., Harpold, D.N., Hartle, R.E., Hunten, D.M., Kasprzak, W.T., Mahaffy, P.R., Owen, T.C., Spencer, N.W. and Way, S.H.: 1996, "The Galileo probe mass spectrometer: Composition of Jupiter's atmosphere", *Science* **272**, 846-849.

Norby, T.: 1990, "Proton conduction in oxides", *Solid State Ionics* **40**, 857-862.

Nowick, A. and Du, Y.: 1995, "High-temperature protonic conductors with pervoskite-related structures", *Solid State Ionics* **77**, 137-146.

O'Nions, R. and Oxbugh, E.: 1983, "Heat and helium in the Earth", *Nature* **306**, 429-431.

Patterson, D., Farley, K. and McInnus, B.: 1997, "Helium isotopic composition of the Tabar-Lihir-Tango-Feni island arc, Papua New Guinea", *Geochim. Cosmochim. Acta* **61**, 2485-2496.

Pedroni, A., Hammerschmidt, K. and Freidricksen, H.: 1999, "He, Ne, Ar, and C isotope systematics of geothermal emanations in the Lesser Antilles Islands Arc", *Geochim. Cosmochim. Acta* **63**, 515-532.

Poirier, J.-P.: 1991, *Introduction to the Physics of the Earth's Interior*, Cambridge University Press, New York, NY, pp. 23-37.

Porcelli, D. and Wasserburg, G.: 1995, "Mass transfer of helium, neon, argon, and xenon through a steady-state upper mantle", *Geochim. Cosmochim. Acta* **59**, 4921-4937.

Rocholl, A., Heusser, E., Kirsten, T., Oehmad, J. and Richter, H.: 1996, "A noble gas profile across a Hawaiian mantle xenolith: Coexisting accidental and cognate noble gases derived from the lithospheric and astenospheric mantle beneath Oahu", *Geochim. Cosmochim. Acta* **60**, 4773-4783.

Titayeva, N.: 1994, *Nuclear Geochemistry*, MIR Publishers, CRC Press, Moscow, Russia, pp. 98-102.

van der Hilst, R. and Karason, H.: 1999, "Compositional heterogeneity in the bottom 1000 kilometers of Earth's mantle: Toward a hybrid convection model", *Science* **283**, 1885-1890.

Wedepohl, K.: 1995, "The composition of the continental crust", *Geochim. Cosmochim. Acta* **59**, 1217-1232.

Yosida, K.: 1996, *Theory of Magnetism*, Springer-Verlag, Berlin/Heidelberg, Germany, pp. 13-15.

Zeigler, J., Biersach, J. and Littmark, U.: 1985, *The Stopping and Range Ions in Solids*, Pergamon Press, New York, NY, 128 pp.

Zhang, Y.: 1998, "The young age of Earth", *Geochim. Cosmochim. Acta* **62**, 3185-3189.

# Ce-Nd-Sr Isotope Systematics of Eucrites and Lunar Rocks

Masaharu Tanimizu and Tsuyoshi Tanaka
*Department of Earth and Planetary Sciences, Graduate School of Science, Nagoya University, Chikusa, 464-8602, JAPAN*
*mash@gcl.eps.nagoya-u.ac.jp*

Abstract: The La-Ce data of Millbillillie are plotted in Figure 2. The age of Millbillillie #2 is 3.7 ± 2.0 Ga. The data for Millbillillie #2 do not fall on the 4.52 Ga reference isochron. There is a possibility that the La-Ce decay system was reset. The fine-grained crystalline clast of Millbillillie #1 made a cluster due to the difficulty in mineral separation. Two different trends between Millbillillie #1 and #2 are also recognized in Figure 2, as well as in the Sm-Nd decay system (Figure 1a) and clearer than it. This fact indicates that Millbillillie is a polymict eucrite, too.

## 1. INTRODUCTION

Long-lived radioactive nuclides are generally used for the age determination of rocks, and the initial isotope ratios provide important information about their origin. In particular, the Sm-Nd dating is essential to discuss the evolution of the earth and planets, as well as the Rb-Sr dating. The characteristic of the Sm-Nd dating is that it is less susceptible to metamorphism and alteration than the other datings, because both Sm and Nd belong to rare earth elements, which usually take the 3+ oxidation state and have similar chemical properties.

A radioactive nuclide, $^{138}$La, decays to $^{138}$Ce and $^{138}$Ba with decay constants of the order of $10^{-12}$/yr. The La-Ce decay system is the other decay system which belongs to rare earth elements. Though the La-Ce decay system was initiated by Tanaka and Masuda (1982), it is difficult to detect the small variation of $^{138}$Ce due to the low abundance of the parent nuclide and its long half-life.

Despite the difficulties, the Ce isotope ratio was measured by several investigators (*e.g.*, Nakamura *et al.*, 1984; Dickin *et al.*, 1987; Makishima and Nakamura, 1991). One of the reasons is that the chemical behavior of Ce is sometimes different from the other rare earth elements because Ce can take the 4+ oxidation state. This results in a Ce anomaly. The Ce anomaly is prominently seen in manganese nodules, cherts and limestone, which are formed in an aquatic environment. A small Ce anomaly is also recognized in some meteorites (*e.g.*, Masuda and Tanaka, 1980) and lunar samples (*e.g.*, Masuda *et al.*, 1972). The Ce isotope ratio must be a key to understand the date and formation mechanism of these Ce anomalies. Especially, it is interesting that some lunar samples have Ce anomalies which are the result of oxidation despite the reduced nature of the moon today. We use the La-Ce decay system to examine the origin of these rocks.

Recently, we developed a Ce isotope measurement using the dynamic multicollector technique. This is the most accurate technique among those used for thermal ionization mass spectrometer (TIMS). Using this technique, the $^{138}$La $\beta$-decay constant was determined and evaluated as 2.32 $\times 10^{-12}$/yr (Tanimizu, 2000).

In this study, La-Ce, Sm-Nd and Rb-Sr decay systems are applied to three lunar rock samples (10017, 14310 and 75015) and two eucrites. Eucrites are considered as the primary basaltic lavas which covered the surface of a proto-planet. The Ce, Nd and Sr isotope ratio of eucrites are also examined to understand the earliest lunar evolution, because the moon and the proto-planet must have experienced an analogous formation process.

## 1.1 Samples and analytical procedure

As eucrite samples, Millbillillie and Camel Donga were examined. Millbillillie contains various different textures (Yamaguchi *et al.*, 1994) and these will be mainly classified into three groups: coarse-grained crystalline clast, fine-grained crystalline clast and matrix. Camel Donga is a unique eucrite in including metallic iron, but its chemical composition, mineralogy and noble gas contents are indistinguishable from other eucrites (Palme *et al.*, 1988). As lunar samples, two high Ti mare basalts 10017 and 75015 and a KREEP basalt 14310 were examined. The presence of a Ce anomaly in 14310 was reported by Masuda *et al.* (1972). We used a part of the sample solutions of 14310 and 75015 which were analyzed by Tanaka *et al.* (1985; 1986).

Two chips of Millbillillie (#1 and #2) and the lunar sample 10017 were separated into minerals with heavy liquids and a hand magnet. Millbillillie #1 consists of the fine-grained crystalline clast and #2 consists mainly of the coarse-grained crystalline clast and includes a vein-like matrix. The vein-like matrix was removed by hand-picking as much as possible after a coarse crushing with an agate mill. Samples were then further crushed and sieved into several fractions. Each fraction was separated into plagioclase ($\rho <$ 2.85; non-magnetic portion) and pyroxene ($\rho >$ 3.33; in the magnetic portion). The remainder of each faction was labeled as "mix" and analyzed, too. The lunar sample 10017 was also crushed and sieved, after a sample was taken for whole rock analysis. Each sieved fraction was separated into plagioclase (2.70 < $\rho$ < 2.85; non-magnetic portion), ilmenite ($\rho >$ 3.33; in the magnetic portion), pyroxene ($\rho >$ 3.33; in the less magnetic portion) and phosphate including fraction (2.85 < $\rho$ < 3.33; non-magnetic portion).

All samples were decomposed by HF, $HClO_4$ and $HNO_3$. The elements: Rb, Sr, La, Ce, Nd and Sm were refined with the conventional column chemistry. The $^{138}Ce/^{142}Ce$, $^{143}Nd/^{144}Nd$, $^{142}Nd/^{144}Nd$ and $^{87}Sr/^{86}Sr$ isotope ratios were measured with a TIMS, VG Sector 54-30. The mass fractionation of each isotope ratio during the measurement was corrected with $^{140}Ce/^{142}Ce$ = 7.941, $^{146}Nd/^{144}Nd$ = 0.7219 and $^{86}Sr/^{88}Sr$ = 0.1194, respectively. The elemental concentrations were determined with the isotope dilution technique. The values of the decay constant of $^{138}La$ to $^{138}Ce$ , $^{138}La$ to $^{138}Ba$, $^{147}Sm$ to $^{143}Nd$ and $^{87}Rb$ to $^{87}Sr$ used in this study were $2.32 \times 10^{-12}$/yr, $4.41 \times 10^{-12}$/yr, $6.54 \times 10^{-12}$/yr and $1.42 \times 10^{-11}$/yr, respectively. All previous isochron ages were recalculated with these values for the following discussion. Details of Ce isotope measurement were described in Tanimizu (2000).

## 2. RESULTS AND DISCUSSION

### 2.1 La-Ce, Sm-Nd and Rb-Sr decay systems in eucrites

The results of eucrites are summarized in Table 1a, 1b and 1c on the following two pages. Following these tables, the $^{147}Sm$-$^{143}Nd$ isochron diagram of eucrites is shown in Figure 1a. The data of Millbillillie #2 define an internal isochron age corresponding to 4.52 ± 0.06 Ga with an initial $^{143}Nd/^{144}Nd$ of 0.50654 ± 0.00009. This initial ratio is the lowest value among those of other eucrites summarized by Faure (1986), though the difference is within the error. The plot of fine fraction of Millbillillie #2 was ignored in this calculation, because it may include small particles of "vein-like matrix". The whole rock Sm-Nd data of Millbillillie were reported by Makishima and Masuda (1993). Their data also suggested its low initial $^{143}Nd/^{144}Nd$ ratio. This may be the result of a slight isotopic heterogeneity in the early solar nebula.

*Table 1a.* La-Ce isotopic result of eucrites

| | La (ppm) | Ce (ppm) | $^{138}La/^{142}Ce$ | $^{138}Ce/^{142}Ce$ |
|---|---|---|---|---|
| Millbillillie #1 | | | | |
| plagioclase (< 325 mesh) | 1.20 | 2.88 | 0.00339 | - |
| pyroxene (150-200 mesh) | 1.62 | 4.36 | 0.00304 | 0.0225840(17) |
| pyroxene (200-325 mesh) | 1.73 | 4.70 | 0.00300 | 0.0225856(6) |
| pyroxene (< 400 mesh) | 1.88 | 5.14 | 0.00298 | 0.0225835(11) |
| mix (325-500 mesh) | 1.95 | 5.10 | 0.00312 | 0.0225863(11) |
| mix (< 400 mesh) | 2.21 | 5.80 | 0.00311 | 0.0225866(13) |
| fine fraction (< 500 mesh) | 3.95 | 10.41 | 0.00310 | 0.0225862(6) |
| Millbillillie #2 | | | | |
| plagioclase (80-150 mesh) | 1.33 | 3.20 | 0.00338 | 0.0225914(14) |
| plagioclase (150-500 mesh) | 1.13 | 2.52 | 0.00367 | 0.0225951(19) |
| pyroxene (80-150 mesh) | 0.795 | 2.55 | 0.00255 | 0.0225869(14) |
| pyroxene (150-400 mesh) | 0.794 | 2.60 | 0.00250 | 0.0225837(9) |
| less magnetic fraction | 0.947 | 3.07 | 0.00252 | 0.0225833(17) |
| fine fraction (< 500 mesh) | 1.96 | 5.71 | 0.00280 | - |
| vein-like matrix | 1.05 | 3.24 | 0.00265 | - |
| Camel Donga | 3.71 | 9.71 | 0.00312 | 0.0225851(27) |

The values in parentheses are error limits and apply to the last digits.
The uncertainty of $^{138}La/^{142}Ce$ is +/- 0.5%.
The average value of a Ce isotope reference, JMC304, is 0.0225889 +/- 0.0000016 ($1\sigma$) for 250 ratios.

*Table 1b.* Sm-Nd isotopic result of eucrites

| | Nd (ppm) | Sm (ppm) | $^{147}Sm/^{144}Nd$ | $^{143}Nd/^{144}Nd$ | $^{142}Nd/^{144}Nd$ |
|---|---|---|---|---|---|
| Millbillillie #1 | | | | | |
| plagioclase (< 325 mesh) | 1.910 | 0.5630 | 0.1781(11) | 0.512045(9) | 1.141786(32) |
| pyroxene (150-200 mesh) | 3.426 | 1.176 | 0.2074(15) | 0.512768(7) | 1.141807(21) |
| pyroxene (200-325 mesh) | 3.694 | 3.773 | 0.2153(14) | 0.513078(7) | 1.141828(23) |
| pyroxene (< 400 mesh) | 4.033 | 1.408 | 0.2109(12) | 0.513035(8) | 1.141835(21) |
| mix (325-500 mesh) | 3.813 | 1.286 | 0.2039(11) | 0.513016(8) | 1.141829(10) |
| mix (<400mesh) | 4.342 | 1.423 | 0.1980(13) | 0.512650(7) | 1.141854(21) |
| mix (<500mesh) | 7.666 | 2.414 | 0.1903(14) | 0.512372(9) | 1.141833(27) |
| Millbillillie #2 | | | | | |
| plagioclase (80-150) | 2.108 | 0.6049 | 0.1734(8) | 0.511725(8) | 1.141818(25) |
| plagioclase (150-500) | 1.612 | 0.4417 | 0.1656(10) | 0.511530(9) | 1.141800(27) |
| pyroxene (80-150 mesh) | 2.451 | 1.001 | 0.2468(13) | 0.513950(7) | 1.141850(16) |
| pyroxene (150-400 mesh) | 2.555 | 1.075 | 0.2541(19) | 0.514149(7) | 1.141880(23) |
| less magnetic fraction | 3.069 | 1.283 | 0.2526(19) | 0.514120(8) | 1.141851(25) |
| fine fraction (< 500) | 4.403 | 1.441 | 0.1977(13) | 0.512558(8) | 1.141801(11) |
| vein-like matrix | 2.525 | 0.9086 | 0.2158(9) | 0.513424(9) | 1.141859(30) |
| Camel Donga | 7.459 | 2.374 | 0.1923(13) | 0.512587(6) | 1.141789(16) |

The values in parentheses are error limits and apply to the last digits.
The average $^{143}Nd/^{144}Nd$ values of Nd isotope references, La Jolla Nd and JNdi-1, are 0.511849 +/- 0.000005 ($1\sigma$) and 0.512104 +/- 0.000003 ($1\sigma$), respectively.
The average $^{142}Nd/^{144}Nd$ values of Nd isotope references are written in text.

*Table 1c.* Rb-Sr isotopic result of eucrites

| | Rb (ppm) | Sr (ppm) | $^{87}Rb/^{86}Sr$ | $^{87}Sr/^{86}Sr$ |
|---|---|---|---|---|
| Millbillillie #1 | | | | |
| plagioclase (< 325 mesh) | 1.35 | 167 | 0.0235 | 0.700430(17) |
| pyroxene (150-200 mesh) | 0.0903 | 19.7 | 0.0132 | 0.699955(15) |
| pyroxene (200-325 mesh) | 0.103 | 21.7 | 0.0137 | 0.700033(14) |
| pyroxene (< 400 mesh) | 0.0641 | 16.7 | 0.0111 | 0.699992(15) |
| mix (325-500 mesh) | 0.517 | 89.3 | 0.0168 | 0.700052(14) |
| mix (< 400 mesh) | 0.572 | 84.1 | 0.0197 | 0.700228(39) |
| fine fraction (< 500 mesh) | 0.696 | 106 | 0.0191 | 0.700185(14) |
| Millbillillie #2 | | | | |
| plagioclase (80-150 mesh) | 0.538 | 170 | 0.00917 | 0.699607(17) |
| plagioclase (150-500 mesh) | 0.537 | 179 | 0.00866 | 0.699573(77) |
| pyroxene (80-150 mesh) | 0.0850 | 11.0 | 0.0224 | 0.700619(15) |
| pyroxene (150-400 mesh) | 0.0885 | 13.2 | 0.0195 | 0.700445(27) |
| less magnetic fraction | 0.139 | 26.5 | 0.0152 | 0.700149(20) |
| fine fraction (< 500 mesh) | 0.422 | 113 | 0.0108 | 0.699765(18) |
| vein-like matrix | 0.240 | 70.4 | 0.00987 | 0.699853(14) |
| Camel Donga | 0.190 | 79.4 | 0.00694 | 0.699452(7) |

The values in parentheses are error limits and apply to the last digits. The uncertainty on the $^{87}Rb/^{86}Sr$ is ± 1%. The average value of a Sr isotope reference, NIST SRM 987, is 0.710250 +/- 0.000011 (1σ).

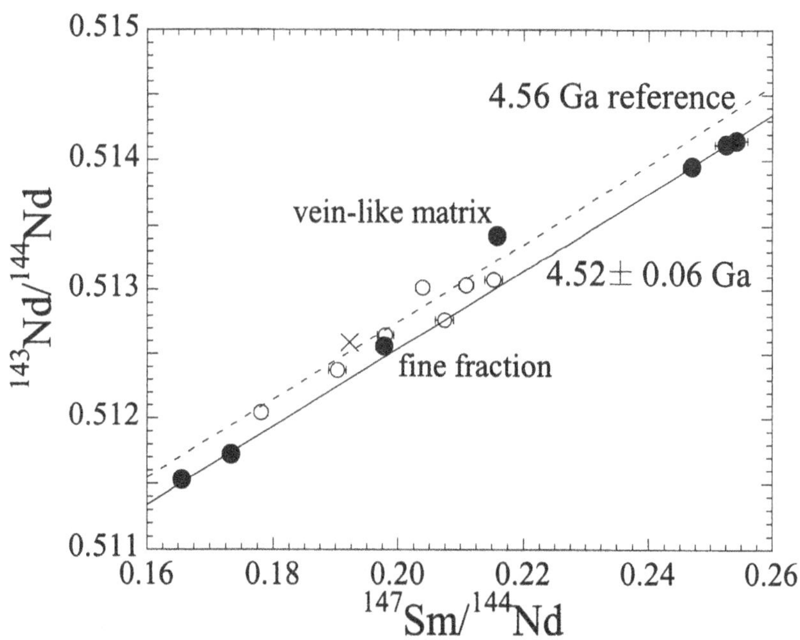


*Figure 1a.* $^{147}Sm$-$^{143}Nd$ isochron diagram of eucrites. Open circles are the fine-grained crystalline clast, Millbillillie #1. Solid circles are the coarse-grained crystalline clast, Millbillillie #2. A cross mark is Camel Donga. The data of Millbillillie #2 defines an internal isochron which corresponds to the age of 4.52 ± 0.06 Ga. The points of the fine fraction and "vein-like matrix" were omitted for the calculation (see text). Those of Millbillillie #1 were scattered above the 4.52 Ga isochron. These two trends indicate that Millbillillie is a polymict eucrite. A dashed line is the 4.56 Ga reference isochron (Makishima and Masuda, 1993).

On the contrary, the plots of the fine-grained crystalline clast, Millbillillie #1, were scattered above the isochron of Millbillillie #2. Millbillillie #1 seems disturbed and partly reset after its primary crystallization. These two trends indicate that Millbillillie is a polymict eucrite. The point of "vein-like matrix" was also plotted above the Millbillillie #2 isochron. This means that the coarse-grained crystalline clast and the "vein-like matrix" have different origins.

The *in situ* decay of $^{146}$Sm to $^{142}$Nd is recognized in Millbillillie #2 (Figure 1b). $^{146}$Sm is one of the extinct nuclides and it decays to $^{142}$Nd with a half-life of $103\times10^6$ yr. The initial $^{146}$Sm/$^{144}$Sm was calculated to be 0.0031 ± 0.0013 from the slope in Figure 1b. This value is almost consistent with those reported by Prinzhofer *et al.* (1992). This $^{146}$Sm/$^{144}$Sm value means that Millbillillie #2 had finally been crystallized, at least before 4.0 Ga.

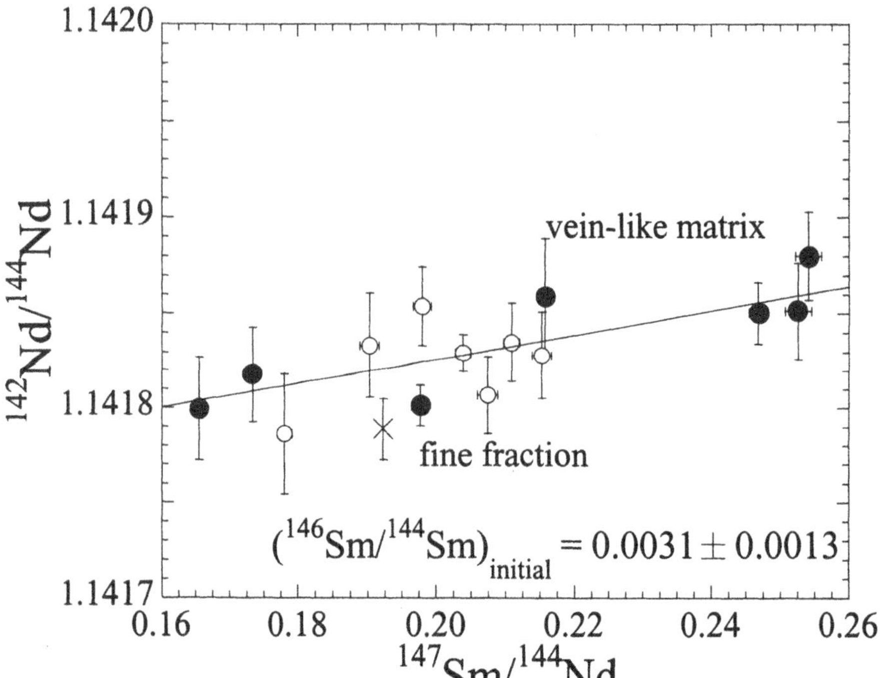


*Figure 1b.* $^{146}$Sm-$^{142}$Nd evolution diagram of eucrites. The symbols are identical with those of Figure 1a. The in situ decay of $^{146}$Sm to $^{142}$Nd was recognized in Millbillillie #2 and the initial $^{146}$Sm/$^{144}$Sm and $^{142}$Nd/$^{144}$Nd ratios were calculated to be 0.0031 ± 0.0013 and 1.1417 ± 0.0006. This $^{146}$Sm/$^{144}$Sm value means that Millbillillie #2 had been generated at 4.0 Ga. The $^{142}$Nd/$^{144}$Nd ratios of La Jolla Nd and JNdi-1 during this study are 1.141832 ± 0.000029 ($1\sigma$, 5 runs) and 1.141835 ± 0.000022 ($1\sigma$, 11 runs), respectively. The ion beam intensity of $^{144}$Nd was set to $0.5\times10^{-11}$ A and the data acquisition time was about three hours.

The La-Ce data of Millbillillie are plotted in Figure 2. The age of Millbillillie #2 is 3.7 ± 2.0 Ga. The data for Millbillillie #2 do not fall on the

4.52 Ga reference isochron. There is a possibility that the La-Ce decay system was reset. The fine-grained crystalline clast of Millbillillie #1 made a cluster due to the difficulty in mineral separation. Two different trends between Millbillillie #1 and #2 are also recognized in Figure 2, as well as in the Sm-Nd decay system (Figure 1a) and clearer than it. This fact indicates that Millbillillie is a polymict eucrite, too.

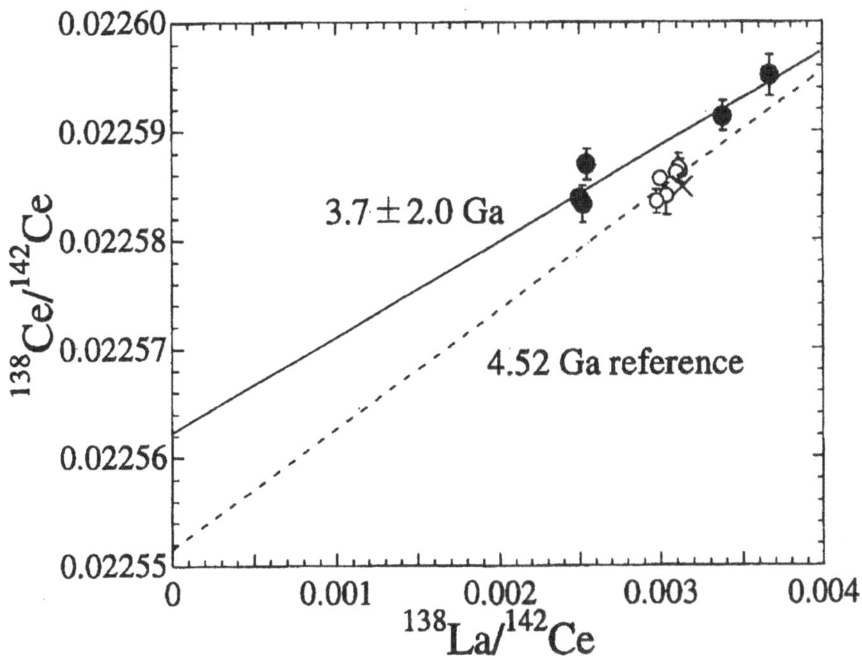


*Figure 2.* La-Ce isochron diagram of eucrites. The La-Ce age of Millbillillie #2 is 3.7 ± 2.0 Ga using the decay constant evaluated by Tanimizu (2000). A large uncertainty is due to the small variation of La/Ce fractionation compared with those of terrestrial samples. Data of Millbillillie #1 are clustering and can not yield a valid age. The dashed line is 4.52 Ga reference (Makishima and Masuda, 1993, adjusted to our JMC304 value). Data of Millbillillie #2 are not on the reference isochron, but those of Millbillillie #1 and Camel Donga were on that. There is a possibility that the La-Ce decay system was reset. Two trends between Millbillillie #1 and #2 are also recognized as well in the Sm-Nd decay system (Figure 1a).

The Rb-Sr isochron diagram of Millbillillie is shown in Figure 3. The minerals from Millbillillie #2 are approximately along the 4.52 Ga reference line, but deviate from it by more than analytical error. This is interpreted as the open-system behavior of Rb during thermal annealing (*e.g.*, Bogard *et al.*, 1993). In fact, Millbillillie is one of the most thermal effected eucrites and has the youngest Ar-Ar age corresponding to 3.55 ± 0.02 Ga (Yamaguchi *et al.*, 1994). Under such an environment, the La-Ce decay may become an open system, though the Sm-Nd decay system is still evolving in a closed system. There is no information about the closure temperature of the La-Ce decay system. More investigation is necessary to know its behavior. The La-Ce dating of eucrites has not been reported until now, except that of an Antarctic eucrite investigated for the weathering origin of its Ce anomaly (Shimizu *et al.*, 1983).

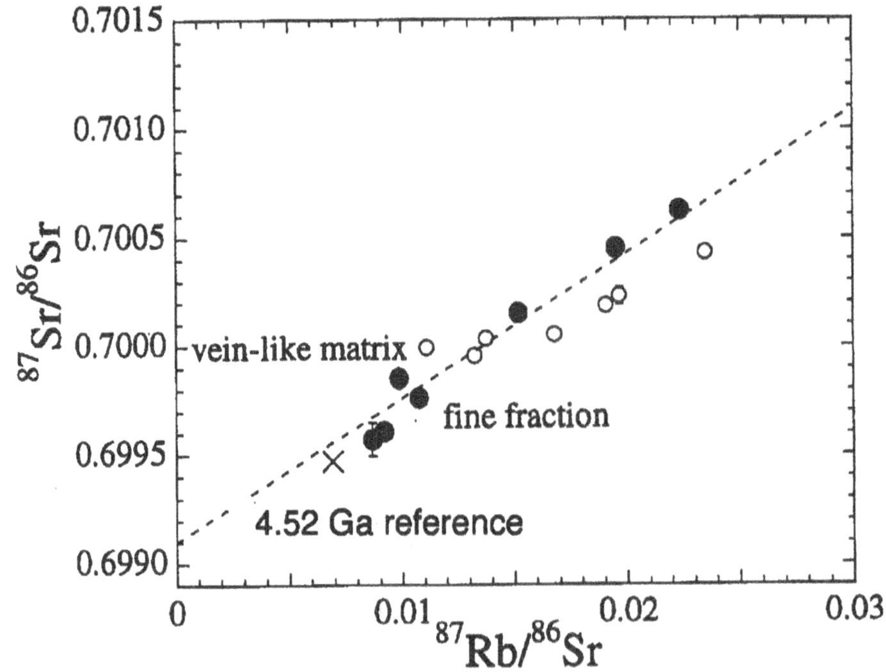


*Figure 3.* Rb-Sr isochron diagram of eucrites. The data of Millbillillie #2 are approximately along the 4.52 Ga reference line, but deviate from it by more than analytical error. This is interpreted as the open system behavior of Rb during thermal annealings. The plots of Millbillillie #1 will make gentler slope than the reference line. Two trends are recognized as well as other decay systems.

## 2.2 Ce-Nd-Sr isotope systematics of lunar rocks

The data of lunar samples are shown on the following three pages as Table 2a, 2b and 2c. Then, the Rb-Sr internal isochron diagrams of 10017 and 14310 are shown in Figure 4a and Figure 4b. The calculated ages, 3.57 ± 0.03 Ga for 10017 and 3.84 ± 0.11 Ga for 14310 are consistent with 3.51 ± 0.05 Ga (Papanastassiou *et al.*, 1970) for 10017 and 3.79 ± 0.04 Ga (Papanastassiou and Wasserburg, 1971) and 3.75 ± 0.07 Ga (Tanaka *et al.*, 1985) for 14310 within the error, respectively. The well defined internal isochrons mean that these rocks are not disturbed by thermal metamorphism or alteration after their Rb-Sr systematics became a closed system. Though the internal isochron age of 75015 is not reported, the range from 3.6 Ga to 3.8 Ga will be acceptable referring to those of the other Apollo 17 rocks (Taylor, 1975).

*Table 2a.* The Rb, Sr and light-REE concentrations in lunar samples

| Sample | Rb (ppm) | Sr (ppm) | La (ppm) | Ce (ppm) | Nd (ppm) | Sm (ppm) |
|---|---|---|---|---|---|---|
| 10017 | | | | | | |
| pyroxene (80-150 mesh) | 4.16 | 83.6 | 19.3 | 57.4 | 46.2 | 16.1 |
| pyroxene (150-255 mesh) | 4.18 | 80.4 | 20.4 | 61.1 | 49.0 | 17.5 |
| $\rho > 3.33$ (255-500 mesh) | 2.15 | 56.3 | 15.3 | 46.6 | 39.1 | 14.6 |
| phosphate (400-500 mesh) | 7.91 | 301 | 29.3 | 85.3 | 65.4 | 22.3 |
| ilmenite (80-150 mesh) | 10.3 | 119 | 44.5 | 130 | 99.5 | 33.8 |
| ilmenite (150-255 mesh) | 9.66 | 109 | 41.9 | 123 | 93.9 | 31.7 |
| whole rock | 5.66 | 162 | 24.8 | 72.8 | 57.1 | 19.8 |
| fine fraction (<500 mesh) | 5.63 | 208 | 24.2 | 71.2 | 55.5 | 19.7 |
| plagioclase | 2.30 | 541 | 6.60 | 18.0 | 12.6 | 3.94 |
| 75015 | 0.606 | 205 | 6.18 | 22.4 | 24.5 | 10.4 |
| 14310 | | | | | | |
| mix (325-500 mesh) | 14.1 | 156 | 79.6 | 212 | 125 | 35.8 |
| fine fraction (<500 mesh) | 13.3 | 207 | 49.4 | 131 | 76.6 | 21.8 |
| pyroxene (325-500 mesh) | 2.77 | 28.2 | 27.4 | 74.5 | 47.4 | 16.7 |
| pyroxene (200-325 mesh) | 21.5 | 121 | 108 | 287 | 173 | 48.9 |
| pyroxene (100-200 mesh) | 23.3 | 209 | 96.4 | 253 | 151 | 42.0 |
| plagioclase (325-500 mesh) | 15.9 | 283 | 36.6 | 93.1 | 53.6 | 14.5 |
| plagioclase (200-325 mesh) | 5.26 | 286 | 16.9 | 42.8 | 23.9 | 6.25 |

*Table 2b.* Isotopic results of lunar samples.

| Sample | $^{87}Rb/^{86}Sr$ | $^{87}Sr/^{86}Sr$. | $^{147}Sm/^{144}Nd$ | $^{143}Nd/^{144}Nd$ | $^{138}La/^{142}Ce$ | $^{138}Ce/^{142}Ce$ |
|---|---|---|---|---|---|---|
| 10017 | | | | | | |
| pyroxene (80-150 mesh) | 0.144 | 0.706816 (16) | 0.2109 (10) | 0.513068 (7) | 0.00274 | 0.0225846 (14) |
| pyroxene (150-255 mesh) | 0.150 | 0.707223 (14) | 0.2161 (11) | 0.513210 (8) | 0.00272 | 0.0225856 (14) |
| $\rho > 3.33$ (255-500 mesh) | 0.111 | 0.705126 (14) | 0.2258 (11) | 0.513440 (7) | 0.00268 | 0.0225831 (14) |
| phosphate (400-500 mesh) | 0.0761 | 0.703406 (14) | 0.2059 (12) | 0.512955 (7) | 0.00281 | 0.0225840 (25) |
| ilmenite (80-150 mesh) | 0.251 | 0.712523 (14) | 0.2052 (12) | 0.512917 (9) | 0.00280 | 0.0225845 (15) |
| ilmenite (150-255 mesh) | 0.257 | 0.712751 (16) | 0.2037 (11) | 0.512920 (8) | 0.00278 | 0.0225844 (15) |

| Sample | $^{87}Rb/^{86}Sr$ | $^{87}Sr/^{86}Sr$. | $^{147}Sm/^{144}Nd$ | $^{143}Nd/^{144}Nd$ | $^{138}La/^{142}Ce$ | $^{138}Ce/^{142}Ce$ |
|---|---|---|---|---|---|---|
| 10017 | | | | | | |
| whole rock | 0.101 | 0.704660 (14) | 0.2099 (9) | 0.513065 (8) | 0.00278 | 0.0225856 (15) |
| fine fraction | 0.0783 | 0.703448 (14) | 0.2147 (16) | 0.513182 (8) | 0.00278 | 0.0225856 (15) |
| (< 500 mesh) | | | | | | |
| plagioclase | 0.0123 | 0.700038 (14) | 0.1884 (10) | 0.512572 (8) | 0.00299 | |
| 75015 | 0.00857 | 0.699695 (13) | 0.2562 (12) | 0.514458 (7) | 0.00225 | 0.0225824 (15) |
| 14310 | | | | | | |
| mix | 0.261 | 0.715171 (16) | 0.1735 (15) | 0.511818 (7) | 0.00307 | 0.0225874 (15) |
| (325-500 mesh) | | | | | | |
| fine fraction | 0.186 | 0.710775 (13) | 0.1720 (12) | 0.511788 (7) | 0.00308 | 0.0225894 (12) |
| (< 500 mesh) | | | | | | |
| pyroxene | 0.284 | 0.716018 (16) | 0.1870 (7) | 0.512239 (7) | 0.00300 | 0.0225879 (14) |
| (325-500 mesh) | | | | | | |
| pyroxene | 0.515 | 0.729211 (19) | 0.1706 (8) | 0.511852 (7) | 0.00308 | 0.0225881 (15) |
| (200-325 mesh) | | | | | | |
| pyroxene | 0.323 | 0.718393 (16) | 0.1679 (8) | 0.511770 (7) | 0.00311 | 0.0225895 (15) |
| (100-200 mesh) | | | | | | |
| plagioclase | 0.162 | 0.709863 (13) | 0.1634 (10) | 0.511619 (7) | 0.00321 | 0.0225877 (16) |
| (325-500 mesh) | | | | | | |
| plagioclase | 0.0533 | 0.703426 (15) | 0.1580 (7) | 0.511533 (8) | 0.00323 | 0.0225897 (16) |
| (200-325 mesh) | | | | | | |

The values in parentheses are error limits and apply to the last digits.
The uncertainties of $^{87}$Rb/86Sr and 138La/142Ce are +/- 1% and +/- 0.5%, respectively.
The average values of isotope references (La Jolla, JNdi-1 and NIST SRM 987) are 0.511854 +/- 0.000003, 0.512108 +/- 0.000005 and 0.710250 +/- 0.000010, respectively.
The average value of a Ce isotope reference, JMC304, is 0.0225925 +/- 0.0000018 for 250 ratios.
Errors are one standard deviation of repeat measurements.

*Table 2c.* Epsilon values of lunar samples at present, 3.6 Ga and 3.8 Ga

| Sample | $\varepsilon Nd_0$ | $\varepsilon Nd_{3.6\ Ga}$ | $\varepsilon Nd_{3.8\ Ga}$ | $\varepsilon Ce_0$ | $\varepsilon Ce_{3.6\ Ga}$ | $\varepsilon Ce_{3.8\ Ga}$ |
|---|---|---|---|---|---|---|
| 10017 | | | | | | |
| pyroxene | 8.38 | 1.6 (0.9) | 1.3 (0.9) | -2.17 (0.62) | -0.94 (0.66) | -0.87 (0.66) |
| (80-150 mesh) | | | | | | |
| pyroxene | 11.16 | 2.0 (1.0) | 1.5 (1.0) | -1.73 (0.62) | -0.42 (0.66) | -0.35 (0.66) |
| (150-255 mesh) | | | | | | |
| $\rho$ > 3.33 | 15.64 | 2.0 (1.0) | 1.2 (1.0) | -2.83 (0.62) | -1.38 (0.66) | -1.30 (0.66) |
| (255-500 mesh) | | | | | | |
| phosphate | 6.18 | 1.8 (1.0) | 1.5 (1.1) | -2.43 (1.11) | -1.45 (1.15) | -1.39 (1.15) |
| (400-500 mesh) | | | | | | |
| ilmenite | 5.44 | 1.3 (1.0) | 1.1 (1.1) | -2.21 (0.66) | -1.18 (0.71) | -1.13 (0.71) |
| (80-150 mesh) | | | | | | |
| ilmenite | 5.49 | 2.1 (1.0) | 1.9 (1.0) | -2.26 (0.66) | -1.17 (0.71) | -1.10 (0.71) |
| (150-255 mesh) | | | | | | |
| whole rock | 8.33 | 2.1 (0.8) | 1.7 (0.9) | -1.73 (0.66) | -0.63 (0.71) | -0.57 (0.71) |

| Sample | $\varepsilon Nd_0$ | $\varepsilon Nd_{3.6\ Ga}$ | $\varepsilon Nd_{3.8\ Ga}$ | $\varepsilon Ce_0$ | $\varepsilon Ce_{3.6\ Ga}$ | $\varepsilon Ce_{3.8\ Ga}$ |
|---|---|---|---|---|---|---|
| 10017 | | | | | | |
| fine fraction (<500 mesh) | 10.62 | 2.1 (1.3) | 1.6 (1.4) | -1.73 (0.66) | -0.63 (0.71) | -0.57 (0.71) |
| plagioclase | -1.29 | 2.4 (0.9) | 2.6 (0.9) | | | |
| 75015 | 35.50 | 7.8 (1.0) | 6.2 (1.1) | -3.14 (0.66) | -0.08 (0.71) | 0.09 (0.71) |
| 14310 | | | | | | |
| mix (325-500 mesh) | -16.01 | -5.4 (1.3) | -4.8 (1.3) | -0.93 (0.66) | -0.93 (0.71) | 0.93 (0.71) |
| fine fraction (<500 mesh) | -16.57 | -5.3 (1.0) | -4.7 (1.1) | -0.04 (0.53) | -0.08 (0.58) | -0.08 (0.58) |
| pyroxene (325-500 mesh) | -7.78 | -3.5 (0.7) | -3.2 (0.7) | -0.71 (0.62) | -0.44 (0.66) | -0.43 (0.66) |
| pyroxene (200-325 mesh) | -15.33 | -3.4 (0.8) | -2.7 (0.8) | -0.62 (0.66) | -0.64 (0.71) | -0.64 (0.71) |
| pyroxene (100-200 mesh) | -16.93 | -3.7 (0.8) | -3.0 (0.8) | 0.00 (0.66) | -0.13 (0.71) | -0.14 (0.71) |
| plagioclase (325-500 mesh) | -19.88 | -4.6 (0.9) | -3.7 (0.9) | -0.80 (0.71) | -1.33 (0.75) | -1.36 (0.75) |
| plagioclase (200-325 mesh) | -21.55 | -3.8 (0.7) | -2.7 (0.7) | 0.09 (0.71) | -0.50 (0.75) | -0.53 (0.75) |

The values in parentheses are $2\sigma$ error limits.
Ce and Nd isotope ratios of CHUR are from Makishima and Masuda (1993). Their Ce isotope ratio of CHUR is adjusted to ours by the comparison of JMC304.

The light rare earth elements (light-REE) patterns of the lunar rocks are shown in Figure 5 and their present $^{138}Ce/^{142}Ce$ and $^{143}Nd/^{144}Nd$ isotope ratios are plotted in Figure 6 in $\varepsilon$Ce and $\varepsilon$Nd values. Essentially, only $\varepsilon$ values of whole rock samples should be plotted, but those of all mineral separates were plotted to reduce the influence of error size owing to mineralogical heterogeneity of small whole rock size.

Figure 6 is called $\varepsilon$Ce - $\varepsilon$Nd diagram and was introduced by Tanaka *et al.* (1987). Its meaning is similar with $\varepsilon$Sr - $\varepsilon$Nd diagram. The $\varepsilon$-notation is the deviation in parts in $10^4$ of isotope ratio of a sample from that in a chondritic uniform reservoir (CHUR). The $\varepsilon$Ce - $\varepsilon$Nd diagram is useful to trace the evolution of light-REE through time.

A schematic figure (Figure 7) explains the relation between the light-REE patterns and the directions of isotopic growth of such samples with time. Lunar sample 10017 and 75015 are depleted in light-REE for 3.6 Ga and sample 14310 has a light-REE enriched pattern (Masuda *et al.*, 1972) for 3.8 Ga. Therefore, the former two samples are expected to be plotted in the second quadrant and the latter one plotted in the fourth.

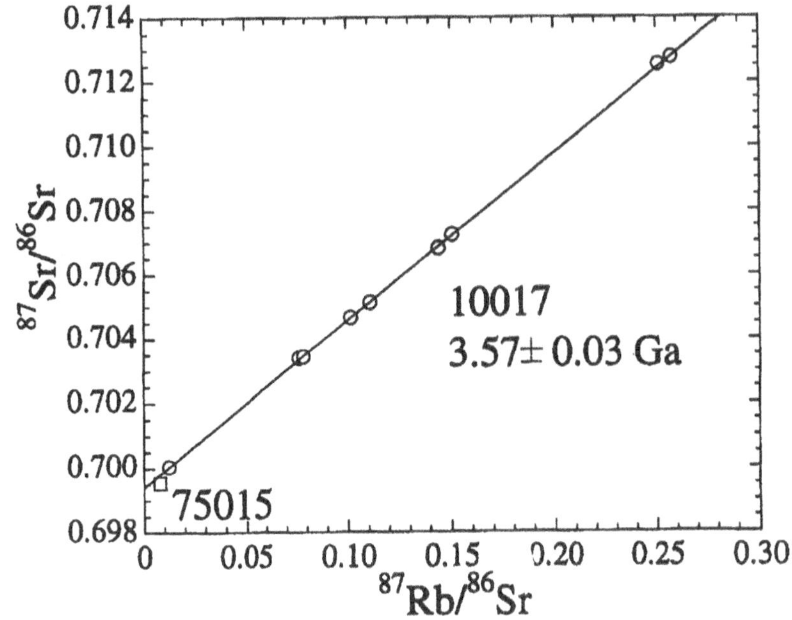


*Figure 4a.* Rb-Sr isochron diagram of lunar sample 10017. The data of sample 10017 define an internal isochron which corresponds to the age of 3.57 ± 0.03 Ga.

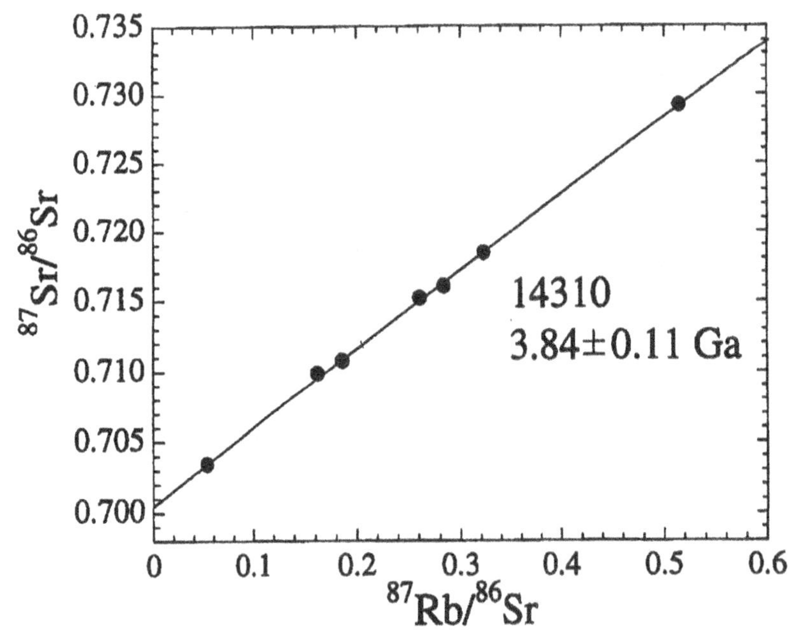


*Figure 4b.* Rb-Sr isochron diagram of lunar sample 14310. The data of sample 14310 define an internal isochron which corresponds to the age of 3.84 ± 0.11 Ga.

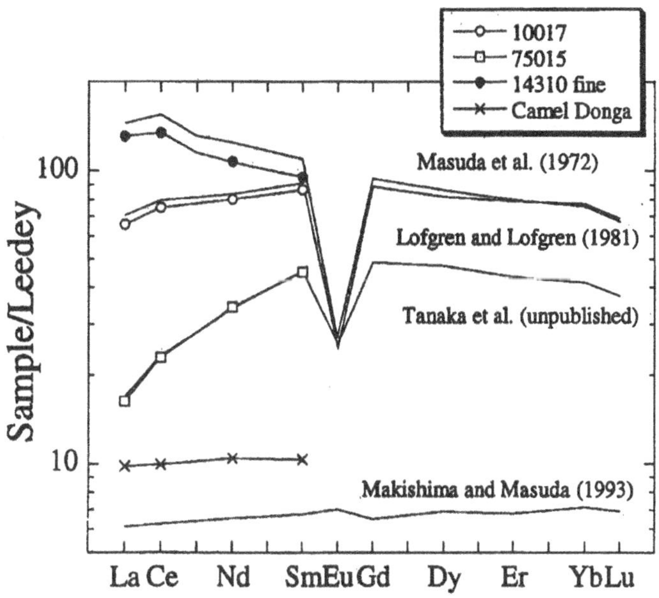


*Figure 5.* Rare earth element patterns of lunar samples and a eucrite, Camel Donga. The concentrations of REE are normalized to Leedey chondrite (Masuda *et al.*, 1973). For the unpublished data Tanaka, Shimizu, Shibata and Masuda, the concentrations of La, Ce, Nd, Sm, Eu, Gd, Dy, Er, Yb and Lu are 6.40, 22.7, 24.3, 10.5, 2.22, 15.1, 18.5, 11.1, 10.4 and 1.45 in ppm.

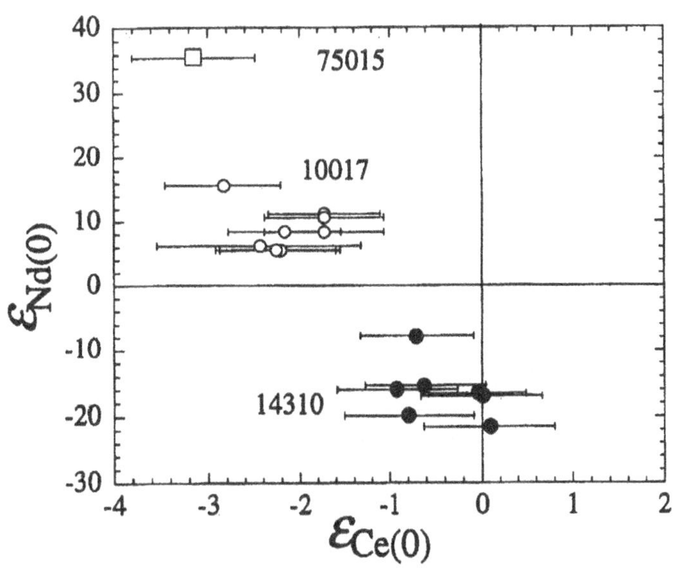


*Figure 6.* εCe-εNd diagram of lunar samples. Open circles and an open square are high Ti mare basalts 10017 and 75015. Solid circles are a KREEP basalt 14310. The εCe and εNd values of CHUR are from Makishima and Masuda (1993, adjusted to our JMC304 value).

The latter sample 14310, however, is in fact plotted in the third quadrant (Figure 6). This is due to the positive Ce anomaly. If the Ce anomaly was caused by some present or recent event, it should be plotted in the fourth quadrant.

Thus, we can conclude that the Ce anomaly of 14310 was not produced recently. The well defined Rb-Sr internal isochron supports this. When was the Ce anomaly produced, then?

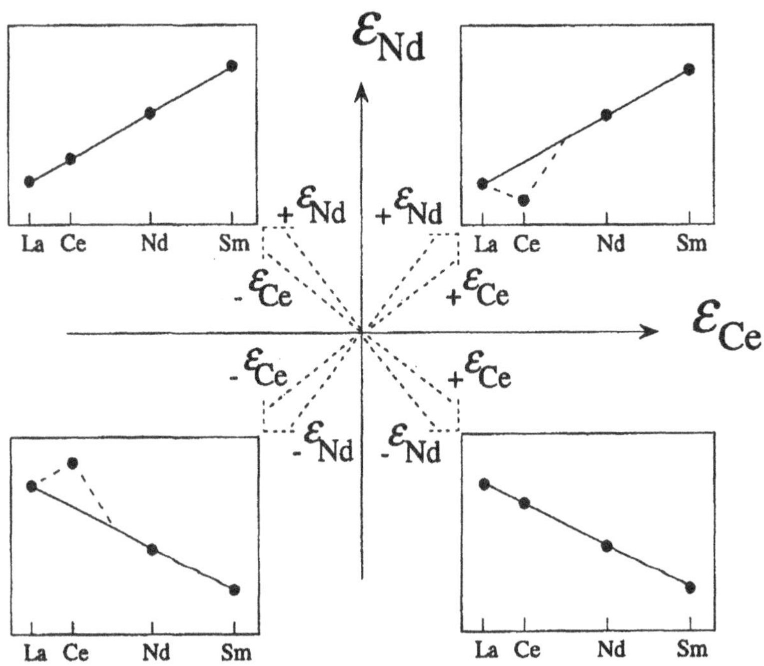


*Figure 7.* A schematic figure which explains the relation between the light-REE pattern of some CHUR-derived samples and the direction of its isotopic growth from CHUR with time.

Now, we focus the discussion on the early period of lunar evolution. Considering the formation of the oldest rock in the lunar sample, the lunar formation age seems to be simultaneous with the other planets. Its elemental abundance is characterized by the loss of volatile elements and enrichment of refractory elements compared with carbonaceous chondrites, but the fractionation among REE would not have occurred. It is believed that there was a widespread melting at a depth of a few hundred km. Proportionally to its cooling, a layered structure was developed by the fractional crystallization. This melting occurred shortly after the formation of the moon. The mare basalts were considered to be generated by the remelting of the layered structure due to radioactive heating. The volcanism continued to around 3.0 Ga (Taylor, 1975).

As the ages of three lunar samples were known, the εCe and εNd values at 3.6 Ga and 3.8 Ga were calculated and plotted in Figures 8a and 8b. In both cases, plots of 10017 and 75015 lie in the second quadrant. This means that the sources of these high Ti basalts are originally depleted in light-REE.

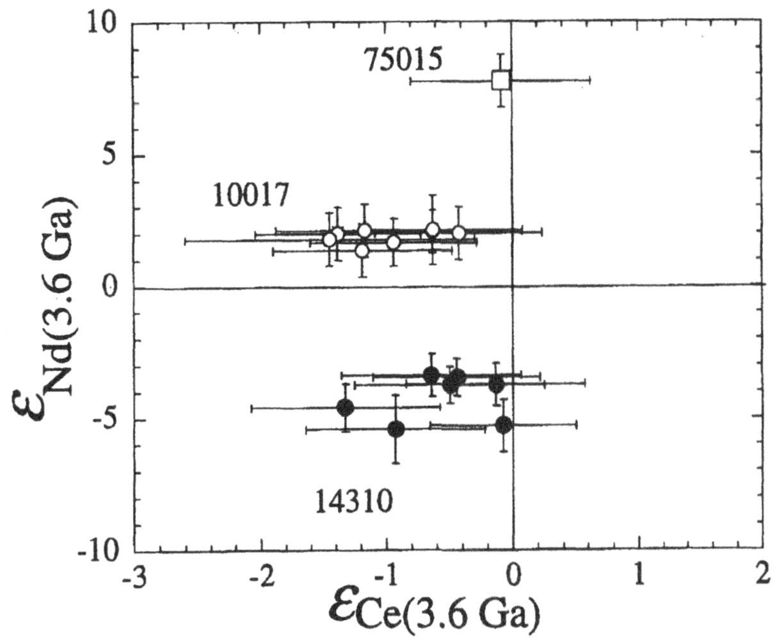


*Figure 8a.* εCe - εNd diagram of the lunar samples at 3.6 Ga.

Now, let us assume the following two things. First, there was no relative fractionation among REE between the melting part and the non-melting residue, when the primary widespread melting occurred. Secondly, the primary melt solidified in a short time compared with the time that passed after the solidification. By assuming these, the εCe and εNd values of lunar samples can be plotted on the point of CHUR (origin of the εCe - εNd diagram) from its formation to the end of the solidification of the primary melt. These assumptions seem to be reasonable. For the former, there is little relative fractionation among REE because of the high degree of melting required to produce such a large size melt. For the latter, the relative partition coefficient between the melt and crystallizing minerals (olivine, orthopyroxene and plagioclase) is almost identical among REE except europium. Eucrites, which are believed to be the basaltic magma that covered the surface of a proto-planet, have flat REE patterns (Consolmagno and Drake, 1977).

By these assumptions, the REE patterns of the magma sources of 10017 and 75015 can be estimated according to the relation between REE patterns and their isotopic growth (Tanaka *et al.*, 1987). Though both 10017 and 75015 are classified as high Ti basalt, the source REE pattern of 75015 is estimated to be more depleted in light-REE than that of 10017. These two rocks must have been produced from different magma sources.

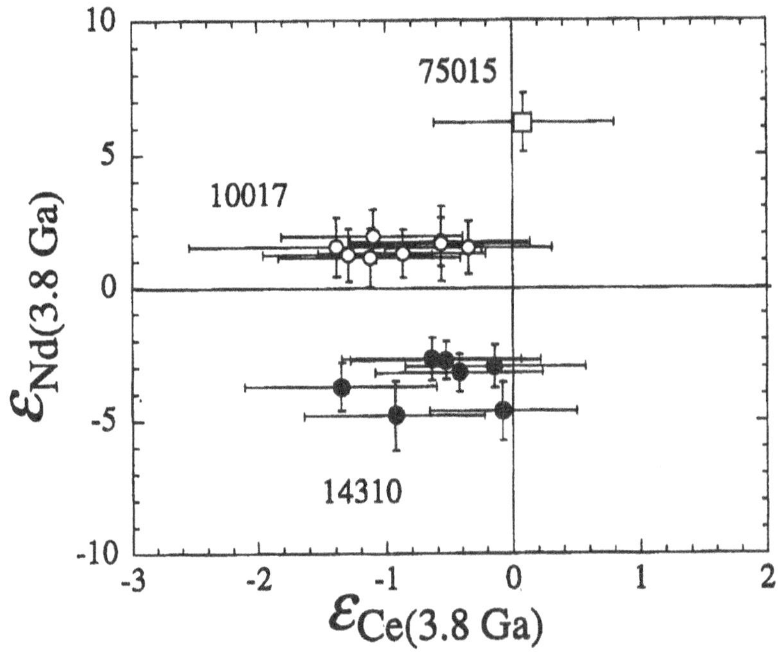


*Figure 8b.* εCe - εNd diagram of the lunar samples at 3.8 Ga.

The magma source of KREEP basalts including sample 14310 is considered as the final residual liquid of the primary melt. The εCe and εNd values of sample 14310 at 3.8 Ga, which was determined above, is still plotted in the third quadrant of Figure 8b. If its Ce anomaly was produced at 3.8 Ga, these ε values should be plotted in the fourth quadrant, because the final solidification part must be enriched in light-REE. The fact that the sample 14310 is plotted in the third quadrant in Figure 8b means the Ce anomaly was produced before 3.8 Ga. Actually, the Rb-Sr internal isochron age of sample 14321 (about 3.9 Ga) is older than that of 14310 and has a Ce anomaly (Masuda *et al.*, 1972). The Ce anomaly in an Antarctic eucrite has been explained by weathering in water or ice (Shimizu *et al.*, 1983). Was there any water on the surface of the ancient moon? Another possibility is that, the final residual liquid of the primary melt is anomalous regarding Ce. For example, a positive Ce anomaly is recognized in zircon in acidic rocks in the earth. Can we apply this phenomenon to lunar rocks, which were formed at low oxygen fugacity?

The Ce anomaly in lunar samples is now widely accepted with the improvement in the precision of REE concentration determination. If there is ice in the lunar polar and it caused Ce anomalies of some lunar samples, we could expect a large Ce anomaly for the rocks in that area. A large Ce anomaly, inversely, may become a pathfinder of water resources in the future. The data used in this paper are preliminary. A more detailed discussion will be possible after more data are obtained. Repeat analyses of the Ce isotope ratio will increase its reliability.

## ACKNOWLEDGMENTS

A part of the eucrite Millbillillie was provided by Mr. Tadashi Sawai. We thank Dr. Akira Yamaguchi for helpful advice and discussion about eucrites. We are indebted to Dr. Cristian Dragusanu for his patience in revising English. This work was supported in part by Research Fellowships of the Japan Society for the Promotion of Science for Young Scientists (10003537) and by Kurata Scholarship.

## REFERENCES

Bogard, D., Nyquist, L., Takeda, H., Mori, H., Aoyama, T., Bansal, B., Wiesmann, H. and Shih, C.-Y.: 1993, "Antarctic polymict eucrite Yamato 792769 and the cratering record on the HED parent body", *Geochim. Cosmochim. Acta* **57**, 2111-2121.

Consolmagno, G.J. and Drake, M.J.: 1977, "Composition and evolution of the eucrite parent body: Evidence from rare earth elements", *Geochim. Cosmochim. Acta* **41**, 1271-1282.

Dickin, A.P., Jones, N.W., Thirlwall, M.F. and Thompson, R.N.: 1987, "A Ce/Nd isotope study of crustal contamination processes affecting Palaeocene magmas in Skye, Northwest Scotland", *Contrib. Mineral. Petrol.* **96**, 455-464.

Faure, G.: 1986, *Principle of Isotope Geology*, John Wiley and Sons, New York, NY, 208 pp.

Lofgren, G.E. and Lofgren, E.M.: 1981, "Catalog of Lunar Mare Basalts Greater Than 40 Grams. Part 1. Major and Trace Chemistry", *NASA Lunar and Planetary Contribution* **438**, 9 pp.

Makishima, A. and Nakamura, E.: 1991, "Precise determination of cerium isotope composition in rock samples", *Chem. Geol. (Isot. Geosci. Sect.)* **94**, 1-11.

Makishima, A. and Masuda, A.: 1993, "Primordial Ce isotopic composition of the solar system", *Chem. Geol.* **106**, 197-205.

Masuda, A., Nakamura, N., Kurasawa, H. and Tanaka, T.: 1972, "Precise determination of rare-earth elements in the Apollo 14 and 15 samples", *Proc. 3rd Lunar Sci. Conf.* **2**, 1307-1313.

Masuda, A., Nakamura, N. and Tanaka, T.: 1973, "Fine structures of mutually normalized rare-earth patterns of chondrites", *Geochim. Cosmochim. Acta* **37**, 239-248.

Masuda, A. and Tanaka, T.: 1980, "Rare earth element distribution in the Melrose-b howardite: Pre-terrestrial negative Ce anomaly", *Earth Planet. Sci. Lett.* **49**, 109-116.

Nakamura, N., Tatsumoto, M. and Ludwig, K.R.: 1984, "Applicability of La-Ce systematics to planetary samples", *J. Geophys. Res.* **89**, B438-B444.

Palme, H., Wlotzka, F., Spettel, B., Dreibus, G. and Weber, H.: 1988, "Camel Donga: A eucrite with high metal content", *Meteoritics* **23**, 49-57.

Papanastassiou, D.A., Wasserburg, G.J. and Burnett, D.S.: 1970, "Rb-Sr ages of lunar rocks from the Sea of Tranquillity", *Earth Planet. Sci. Lett.* **8**, 1-19.

Papanastassiou, D.A. and Wasserburg, G.J.: 1971, "Rb-Sr ages of igneous rocks from the Apollo 14 mission and the age of the Fra Mauro formation", *Earth Planet. Sci. Lett.* **12**, 36-48.

Prinzhofer, A., Papanastassiou, D.A. and Wasserburg, G.J.: 1992, "Samarium-neodymium evolution of meteorites", *Geochim. Cosmochim. Acta* **56**, 797-815.

Shimizu, H., Masuda, A. and Tanaka, T.: 1983, "Ce anomaly in REE pattern of Antarctic eucrite", *Proc. 8th Symp. Antarctic Meteorites*, 341-348.

Tanaka, T. and Masuda, A.: 1982, "The La-Ce geochronometer: A new dating method", *Nature* **300**, 515-518.

Tanaka, T., Shimizu, H., Shibata, K. and Masuda, A.: 1985, "Positive cerium anomaly of lunar 14310: Examination by $^{138}Ce/^{142}Ce$", *Lunar Planet. Sci.* **XVI**, 847-848.

Tanaka, T., Shimizu, H., Shibata, K. and Masuda, A.: 1986, "Water in ancient moon? Possible oxic alteration of 14310", *Lunar Planet. Sci.* **XVII**, 867-868.

Tanaka, T., Shimizu, H., Kawata, Y. and Masuda, A.: 1987, "Combined La-Ce and Sm-Nd isotope systematics in petrogenic studies", *Nature* **327**, 113-117.

Tanimizu, M.: 2000, "Geochronological determination of $^{138}La$ $\beta$-decay constant and its evaluation", submitted to *Phys. Rev. C.*

Taylor, S.R.: 1975, *Lunar Science: A Post-Apollo View*, Pergamon Press, Oxford, UK, 372 pp.

Yamaguchi, A., Takeda, H., Bogard, D.D. and Garrison, D.: 1994, "Textual variations and impact history of the Millbillillie eucrite", *Meteoritics* **29**, 237-245.

PART VI

# THE ORIGIN OF THE SOLAR SYSTEM

# Abundance of $^{182}$Hf and the Supernova Model of the Solar System

S. Ramadurai
*Astrophysics Group, Tata Institute of Fundamental Research, Homi Bhabha Road, Navy Nagar, Colaba, Mumbai 400 005, INDIA*
*durai@tifr.res.in*

Abstract: The possibility of the actinides and short-lived non-actinides arising from diverse supernova sources is suggested by Wasserburg *et al.* (1996) based on the inferred abundances of $^{182}$Hf and $^{129}$I in the primitive solar nebula. But the addition of neutrino induced nucleosynthesis of $^{182}$Hf in the supernova will modify the need for multiple sources for the short- and long-lived radioactive species. It is shown that a modified version of the Single Supernova Origin of the Solar System Model of Manuel and Sabu (1975) is able to account for the extinct short-lived as well as long-lived radioactive species in the primitive solar nebula, if one takes into account all the nucleosynthetic processes properly.

## 1. INTRODUCTION

The discovery that $^{182}$Hf was present live in the early solar system with $^{182}$Hf/$^{180}$Hf ~ 2.8 x $10^{-4}$ (See Harper and Jacobsen, 1996 for a detailed review) has opened up a great puzzle about the r-process nucleosynthesis yields. Following the suggestion of Cameron (1993), Wasserburg *et al.* (1995) investigated the AGB sources and found that the $^{182}$Hf yield predicted falls short by nearly a factor of 100, if the correct yields of other short lived species like $^{26}$Al, $^{107}$Pd, $^{41}$Ca and $^{60}$Fe are supposed to be produced by the same process. This is due to the low neutron densities required to produce branchings in the main s-component. Wasserburg *et al.* (1996) went on to show that $^{182}$Hf is an r-process product and that uniform production of actinides and $^{182}$Hf over the history of the galaxy yields self-consistent results. But the low $(^{129}I/^{127}I)_0$ ratio of $10^{-4}$ is then a puzzle. So Wasserberg *et al.* (1996) suggested diverse r-process yields from supernovae. Ramadurai (1993) had suggested the possibility of significant yields of $^{182}$Hf from the neutrino induced nucleosynthetic process in a supernova. Subsequently Qian *et al.* (1998) have done a detailed calculation of r-process yields including the neutrino induced reactions. Recently, the astrophysical r-process has been studied by Freiburghaus *et al.* (1999) in the framework of neutron densities and temperature using the adiabatic expansion model of the supernova explosion. The main aim of the present investigation is to show that if one takes into account the neutrino induced nucleosynthesis, then for the same flux of neutrinos expected from a supernova explosion, observed yields of the p-process nuclides like $^{146}$Sm and $^{180}$Ta are the result in addition to the contribution of $^{182}$Hf. This brings us back to the question of whether a single supernova can account for the observed diverse isotopic anomalies. The answer is that a modified version by Ramadurai (this volume) of the single supernova hypothesis (Manuel and Sabu, 1975) will be in accord with all the extinct radioactivities.

## 2. NEUTRINO INDUCED NUCLEOSYNTHESIS

The importance of neutrinos in the supernova phenomenon has been emphasised by Colgate and White (1963) from the dynamical point of view. The earliest serious attempt to stress the importance of neutrino processes for nucleosynthesis is by the Russian group (Domagatskii and Nadyozhin, 1977). Detection of neutrinos from SN1987A revived interest, with Epstein *et al.* (1988) pointing out the r-process yields will be modified substantially with neutrino processes. A detailed investigation of this suggestion was taken up by Woosley *et al.* (1990), who showed its importance in producing several nuclei attributed to various processes like the p-process, cosmic ray spallation, etc. It was shown by Ramadurai (1993) that even a nuclide like $^{182}$Hf, which is not shielded, would have a significant contribution from the neutrino process. Woosley *et al.* (1990) conclude from their detailed calculations that for the neutrino luminosity observed for SN1987A, one is able to get the abundance of $^{180}$Ta and $^{146}$Sm to match with solar values. For the same flux, the $^{182}$Hf yield is given by the following:

$$R_\nu = n_{AZ}\frac{L_\nu}{4\pi r^2\langle E_\nu\rangle}\langle\sigma v_{AZ}\rangle$$

where $R_\nu$ is the reaction rate, $n_{AZ}$ is the number density of target nuclei, $L_\nu$ is the total energy carried by neutrinos of a given type per unit time, $\langle E_\nu \rangle$ being the mean energy of individual neutrinos; r is the distance of the neutrino irradiated matter from the centre of the collapsing core. Here the neutrino excitation crosssection $\langle \sigma v_{AZ} \rangle$ has been averaged over the neutrino energy spectrum. $\delta n_{AZ}$, the number of the excited target nuclei is given by $R_\nu$ from which we get

$$\frac{\delta n_{AZ}}{n_{AZ}} = \frac{L_\nu}{4\pi r^2 \langle E_\nu \rangle} \langle \sigma v_{AZ} \rangle$$

The ratio of the resultant nuclei of interest can be obtained by multiplying the result by the α branching ratio. In the case of hafnium and tungsten, the total Hf to the $^{186}$W ratio happens to be 2 (Cameron, 1982). Hence the ratio of $^{182}$Hf/Hf is given as

$$\frac{N_{^{182}Hf}}{N_{Hf}} = \frac{\delta n_{^{186}W}}{n_{^{186}W}} \cdot b_\alpha \cdot 2n_{^{186}W} \approx 1.86x10^{-4}$$

which is nearly the same as the observed upper limit (Harper and Jacobsen, 1996). Thus the supernova is able to produce enough $^{182}$Hf. The detailed analysis of Qian *et al.* (1998) of the r-process yields including the neutrino process bear witness to the correctness of this statement. But it should be emphasised that what Qian *et al.* were concerned with mainly were r-process yields, and they showed that provided one takes as free parameters the neutrino luminosity at the irradiation point given by $L_{\nu 51}/r^2$, mean decay lifetime $\tau$, and the neutron to seed ratio, denoted by n/s, one is able to get a consistent picture of all the yields including actinides. The problem of $^{129}$I is solved by choosing the parameters for the low frequency supernovae. Actually the proper adjustment of $L_\nu/r^2$ will be able to yield the correct ratio for both $^{129}$I and $^{182}$Hf. We have already shown that the p-process nuclides too are easily produced in the supernova hypothesis. Thus from the nucleosynthetic point of view the yield from a single supernova is in good agreement with the solar system values. However one has to emphasise that though by adjusting the various parameters we are able to get agreement with observations of the p- and r-process yields, the s-process yields pose a problem, as these strongly indicate the presence of AGB products (See Goswami, this volume). Now this brings us to the question of the actual scenario of the solar system formation, as the nucleosynthetic yields are only one aspect of the many processes involved in the formation of the solar system.

## 3. SUPERNOVA MODEL OF THE SOLAR SYSTEM FORMATION

Based on the observed correlation of the noble gas isotopic anomalies, Manuel and Sabu (1975) suggested that the solar system is the result of the products of a single supernova. Essentially their argument was based on the simplistic interpretation of the Burbidge *et al.* (1957) r- and p-process yields from a supernova combined with the abundance estimates of the other nuclei. The observation of several isotopic anomalies in individual grains brought into focus the sheer impossibility of having a single supernova providing answers to all the isotopic anomalies observed in meteorites. But as can be seen from the discussion of the case of $^{182}$Hf, there is a very strong case for the presence of yields from a single supernova in the solar system. Hence a modified version of the single supernova hypothesis has been put forward by Ramadurai (this volume). In this model, the solar system formed out of the gas present in the interstellar clouds, which has been polluted by heavy elements since the formation of our galaxy until the solar system started to form. The star formation episode in our neighbourhood is triggered by the supernova explosion providing the necessary thrust to the gas, which has been already accreting over several billion years onto the primordial blackhole of about $10^{-18}M_{0}$. When the supernova went off, the entire mass of the gas was thrown with speeds much larger than the sound speed. This initially dispersed the material, but the blackhole core serves as a point of accretion and the supernova dispersed gas falls back onto the tiny core to make the sun and the other planets. Thus one is able to get the advantage of the yields from a single supernova forming the dominant component, thus in agreement with the observations of all the isotopic anomalies, both short- and long-lived radioactive isotopes. Since the gas had been polluted by the grains from the earlier generations of stars, the yields from the other processes will be present. The detailed dynamics of this scenario, though not completely worked out, is outlined in Ramadurai (this volume). With this scenario, it is easy to overcome the problem of Hf and I too. The single supernova can bring in the $^{129}$I and the neutrino irradiated part will bring in the $^{182}$Hf, as outlined in the detailed model of the supernova origin of the solar system.

## 4. CONCLUSION

The supernova r- and p-process yields including the neutrino captures is able to account for the short- and long-lived radioactive species present in the early solar system. This depends on three free parameters, the neutrino luminosity at the irradiation point, the decay time and the neutron to seed nuclei ratio. For a set of parameters, there is good agreement among all the yields. But the hypothesis of the single supernova origin has to be modified to say that the dominant component in the solar system material is from a single supernova, thus not excluding yields from the other sources totally.

## ACKNOWLEDGEMENTS

The author wishes to thank the Foundation for Chemical Research, Inc., the ACS Division of Nuclear Chemistry & Technology, the University of Missouri-Rolla and Professors Paul Kuroda and Oliver Manuel for financial assistance which made it possible for the author to attend the meeting.

## REFERENCES

Burbidge, E.M., Burbidge, G.R., Fowler, W.A. and Hoyle, F.: 1957, "Synthesis of elements in stars", *Rev. Mod. Phys.* **29**, 547-650.

Cameron, A.G.W.: 1982, "Elemental and nuclidic abundances in the solar system", in *Essays in Nuclear Astrophysics*, eds., Barnes, C.A., Clayton, D.D. and Schramm, D.N., Cambridge Univ. Press, Cambridge, UK, pp. 23-43.

Cameron, A.G.W.: 1993, "Nucleosynthesis and star formation", in *Protostars and Planets III*, eds., Levy, E.H. and Lunine, J.I., Univ. Arizona Press, Tucson, AZ, pp. 47-74.

Colgate, S.A. and White, R.H.: 1963, "Cosmic rays from large supernovae", in *Proc. Int. Conf. Cosmic Rays*, Jaipur, India, eds., Daniel, R.R., Lavakare, P.J., Menon, M.G.K., Naranan, S., Nerurkar, N.W., Pal, Yash and Sreekantan., B.V., Tata Institute of Fundamental Research, Bombay, India, vol. **3**, pp. 335-359.

Domagatskii, G.V. and Nadyozhin, D.K.: 1977, "Neutrino induced production of bypassed elements", *Mon. Not. Royal Astron. Soc.* **178**, 33P-35P.

Epstein, R.I., Colgate, S.A. and Haxton, W.C.: 1988, "Neutrino induced r-process nucleosynthesis", *Phys. Rev. Lett.* **61**, 2038-2041.

Freiburghaus, C., Rembges, J.-F., Rauscher, T., Kolbe, E., Thielemann, F.-K., Kratz, K.-L, Pfeiffer, B. and Cowan, J.J.: 1999, "The astrophysical r-process: A comparison of calculations following the adiabatic expansion with classical calculations based on neutron densities and temperatures", *Ap. J.* **516**, 381-398.

Harper, C.L. and Jacobsen, S.B.: 1996, "Evidence for $^{182}$Hf in the early solar system and constraints on the timescale of terrestrial accretion and core formation", *Geochim. Cosmochim. Acta* **60**, 1131-1153.

Manuel, O.K. and Sabu, D.D.: 1975, "Elemental and isotopic inhomogeneities in noble gases: The case for local element synthesis of the chemical elements", *Trans. Missouri Acad. Sci.* **9**, 104-122.

Qian, Y.Z., Vogel, P. and Wasserburg, G.J.: 1998, "Diverse supernova sources for the r-process", *Ap. J.* **494**, 285-296.

Ramadurai, S.: 1993, "Neutrino induced nucleosynthesis of $^{182}$Hf", in *Origin and Evolution of the Elements*, eds., Prantzos, N., Vangioni-Flam, E. and Casse, M., Cambridge Univ. Press, Cambridge, UK, pp. 457-464.

Wasserburg, G.J., Gallino, R., Busso, M., Goswami, J.N. and Raiteri, C.M.: 1995, "Injection of freshly synthesized $^{41}Ca$ in the early solar nebula by an asymptotic giant branch star", *Ap. J. Lett.* **440**, L101-L104.

Wasserburg, G.J., Busso, M. and Gallino, R.: 1996, "Abundances of actinides and short-lived non-actinides in the interstellar medium: Diverse supernova sources for the r-process". *Ap. J. Lett.* **466**, L109-L113.

Woosley, S.E., Hartmann, D.H., Hoffman, R.D. and Haxton, W.C.: 1990, "The ν-process", *Ap. J.* **356**, 272-301.

*Note added in proof: It has been pointed out by Kratz that the neutrino-excitation of nuclei may not be effective once they take part in the momentum transfer to the outer layers to cause the supernova explosion. This question is under active investigation by the author at present, as it is not immediately clear that the neutrinos are not left with enough energy to cause the excitation of the nuclei following the explosion process.*

# Binary Origin of Solar System

M. N. Vahia
*Tata Institute of Fundamental Research, Homi Bhabha Road, Mumbai 400 005, INDIA; vahia@tifr.res.in*

Abstract: We investigate the scenario that the solar system may have originated in a binary whose companion exploded into a supernova. We estimate the angular momentum distribution between the sun and an accretion disk and show that the disk will have a much larger angular momentum. We show that such a disk would be unstable against convection once its density exceeds $10^{15}$ p/cm$^3$. For reasonable time scales of $10^3$ years, such a disk would have convective motions up to about 5 AU. Under such conditions, diffusion and convection could combine to give the solar system its presently observed structure in terms of composition, mass and angular momentum distribution.

## 1. INTRODUCTION

The origin of the solar system has been studied from a variety of different perspectives (*cf.*, Taylor, 1992; Lewis, 1995; Kaula, 1986 for comprehensive reviews). It is conventionally assumed that the solar system was born along with the sun in a proto solar system cloud that condensed to form the sun and the planets. This common origin is assumed in view of the fact that the composition of the solar photospheric material is, in general, similar to that of the planetary material with some differences. However, such models have several problems requiring external triggers to form the collapse and the elemental abundance characteristics, as well as dynamical parameters, are difficult to define. Also, the abundance anomalies and parameters such as the relative angular momentum of the sun and planets are difficult to reconcile under a single source origin.

*Origin of Elements in the Solar System: Implications of Post-1957 Observations.*
Edited by O. Manuel, Kluwer Academic/Plenum Publishers, 2000. 

Here we investigate the possibility that the proto-solar system cloud that resulted in the planets was formed *after* the sun had been ignited. We follow a scenario that the sun was in a binary with a massive companion that exploded into a supernova. Under such a condition, the sun would have been subjected to a large acceleration force. It would also trap some of the remnant material of the supernova that would form a thin disk around the sun. Hoyle (1945) had shown that such a scenario would account for the large angular momentum of the planets compared to the sun. We calculate the angular momentum and disk dynamics for such a scenario. Shu *et al.* (1997) have discussed the possible meteoritic evolution in such an accretion disk.

## 2. MODEL CALCULATIONS

Some of the properties of such a cloud are given in Tables 1a and 1b. The distances at which the parameters are calculated are taken to coincide with planets and have been extended till the Oort's cloud. Columns 3 to 5 are derived from Newtonian Gravity (We take the formulation from Harwitt, 1988). We discuss the other parameters below.

*Table 1.* Properties of the proto solar system disk

| Planetary | Nebular | Proto-planet | Thermal | Conv. | Conv. | Density | Diffusion | Diffusion |
|---|---|---|---|---|---|---|---|---|
| Mercury | $1.29 \times 10^4$ | $4.48 \times 10^2$ | $3.34 \times 10^3$ | $8.84 \times 10^1$ | $2.91 \times 10^4$ | $2.98 \times 10^8$ | $6 \times 10^{12}$ | $6 \times 10^1$ |
| Venus | $6.92 \times 10^3$ | $3.28 \times 10^2$ | $2.86 \times 10^3$ | $3.46 \times 10^1$ | $4.65 \times 10^4$ | $8.55 \times 10^7$ | $2 \times 10^{13}$ | $2 \times 10^2$ |
| Earth | $5.00 \times 10^3$ | $2.79 \times 10^2$ | $2.64 \times 10^3$ | $2.13 \times 10^1$ | $5.93 \times 10^4$ | $4.47 \times 10^7$ | $4 \times 10^{13}$ | $5 \times 10^2$ |
| Mars | $3.28 \times 10^3$ | $2.26 \times 10^2$ | $2.37 \times 10^3$ | $1.13 \times 10^1$ | $8.13 \times 10^4$ | $1.93 \times 10^7$ | $1 \times 10^{14}$ | $1 \times 10^3$ |
| Jupiter | $9.61 \times 10^2$ | $1.22 \times 10^2$ | $1.75 \times 10^3$ | $1.79 \times 10^0$ | $2.04 \times 10^5$ | $1.65 \times 10^6$ | $1 \times 10^{15}$ | $2 \times 10^4$ |
| Saturn | $5.24 \times 10^2$ | $9.02 \times 10^1$ | $1.50 \times 10^3$ | $7.23 \times 10^{-1}$ | $3.22 \times 10^5$ | $4.91 \times 10^5$ | $4 \times 10^{15}$ | $8 \times 10^4$ |
| Uranus | $2.61 \times 10^2$ | $6.36 \times 10^1$ | $1.26 \times 10^3$ | $2.53 \times 10^{-1}$ | $5.43 \times 10^5$ | $1.21 \times 10^5$ | $2 \times 10^{16}$ | $4 \times 10^5$ |
| Neptune | $1.66 \times 10^2$ | $5.08 \times 10^1$ | $1.13 \times 10^3$ | $1.29 \times 10^{-1}$ | $7.61 \times 10^5$ | $4.94 \times 10^4$ | $4 \times 10^{16}$ | $1 \times 10^6$ |
| Pluto | $1.27 \times 10^2$ | $4.43 \times 10^1$ | $1.05 \times 10^3$ | $8.57 \times 10^{-2}$ | $9.34 \times 10^5$ | $2.86 \times 10^4$ | $7 \times 10^{16}$ | $2 \times 10^6$ |
| | $5.00 \times 10^1$ | $2.79 \times 10^1$ | $8.33 \times 10^2$ | $2.13 \times 10^{-2}$ | $1.87 \times 10^6$ | $4.47 \times 10^3$ | $4 \times 10^{17}$ | $2 \times 10^7$ |
| | $1.00 \times 10^1$ | $1.25 \times 10^1$ | $5.57 \times 10^2$ | $1.91 \times 10^{-3}$ | $6.27 \times 10^6$ | $1.79 \times 10^2$ | $1 \times 10^{19}$ | $6 \times 10^8$ |
| | $5.00 \times 10^0$ | $8.81 \times 10^0$ | $4.69 \times 10^2$ | $6.74 \times 10^{-4}$ | $1.05 \times 10^7$ | $4.47 \times 10^1$ | $4 \times 10^{19}$ | $3 \times 10^9$ |
| Oort's Cloud | $2.50 \times 10^{-1}$ | $1.97 \times 10^0$ | $2.22 \times 10^2$ | $7.53 \times 10^{-6}$ | $9.97 \times 10^7$ | $1.12 \times 10^{-1}$ | $2 \times 10^{22}$ | $2 \times 10^{12}$ |
| | $1.00 \times 10^{-1}$ | $1.25 \times 10^0$ | $1.76 \times 10^2$ | $1.91 \times 10^{-6}$ | $1.98 \times 10^8$ | $1.79 \times 10^{-2}$ | $1 \times 10^{23}$ | $2 \times 10^{13}$ |

### 2.1 The scenario

About half of all stars in the galaxy are believed to be born in binaries (Harwitt, 1988). We therefore investigate the possibility that the sun may also have been born in a binary with a massive star. Such massive stars would have lifetime less than a billion years and would explode into a supernova leaving behind a large amount of r-process material. The energy release from such a supernova would disrupt the binary and give the sun a velocity with respect to the local rest frame. The sun has a velocity of about 16 km/s in this local rest frame that rotates with the spiral arm (Frisch, 1995). If this velocity of the sun is a remnant of such an explosion, we can determine several parameters of the residuals.

On the basis of these assumptions, we will argue that the planetary angular momentum would be much larger than that of the sun. Such a disk would also have a large temperature gradient making it convectively unstable. We estimate the convection time scales in the disk and diffusion time-scales elsewhere and show that they will be similar. Its size would also be less than 500 AU due to thermal motions. Such a scenario, with an inner convective disk up to about 5 AU and an outer stable disk would give rise to planets of different types of composition inside and outside 5 AU.

### 2.1.1 Binary parameters

The sun has a velocity of about 20 km/s in the local rest frame. If we assume that this velocity was a result of the acceleration from a supernova disruption, the sun would have gained a kinetic energy of $10^{46}$ ergs. This is typically $10^{-5}$ of a supernova kinetic energy of $10^{52}$ ergs. Hence, the sun must have subtended an effective angular size of $10^{-5}$ of the total sky area of 40,000 deg$^2$. The binary separation therefore must be of the order of 1 AU.

### 2.1.2 Angular Momentum

One of the outstanding problems of the models for solar system origin is that the planets have much larger angular momentum than the sun (Taylor, 1992). The angular momentum of the sun is several orders of magnitude less than that of the planets. If the solar system originated from the material of an exploded binary, the companion's initial velocity and distance would decide the angular momentum of the material captured by the sun. If we assume that 0.1 $M_{sun}$ of material moving with $10^8$ cm/s was captured at an average distance of 100 $R_{sun}$, the angular momentum of the matter trapped by the sun would be about $10^{53}$ gm cm$^2$/s. This is about $10^5$ times larger than the angular momentum of the sun 1.9 x $10^{48}$ gm cm$^2$/s (Pijpers, 1998).

### 2.1.3 Convective instability

If the material from the binary settles down into an accretion disk, it would be possible to calculate the dynamics of such a disk. Assuming the disk mass to be 0.1 $M_{sun}$ spread over a distance of 100 AU, we get the mean density to be about $10^{15}$ p/cm$^3$ for a disk thickness of 0.1 $R_{sun}$. In order to investigate the stability of the disk from solar radiation, we define the maximum luminosity $L_{max}$ that can pass through a region without inducing convective instability (Bowers and Deeming, 1997a,b). $L_{max}$ (in ergs/s) for a disk is given by

$$L_{max} = 1.23\ 10^{-18}\ \mu\ (M_{sun}/\kappa)\ (T^3/\rho) \tag{1}$$

Here $\kappa$ is the opacity, $\rho$ is the density and $\mu$ is the mean molecular weight. For the sun, $L_{sun} = 4 \times 10^{33}$ erg/s. We therefore get that the disk will be convective for $\rho > 10^{15}$ p/cm$^3$. Hence, as the disk begins to settle in the gravitational field of the sun, the adiabatic gradient will fall more rapidly than the temperature gradient and the disk will be *convectively unstable.*

### 2.1.4 Other properties

We derive the temperature profile assuming a free fall characteristic for the gas. We define the following boundary conditions. We take $T_0 = 2\ 10^6$ K at $r_0 = 0.005$ AU (~ 1.5 $R_{sun}$). If we assume that the temperature falls as C/r (where r is the radial distance from the sun and C is a constant), the temperature of the disk will reach 100 K by about 100 AU. On the basis of this boundary condition, we calculate the value C. We take the temperature to be constant beyond this point. Under such a condition, the thermal velocity of the particles given by

$$V_{thermal} = \sqrt{3kT/m_p} = \sqrt{3kC/m_p r} \tag{2}$$

Here $k$ is the Boltzmann's Constant, T the temperature, and $m_p$ the mass of the proton. This is given in columns 6 and 7 of Table 1a. By about 500 AU, the thermal velocity will exceed the escape velocity of the particles and the accretion disk will be self limited to about 500 AU. This value would be smaller once buoyancy due to magnetic fields and other effects are considered.

We next calculate the time scales of convection. The following relation gives the convective acceleration (Kaula, 1986). $d^2r/dt^2 = M(r)G/r^2\ dT/T$. From this the convective time scale can be calculated as

$$\tau_{convection} = \sqrt{2Rr2/GM\ dT/T} \tag{3}$$

where $R$ is the size of the convective cell. In order to evaluate this equation, we take $R = 0.25$ AU and $M = M_{sun}$. As can be seen in columns 2 and 3 of

Table 1b, for regions up to Mars these time scales are of the order of $10^3$ years.

### 2.1.5 Diffusion time scales

In the regions with particle density < $10^{15}$ p/cm$^3$, diffusive mixing will be most important. Diffusive velocity and particle density can be used to calculate the diffusion time scales. The diffusion mean free path ($\lambda$) is calculated using the formula (Longair, 1992).

$$\lambda = \left(\sqrt{2\pi\rho d2}\right)^{-1} \tag{4}$$

where $\rho$ is the number density of particles and *d* is the particle diameter. The diffusion time scales are given by

$$\tau_{diffusion} = \lambda / 3\, V_{thermal} \tag{5}$$

We calculate the $\rho$ assuming that it also falls as $r^{-1}$. The value of the constant of proportionality is derived assuming $\rho_0 = 10^7$ p/cm$^3$ at $r_0 = 0.005$ AU. This is given in column 4 of Table 1b. The diffusive time scales are given in columns 5 and 6 of Table 1b. The diffusive time scales will be of the order of $10^3$ years for accretion into the inner planets and will be sensitive to the particle mass. The gravitational acceleration, which is sensitive to particle mass, would increase the accumulation rate of heavy material into the inner regions of the solar system making them substantially metal rich.

Consequently, there would be both a convective mixing as well as a diffusive drift of the particles into the inner Heliosphere. The result of this mixing will be a collection of heavier elements into the inner solar system, which would be cooled very rapidly to form granules. This limits the effective convective cell size to about 5 AU since the convective time scales beyond this size would be much larger than the cooling time scales limiting the mixing of the material. The region up to 5 AU therefore would be enriched in heavier elements due to a mixture of diffusion and convection while outer regions will be more gaseous. This would have significant effect on planetary condensations once the nebula is thin enough to be optically transparent.

### 2.1.6 Dynamical parameters

We next consider the dynamical parameters. The basic dynamics of the disk are parameterized by three parameters ($\alpha,\beta,\delta$), namely, the ratio of thermal, rotational and magnetic energy to the gravitational energy of the system (Kaula, 1986). Detailed simulation studies have shown that $\alpha$ is the most important of these parameters. For high values of $\alpha$, fragmentation of the proto solar system disk will be suppressed. As $\alpha$ falls below 0.4, the disk will begin to fragment into rings. Once $\alpha$ falls below 0.1 the fragments become unstable to collapse and simulations of such systems show that about 20 percent of the angular momentum is converted into the spin of the planets. Under the standard scenario however, it would take considerable time for $\alpha$ to change significantly. However, under the assumption of a cooling disk in the presence of the sun, such a transition would be much more rapid.

The Oort's cloud of minor bodies exists at about 20,000 AU. Recent measurements of the Heliopause clearly indicate that the zone of influence of the solar wind is within about 100 AU, where the interstellar matter stops cosmic rays from falling in. Interstellar dust however, is a major component of matter at 5 AU. It is therefore not possible for *solar* material to have been spread over such a large distance of Oort's cloud. However, if the Oort's cloud is the outer edge of the supernova that formed the proto solar system cloud, such a material can move with the sun.

## 2.2 Cosmochemistry

We next discuss the evidence from cosmochemistry. Shu *et al.* (1997) have discussed the meteoritic observations in the light of the suggestion that the solar system may have condensed from a disk around the sun. The most notable feature of composition around the sun is that the inner planets are richer in metals compared to the outer planets. Also, there is a marked dual structure of mass composition with the earth being the heaviest of the inner planets. The outer planets begin with the heaviest member Jupiter and then taper down to Pluto, which is roughly of 1 earth mass. If convection were effective to about 5 AU as suggested by calculations above, the inner planets would be mixed by convection and enriched by diffusion, while the outer planets are poorly heated and collect gaseous material. Qualitatively, this is what is observed. It is interesting that all stars around which planets have been found are unusually metal rich (Lewis, 1995). This probably indicates that the most probable way that planets may condense around a star is by binary explosion, which will enrich the surviving star's atmosphere with high Z elements of varying amounts depending on proximity of the original binary partner. Hence, if this suggestion is accurate, the closer the planetary component, the higher should be the metalicity of the parent star.

Studies of the composition show interesting abundance patterns. The first of these features is the similarity in the universal abundance, planetary abundance and the solar photospheric abundance. This would arise as a direct consequence of the material of the supernova remnant being gravitationally pulled onto the solar surface. Detailed analysis of the chondritic material show that these are a mixture of both, s- and r-process material as well as material that has been suddenly quenched to temperatures below 1500K from much higher temperatures (Taylor, 1992; Lewis, 1995). We have calculated the convection time scales (Table 1a,b) for material transferred across the disk. These are of the order of a few months and hence any chondritic material which is transferred by convection would be quenched as rapidly forming grains, a certain fraction of which will be reheated as it returns to the inner regions.

## 3. CONCLUSIONS

In conclusion therefore, we propose that there is considerable merit in evaluating a scenario where the solar system was formed *later* than the sun from the residue of a supernova explosion of a companion of the sun. Such a scenario would explain some of the aspects of the observed features of the solar system.

## REFERENCES

Bowers R. and Deeming T.: 1997a, *Astrophysics I. Stars,* Jones and Bartlett Publishers, Sudbury, MA, 343 pp.

Bowers R. and Deeming T.: 1997b, *Astrophysics II. Interstellar Matter and Galaxies,* Jones and Bartlett Publishers, Sudbury, MA, 588 pp.

Frisch, P.C.: 1995, "Characteristics of nearby interstellar matter", *Space Sci. Rev.* **72**, 499-592.

Gonzalez G.: 1998, "Extra solar planets and Eti", *Astron. Geophys.* **39**, 8.

Harwitt M.: 1988, *Astrophysical Concepts*, Springer Verlag, New York, NY, 561 pp.

Hoyle F: 1945, "Note on the origin of the solar system", *Mon. Not. Royal Astron. Soc.* **105**, 175-178

Kaula W.M.: 1986, "Formation of the sun and its planets", in *Physics of the Sun,* vol. **III**, eds., Sturrock, P.A., Holzer, T.E., Mihalas, D.M. and Ulrich, R.K., D. Reidel Publishing Company, Dordretch, Holland, pp. 1-32

Lewis, J.S.: 1995, *Physics and Chemistry of the Solar System*, Academic Press, San Diego, CA, pp. 66-128.

Longair, M.S.: 1992, *High Energy Astrophysics*, Second Edition, Cambridge University Press, Cambridge, UK, 393 pp.

Pijpers, F.P.: 1998, "Helioseismic determination of the solar gravitational quadrupole-moment", *Mon. Not. Royal Astron. Soc.* **297**, L76-L80.

Shu, F. H., Shang, H., Glassgold, A. E. and Lee, T.: 1997, "X-rays and fluctuating x-winds from protostars", *Science* **277**, 1475-1479.

Taylor S.R.: 1992, *Solar System Evolution: A New Perspective*, Cambridge University Press, Cambridge, UK, 307 pp.

# Origin of Elements in the Solar System

O. Manuel
*Chemistry Department, University of Missouri, Rolla, MO 65401 USA*
*om@umr.edu*

*"If our inconceivably ancient Universe even had any beginning, the conditions determining that beginning must even now be engraved in the atomic weights."* Theodore W. Richards (1919)

Abstract: The solar system is chemically and isotopically heterogeneous. The earth contains only 0.0003% of the mass of the solar system, but the abundance pattern of non-radiogenic isotopes for each terrestrial element has been defined as "normal".

The outer planets consist mostly of light elements like H, He and C. The inner planets are rich in heavy elements like Fe and S. Isotopic irregularities are closely linked with these chemical differences in planets, as well as in the primary minerals of chondritic meteorites.

Chondrites are heterogeneous, agglomerate rocks from the asteroid belt that separates the two types of planets. They contain troilite (FeS) inclusions with isotopically "normal" Xe, like that found in the inner planets. Chondrites also contain diamond inclusions (C) with abundant He and "strange" Xe enriched in isotopes from the r- and p-processes. The Galileo probe found similar r-products in the Xe isotopes of Jupiter, a planet rich in He and C. The sun is a mixture of the chemically and isotopically distinct components found in its planetary system. Inter-linked chemical and isotopic irregularities, short-lived radioactivities and other post-1957 observations are used here to evaluate the two most conflicting opinions on the origin of elements in the solar system.

The first of these is that the elements in the solar system originated via remote element synthesis (RES). The RES model is the modern version of the classic nebular model postulated by Kant and Laplace over 200 years ago for the origin of the solar system. It is a natural extension of the cosmological view of element synthesis. According to the nebular RES view, products of nuclear reactions collected from multiple stellar sources over vast regions of space and produced a well-mixed protosolar nebula having approximately the elemental composition of the sun's photosphere and isotopic ratios of carbonaceous chondritic meteorites. The sun formed as a fully convective, homogeneous protostar. Elements in the planetary system were subsequently redistributed to produce the solar system's current chemical gradients. This nebular RES model was modified in the late 1970s and early 1980s to try to explain the occurrence of isotopic anomalies and decay products of short-lived nuclides by the addition of very small amounts of exotic nucleogenetic material, which either had been injected from nearby stars or survived as interstellar grains that became embedded in meteorites.

An alternative explanation is that the elements in our solar system were produced by local element synthesis (LES). The LES model is the latest version of the classic catastrophic model postulated by Buffon over 250 years ago. This asserts that the entire solar system formed directly from poorly-mixed debris of a spinning star, concentric with the present sun, which exploded axially as a supernova (SN). The sun formed on the SN core and the chemical gradients in the planetary system were inherited from the parent SN. The outer planets formed mostly from the light elements in the outer SN layers. Iron meteorites and the cores of the terrestrial planets formed in a central, iron-rich region of the SN debris. These planet cores were subsequently overlaid as lighter material from other SN layers fell toward the condensing sun. Diffusion enriched lighter nuclei at the evolved sun's surface making it appear to be composed of light elements such as H and He. Lyttleton, Hoyle, and Brown earlier suggested other versions of LES to explain the distribution of angular momentum in the solar system.

Reynolds' mass spectrometer began generating empirical challenges to the nebular RES model soon after publication of the classical papers by Burbidge *et al.* (1957) and Cameron (1957). Data from space probes and ion-probe mass spectrometers have made the RES model even less attractive. The catastrophic LES model now explains the maximum number of phenomena (chemical and isotopic irregularities, decay products of short-lived radioactivities, and observations on planets, the sun and other stars) with the minimum number of postulates, although this implies that elemental abundances for the sun are similar to those of the inner planets and that nuclear evolution is much more advanced for elements in the sun's interior than for elements in the solar photosphere or in the Jovian planets. The most abundant elements for the bulk sun are the same ones concluded 80+ years ago from early analyses of meteorites (Harkins, 1917). They are: Fe, Ni, O, Si, S, Mg, and Ca. Others have reached somewhat similar conclusions about the sun from efforts to find a unique solution to the solar structure equations (Rouse, 1975), to explain the low flux of solar neutrinos (Hoyle, 1975), and to explain isotopic anomalies in meteorites (Lavrukhina, 1980; Lavrukhina and Kuznetsova, 1982).

# 1. INTRODUCTION

This paper brings together and evaluates the observations that provide meaningful information on the origin of elements in the solar system, with emphasis on new data not available to Burbidge *et al.* (1957) and Cameron (1957) in writing their classical papers on stellar nucleosynthesis. It is not intended that this work will be an exhaustive review of all the earlier research on the subject. The number of authors who have contributed to the knowledge of the origin and abundances of the elements is too great to be included in a review paper of reasonable length. The small book by Kuroda (1982) gives an excellent historical review of scientific developments on the origin of chemical elements, and includes the opening quote by Theodore Richards on receiving the Nobel Prize in Chemistry in 1919.

At the onset it is useful to acknowledge that most of the material in the solar system lies hidden in the sun's interior. For the small fraction of material in the planetary system, the chemical composition varies with radial distance from the sun, and the individual planets themselves consist of layers of different compositions. Thus, (a) only a small fraction of material in the solar system is available for study, and (b) there are steep chemical gradients in the accessible material.

In spite of this difficulty, recent measurements have provided much better experimental data with which to evaluate the models proposed for the origin of elements in the solar system than for those in any other part of the universe. Therefore it is logical to focus on, and to attempt to explain, the system for which one has the most information. There is not enough detailed experimental data available to decide the origin of elements elsewhere in the universe.

Most authors assume that local abundances of elements and isotopes were generally representative of those in the galaxy, and perhaps beyond when the protosolar nebula formed. Thus, "universal" and "cosmic" are frequently substituted for "solar" as the adjective describing these abundances. If the sun and other stars developed from homogeneous material which had been collected from multiple stellar sources over vast regions of space, then "universal" or "cosmic" abundances may be the same as "solar" ones. Indeed, that is one of two models proposed to explain the origin of elements in the solar system. This model is known as remote element synthesis (RES). An alternative model of local element synthesis (LES) was proposed in the 1970s. These are summarized below.

1. *Remote element synthesis (RES).* According to this popular view (Wood, 1999), products of nuclear reactions gravitated from multiple stellar sources over vast regions of space to produce a dense interstellar cloud of gas and dust such as the Orion Nebula. Simultaneous contraction and

fragmentation of this interstellar cloud occurred because, "As the cloud contracted, successively smaller portions of it became gravitationally unstable and begin to contract independently." (Wood, 1978, p. 327). Figure 1 is a typical text-book illustration of the formation of the solar system from one of these small, contracting sub-units (Zeilik, 1982, Figure 12.21).

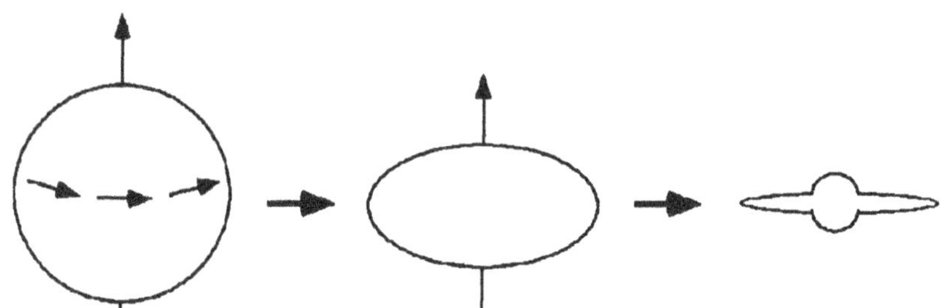


*Figure 1.* Schematic drawing of events for the formation of the solar system from a contracting fragment of an interstellar cloud with angular momentum. This is the nebular RES model.

On the left is the well-mixed protosolar nebula having approximately the elemental composition of the sun's photosphere and the isotopic composition of carbonaceous chondritic meteorites (*e.g.*, Arnett, 1996). Angular momentum is conserved as this fragment of interstellar cloud contracts, thus causing it to rotate faster and to take on a disk-like shape. The sun forms in the center as a fully convective, homogeneous protostar. Then, elements redistribute to produce the solar system's current chemical gradients. The RES model is an outgrowth of the classic nebular model postulated by Kant and Laplace over 200 years ago for the origin of the solar system (Miyake, 1965). The RES model was recently modified to explain the occurrence of isotopic anomalies and the decay products of short lived nuclides in meteorites by the addition of about 0.0001 parts of exotic nucleogenetic material from a separate stellar source (Fowler, 1984; Wasserburg, 1987; Anders and Zinner, 1993; Cameron, 1995; Zinner, 1997). This may have been injected from a nearby supernova (Manuel *et al.*, 1972; Cameron and Truran, 1977) or survived within interstellar grains of dust that had been in the early protosolar nebula and became embedded in meteorites like microscopic time capsules of exotic material (Black, 1972; R. Clayton *et al.*, 1973; D. Clayton, 1975a,b; Ott, this volume).

2. *Local element synthesis (LES).* According to this view, the solar system formed from locally-produced stellar debris. The LES model is an

outgrowth of the classic catastrophic model postulated by Buffon in 1745 and later refined by Chamberlin and Moulton and by Jeans early in this century (Miyake, 1965). Manuel and Sabu (1975, 1977) asserted that the LES model also explains the occurrence of isotopic anomalies, inter-linked chemical and isotopic irregularities, and the decay products of short-lived nuclides. They suggested that the entire solar system formed directly from chemically and isotopically heterogeneous debris of a spinning star, concentric with the present sun. Figure 2 shows their mechanism for the formation of the solar system.

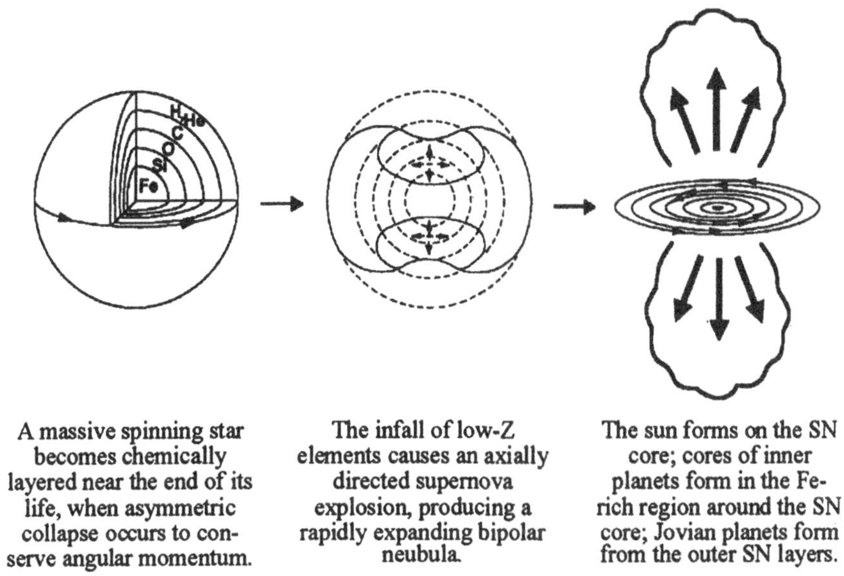


*Figure 2.* Schematic drawing of events for the formation of the solar system and the LES model (Manuel *et al.*, 1998a).

According to this view, the sun formed on the supernova (SN) core and the outer planets developed mostly from the light elements in the outer SN layers. Iron meteorites and cores of the terrestrial planets formed in a central, Fe-rich region of the SN debris. The sun and the inner planets accreted in layers, as lighter material from other SN layers fell toward the forming sun. Diffusion continued to enrich the lighter elements and the lighter isotopes of individual elements at the evolved sun's surface (Manuel and Hwaung, 1983). According to the LES model, chemical gradients were inherited from the parent SN, and these decreased with time from natural mixing. Three earlier versions of the LES model suggested that a spinning star made the solar system (Lyttleton, 1941), that debris from an explosion of the sun's binary companion formed the planetary system (Hoyle, 1944), and that the SN shell ejected by an explosion fragmented to form many separate solar systems (Brown, 1970).

Most observations available in 1957 seemed to favor the RES model. Recent reviews on element synthesis (Arnett, 1996; Wallerstein *et al.*, 1997) and almost all the papers presented at a recent St. Louis symposium on isotopic anomalies and short-lived nuclides in meteorites (Bernatowicz and Zinner, 1997) also conclude that the latest observations are consistent with the concept of RES in multiple stellar sources. For example, in the overview, Zinner (1997, p. 3) states that "*It was realized early on that the solar system not only is a very homogeneous mixture of material from many different stellar sources but that these stars themselves incorporated the debris of previous generations of stars ("galactic chemical evolution").*" This homogeneous mixture of the RES model is illustrated in the left frame of Figure 1. A follow-up news report, begins, "*Amazingly, individual grains of dust from stars that existed before the sun was born have made their way to Earth in meteorites.*" (Bernatowicz and Walker, 1997, p. 26). These illustrate the continued dominance of the RES model in the space sciences community.

However, one point that has troubled scientists for decades (*e.g.*, Lyttleton, 1941; Krat, 1952; Urey, 1956; Huang, 1957) concerns the production of binary, multiple, and planetary systems by the simple condensation of gas and dust with large angular momentum. In spite of this concern, variant ideas about the origin of elements in the solar system have encountered stiff resistance, perhaps because the 1957 papers made such brilliant advancements in noting the connection between abundances of nuclides and their syntheses by reasonable nuclear reactions at each stage of stellar evolution, up to the final supernova shown in Figure 2.

A large amount of new data on the solar system has been reported since 1957. Isotopic ratios in the solar wind and in solar flares, the neutrino flux, and helioseismology have given new insight into the sun's interior. Analyses on samples of meteorites, the moon, Jupiter, Mars, and material from the Earth's interior have revealed the decay products of short-lived nuclides and evidence that elements, and even the isotopes of individual elements, were poorly mixed at the birth of the solar system. The Hubble telescope, pulsar planets, and SN1987A have given new insight into the birth and death of stars. These post-1957 observations can be compared with predictions of the RES and LES models to evaluate their merits.

In many cases, the nebular RES model (Figure 1) and the catastrophic LES model (Figure 2) require or predict specific trends that are different from one another. A few examples are below.

(a) According to the RES model, that stable isotope which has less nuclear binding energy than any other, $^{1}H$, accounts for about 90% of all atoms in the solar system; the LES model claims that most of the solar system came from the hot interior of a supernova where nuclei with much greater nuclear stability, *e.g.*, $^{56}Fe$, are more abundantly formed.

(b) According to the RES model, chemical heterogeneities in the solar nebula increased with time, whereas for the LES model chemical and isotopic diversity clearly decreased with time.

(c) The RES model requires that major chemical heterogeneities in the planetary system were produced by variations in P, T, oxygen fugacity, etc.; the LES model requires that these heterogeneities were inherited from the parent star.

(d) The RES model predicts no linkage between elemental and isotopic compositions; the LES model predicts that some elemental and isotopic irregularities of the supernova remained coupled, not only in microscopic grains of meteorites, but also in planetary-sized objects.

(e) Excesses and depletions of the same isotope are difficult to explain by the RES model, usually requiring separate stellar sources; excesses and depletions of the same isotope are predicted in the LES model because "normal" isotopic ratios are produced by mixing these.

(f) The RES model predicts no coupling of nucleogenetic isotopic anomalies with the decay products of short-lived radioactivities; the LES model predicts that these will be coupled because natural mixing reduced the levels of nucleogenetic isotopic anomalies while short-lived nuclides in the solar nebula decayed away (Figure 2).

(g) The RES model predicts similar elemental and isotopic compositions for the sun and Jupiter (Figure 1); the LES model predicts that the compositions of these two largest bodies in the solar system will differ because they consist of material from different regions of the supernova that gave birth to the solar system (Figure 2).

In what follows, abundance estimates are reviewed and the literature on the origin of elements is evaluated. An extensive review is given of post-1957 observations, including trends in composition, tracer isotopes, inter-linked chemical and isotopic heterogeneities in meteorites and in planets, decay products of short-lived nuclides, isotopic anomaly patterns, hetero-geneities in the sun, and pertinent astronomical and geological observations. Although the review of literature on the origin of the elements follows the review of abundance estimates, the abundance and origin of the elements intertwine. Accordingly, the different models of origin, both RES and LES, each have their own abundance counterparts.

## 1.1 ABUNDANCES OF ELEMENTS

A great deal of dedicated labor by a large group of scientists has been expended in collecting elemental and isotopic abundance data. Readers are referred to reviews on this subject in the books and articles cited below and to the articles by Ebihara *et al.* (this volume) and Grevesse and Sauval (this volume). Readers should also note especially the excellent summaries on the sources of abundance data in Chapters 2 and 3 of Kuroda (1982), Section II of Trimble (1975), Chapter 2 of Arnett (1996), and Trimble (1996).

In their classical papers on the synthesis of elements in stars, both Cameron (1957) and Burbidge *et al.* (1957) relied heavily on the Suess and Urey (1956) compilation of data for abundances of the elements. The data by Suess and Urey data are still widely cited. The scientific community has adopted only minor changes in elemental abundance estimates over the past 40+ years (*e.g.*, Cameron, 1968, 1973, 1982; Palme *et al.*, 1981; Anders and Ebihara, 1982; Anders, 1988; Anders and Grevesse, 1989; Grevesse and Sauval, this volume), in spite of rapid advancements in analytical instrumentation and space exploration that have increased the availability of material from various parts of the solar system.

However, estimates of the abundances of the elements underwent radical changes during the preceding 40 year period, between 1917 and 1957. That change came primarily from assuming that abundances of elements in the solar photosphere are representative of those in the interior of the sun (Goldschmidt, 1938). For reasons given below, there may be reason to question whether some of the changes were improvements.

The pioneer in building a reliable data base for elemental abundance estimates, William Harkins of the University of Chicago, was also a leader in recognizing a possible relationship between elemental abundances and nuclear stability: "*...the more stable atoms should be the more abundantly formed...*" (Harkins, 1917, p. 859). He warned that elemental abundances in the Earth's crust and the sun's gaseous envelope may not represent the overall composition of these bodies.

After cautioning against use of surface abundances, he pointed out that, "*There is, however, material available of which accurate quantitative analyses can be made, and which falls upon the earth's surface from space. The bodies which fall are called meteorites, and no matter what theory of their origin is adopted, it is evident that this material comes from much more varied sources than the rocks on the surface of the earth. In any event, it seems probable that the meteorites represent more accurately the average composition of material at the stage of evolution corresponding to the earth than does the very limited part of earth's material to which we have access.*" (Harkins, 1917, p. 861).

Harkins therefore used chemical analyses on 318 iron meteorites and 125 stone meteorites to arrive at his conclusion that the first seven elements in order of abundance consist of Fe, O, Ni, Si, Mg, S, and Ca. He stated that, "*...not only do all these elements have even atomic numbers, but in addition* *they make up 98.6% of the material of the meteorites.*" (Harkins 1917, p. 862). Fe alone accounts for 72% of these elements.

Harkins' abundance estimates clearly reflected the enhanced nuclear stability of elements with even atomic numbers and the occurrence of maximum nuclear stability for Fe (Mayer, 1948), Harkins' most abundant element. When Harkins (1917) considered only the elements in stone meteorites, he found the same seven most abundant elements, but their order changed slightly to O, Fe, Si, Mg, S, Ca, and Ni.

The positions of Fe, Ni and S relative to those of O, Si, Mg, Ca and Ca depend on the ratio assumed for metal: sulfide: silicate phases. Noddack and Noddack (1930) assumed a ratio of (iron: troilite: stone) = 68: 9.8: 100 and used the chemical analyses of iron meteorites, troilite inclusions (FeS), and stone meteorites to conclude the same seven most abundant elements, in slightly different order, O, Fe, Si, Mg, Ni, S and Ca.

Most later authors (Noddack and Noddack, 1934; Goldschmidt, 1938; Urey, 1952) assumed a lower ratio of (metal : silicate), except Brown (1949). Brown assumed that the Earth's core and mantle have the compositions of iron and stone meteorites, respectively. He used the location of the seismic discontinuity for the Earth's mantle/core boundary to obtain a value of (metal : silicate) = 67 : 100 (Brown, 1949).

Goldschmidt (1938, 1958) suggested that chondritic meteorites may best represent the abundances of nonvolatile elements and that stellar spectra may estimate the abundances of volatile elements. This assumes the validity of the RES model, in which an initially homogeneous cloud formed the sun and its planets. If so, then "*The enormous difference in chemical composition of the giant planets, the earth-like planets, and meteorites, is quite natural when examined from the point of view of cosmic evolution.*" (Goldschmidt, 1958, p. 70). He continued: "*Thus the giant planets, the earth-like planets, planetoids and meteorites differ in that only the giant planets have been able to retain much of their original share of hydrogen (and probably helium) while all minor and miniature 'autonomous' bodies have lost their hydrogen and helium...*" (Goldschmidt, 1958, p. 71).

The elemental abundance estimates by Urey (1952) were also based on the abundances of the light, volatile elements in stellar spectra and on the abundances of nonvolatile elements in meteorite phases weighed, "*...in such a way as to secure a mixture having the density of the moon.*" (Urey, 1952, p. 230). Suess and Urey (1956) did not accept the (metal : silicate) ratio recommended by Brown (1949), adopting instead the suggestion by Urey (1952) that fractionation occurred during the formation of the terrestrial planets, thus, "*...separating metal from silicate in such a way that the silicate was lost preferentially. Hence, the ratio of core to mantle of the earth cannot serve as a basis for an estimate of the respective cosmic ratios.*"

(Suess and Urey, 1956, p. 56). Table 1 compares estimates by Harkins (1917) and Suess and Urey (1956) for the seven most abundant elements.

*Table 1.* Examples of pre-1957 estimates of the solar system's most abundant elements

| Elements | Harkins (1917) | Suess and Urey (1956) |
|---|---|---|
| The Most Abundant | Iron (Fe), element 26 | Hydrogen (H), element 1 |
| 2nd Most Abundant | Oxygen (O), element 8 | Helium (He), element 2 |
| 3rd Most Abundant | Nickel (Ni), element 28 | Oxygen (O), element 8 |
| 4th Most Abundant | Silicon (Si), element 14 | Neon (Ne), element 10 |
| 5th Most Abundant | Magnesium (Mg), element 12 | Nitrogen (N), element 7 |
| 6th Most Abundant | Sulphur (S), element 16 | Carbon (C), element 6 |
| 7th Most Abundant | Calcium (Ca), element 20 | Silicon (S), element 14 |

Iron (Fe) is the most abundant element in Harkins' estimate. This consists mostly of $^{56}$Fe, the decay product of a doubly-magic nuclide, $^{56}$Ni. $^{56}$Fe is the most stable of all nuclides, i.e., it has minimum mass per nucleon. The other heavy element on Harkins' list, nickel (Ni) has two stable isotopes, $^{62}$Ni and $^{60}$Ni, with the second and third lightest mass per nucleon. All seven of Harkins' elements exhibit the enhanced nuclear stability associated with even values of atomic numbers, and the five lightest elements on his list consist mostly of nuclides ($^{16}$O, $^{24}$Mg, $^{28}$Si, $^{32}$S and $^{40}$Ca) that have the heightened nuclear stability associated with multiples of the $^{4}$He nuclide.

If the elemental abundances shown in the center column of Table 1 are wrong, then it is a remarkable coincidence that Harkins' mistake in 1917 yielded an abundance pattern of elements that would so closely follow the nuclear properties that were to be discovered years later. Since nuclear evolution is most advanced in the central region of an evolved star (Burbidge *et al.,* 1957), elements that are produced deep in the stellar interior (Figure 2) will display the relationship that Harkins (1917) found between elemental abundances and nuclear stability.

According to the tabulation by Suess and Urey (1956), the first seven elements in order of abundance are H, He, O, Ne, N, C, and Si. Their most abundant element, hydrogen (H), consists mostly of $^{1}$H. This has the highest mass per nucleon (and lowest nuclear binding energy) of any stable nuclide. The five elements with even atomic numbers, He, O, Ne, C, and Si, account for only 7% of these seven most abundant elements; the two with odd atomic numbers, H and N, account for the other 93%.

The sharp contrast between the abundance data of Suess and Urey (1956) and those that reflect nuclear stability (Harkins, 1917) results directly from the assumption by Suess and Urey that abundances of elements in the solar photosphere are representative of those in the sun's interior. This assumption remains in vogue today, although for such assumption, Arnett (1996, p. 10) notes that the most abundant nuclei, "*...are not the most tightly* *bound*", and he later asks, "*...how can the heavier yet more tightly bound nuclei have remained so rare?*" (Arnett, 1996, p. 12).

The solution to this paradox came from signs that internal diffusion in the sun segregates elements and isotopes within it (Manuel and Hwaung, 1983; MacElroy and Manuel, 1986; Bahcall *et al.,* 1995, 1997), as will be illustrated in Table 2 of the next section. It will also be shown that divergent views about the origin of elements in the solar system, whether based on the RES model or the LES model, closely parallel the differing opinions about the internal composition of the sun.

## 1.2 ORIGIN OF ELEMENTS

Most of the early papers about the origin of elements dealt with primordial or cosmological, rather than stellar, synthesis. This concept predates that of stellar synthesis and is by definition remote element synthesis, RES. The possibility of catastrophic local element synthesis, LES, was not considered by Cameron (1957) or Burbidge *et al.* (1957), although one of the authors had earlier suggested that the sun's planetary system formed directly from SN debris (Hoyle, 1944, 1945, 1946a). Thus, nebular RES was tacitly assumed by default in 1957.

### 1.2.1 1957 Views on Element Synthesis

At the time of publication of the papers by Cameron (1957) and Burbidge *et al.* (1957) on stellar nucleosynthesis, there existed two competing ideas of cosmological element synthesis. The *yelm* theory stated that the universe was created at a particular instant of time and expanded ever since. The other idea, the steady state theory, assumed that the universe had no beginning and that matter is created throughout space at the rate that it disappears at the boundary of the expanding universe (Bondi and Gold, 1948; Hoyle, 1948).

Burbidge *et al.* (1957) criticized primordial theories, "*....which demand matter in a particular primordial state for which we have no evidence...*" (Burbidge *et al.*, 1957, p. 550), but they did not explicitly endorse the steady state concept. Cameron (1957) and Burbidge *et al.* (1957) instead focused attention on nuclear reactions that convert H into heavier elements during normal stages of stellar evolution and on mechanisms that eject these nucleosynthesis products back into the interstellar medium.

Both Cameron (1957) and Burbidge *et al.* (1957) used abundance estimates based mostly on elements in the solar system (Suess and Urey, 1956), and both referred to Hoyle's earlier work on thermonuclear reactions in stars (Hoyle, 1946b). But neither mentioned the possibility that elements in the solar system might have been produced locally, LES, nor the suggestion that our planetary system may have formed from debris of the sun's binary companion (Hoyle, 1944, 1945, 1946a).

Burbidge *et al.* (1957) and Cameron (1957) instead laid the foundation for the currently popular concepts of remote element synthesis, RES (Wood, 1978, 1999; Anders and Zinner, 1993; Cameron, 1995; Zinner, 1997) and chemical evolution of the galaxy (Trimble, this volume and 1988; Timmes *et al.*, 1995; Zinner, 1997) by statements such as, "*It is tempting to believe that our galaxy may have been originally composed entirely of hydrogen.*" (Cameron, 1957, p. 203) and, "*Since stars eject the products of nuclear synthesis into the interstellar gas it seems highly probable that only the 'first' stars can have consisted of pure hydrogen.*" (Burbidge *et al.*, 1957, p. 569).

However, that same year Huang (1957) speculated that binary and multiple stars and those with planetary systems may have formed on "prestellar objects"; an idea that had previously been advanced by Krat (1952) and Urey (1956). The catastrophic LES model (Figure 2) is consistent with both i) stellar nucleosynthesis to a final supernova (Burbidge *et al.*, 1957), and ii) the formation of a secondary star and planetary system on the SN core and stellar debris (Huang, 1957).

Burbidge *et al.* (1957) even recognized trends in the abundance data which suggested LES for material in the solar system. Examples are the introductory and concluding statements from Section XII. D. of Burbidge *et al.* (1957, p. 639): "*As mentioned in Sec. VII, the forms of the back sides of the abundance peaks of the r-process isotopes might suggest that the conditions obtaining in a single supernova were responsible for their synthesis.*", and "*It does not appear unreasonable from this point of view, therefore, that a single supernova has been responsible for all of the material built by the r-process currently present in the solar system.*"

Since publication of the classical papers by Burbidge *et al.* (1957) and Cameron (1957), space studies have become popular and important new information has been obtained from analyses of samples from other parts of the solar system and from astronomical observations on the sun and other stars. Results from sample analyses raised questions about the validity of the standard solar model and provided evidence of chemical and isotopic heterogeneities and decay products of short-lived nuclides at the birth of the solar system. Results from astronomical observations raised additional questions about the composition of the sun's interior and revealed asymmetric ejections of poorly-mixed material from dying stars.

The following section will briefly show how a newly developed instrument began finding many apparent exceptions to the nebular RES model for the origin of the solar system soon after publication of the papers by Burbidge *et al.* (1957) and Cameron (1957).

### 1.2.2 Empirical Challenges to the Nebular RES Model for the Solar System

The idea of remote element synthesis (RES) seemed both attractive and natural in 1957 as attention first shifted from cosmological to stellar synthesis. However, the RES concept did not anticipate later findings that the solar system formed from material containing chemical and isotopic irregularities from nucleosynthesis, in addition to short-lived radioactivities.

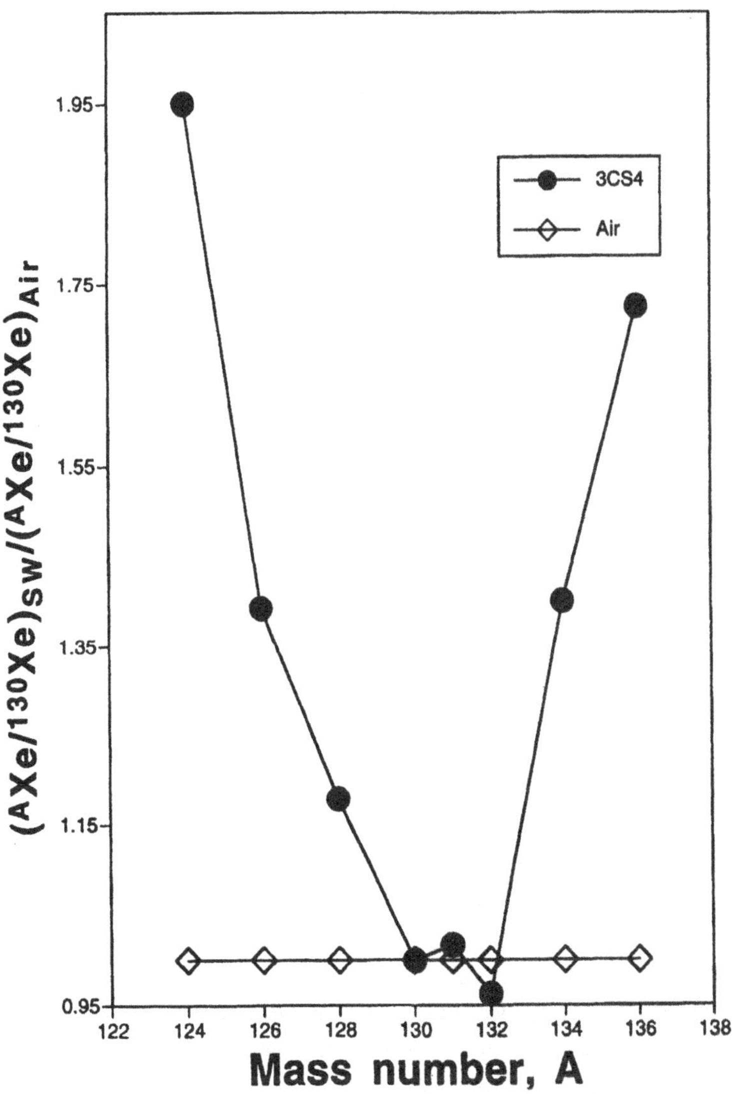


*Figure 3.* A comparison of isotopically "normal" xenon in air (◊) with the isotopically "strange" xenon in the 3CS4 mineral separate of the Allende carbonaceous chondrite (●).

One year before publication of the 1957 papers on stellar synthesis, Reynolds (1956) reported the development of a stable isotope mass spectrometer for noble gases that uncovered many of these features of the early solar system. The new mass spectrometer first uncovered the presence of radiogenic $^{129}$Xe, the decay product of extinct $^{129}$I, in the Richardton meteorite (Reynolds, 1960a). The instrument also exposed this decay product embedded in xenon with a primordial isotopic anomaly pattern for the other eight isotopes, $^{124\text{-}136}$Xe (Reynolds, 1960b).

Reynolds' instrument later showed that xenon in meteorites contains excesses of $^{134,136}$Xe and $^{124,126}$Xe made by the r- and p-processes in a supernova explosion (Manuel *et al.*, 1972). These r- and p-products of xenon occur only in a meteorite phase with abundant primordial He (Sabu and Manuel, 1980), as discovered by a group using the Reynolds' mass spectrometer at the University of Chicago (Lewis *et al.*, 1975). They also found that another meteorite phase trapped a complementary xenon component with excess $^{128\text{-}132}$Xe from the s-process of nucleosynthesis, before a star reaches the final supernova stage (Srinivasan and Anders, 1978).

Figure 3 compares isotopic abundances of "strange" xenon in mineral separate 3CS4 of the Allende carbonaceous chondrite (Lewis *et al.*, 1975) with those of Xe in air. The light, neutron-poor isotopes, $^{124,126}$Xe, and the heavy, neutron-rich isotopes, $^{134,136}$Xe, are enriched by factors of $\approx$ 40-95%. Other heavy elements trapped with isotopically "strange" xenon also display excesses of isotopes produced in a supernova explosion (Maas *et al.*, this volume; Ballad, *et al.*, 1979; Oliver *et al.*, 1981), but the light elements (He, C, Ne) trapped there are isotopically "normal".

Figure 3 shows one of the unexpected discoveries made possible by the use of Reynolds' (1956) high-sensitivity mass spectrometer. Such large isotopic anomalies from nucleosynthesis, the close association of the "strange" xenon with primordial helium, and the decay products of short-lived nuclides in meteorites first prompted Manuel and Sabu (1975, 1977) to propose local elements synthesis, LES.

### 1.2.3 The Rise of the Catastrophic LES Model for the Solar System

The nebular RES model shown in Figure 1 offers no explanation for the reciprocal isotopic anomalies first reported for xenon in meteorites (excess r- and p-products in some minerals and excess s-products in others) and since found in several other heavy elements (Begemann, 1993). Nor does RES explain the close association of isotopically "normal" light elements with isotopically "strange" heavy elements.

The LES model shown in Figure 2 was developed to explain why essentially all primordial He in meteorites is trapped with isotopically "strange" Xe, Kr, and Ar, but none with isotopically "normal" Xe, Kr, and Ar (Manuel and Sabu, 1975, 1977; Sabu and Manuel, 1980). This feature of noble gases is illustrated in Figure 4 for mineral separates of the Allende carbonaceous chondrite (Lewis *et al.*, 1975).

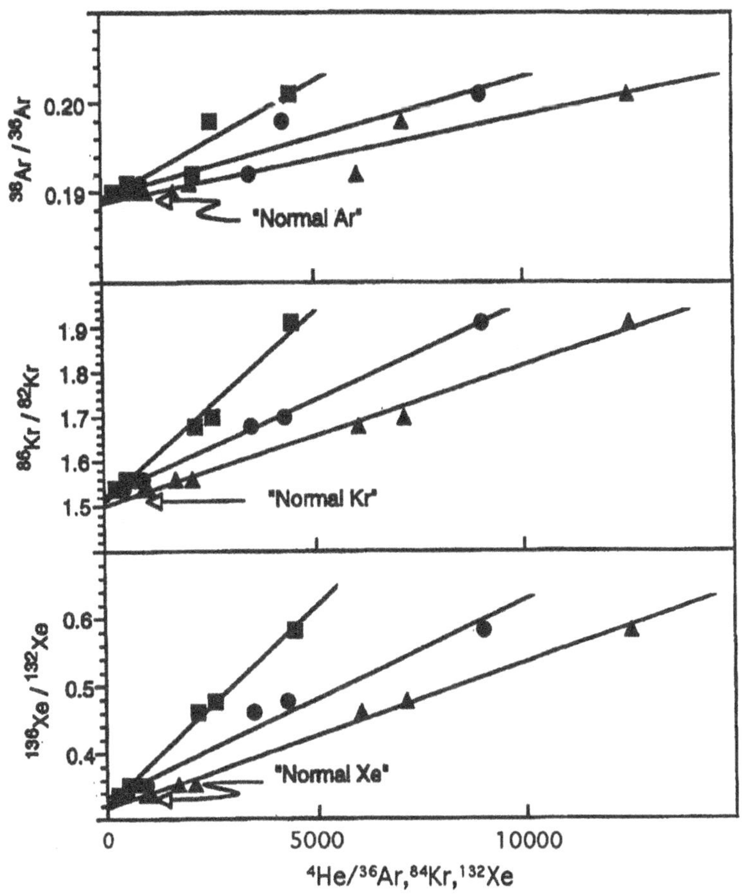


*Figure 4.* The close coupling of primordial He and isotopically "strange" Ar, Kr and Xe. Isotopically "normal" Ar, Kr and Xe are at the intercepts, where the abundance of primordial He vanishes. (Data from Lewis *et al.*, 1975). ▲ represents values of ($^4He/^{132}Xe$) x 0.15; ● represents values of ($^4He/^{84}Kr$) x 0.08; ■ represents values of ($^4He/^{36}Ar$) x 5.

To explain the correlation of elemental and isotopic irregularities shown in Figure 4, Manuel and Sabu (1975, 1977) abandoned earlier suggestions that the "strange" Xe might be a fission product of a superheavy element (Srinivasan *et al.*, 1969) or material added to the solar system from a supernova (Manuel *et al.*, 1972). They proposed that the entire solar system instead formed directly from chemically and isotopically heterogeneous debris of a single supernova, as illustrated in Figure 2.

According to their view, isotopically "strange" Xe, Kr, and Ar are closely associated with abundant He because these elements came from the outer SN layers that also produced the giant Jovian planets. Likewise, isotopically "normal" Xe, Kr, and Ar came from the stellar interior where fusion reactions had consumed light elements such as He to produce heavier elements like those found in iron meteorites and in the rocky, iron-rich, terrestrial planets. Thus, the correlations of isotopic ratios of heavy noble gases with elemental abundances of He (Figure 4) reflect the fact that some minerals formed in the outer layers, and other phases formed in the inner layers, of the SN that produced the solar system (Figure 2).

The possibility of producing the solar system directly from debris of a local star had been proposed earlier to explain its distribution of angular momentum. Lyttleton (1941) imagined that the solar system was produced by the break-up of a single star due to rotational instability. Hoyle (1944, 1945, 1946a) envisioned that the early sun had a larger, binary companion that exploded as a nova or a supernova. A small fraction of the material thrown off by the exploding star became the sun's planetary system. Brown (this volume; 1970-1991) and Brown and Gritzo (1986) focused on the dynamics of SN shell fragmentation to produce solar systems.

The LES model shown in Figure 2 was developed to explain interlinked chemical and isotopic inhomogeneities and decay products from short-lived radioactivities in meteorites (Oliver *et al.*, 1981). However, it also explains the distribution of angular momentum in the solar system.

Support for the LES model shown in Figure 2 came from who found that the abundances of short-lived nuclides in meteorites are consistent with a single stellar source (Kuroda and Myers, 1996, 1997; Sahijpal *et al.*, 1998; Goswami, this volume) and from the efforts of Rouse in developing equations for the calculation of solar structure without the assumption of a constant mean molecular weight (Rouse, 1964, 1969), presenting calculations showing a higher mean molecular weight in the sun's core than indicated by the standard solar model (Rouse, 1975, 1983), and demonstrating that new data on the solar neutrino flux (Rouse, this volume; 1995) are consistent with the presence of a high-Z, iron-like core in the sun.

Other support for the occurrence of an event such as that depicted in Figure 2 came from observations of bipolar nebulae with the Hubble telescope (Frank, this volume; 1997). Frank's (1997, p. 38) idealized drawing for the formation of bipolar nebulae is remarkably similar to the diagram shown by Oliver *et al.* (1981) for the origin of the solar system (Figure 2).

Finally, it should be mentioned that Reynolds' high sensitivity mass spectrometer (Reynolds, 1956) made possible the detection of fission products from $^{244}$Pu in meteorites (Rowe and Kuroda, 1965). This proved the occurrence of a supernova and provided a way to date the time of the supernova event. Kuroda and Myers (1996) combined $^{244}Pu$-$^{136}Xe$ dating with U-Pb dating to show that the SN event occurred 5 Gyr ago. They also showed that the $^{26}Al$-$^{26}Mg$ dates of SiC in meteorites indicate that the larger grains formed 1-2 Myr after the SN event (Kuroda and Myers, 1997).

Thus many features of the catastrophic LES model were confirmed. However, one obvious problem remained. Since light elements like H and He are totally consumed from the interior of an evolved star, the presence of an H,He-rich star at the center of the solar system seemed inconsistent with the suggestion that the sun formed in the manner illustrated in Figure 2.

Data from high sensitivity mass spectrometers (Reynolds, 1956) also solved this enigma by exposing differences in the isotopic compositions of noble gases in the solar wind (SW) and those in the two major primordial noble gas components in the planetary system.

Component X contains isotopically normal He and Ne and isotopically "strange" Xe, Kr and Ar. It is on the right side of Figure 4. Component Y contains isotopically "normal" Ar, Kr and Xe, with little or no He or Ne. It is on the left side of Figure 4. Mixtures of these two dominant types of primordial noble gases (Sabu and Manuel, 1980) produce the correlations shown in Figure 4 for Allende mineral separates (Lewis *et al.*, 1975).

By assuming that noble gases in the sun are a mixture of these two primordial noble gas components, Manuel and Hwaung (1983) were able to identify a systematic enrichment of lighter mass isotopes in all five noble gases of the solar wind (SW) from diffusion inside the sun that selectively moves lighter mass particles to the solar surface. When solar photospheric abundances were corrected for this diffusive mass-fractionation, the seven most abundant elements in the bulk sun were found to be: Fe, Ni, O, Si, S, Mg, and Ca, in decreasing order (See other paper by Manuel, this volume).

Table 2 compares the seven most abundant elements found by Manuel and Hwaung (1983) with those from the recent estimate by Grevesse and Sauval (this volume). The most noteworthy feature of Table 2 is its remarkable similarity to the two lists of elements in Table 1. Another noteworthy feature is the discord between the list of elements in the middle and the right columns of each table.

*Table 2.* Examples of post-1957 estimates of the solar system's most abundant elements

| Elements | *Manuel and Hwaung (1983) | Grevesse and Sauval (this volume) |
|---|---|---|
| The Most Abundant | Iron (Fe), element 26 | Hydrogen (H), element 1 |
| 2nd Most Abundant | Nickel (Ni), element 28 | Helium (He), element 2 |
| 3rd Most Abundant | Oxygen (O), element 8 | Oxygen (O), element 8 |
| 4th Most Abundant | Silicon (Si), element 14 | Carbon (C), element 6 |
| 5th Most Abundant | Sulphur (S), element 16 | Neon (Ne), element 10 |
| 6th Most Abundant | Magnesium (Mg), element 12 | Nitrogen (N), element 7 |
| 7th Most Abundant | Calcium (Ca), element 20 | Magnesium (Mg), element 12 |

*Photospheric abundances of Ross and Aller (1976) corrected for diffusive mass fractionation.

In terms of nuclear physics, the most abundant elements in the middle and right columns of Tables 1 and 2 are opposites: One claims that the most abundant nuclide is $^{56}$Fe, the ash of thermonuclear reactions at equilibrium. The other claims that the most abundant one is $^{1}$H, the stable nuclide that releases the greatest energy per nucleon in thermonuclear reactions. Nuclear evolution is highly advanced in the first case; it has hardly begun in the second.

Slight differences in the order of elements in the right columns of Table 1 and Table 2 reflect the small changes in mainstream elemental abundance estimates over the past 43 years (*e.g.*, Suess and Urey, 1956; Grevesse and Sauval, this volume). In addition to a slight change in order, silicon has been replaced with magnesium as the seventh most abundant element.

However, the agreement of the list of elements in the middle columns of Table 1 and Table 2 is even more remarkable. Harkins (1917) and Manuel and Hwaung (1983) considered completely different measurements in concluding that Fe, O, Ni, Si, Mg, S and Ca are the seven most abundant elements. Harkins used analyses of 443 meteorites to arrive at this conclusion. Manuel and Hwaung combined analyses of the photospheric spectrum by Ross and Aller (1976) with isotopic analyses of noble gases in the solar wind (Eberhardt *et al.*, 1972) to conclude the same seven elements, in only slightly different order. The question is whether this agreement might be happenstance.

Eighty-three (83) elements remain in the solar system, two long-lived actinide elements ($_{90}$Th and $_{92}$U) plus 81 stable ones in the range of $_{1}$H-$_{83}$Bi. If each element has an equal chance of being selected, then the random chance of selecting Harkins' set of seven elements is (7!)(83-7)!/(83!), i.e., there is one chance in 4,000,000,000 for meaningless agreement between the compilations of Harkins (1917) and Manuel and Hwaung (1983). Since these are trace elements in the solar photosphere, the chance for a meaningless selection of these elements from the photosphere is even less if atomic abundances in the photosphere determine the selection probability for an element.

Since nuclear evolution is more advanced in the central region of an evolved star (Burbidge *et al.*, 1957), it should come as no surprise that the most abundant elements in the sun are those produced in the deep interior of a star (See Figure 2). In fact, so much $^{56}$Co was produced in SN1987A that the decay of this $^{56}$Fe-precursor dominated the light curve for about 120-900 days after the supernova explosion (Chevalier, this volume).

In Section 2 below, the tracer isotope technique is used to determine the origin of chemical and isotopic diversity in the solar system, from the nanometer scale (nm) of the tiniest meteorite grains to the astronomical unit scale (AU) of interplanetary distances. Some readers may want to skip Section 2 and use the list of conclusions in Section 3 to select portions of Section 2 for further study.

## 2. DETAILS OF OTHER POST-1957 OBSERVATIONS

Even in 1957, scientists knew that the planetary system was heterogeneous on a macroscopic scale (Urey, 1952). Since then, it has been found that meteorites, the oldest known solids in the solar system, display both microscopic- and macroscopic-scale irregularities in isotopic as well as in elemental compositions. These findings do not agree with predictions of the nebular model of Kant and Laplace (Figure 1) because "*The classical picture of the pre-solar nebula is that of a hot, well-mixed cloud of chemically and isotopically uniform composition.*" (Begemann, 1980, p. 1309).

Exactly what information do these heterogeneities contain about the origin of local elements? Can they be explained by reasonable processes acting on a well-mixed cloud of "cosmic" composition (Figure 1), as indicated by the nebular RES model (Wood, 1999)? Or are they remnants of extremely heterogeneous debris from the supernova (Figure 2) that gave birth to the solar system according to the catastrophic LES model (Manuel and Sabu, 1975, 1977)?

Can less drastic starting material explain the data, such as a rather uniform cloud of interstellar gas and dust but for grains of dust that survived conditions in the presolar nebula and became embedded in meteorites (Black, 1972; R. Clayton *et al.*, 1973; D. Clayton, 1975a, b, 1982; Clayton and Hoyle, 1976)? What significance might these observations have for the injection of exotic material from a nearby supernova explosion, *i.e.*, from a blend of RES methodology with a small contribution from LES methodology (Cameron and Truran, 1977; Goswami, this volume)? Regardless of the mechanism, can the observations be explained if 99.99% of the starting material had normal isotopic composition and only 0.01% had exotic nucleogenetic composition (Fowler, 1984; Wasserburg, 1987)?

Differences in the natural abundances of isotopes allow the "tracer isotope technique" to answer these questions. To illustrate the power of tracer isotopes, suppose there are three samples that consist of the same elements but the isotopic abundances of one of the elements is different in each sample. An example might be samples A, B and C of hydrogen bromide (HBr) with $(^{79}Br/^{81}Br)_A = 1.0$, $(^{79}Br/^{81}Br)_B = 2.0$, and $(^{79}Br/^{81}Br)_C = 0.50$. Each of these three compounds has an isotopic "fingerprint" that allows isotopic analyses to identify possible parent sources and reaction products. Further, tracer isotopes constrain the production of one sample from another. For example, sample A might be a mixture of samples B and C, but obviously sample B could not be a mixture of A and C, nor could sample C be a mixture of A and B.

After a brief review of chemical gradients in the solar system, isotopic irregularities and the tracer isotope technique will be used to decide the roles of RES and LES in producing the elements in the solar system.

## 2.1 TRENDS IN ELEMENTAL COMPOSITION

In addition to new discoveries since 1957, remarkable patterns appear in the abundance distribution of elements in the solar system that may, or may not, have significance. A logical question might be why is H, which is the first element in the periodic table and the lightest of all elements, also the most abundant element in the solar photosphere? Likewise, one might ask why is He, which is the second element in the periodic table and the second lightest of all elements, also the next most abundant element in the solar photosphere? Are these coincidences?

Likewise, one might question the remarkable similarity in the distribution of elements in highly evolved, massive stars, *e.g.*, Figure 46 of Taylor (1972) and Figure 20.44 of Zeilik (1982), as well as in the interior of the Earth, *e.g.*, Figure 9 of Taylor (1972) and Figure 8.2 of Zeilik (1982), and in the planetary system that orbits the sun, for example Figure 1 and Table 1.1 of Miyake (1965). In each case, the central region has abundant even-Z elements with maximum nuclear stability, such as Fe and Ni (Z = 26 and 28). The outer region contains higher abundances of light elements of low atomic number and relatively low nuclear binding energy, such as H and C (Z = 1 and 6). The intermediate region contains elements of intermediate atomic number and nuclear binding energy, such as Mg and Si (Z = 12 and 14).

### 2.1.1 Trends in Composition of the Earth

Terrestrial planets have obviously undergone some metamorphic processing to form the crust and atmosphere, and the chemical boundaries are not sharply defined. Oxygen (Z = 8), for example, is abundant at the surface and in the mantle of the Earth, but not in the core where metallic Fe and Ni are dominant. The Earth's inner core is generally believed to consist almost entirely of Fe and Ni, but its outer core may also contain lighter elements such as S (Z = 16), in addition to Fe,Ni (J. Lewis, 1974; Wood, 1978; Hou *et al.*, 1993). Chapter III of Miyake (1965) contains a superb review of the early theories proposed to explain the Earth's layered structure.

It is widely assumed that low abundances of light elements in the Earth and other terrestrial planets were caused by heating and loss of volatiles and that their layered internal structures were caused by gravitational sinking of high-density Fe, Ni to the core (Urey, 1952, 1954; Miyake, 1965). Grossman (1972), Grossman and Larimer (1974), and Saxena and Eriksson

(1983) reviewed the early literature, and Yoneda and Grossman (1995) and Larimer (this volume) present more recent work on the possibility of producing meteorites and Earth-like objects from a gas initially having the composition of the solar photosphere if pressure, oxygen fugacity, and composition of the gas phase are considered as variables.

Eucken (1944) also envisioned that the solar system formed from a primitive nebula rich in H and He, but he suggested that the Earth may have developed in layers, beginning with the accretion of Fe that condensed first from a high temperature, high pressure, central region of the nebula. Eucken's proposal generated considerable interest in the possibility of both equilibrium condensation and heterogeneous accretion of the Earth.

Turekian and Clark (1969) and Vinogradov (1975) endorsed the concept of heterogeneous accretion of the Earth. However, these authors assumed that an inner Fe,Ni-rich region of the solar nebula was produced by a combination of condensation and redistribution of material from an initially homogeneous solar nebula. According to their model, "*The iron body that is now the Earth's core formed by accumulation of the condensed iron-nickel in the vicinity of its orbit. It then served as the nucleus upon which the silicate mantle was deposited, and the mantle in turn shielded the core from subsequent reaction with $H_2S$ and $H_2O$ to form sulfides and oxides.*" (Turekian and Clark, 1969, p. 347). Vinogradov (1975) suggested that identical processes formed iron meteorites and the Earth's core. Alexander (this volume) shows that some iron meteorites were not produced by planetary differentiation, and Qi-Lu and Masuda (this volume) show that some iron meteorites retain a primary nature, including isotopic anomalies from nucleosynthesis.

### 2.1.2 Trends in Composition of Meteorites

Meteorites are the oldest known solids in the solar system. Thus trends in their chemical constituents may indicate the origin of compositional differences across the solar system and in planetary bodies. Their point of origin is believed to be the asteroid belt, which marks the divide between the smaller, high density, rocky, iron-rich, inner planets and the giant, low density, gaseous, outer planets. According to the nebular RES view of the early solar nebula (Figure 1), the asteroid belt is, "*... evidently a transition region between the hot, inner solar system, where high temperature condensates accreted into small bodies and eventually into rocky planets, and the cold, outer solar system where volatile ices condensed along with high- and low-temperature minerals.*" (Chapman, 1998, p. 16).

According to the catastrophic LES view of the early solar nebula (Figure 2), the asteroid belt is also the boundary that separates the condensation products of the inner supernova layer, in which the dense, iron-rich, rocky planets were made, and the outer supernova layer, in which the low density, gaseous, giant planets were made. The differences seen in the elemental and isotopic compositions of mineral separates of meteorites (Figures 3 and 4) are consistent with their origin at this major boundary.

Meteorites display many of the same macroscopic differences in composition that are recognized in the Earth, in the planetary system, and in models of highly evolved massive stars. There are three main classes of meteorites: Iron, stony iron, and stone. Iron meteorites consist mostly of metallic Fe,Ni, as does the Earth's core. Stony irons contain separate regions of metallic Fe,Ni and ferromagnesium silicates. Stone meteorites consist of two types: chondrites and achondrites. About 90% of the stone meteorites are extremely heterogeneous, agglomerate rocks known as chondrites. The extent of their heterogeneity is well illustrated in Figure 4 of Goswami (this volume), Figures 34-38 of Mason (1962), Figures 11 and 18 of Wood (1978), and in Figures 2.1 and 3.1 of Dodd (1981).

Vinogradov (1975) noted that the blebs of Fe,Ni alloy in stone meteorites exhibit the same polymorphic varieties that are seen in iron meteorites, such as kamacite with up to 6% Ni and taenite with up to 50% Ni. Thus the process that formed the Fe,Ni inclusions of stone meteorites not only formed the iron meteorites but probably even the larger iron cores of the inner planets. The remaining 10% of the stone meteorites are achondrites. They are more highly differentiated than chondrites and contain no chondrules, the aerodynamically-shaped, glassy, rapidly-quenched droplets of silicates, which are designated by the name chondrite.

Currently, the relative abundances of the three types of meteorites in observed falls are 3% irons, 1% stony irons, and 96% stones, with chondrites:achondrites ≈ 10:1 (Dodd, 1981). The composition of the Earth's core may indicate that the irons were dominant in the accretion process when the cores of the terrestrial planets formed. (Vinogradov, 1975).

Olbers (1803) suggested that asteroids and meteorites are pieces of a disrupted planet between Mars and Jupiter. Olbers' suggestion is still in vogue, except that the asteroids of different composition (Anders, 1964; Dodd, 1981) are now assumed to be the parent bodies for the meteorites instead of one planetary object. Like the planets, "*...the distribution of compositional types among the asteroids is found to vary systematically with heliocentric distance.*" (Gradie and Tedesco, 1982, p. 1405).

It has generally been assumed that meteorites were derived from material having "cosmic" elemental abundances (Mason, 1962) in accordance with the nebular RES model, and that carbonaceous chondrites represent the most primitive accretion products. However, the largest isotopic anomalies are found in carbonaceous chondrites. Even if the chemical irregularities in meteorites arose from large differences in temperature, pressure, and oxygen fugacity, this leaves the isotopic irregularities unexplained. Further, it is not easy to understand how planetary layers of various composition formed from material with "cosmic" abundances or by accreting chondritic meteorites. For example, Wood (1978, p. 344) recognized differences in the oxidation state of Fe and in the thermal histories of the tiny grains of material in a carbonaceous chondrite and remarked: "*Thus, even on a millimeter scale, this aggregation, which we believe to be very similar to the raw material the planets are made of, is a disequilibrium mixture of high- and low-temperature components. If the Earth was put together out of material like this, one might logically ask why the metal didn't react with the ferric iron until one of them was used up, and how did we end up with iron in three states of oxidation in the Earth?*"

Chemical, isotopic, textural and/or mineralogical differences are not limited to primitive chondrites. Some achondrites, as well as chondrites, are breccias. The Erevan meteorite is part of a group of achondrites, which closely resemble the lunar soil in texture (Dodd, 1981). A carbonaceous chondrite clast in this achondrite contains a fragment of phosphorus-rich sulfides that is believed to have condensed from the solar nebula and to be a primary carrier of P and K in carbonaceous chondrites (Nazarov, 1994). Recently, Nazarov *et al.* (1998) showed that this P-rich sulfide fragment is enriched in elements produced by the s-process, with an abundance pattern for refractory elements that is similar to the one reported in "presolar" SiC grains (Amari *et al.*, 1995; Lodders and Fegley, 1995).

As noted earlier, the tracer isotope technique can be used to decide if chemical and isotopic irregularities seen across the planetary system and in individual grains of meteorites are products of condensation and segregation of an interstellar cloud according to the nebular RES model (Figure 1), or that coupled with a small component of exotic nucleogenetic material (Fowler, 1984; Wasserburg, 1987), or residual heterogeneities from the stellar debris that gave birth to the solar system (Figure 2).

## 2.2 TRACER ISOTOPES IN THE SOLAR SYSTEM

The tracer isotope technique can be used to decide the merits of the nebular RES model (Figure 1) versus the catastrophic LES model (Figure 2). One wants to determine whether the bulk solar system (99.99%) was derived from homogeneous and isotopically "normal" ingredients according to the latest RES model with only about 0.01% of exotic nucleogenetic material with "strange" isotopic ratios added (Fowler, 1984; Wasserburg, 1987), or if the entire solar system was derived from chemically and isotopically heterogeneous supernova debris, according to the LES model (Manuel *et al.*,

1998a). If the solar system (Lyttleton, 1941; Brown, 1970; Ramadurai, this volume) or some part of it (Hoyle, 1944; Vahia, this volume) formed from stellar debris by some other mechanism, then elemental and isotopic variations might be somewhat like those expected for the catastrophic LES model, provided that there is a viable mechanism such as that shown in Figure 2 to preserve these stellar-produced gradients as the solar system developed.

A word of caution is in order before proceeding to use the tracer isotope technique to decide the origin of chemical and isotopic heterogeneities in the solar system. In the following discussion of tracer isotopes in the solar system, it is assumed that chemically produced isotopic effects arise solely because of differences in mass. An exception to this general rule has been demonstrated for the production of ozone with an electrical discharge of molecular oxygen (Thiemens and Heidenreich, 1983) and for the production of molecular oxygen with an electrical discharge of carbon dioxide (Heidenreich and Thiemens, 1985). However, it has not been shown that any of the "strange" isotopic ratios observed in meteorites came from electrical discharges, and it is highly unlikely that such a process produced a significant fraction of the isotopic anomalies that match the patterns expected from the accumulation of the decay products of extinct radionuclides and the presence of poorly mixed products of the nucleosynthesis reactions that made the elements (Begemann, 1980).

### 2.2.1 HETEROGENEITIES IN THE PLANETARY SYSTEM

Systematic gradients in the chemical composition of the sun's planetary system were considered above in Section 2.1. The composition varies with radial distance from the sun: Light elements with low nuclear binding energy (H, C) are abundant in the outer planets, such as Jupiter; elements with maximum nuclear stability (Fe, Ni) are abundant in the inner planets, such as Mercury; and intermediate weight elements of intermediate nuclear binding energy (Mg, Si) are abundant in intermediate planets, such as Mars. The tracer isotope technique can determine if the these chemical gradients in the planetary system are better explained by the catastrophic LES or the nebular RES models, or some hybrid of the two.

#### 2.2.1.1 Inter-linked Chemical and Isotopic Diversities

In the following three sections, the tracer isotope technique will be used to answer the following questions: 1) Were chemical and isotopic heterogeneities initially linked? 2) Are isotopic variations still linked with chemical heterogeneities, and if so are they coupled over planetary distances? Did isotopic and chemical heterogeneities decrease with time?

**2.2.1.1.1 Primordial Coupling of Isotopic and Chemical Diversities.**

Figure 4 shows the first experimental evidence that elemental abundances of a light element, He, are closely coupled with isotopic abundances of its heavy congeners, Ar, Kr, and Xe, in mineral separates of the Allende carbonaceous chondrite. These data from R. Lewis *et al.* (1975) formed the basis for the original suggestion that the entire solar system formed directly from debris of a single supernova (Manuel and Sabu, 1975; 1977). According to this model, isotopically "strange" Ar, Kr, and Xe had come from the outer layers of a supernova where fusile elements (H, He, C) like those in the outer planets are still abundant. In meteorites these isotopically "strange" heavy noble gases are trapped with abundant He and Ne in nanometer sized diamonds (Lewis and Anders, 1988), which are plausible vapor condensation products from material rich in H, He, and C (Shindo *et al.*, 1985).

Isotopically "normal" Ar, Kr, and Xe, such as seen in the inner part of the solar system, are at the intercepts of the correlations shown in Figure 4, where elemental abundances of He vanish. This noble gas component may be from the star's hot interior, where He was possibly destroyed by fusion to make Fe, S, and other abundant elements in the inner planets.

Correlations between elemental abundances of primordial He with isotopic excesses of $^{38}$Ar, $^{86}$Kr, and $^{136}$Xe are exactly those expected if isotopically anomalous Ar, Kr, and Xe were initially associated with an ingredient of the solar nebula that contained all of its He. In other words, these correlations are to be expected if the solar system had been formed from poorly-mixed supernova debris (Figure 2). Because chemical differences between noble gases are minimal, this evidence for a primordial linkage between the elemental abundances of a light, fusile element and isotopic enrichments of heavy elements apparently survived the condensation and chemical reactions which occurred between the time of the nucleosynthesis and the formation of the planetary solids.

The close coupling of primordial He with isotopically "strange" Ar, Kr, and Xe, as shown in Figure 4, is a common property of the noble gases in meteorites (Sabu and Manuel, 1980). This property is consistent with the predictions of the catastrophic LES model for the entire solar system. It is unexplained by the nebular RES model and is generally overlooked by its supporters. Because RES posits that He is the second most abundant element, the results cannot be explained by injecting 0.0001 parts of exotic nucleogenetic material (Fowler, 1984; Wasserburg, 1987).

Chemical and physical methods used to isolate chemically-distinct phases from primitive meteorites with different nucleogenetic components is another obvious, but seemingly overlooked, indication for primordial coupling of chemical and isotopic irregularities in the elements that formed meteorites. Lee *et al.* (1996a,b, 1997) and Manuel *et al.* (1998a) cite other examples of inter-linked chemical and isotopic irregularities in the early solar nebula, as expected from LES, such as coupling of the chemical classification of meteorites with their levels of excess $^{16}O$ (R. Clayton *et al.*, 1976), the linking of the FeS in diverse meteorites with the Xe of terrestrial isotopic composition (Lee *et al.*, 1996a,b), and the connecting of SiC inclusions with the s-products of nucleosynthesis (Huss and Lewis, 1995).

**2.2.1.1.2 Planetary Coupling of Isotopic and Chemical Diversity.**

In 1960 the first hint of planetary-scale isotopic diversity was found in primordial Xe (Reynolds, 1960a,b). This work as well as many later analyses in laboratories around the globe confirmed that the isotopic composition of primordial Xe in the Earth differs from that in carbonaceous chondrites that formed farther away from the sun (*e.g.*, Eugster *et al.*, 1967). These analyses ascertained the residual link of isotopic heterogeneity with the radial chemical gradient of the planetary system.

Oxygen further ascertained the residual coupling of chemical and isotopic heterogeneities over planetary distances when R. Clayton *et al.* (1976) demonstrated that fractional enrichments of monoisotopic $^{16}O$ can be used to group meteorites and planets into at least six distinct categories. Meteorites or planets in one category cannot be made by differentiating objects from another category. The fractional level of $^{16}O$ in the Earth and the Moon is like that in achondrites, stony-iron meteorites, and enstatite chondrites. Their level of $^{16}O$ is unlike that in carbonaceous and ordinary chondrites and in urelites.

In 1972, it was shown that the type of Xe trapped in average carbonaceous chondrites, AVCC Xe, is a mixture of mass fractionated terrestrial Xe and the isotopically "strange" Xe, which is released from carbonaceous chondrites at 600°-1000°C (Manuel *et al.*, 1972). This "strange" Xe was trapped in nanometer-sized diamonds (Lewis and Anders, 1988) with abundant primordial He, but isotopically "normal" Xe was trapped with little or no He, as shown in Figure 4 at the intercept where He vanishes.

Residual linkage of isotopically "strange" Xe with primordial He (Figure 4) was recently confirmed with the release of data from the 1995 Galileo probe, which penetrated the He-rich atmosphere of Jupiter. In spite of large experimental errors, the value of the $^{136}Xe/^{134}Xe$ ratio in Jupiter is distinctly larger than that in the solar wind and approximately equal to the value seen in Xe extracted from the Allende mineral separate 3CS4 (Windler, this volume; Manuel *et al.*, 1998b,c). Xenon from this same mineral separate is shown as an example of isotopically "strange" Xe in Figure 3, and it is represented by the data points with the largest enrichment of $^{136}Xe$ in Figure 4. The presence of isotopically "strange" Xe in the He-rich atmosphere of Jupiter is another example that nanometer-scale coupling of chemical and isotopic components in meteorites has survived as large-scale gradients in the chemical and isotopic compositions of the planetary system.

Primordial Xe of the type found on Earth is also dominant in Mars and in the troilite (FeS) of diverse meteorites, including the troilite separates from the very primitive, heterogeneous, Allende carbonaceous chondrite (Hwaung and Manuel, 1982; Swindle *et al.*, 1995; Lee *et al.*, 1996a,b). The presence today of terrestrial-type Xe in meteorite inclusions of FeS and in planets that are rich in Fe and S further illustrates that planets are composed of material with the elemental and isotopic heterogeneities seen on a much smaller scale in meteorites.

The above examples illustrate that isotopic differences are still linked with residual chemical gradients in the planetary system as expected from the catastrophic LES model (Figure 2). However, they are unexplained by the nebular RES model, even with an injection of 0.0001 parts of exotic nucleogenetic material (Fowler, 1984; Wasserburg, 1987). Nevertheless, the examples may be consistent with other suggestions that the solar system formed from stellar debris (Lyttleton, 1941; Hoyle, 1944; Brown, 1970; Vahia, this volume; Ramadurai, this volume) provided that there is a mechanism to preserve inter-linked chemical and isotopic gradients over planetary distances in forming the sun's planetary system.

#### 2.2.1.1.3 Coupling of Isotopic Anomalies with Short-Lived Nuclides.

If the LES model is used to postulate that the elements in the planetary system came from heterogeneous supernova ejecta, as illustrated in Figure 2, natural mixing would be expected to diminish the isotopic anomalies while natural decay would decrease the abundance of short-lived nuclides. Thus, the amount of extinct radioactivity might be coupled with the magnitude of the isotopic anomalies.

Linkage of isotopic anomalies with extinct radioactivity was first observed when Reynolds (1960a) found the decay product of extinct $^{129}$I embedded in Xe with a general isotopic anomaly pattern across all nine Xe isotopes (Reynolds, 1960b). The decay product of extinct $^{129}$I ($t_{1/2}$ = 16 My) might be used to date the supernova event, except that (a) $^{129}$I can be produced by nuclear reactions other than the r-process of a supernova, and (b) iodine is a volatile element that is unlikely to condense into the solid phase until long after the supernova event.

The decay product of extinct $^{244}$Pu ($t_{1/2}$ = 82 My) is better suited to date the last supernova event because (a) $^{244}$Pu cannot be made by any nuclear reaction other than the r-process, and (b) $^{244}$Pu became part of meteorite minerals that trapped other actinide elements like U. Thus, the $^{244}$Pu-$^{136}$Xe and the U-Pb chronometers can be cross-calibrated by using U-Pb dating to determine the age of actinide-rich meteorite minerals and $^{244}$Pu-$^{136}$Xe dating to determine how much earlier a supernova produced the $^{244}$Pu (Figure 5).

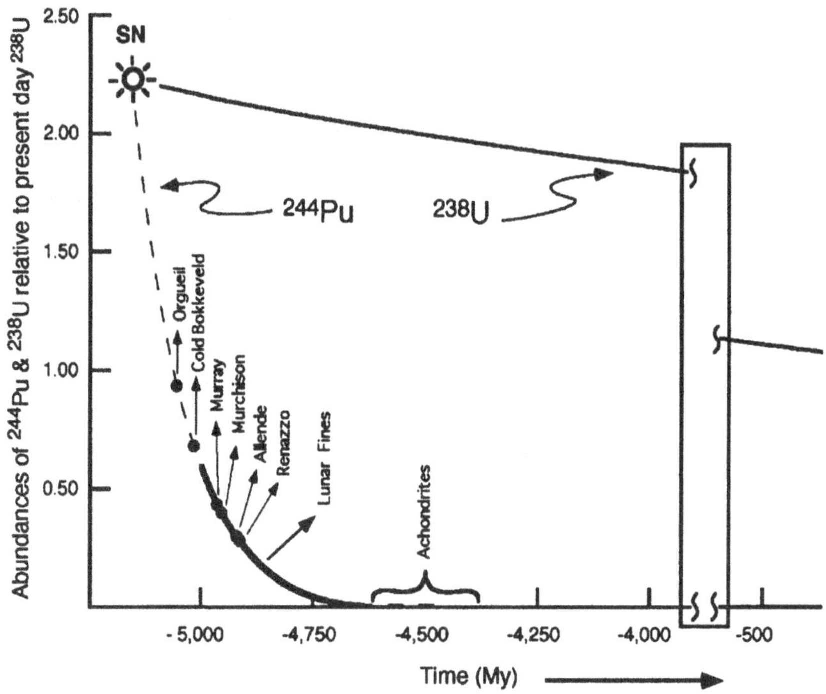


*Figure 5.* A supernova produced $^{244}Pu$ about 5000 My ago, and carbonaceous chondrites started to retain its gaseous fission product about 100 My later (Manuel *et al.*, 1998a).

As illustrated in Figure 5, Kuroda and Myers (1996) used this technique to show that (a) a supernova event about 5000 My ago produced the $^{244}Pu$ that was trapped in meteorites, (b) the carbonaceous chondrites started to retain excess $^{136}Xe$ from the spontaneous fission of $^{244}Pu$ about 100 My later, and (c) the more highly differentiated achondrites started to retain this gaseous fission product about 500 My after the supernova. Manuel *et al.* (1998a) presented this chronology in the manner shown in Figure 5.

Much larger isotopic anomalies have been found routinely in carbonaceous chondrites than in achondrites. Thus, the chronology of events outlined in Figure 5 provide further evidence for a link between extinct radioactivities and isotopic anomalies (Reynolds, 1960a,b). However, that early work on bulk meteorites could not yield the microscopic detail that is now available with ion microprobe mass spectrometers. These have revealed extinct radioactivity linked with isotopic anomalies in grains with the same physical properties exhibited by other high-temperature condensation products.

Five X-type grains of SiC, which Amari *et al.* (1992) recovered from the Murchison carbonaceous chondrite, best depict these properties. They are physically larger, and the initial $^{26}Al/^{27}Al$ ratios in these grains are 1-2 orders of magnitude more than those in other SiC grains from the Murchison chondrite. The grains formed within 1-2 My of a supernova explosion, over a time period in which the $^{26}Al/^{27}Al$ ratio decayed from a value of 0.60 to 0.10 (Kuroda and Myers, 1997; Manuel *et al.*, 1998a). Amari *et al.* (1992, p. L43) reported that the grains, "... *exhibit extremely exotic isotopic compositions, distinct from the majority of the SiC grains, ...*" As expected of early condensate from heterogeneous supernova debris, large isotopic anomalies have been found in other elements, including C, N, Si, Ti, and Ca.

Kuroda and Meyers (1997) noted that the levels of extinct radioactivity in these SiC grains decrease with particle size, as has also been found in fallout particles from nuclear weapons testing. Apparently, higher values of the $^{26}Al/^{27}Al$ ratio occur in SiC grains that started to nucleate early and grew larger; lower values of the $^{26}Al/^{27}Al$ ratio occur in smaller grains that started their growth later. Figure 6 (Manuel *et al.*, 1998a) illustrates how the $^{26}Al/^{27}Al$ ratio in these SiC grains varies with the particle size.

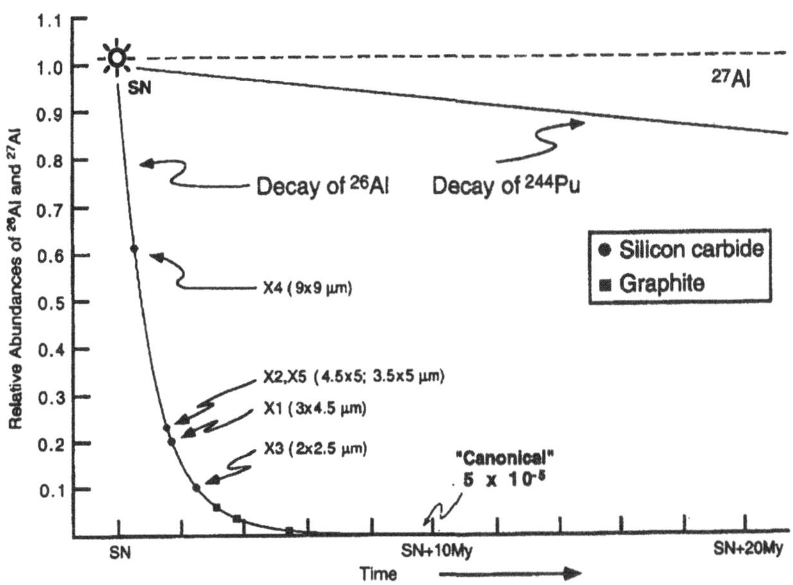


*Figure 6.* High levels of radioactive $^{26}Al$ and large isotopic anomalies were trapped in meteorite grains of silicon carbide (Amari *et al.*, 1992) and graphite (Zinner *et al.*, 1991) that formed within 1-6 My of the supernova event. Reprinted from Manuel *et al.* (1998a).

Figure 6 shows that larger grains of SiC formed within 1-2 Myr of the supernova event and graphite grains formed within 3-6 Myr. One SiC grain contains a large excess of $^{44}Ca$ (Amari *et al.*, 1992) from the decay of $^{44}Ti$, a supernova-product with a 59-year half-life (Ahmad *et al.*, this volume). This grain may have formed on an even more refractory, Ti-rich particle that condensed even before the complete decay of $^{44}Ti$.

Thus high abundances of short-lived nuclides, such as $^{26}$Al, were initially coupled with large isotopic anomalies in other elements, such as C, N, Si, Ti, and Ca. These anomalies are found in grains with the same physical properties exhibited by other high-temperature condensates. These findings imply that heterogeneities in the solar nebula decreased with time, as expected from application of the catastrophic LES model (Figure 2) but not from that of the nebular RES model (Figure 1).

The RES model, which implies a supply of elements from multiple stellar sources, even with a late injection of exotic material, explains neither the observed linkage of extinct radioactivity with isotopic anomalies nor the coupling of these isotopic properties of the elements with the physical size of the host grains. Further, Figure 6 shows no evidence for a "canonical" value of $^{26}$Al/$^{27}$Al = 5 x $10^{-5}$ (Podosek and Swindle, 1988), as might be produced from an injection of $10^{-4}$ parts of exotic nucleogenetic material into an otherwise homogeneous solar nebula (Fowler, 1984; Wasserburg, 1987).

#### 2.2.1.2 The Relation of "Strange" and "Normal"

The LES model (Figure 2) requires that there be isotopic excesses as well as depletions in nucleogenetic isotopic anomalies so that elements of "normal" isotopic composition can be made by mixing two or more isotopically "strange" components. This is not expected for the RES model (Figure 1), nor for RES plus an added exotic nucleogenetic component. Different alien sources are required to explain excesses and depletions of the same isotope for the RES model, and the probability of alien stellar sources decreases as their number increases. Thus, the tracer isotope technique can determine whether there is a relationship between the isotopic compositions of "strange" and "normal" components, as expected of LES, or if the "strange" isotopic components are indeed totally alien to those of "normal" material in the solar system, as expected of RES.

Complementary enrichments and depletions in isotopic ratios were first noticed in Xe. One component is enriched in light isotopes, $^{124,126}$Xe, from the p-process and in heavy isotopes, $^{134,136}$Xe, from the r-process (Manuel *et al.*, 1972). A complimentary component was later discovered with an excess of the intermediate mass isotopes, $^{128\text{-}132}$Xe, from the s-process of nucleosynthesis (Srinivasan and Anders, 1978).

Oliver *et al.* (1981) noticed positive and negative isotopic anomalies in U, Ba, Ca, Xe, and Te, which are unexpected from the addition of alien nucleogenetic material. On the basis of their observation, they commented that, "*The existence of positive and negative isotopic anomalies are expected, however, if the solar system formed directly from unmixed stellar debris that contained products of the different nucleosynthesis reactions that collectively produced all of the isotopes of each element.*"

Begemann (1993) found the most compelling evidence of complementary enrichment and depletion of isotopes when he compared the elements, Ba, Nd, and Sm in the inclusion EK-1-4-1 of the Allende carbonaceous chondrite with the same elements in the grains of SiC from the Murchison carbonaceous chondrite. In the inclusion EK-1-4-1 of the Allende meteorite, these three elements were enriched in isotopes that were made by the r- and p-processes of nucleosynthesis. In the SiC grains from the Murchison meteorite these same isotopes were depleted. The relative depletions in the SiC grains of Murchison are in the same proportion as the enrichments in the inclusion EK-1-4-1 of Allende. In other words, the isotopic anomaly patterns are "mirror-images," as illustrated in Figure 7 for any element with seven stable isotopes.

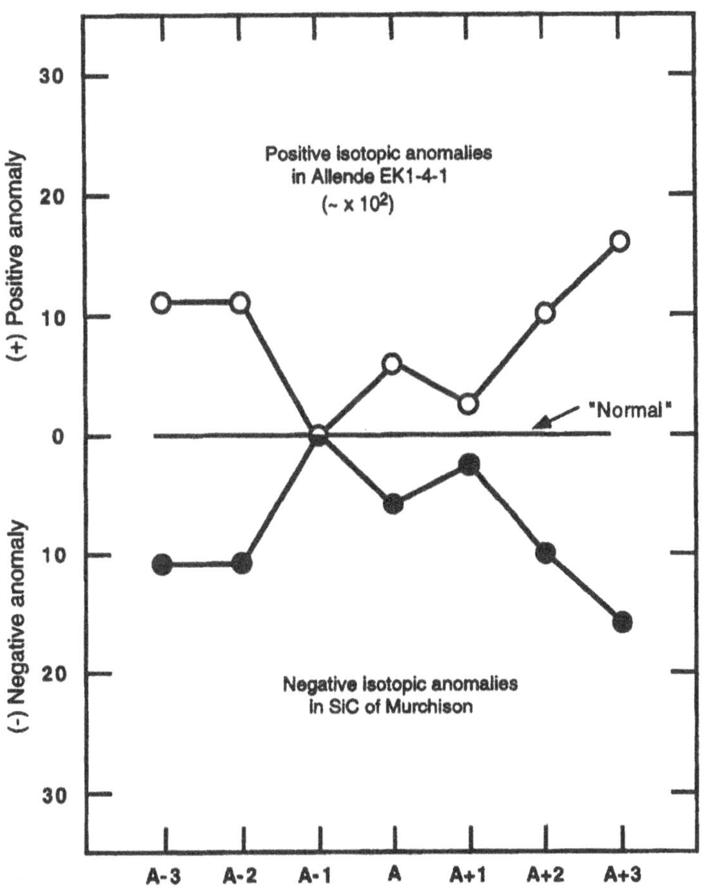


*Figure 7.* An illustration of "mirror-image" (+ and -) isotopic anomaly patterns seen in Ba, Nd and Sm. These elements are enriched in s-products in SiC inclusions of the Murchison meteorite; these same elements are enriched in r- and p-products in inclusion EK-1-4-1 of the Allende meteorite (Begemann, 1993). This figure is from Manuel *et al.* (1998a).

Individual plots for the enrichments and the depletions of stable isotopes of Ba, Nd, and Sm in inclusions from both of these meteorites are shown in Figure 2 of Begemann (1993) and in Figure 3 of Zinner (1997). In commenting on the pattern displayed in Figure 7, Begemann (1993) stated that, "*For Ba, Nd and Sm the surprisingly good (anti)-match of the anomalies pattern in MuSiC and Allende FUN inclusions appears too good to be entirely serendipitous, the implication being that the s-Ba, Nd, Sm in SiC from Murchison are quantitatively complementary to the excess of r-Ba, Nd and Sm in Allende inclusions.*"

The data in Figure 7 demonstrate a clear relationship between "strange" and "normal" isotopic components, as expected from use of the catastrophic LES model (Figure 2). It is of course also possible that the isotopic anomaly patterns shown in Figure 7 came from multiple injections of exotic nucleogenetic material into a cloud of interstellar material. In that case, it is entirely serendipitous that two extraneous stellar sources would generate "mirror-image" anomaly patterns, which happen to sum to the isotopic composition of "normal" solar system material.

### 2.2.1.3 Scale of Chemical and Isotopic Variations

The scale of chemical and isotopic variations can also be used to decide if the observations fit the nebular RES model (Figure 1) or the catastrophic LES model (Figure 2). If the elements in the solar system were produced by the RES model shown in Figure 1 and the isotopically anomalous material represents only about 0.01% of the bulk (Fowler, 1984; Wasserburg, 1987), then planets or other large objects (*e.g.*, those representing ≈ 0.1-1.0% of the solar system) will not display the same isotopic anomalies that are seen on the small scale represented by individual grains in meteorites.

However, if the elements in the solar system were produced from heterogeneous supernova debris, as suggested by the LES model (Figure 2), then the coupling of chemical and isotopic irregularities will be less dependent on the size of the objects, except for differences in the times required to form large and small objects and in the quantities of material averaged together in large ones. In other words, for the LES model the same linkage between elemental and isotopic abundances might be expected in microscopic grains of meteorites and in planetary-sized objects.

Thus for the solar system, the LES model suggests that the isotopic compositions of the elements, which formed the sun and the planets, may not only differ, but be part of the isotopic irregularities retained in tiny grains of material in the meteorites as well. It has already been shown in Section 2.2.1.1.2. that there are planetary-scale variations in the isotopic ratios of some elements. This was further demonstrated by the recent finding that the $^{136}Xe/^{134}Xe$ ratio in Jupiter (Windler, this volume; Manuel *et al.*, 1998b,c) is indistinguishable from isotopically "strange" Xe in the Allende mineral separate 3CS4, in which $^{136}Xe/^{134}Xe$ = 1.04 (Lewis *et al.*, 1975), but distinctly larger than that in the solar wind, where $^{136}Xe/^{134}Xe$ = 0.80 (Kaiser, 1972).

The sun and Jupiter are the two largest bodies in the solar system, yet this disclosure confirms differences in the isotopic abundances of the elements that formed them. This lends further support to the catastrophic LES model for explaining either the presence of all of the elements in the solar system (Lyttleton, 1941; Brown, 1970; Manuel and Sabu, 1975, 1977; Ramadurai, this volume) or at least those in the planetary system (Hoyle, 1944; Vahia, this volume). The nebular RES model (Figure 1), even with the addition of a 0.01% of exotic nucleogenetic material, does not explain how the sun and Jupiter might have formed from isotopically distinct reservoirs.

The two heaviest isotopes of Xe, $^{136}Xe$ and $^{134}Xe$ are generally believed to be produced solely by the r-process of nucleosynthesis (Burbidge *et al.*, 1957). Since the fractional enrichment of $^{136}Xe$ is greater than that of $^{134}Xe$ in "strange" Xe (Manuel *et al.*, 1972), some have suggested that the "strange" Xe represents an exotic r-process (Heymann and Dziczkaniec, 1979; Clayton, 1989; Howard *et al.*, 1992). However, Ott (1996) has shown that the standard r-process can explain the isotopic ratios in "strange" Xe if there was a separation of the stable Xe isotopes from their radioactive precursors within a few hours of their production. Thus, there is no need for an *exotic* nucleogenetic component in Xe. The anomalously high abundances of $^{136}Xe$ in "strange" Xe of meteorites may be the counterpart to large excesses of $^{132}Xe$ in "strange" Xe of terrestrial samples (Kuroda, 1960; Bennett and Manuel, 1970).

### 2.2.2 HETEROGENEITIES IN THE SUN

Isotopic ratios in the solar wind and in solar flares can be used to determine the internal composition of the sun. The nebular RES model (Figure 1) and the standard solar model imply that the solar system formed from elements that have the same composition seen on the solar surface. The catastrophic LES model (Figure 2) indicates that the sun grew heterogeneously and that its interior may have a different composition than its surface (Manuel *et al.*, 1998a). Several years ago Hoyle (1975) also proposed an Fe-rich solar core to account for the low flux of solar neutrinos, stating that, "*The exterior of the sun, comprising 50 percent or more of the present solar mass, is taken to have been added 4.7 x $10^9$ years ago.*" (Hoyle, 1975, p. L127).

#### 2.2.2.1 Elemental Composition

The isotopic compositions of the elements in the solar wind and in solar flares can be used to determine whether diffusion in the sun enriches the light elements at its surface. If diffusion enriches lighter nuclei at the surface, the effects may be seen in isotopic abundances of elements there.

It has long been known that thermal diffusion can separate particles on the basis of mass. For an un-ionized gas, Chapman and Cowling (1952, p. 255) noticed that, "*...the heavier molecules tend to diffuse into the cooler regions...*" This may describe matter at the solar radius with an effective temperature of about 5800 K and a mass density of about 2.7 x $10^{-7}$ g/cc (Rouse, 1998). Inward the degree of ionization increases and thermal diffusion may enrich lighter nuclei at stellar surfaces. Heavier ions tend to diffuse towards the warmer region of an ionized gas. Chapman and Cowling (1952) noted that, "*This must happen in the sun and the stars, where thermal diffusion will assist pressure diffusion in concentrating the heavier nuclei towards the hot central regions.*" (Chapman and Cowling, 1952, p. 255).

The first report on the isotopic composition of Xe in the solar wind stated that, "*... terrestrial atmospheric and solar-type xenon may be related to each other by a strong mass fractionating process ...*" (Marti, 1969, p. 1265). Compared to terrestrial Xe, solar wind Xe is enriched in the lighter mass isotopes by approximately 4% per amu (Boulos and Manuel, 1971; Kaiser, 1972), although, as much as 7-8% of $^{136}$Xe in the solar wind may be the isotopically "strange" Xe seen in meteorites (Sabu and Manuel, 1976a; Pepin *et al.*, 1995).

Figure 8 shows the fractionation relationship between solar wind and fission-free terrestrial Xe (Sabu and Manuel, 1976a). Since air contains only about 1/10th of the Earth's total Xe (Canalas *et al.*, 1968; Hennecke and Manuel, 1975), the correction for fission in Figure 8 amounts to only 1.8% at $^{136}$Xe. The fission correction is significantly smaller for $^{131\text{-}134}$Xe.

The fission correction to atmospheric xenon in Figure 8 is based on the initial inventory of $^{244}$Pu that would have been associated with the uranium content of the Earth's crust and upper mantle (Becker *et al.*, 1968). Isotopic ratios for solar wind Xe are averages of the three separate analyses of lunar soils cited by Sabu and Manuel (1976a). The effects of mass fractionation (MF), which selectively enriches the lighter mass isotopes by 3.5% per amu, are shown in Figure 8 by the sloping MF-line. Positive deviations from the MF-line at $^{124,126}$Xe and $^{134,136}$Xe suggest the presence of a small component of isotopically "strange" Xe in the sun (Sabu and Manuel, 1976a; Pepin *et al.*, 1995).

As indicated in Section 1.2.3, Manuel and Hwaung (1983) identified a systematic enrichment of lighter mass isotopes across all five of the noble gases in the solar wind. When solar photospheric abundances were corrected for this fractionation, the seven most abundant elements in the bulk sun were found to be: Fe, Ni, O, Si, S, Mg, and Ca.

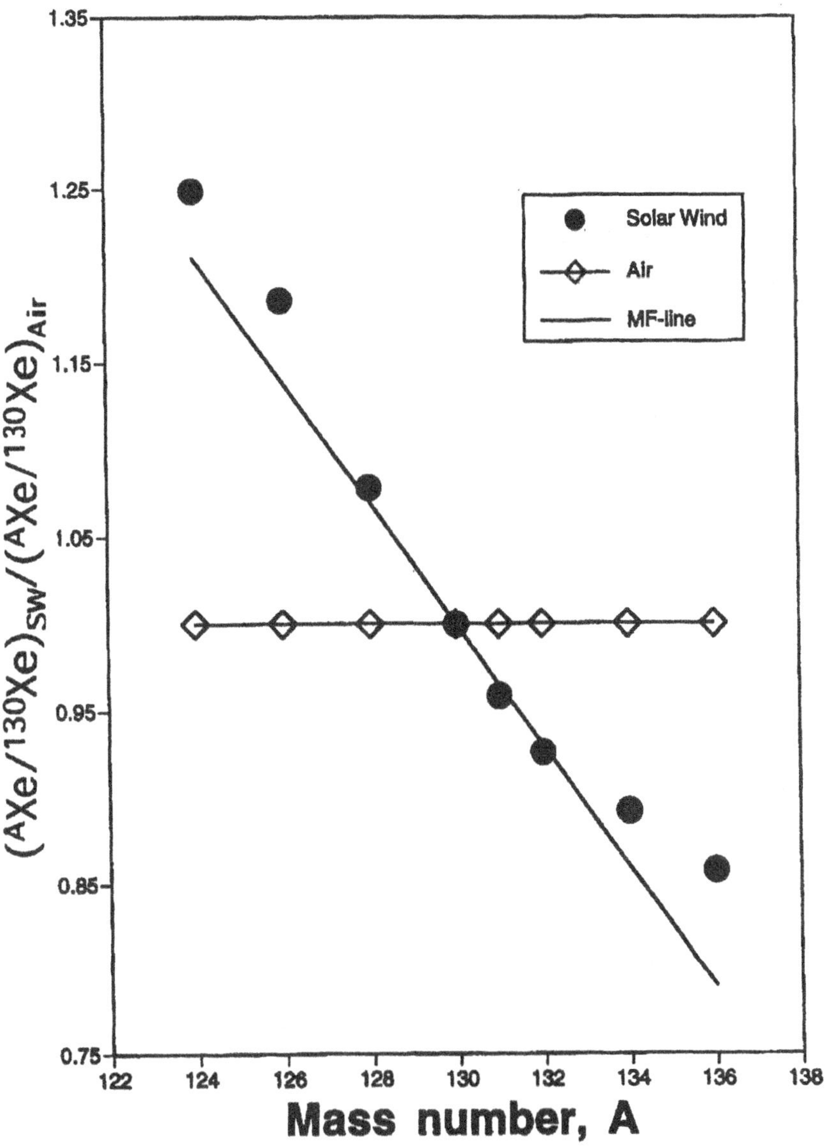


*Figure 8.* An illustration of the mass fractionation pattern that relates the isotopic composition of SW and terrestrial Xe. The sloping line shows the effects of mass fractionation. Deviations from the fractionation line at $^{124,126}$Xe and $^{134,136}$Xe are due to the presence of a small component of isotopically "strange" Xe in the sun (Sabu and Manuel, 1976a).

Acknowledging that their conclusion about solar abundances might "*seem extreme and controversial*", Manuel and Hwaung (1983, pp. 219-220) proposed three tests for their conclusions based on future measurements of:

1. The isotopic composition of a refractory element like Mg in the solar wind;

2. The solar neutrino flux using detectors other than $^{37}Cl$;
3. The isotopic composition of Xe in Jupiter.

All three of the measurements have now been made and, although the quality of the data is less than ideal, the results are affirmative for all three tests (Manuel, 1998).

Grevesse and Sauval (this volume) note that solar modelers will have difficulty in computing a shining sun if Fe, Ni, O, etc., are its most abundant elements. Modelers have also been unsuccessful in explaining other observations, *e.g.*, the solar neutrino flux (See papers by Guzik and Neuforge, Ramadurai, and Rouse in this volume and references therein). The standard solar model itself may be changing; compare for example the definition of the standard solar model given by Dar and Shaviv (1996, p. 935) with recent comments about the possibility of diffusion and elemental segregation in the sun (Bahcall *et al.*, 1997; Grevesse and Sauval, this volume).

Thus, the most abundant element in the interior of the sun is probably iron, rather than hydrogen. *I.e.*, the sun's most abundant nuclide is $^{56}Fe$, the one with minimum mass per nucleon, rather than $^{1}H$, the stable nuclide with the lowest nuclear binding energy and highest mass per nucleon.

There is, however, one observation which seems to favor the standard solar model and the mainstream elemental abundances recommended by Goldschmidt (1938), Suess and Urey (1956) and others. This is the remarkable similarity between the abundances of non-volatile elements in CI-type carbonaceous chondrites and those in the solar photosphere (*e.g.*, Ebihara *et al.*, this volume; Grevesse and Nauval, this volume). This is an unlikely coincidence. If CI carbonaceous chondrites formed from material in the outer part of a supernova, as suggested by the catastrophic LES model (Figure 2), the exceptional likeness between the abundances of non-volatile elements in carbonaceous chondrites and those in the solar photosphere suggests the possibility of chemical layering, not only in the sun, but also in the sun's presursor star (Figure 2) and perhaps in other stars.

#### 2.2.2.2 Isotopic Composition

A companion paper (Manuel, this volume) shows that the energetic events which produce solar flares also disrupt diffusion in the sun and cause shifts in the isotopic abundance ratios of He, Ne, Mg and Ar. Thermonuclear fusion near the sun's surface has increased the $^{15}N/^{14}N$ ratio in the solar wind over geologic time (Kerridge, 1975), *via* the $^{14}N(^{1}H, \beta^{+}\nu)^{15}N$ reaction suggested by Bethe (1939). Solar flares bring up bubbles of less mass-fractionated He, Ne, Mg and Ar to the solar surface. Flares also dredge up nitrogen with less of the solar proton-capture product, $^{15}N$. These observations are consistent with the catastrophic LES model (Figure 2) for the origin of our elements. For the conventional, nebular RES view and the standard solar model (Figure 1), "*Many explanations have been advanced to* *account for the long term variation in $^{15}N/^{14}N$ ...., but none has yet been accepted as satisfactory.*" (Kim *et al.*, 1995, p. 383).

Hwaung and Manuel (1982) identified, and Lee *et al.* (1996a,b; 1997) confirmed, the close association of terrestrial-type Xe with meteorite inclusions and with planets rich in Fe and S. Thus, the presence of terrestrial-type Xe in the solar wind affirms that Fe and S are abundant in the interior of the sun, although this is accompanied by high concentrations of He and other light elements that diffusion has enriched at the sun's surface. In this regard it is noteworthy that Harkins (1917) and Manuel and Sabu (1983) both concluded that Fe and S are among the more abundant elements in the solar system.

Thus, the experimental observations cited above indicate that the sun itself is heterogeneous, because:

(a) Diffusion inside the sun enriches the lighter elements and the lighter isotopes of individual elements at the solar surface, but the diffusive mass fractionation process is partially disrupted for particles ejected by solar flares (Manuel, this volume);

(b) The $^{14}N(^{1}H, \beta^{+}\nu)^{15}N$ reaction suggested by Bethe (1939) has increased the $^{15}N/^{14}N$ ratio in the solar wind, but nitrogen ejected by solar flares contains less excess $^{15}N$ from this reaction;

(c) The presence of terrestrial-type Xe in the solar wind and the close association of this type Xe with Fe and S in meteorites and planets reinforce the claim that Fe and S are major constituents of the bulk sun although they are only trace elements in the photosphere.

## 2.3 ASTRONOMICAL AND GEOLOGICAL OBSERVATIONS

### 2.3.1 SUPERNOVA EXPLOSIONS

In 1976, an illustration of the formation of the solar system from supernova debris (Figure 2) was presented at the Spring AGU meeting in Washington, D. C. (Sabu and Manuel, 1976b). Although it was widely believed at that time that it was impossible to form a planetary system directly from supernova debris, a few years earlier, LeBlanc and Wilson (1970) had in fact calculated that the explosion of a rotating star might produce oppositely-directed, axial jets of material.

Later observations on Cassiopeia A (Chevalier and Kirshner, 1979), on SN1987A (Chevalier, this volume; 1992-1996), and observations on more distant dying stars with the Hubble telescope, *e.g.*, the photographs of bipolar nebulae (Frank, this volume; 1997), clearly demonstrate that supernova debris can be ejected axially and leave a spatial record of chemical and isotopic inhomogeneities from the parent star. These observations offer an obvious explanation for the origin of inter-linked elemental and isotopic irregularites that are observed in the sun and its planetary system.

### 2.3.2 FORMATION OF PLANETARY SYSTEMS

Despite the widespread belief that the earliest planetary solids formed about 4,550 My ago, in a paper included in this volume Kuroda and Myers have shown many examples of radiogenic/radioactive ratios of nuclides in terrestrial and lunar samples that are only expected in samples that are older than this. Such measurements should not be rejected as necessarily indicating experimental error or the transport of the parent or daughter into/out of the sample. Some refractory grains of meteorites condensed 1-2 My after a supernova (Figure 6), and combined $^{244}$Pu-$^{136}$Xe and U-Pb dating indicate that the supernova event itself was 5000 My ago (Figure 5).

Recent astronomical observations show that supernova debris can form planetary systems: Wolszczan and Frail (1992) discovered and Wolszczan (1994) confirmed the first planetary system that lies beyond the Earth's solar system. This system orbits the collapsed supernova core, which produced the millisecond pulsar, PSR 1257 + 12. The planets within the system most likely formed directly from the supernova's debris: Previously formed planets around the parent star would not have survived the explosion.

More interestingly, small Earth-like planets composed of dense material representing elements in an advanced state of nuclear evolution, were reported within 0.5 AU of the pulsar (Wolszczan, 1994). Such material occurs in the four terrestrial planets close to the sun and near the center of highly evolved massive stars (Figure 46 of Taylor, 1972). These findings certainly suggest that planetary systems may form directly from supernova debris and even inherit chemical heterogeneities from the parent star.

Lin *et al.* (1991) had indicated that a planetary system could be produced around the collapsed core of a supernova from a rotationally-supported disk of material that may fall back after a stellar explosion, and Manuel *et al.* (1998a) showed that the planetary system around the pulsar PSR 1257+12, could have been made in this manner. The question is whether the Earth and the rest of the sun's planetary system may have been formed this way.

As noted earlier, Eucken (1944), Turekian and Clark (1969), and Vinogradov (1975) independently suggested that the Earth's iron core formed first and was then layered by a mantle of silicates. The concept of heterogeneous accretion of the Earth is still debated (Palme and O'Neill, 1996), but the tracer isotope technique effectively rules out the possibility of the Earth's internal structure being produced by geochemical differentiation.

Manuel and Sabu (1981) and Allègre *et al.* (1983) independently concluded from isotopic measurements of terrestrial noble gases that the Earth's mantle consists of two basic layers. The upper mantle was exhaustively degassed in forming the Earth's crust and atmosphere, "*...within the first 200 My of the Earth's history.*" (Manuel and Sabu, 1981, p. 260). Regarding the upper depleted mantle and the lower primitive one, Allègre *et al.* (1983, p. 766) stated that "*... these two reservoirs must have been separated for at least 4,400 Myr...*".

Both of the above research teams concluded that the lower mantle has retained primordial $^3$He. Manuel and Sabu (1981, p. 260) also posited that, "*The primitive nature of the lower mantle rules out the formation of the iron core by partial melting and geochemical differentiation.*" In view of the high mobility of $^3$He, it is difficult to imagine a process that would extract iron, for transport into the Earth's core, and leave $^3$He in the lower mantle surrounding that core.

However, Lee and Halliday (1995, 1996, 1997) and Halliday *et al.* (1996) used the dating method based on $^{182}$Hf-$^{182}$W ($t_{1/2}$ = 9 My) to conclude that the Earth's core formed late in the genesis of the planet, after a delay of about 50 Myr when most of the $^{182}$Hf had decayed away. Their conclusion was based on the fact that Hf is a lithophile element but W is a moderately siderophile element and that the isotopic composition of W in the bulk terrestrial silicates is identical to that of the W in chondritic meteorites. It should be pointed out that their data show equal "deficits" of $^{182}$W in the metal phase of ordinary chondrites and iron meteorites and equal "enrichments" of $^{182}$W in chondrites and bulk terrestrial silicates. It is thus possible that W isotopes provide no details about core formation but give additional evidence of heterogeneous accretion of the Earth.

Analyses of supposedly "differentiated" iron meteorites also indicate that some formed from chemically and isotopically heterogeneous material rather than by geochemical differentiation (See Alexander, this volume; Qi-Lu and Masuda, this volume). Alexander and Manuel (1967) detected a higher concentration of radiogenic $^{129}$Xe, which is the decay product of extinct $^{129}$I, in a graphite inclusion of the Canyon Diablo iron meteorite than had been found previously in all but two stone meteorites. Niemeyer (1979) determined that the $^{129}$I-$^{129}$Xe formation times of silicate inclusions in iron meteorites closely parallel those of "primitive" chondritic stone meteorites, but the $^{129}$I-$^{129}$Xe formation time of a troilite (FeS) inclusion predates that of both chondrites and silicate inclusions of iron meteorites. He concluded that, "*...this troilite formed in a different nebular region from the silicate and metal, and was later mechanically mixed with these other phases.*" (Niemeyer, 1979, p. 843). Much earlier, Urey (1952) had noted that the spherical shape of troilite nodules in iron meteorites, as opposed to the flattened shape expected from large overburden pressures, might indicate formation in small bodies.

Downing and Manuel (1982) observed large variations in the ratio of spallation-produced noble gas isotopes, $^{21}Ne/^{38}Ar$, in the Canyon Diablo iron meteorite. They concluded that these variations could not be explained by differences in depth or target element chemistry within the parent body. Similar variations had been reported earlier in mineral separates of the Odessa, Toluca, and Grant iron meteorites, leading Downing and Manuel (1982) to conclude that some of the different phases of iron meteorites had existed before compaction and been pre-irradiated.

Niemeyer's conclusion about the formation of troilite in a distinctive nebular region was further confirmed by later investigations, which showed that there is a characteristic isotopic composition of primordial Xe in the troilite inclusions of iron and stone meteorites (Hwaung and Manuel, 1982; Lee *et al.*, 1996a,b; 1997; Lee and Manuel, 1996). This same form of primordial Xe occurs not only in the FeS inclusions of diverse meteorites but in the terrestrial planets that are rich in these elements. This observation effectively rules out geochemical differentiation of initially homogeneous material to produce a reservoir in the solar system with abundant Fe and S.

Masuda and Qi-Lu (1998) recently found additional evidence that iron meteorites did not form from the differentiation of isotopically homogeneous solar material. Their measurements on the isotopic composition of Mo in six meteorites revealed an isotopic heterogeneity from nucleosynthesis. They concluded, quite correctly, that, "*It is obvious that the bulk iron meteorite retains a primary nature.*" Of course, isotopic anomalies in iron meteorites are small, when measured relative to terrestrial standards. Much larger isotopic anomalies of Mo are to be found in material that formed in the outer part of the solar system, *e.g.*, Nicolussi *et al.* (1998).

Thus, there is overwhelming geochemical evidence that the solar system formed from chemically and isotopically heterogeneous material and that the major chemical differences within the Earth and across the solar system were inherited. Further, there are astronomical observations which confirm that planetary systems can form directly from supernova debris. Conversely, there is no corroborative evidence that any of the so-called "presolar" grains traversed interstellar distances before becoming embedded in meteorites. In fact, irons constitute the only class of meteorites which contain material that has been exposed to cosmic rays for the long time period needed for a journey across interstellar distances, but iron meteorites display the smallest isotopic irregularities from nucleosynthesis when measured relative to terrestrial standards.

### 2.3.3 OBSERVATIONS OF THE SUN

Several different groups have suggested that modern helioseismic observations may indicate diffusion, limited mixing, and layering in the chemical composition of the sun (Rouse, 1985; Bahcall *et al.*, 1995, 1997; Basu, 1997). Further, a recent paper by Antia and Chitre (1998) suggests that various elemental abundance profiles inside the sun are consistent with modern helioseismic observations.

Many other observations of the sun indicate that the composition of its photosphere does not represent the solar interior. It has been suggested, for example, that the solar neutrino puzzle (Bahcall and Davis, 1976) may indicate chemical layering in the sun, with more high-Z material in the core than at the surface (Rouse, 1975; Wheeler and Cameron, 1975), perhaps even an iron-rich core (Hoyle, 1975; Rouse, 1975, 1983, 1995) or a central black hole (Clayton *et al.*, 1975; Ramadurai, this volume). However, solar neutrinos are only part of a unison of observations indicating that the interior of the sun is unlike its surface. Rather than repeat here information that has already been given, the reader is referred to Table 3, which summarizes the observations.

*Table 3.* Hints on the composition of the solar interior

| Problem | Solution | Ref. |
|---|---|---|
| Chemical disequilibrium of iron and high abundances of siderophile elements in Earth's mantle | Accretion of the iron core in a central, Fe-rich region of the protosolar nebula | a,b |
| The low flux of solar neutrinos | The sun's core is also Fe-rich | c |
| A unique solution to the equations for solar structure | Requires high mass particles like Fe in the core of the sun | d |
| Enriched light isotopes of He, Ne, Ar, Kr and Xe in the solar wind | Diffusion in the sun enriches light-weight nuclei at the solar surface | e |
| Light-weight isotopes are less enriched in solar flare particles | Energetic events disrupt diffusion; bring up less fractionated material | f,g |
| Terrestrial-type Xe in the solar wind; "strange" Xe in Jupiter | Tracer isotopes indicate abundant Fe & S in the sun; abundant C & He in Jupiter | e,h,i |
| High photospheric abundances of beryllium | Only superficial mixing occurs in the solar photosphere | j |
| Strange isotope ratios of N and Li in the solar wind | Thermonuclear reactions occur in the outer part of the sun | k,l |

a. Turekian and Clark (1969)
b. Vinogradov (1975)
c. Hoyle (1975)
d. Rouse (1975)
e. Manuel and Hwaung (1983)
f. Rao *et al.* (1991)
g Manuel and Ragland (1997)
h. Kaiser (1972)
i. Manuel *et al.* (1998a,b)
j. Balachandran and Bell (1998)
k. Manuel (1999)
l. Chaussidon and Robert (1999)

Of course, no "direct" observations of material in the sun's interior exist, but close scrutiny of the available measurements, such as those listed in Table 1, offers hints that the standard solar model (Dar and Shaviv, 1996) may have flaws and that the composition of the solar interior differs from the photosphere.

In summary, the information in Table 2 indicates that the standard solar model does not provide a unique solution for the structure of the sun and the solar neutrino flux suggests problems with the model. Observations since 1957 have led several to suggest an iron-rich solar core. This does not require segregation of the sun over the past 4-5 Gyr. The core of the sun probably formed from the same Fe-rich reservoir of elements that produced the iron meteorites and the cores of terrestrial planets in the inner part of the protosolar nebula.

These findings are consistent with predictions of the catastrophic LES model (Figure 2) for the origin of the solar system. The companion paper (Manuel, this volume) provides additional evidence for drastic chemical layering in the sun and for fusion reactions near its surface. These observations are unexplained by the nebular RES model (Figure 1) for the origin of the solar system.

## 3. CONCLUSIONS

The conclusions of this study, in the approximate order of the paper, are below:

1. Ideas about the abundances of elements and their origin intertwine. If Harkins (1917) was correct in using analyses of meteorites to conclude that the seven most abundant elements in the solar system are Fe, O, Ni, Si, Mg, S and Ca, then elemental abundances in the solar system are closely linked with the stability of their nuclei. Such elements are produced in the interior of highly evolved stars. Thus Harkins' abundance pattern favors the catastrophic LES model shown in Figure 2. Burbidge *et al.* (1957) and Cameron (1957) relied on the abundance table of Suess and Urey (1956), which is based more on abundances in the solar photosphere. For these abundances, nuclear evolution is much less advanced and the most abundant

nuclide, $^{1}H$, is the one with minimum nuclear binding energy. The Suess-Urey abundance pattern favors the nebular RES model shown in Figure 1. Table 1 compares these two sets of abundances.

A new high-sensitivity mass spectrometer reported the year before the classical papers by Burbidge *et al.* (1957) and Cameron (1957) provided the following empirical challenges to the nebular RES model:

(a) Amounts of the decay products of $^{129}I$ and other short-lived nuclides in meteorites meant that there was too little time to produce a well-mixed nebula between nucleosynthesis and the formation of meteorite grains.

(b) By 1960, an anomalous abundance pattern across all nine stable xenon isotopes, $^{124-136}Xe$, in chondritic meteorites hinted that atomic weights in the solar system might vary across planetary distances.

(c) In the 1970s, monoisotopic $^{16}O$ and excesses and depletions of the isotopes of heavy elements made by the r-, p-, and s-processes were found in grains of meteorites. This implied that different nucleogenetic components of a given element were not mixed before condensation. Examples are shown in Figure 3 and in Figure 7.

(d) In 1975 it was shown that elemental abundances of He are closely linked with isotopic abundances of Ar, Kr and Xe in meteorites (Figure 4). This linkage between chemical abundances of a light element and isotopic abundances of heavy elements is expected in poorly mixed debris of a supernova (Figure 2).

(e) In the early 1980s, lighter mass isotopes of elements were shown to be systematically enriched in the solar wind. The empirical fractionation pattern is that expected if diffusion in the sun enriches the lighter elements and the lighter isotopes of individual elements at the solar surface.

(f) When solar photospheric abundances were corrected for diffusive mass fractionation, the seven most abundant elements in the bulk sun turned out to be the same ones that Harkins (1917) had concluded from analyses of meteorites. These results are shown in Table 2. The chance of producing this agreement by random selection is about $\approx 2 \times 10^{-10}$

(g) Across the planetary system, the abundance of Fe increases towards the sun. When corrected for diffusion, the composition of the bulk sun itself fits this elemental abundance trend.

2. The tracer isotope technique was used to determine the origin of microscopic- and macroscopic-scale variations in the chemical and isotopic compositions of meteorites, planets and the sun. The results indicate that:

(a) There was a primordial coupling of isotopic and chemical irregularities in the material that first formed solids in the solar system.

(b) This initial linkage of chemically and isotopically distinct components included trace as well as major elements and compounds, *e.g.*, He, C, O, Ne, Xe, SiC, FeS, and even different chemical classes of meteorites.

(c) The coupling of chemically and isotopically discrete members has been preserved across the planetary system and across nanometer distances in the diverse, agglomerate rocks we call chondritic meteorites.

(d) Combined U-Pb and $^{244}$Pu-$^{136}$Xe dating shows that a supernova produced the solar system's $^{244}$Pu about 5000 My ago (Figure 5).

(e) $^{129}$I-$^{129}$Xe and $^{244}$Pu-$^{136}$Xe dating of bulk meteorites suggested a link between extinct radionuclides and general isotopic anomalies. $^{26}$Al-$^{26}$Mg dating of individual meteorite grains confirmed the coupling of larger isotopic anomalies with shorter-lived nuclides in grains of SiC that started to nucleate earlier and grew larger (Figure 6).

(f) The reciprocal isotopic anomaly patterns in meteorite grains fit those expected from poorly-mixed products of the same nuclear reactions that made isotopically "normal" elements in the solar system (Figure 7).

(g) The Galileo probe into Jupiter's helium-rich atmosphere revealed a tracer isotope ratio there, $^{136}$Xe/$^{134}$Xe, that is unlike that in the solar wind, but indistinguishable from the isotopically "strange" xenon that is trapped in helium-rich meteorite minerals.

(h) Xenon in the sun appears to be a mixture of the "strange" and "normal" isotopic components found in its planetary system (Figure 8), consisting mostly of the "normal" xenon that occurs in Fe, S-rich planets and meteorite inclusions.

(i) When solar photospheric abundances are corrected for the effects of diffusion in the sun (Figure 8), the sun's most abundant nuclide is the ash of thermonuclear reactions at equilibrium, $^{56}$Fe, instead of $^{1}$H, the stable nuclide that is the best fuel for these thermonuclear reactions.

(j) The remarkable similarity between the abundances of non-volatile elements in CI-carbonaceous chondrites and those in the solar photosphere may indicate chemical layering, not only in the sun, but also in the sun's presursor star (Figure 2) and probably in other stars.

(k) The isotopic ratios of many elements in the solar wind indicate that diffusion occurs inside the sun, but material ejected in solar flares is less mass-fractionated (See Figure 1 and Table 1 of Manuel, this volume).

(l) The $^{14}$N($^{1}$H, $\beta^{+}\nu$)$^{15}$N portion of the CNO cycle has increased the $^{15}$N/$^{14}$N ratio over geologic time near the surface of the sun, but nitrogen ejected by solar flares contains less $^{15}$N from this proton-capture reaction.

(m) The presence of terrestrial-type Xe as the dominant Xe component in the sun (Figure 8), and its close association with Fe and S in meteorite minerals and planets, reinforces the claim that Fe and S are major, rather than trace, elements in the bulk sun.

(n) Post-1957 observations in astronomy and geology do not support the concept of nebular RES over catastrophic LES.

These facts strongly favor the catastrophic LES model illustrated in Figure 2 over the nebular RES model (Figure 1), even when the injection of exotic nucleogenetic material has been posited (Fowler, 1984; Wasserburg,

1987). The LES model is further supported by Lyttleton's initial proposal that the solar system was produced by a rotating star (Lyttleton, 1941), by the solar models of Rouse (this volume, 1964-1995), by the chronologies determined from the decay products of short-lived nuclides (Kuroda and Meyers, this volume, 1996, 1997; Sahijpal *et al.*, 1998; Goswami, this volume), and by the fact that the different nuclear reactions, which made the elements in the solar system at various stages of stellar evolution (Cameron,1957; Burbidge *et al.* 1957), did not completely mix and remain engraved as "strange" and "normal" atomic weights of the elements.

The conclusions of this study are consistent with the suggestions by Buffon, Chamberlin and Moulton, and Jeans (Miyake, 1965) that a catastrophic event produced the solar system, rather than the gravitational collapse of a nebula as proposed by Kant and Laplace (Miyake, 1965).

The opposing view, remote element synthesis (RES) and the classic nebular model of Kant and Laplace, has been assumed in most review articles and books (Clayton, 1968; Trimble, 1975, 1982, 1983, 1988, 1991; Kuroda, 1982; Barnes *et al.*, 1982; Cameron, 1984; Tayler, 1988; Pagel, 1988; Wheeler *et al.*, 1989; Arnett, 1996; Wallerstein *et al.*, 1997; Wood, 1999), symposia (Reeves, 1972; Milligan, 1978; Mathews, 1988; Prantos *et al.*, 1993; Bernatowicz and Zinner, 1997; Goswami *et al.*, 1997), and award lectures (Fowler, 1984; Wasserburg, 1987; Cameron, 1995) on the origin of elements. This assumption is implicit in titles like the "Origin of Elements" without any stipulation of the specific batch of elements being considered.

Additional studies of the sun and isotopic analyses of material from other regions of the solar system should resolve these differences of opinion within the next few years and give us a broadly based consensus on the origin of elements in the solar system.

*Postscript* — The words of three distinguished scientists encouraged the author in writing this review. First, the assertion by Edward Anders that, "*It is a well-known maxim underlying all of physical sciences: that one should strive to explain the maximum number of phenomena by the minimum number of arbitrary postulates*" (Anders, 1964, p. 587). Another was Albert Einstein's statement that, "*The right to search for truth implies also a duty. One must not conceal any part of what one has recognized to be true*" (Broad, 1978, p. 951). Finally, sage advice from Sir Fred Hoyle (1982, p. 9), "*As I say, I think we should have persisted, despite all the catcalls and brick-bats, the shaking heads and the curled lips that one so frequently encounters when presenting ideas outside the mainstream of the search for knowledge.*"

The abundance pattern of the various atomic species contains a record of their origin (Richards, 1919), but that record could not be deciphered until the composition of the sun was resolved. Although the conclusions reached here are unpopular, I am confident future generations will find that Figure 2 is a reasonable explanation for the initial and present distribution of nuclides in the solar system, including high abundances of light elements like hydrogen and helium at the sun's surface and an iron-rich interior with an abundance peak at $^{56}Fe$.

## ACKNOWLEDGEMENTS

My wife, Caroline, and all my children have shown great patience and kindness as I ignored other responsibilities in order to pursue my lifelong obsession to understand the origin of the solar system. I am also grateful to Professors Paul K. Kuroda and John H. Reynolds for introducing me to this exciting field, to John Koenig, Daniel Armstrong, Carl A. Rouse, Ken Windler, and V. A. Samaranayake for helpful comments on an earlier version of this manuscript, to the undergraduate students, Ken Windler, Adam Nolte, Dan Ragland, Lucie Johannes, Joshua Zirbel and Cara Lietz who helped decipher raw data from the mass spectrometer on the Galileo probe into Jupiter's atmosphere, to Dr. Daniel S. Goldin of NASA for releasing those data, and to Bill James for accepting a heavier teaching responsibility so I would have time to research and write this review.

## DEDICATION

This paper is dedicated to Bo Lozoff, founder of the Human Kindness Foundation of Durham, NC and to Gerry D. Reece (LWP#179321), Robert Driscoll (CP#33), Rufus Ervin (CP#136) and others who spend their lives in prison because society does not yet acknowledged that we are all basically the same, except that the behavior of individuals is determined by cause and effect (Menninger, 1968; Humes, 1999), just as the origin of the elements is engraved in their atomic weights.

## REFERENCES

Alexander, E.C., Jr. and Manuel, O.K.: 1967, "Isotopic anomalies of krypton and xenon in Canyon Diablo graphite", *Earth Planet Sci. Lett.* **2**, 220-224.

Allègre, C.J., Staudacher, T., Sarda, P. and Kurz, M. : 1983, "Constraints on evolution of Earth's mantle from rare gas systematics", *Nature* **303**, 762-766.

Amari, S., Hoppe, P., Zinner, E. and Lewis, R.S. :1992, Interstellar SiC with unusual isotopic compositions: Grains from a supernova?", *Ap. J.* **394**, L43-L46.

Amari, S., Hoppe, P., Zinner, E. and Lewis, R.S. :1995, "Trace-element concentrations in single circumstellar silicon carbide grains from the Murchison meteorite", *Meteoritics* **30**, 679-693.

Anders, E.: 1964, "Origin, age, and composition of meteorites", *Space Science Reviews* **3**, 583-714.

Anders, E.: 1988, "Solar-system abundances of the elements", *in Origin and Distribution of the Elements*, ed., Mathews, G.J., World Scientific, Singapore, pp. 349-353.

Anders, E. and Ebihara, M.: 1982, "Solar-system abundances of the elements", *Geochim. Cosmochim. Acta* **46**, 2363-2380.

Anders, E. and Grevesse, N.: 1989, "Abundances of the elements: Meteoritic and solar", *Geochim. Cosmochim. Acta* **53**, 197-214.

Anders, E. and Zinner, E.: 1993, "Interstellar grains in primitive meteorites: Diamond, silicon carbide, and graphite", *Meteoritics* **28**, 490-514.

Antia, H.M. and Chitre, S.M.: 1998, "Determination of temperature and chemical composition profiles in the solar interior from seismic models", *Astron. Astrophys.* **339**, 239-251.

Arnett, D.: 1996, *Supernova and Nucleosynthesis: An Investigation of the History of Matter from the Big Bang to the Present*, Princeton University Press, Princeton, N.J., 598 pp.

Bahcall, J.N. and Davis, R.: 1976, "Solar neutrinos: A scientific puzzle", *Science* **191**, 264-267.

Bahcall, J.N., Pinsonneault, M.H., and Wasserburg, G.J.: 1995, "Solar models with helium and heavy-element diffusion", *Rev. Mod. Phys.* **67**, 781-808.

Bahcall, J.N., Pinsonneault, M.H., Basu, S. and Christensen-Dalsgaard: 1997, "Are standard solar models reliable?", *Phys. Rev. Lett.* **78**, 171-174.

Balachandran, S.C. and Bell, R.A.: 1998, "Shallow mixing in the solar photosphere inferred from revised beryllium abundances", *Nature* **392**, 791-793.

Ballad, R.V., Oliver, L.L., Downing, R.G. and Manuel, O.K.: 1979, "Isotopes of tellurium, xenon, and krypton in Allende meteorite retain record of nucleosynthesis", *Nature* **277**, 615-620.

Barnes, C.A., Clayton, D.D. and Schremm, D.N, eds.: 1982, *Essays in Nuclear Astrophysics*, Cambridge University Press, Cambridge, UK, 466 pp.

Basu, S.: 1997, "Seismology of the base of the solar convection zone", *Mont. Not. R. Astron. Soc.* **288**, 572-584.

Becker, V., Bennett, J.H. and Manuel, O.K.: 1968, "Iodine and uranium in ultrabasic rocks and carbonatites", *Earth Planet Sci. Lett.* **4**, 357-362.

Begemann, F.: 1980, "Isotopic anomalies in meteorites", *Rep. Prog. Phys.* **43**, 1309-1356.

Begemann, F.: 1993, "Isotopic abundance anomalies and the early solar system" in *Origin and Evolution of the Elements*, ed., Prantos, N., Vangioni-Flam, E. and Cassé, M., Cambridge University Press, Cambridge, UK, pp. 518-527.

Bennett, G. A. and Manuel, O. K.: 1970, "Xenon in natural gases", *Geochim. Cosmochim. Acta* **34**, 593-610.

Bernatowicz, T.J. and Walker, R.: 1997, "Ancient stardust in the laboratory", *Physics Today* **50**, 26-32.

Bernatowicz, T.J. and Zinner, E.: 1997, *AIP Conference Proceeding 402: Astrophysical Implications of the Laboratory Study of Presolar Materials, American Institute of Physics*, Woodbury, N.Y., 750 pp.

Bethe, H.: 1939, "Energy production in stars", *Phys. Rev.* **55**, 103.

Black, D.C.: 1972, "On the origins of trapped helium, neon and argon isotopic variations in meteorites-II. Carbonaceous meteorites", *Geochim. Cosmochim. Acta* **36**, 377-394.

Bondi, H. and Gold, T.: 1948, "The steady-state theory of the expanding universe", *Monthly Notices Roy. Astron. Soc.* **108**, 252-270.

Boulos, M.S. and Manuel, O.K.: 1971, "The xenon record of extinct radioactivities in the earth", *Science* **174**, 1334-1336.

Broad, W.J.: 1978, "Statue on the mall to hail Einstein's 100th", *Science* **202**, 951.

Brown, H.: 1949, "A table of relative abundances of nuclear species", *Rev. Mod. Phys.* **21**, 625-634.

Brown, W.K.: 1970, "A model for formation of solar systems from massive supernova fragments", Los Alamos Scientific Laboratory report LA 4343, 28 pp.

Brown, W.K.: 1971, "A solar system formation model based on supernovae shell fragmentation", *Icarus* **15**, 120-134.

Brown, W.K.: 1987a, "High explosive simulations of supernovae and the supernovae shell fragmentation model of solar system formation", Los Alamos National Laboratory report LA-11005, 10 pp.

Brown, W.K.: 1987b, "Possible mass distributions in the nebulae of other solar systems", *Earth, Moon and Planets* **37**, 225-239.

Brown, W.K.: 1987c, "High explosive simulation of supernovae", *Pub. Astro. Soc. Pacific* **99**, 858-861.

Brown, W.K.: 1991, "The supernovae as a genesis site of solar systems", *Speculat. Sci. Tech.* **15**, 149-160.

Brown, W.K. and Gritzo, L.A.: 1986, "The supernovae fragmentation model of solar system formation", Astrophys. Space Sci. 123, 161-181.

Burbidge, E.M., Burbidge, G.R., Fowler, W.A. and Hoyle, F.: 1957, "Synthesis of the elements in stars", *Rev. Mod. Phys.* **29**, 547-650.

Cameron, A.G.W.: 1957, "Nuclear reactions in stars and nucleogenesis", *Publ. Astron. Soc. Pac.* **69**, 201-222. See also "Stellar evolution, nuclear astrophysics and nucleosynthesis", *Chalk River Report CRL-41*, Atomic Energy of Canada, Ltd.

Cameron, A.G.W.: 1968, "A new table of abundances of the elements in the solar system", in *Origin and Distribution of the Elements*, ed., Ahrens, L.H., Pergamon Press, pp. 125-143.

Cameron, A.G.W.: 1973, "Abundances of the elements in the solar system", *Space Sci. Rev.* **15**, 121-146.

Cameron, A.G.W.: 1982, "Elemental and nuclidic abundances in the solar system", in *Essays in Nuclear Astrophysics*, eds., Barnes, C.A., Clayton, D.D. and Schramm, D.N., Cambridge University Press, Cambridge, UK, pp. 23-43.

Cameron, A.G.W.: 1984, "Star formation and extinct radioactivities", *Icarus* **60**, 416-427.

Cameron, A.G.W.: 1995, "The first ten million years in the solar nebula", *Meteoritics* **30**, 133-161.

Cameron, A.G.W. and Truran, J.W.: 1977, "The supernova trigger for formation of the solar system", *Icarus* **30**, 447-461.

Canalas, R.A., Alexander, E.C., Jr. and Manuel, O.K.: 1968, "Terrestrial abundance of noble gases", *J. Geophys. Res.* **73**, 3331-3334.

Chapman, C.R.: 1998, "Solar-system - 2 shades beyond Neptune", *Nature* **392**, 16-17.

Chaussidon, M. and Robert, F.: 1999, "Lithium nucleosynthesis in the sun inferred from the solar-wind $^{7}Li/^{6}Li$ ratio", *Nature* **402**, 270-273.

Chapman, S. and Cowling, T.G.: 1952, "An a account of the kinetic theory of viscosity, thermal conduction, and diffusion in gases", in *The Mathematical Theory of Nonuniform Gases*, Cambridge University Press, Cambridge, UK, p. 255.

Chevalier, R.A.: 1992, "Supernova 1987A at five years of age", *Nature* **355**, 691-696.

Chevalier, R.A.: 1996, "Shocking supernova tales", *Bull. Am. Astron. Soc.* **28**, 1273.

Chevalier, R.A. and Kirshner, R.P.: 1979, "Abundance inhomogeneities in the Cassiopeia A supernova remnant", *Ap. J.* **233**, 154-162.

Clayton, D.D.: 1968, *Principles of Stellar Evolution and Nucleosynthesis*, University of Chicago Press, Chicago, IL., 612 pp.

Clayton, D.D.: 1975a, "Extinct radioactivities: Trapped residuals of presolar grains", *Ap. J.* **199**, 765-769.

Clayton, D.D.: 1975b, "$^{22}$Na, Ne-E, extinct radioactive anomalies and unsupported 40Ar", *Nature* **257**, 36-37.

Clayton, D.D.:1982, "Cosmic chemical memory: a new astronomy", *Q. Jl R. Astr. Soc.* **23**, 174-212.

Clayton, D.D.: 1989, "Origin of heavy xenon in meteoritic diamonds", *Ap. J.* **340**, 613-619.

Clayton, D.D. and Hoyle, F.: 1976, "Grains of anomalous isotopic composition from novae", *Ap. J.* **203**, 490-496.

Clayton, D.D., Newman, M.J. and Talbot, R.J., Jr.: 1975, "Solar models of low neutrino-counting rate: The central black hole", *Ap. J.* **201**, 489-493.

Clayton, R.N., Grossman, L. and Mayeda, T.K.: 1973, "A component of primitive nuclear composition in carbonaceous meteorites", *Science* **182**, 485-488.

Clayton, R.N., Onuma, N. and Mayeda, T.K.: 1976, "A classification of meteorites based on oxygen isotopes", *Earth Planet Sci. Lett.* **30**, 10-18.

Dar, A. and Shaviv, G.: 1996, "Standard solar neutrinos", *Ap. J.* **468**, 933-946.

Dodd, R.T.: 1981, *Meteorites: A petrologic-chemical synthesis*, Cambridge University Press, Cambridge, UK, 368 pp.

Downing, R.G. and Manuel, O.K.: 1982, "Composition of the noble gases in Canyon Diablo", Geochem. J. **16**, 157-178.

Eberhardt, P., Geiss, J., Graf, H., Grögler, N., Mendia, M.D., Mörgeli, M., Schwaller, H., Stettler, A., Krähenbühl, U. and von Guten, H.R.: 1972, "Trapped solar wind noble gases in Apollo 12 lunar fines 12001 and Apollo 11 breccia 10046", *Proc. 3rd Lunar Sci. Conf.* **2**, 1821-1856.

Eucken, A.: 1944, "Physikalisch-chemische betrachtungen über die früheste Entwicklungs-geschichte der Erde", *Nachr. d. Akad. d. Wiss. in Göttingen, Math-Phys.* **1**, 1-25.

Eugster, O., Eberhardt, P. and Geiss, J.: 1967, "Krypton and xenon isotopic composition in three carbonaceous chondrites", *Earth Planet. Sci. Lett.* **3**, 249-257.

Fowler, W.A.: 1984, "The quest for the origin of the elements", *Science* **226**, 922-935.

Frank, A.: 1997, "Blowing cosmic bubbles", *Astronomy* 25, 36-43.

Goldschmidt, V.M.: 1938, Geochemische Verteilungsgestze der Elemente. IX. Die Mengenverhältnisse der Elemente und der Atom-Arten, *Skrifter Norske Videnskaps-Akad., Oslo I Math.-Naturv. Klasse.* no. 4, 148 pp.

Goldschmidt, V.M.: 1958, *Geochemistry*, ed., Muir, A., Oxford University Press, London, UK, 730 pp.

Goswami, J.N., Sahijpal, S. and Chakrabarty, P., eds.: 1997, *International Conference on Isotopes in the Solar System, Abstracts*, Physical Research Laboratory, Ahmedabad 380009, India., 218 pp.

Gradie, J. and Tedesco, E.: 1982, "Compositional structure of the asteroid belt", *Science* **216**, 1405-1407.

Grossman, L.: 1972, "Condensation in the primitive solar nebula", *Geochim. Cosmochim. Acta* **36**, 597-619.

Grossman, L. and Larimer, J.W.: 1974, "Early chemical history of the solar system", *Rev. Geophys. Space Phys.* 12, 71-101.

Halliday, A., Rehkämper, M., Lee, D.-C. and Yi, W.: 1996, "Early evolution of the Earth and Moon: new constraints from Hf-W isotope geochemistry", *Earth Planet Sci. Lett.* **142**, 75-89.

Harkins, W.D.: 1917, "The evolution of the elements and the stability of complex atoms", *J. Am. Chem. Soc.* **39**, 856-879.

Heidenreich, J.E. III and Thiemens, M.H.: 1985, "The non-mass-dependent oxygen isotope effect in electro-dissociation of carbon dioxide: A step toward understanding NoMaD chemistry", *Geochim. Cosmochim. Acta* **49**, 1303-1306.

Hennecke, E.W. and Manuel, O.K.: 1975, "Noble gases in an Hawaiian xenolith", *Nature* **257**, 778-779.

Heymann, D. and Dziczkaniec, M.: 1979, "Isotopic compositions of xenon from explosive carbon burning: A global look", *Proc. Lunar and Planet. Sci. Conf.* **X**, 549-551.

Hou, W., Ouyang, Z.-Y., Xie, H.-S., Zhang, Y.-M., Xu, H.-G. and Zhou, Y.-L.: 1993, "The melting-differentiation of chondrite and the initial evolution of the earth", *Sci. China* **B36**, 121-128.

Howard, W.M., Meyer, B.S. and Clayton, D.D.: 1992, "Heavy-element abundances from a neutron burst that produces Xe-H", *Meteoritics* **27**, 404-412.

Hoyle, F.: 1944, "On the origin of the solar system", *Proc. Camb. Phil. Soc.* **40**, 256-258.

Hoyle, F.: 1945, "Note on the origin of the solar system", *Monthly Notices Roy. Astron. Soc.* **105**, 175-178.

Hoyle, F.: 1946a, "On the condensation of the planets", *Monthly Notices Roy. Astron.* Soc. **106**, 406-422.

Hoyle, F. 1946b, "The synthesis of the elements from hydrogen", *Monthly Notices Roy. Astron. Soc.* **106**, 343-383.

Hoyle, F.: 1948, "A new model for the expanding universe", *Monthly Notices Roy. Astron. Soc.* **108**, 372-382.

Hoyle, F.: 1975, "A solar model with low neutrino emission", *Ap. J.* **197**, L127-L131.

Hoyle, F.: 1982, "Two decates of collaboration with Willy Fowler" *in Essays in Nuclear Astrophysics*, eds., Barnes, C.A., Clayton, D.D. and Schramm, D.N., Cambridge University Press, Cambridge, UK, pp. 1-9.

Huang, S.-S.: 1957, "A nuclear-accretion theory of star formation", *Astron. Soc. Pacific* **69**, 427-430.

Humes, E.: 1999, *Mean Justice*, Simon and Schuster, Inc., New York, N.Y., 672 pp.

Huss, G.R. and Lewis, R.S.: 1995, "Presolar diamond, SiC and graphite in primitive chondrites: Abundances as a function of meteorite class and petrologic type", *Geochim. Cosmochim. Acta* **59**, 115-160.

Hwaung, G and Manuel, O.K.: 1982, "Terrestrial-type xenon in meteoritic troilite", *Nature* **299**, 807-810.

Kaiser, W.A.: 1972, "Rare gas studies in Luna 16-G-7 fines by stepwise heating technique", *Earth Planet Sci. Lett.* **13**, 387-399.

Kerridge, J.F.: 1975, "Solar nitrogen: Evidence for a secular increase in the ratio of nitrogen-15 to nitrogen-14", *Science* **188**, 162-164.

Kerridge, J.F.: 1993, "Long term compositional variation in solar corpuscular radiation: Evidence from nitrogen isotopes in lunar regolith", *Rev. Geophys.* **31**, 423-437.

Kim, J.S., Kim, Y., Marti, K. and Kerridge, J.F.: 1995, "Nitrogen isotope abundances in the recent solar wind", *Nature* **375**, 383-385.

Krat, V.A.: 1952, *Problems in Cosmology*, vol. **1**, Academy of Science, Moscow, USSR, p. 34.

Kuroda, P. K.: 1960, "Nuclear fission in the early history of the Earth", *Nature* **187**, 36-38.

Kuroda, P.K.: 1982, *The Origin of the Chemical Elements and the Oklo Phenomenon*, Springer-Verlag, New York, N.Y., 165 pp.

Kuroda, P.K. and Myers, W.A.: 1996, "Iodine-129 and plutonium-244 in the early solar system", *Radiochim. Acta* **77**, 15-20.

Kuroda, P.K. and Myers, W.A.: 1997, "Aluminum-26 in the early solar system", *J. Radioanal. Nucl. Chem.* **211**, 539-555.

Lavrukhina, A.K.: 1980, "On the nature of isotopic anomalies in meteorites", *Nukleonika* **25**, 1495-1515.

Lavrukhina, A.K. and Kuznetsova, R.I.: 1982, "Irradiation effects at the supernova stage in isotopic anomalies", *Lunar Planet. Sci.* **XIII**, 425-426.

LeBlanc, J.M. and Wilson, J.R.: 1970, "A numerical example of the collapse of a rotating magnetized star", *Ap. J.* **161**, 541-551.

Lee, D.-C. and Halliday, A.N.: 1995, "Hafnium-tungsten chronometry and the timing of terrestrial core formation", *Nature* **378**, 771-774.

Lee, D.-C. and Halliday, A.N.: 1996, "Hf-W Isotopic evidence for rapid accretion and differentiation in the early Solar System", *Science* **274**, 1876-1879.

Lee, D.-C. and Halliday, A.N.: 1997, "Core formation on Mars and differentiated asteroids", *Nature* **388**, 854-857.

Lee, J.T., Li, B. and Manuel, O.K.: 1996a, "Terrestrial-type xenon in sulfides of the Allende meteorite", *Geochem. J.* **30**, 17-30.

Lee, J.T., Li, B. and Manuel, O.K.: 1996b, "Xenon isotope record of nucleosynthesis and the early solar system", *Chinese Sci. Bull.* **41**, 1778-1782.

Lee, J.T., Li, B. and Manuel, O.K.: 1997, "On the signature of local element synthesis", *Comments Astrophys.* **18**, 335-345.

Lee, J.T. and Manuel, O.K.: 1996, "On the isotopic composition of primordial xenon in meteoritic troilite and the origin of the chemical elements", *Proc. Lunar Planet. Sci. Conf.* **XXVII**, 738a-738b.

Lewis, J.S.: 1974, "The chemistry of the solar system", *Scientific American* **230**, 50-65.

Lewis, R.S. and Anders, E.: 1988, "Xenon-HL in diamonds from the Allende meteorite - composite nature", *Proc. Lunar Planet. Sci. Conf.* **19**, 679-680.

Lewis, R. S., Srinivasan, B. and Anders, E.: 1975, "Host phase of a strange xenon component in Allende", *Science* **190**, 1251-1262.

Lin, D.N.C., Woosley, S.E., Bodenheimer, P.H.: 1991, "Formation of a planet orbiting pulsar 1829-10 from the debris of a supernova explosion", *Nature* **353**, 827-831.

Lodders, K. and Fegley, B., Jr.: 1995, "The origin of circumstellar silicone carbide grains found in meteorites", *Meteoritics and Planet. Sci.* **30**, 661-678.

Lyttleton, R.A.: 1941, "On the origin of the solar system", *Mont. Not. Roy. Astron. Soc.* **101**, 216-226.

MacElroy, J.M.D. and Manuel, O.K.: 1986, "Can intrasolar diffusion contribute isotope to anomalies in the solar wind?" *J. Geophys, Res.* **91**, D473-D482.

Manuel, O.: 1998, "Isotopic ratios in Jupiter confirm intrasolar diffusion", *Meteoritics and Planet. Sci.* **33**, A97.

Manuel, O.: 1999, Abstract 60, "Proton capture on $^{14}$N generates excess $^{15}$N in the solar wind", *Annual Meeting of the Missouri Academy of Sciences*, Cape Girardeau, MO, *Missouri Acad. Sci. Bull.* **27**, no. 4, p. 14.

Manuel, O.K., Hennecke, E.W. and Sabu, D.D.: 1972, "Xenon in carbonaceous chondrites", *Nature* **240**, 99-101.

Manuel, O.K. and Hwaung, G.: 1983, "Solar abundances of the elements", *Meteoritics* 18, 209-222.

Manuel, O.K., Lee, J.T., Ragland, D.E., MacElroy, J.M.D., Li, Bin and Brown, W.K.: 1998a, "Origin of the solar system and its elements", *J. Radio Anal. Nucl. Chem.* **238**, 213-225.

Manuel, O, Ragland, D., Windler, K., Zirbel, J., Johannes, L. and Nolte, A.: 1998b, "Strange isotope ratios in Jupiter", *Bull. Am. Astron. Soc.* **30**, 852-853.

Manuel, O., Windler, K., Nolte, A., Johannes, L., Zirbel, J. and Ragland, D.: 1998c, "Strange xenon in Jupiter", *J. Radioanal.Nucl. Chem.* **238**, nos. 119-121.

Manuel O.K. and Ragland D.E.: 1997, Abstract 281, "Diffusive mass fractionation effects across the isotopes of noble gases and magnesium in the solar wind and in solar flares", *1997 Midwest Regional Meeting of the American Chemical Society*, Osage Beach, MO, USA.

Manuel, O.K. and Sabu, D.D.: 1975, "Elemental and isotopic inhomogeneities in noble gases: The case for local synthesis of the chemical elements", *Trans. Missouri Acad. Sci.* **9**, 104-122.

Manuel, O.K. and Sabu, D.D.: 1977, "Strange xenon, extinct superheavy elements and the solar neutrino puzzle", *Science* **195**, 208-209.

Manuel, O.K. and Sabu, D.D.: 1981, "The noble gas record of the terrestrial planets", *Geochem. J.* **15**, 245-267.

Marti, K.: 1969, "Solar type xenon: A new isotopic composition of xenon in the Pesyanoe meteorite", *Science* **166**, 1263-1265.

Mason, B.: 1962, *Meteorites*, John Wiley & Sons, New York, NY, 274 pp.

Masuda, A. and Qi-Lu: 1998, "Isotopic composition of molybdenum in iron meteorites viewed from nucleosynthesis", *Meteoritics Planet. Sci.* **33**, A99.

Mathews, G.J.: 1988, *Origin and Distribution of the Elements, Proceedings* of a 1987 ACS symposium, World Scientific, Singapore, 767 pp.

Mayer, M.G.: 1948, "On closed shells in nuclei", *Phys. Rev.* **74**, 235-239.

Menninger, Karl A.: 1968, *The Crime of Punishment*, Viking Press, New York, NY, 305 pp.

Milligan, W.O., ed.: 1978, Proceedings of the *Robert A Welch Foundation Conferences on Chemical Research, XXI. Cosmochemistry*, The Robert A. Welch Foundation, Houston, TX, 397 pp.

Miyake, Y.: 1965, *Elements of Geochemistry*, Maruzen Co., Tokyo, Japan, 475 pp.

Nazarov, M.A., Hoppe, P., Brandstaetter, F. and Kurat, G.: 1998, "Presolar trace element signature in P-rich sulfide from a CM chondrite clast in the Erevan howardite", *Lunar Planet. Sci. Conf.* **XXIX**, 1596-1597.

Nazarov, M.A.: 1994, "P-rich sulfide, barringerite, and other phases in carbonaceous clasts of the Erevan howardite", *Lunar Planet. Sci. Conf.* **XXV**, 979-980.

Nicolussi, G.K., Pellin, M.J., Lewis, R.S., Davis, A.M., Amari, S. and Clayton, R.N.: 1998, "Molybdenum isotopic composition of individual presolar silicon carbide grains from the Murchison meteorite", *Geochim. Cosmochim. Acta* **62**, 1093-1104.

Niemeyer, S.: 1979, "I-Xe dating of silicate and troilite from IAB iron meteorites", *Geochim. Cosmochim. Acta* **43**, 843-860.

Noddack, I. and Noddack, W.: 1930, "Die Häufigkeit der chemischen Elemente", *Naturwissenschaften* **18**, 757-764.

Noddack, I. and Noddack, W.: 1934, "Die geochemischen Verteilungkoeffizienten der Elemente", *Svensk Kemisk Tidskrift* **XLVI**, 173-201.

Olbers, W.: 1803, "Uber die vom himmel gefellenen steine", *Ann. Phys.* **14**, 38-45.

Oliver, L.L., Ballad, R.V., Richardson, J.F. and Manuel, O.K.: 1981, "Isotopically anomalous tellurium in Allende: Another relic of local element synthesis", *J. Inorg. Nucl. Chem.* **43**, 2207-2216.

Ott, U.: 1996, "Interstellar diamond xenon and timescales of supernova ejecta", *Ap. J.* **463**, 344-348.

Pagel, B.E.J.: 1988, "The origin and distribution of the elements", in *Origin and Distribution of the Elements*, ed., Mathews, G.J., World Scientific, Singapore, pp. 253-271.

Palme, H. and O'Neill, H.St.C.: 1996, "Formation of the earth's core", *Geochim. Cosmochim. Acta* **60**, 1105-1108.

Palme, H., Suess, H.E. and Zeh, H.D.: 1981, "Abundances of the elements in the solar system", in *Landolt-Börnstein Numerical Data and Functional Relationships in Science and Technology* **2**, 257-272.

Pepin, R.O., Becker, R.H., and Rider, P.E.: 1995, "Xenon and krypton isotopes in extraterrestrial regolith soils and in the solar wind", *Geochim. Cosmochim. Acta* **59**, 4997-5022 .

Podosek, F.A. and Swindle, T.D.: 1988, "Extinct radionuclides", in *Meteorites and the Early Solar System,* eds., Kerridge, J.F. And Matthews, M.S., University of Arizona Press, Tuscon, AZ, pp. 1093-1146-.

Prantos, N., Vangioni-Flam, E. and Cassé, M., eds.: 1993, *Origin and Evolution of the Elements*, Cambridge University Press, Cambridge, UK, 545 pp.

Rao, M.N., Garrison D.H., Bogard D.D., Badhwar G. and Murali A.V.: 1991, "Composition of solar flare noble gases preserved in meteorite parent body regolith" *J. Geophys. Res.* **96**, 19.321-19.330.

Reeves, H., ed.: 1972, *Symposium on the Origin of the Solar System*, Centre National de la Recherche Scientifique, Paris, France, 379 pp.

Reynolds, J.H.: 1956, "High sensitivity mass spectrometer for noble gas analysis", *Rev. Sci. Instruments* **27**, 928-934.

Reynolds, J.H.: 1960a, "Determination of the age of the elements", *Phys. Rev. Lett.* **4**, 8-10.

Reynolds, J.H.: 1960b, "Isotopic composition of primordial xenon", *Phys. Rev. Lett.* **4**, 351-354.

Richards, T.W.: 1919, "Atomic Weights" Nobel lecture, December 6, 1919, *Nobel Lectures in Chemistry*, pp. 280-292.

Ross, J.E. and Aller, L.H.: 1976, "The chemical composition of the Sun", *Science* **191**, 1223-1229.

Rouse, C.A.: 1964, "Calculation of stellar structure using an ionization equilibrium equation of state", *University of California, UCRL Report 7820-T.*

Rouse, C.A.: 1969, "Calculation of stellar structure", in *Progress in High Temperature Physics and Chemistry*, vol. **2**, ed. Rouse, C.A., Pergamon Press, Oxford, UK, pp. 97-126.

Rouse, C.A.: 1975, "A solar neutrino loophole: Standard solar models", *Astron. Astrophys.* **44**, 237-240.

Rouse, C.A.: 1983, "Calculation of stellar structure. III. Solar models that satisfy the necessary conditions for a unique solution to the stellar structure equations", *Astron. Astrophys.* **126**, 102-110.

Rouse, C.A.: 1985, "Evidence for a small, high-Z, iron-like solar core", *Astron. Astrophys.* 149, 65-72.

Rouse, C.A.: 1995, "Calculation of solar structure IV. Results using a detailed energy generation subroutine", *Astron. Astrophys.* **304**, 431-439.

Rouse, C.A.: 1998, personal communication.

Rowe, M.W. and Kuroda, P.K.: 1965, "Fissiogenic xenon from the Pasamonte meteorite", *J. Geophys. Res.* **70**, 709-714.

Sabu, D.D. and Manuel, O.K.: 1976a, "Xenon record of the early solar system", *Nature* **262**, 28-32.

Sabu, D.D. and Manuel, O.K.: 1976b, "The xenon record of element synthesis", *Eos* **57**, 278.

Sabu, D.D. and Manuel, O.K.: 1980, "Noble gas anomalies and synthesis of the chemical elements", *Meteoritics* **15**, 117-138.

Sahijpal, S., Goswami, J.N., Davis, A.M., Grossman, L. and Lewis, R.S.: 1998, "A stellar origin for the short-lived nuclides in the early Solar System", *Nature* **391**, 559-561.

Saxena, S.K. and Eriksson, G.: 1983, "High temperature phase equilibria in a solar-composition gas", *Geochim. Cosmochim Acta* **47**, 1865-1874

Shindo, H., Miyamoto, M., Matsuda, J. and Ito, K.: 1985, "Vapor deposition of diamond from methane-hydrogen mixture and its bearing on the origin of diamond in ureilite: A preliminary report", *Meteoritics* **20**, 754.

Srinivasan, B., Alexander, E.C., Jr., Manuel, O.K. and Troutner, D.E.: 1969, "Xenon and krypton from the spontaneous fission of californium 252", *Phys. Rev.* **179**, 1166-1169.

Srinivasan, B. and Anders, E.: 1978, "Noble gases in the Murchison meteorite: Possible relics of s-process nucleosynthesis" *Science* **201**, 51-56.

Swindle, T.D., Grier, J.A. and Burkland, M.K..: 1995, "Noble gases in ortho pyroxenite AlH84001: A different kind of martian meteorite with an atmospheric signature", *Geochim. Cosmochim. Acta* **59**, 793-801.

Suess, H.E. and Urey, H.C.: 1956, "Abundances of the elements", *Rev. Mod. Phys.* **28**, 53-74.

Tayler, R.J.: 1972, *The Origin of the Chemical Elements*, Wykeham Pub. Ltd., London, 169 pp.

Tayler, R.J.: 1988, "Nucleosynthesis and the origin of the elements", *Philos. Trans. R. Soc. London*, A **325**, 391-403.

Thiemens, M.H. and Heidenreich J.E. III: 1983, "The mass-independent fractionation of oxygen: A novel isotope effect and its possible cosmochemical implications", *Science* **219**, 1073-1075.

Timmes, F.X., Woosley, S.E., and Weaver, T.A.: 1995, "Galactic chemical evolution: Hydrogen through zinc", Astrophys. J. *Suppl.* **98**, 617-658.

Trimble, V.: 1975, "The origin and abundances of the chemical elements", *Rev. Mod. Phys.* **47**, 877-976.

Trimble, V.: 1982, "Supernovae. Part I: the events" *Rev. Mod. Phys.* **54**, 1183-1224.

Trimble, V.: 1983, "Supernovae. Part II: the aftermath" *Rev. Mod. Phys.* **55**, 511-563.

Trimble, V.: 1988, "Galactic chemical evolution: Perspectives and prospects", in *Origin and Distribution of the Elements, Proceedings of a 1987 ACS symposium*, ed., Mathews, G.J., *World Scientific*, Singapore, pp. 163-175.

Trimble, V.: 1991, "The origin and abundances of the chemical elements revisited", *Astron. Astrophys. Rev.* **3**, 1-46.

Trimble, V.: 1996, "Cosmic abundances: Past, present and future", *ASP Conf. Series* **99**, eds., Holt, S.S. and Sonneborn, G., pp. 3-35.

Turekian, K.K. and Clarke, S.P.: 1969, "Inhomogeneous accumulation of the earth from the primitive solar nebula", *Earth Planet Sci. Lett.* **6**, 346-348.

Urey, H.C.: 1952, " *The Planets, Their Origin and Development*, Yale University Press, New Haven, Connecticut, 245 pp.

Urey, H.C.: 1954, "On the dissociation of gas and volatilized elements from proto-planets", *Ap. J. Suppl.* **1**, 147-174.

Urey, H.C.: 1956, "Diamonds, meteorites, and the origin of the solar system", *Ap. J.* **124**, 623-637.

Vinogradov, A.P.: 1975, "Formation of the metal cores of the planets", *Geokhimiya* **10**, 1427-1431.

Wallerstein, G., Iben, I., Parker, P., Boesgaard, A.M., Hale, G.M., Champagne, A.E., Barnes, C.A., Käppler, F., Smith, V.V., Hoffman, R.D., Timmes, F.X., Sneden, C., Boyd, R.N., Meyer, B.S. and Lambert, D.L.:1997, "Synthesis of elements in stars: forty years of progress", *Rev .Mod. Phys.* **69**, 995-1084.

Wasserburg, G.J.: 1987, "Isotopic abundances: inferences on solar system and planetary evolution", *Earth Planet. Sci. Lett.* **86**, 129-173.

Wheeler, J.C. and Cameron, A.G.W.: 1975, "The effect of primordial hydrogen/helium fractionation on the solar neutrino flux", *Ap. J.* **196**, 601-605.

Wheeler, J.C., Sneden, C. and Truran, J.W.: 1989, "Abundance ratios as a function of metallicity", *Annu. Rev. Astro. Astrophys.* **27**, 279-349.

Wolszczan, A.: 1994, "Confirmation of earth-mass planets orbiting the millisecond pulsar PSR B 1257+12", *Science* **264**, 538-542.

Wolszczan, A. and Frail, D.A.: 1992, "A planetary system around the millisecond pulsar PSR1257+12 *Nature* **355**, 145-147.

Wood, J.: 1978, "Ancient chemistry and the formation of the planets", in *Proceedings of the Robert A Welch Foundation Conferences on Chemical Research*, **XXI**. Cosmochemistry, ed., Milligan, W.O., The Robert A. Welch Foundation, Houston, TX, pp. 323-362.

Wood, John A.: 1999, "Forging the planets", *Sky & Telescope* **97**, 36-48.

Yoneda, S. and Grossman, L.: 1995, "Condensation of CaO-MgO-Al2O3-SiO2 liquid from cosmic gases *Geochim. Cosmochim. Acta* **59**, 3413-3444.

Zeilik, M.: 1982, *Astronomy: The Evolving Universe*, Harper & Row, New York, NY, 623 pp.

Zinner, E., Amari, S., Anders, E. and Lewis, R.S.: 1991, "Large amounts of extinct $^{26}$Al in interstellar grains from the Murchison carbonaceous chondrite", *Nature* **349**, 51-54.

Zinner, E.: 1997, "Presolar material in meteorites: an overview", *AIP Conference Proceeding 402: Astrophysical Implications of the Laboratory Study of Presolar Materials*, eds., Bernatowicz, T.J. and Zinner, E., American Institute of Physics, Woodbury, NY, pp. 3-26.

# Author Index

AHMAD, I., 203
ALEXANDER, JR., E. CALVIN, 401
ANDERS, E., 361
ANTHONY, D. W., 51
ARLANDINI, C., 93
ARMBRUSTER, P., 35
BERNAS, MONIQUE, 71
BILDSTEN, L., 153
BROWN, WILBUR K., 225
BROWNE, EDGARDO, 211
BURBIDGE, G., 167
CHEVALIER, ROGER A., 217
CUMMING, A., 153
D'AURIA, J. M., 51, 63
DE LAETER, J. R., 361
DRAGON COLLABORATION, 63
EBIHARA, M., 289
FOWLER, M. M., 103
FRANK, ADAM, 241
GHIORSO, ALBERT, 3
GIESEN, U., 51
GOSWAMI, JITENDRA NATH, 407
GREENE, J. P., 203
GREGORICH, K. E., 21
GREVESSE, N., 261
GUZIK, JOYCE ANN, 301
HAIGHT, R. C., 103
HASHIMOTO, Y., 511
HOFFMAN, ROBERT D., 143
HUSS, G. R., 361
HWAUNG, GOLDEN, 345
KÄPPELER, F., 93, 103
KOEHLER, P., 103
KRATZ, K.-L., 111, 119
KURODA, P. K., 431
KUTSCHERA, W., 203
LAL, D., 351
LEWIS, R. S., 361
LHERSONNEAU, G., 111
LIETZ, CARA, 529
LOFY, P. A., 51
LOSS, R. D., 361
LU, QI, 385
LUGMAIR, G. W., 361
MAAS, R., 361
MANTICA, P. F., 51
MANUEL, O., 279, 589
MASUDA, AKIMASA, 385
MILLER, G. G., 103
MÖLLER, P., 119
MORRISSEY, D. J., 51
MYERS, W. A., 431
NAKANISHI, T., 511
NEUFORGE, CORINNE, 301
NINOV, V., 21
NOLTE, ADAM, 529
NORMAN, ERIC B., 211
OTT, ULRICH, 361, 369
OURA, Y., 289
OZAKI, H., 289
PALMER, P. D., 103
PAUL, M., 203
PFEIFFER, B., 111, 119
PRISCIANDARO, J. I., 51
RAMADURAI, S., 253, 575
RAUSCHER, THOMAS, 143
ROSMAN, K. J. R., 361
ROUSE, CARL A., 317
ROWE, MARVIN W., 501
RUNDBERG, ROBERT S., 103
SAKAMOTO, K., 511

SAUVAL, A. J., 261
SCHATZ, H., 153
SHINOTSUKA, K., 289
STEINER, M., 51
SORLIN, O., 81
TANAKA, TSUYOSHI, 555
TANIMIZU, MASAHARU, 555
THIELEMANN, FREIEDRICH-KARL, 119, 143
TRIMBLE, VIRGINIA, 175
ULLMAN, J. L., 103
VAHIA, M. N., 351, 581
VIOLA, V. E., 189
VOSS, F., 93
WARD, THOMAS E., 545
WIESCHER, M., 153
WILHELMY, J. B., 103
WINDLER, K., 519
WISSHAK, K., 93
WOOSLEY, STANFORD E., 143

**SPRINGER NATURE**

# GPSR Compliance

*The European Union's (EU) General Product Safety Regulation (GPSR) is a set of rules that requires consumer products to be safe and our obligations to ensure this.*

*If you have any concerns about our products, you can contact us on ProductSafety@springernature.com*

In case Publisher is established outside the EU, the EU authorized representative is:

Springer Nature Customer Service Center GmbH
Europaplatz 3
69115 Heidelberg, Germany

The manufacturer's authorised representative in the EU is Springer Nature Customer Service Centre GmbH, Europaplatz 3, 69115 Heidelberg, Germany. If you have any concerns regarding our products, please contact ProductSafety@springernature.com

Printed and bound by CPI Group (UK) Ltd, Croydon, CR0 4YY
22/07/2026
02174140-0004